TRAFFIC ENGINEERING
AND
TRANSPORT PLANNING

L.R. KADIYALI
B.E. (Hons.) (*Mumbai*)
P.G. Dip. in Highway and Traffic Engg.
(Newcastle–Upon Tyne).
Ph. D. (Kakatiya University)
Formerly Chief Engineer (Road Wing)
Ministry of Shipping and Transport
New Delhi
and
Study Director, Road User Cost Study
Central Road Research Institute
New Delhi

KHANNA PUBLISHERS®
Investing in Learning®
Operational Office :
4575/15, Onkar House, Opp. Happy School,
Ground Floor, Daryaganj, New Delhi 110 002
Phones : 011-45033819 • *Mob.* 09811541460
email : contactus@khannapublishers.in

Published by :

Romesh Chander Khanna & Vineet Khanna
for **KHANNA PUBLISHERS**
2-B, Nath Market, Nai Sarak,
Delhi-110006 (India)

Visit us at : www.khannapublishers.in

Hologram & Description

To all readers of our books, to prevent yourself from being frauded by pirated books. Please make sure that there is an Hologram on the cover of our books with the below specifications. If you find any book without the Hologram or Description, please mail us at **khannapublishers@yahoo.in**.

Thanking you

ISBN No. : 978-81-7409-220-5

First Edition : ***1999***
Ninth Edition : ***2017***
Second Reprint : 2025

Price Rs. 639.00

Dedication:

The book is dedicated to the loving memory of my father, late Shri K. Rama Bhat, who was a source of inspiration to me in the development of my career.

Preface

The spectacular growth of the automobile as one of the most convenient modes of travel has brought in its wake frustrating problems of parking, accidents, delay, congestion and environmental degradation. It is ho longer sufficient to build roads of adequate structural strength to cater to the needs of vehicles, but it is equally important that the safe, efficient and comfortable movement of traffic and other road user is ensured on these roads. This need has given birth to new branch of engineering, known as Traffic Engineering. The subject is growing in important as the vehicle population increases, and the attendant problems become alarming.

Considering the relevance of the subject to our times, the author has pleasure in offering this book to graduate and postgraduate students in the discipline of Highway Engineering and Traffic Engineering, Professional Highway Engineers and Traffic Transportation Engineers and Town Planners in India. In fact, Traffic Engineering has a multi-disciplinary coverage, and includes engineers, planners, statisticians, the police, administrators and so on. That the interest in the subject is growing in our country too is obvious from the fact that many universities in our country are offering postgraduate courses in Highway and Traffic Engineering. Even at the undergraduate level, the subject of Highway Engineering as now taught contains substantial content of Traffic Engineering. It is hoped the present book will fill a great void in the availability of suitable textbooks to such students. At the same time, the book has been written with larger need in view, *viz.* that of the practising and professional Highway and Traffic Engineers who has probably missed the requisite university level specialization in Traffic Engineering, but has to all the same deal with day-to-day problems pertaining to Traffic Engineering.

The treatment of the text has been given with a bias towards traffic conditions obtaining in India, and the Indian approach for dealing with the situation. Though Traffic Engineeing is yet in its infacy in this country, the time is most appropriate now to apply sound traffic control, regulatory and management solutions to the problems. Already, the accident situation in the country is alarming, and if matters have not to be allowed to degenerate further, there is a great need for qualified traffic engineers to handle the situation. Although a bias towards Indian conditions and solutions is clearly discernible in the text, the author has included the latest thinking, trends and research abroad so that the inequisitive reader is aware of the developments abroad and, in the absence of guidance from Indian practice, can turn to practices elsewhere.

The author has benefited from extensive published literature on the subject and has included figures, tables and formulae contained in them and taken due pains to give reference to the sources. The help received is gratefully acknowledged. However, since the subject has evolved through the work and contributions of countless over the past fifty years, lot of ideas have been assimilated so thoroughly in the present day knowledge of the subject that it is now impossible to trace them to their originators. Perhaps it is not necessary to do that either, and the very fact that the subject has grown to its present interesting state is a fitting salute to all the past contributors.

In its initial period of evolution, the subject of Traffic Engineering merely confined itself to deal with day-to-day problems concerning traffic met with in cities. Traffic signals, design of intersections, traffic management and regulatory measures and such other simple remedies were all that were needed to grapple with the problems. But no longer are these the entire field of the subject. Basic issues as to how traffic generates and how it interacts with land-use and planning now seem to be equally important. Transport Planning, which deals with these issues has itself become a major subject. Any book on Traffic Engineering will be incomplete if an elementary treatment of Transport Planning is omitted. This is the explanation for the title of the present book. Other important areas of interest are the transport economics and the impact of traffic on the environment. These have also been introduced to the reader in the book.

It is hardly necessary to emphasise the importance of Statistics in Traffic Analysis. In fact, the subject of Statistics has many applications in all branches of engineering. It is high time this is recognised and the universities include this subject in the curricula of studies at the undergraduate level. In this book a very elementary introduction to this subject is given, but the reader is advised to refer to more exhaustive textbook on the subject for better understanding.

The author would be grateful for suggestions for improvement and pointing out errors that might have inadvertantly crept in the text. Efforts would be made to incorporate the suggestions in future editions.

Acknowledgements are due to my daughter, Vasudha, for painstakingly typing the manuscript. The author wishes to thank the Director General (Road Development) and Additional Secretary Ministry of Shipping and Transport (Roads Wing), Government of India for permission to publish this book.

L.R. KADIYALI

Foreward to the Revised Edition

Since the last revision of the book, many developments have taken place in India and abroad, and the author has included reference to these.

In the text, there are frequent reference to Indian rupees. For the benefit of internation readers, the following may be noted :

1 crore rupees	$= 1 \times 10^7$ rupees
1 lakh rupees	$= 1 \times 10^5$ rupees
1 US $	= ₹ 65 Indian rupees (2018)

India follows the "Keep to the left rule" and this may be noted when referring to diagrams and text.

L.R. KADIYALI

Foreward to the Revised Edition

Since the last revision of the book, many developments have taken place in India [illegible] and references to these.

In the text there are frequent reference to Indian rupees. For the benefit of international readers the following may be noted:

1 [illegible] rupees = 1 × 10[illegible] rupees

1 lakh rupees = 1 × 10[illegible] rupees

1 US $ = [illegible] Indian rupees (2018)

India follows the 'Keep to the left' rule and this may be noted when referring to diagrams and text.

L.R. KADIYALI

Contents

SECTION 7. TRAFFIC REGULATIONS

SECTION 1

INTRODUCTION

1. Traffic Engineering Administration and Functions

1.1. Definition

The Institute of Traffic Engineers defines traffic engineering as "that phase of engineering which deals with planning, geometric design and traffic operations of roads and streets and highways, their networks, terminals, abutting lands, relationships with other modes of transportation for the achievement of safe, efficient and convenient movement of persons and goods." (Ref. 1) The definition contained in the Glossary on Traffic Engineering Terms of the PIARC (Ref. 2) is :

"That phase of engineering which deals with the planning and geometric design of streets, highways and abutting lands, and with traffic operation thereon, as their use is related to the safe, convenient and economic transportation of persons and goods."

In addition to the qualifications such as safe, economic, convenient, comfortable and efficient, it may also be appropriate to add the term *"environmentally sustainable"* in the present threat of global warming and greenhouse gases generated by transport system.

The above definitions adequately illustrate the scope of the subject.

1.2. Growth of the Subject of Traffic Engineering

Traffic engineering is a comparatively new branch of engineering and has grown with the increase in traffic in recent years. As vehicular traffic began to increase, the congestion on the streets began to hamper the safe and efficient movement of traffic. More and more accidents were caused, and serious problems of parking and environmental pollution began to be felt. It was, therefore, necessary to give increasing attention to the operational characteristics of highway transportation and study the need for better geometric design, capacity, intersections, traffic regulations, signals, traffic signs, roadway markings, parking facilities, design of bus stands and truck terminals and street lighting. The above specialised needs were sought to be met with by the services of the Traffic Engineer.

Traffic surveys were made in a number of places in the United States, and much of this pioneering work is due to W.P. Eno, through whose endowment the Eno Foundations, Saugatuck is carrying on useful work on traffic research. The now familiar three colour light signals made their appearance in 1918. The Institute of Traffic Engineers was founded in 1931 and with this the profession was officially established and defined (Ref. 3).

Highway Engineering and Traffic Engineering are related subjects, and the latter can be deemed to be an offshoot of the former. In the United States, Traffic Engineering has been now recognized as a specialised branch. A number of universities offer graduate and post-graduate courses in Traffic Engineering. In European countries, Traffic Engineering is now becoming increasingly important and the need for a Traffic Engineering profession has been keenly felt (Ref. 4). In India, Traffic Engineering is still very much an adjunct of Highway Engineering, though the problems facing the urban centres in traffic and transportation are increasingly becoming more and more formidable, justifying the application of specialised techniques and knowledge. It will not be long before that the Traffic Engineer will be recognised for the useful role he has to play in better management of the traffic in our cities and towns. Some Universities in India are now offering post-graduate courses in Traffic Engineering.

1.3. Functions

The functions of a Traffic Engineer include the following :

1.3.1. Collection, analysis and interpretation of data pertaining to traffic. One of the important functions of a traffic engineer is to organise and implement various surveys and studies aimed at collection of data pertaining to traffic characteristics. Such studies include (*i*) origin and destination survey, (*ii*) volume counts, (*iii*) speed, travel time and delay measurements, (*iv*) accident statistics, (*v*) parking characteristics, (*vi*) pedestrian behaviour and use of streets, (*vii*) capacity studies and (*viii*) economic loss caused by inferior traffic facilities.

The data collected by the above studies are analysed by the traffic engineer and interpreted to take advantage of the observed regularities. Accurate understanding of the scientific phenomena behind these regularities enables the traffic engineer to select appropriate solutions to problems.

1.3.2. Traffic and transportation planning. The traffic engineer is concerned with the preparation of traffic and transportation plans to ensure a safe, orderly and fully integrated transportation system. This phase of activity concerns itself with the relation of land use with transportation and study of travel characteristics. Based on the analysis of the results of such a study, mathematical models are formulated to predict how a system will behave under a given set of conditions. Alternative solutions for the development of the street system are then thought out and are evaluated for their comparative merits and demerits. The optimal solution is then selected and implemented.

1.3.3. Traffic design. This part of the traffic engineer's function concerns itself with geometric design of highways and streets, intersection design, schemes for grade separated inter-changes, design of off-street and on-street parking facilities and design of terminal facilities for trucks and buses. In the field of geometric design, the functions of the highway engineer and the traffic engineer overlap and there is need for closer co-operation and consultation between the two.

1.3.4. Measures for operation of traffic. For efficient and safe operation of traffic, the traffic engineer has to take recourse to a number of measures such as :

(*i*) Legislation and enforcement measures for regulating the driver, vehicle and road users like the pedestrians, cyclists and motorcyclists.

(*ii*) Management measures such as one-way streets, prohibited turnings at junctions and tidal flow arrangements, exclusive bus lanes etc. with a view to get the maximum use out of the available street facilities.

(*iii*) Measures for regulation of parking of vehicles.

(*iv*) Traffic control devices such as traffic signs, traffic signals, pavement markings and channelization techniques to guide and secure the safe and efficient flow of traffic.

1.3.5. Administration. One of the functions of the traffic engineer is to organise and administer the various programmes intended to secure safe and efficient traffic in towns and cities. In this field he is closely associated with the legal and administrative framework of the city department dealing with education, legislation and enforcement measures.

1.4. Organisation of the Traffic Engineering Department

With the growth of the vehicular traffic on the roads, it is increasingly realised that the organizational setup designed to deal with the attendant problems of growing traffic has to be a separate unit under competent traffic engineers. The need for such a unit exists at the national, state and city level.

Table 1.1. Organisation of the State Traffic Engineering Unit in U.S.A.

(*Source* : Ref. 5)

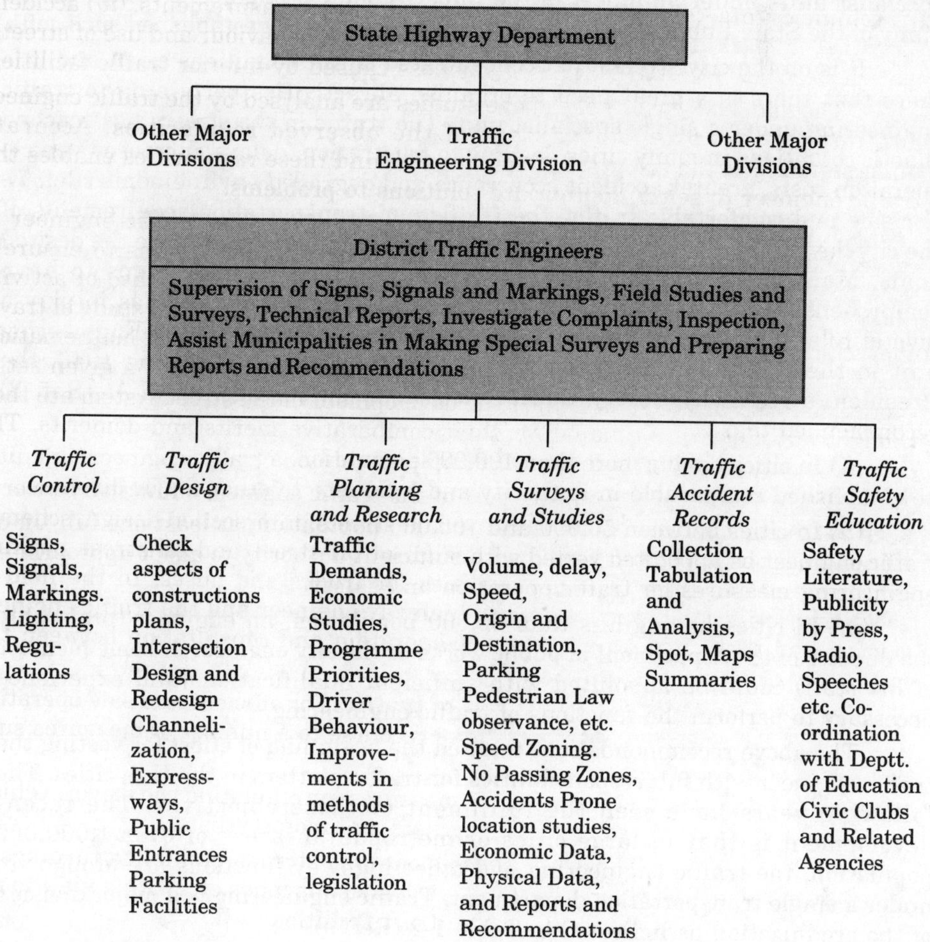

In India, the Central Government have the direct responsibility for the construction and maintenance of National Highways. The responsibility requires that data on flow, speed, capacity, cost of Highways be collected in a systematic way, analysed and interpreted. This will facilitate drawing up plans for the orderly and regulated development of the highways and evolve appropriate geometric standards for designing the facilities. This need has been recognised by the Roads Wing of the Ministry of Transport and Shipping and a separate cell under a Chief Engineer established some years ago. However, this cell now is being looked after by the Chief Engineer (Research and Standardisation).

In the United States of America, the pace-setter in Traffic Engineering, the Federal Department of Transportation contains a number of organizations dealing exclusively with highway safety and urban transportation. In U.K., the Department of the Environment has on its staff traffic engineering specialists.

At the State level, a separate traffic engineering unit exists in the United States. Table 1.1 indicates the several functions assumed by the traffic engineering unit at the State level in the United States (Ref. 5).

In U.K., Traffic Engineering units exist at the county level. In India, such units exist only in some States. Recognising this deficiency, the Indian Roads Congress had made recommendations to the various States to set up such specialist units under an officer of the Superintending Engineer's rank (Ref. 6). Many of the State Public Works Departments have already established such cells.

It is on the city streets that concentrated traffic conditions exist, and it is there that there is a great need to organise all activities pertaining to traffic engineering under a single specialist unit. The traffic in the streets has reached chaotic conditions in many cities, leading to frustration, delays, increased vehicle operation costs, greater accident occurrence and degraded environments (Ref. 7). For safe and comfortable traffic, the traffic and transportation requirements of the city have to be assessed and plans have to be formulated to achieve the desired goals. Many metropolitan cities are already taking suitable steps to have comprehensive transportation studies where the traffic engineer has to play a pivotal role. With this background, the need for a separate Traffic Engineering unit in the City Administration set-up is paramount. As early as 1946, the President's Highway Safety Conference on Engineering in U.S.A. (Ref. 8) recommended that :

(*i*) In cities having more than 100,000 population, a traffic engineering unit be established comparable in authority and influence to other major divisions.

(*ii*) In cities between 50,000 and 100,000 population, at least one full-time traffic engineer be appointed vested with sufficient authority to adopt appropriate engineering measures for traffic operation and safety.

(*iii*) In cities having less than 50,000 population, an engineer (preferably the director of the department of public works or the city engineer or some member of his staff) could be appointed with sufficient qualifications and experience necessary to perform the functions of traffic engineering.

The above recommendations marked the beginning of effort for vesting the traffic engineer with full responsibilities for traffic matters in the U.S. cities. The following years have seen the fulfilment of these objectives. The recent development is that in large cities or metropolitan areas of over 1,000,000 population, the traffic engineering and other highway functions are brought in under a single transportation department. Traffic engineering is a major division of the organization as indicated in Table 1-2 (Ref. 9).

Table 1.2. Traffic Engineering Administration in a Department of Transportation (*Source* : Ref. 9)

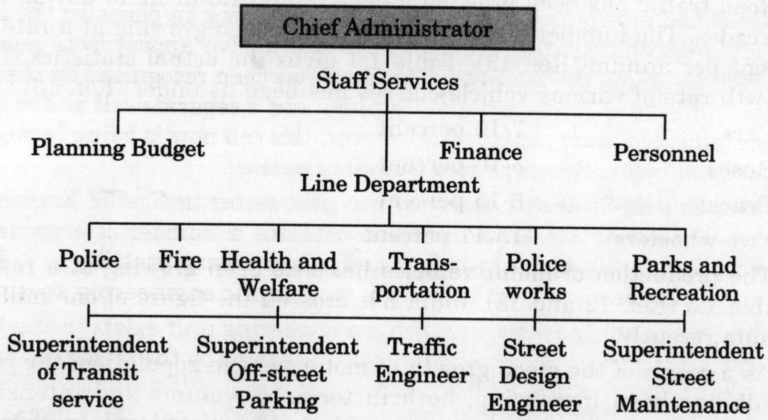

The Institute of Traffic Engineers now recommends that the above system be adopted for cities with a population more than 750,000. (Ref. 9). For cities with a population of between 250,000 and 750,000, the Institute recommends a separate major city department, dealing with streets and traffic, and having on its roles several professional engineers and a fairly large supporting staff. For cities between 50,000 and 250,000 a full-time traffic engineer and staff is deemed to be necessary. For cities with a population smaller than 50,000, the Institute reckons that a full-time engineer or staff for traffic functions may not be justified, the city engineer himself assuming responsibility for the overall supervision of traffic engineering activities. In cities having a separate Traffic Engineering Department, the organizational structure and functions can be as shown in Table 1.3 (Ref. 10).

At the city level, the traffic departments in India are managed by high ranking police officers, with traffic engineers playing very minor roles in the organizations (Ref. 11). This is a very unsatisfactory arrangement and needs to be ended soon. Separate Traffic Engineering cells are needed answer to tackle the traffic problems baffling Indian cities.

Table 1.3. Organisational Structure for the Traffic Engineering Department in City (*Source* : Ref. 10)

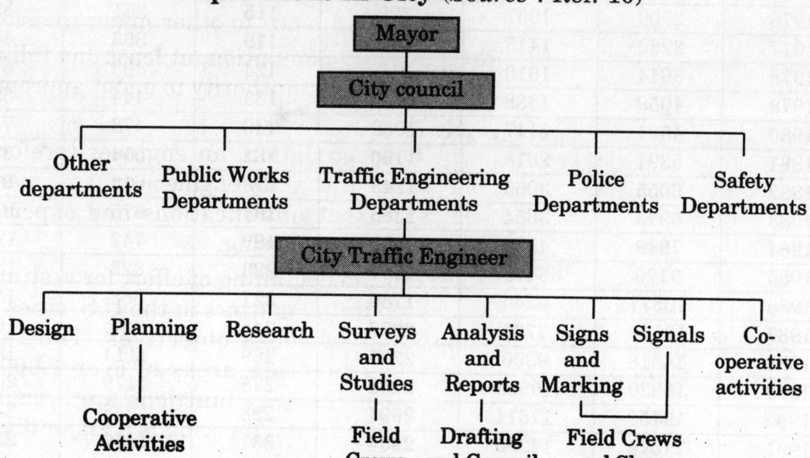

1.5. Importance of Traffic Engineering under Indian Conditions

Road traffic has been growing a very rapid rate in India during the past three decades. The number of motor vehicles has been growing at a rate around 10 percent per annum (Ref. 13). Table 1.4 gives the actual statistics (Ref. 12). The growth rate of various vehicle classes has been as under (Ref. 13) :

Cars : 7-10 percent
Buses : 5-10 percent
Trucks : 6-15 percent
Two-wheelers : 15-17 percent.

The production of motor vehicles has also been growing at a rapid rate, vide Table 1.5 (Ref. 12 and 15). India has crossed the figure of one million cars per annum recently.

As a result of the steep growth of motor vehicle population, the traffic on the roads has been increasing, both in terms of volume and intensity. The investments on roads have not kept pace with the growth of traffic, with the result there is severe congestion on the roads. Speeds are low and vehicle operating costs are high. Accident rate is usually high.

Traffic engineering measures provide a valuable tool to understand the problems and evolve suitable measures to overcome the deficiencies.

Table 1.4. Total Registered Motor Vehicles in India – 1951-93 (Source : Ref. 12,14) (in Thousands)

Year (as on 31st March)	*All Vehicles*	*Two wheelers*	*Car, jeep and Taxis*	*Buses*	*Goods Vehicles*	*Others**
1951	306	27	159	34	82	4
1956	426	41	203	47	119	16
1961	665	88	310	57	168	42
1966	1099	226	456	73	259	85
1971	1865	576	682	94	343	170
1972	2045	656	740	100	364	185
1973	2109	734	709	95	308	263
1974	2327	838	768	105	323	293
1975	2472	946	766	114	335	311
1976	2700	1057	779	115	351	398
1977	3260	1415	878	119	383	465
1978	3614	1618	919	124	403	550
1979	4059	1888	996	133	444	598
1980	4521	2117	1059	140	473	732
1981	5391	2618	1160	162	554	897
1982	6055	3065	1243	173	613	961
1983	6973	3654	1385	185	675	1074
1984	7949	4351	1455	199	742	1202
1985	9170	5179	1607	223	822	1339
1986	10577	6245	1780	227	863	1462
1987	12618	7739	2007	245	984	1643
1988	14818	9300	2295	269	1114	1840
1989	16920	10965	2486	278	1179	2012
1990	19152	12611	2694	298	1238	2311
1991	21374	14200	2954	331	1356	2533

1992	23507	15661	3205	358	1514	2769
1993	25299	17026	3330	381	1599	2963
1994	26464	17936	3446	390	1791	2901
1995	30295	20831	3841	423	1938	2770
1996	33786	23252	4204	449	2031	3850
1997	37332	25729	4672	484	2343	4104
1998	41368	28642	5138	538	2536	4514
1999	44875	31328	5556	540	2554	4897
2000	48857	34118	6143	562	2715	5319
2001	54991	38556	7058	634	2948	5795
2002	58924	41581	7613	635	2974	6121
2003 (P)	67033	47525	8619	727	3488	6674

Includes Omni buses.

(P) : Provisional

*Others included tractors, trailors, three wheelers (passenger vehicles) and other miscellaneous vehicles which are not separately classified.

Table 1.5. Production of Motor Vehicles in India (1950-93) (Source : Ref. 12 and 15) (in Thousands)

Year (as on 31st Dec)	*Jeeps and Utility Vehicles*	*Car*	*Commercial Vehicles*	*Two Wheelers*				*Three wheelers*	*Tractors*
				Scooters	*Mopeds*	*Motor cycles*	*Total (Col.5-7)*		
1	**2**	**3**	**4**	**5**	**6**	**7**	**8**	**9**	**10**
1950	–	2.2	1.9	–	–	–	–	–	–
1955	2.9	10.0	9.3	0.6	–	0.4	1.0	–	–
1960	5.5	19.1	27.5	12.0	0.9	4.0	16.9	0.5	0.1
1965	10.5	24.8	37. 4	20.3	7.4	21.4	49.1	1.9	5.6
1970	9.3	35.2	41.0	58.4	11.7	42.9	113.0	4.2	19.9
1975	8.1	20.1	43.0	101.8	36.2	69.7	207.7	12.2	32.4
1980	15.1	30.5	68.3	209.9	106.1	101.6	417.6	26.5	67.1
1981	17.0	42.1	89.8	202.9	185.4	110.8	499.1	24.8	82.5
1982	19.0	42.7	90.8	250.7	212.6	130.0	593.3	30.6	67.6
1983	21.7	45.1	87.4	273.9	329.1	156.2	759.2	37.7	71.5
1984	22.2	64.0	94.7	297.3	377.0	175.3	849.6	41.8	78.9
1985	26.9	102.5	99.7	422.3	455.3	248.0	1125.6	49.3	78.3
1986	27.8	116.0	93.6	588.1	449.8	314.7	1352.6	53.1	74.7
1987	32.4	148.5	108.1	625.6	475.6	300.6	1401.8	60.2	63.8
1988	35.1	159.9	117.3	658.9	504.2	411.6	1574.7	69.1	102.4
1989	42.4	177.2	117.6	848.5	481.1	420.8	1750.4	83.2	117.2
1990	41.9	176.6	145.6	968.4	428.6	478.5	1875.5	95.5	128.8
1991	30.4	178.9	145.7	778.6	398.1	424.6	1601.3	79.4	152.1
1992	38.2	153.9	128.1	684.3	405.9	387.0	1477.2	66.4	153.2
1993	44.3	199.6	127.8	793.8	451.4	428.5	1673.7	77.5	135.1
2002	169.4	500.3	162.5	937.5	427.5	2906.3	4271.3	212.7	NA
2003	165.9	557.4	203.7	848.4	351.6	3876.2	5076.2	276.7	NA
2004	207.0	782.6	275.0	935.3	332.3	4355.2	5622.8	356.2	NA
2005	249.3	960.4	353.7	987.5	348.4	5193.9	6529.8	374.4	NA
2006	263.0	1045.9	391.1	1020.0	379.6	6201.2	7600.8	434.4	NA

NA : Not Available.

I. MULTIPLE CHOICE QUESTIONS

1. Traffic engineering only includes
   (*a*) analysis of traffic characteristics
   (*b*) design and application of control devices
   (*c*) planning and control of regulatory measures
   (*d*) traffic operation
   (*e*) all the above.
2. Which set of traffic studies is needed for functional design as well as for 'highway capacity' design?
   (*a*) origin and destination studies (*b*) parking and accident studies
   (*c*) speed and volume studies (*d*) axle load studies
3. A traffic engineering unit to be established if a city having
   (*a*) more than 100,000 populations (*b*) more than 50,000 populations
   (*c*) more than 10,000 populations (*d*) none of these
4. Which of the following is known as design capacity?
   (*a*) basic capacity (*b*) theoretical capacity
   (*c*) possible capacity (*d*) practical capacity
5. For the movement of vehicles at an intersection of two roads, without any interference, the type of grade separator generally preferred to, is
   (*a*) delta (*b*) trumpet
   (*c*) diamond interchange (*d*) clover leaf.
6. If the ruling gradient on any highway is 3%, the gradient provided on the curve of 300 metre radius, is
   (*a*) 2.00% (*b*) 2.25%
   (*c*) 2.50% (*d*) 2.75%
7. As per recommendations of I.R.C., traffic volume study is carried out for rural roads for 7 days continuously during
   (*a*) harvesting (*b*) lean season
   (*c*) harvesting and lean season (*d*) none of these.
8. The minimum ratio of the radii of two circular curves of a compound curve, is kept
   (*a*) 1.25 (*b*) 1.5
   (*c*) 1.75 (*d*) 2.0
9. Design of horizontal and vertical alignments, super elevation, sight distance and grades, is worst affected by
   (*a*) width of the vehicle (*b*) length of the vehicle
   (*c*) height of the vehicle (*d*) speed of the vehicle
10. The growth rate of buses in the country is
    (*a*) 15-17% (*b*) 7-10%
    (*c*) 5-10% (*d*) None of these
11. The number of motor vehicles has been growing at a rate around
    (*a*) 10% per annum (*b*) 20% per annum
    (*c*) 15% per annum (*d*) None of these

Answers

1. (*e*)	2. (*c*)	3. (*a*)	4. (*a*)	5. (*d*)
6. (*d*)	7. (*c*)	8. (*b*)	9. (*d*)	10. (*c*)
11. (*a*)				

II. QUESTIONS AND ANSWERS

Q. 1. Define traffic engineering.

Ans. Traffic engineering is the branch of engineering which deals with the improvement of traffic performance of the road networks and terminds focuses on planning, geometric design and traffic operations of roads and streets and highways, their networks, terminals, abutting lands, relationships with other modes of transportation for the achievement of safe, efficient and convenient movement of persons and goods.

Q. 2. What are the functions of traffic engineer?

Ans. A traffic engineer performs following functions:

1. He train and educates staff on criteria and standards of section. He provides engineering guidance to staff and makes engineering recommendations to county officials and other departments.
2. He developes, monitors, and controls the section's budget.
3. He coordinates the section's activities with other sections, departments, and jurisdictions.
4. Maintaines regular, predictable and punctual attendance during regularly scheduled work hours at assigned worksite.
5. He meets travel requirements of the position.
6. Performs the physical requirements of the position; work within the established working conditions of the position.
7. He work a flexible schedule, which may include evenings, weekends, holidays.

Q. 3. What do you mean by traffic congestion?

Ans. Traffic congestion is a condition on transport networks that occurs when the use of transport increases. It is characterized by slower speeds, longer trip times, and increased vehicular queueing. As demand approaches the capacity of a road (or of the intersections along the road), extreme traffic congestion sets in.

Q. 4. What measures a traffic engineer has to take?

Ans. A traffic engineer has to take the following measures:

- Legislation and enforcement measures for regulating the driver, vehicle and road users like the pedestrians, cyclists and motor-cyclists.
- Management measures such as one-way streets, prohibited turning at junctions and tidal flow arrangements, exclusive bus lanes etc. with a view to get the maximum use out of the available street facilities.
- Measures for regulation of parking of vehicles.
- Traffic control devices such as traffic signs, traffic signals, pavement markings and channelization techniques to guide and secure the safe and efficient flow of traffic

Q. 5 What is the importance of traffic engineering?

Ans. Traffic engineering uses engineering techniques to achieve the safe and efficient movement of people and goods. It focuses mainly on research and construction of the immobile infrastructure necessary for this movement, such as roads, railway tracks, bridges, traffic signs and traffic lights. It manages the traffic density by limiting the rate that vehicles enter the highway during peak periods can keep both speeds and lane flows at bottlenecks high. Ramp meters, signals on entrance ramps that control the rate at which vehicles are allowed to enter the mainline facility, provide this function (at the expense of increased delay for those waiting at the ramps).

Q. 6. What the various objectives of traffic engineering?

(B.E. Exam. 2010)

Ans. A major objective of traffic engineering is to minimize or eliminate high-loss situations. In particular, the number of rejected messages or failed call attempts should be as close to zero as possible. Another goal of traffic engineering is to balance the QoS (Quality of Service) against the cost of operating and maintaining the network.

It is a method of optimizing the performance of a telecommunications network by dynamically analyzing, predicting and regulating the behavior of data transmitted over that network. Traffic engineering is also known as tele traffic engineering and traffic management. The techniques of traffic engineering can be applied to networks of all kinds, including the PSTN (public switched telephone network), LANs (local area networks), WANs (wide area networks), cellular telephone networks, proprietary business and the Internet.

III. REVIEW QUESTIONS

1. What is the scope of traffic engineering?
2. Briefly explain the growth of traffic engineering.
3. Discuss the important functions of traffic engineering.
4. Describe the road traffic in India.

REFERENCES

1. Institute of Traffic Engineers, Year Book, 1970, Washington.
2. Dictionary of Traffic Engineering Terms, Permanent International Association of Road Congress, Paris.
3. U.K. Evans (Editor). Traffic Engineering Handbook, Institute of Traffic Engineers, New Haven, 1950.
4. Traffic Engineering and Control in the U.S.A., European Productivity Agency, Organization for European Economic Co-operation, Paris, 1955.
5. Traffic Engineering Functions and Administration, Publication Number 100, Public Administration Service, Chicago, 1948.
6. Need for Traffic Engineering Cells in the Central and State Highway Departments, Editorial, Indian Highways, Vol. 1, No. 6, New Delhi, 1973.
7. Traffic in Towns, Buchanan Report, Her Majesty's Stationery Office, London, 1962.
8. Report of Committee on Engineering, The President's Highway Safety Conference, U.S. Government Printing Office, Washington, 1946.
9. An Introduction to Transportation Engineering, Institute of Traffic Engineers, New Haven, 1968.

10. T.M. Maston, W.S. Smith and F.W. Hurd, Traffic Engineering, MacGraw Hill Book Co., New York, 1955.
11. M.V. Prakash Rao, Place of Traffic Engineering in Various Organizations, Indian Highway, Vol. 8, New Delhi, 1974.
12. Motor Transport Statistics, Ministry of Surface Transport, Transport Research Division, New Delhi. (From Internet)
13. Kadiyali, L.R. Road, Transport Demand Forecast for 2000 A.D., Vol. 48-3, Journal of the Indian Roads Congress, New Delhi, 1987.
14. I.R.T.D.A. Newsletter, November, 1995, Mumbai.
15. Society of Indian Automobile Manufactures, through Internet, 2006.

2. The Road User and the Vehicle

2.1. Introduction

The traffic engineer has to design and operate the traffic facilities which will be used by pedestrians, cyclists and motorists. A proper understanding of the behaviour of the road user and the vehicle characteristics is necessary for this purpose. The human factors that govern the behaviour of the road user can offer many clues in the design of a traffic facility. A knowledge of how this behaviour is influenced by various external conditions will be useful for the same purpose. The characteristics of the vehicle will determine the design of the geometric elements of the road and in understanding the safety aspects.

2.2. Human Factors Governing Road User Behaviour

2.2.1. The human body as a complex system exhibiting reactions to external stimuli

The aspects of human behaviour having a bearing on traffic can be broadly considered under two groups : (*i*) physiological and (*ii*) psychological. Under the physiological group, the following factors may be considered : (*a*) vision, (*b*) hearing. Under the psychological factors, the perception, intellection, emotion and volition need to be considered. These are discussed in detail below.

2.2.2. Vision

Vision is one of the important factors that affects almost all aspects of highway design and safety. The human eye is the sensory organ that enables one to see and evaluate the size, shape and colour of objects and estimate distances and speeds of bodies. The light rays from the object get refracted through the lens and are brought to focus on the retina. The image formed on the retina is transmitted to the brain through the optic nerve which perceives the object.

The zone of acute vision (visual acuity) is formed by a cone whose angle is 3° about the centre of the retina. This signifies that for very distant vision, the objects should be within this narrow cone for satisfactory perception. However, the vision is still satisfactory when the angle of the cone is upto 10° or 12°. This is important when locating traffic signs and signals.

Peripheral vision deals with the total visual field for the two eyes. Within this field the eyes are able to see objects, but without clear details and colour. For example, a movement or a bright light can alert a driver in the zone of peripheral vision. The angle of peripheral vision is about 160° in the horizontal direction and 115° in the vertical direction. If the detailed attention of the eye is needed, the driver turns his head or eyes so that the object now comes within the cone of clear vision. The cone of peripheral vision also depends upon speed. The angle of the cone falls down from about 110° at 30 K.P.H. speed to about 40° at 100 K.P.H. speed.

Colour vision is important for discerning the traffic lights and colour schemes in traffic signs. But colour blindness need not be of serious concern since drivers can learn other means of recognising signs and signals. For example, the relative position of lights in a signal head being fixed, a driver can without difficulty recognise the proper signals.

The ability of the driver's eyes to adapt to glare due to headlights or to variations in the lighting conditions (entrance to tunnels, exit from tunnels) is an important factor. The age of the driver and the wearing of eye glasses determine the quickness with which he adapts to these changed conditions. The glare recovery time varies from 3 to 6 seconds.

The ability to judge the depth and distance of an object stereo scopically and its speed is important to the road user. To the predestrian waiting for a gap in the stream, the true assessment of the speed of the on coming vehicle and its location is vital. To the driver moving at a speed, the judgment of speeds of other vehicles is needed to facilitate various manoeuvres.

2.2.3. Hearing

Hearing is an aid to the road user which can at times be very vital. The sound of a horn or the sound of the nearing vehicle itself can alert a pedestrian to safety. Elderly persons with falling eyesight can perceive better through hearing than through seeing. Defective hearing is, however, not a very serious handicap.

2.2.4. Perception, intellection, emotion and volition

The important psychological characteristic of road user concerns perception, intellection, emotion and volition, abbreviated as PIEV and the time taken for these processes is known as PIEV time.

Perception is the process of perceiving the sensations received through the eyes, ears, nervous system and the brain. The exact time required for this is dependent upon the individual's psychological and physiological build-up.

Intellection is the identification of the stimuli by the development of new thoughts and ideas. It is slightly different from simple recognition by past experiences, which is part of the normal perception process. When a person perceives certain stimuli, new thoughts and ideas may form leading to better understanding of the stimuli.

Emotion is the personal trait of the individual that governs his decision making process, after the perception and intellection of the stimuli.

Volition is the will to react to a situation.

The reaction time that elapses before a road user perceives a danger and decides to take action is an important design consideration. For example, in the design of sight distances, the perception and brake reaction time comes into the picture. The perception time is the time required for a driver to come to the realisation that brakes must be applied. The brake reaction time is that time lag between the perception of danger and the effective application of the brakes. AASHTO Policy (Ref. 1) is to assume that the total perception and brake reaction time is 2.5 seconds. This total time may be considered to be the PIEV time. The Indian Roads Congress (Ref. 8) adopts the same value.

A detailed discussion about the factors that influence the behaviour of drivers and pedestrians is given in Chapter 18 dealing with Road Accidents.

2.3. Power Performance of Vehicles

2.3.1. A knowledge of the power performance of a vehicle is necessary to determine the vehicle running costs and the geometric design elements like grades.

2.3.2. Resistance to motion of a vehicle

The power developed by the engine (P_p) should be sufficient to overcome all resistance to motion at the desired speed and to accelerate at any desired rate to the desired speed. The following forces have to be overcome for this purpose (Ref. Fig. 2.1) :

1. Rolling resistance. (P_f) 2. Air resistance. (P_a)
3. Grade resistance. (P_i)
4. Inertia forces during acceleration and deceleration. (P_j)
5. Transmission losses.

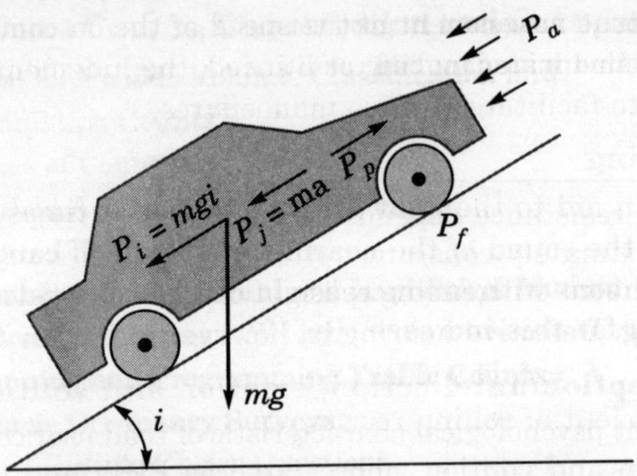

Fig. 2.1. Forces acting on a vehicle.

2.3.3. Rolling resistance

When the vehicle wheels roll over the road surface, the irregularities and the roughness of the surface cause of deformation of the tyres. The road surface itself may undergo deformations. Shocks and impacts are caused by such a motion and these hinder rolling motion of the wheels. The rolling resistance varies with the type of surfacing. The values given in Table 2.1, are typical.

Table 2.1. Values of Coefficient of Rolling Resistance
(*Source* : Ref. 2)

Type of surfacing	*Coefficient of rolling resistance*
(*a*) Cement concrete and asphalt surfacing.	0.01 to 0.02
(*b*) Road with smooth chippings or gravel surface, treated with bituminous binders.	0.02 to 0.025
(c) Chippings or gravel surfacings, not treated with binder, having small pot-holes.	0.03 to 0.04
(*d*) Cobblestone pavement.	0.04 to 0.05
(e) Earth road, smooth, dry and compact.	0.03 to 0.06
(*f*) Ploughed field, saturated and swampy ground, loose sand.	0.15 to 0.30 and over

Research carried out in India has yielded the following values of rolling resistance (Ref. 3), Table 2.2.

The rolling resistance is given by

$$P_f = mfg \qquad \text{...(2.1)}$$

where, m = mass of the vehicle in kg

f = co-efficient of rolling resistance

P_f = rolling resistance in N

g = acceleration due to gravity in m/sec^2

Table 2.2. Values of Rolling Resistance Co-efficient from Indian Studies (*Source* : Ref. 3)

	Type of surface	*Rolling resistance co-efficient*
1.	Asphaltic concrete	0.01
2.	Premixed carpet in good condition	0.016
3.	Premixed carpet in bad condition	0.022
4.	Water-bound macadam in good condition	0.025
5.	Water-bound macadam bad condition	0.037
6.	Gravel	0.046
7.	Earth	0.055

The rolling resistance depends on the speed of the vehicles also. Though its value is approximately constant upto a speed of about 50 K.P.H., at higher values of speed the co-efficient increases in value. The following approximate equation accounts for this increase (Ref. 2) :

$$f_v = f_0\,[1 + 0.01\,(V - 50)] \qquad \text{...(2.2)}$$

where, f_v = co-efficient of rolling resistance at speed V.

V = speed in K.P.H.

f_0 = co-efficient of rolling resistance, assumed constant upto a speed of 50 K.P.H. and can be taken from Table 2.1.

2.3.4. Air resistance

When a vehicle is in motion, air resists it in the following ways :

(*i*) Since air has density, it exerts a reaction pressure against the front of the vehicle when it moves at speed.

(*ii*) The friction of air against the sides of the vehicle body causes resistance.

(*iii*) The eddying of the air stream behind the vehicle, under the body and around the wheels causes power loss.

(*iv*) The flow of air through the vehicle for ventilating and cooling causes resistance to motion.

The following formula can be used to determine the air resistance, P_a (Ref. 3) :

$$P_a = C_a \cdot Av^2 \qquad \text{...(2.3)}$$

where, P_a = air resistance in N

A = projected front area of the vehicle in sq. metres on a plane at right angles to the direction of motion; for Indian vehicles, it can be taken from Table 2.3.

v = speed of the vehicle relative to air in m/sec.

C_a = Co-efficient of air resistance, having values set forth in Table 2.3.
g = Acceleration due to gravity, 9.81 m/sec².

Table 2.3. Values of Frontal Area and Co-efficient of Air Resistance (*Source* : Ref. 3, 7)

	Type of Vehicle	*Frontal area (m^2)*	*Mass (kg)*	*Co-efficient of resistance C_a (kg/m^2)*
1.	Premier Padmini car	1.63	1065	0.42
2.	Ambassador car	2.15	1365	0.39
3.	Jeep	2.38	1200	0.37
4.	Tata truck	5.37	6120	0.48
5.	Ashok Leyland truck	5.37	8125	0.48
6.	Maruti car	1.54	880	0.40

2.3.5. Grade resistance

When a vehicle which was moving on a level stretch at a particular speed has to move up an incline, additional work has to be done in keeping the vehicle at the same speed as in the level stretch. The additional work is equal to the work that will be needed to lift the vehicle through a height represented by the inclination. If the horizontal distance is unity (*i.e.*, 1 metre), and the slope is i per cent, the rise will be $\frac{i}{100}$ m. If the mass of the vehicle is m kg, the additional force to move the vehicle up the incline, P_i, is given by,

$$\pm P_i = \frac{m.i.g}{100} \qquad ...(2.4)$$

It may be noted that if the slope becomes downward, i becomes –ve, and P_i also becomes –ve, representing a reduction in the force to move the vehicle.

2.3.6. Inertia forces during acceleration and deceleration

When the speed of a moving vehicle needs to be increased some additional power is needed to accelerate. Similarly if the vehicle has to gather a desired speed from a stopped position, additional force is needed to accelerate. The additional force P_j is given by

Force = Mass × Acceleration

$$\text{Hence,} \quad (\pm P_j) = ma = m \cdot \frac{dv}{dt} \qquad ...(2.5)$$

where, p_j = Force to accelerate, N
m = Mass of the vehicle, kg
a = Average acceleration of the vehicle, m/sec² = dv/dt

The value of P_j will be +ve if the vehicle is to accelerate and –ve if the vehicle is to decelerate.

2.3.7. Transmission losses

Losses in power occur to the mode of power transmission (clutch or automatic fluid coupling) from the engine to the gear system and in the gear system itself. The vehicle has a system of gears such that the speed of the vehicle can be altered relative to the engine speed. At the start of the vehicle, high power is needed but at low speed. Similarly, a high engine power is needed while climbing uphill, which is accomplished at a lower road speed than when driving at a level stretch. These manoeuvres are made at the lowest gears. For movement along a good road where the resistance to motion will be small, a high gear will tend to be used. The highest forward gear will generally be 1 : 1, representing direct drive, A further gear reduction is made at the rear axle. The total effect of all the above is to consume about 10 – 15 per cent, of the engine power, which may be as high as 25 per cent in case of trucks in their lowest gear (Ref. 4).

2.3.8. Power requirements of the vehicle

The mechanical power developed by the engine is transmitted to the driving wheels by the transmission system. The torque developed at the flywheel is converted to a torque at the rear axle and the following equation holds good :

Rear-axle torque, $T_a = kT_cG_tG_a$ ...(2.6)

where, T_a = rear-axle torque

k = efficiency of the transmission system, which takes into account the loss of power due to overcoming the resistance of all mechanism between the engine and the driving wheels, and can be taken to be about 0.85 to 0.90

T_c = engine torque at the fly-wheel

G_t = transmission gear ratio

G_a = rear-axle gear ratio.

The rear axle torque imparts a tractive force P_p at the contact of the wheel and the road. This tractive force also known as *tyrerim pull*, is given by the following equation.

$$P_p = \frac{\text{Rear axle torque}}{\text{Radius of the rolling drive tyre}}$$

$$= \frac{kT_e G_t G_a}{r_w} \quad \text{...(2.7)}$$

r_w is related to the radius of the tyre r_0 by the following formula, Fig. 2.2.

$$r_w = \lambda r_0 \quad \text{...(2.8)}$$

where, λ = tyre deformation factor will have a value of 0.945 – 0.950 for high pressure air tyres and 0.930 – 0.935 for low-pressure tyres, on hard surfaces (Ref. 2).

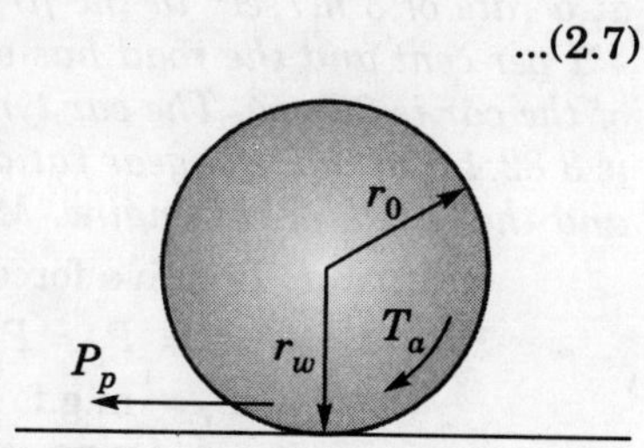

Fig. 2.2. Tractive effort and the driving wheel.

The horse-power corresponding to the tractive effort P_p when the vehicle moves at a speed of v m/sec is

Power output = $P_p v$ ...(2.9)

But $v = \frac{V \times 1000}{3600} = \frac{V}{3.6}$, V being speed in K.P.H. ...(2.10)

$\therefore$ Power output = $\frac{P_p \times V}{3.6}$ ...(2.11)

Also $$V = \frac{2\pi r_w n}{60\, G_t\, G_a} \qquad ...(2.12)$$

where n is the engine speed in R.P.M. Substituting Eq. (2.12) in (2.10)

$$V = \frac{0.377\, r_w n}{G_t\, G_a} \qquad ...(2.13)$$

Substituting Eq. (2.13) in (2.11),

Power output $$= \frac{P_v \times 0.377 \times r_w n}{G_t\, G_a}$$

Engine power (in Watts)

$$= \frac{P_v \times 0.377 \times r_w \times n}{G_t \times G_a \times k}$$

where, k = transmission efficiency

Engine horse-power (metric)

$$= \frac{\text{Engine power (in watts)}}{735} \qquad ...(2.14)$$

It may be noted that the tractive effort developed at the wheels should be equal to the resistance to be overcome.

P_p = Rolling resistance + Air resistance + Grade resistance + Inertia forces due to acceleration and deceleration

$$= P_f \pm P_a \pm P_i \pm P_j \qquad ...(2.15)$$

The following example will illustrate the use of the formulae developed above.

Problem 2.1. *A passenger car weighing 2 tonnes is required to accelerate at a rate of 3 m/sec² in the first gear from a speed of 10 K.P.H. The gradient is + 1 per cent and the road has a black topped surface. The frontal projection area of the car is 2.0 m². The car tyres have radius of 0.33 m. The rear axle gear ratio is 3.82:1 and the first gear ratio is 2.78:1. Calculate the engine horsepower needed and the speed, of the engine. Make suitable assumptions.*

Solution. Tractive force needed,

$= P_p = P_f \pm P_a \pm P_i \pm P_j$

P_f = m.g.f. (Equation 2.1)

Assume $f = 0.02$

$P_f = 2000 \times 9.81 \times 0.02 = 392.4$ N

$P_a = C_a.A.v^2$ (Equation 2.3)

$C_a = 0.39$ (Table 2.3)

$A = 2.15\ m^2$

V = 10 K.M.P.H initially, and increased to 20 K.M.P.H. eventually. So, assume average value of 15 K.M.P.H.

$$P_a = 0.39 \times 2.15 \times \frac{15}{3.6} \times \frac{15}{3.6} = 14.5 \text{ N}$$

Pi = m.g.i. (Equation 2.4)

$= 2000 \times 9.81 \times 1/100 = 196.2$ N

$$P_j = m.a = 2000 \times 3 = 6000 \text{ N}$$
$$P_p = 392.4 + 14.5 + 196.2 + 6000 = 6603.1 \text{ N}$$

$$\text{Power output} = P_p \times v = P_p \times \frac{V}{3.6} = 6603.1 \times \frac{10}{3.6}$$
($\therefore$ V = 10 K.M.P.H. initially)

$$= 18341.9 \text{ W} = \frac{18341.9}{735} \text{ hp} = 24.95 \text{ hp.}$$

Assume a transmission efficiency of 0.90, the engine horse-power

$$V = \frac{24.95}{0.9} = 27.72 \text{ hp} = \frac{0.377\, r_w n}{G_t\, G_a} \quad \text{(Equation 2.13)}$$
$$r_w = \lambda r_0$$
$$= 0.935 \times 0.33, \text{ assuming } \lambda = 0.935 = 0.308 \text{ m}$$
$$n = \frac{10 \times 2.78 \times 3.82}{0.377 \times 0.308} = 915 \text{ R.P.M.}$$

2.4. Other Vehicle Characteristics

2.4.1. The other vehicle characteristics that are of concern to the traffic engineer are :

(*i*) Dimensions and weight (*ii*) Turning capability
(*iii*) Braking system (*iv*) Acceleration and deceleration.
(*v*) Vehicle lighting system. (*vi*) Features of the vehicle body.
(*vii*) Tyres etc.

2.4.2. The dimensions and weight of road vehicles and their turning capability have an important influence on the geometric design of roads. They have been discussed in Chapter 11 on Geometric Design.

2.4.3. Ths braking system of a vehicle is important from the point of view of safety. The safe stopping distance is composed of the distance travelled by the vehicle during the perception and brake reaction time and the distance required to stop the vehicle after the brakes are applied. If *f* is the coefficient of friction between the tyre and the pavement, the braking distance is given by the formula (Ref. 8) :

$$d = \frac{V^2}{254 f} \quad \text{...(2.16)}$$

where, d = braking distance, metres
V = speed in K.P.H.
f = coefficient of friction between the tyre and the pavement.

Values of *f* usually adopted in geometric design are discussed in Chapter 11. The braking capability of a vehicle influences the skidding that can take place where brakes are applied suddenly. Because of its close connection with road accidents, skidding is discussed under Chapter 18.

2.4.4. Acceleration characteristics of a vehicle need to be understood when designing the intersection elements (acceleration lane etc.) and overtaking sight distance. Acceleration rate is governed by the vehicle transmission system, weight and horsepower. The acceleration rate also varies with speed, being high at lower speeds and low at higher speeds. The following are the ranges in values of acceleration rates of different types of vehicles :

Medium passenger cars : 3 – 8 K.P.H. per sec.
Trucks and buses : 1 – 4 K.P.H. per sec.

2.4.5. When driver applies the brakes, the vehicle decelerates.

The deceleration rate that a driver utilises is generally less than that which a vehicle is capable of developing. Only in an emergency does the driver attempt to fully utilise the maximum deceleration. The maximum deceleration is related to the co-efficient of friction between the tyre and the pavement in the following way :

The force required to decelerate a vehicle is given by :

$$F = md \quad ...(2.17)$$

where, F = force required to decelerate, N
m = mass of the vehicle, kg
d = deceleration in m/sec^2.

Rearranging the terms,

$$d = \frac{F}{m} \quad ...(2.18)$$

But $\frac{F}{mg} = f.$

f being the co-efficient of friction.

$$d = f \cdot g \quad ...(2.19)$$
$$= 9.81 f \quad ...(2.20)$$

Table 2.4 gives some observed deceleration rates.

Table 2.4. Deceleration Rates from 115 K.P.H.
(*Adapted from* Ref. 5)

	Deceleration	
	(m/sec^2)	*(f)*
Comfortable to passengers, preferred by driver	2.62	0.27
Undesirable but not alarming to passengers; driver-would rather not use	3.39	0.34
Severe and uncomfortable to passengers; slides objects off seats–driver classifiess as emergency stop.	4.26	0.43

2.4.6. The vehicle lighting system, other features of the body and tyres are discussed in Chapter 18 because of their close connection with road accidents.

2.5. Travelling Vehicle is Suddenly Stopped

When the vehicle which is travelling at a particular speed is suddenly allowed to coast by switching off the engine and putting the gear to neutral, deceleration is caused. From the force equation (Eq. 2.15).

$$P_p = P_f \pm P_a \pm P_i \pm P_j$$

Since $P_p = 0$, and i (gradient) = 0,

$$\therefore \quad P_j = P_a + P_f$$

$$\therefore \quad m\frac{dv}{dt} = C_a.A.v^2 + m.f.g.$$

$$\therefore \quad \frac{dv}{dt} = \frac{C_v.A.v^2}{m} + f.g \quad ...(2.21)$$

The following example illustrates the use of the formula.

Problem 2.2. *An Ambassador car travelling at a speed of 60 K.M.P.H. on a level W.B.M. road in good condition is suddenly allowed to coast by switching off the engine and putting the gear in neutral. What is the deceleration caused ?*

Solution. From Eq. 2.21,

$$\frac{dv}{dt} = \frac{C_v.A.v^2}{m} + f.g$$

For an Ambassador car, (Table 2.3)

$$C_a = 0.39 \text{ kg/m}^3$$
$$A = 2.15 \text{ m}^2$$
$$m = 1365 \text{ kg.}$$

For a W.B.M. road,

$$f = 0.025 \text{ (Table 2.2)}$$

$$\therefore \quad \frac{dv}{dt} = \frac{0.39 \times 2.15}{1365} \times \left(\frac{60}{3.6}\right)^2 + 0.025 \times 9.81$$

$$= 0.17 + 0.25 = 0.42 \text{ m/sec}^2$$

2.6. Characteristics of Slow Moving Traffic in India

The slow moving traffic in India consists of cycles, cycle rickshaws, bullock carts and horse carts.

An understanding of some of the characteristics of these vehicle types is essential for traffic engineer.

The dimensions of the slow moving vehicles are given in Table below.

Table 2.5. Dimensions of Slow Moving Vehicles

(*Source* : Ref. 6)

S. No.	*Vehicle Type*	*Length (m)*
1.	Cycle	1.91
2.	Cycle rickshaw	2.59
3.	Bullock cart	5.87
4.	Horse cart	4.11

The speeds of the vehicles are given in Table below :

Table 2.6. Mean Speeds of Slow Moving Vehicles

(*Source* : Ref. 6)

S. No.	*Vehicle Type*	*Mean Speed (K.M.P.H.)*	*Standard Deviation (K.M.P.H.)*
1.	Cycles	15.63	2.87
2.	Bullock cart	3.78	0.65
3.	Horse cart	12.24	3.42

The average headway between slow moving vehicles when moving in groups is as below :

Headway between cycles : 2 metres.

Headway between bullock carts : 1 metre.

Headway between horse carts : 8 metres.

The space occupied by cycles, when moving in groups, across the pavement is given below :

Table 2.7. Lateral Space Occupied by Cycle Across the Pavement (*Source* : Ref. 6)

Group size (Number of cycles)	*Lateral space occupied by one cycle (metre)*
1	1
2	0.9
3	0.8
4	0.7
5, 6 and 7	0.6

The speed of a cyclist when crossing an intersection from a stopped position is about 13 km/hr.

I. MULTIPLE CHOICE QUESTIONS

1. The effect of having excess camber is
   (*a*) hard steering
   (*b*) excessive steering alignment torque
   (*c*) too much traction
   (*d*) uneven tyre wear
2. In automobiles G.V.W. refers to
   (*a*) gross vehicle weight (*b*) gross vehicle width
   (*c*) gross vehicle wheel base (*d*) gross vehicle wheel track
3. When two roads with two-lane, two-way traffic cross at an uncontrolled intersection, the total number of potential major conflict points would be
   (*a*) 32 (*b*) 24 (*c*) 16 (*d*) 4 (*IES 1999*)
4. When a vehicle is involved in an accident causing injury to any person
   (*a*) Take the vehicle to the nearest police station and report the accident
   (*b*) Stop the vehicle and report to the police station
   (*c*) Take all reasonable steps to secure medical attention to the injured and report to the nearest police station within 24 hours
5. On a road designated as one way
   (*a*) parking is prohibited (*b*) overtaking is prohibited
   (*c*) should not drive in reverse gear
6. You can overtake a vehicle in front
   (*a*) through the right side of that vehicle
   (*b*) through the left side
   (*c*) through the left side, if the road is wide
7. When a vehicle approaches an unguarded railway level crossing, before crossing it, the driver shall

(*a*) stop the vehicle on the left side of the road, get down from the vehicle, go to the railway track, and

(*b*) ensure that no train or trolley is coming from either side

(*c*) sound horn and cross the track as fast as possible

(*d*) Wait till the train passes

8. How can you distinguish a transport vehicle.

(*a*) by looking at the tyre size

(*b*) by colour of the vehicle

(*c*) by looking at the number plate of the vehicle

9. In a road without footpath, the pedestrians

(*a*) should walk on the left side of the road

(*b*) should walk on the right side of the road

(*c*) may walk on either side of the road

10. In traffic moving two lanes in each direction, the car directly ahead of you is waiting to make a right turn. You may

(*a*) pass the vehicle from right side

(*b*) pass the vehicle from left side

(*c*) do not pass the vehicle till the vehicle makes the right turn

11. When passing a procession, body of troops, or men at work, we should

(*a*) proceed at regular speed (*b*) stop

(*c*) proceed at not more than 25 km/hour and carefully

12. Which are the vehicles which have got free passage

(*a*) motor cycles with pillion riders

(*b*) private cars (*c*) emergency vehicles

13. To stop the motor cycles

(*a*) apply the back brake first and after slowing down apply front brake

(*b*) apply front brake first and after slowing down apply back brake

(*c*) apply the front and back brakes simultaneously

14. As you approach a set of traffic lights at an intersection, they change from green to yellow. You must

(*a*) stop before the lights, unless you are so close that sudden braking might cause an accident

(*b*) drive through the intersection without accelerating

(*c*) accelerate to clear the intersection before the lights change to red

15. If you refuse to take a breath test and are convicted for this, you could be disqualified from driving

(*a*) for no more than five years (*b*) for more than five years

(*c*) for at least one year

16. When you apply the brakes, it takes at least 20 metres to bring your car to a stop on a road from 50 km/h. If you double your speed to 100 km/h, what minimum distance will it take to stop your car when you apply the brakes on the same road surface?

(*a*) 30 metres (*b*) 80 metres (*c*) 40 metres (*d*) 20 metres

17. Providing they are not breaking the speed limit, what is the maximum speed for learner drivers who are not in a driving school vehicle fitted with dual brake controls?
(*a*) 60 km/h (*b*) 100 km/h (*c*) 80 km/h

18. When passing a cyclist, what are the required minimum distances that must be adhered to?
(*a*) at least 1.7 meters for any speed limit
(*b*) at least 1 meter when the speed limit is 60 km/h or under and at least 1.5 meters for a speed limit over 60 km/h
(*c*) Any distance as long as it is safe

19. Where there is parallel kerbside parking, are you allowed to double park alongside a parked vehicle?
(*a*) No, not at any time (*b*) Yes, if not obstructing traffic
(*c*) Yes, if delivering goods

20. What is the speed limit at a children's crossing (Koala crossing) when the yellow flashing lights are operating?
(*a*) 40 km/h (*b*) 60 km/h (*c*) 25 km/h

21. You are waiting to turn right from behind the stop line at an intersection. You are faced with a green light and a red right turn arrow. The red arrow light goes out. What should you do?
(*a*) wait behind the stop line until the green light changes to yellow and then move into the intersection and complete the right turn, if it is safe.
(*b*) wait behind the stop line and until the green arrow appears again before you complete the right turn.
(*c*) enter the intersection and wait until it is safe to complete your right turn.

Answers

1. (*d*)	**2.** (*a*)	**3.** (*b*)	**4.** (*c*)	**5.** (*c*)
6. (*a*)	**7.** (*a*)	**8.** (*c*)	**9.** (*b*)	**10.** (*b*)
11. (*c*)	**12.** (*c*)	**13.** (*c*)	**14.** (*a*)	**15.** (*c*)
16. (*b*)	**17.** (*a*)	**18.** (*b*)	**19.** (*a*)	**20.** (*c*)
21. (*b*)				

Explanation

3. (*b*) Point of potential conflicts depends on the number of lanes an intersecting lanes. For two way traffic as a right angled road inter-section the conflict points are it whereas for two way traffic on T-intersection the conflict point are 18 only.

II. SOLVED QUESTIONS AND ANSWERS

Q. 1. What is hearing?

Ans. Hearing is an aid to the road user which can at time be very vital. The sound of a horn or the sound of the nearing vehicle itself can alert a pedestrian to safety.

Q. 2. What do you mean by PIEV time?

Ans. The **PIEV** is important psychological characteristics of road user which concerns perception, intellection, emotion and volition. It is abbreviated as **PIEV**. The time taken for these processes is known as **PIEV** time.

Q. 3. What is perception?

Ans. Perception is the process of perceiving the sensations received through the eyes, ears, nervous system and the brain. The exact time required for this is dependent upon the individuals psychological and physiological build-up.

Q. 4. What is volition?

Ans. Volition is the will to react to a situation.

Q 5. What do you mean by rolling resistances?

Ans. The force that resists the motion when a body rolls on a surface is called the rolling resistance or the rolling friction.

Q. 6. How can the rolling resistance be expressed?

Ans. The *rolling resistance* can be expressed as:

$$F_r = c\ W \qquad ...(1)$$

where, F_r = rolling resistance or rolling friction (N, lb_r)

c = rolling resistance co-efficient – dimensionless (co-efficient of rolling friction – CRF)

$W = mg$ = normal force – weight of the body (N, lb_f)

m = mass of body (kg, lb)

g = acceleration of gravity (9.81 m/s^2, 32.174 ft/s^2)

The rolling resistance can alternatively be expressed as

$$F_r = c_1\ W/r \qquad ...(2)$$

where, c_1 = rolling resistance co-efficient with dimension length (co-efficient of rolling friction) (mm, in)

r = radius of wheel (mm, in)

Q. 7. What is air resistance?

Ans. When a vehicle is in motion, air resists it in the following ways;

(*i*) since air has density, it exerts a reaction pressures against the front to the vehicle.

Q. 8. What are the driving tips for city driving in slow traffic?

Ans. Driving in slow moving city traffic can be quite a workout session. City driving puts lots of stresses on a driver as he has to constantly be mindful of his environment and frequently keep making adjustments to his driving style. The slow moving traffic in India consists of cycles, cycle rickshaws, bullocks carts and horse carts.

Most Indian cars are all manual transmission variants, with automatics making up a very minute population of cars – less than 0.01% of the vehicular density in a large city like New Delhi. In such a scenario, CarToq offers you some tips on how to handle the stresses of city traffic.

Q. 9. Why did the licence weight definition change?

Ans. Changing the definition of licence weight greatly simplified administration for both industry and government agencies.

Q. 10. Why a tyre has some resistance to the motion of a bike as it rolls over the ground.

Ans. Anything on wheels experiences rolling resistance because there is a friction, but anything with the wheel, where the wheel touches the ground, it deforms slightly. It's got a flat section near the ground, and as you roll it over, that flat section moves around the tyre, and so you are constantly having to change the shape of the tyre. So that gives you an extra resistive force, and you notice that, if you leave your bicycle tyres to go flat, it's much harder to pedal than if they are well pumped up and quite firm. That's rolling resistance is what you are feeling.

Q. 11. What is transmission losses?

Ans. Transmission loss is defined as the difference between the power incident on a duct acoustic device (muffler) and that transmitted downstream into an anechoic termination.

Q. 12. Is there any change to how high productivity vehicles are dealt with?

Ans. Vehicles that are used consistently to carry overweight loads have the option of having a special "H vehicle" type RUC licence. This enables them to carry weights up to the maximum under a high productivity vehicle permit at all times. Vehicles that only operate overweight occasionally, or carry varying overweight loads, have the option of purchasing additional RUC licences as required to cover specific journeys.

III. REVIEW QUESTIONS

1. What are the human factors which govern road user behaviour?
2. What do you mean by rolling resistance?
3. Discuss the power requirement of the vehicle?
4. Discuss the characteristics of vehicle.
5. What are the characteristics of slow moving traffic in India?

REFERENCES

1. A Policy on Geometric Design of Highways and Streets. American Association of State Highway and Transportation Officials, Washington, 1994.
2. Babkov, V. and M. Zamakhayev, Highway Engineering, MIR Publishers, Moscow, 1967.
3. Kadiyali, L.R., *et al*, Rolling resistance and air resistance factors for Indian road and vehicle conditions, Highways, Research Bulletin 19, Indian Roads Congress, New Delhi, 1982.
4. Winfrey, R., Economic Analysis for Highways, International Textbook Company, Scranton, 1969.
5. Wilson, E.E., Deceleration Distances for High Speed Vehicles, Proceedings, Highway Research Board, Vol. 20, 1940, Washington.
6. Kadiyali, *et al*, Some characteristics of slow moving traffic on Indian Road, RUCS Technical Paper No. 47, Eleventh Quarterly Report, Road User Cost Study, Central Road Research Institute, New Delhi, 1980.
7. Study for updating Road User Cost Data, Ministry of Surface Transport, Report prepared by Dr. L.R. Kadiyali and Associates, New Delhi, 1992.
8. Recommended Practice for Sight Distances on Rural Highways, IRC : 66—1976, Indian Roads Congress, New Delhi, 1976.

SECTION 2

TRAFFIC SURVEYS

3. Speed, Journey Time and Delay Surveys

3.1. Introduction

Speed is one of the most important characteristics of traffic and its measurement is a frequent necessity in traffic engineering studies. Speed is the rate of movement of traffic or of specified components of traffic and is expressed in metric units in kilometres per hour (K.P.H.). First of all it is necessary to know the different types of speed measurement met with in traffic engineering and their exact meaning.

Spot speed is the instantaneous speed of a vehicle at a specified location.

Running speed is the average speed maintained by a vehicle over a given course while the vehicle is in motion. It is significant to note the clause 'while the vehicle is in motion', because the running speed is obtained by dividing the length of the course by the time the vehicle is in motion, *i.e.,* by the running time, which excludes that part of the journey time when the vehicle suffers delay. Thus,

$$\text{Running speed} = \frac{\text{Length of course}}{\text{Running time}} = \frac{\text{Length of course}}{\text{Journey time} - \text{delay}}$$

Journey speed, also known as overall travel speed, is the effective speed of a vehicle between two points, and is the distance between two points divided by the total time taken by the vehicle to complete the journey, including all delays incurred enroute. Thus :

$$\text{Journey speed} = \frac{\text{Distance}}{\text{Total journey time (including delays)}}$$

Time-mean speed is the average of the speed measurements at one point in space over a period of the time. It is the average of a number of spot speed measurements.

Space-mean speed is the average of the speed measurements at an instant of time over a space.

The relation between time-mean speed and space-mean speed discussed in a later Chapter (Chapter 9).

3.2. Use of Speed, Journey Time and Delay Studies

Speed measurements are needed for a number of purposes given below :

3.2.1. Spot speeds

(*i*) For geometric design of roads, it is necessary to have a realistic estimate of the speeds at which vehicles travel. Based on the speed studies, the design speed can be selected, and other geometric elements of design such as horizontal curvature, vertical profile, sight distances and super elevation can be determined.

(*ii*) For regulation and control of traffic operations, spot speed studies are needed. They enable the safe speed limits to be established and speed zoning to be determined. Traffic signal designs require speed data. Location and size of traffic signs are dependent upon speed.

(*iii*) For analysing the causes of accidents and identifying any relation between speed and accidents, spot speed data are needed.

(*iv*) For before and after studies of road improvement schemes it is necessary to have spot speed data, enabling a meaningful analysis to be carried out of the effect of improvements.

(*v*) For determining the problems of congestion on roads and relating capacity with speeds, accurate speed data are useful.

3.2.2. Journey speeds and delays

(*i*) The cost of a journey depends upon the speed at which it is made. In all highway economic studies, journey speeds and delays are highly important.

(*ii*) Journey-time studies on a road network in a town are useful to evaluate congestion, capacity, level of service and the need for improvements.

(*iii*) In transportation planning studies, the determination of the travel time is necessary for carrying out the trip assignment. Also, travel time and delays are some of the factors affecting modal choice.

(*iv*) Before-and-after studies pertaining to journey time are useful for assessing the effectiveness of improvement measures.

(*v*) Delay studies at intersections provide data for the design and installation of the appropriate traffic control device.

3.3. Methods of Measuring Spot Speeds

3.3.1. The methods available for measuring spot speeds can be grouped as under :

(*a*) Those that require observation of the time taken by a vehicle to cover a known distance.

(*b*) Radar speedmeter which automatically records the instantaneous speed.

(c) Photographic method.

Methods under (*a*) above can be further classified into :

(*i*) those in which vehicles are timed over a long distance (long base), and

(*ii*) those in which vehicles are timed over a short distance (short base).

The long-base methods commonly used are :

(*i*) Direct timing procedure. (*ii*) Enoscope.

(*iii*) Pressure contact tubes.

The short-base method uses vehicle detectors over a short distance.

3.3.2. In all the above methods, there are some general considerations for selecting the site of the survey. They are :

(*i*) The location at which measurements are taken is governed by the specific purpose for which the data are required. For example, if the data is likely to be used to analyse the accident pattern, the survey will be needed at high accident frequency locations.

(*ii*) The site selected should reduce to a minimum the influence of the study team and equipment upon the vehicle speeds. For this purpose, the study team and the equipment should be located as inconspicuously as possible.

(*iii*) Generally straight, level and open sections of highways should be selected so as to minimise the influence of geometric features of the highways, road-side development and intersections.

3.3.3. Recommended base lengths

The following base lengths for the long-base methods are adapted from the recommendations of the Traffic Engineering Hand-book (Ref. 1) :

Table 3.1. Recommended Base Lengths for Spot Speed Survey

(*Source* : Adapted from Ref. 1)

Average Speed of Traffic Stream (K.P.H.)	*Base length (metres)*
Less than 40	27
40 to 65	54
Greater than 65	81

3.4. Direct-Timing Procedure for Spot Speed Determination

This is one of the simplest of methods for spot speed determination. Two reference points are marked on the pavement at a suitable distance apart and an observer starts and stops and accurate stop-watch as a vehicle crosses these two marks. From the known distance and the measured time intervals speeds are calculated. Skilled observers can read a stop-watch to an accuracy of 0.2 second. If the observer stations himself inconspicuously, the speed readings are not influenced by driver reaction. The disadvantage with this method is that large errors are likely to be introduced because of the parallax effect.

A simple variation of this method is to station two observers, one at each reference point. The observer standing at the reference point which the vehicles pass first, signals that a vehicle to be timed is passing the point and the second observer then starts a stopwatch. The second observer stops the stop watch when he observes the same vehicle passing the reference point. The disadvantage with this method is that it involves the reaction time of two individual observers.

3.5. Enoscope

A simple device called Enoscope eliminates the parallax effect that creeps in when the direct readings are taken by one observer. This device, also known as the Mirror-box, is an *L*-shaped box, open at both ends, with a mirror set at a 45-degree angle to the arms of the instrument, as shown in Fig. 3.1.

The instrument bends the line of sight of the observer so that it is perpendicular to the path of the vehicle. The method can be used with one enoscope or with two enoscopes. If one enoscope is used, the instrument is placed directly opposite the first reference point and the observer stations himself at the other reference point, as shown in Fig. 3.2.

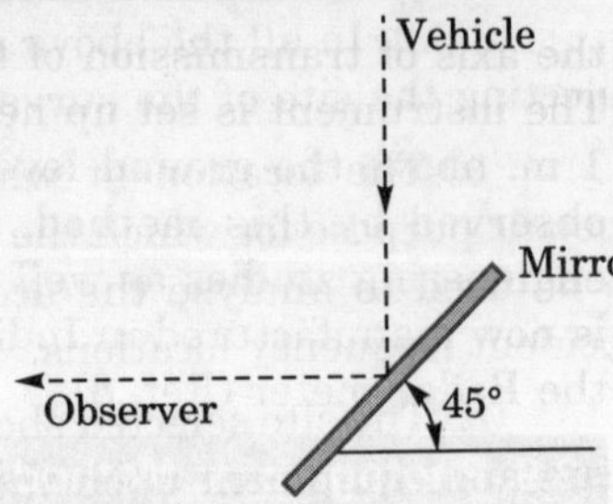

Fig. 3.1. Enoscope.

The stop-watch is started as soon as the vehicle passes the first reference point and is stopped as soon as it passes the observer. If two enoscopes are used, the observer stations himself mid-way between the two reference points and starts the stop-watch as soon as a vehicle crosses the first reference point and stops the stop-watch when It crosses the second reference point.

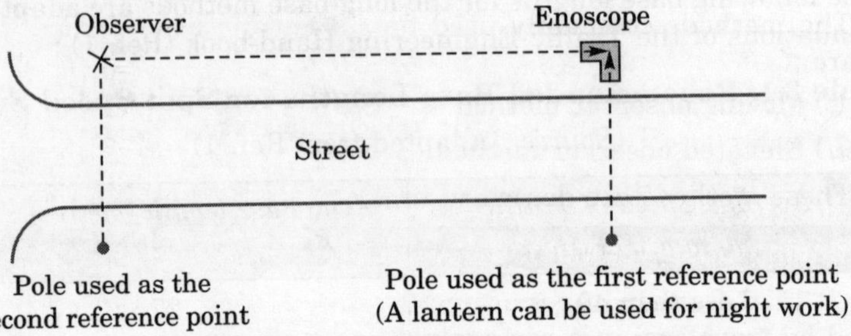

Fig. 3.2. Use of one enoscope.

3.6. Pressure Contact Tubes

In this method, detectors (usually pneumatic tubes) are used to indicate the time of entering and leaving the base length. When a vehicle passes over the tube laid at the first reference point, an air impulse is sent, which activates an electromagnetically controlled stop-watch in the hands of the observer. When the vehicle passes over the second tube, the stop watch automatically stops. The reading is noted by the observer. Alternatively, the readings can be recorded by automatic data recorders;. The disadvantage with the pressure contact tubes is that they are easily seen by drivers and this may affect their behaviour.

3.7. Short-Base Methods for Determining Spot Speeds

With the short-base methods, the base length for measuring the speeds is very short, say about 2 m. The instruments are electronic and are used in conjunction with pneumatic tubes or electric detectors laid across the pavement. Recording can be manual, the instantaneous speeds being noted by the observer or can be automatic with an electronic system.

3.8. Radar Speed Meters

These meters work on the Doppler principle that the speed of a moving body is proportional to the change in the frequency between the radio wave transmitted to the moving body and the radio wave received back. The instrument directly measures the speed, and an accuracy of at least ± 1.5 to 3 K.P.H. is possible (Ref. 2). The instrument is battery operated and is portable. The speed meter is so kept that the angle between the direction of travel of the vehicle and the axis of transmission of the radio wave is as low as possible, say within 20°. The instrument is set up near the edge of the carriage way at a height of about 1 m, above the ground level. The speeds of vehicles in both directions can be observed by this method. The device has been extensively used for traffic engineering studies as well as for enforcement by traffic police. The instrument is now manufactured in India. Some important speed studies in India have used the Radar meter (Ref. 8).

3.9. Photographic Method and Video Camera Method

Time-lapse camera photography has been used successfully to determine the speeds of vehicles accurately in crowded streets. According to this method, photographs are taken as fixed intervals of time (say one second per frame) on a special camera. By projecting the film on a screen, the passage of any vehicle can be traced with reference to time. Images by video cameras can also be used.

3.10. Methods for Measurement of Running Speed and Journey Speed

The methods commonly used for measuring running speed and journey speed are :

(*i*) Moving observer method (*ii*) Registration number method

(*iii*) Elevated observer method.

These methods are described in detail in the following sections.

3.11. Moving Observer Method

3.11.1. General procedure. In this method, the speed and flow can be obtained by travelling in a car against and with the flow, and noting down the journey time, the number of vehicles met with from the opposite direction, and number of vehicles overtaking the test vehicle (Ref. 2–6). The theory behind the method will first be described before the detailed field procedure is outlined.

3.11.2. Theory behind the method. Consider a stream of traffic moving along a section of road, of length *l,* in such a way that the average number of vehicles passing through the section per unit time is constant.

The stream can be regarded as consisting of flows q_1 moving with speed V_1, q_2 moving with speed V_2, etc.

Suppose an observer travels with the stream at speed v_w, and against the stream at speed v_a. Then the flows relative to him of vehicles with flow q_1 and speed v_1 are :

$$= \frac{q_1 (v_1 - v_w)}{v_1} \quad \text{and} \quad \frac{q_1 (v_1 - v_a)}{v_1}$$

respectively.

If, t_w = journey time, l/v_w corresponding to speed v_w,

t_a = journey time, l/v_a corresponding to speed v_a,

x_1 = number of vehicles with speed v_1 met by the observer when travelling against the stream,

y_1 = number of vehicles with speed v_1 which overtake the observer when travelling with the stream minus the number he overtakes,

t_1 = journey time corresponding to speed $v_1 = l/v_1$

Then $x_1 = q_1 (t_a + t_1)$ ...(3.1)

$y_1 = q_1 (t_w - t_1)$ ...(3.2)

Similar results hold good for all speeds, so that

q = flow of all vehicles in the stream

$= q_1 + q_2 + ...$

x = total number of vehicles met in the section when travelling against the stream.

$= x_1 + x_2 ......$

y = number of vehicles overtaking the observer minus the number he overtakes when travelling with the stream,

$= y_1 + y_2 ......$

$\bar{t}$ = mean journey time of all vehicles in the stream

$$= \frac{(q_1 t_1 + q_2 t_2 + ...)}{q}$$

Then summing up,

$$x = x_1 + x_2 + ...$$

$$= q_1 (t_a + t_1) + q_2 (t_a + t_2) ......$$

$$= (q_1 + q_2 + ...) (t_a + q_1 t_1 + q_2 t_2 ......)$$

$$= q t_a + q \bar{t} = q (\bar{t} + t_a) \quad ...(3.3)$$

$$y = y_1 + y_2 + ... = q_1 (t_w - t_1) + q_2 (t_w - t_2) + ...$$

$$= (q_1 + q_2 + ...) t_w - (q_1 t_1 + q_2 t_2 ......)$$

$$= q t_w - q \bar{t} = q(t_w - \bar{t}) \quad ...(3.4)$$

Therefore, $x + y = q(t_a + t) + q(t_w - t)$

$$= q(t_a + t_w) \quad ...(3.5)$$

$$\therefore \quad q = \frac{(x + y)}{(t_a + t_w)} \quad ...(3.6)$$

and, $$\bar{t} = t_w - \frac{y}{q} \quad ...(3.7)$$

$$\text{The mean journey speed} = \frac{l}{t} \quad ...(3.8)$$

and this is the space mean speed.

In the above formulae, it is useful to affix subscripts n and s, signifying north bound and south bound traffic respectively.

Thus, $$q_n = \frac{x_s + y_n}{t_s + t_n} \quad ...(3.9)$$

where, q_n = flow in the north bound direction

x_s = opposing traffic count of vehicles met when the test car was travelling south

y_n = number of vehicles overtaking the test car minus the number overtaken by the test car, when the test car is travelling north.

Also $$\bar{t}_n = t_n - \frac{y_n}{q_n} \quad ...(3.10)$$

where $\bar{t}_n$ = mean journey time in the north bound direction

t_n = journey time when the test car travelled in the north bound direction

and y_n, q_n have the meaning as above.

3.11.3. Detailed field procedure. A small, preferably even, number of test cars is required usually two-each car carrying a driver and three observers. One observer in the car counts opposing traffic, using hand tallies. Another observer carries a recording board on which a watch is mounted. The recording board carries a journey log prepared in advance, on which the observer records the totals from the hand tallies and times at predetermined points en-route, together with the times of stopping and starting at intersections. It is desirable to have two stop watches, one for recording the continuous time at different points enroute, and the other for measuring the individual stopped time delays. The

Route Plan : A–B–C–D–E–F–G–H–I–J–K–L–H–M–N–O–H–G–F–D–C–P–Q–R–S–P–O–F.

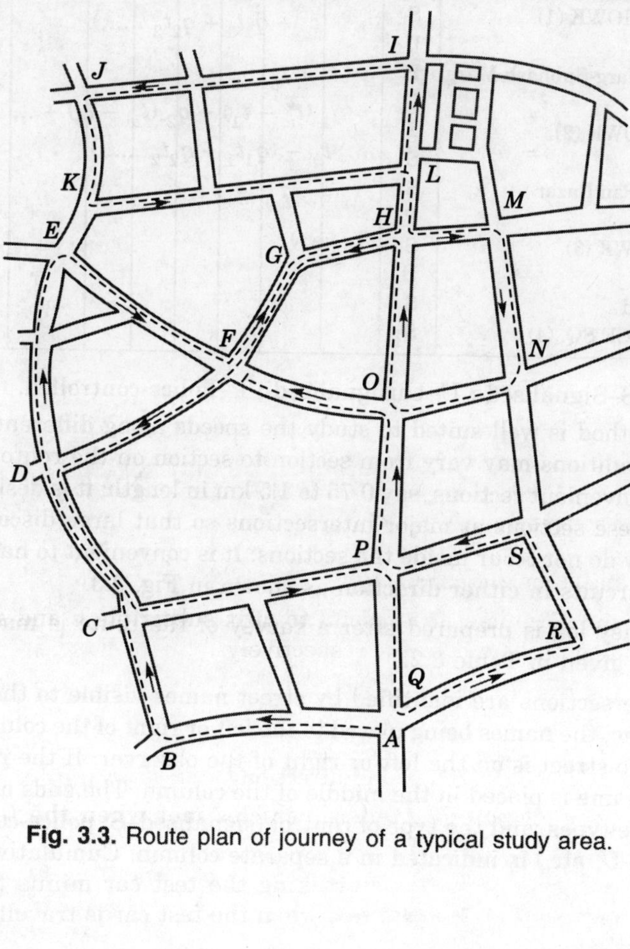

Fig. 3.3. Route plan of journey of a typical study area.

second watch can accumulate the delay time as the observer operates buttons. A third observer records the number of overtaking and overtaken vehicles, and if required, the number of parked vehicles. If only two observers are available, the driver is instructed to overtake as many vehicles as do overtake him and no record need be kept of those vehicles.

Table 3.2. A Typical Journey Log for the Moving Observer Method of Speed and Volume Study

Day	Vehicle No.	Route
Data		Recorder
Time of start	Driver	Observer

Timing point intersection	*Types of control*	*Journey time*	*Stopped time*	*Opposing traffic*				*Same direction traffic*		*Parking*
				Cars	*Trucks*	*Buses*	*Cycles*	*Overtaking vehicles*	*Vehicles passed by the test car*	
GANDHI CHOWK (1)	S									
Bank St.	U									
Saraswati Marg-Subhash Marg	P									
Tilak St.	U									
PATEL CHOWK (2)	S									
School Lane	S									
Market St.-Pan Bazar	S									
Vinay Marg	P									
PANT CHOWK (3)	S									
Purvi Marg	U									
College Road	S									
POST OFFICE SQ. (4)	S									

Note : S–Signalised ; U–Unsignalised ; P–Police-controlled.

The method is well suited to study the speeds along different roads in an area. Since conditions may vary from section to section on the route, the route is divided into convenient sections, say 0.75 to 1.5 km in length; it is desirable to have the ends of these sections at major intersections so that large discontinuities in speed and flow do not occur inside the sections. It is convenient to have the routes to be closed circuits in either direction as shown in Fig. 3.3.

A journey log is prepared after a survey of the route is made. A typical journey log is given in Table 3.2.

All intersections are identified by street names visible to the recorder as he travels along, the names being placed to the left or right of the column according to whether the street is on the left or right of the observer. If the route is along the road, its name is placed in the middle of the column. The ends of sections are shown in large types, and the type of control (signalised–*S*, police-controlled —*P*, uncontrolled –*U*, etc.) is indicated in a separate column. Cumulative tally totals and watch times of stopping and starting at all intersections are recorded. The distance is measured as recorded by the speedometer or from Survey Maps.

It is desirable that twelve to sixteen runs in each direction along the route be made and the results averaged out so as to arrive at an accurate estimate of the speed and flow.

3.11.4. Advantages of the method

(*i*) The method gives an unbiased estimate of the flow. Random errors can, however, occur due to observer's errors and random fluctuations in flow, but these are not serious under normal conditions.

(*ii*) As compared to the stationary observer method, the moving observer method is equivalent to a stationary count over twice the single journey time. Hence it is economical in manpower.

(*iii*) It enables data on speed and flow to be collected at the same time. This is particularly advantageous when analysing the relations between the two.

(*iv*) It gives mean values of flow and speed over a section, rather than at a point. Thus it gives directly the space mean speed, whereas spot speed studies gives the time mean speed.

(*v*) It gives additional information on stops at intersections, delays, parked vehicles etc.

3.12. Illustrative Problems

Problem 1. *The following tables give the particulars collected for a section of road 0.7 km long during the course of a moving observer study :*

Table 3.3

Journey : North bound

Run No.	*Journey Time (minutes)*	*Stopped Time (minutes)*	*Vehicles met with in the opposing direction*			*Vehicles in the same direction*	
			Car	*Bus*	*Truck*	*Overtaking Vehicles*	*Overtaken Vehicles*
1.	1.01	0.04	11	0	5	1	0
2.	0.92	0.10	13	0	0	2	1
3.	0.77	0.08	19	2	11	1	1
4.	0.03	0.14	14	2	4	1	0
5.	0.84	0.08	2	0	11	0	1
6.	1.06	0.13	19	1	7	2	1
Total	5.63	0.57	78	5	38	7	4
			PCU = 78 + 5 × 3 + 38 × 2 = 169				
Av.	0.94	0.10	$\frac{169}{6}$ = 28			1.2	0.7

Table 3.4

Journey : North bound

Run No.	Journey Time (minutes)	Stopped Time (minutes)	Vehicles met with in the opposing direction			Vehicles in the same direction	
			Car	Bus	Truck	Overtaking Vehicles	Overtaken Vehicles
1.	1.00	0.05	10	0	2	1	1
2.	0.87	0.07	2	0	3	1	0
3.	1.20	0.11	23	1	6	2	1
4.	1.18	0.12	7	0	1	2	0
5.	1.06	0.09	8	0	1	1	1
6.	1.02	0.10	11	0	8	2	0
Total	6.33	0.54	61	1	21	9	3
			PCU = 61 + 1 × 3 + 21 × 2 = 106				
Av.	1.06	0.09	$\frac{109}{6} = 18$			1.5	0.5

Calculate the flow in PCU per hour in both directions of traffic assuming an equivalency factor of one per car, 3 for bus and 2 for trucks. Calculate the journey speed and running speed.

Solution. The conversion of the traffic into PUCs is done in the above tables.

Using the formula, $q = \dfrac{x + y}{(t_a + t_w)}$.

and using the suffix n for north bound and s for south bound traffic,

$$q_n = \frac{(x_s + y_n)}{(t_a + t_w)}$$

$$= \frac{18 + (1.2 - 0.7)}{1.06 + 0.94} = \frac{18 + 0.5}{2.00}$$

$$= 9.25 \text{ PCUs/min} = 555 \text{ PCUs/hour}$$

$$q_s = \frac{x_n + y_s}{(t_n + t_s)} = \frac{28 + (1.5 - 0.5)}{2.0}$$

$$= \frac{28 + 1}{2} = 14.50 \text{ PCUs/min}$$

$$= 870 \text{ PCUs/hour}$$

$$\bar{t}_n = t_n - \frac{y_n}{q_n} = 0.94 - \frac{(1.2 - 0.7)}{9.25}$$

$$= 0.94 - \frac{0.5}{9.25} = 0.94\text{–}0.05 = 0.89 \text{ min.}$$

∴ Mean journey speed in north bound direction

$$= \frac{0.7 \times 60}{0.89} = 47 \text{ K.P.H.}$$

$$\bar{t}_s = t_s - \frac{y_s}{q_s} = 1.06 - \frac{1.5 - 0.5}{14.5}$$

$$= 1.06 - 0.07 = 0.99 \text{ min.}$$

∴ Mean journey speed in south bound direction

$$= \frac{0.70}{0.99} \times 60 = 42.5 \text{ K.P.H.}$$

∴ Mean running time in north bound direction

$\bar{t}_s$, Stopped time = 0.89 – 0.10 = 0.79 min.

∴ Running speed in north bound direction

$$= \frac{0.70}{0.79} \times 60 = 53 \text{ K.P.H.}$$

Mean running time in south bound direction

$\bar{t}_s$ – Stopped time = 0.99 – 0.09 = 0.90 min

Running speed in south bound direction

$$= \frac{0.70}{0.90} \times 60 = 46.7 \text{ K.P.H.}$$

Problem 2. *ABCD is a street running north to south comprising of 3 adjacent sections. Table 3.5 below gives the length of each section :*

Table 3.5

Section	*Length (m)*
AB	*500*
B	*Intersection*
BC	*600*
C	*Intersection*
CD	*450*

Tables 3.6 and 3.7 give the details of observations (average of test runs) by the moving observer team :

Table 3.6. Moving Observer Travelling South

Section	*Time taken to traverse (sec.)*	*Vehicles met with in opposing direction*	*Vehicles in the same direction*	
			Vehicles overtaking	*Vehicles overtaken*
AB	152	25	4	3
B	15	–	–	–
BC	172	31	2	2
C	10	–	–	–
CD	138	28	3	4

Table 3.7. Moving Observer Travelling North

Section	*Time taken to traverse (secs.)*	*Vehicles met with in opposing direction*	*Vehicles in the same direction*	
			Vehicles overtaking	*Vehicles overtaken*
DC	132	50	2	1
C	12	–	–	–
CB	160	55	3	4
B	18	–	–	–
BA	145	45	1	1

Calculate the flows and running speeds in each section in both directions and the overall journey speeds in each direction.

Solution. Flows in north bound direction

$$q_n \text{ in section } BA, = \frac{x_s + y_n}{t_s + t_n} = \frac{25 + (1-1)}{152 + 145} \times 3600 \text{ veh/hour}$$

$$= \frac{25}{297} \times 3600 = 303 \text{ veh/hour}$$

$$q_n \text{ in section } CB, = \frac{31 + (3-4)}{172 + 160} \times 3600 \text{ veh/hour}$$

$$= \frac{30}{332} \times 3600 \text{ veh/hour} = 326 \text{ veh/hour}$$

$$q_n \text{ in section } DC, = \frac{28 + (2-1)}{138 + 132} \times 3600 \text{ veh/hour}$$

$$= \frac{29}{270} \times 3600 \text{ veh/hour} = 387 \text{ veh/hour.}$$

Flows in south bound direction

$$q_s \text{ in section } AB, = \frac{45 + (4-3)}{152 + 145} \times 3600 \text{ veh/hour}$$

$$= \frac{46}{297} \times 3600 \text{ veh/hour} = 557 \text{ veh/hour.}$$

$$q_s \text{ in section } BC, = \frac{55 + (2-2)}{172 + 160} \times 3600 \text{ veh/hour}$$

$$= \frac{55}{332} \times 3600 \text{ veh/hour} = 596 \text{ veh/hour.}$$

$$q_s \text{ in section } CD, = \frac{50 + (3-4)}{138 + 132} \times 3600 \text{ veh/hour}$$

$$= \frac{49}{270} \times 3600 \text{ veh/hour} = 654 \text{ veh/hour}$$

Running speeds, $\bar{t}_n = t_n - \dfrac{y_n}{q_n}$

$$\bar{t}_n \text{ in section } BA, = 145 - \frac{(1-1)\times 3600}{303} = 145 \text{ seconds.}$$

$\therefore$ **Running speed of north bound traffic in section** BA

$$= \frac{500}{145} \times \frac{3600}{1000} = 12.4 \text{ K.P.H.}$$

$$\bar{t}_n \text{ in section } CB = 160 - \frac{(3-4)}{326} \times 3600 = 160 + 11$$

$$= 171 \text{ seconds.}$$

$\therefore$ **Running speed of north bound traffic in section** CB

$$= \frac{600}{171} \times \frac{3600}{1000} = 12.6 \text{ K.P.H.}$$

$$\bar{t}_n \text{ in section } CB = 160 - \frac{(3-4)}{326} \times 3600 = 160 + 11$$

$$= 122.7 \text{ seconds.}$$

$\therefore$ **Running speed of north bound traffic in section** DC

$$= \frac{450}{122.7} \times \frac{3600}{1000} = 13.2 \text{ K.P.H.}$$

$$\bar{t}_n \text{ in section } AB = 152 - \frac{(4-3)\times 3600}{557} = 152 - 6.5$$

$$= 145.5 \text{ seconds.}$$

$\therefore$ **Running speed of south bound traffic in section** AB

$$= \frac{500}{145.5} \times \frac{3600}{1000} = 12.4 \text{ K.P.H.}$$

$$\bar{t}_s \text{ in section } BC = 172 - \frac{(2-2)\times 3600}{596} = 172 \text{ seconds.}$$

$\therefore$ **Running speed of south bound traffic in section** BC

$$= \frac{600}{172} \times \frac{3600}{1000} = 12.6 \text{ K.P.H.}$$

$$\bar{t}_n \text{ in section } CD = 138 - \frac{(3-4)\times 3600}{654} = 138 + 5.5$$

$$= 143.5 \text{ seconds.}$$

$\therefore$ **Running speed of south bound traffic in section** CD

$$= \frac{45}{143.5} \times \frac{3600}{1000} = 11.3 \text{ K.P.H.}$$

Journey Speeds

Journey time when travelling south from A **to** D

$$= 145.5 + 172 + 143.5 + 15 + 10 = 486 \text{ seconds.}$$

Journey speed in south bound direction

$$= \frac{(500 + 600 + 450)}{486} \times \frac{3600}{1000}$$

$$= \frac{1550}{486} \times \frac{3600}{1000} = 11.5 \text{ K.P.H.}$$

Journey time when travelling north from D to A

$$= 145 + 171 + 122.7 + 12 + 18 = 468.7 \text{ seconds.}$$

∴ Journey speed in northwards direction

$$= \frac{1550}{468.7} \times \frac{3600}{1000} = 11.9 \text{ K.P.H.}$$

3.13. Registration Number Method

3.13.1. In this method observers are stationed at the ends of a measuring section, about 0.5 – 1.00 km long (Ref. 9). The time and registration number of the vehicles entering and leaving the section are noted using synchronized watches. By matching the registration numbers later, the journey times and speed can be determined. If the traffic is too heavy for being recorded, some method of sampling can be adopted. For instance, noting down only numbers ending with an odd number will yield a 50 percent sample. The vehicle classification can also be noted in this method thus yielding this additional information. The accuracy that is possible in recording time is about + 0.01 minutes, which then gives an accuracy of 1 to 2 percent for journeys lasting about one minute. The watches are supported on the recording board and are allowed to run continuously. Watches divided into tenths and hundredths of a minute are desirable as they facilitate subsequent analysis. A duration of more than two hours is likely to cause undue strain on the observers. Two observers are normally employed for each direction of travel, one for noting the registration number and calling the same, the other for recording it alongwith the time. Two observers can record at the rate of about 300 vehicles per hour (Ref. 7).

3.13.2. The advantage of the method is that no sophisticated instruments are needed except stop watches. The analysis is however, laborious and time consuming, but can be rendered easier by the help of computers. The method can only be used on highway sections having minor or no intersections, since the vehicles may enter, leave or stop within the section having intersections. Hence the method is generally suitable for rural highway sections. It fails to yield data on delays, and hence records journey speeds only and not running speeds.

3.14. Elevated Observer Method

In this method, the observers stationed on top of an elevated building select vehicles at random and follow their course along the road, noting the time of entering the test section, duration and nature of delays suffered and the time of leaving. The test section has to be short, such as a street in the central area of a city.

3.15. Presentation of Travel Time and Journey Speed Data

The journey speed and running speed data pertaining to a road network in an area can be conveniently represented in the graphical form shown in Fig. 3.4.

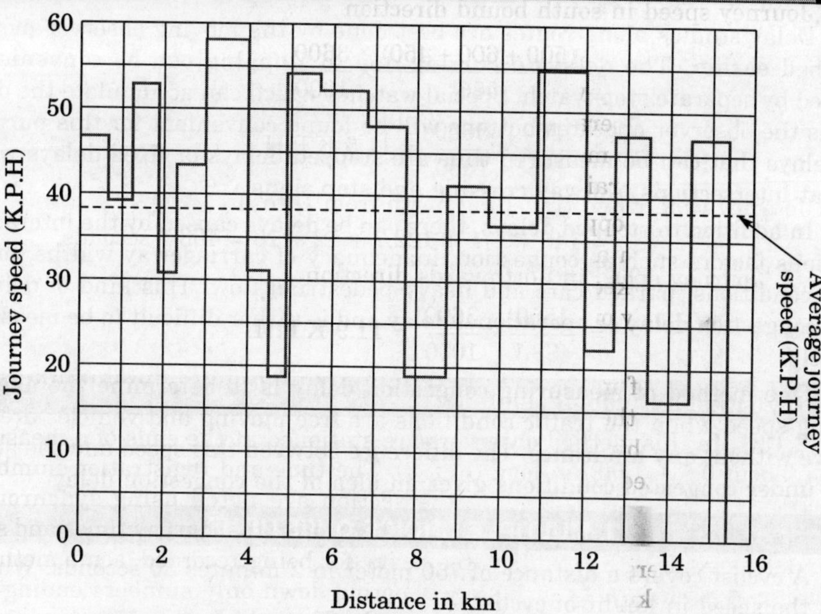

Fig. 3.4. Graphical presentation of journey speed data.

The actual time taken for journeys to various points from a common point of origin or destination can be depicted in a time-zone map. In this map, the time taken is shown in the form of contours, shown in Fig. 3.5, and known as Isochrones.

3.16. Delay Studies

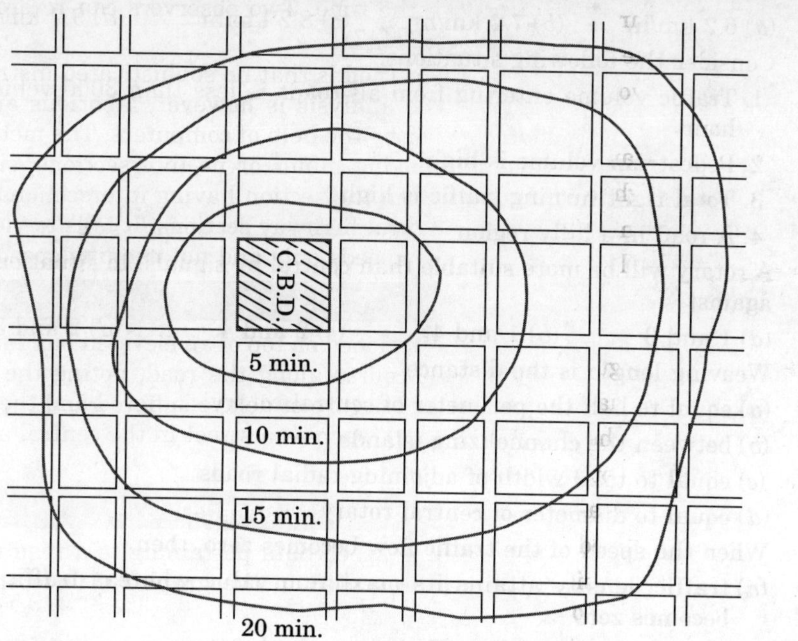

Fig. 3.5. Time zone map around a C.B.D.

Delay studies along routes are best done by the moving observer method described earlier. The delays occurring due to stopping can be conveniently recorded by separate stop-watch. Special watches which can accumulate the delay time as the observer operates buttons will be found convenient for this purpose. The delays that can be measured thus are stopped delays or fixed delays which occur at intersections, railway crossing and stop signs.

In addition to stopped delays, there can be delays caused by the interaction of various factors such as congestion, inadequacy of carriageway widths, mixed traffic conditions, parked cars and heavy pedestrian flow. This kind of delay is called congestion delay or operational delay and is rather difficult to be measured precisely.

One method of measuring congestion delay is to determine the average running speed when the traffic conditions are free moving and vehicles are free to move without any hindrance. The difference between this speed and the actual speed under congested conditions gives an idea of the congestion delay.

I. MULTIPLE CHOICE QUESTIONS

1. A cyclist covers a distance of 750 meter in 2 minutes 30 seconds. What is the speed in km/hr of cyclist

(*a*) 16 km/hr (*b*) 17 km/hr (*c*) 18 km/hr (*d*) 19 km/hr

2. A car moves at 80 km/hr. What is the speed of the car in meters per second?

(*a*) $20\frac{2}{9}$ m sec (*b*) $22\frac{2}{9}$ m sec (*c*) $24\frac{2}{9}$ m sec (*d*) $26\frac{2}{9}$ m sec

3. An athlete runs 200 meters in 24 seconds. His speed is?

(*a*) 10 km/hr (*b*) 17 k/hr (*c*) 27 km/hr (*d*) 30 km/hr

4. A person crosses a 600 meter long street in 5 minutes. What is the speed in km/hr?

(*a*) 6.2 km/hr (*b*) 7.2 km/hr (*c*) 8.2 km/hr (*d*) 9.2 km/hr

5. Consider the following situations:

1. Traffic volume entering from all roads is less than 3000 vehicles per hour
2. Pedestrian volume is high
3. Total right turning traffic is high
4. A road in a hilly region

A rotary will be more suitable than control by signals, in situations listed against

(*a*) 1 and 3 (*b*) 1 and 4 (*c*) 2 and 4 (*d*) 2 and 3

6. Weaving length is the distance

(*a*) equal to half the perimeter of central rotary

(*b*) between the channelizing islands

(*c*) equal to total width of adjoining radial roads

(*d*) equal to diameter of central rotary

7. When the speed of the traffic flow becomes zero, then

(*a*) traffic density attains its maximum value whereas traffic volume becomes zero

(*b*) traffic density and traffic volume both attain respective maximum values, (*c*) traffic density and traffic volume both become zero

(*d*) traffic density becomes zero whereas traffic volume attains its maximum value

Answers

1. (*c*)	**2.** (*b*)	**3.** (*d*)	**4.** (*b*)	**5.** (*d*)
6. (*b*)	**7.** (*a*)			

II. QUESTIONS AND ANSWERS

Q. 1. What is speed?

Ans. Speed is one of the most important characteristics of traffic and its measurement is a frequent necessity in traffic engineering studies. Speed is the rate of movement of traffic or of specified components of traffic and is expressed in metric units in kilometers per hour (K.P.H).

Q. 2. What is spot speed?

Ans. Spot speed is the instantaneous speed of a vehicle at a specified location. Spot speed can be used to design the geometry of road like horizontal and vertical curves, superelevation etc. Location and size of signs, design of signals, safe speed, and speed zone determination, require the spot speed data. Accident analysis, road maintenance, and congestion are the modern fields of traffic engineer, which uses spot speed data as the basic input. Spot speed can be measured using an enoscope, pressure contact tubes or direct timing procedure or radar speedometer or by time-lapse photographic methods. It can be determined by speeds extracted from video images by recording the distance travelling by all vehicles between a particular pair of frames.

Q. 3. What is running speed?

Ans. Running speed is the average speed maintained by a vehicle over a given course while the vehicle is in motion. It is significant to note the clause 'while the vehicle is *n* motion', because the running speed is obtained by dividing the length of the course by the time the vehicle is *n* motion, *i.e.,* by the running time, which excludes that path of the journey time when the vehicle suffers delay.

Q. 4. What is meant by journey speed?

Ans. Journey speed, also known as overall travel speed, is the effective speed of a vehicle between two points and is the distance between two points divided by the total time taken by the vehicle to complete the journey, including all delays incurred en-route.

$$\text{Thus journey speed} = \frac{\text{Distance}}{\text{Total journey time (including delays)}}$$

Q. 5. Define time-mean speed.

Ans. Time-mean speed is the average of the speed measurements at one point in space over a period of the time. It is the average of a number of spot speed measurement.

Q. 6. What is space speed?

Ans. Space-mean speed is the average of the speed measurements at an instant of time over a space.

Q. 7. How journey speeds and delays are related to each other?

Ans. Because

(*i*) The cost of a journey depends upon the speed at which it is made. In all highway economic studies, journey speeds and delays are highly important.

(*ii*) Journey-time studies on a road net-work in a town are useful to evaluate congestion, capacity, level of service and the need for improvements.

(*iii*) In transportation planning studies, the determination of the travel time is necessary for carrying out the trip assignment. Also, travel time and delays are some of the factors affecting modal choice.

(*iv*) Before and after studies pertaining to journey time are useful for assessing the effectiveness of improvement measures.

Q. 8. What is speed and travel time?

Ans. Speed and travel time are the most commonly used indicators of performance for traffic facilities and networks. Delays are often used to measure the performance of traffic flow at intersections.

Q. 9. Explain travel time study.

Ans. Travel time is the elapsed time it takes for a vehicle to traverse a given segment of a street. Travel time studies provide the necessary data to determine the average travel time. Combined with the length of the corridor under study, this data can be used to produce average travel speed. Travel time and delay are two of the principal measures of roadway system performance used by traffic engineers, planners and analysts. Since vehicle speed is directly related to travel time and delay, it is also an appropriate measure-of-performance to evaluate traffic systems.

Q. 10. What is delay?

Ans. Delay is defined as an extra time spent by drivers against their expectation. Delay can have many forms depending on different locations. A study made to provide information concerning the amount, cause, location, duration and frequency of delay as well as travel time and similar value. The time lost by traffic due to traffic friction and traffic control device is called delay.

Q. 11. What are the types of delays?

Ans. There are following types of delay:

Congestion delay. Congestion delay is the delay caused by the constricting or slowing down effect of overloaded intersections, inadequate carriageway widths, parked cars, crowded pavement and similar factor.

Fixed delay. The delay to which a vehicle is subjected regardless of the amount of traffic volume and interference present.

Operational delay. The delay caused by interference from other component of the traffic stream. Examples include time lost while waiting for a gap in a conflicting traffic stream, or resulting from congestion, parking maneuvers, pedestrians, and turning movement.

Stopped delay. The time a vehicle is not moving.

Travel time delay. The difference between the actual time required to traverse a section of street or highway and the time corresponding to the average speed of traffic under uncongested condition. It includes acceleration and deceleration delay in addition to stopped delay.

Approach delay. Travel time delay encountered to an approach to an intersection.

Q. 12. What is the purpose of travel time and delay studies?

Ans. The purpose of a Travel Time and Delay Study is to evaluate the quality of traffic movement along a route and determine the locations, types, and extent of traffic delays by using a moving test vehicle.

This study method can be used to compare operational conditions before and after roadway or intersection improvements have been made. It can also be used as a tool to assist in prioritizing projects by comparing the magnitude of the operational deficiencies (such as delays and stops) for each project under consideration.

Q. 13. What are the four basic methods of collecting and processing license plates?

Ans. Four basic methods of collecting and processing license plates normally considered are:

Manual. Collecting license plates via pen and paper or audio tape recorders and manually entering license plates and arrival times into a computer.

Portable computer. Collecting license plates in the field using portable computers that automatically provide an arrival time stamp.

Video with manual transcription. Collecting license plates in the field using video cameras or camcorders and manually transcribing license plates using human observers.

Video with character recognition. Collecting license plates in the field using video, and then automatically transcribing license plates and arrival times into a computer using computerized license plate character recognition.

Q. 14. What is photographic method?

Ans. This method is primarily a research tool, it is useful in studies of interrelationship of several factors such as spacing, speeds, lane usage, acceleration rates, merging and crossing maneuvers, and delays at intersections. This method is applicable to a short test section only.

Q. 15. Explain interview method.

Ans. This method may be useful where a large amount of material is needed in a minimum of time and at little expense for field observation. Usually the employees of a farm or establishment are asked to record their travel time to and from work on a particular day.

Q. 16. What is cycle-based method?

Ans. Cycle-based method is applicable to all under saturated signalized intersections. For over-saturated conditions, queue buildup normally makes the method impractical. The method described here is applicable to situations in which the average maximum queue per cycle is no more than about 20 to 25 veh/ln.

When queues are long or the demand to capacity ratio is near 1.0, care must be taken to continue the vehicle-in-queue count past the end of the arrival count period, vehicles that arrived during the survey period until all of them have exited the intersection.

REFERENCES

1. Traffic Engineering Handbook, Institute of Traffic Engineers, 1965.
2. Research on Road Traffic, Road Research Laboratory, (U.K.) Her Majesty's Stationery Office, London, 1965.

3. Wardrop, J.G. and G. Charlesworth, A Method of Estimating Speed and Flow of Traffic from a Moving Vehicle, Proceedings of the Institution of Civil Engineers, 3, Part 2, 1954, London.
4. Mortimer W.J., Moving Vehicle, Method of Estimating Traffic Volumes and Speeds, Traffic Engineering, 28 (12) 1956.
5. Blensly, R.C. Moving Vehicle Method of Estimating Traffic Volumes, Traffic Engineering, 27 (3), 1956.
6. Williams, T.E.H., and Emerson, J., Traffic Volumes, Journey Time and Speed by Moving-Observer Method, Traffic Engineering and Control, 3 (3) 1961.
7. O'Flaherty, C.A. Highways and Traffic, Vol. I, Edward Arnold, London, 1974.
8. Road User Cost Study in India, Final Report, Vol. VI, Instrumentation, Central Road Research Institute, New Delhi, 1982.
9. Study for updating Road User Cost Data in India, Final Report prepared for the Ministry of Surface Transport, Dr. L.R. Kadiyali and Associates, New Delhi, 1992.

4. Vehicle Volume Counts, Classification and Occupancy

4.1. Need for Vehicle Volume, Classification and Occupation Counts

One of the fundamental measures of traffic on a road system is the volume of traffic using the road in a given interval of time. It is also termed as flow and it is expressed in vehicles per hour or vehicles per day. When the traffic is composed of a number of types of vehicles, it is the normal practice to convert the flow into equivalent passenger car unit (PCUs), by using certain equivalency factors. The flow is then expressed as PCUs per hour or PCUs per day.

A knowledge of the vehicular volume using a road network is important for understanding the efficiency at which the system works at present and the general quality of service offered to the road users. Knowing the flow characteristics one can easily determine whether a particular section of the road is handling traffic much above or below its capacity. If the traffic is heavy, the road suffers from congestion with consequent loss in journey speeds. Lower speeds cause economic loss to the community due to time lost by the occupants of the vehicles and the higher operational cost of vehicles. Congestion also leads to traffic hazards. Volume counts are, therefore, indicators of the need to improve the transport facilities and are an invaluable tool in the hands of a transport planner. They enable him to draw up schemes for improvement of roads based on a system of relative priorities and to allocate the scarce economic resources most advantageously. They provide an equitable basis for the selection of pavement, shoulder and bridge widths.

If traffic flow data are available over the past number of years, the rate at which traffic flow has increased in the past can be easily determined. Extrapolating the past trend into the future, a reasonable indication of the future rate of growth of traffic is made possible. Traffic forecasting is an important step in the transportation planning process.

If the average annual flow and the length of the highway are known, the annual vehicle miles of travel can easily be computed. This is of use in studies pertaining to accident rates. The structural design of the road pavement requires data pertaining to the number of commercial vehicles using a road. Though more accurate analysis demands data pertaining to the axle load spectrum and repetitions, the number of commercial vehicles using the road will itself be a good guide in pavement design.

The programming of the maintenance needs of a highway is often based on the traffic using the road. As an example, the current practice in the Roads Wing of the Government of India may be cited. The requirement of the renewal programme of the road surfaces as part of normal maintenance is assessed by this organization on the basis of the number of commercial vehicles per day. (Ref. 1).

Traffic regulatory and control systems are designed on the basis of accurate vehicle flow data. The design of signals and road junctions is possible only if, among other things, the vehicle flow data are available.

For evaluating the financial viability of privately financed toll roads, the important consideration is the volume of traffic-both present and future. Volume count data are, therefore, collected very carefully for such projects.

The traffic using a road is composed of a variety of vehicles ranging from the simple pedal cycles to the motor car and the heavy commerical vehicles. Each type having an influence on the performance of the road in its own way, a simple volume count, without classifying the vehicles into distinct types, is of limited use. It is, therefore, the normal practice to classify the vehicles into distinct types when doing the volume counting.

The ultimate aim of travel is to transport men and goods. The number of people involved in travel is an important measure in transportation planning. When it is desired to find out the number of people travelling in vehicles, an estimate of the average occupancy becomes necessary. This data can be easily gathered by noting the number of occupants in each vehicle as the volume count is being taken. The average occupancy thus determined can be used for a number of purposes. For instance, if the economic analysis of a highway project is being carried out, it becomes necessary to evaluate the loss suffered by vehicle occupants due to delays in terms of monetary value. If the vehicle count is available, and the average occupancy of each type of vehicle is known, the total number of persons travelling can be easily found out. The delay caused in terms of man-hours and in terms of monetary value can then be determined.

4.2. Types of Counts

4.2.1. Levels of measurement of flow

The three common levels of measurement of vehicle flow are :

(*i*) Average annual flow, expressed in vehicles per year.

(*ii*) Annual Average Daily Traffic (AADT), expressed in vehicles per day.

(*iii*) Hourly flow, expressed in vehicles per hour.

The need for expressing the flow in the above different levels arises from the variations in the pattern of volume over different time bases.

The average annual flow measurement is useful for estimating the total travel on a road system and for determining the gross annual revenues from road users. In accident statistics the total annual vehicle-kilometres forms the usual basis for comparison.

The A.A.D.T., which is $\frac{1}{365}$ th of the total annual flow, is a common measure of flow utilized in geometric standards for high-ways, improvement of existing facilities and standards for pavement design and maintenance. If the flow is not measured for all the 365 days, but only for a few days the average flow is known by the term ADT, Average Daily Traffic.

Short term variations occurring in the course of a day, and especially the peaking situations in the morning and evening rush hours are needed for design of traffic control systems and complicated intersections.

4.2.2. Short term and long term counts

The duration of the counts depends upon the purpose for which the data are needed and the financial and man-power resources at the command of the traffic engineer. Sometimes it is only necessary to measure the flow for a short term, say an hour; at other times the flow may be measured for an intermediate period such as a full day of twenty-four hours ; in some situations, the count may extend for a duration of a full week; and lastly, the count may be a continuous and regular affair. The examples of situations where the above types of counts are made are given in Table 4.1.

4.3. Method Available for Traffic Counts

The methods available for traffic counts are listed below :

(*i*) Manual methods.

(*ii*) Combination of manual and mechanical methods.

(*iii*) Automatic devices.

(*iv*) Moving observer method.

(*v*) Photographic methods.

Manual methods use field personnel to count and classify traffic flowing past a fixed point. Automatic devices enable a count of traffic to be taken at any given location and a record to be kept of the count. A combination of manual and mechanical methods involves the services of field personnel who operate mechanical devices to count and record the arrival of vehicles at any given point across a road. The moving observer method is a special traffic engineering technique which results in the collection of data on the flow and speed of traffic, travel time, delay at junctions and parking and has been considered in detail in Chapter 3. Photographic methods for measuring flow and other traffic characteristics are becoming powerful tools in the hands of the traffic engineer and are dealt in Chapter 7.

Table 4.1. Types of Vehicle Volume Counts and their Uses

Type of counts

Short-term counts, say, for one hour or two hours	Counts for a full day	Counts for a full week	Continuous counts
Purpose and use	**Purpose and use**	**Purpose and use**	**Purpose and use**
1. To determine the flow or traffic in the peak hour.	1. To determine hourly fluctuation of flow.	1. To determine the hourly and daily fluctuation of flow.	1. To determine the fluctuation of flow daily, weekly, seasonally and yearly.
2. Used in measuring the saturation flow at signalised intersections.	2. Used in intersection counts.	2. Used generally in developing countries as the only means of traffic census on non-urban high-ways.	2. To determine the annual rate of growth of traffic.

3. Used in intersection counts during the morning and evening peaks.	3. Used in cordon line and screen line counts as part of transportation survey.	3. Used in cordon line and screen line counts as part of urban transportation survey.	3. Used generally in developed countries at selected number of stations for a continuous monitoring of traffic flow in urban or non-urban locations.

The advantages of manual methods and situations where these are to be preferred are :

(*i*) Details such as vehicle classification and number of occupants can be easily obtained. With automatic devices these data are unfortunately lacking, and hence automatic counting should be supported by manual counts.

(*ii*) The data can be collected giving the breakdown of traffic in each direction of travel.

(*iii*) Specific vehicular movements such as left-turns, right-turns, straight-aheads etc. at a junction can be noted and recorded.

(*iv*) Manual methods enable any unusual conditions obtaining at the time of the count to be recorded. This will help in understanding and analysing the traffic characteristics. Such unusual conditions can be adverse weather conditions, traffic breakdowns, temporary closure of any lane of the highway for maintenance operations etc.

(*v*) In developing countries, sophisticated automatic devices are not indigenously produced. On the other hand, manpower for counting is available comparatively cheaply.

(*vi*) Pilferage and vandalism often prevent the use of costly equipment in remote rural areas, and in such cases manual methods are the only solution.

(*vii*) Even if automatic devices are used, it is often necessary to check the accuracy of these devices periodically and manual methods serve this purpose.

(*viii*) Data accumulated by manual methods are easy to analyse.

(*ix*) Manual methods are suitable for short-term and non-continuous counts.

The advantages of automatic devices and situations dictating their use are :

(*i*) Where a continuous and regular record of traffic flow is needed, the only alternative is to install automatic counters.

(*ii*) Their use has been very wide-spread in the developed countries, where traffic engineering and transportation planning have attained a high degree of perfection. These are now used in India too.

(*iii*) If counts of remote areas are needed, automatic devices are useful.

(*iv*) Automatic devices usually yield data which are amenable to easy analysis by computers.

4.4. Manual Methods

4.4.1. Number of observers

The number of observers needed to count the vehicles depends upon the number of lanes in the highway on which the count is to be taken and the type of information desired. The indications in Table 4.2 can be used as rough guides.

Table 4.2. Number of Observers Needed for Manual Volume Counts

Road features and counting requirements	*No. of vehicles per hour that can be counted by one trained observer*
1. 2–lane two-way road, with separate observers for each direction; vehicles to be counted and classified.	500 vehicles per hour in one direction.
2. 2–lane two-way road, with one observer for both the directions; vehicles to be counted and classified for each direction separately.	200 vehicles per hour in both directions.
3. 2–lane two-way road, with one observer for both directions; vehicles to be simply counted with no requirement for classification and posting into separate direction.	800 vehicles per hour both directions.

It is perhaps more desirable to record traffic in each direction of travel separately and post separate observers for each direction. In fact, the Indian Roads Congress Standard (Ref. 3) deems this necessary. Enumerators should be literate persons with preferably middle or matriculation level qualifications and can be trained suitably for the purpose. For all-day counts, work in three shifts of 8 hours each could be organised. A separate observer is needed if the occupancy count is to be made.

4.4.2. Equipment needed

The following equipment is needed :

(*i*) A watch

(*ii*) Pencils, eraser and pencil sharpener

(*iii*) Supply of blank field data sheets

(*iv*) Clip board.

An additional equipment which is very handy is the multibank hand tally counter. If five types of vehicles are to be counted the hand tally should have five counters, with a label stuck on each pressing knob indicating the vehicle type it signifies. The pressing of the knob operates the counter and records the vehicle. The numbers indicated on the counters are periodically transferred to the forms.

4.4.3. Field data sheets and summary sheets

When multibank hand tally counters are not available, the data is recorded conveniently by the five-dash system, whereby vertical strokes are entered for the first four vehicles, followed by an oblique stroke for the fifth vehicle so as to depict a total of five.

The field data sheet prescribed by the Indian Roads Congress Standard (Ref. 3) is depicted in Table 4.3. This form is intended to last for four hours, but if the hourly flow is large, one form may be needed for one hour.

The data can be summarised for each hour of the day in a form prescribed by the Indian Roads Congress Standard in Table 4.4. This form itself can be used for transferring the totals from multi-bank hand tallies, if these are used instead of the five-dash system.

Table 4.3. Field Data Sheet for Traffic Census (*Source* : Ref. 3)

Date and day of the week..................
Direction of traffic.............................. From.............To.............

Road Classification..........
Kilometrage/mileage.......
Route number (if any).....
District...........State........

Type of vehicles (1) / *Hour of count*	*Cars, jeeps, vans, three wheelers* (2)	*Buses* (3)	*Trucks* (4)	*Motorcycles and scooters* (5)	*Animal drawn vehicle*** (6)	*Cycles* (7)	*Others (specify)* (8)	*Remarks, including weather conditions* (9)
From—Hours* To—Hours								
Hourly Total								
From—Hours* To—Hours								
Hourly Total								
From—Hours* To—Hours								
Hourly Total								
From—Hours* To—Hours								
Hourly Total								

Notes: 1 Record traffic volume in columns 2 to 8 by making tallies in the form of vertical strokes for first four vehicles and drawing an oblique stroke for every 5th as shown within brackets (~~||||~~)

2. Some roads carry appreciable volume of other traffic like cycle rickshaws. Record the volume of such vehicles in column 8 after specifying the vehicle type.

* The hour of count should be entered before the start of the enumeration. 7 P.M. hours should be recorded after adding 12 to the actual hour, for example, P.M. should be recorded as 14.00 hours.

** If felt necessary by Highway Authority, this column could be subdivided into two for recording the volume of'pneumatic-tyred' and 'iron-tyred' vehicle separately.

Name and signature of enu-merator with date ......

Name and signature of supervisor with date ......

Table 4.4. Daily Summary Sheet for Traffic Census (*Source* : Ref. 3)

From..........Hrson*........To..........Hrson*
Direction of traffic : From........To........(UP)
From........To........(Down)

Road Classification..........
Kilometrage/mileage.......
Route number (if any).....
District............State........

Count Hour	*Cars, jeeps, Vans, three-wheelers etc.*		*Buses*		*Trucks*		*Motor cycle, and scooters*		*Total fast*		*Total col. 10 & 11*	*Animal drawn vehcles*		*Cycles*		*Others (specify)*		*Total slow*		*Total col. 19 & 20*	*Remarks*
(1)	*Up* (2)	*Down* (3)	*Up* (4)	*Down* (5)	*Up* (6)	*Down* (7)	*Up* (8)	*Down* (9)	*Up* (10)	*Down* (11)	(12)	*Up* (13)	*Down* (14)	*Up* (15)	*Down* (16)	*Up* (17)	*Down* (18)	*Up* (19)	*Down* (20)	(21)	(22)
0600—0700																					
0700—0800																					
0800—0900																					
0900—1000																					
1000—1100																					
1100—1200																					
1200—1300																					
1300—1400																					
1400—1500																					
1500—1600																					
1600—1700																					
1700—1800																					
1800—1900																					
1900—1900																					
2000—2100																					
2100—2200																					
2200—2300																					
2300—2400																					
0000—0100																					
0100—0200																					
0200—0300																					
0400—0500																					
0500—0600																					
Total																					
Total Up and Down for vehicle type																					

***Enter Date and Day of week**

Name and Signature of Supervisor with date ....

The data for a period of a week, as is the recommended practice in the Indian Roads Congress Standard, can be summarised in the form given in Table 4.5.

4.4.4. Manual counts at intersections

The field data sheets can be modified to suit the particular requirements of any intersection. Traffic enumerators need to be posted on each arm of the intersection. At a four arm intersection, the count at each arm of the traffic entering the intersection can be broken down into three categories, *viz.*, left turning, right turning and straight ahead traffic. The field data sheet can be modified to suit, as in Table 4.6. For heavy flows, a multi-bank hand counter will be found useful.

4.5. Combination of Manual and Mechanical Method

An example of a combination of manual and mechanical methods is the multiple pen recorder. A chart moves continuously at the speed of a clock. Different pens record the occurrence of different events on the chart. The actuation of the event recorder pen is by pressing the electric switch associated with each pen recorder. For instance, a particular switch may be pressed whenever a particular class of vehicles (cars, buses trucks etc.) arrives and this operates the pen on the channel which can be identified with the arrival of a particular class of vehicle. The advantages of an event recorder of this type are:

(*i*) A permanent record is kept of the arrival of each class of vehicle. The classification and vehicle count are performed simultaneously.

(*ii*) Additional information such as time-headways between successive vehicles and the arrivals per unit time become available (see Chapter 26).

The analysis of data accumulated by the multiple pen recorder is time consuming.

4.6. Automatic Devices

4.6.1. Essential parts of automatic devices

The automatic devices essentially consist of an equipment for detecting the passage or presence of a vehicle (called detector or sensor) and another for recording the count. The sensor usually transmits some form of electric impulse which activates the accumulating register or recording chart.

4.6.2. Sensors (detectors)

Sensors operated on several different principles are available. They are :

(*i*) **Pneumatic tube (road tube).** A flexible tube with one end sealed is clamped to the road surface at right angles to the pavement. The other end of the tube is connected to a diaphragm actuated switch. When an axle of a vehicle crosses the tube, a volume of air gets displaced thus creating a pressure which instantaneously closes the electrical contact through the switch. Two such contacts result in one count being registered, thus representing the two axles in a vehicle. Inaccuracies are caused when vehicles with more than two axles are present in the traffic stream in appreciable numbers. Because of their simplicity and their cheapness, pneumatic tube sensors are very popular. Some difficulty may be caused in fixing them to gravel surfaces and they have the additional drawback that they are easily pilfered by vandals. They are likely to be damaged by crawler tractors, tyre chains, snow ploughs and similar equipment. They cannot detect vehicles by lanes.

Table 4.5. Weekly Summary Sheet of Traffic Census

From................ To................
(Date) (Date)
Details of census point...............

Road classification..........
Kilometrage/mileage.......
Route number (if any).....
District............State........

Period				Fast vehicles					Slow vehicles				
From		To		Cars, jeeps, Vans, three-wheelers	Buses	Trucks	Motor cycle, and scooters	Total (Col 5 to 8)	Animal drawn vehcles	Cycles	Others (specify)	Total (Col. 10 to 12)	Remarks
Date	Hour	Date	Hour										
(1)	(2)	(3)	(4)	(5)	(6)	(7)	(8)	(9)	(10)	(11)	(12)	(13)	(14)
Total for the week													
Average daily traffic for the week													

Name and Signature of Supervisor with date ....

Table 4.6. Data Sheet for Intersection Volume Count.

Day and date of the week...................

Period from............Hours to.............Hours.............

Traffic entering intersection from (Name of Approach road) and travelling (Compass direction)

Types of vehicles / *Direction of flow*	*Cars, jeeps, vans, three-wheelers*	*Buses*	*Trucks*	*Motor-cycles and scooters*	*Animal drawn vehicles*	*Cycles*	*Others (specify)*	*Remarks, including weather conditions*
(1)	*(2)*	*(3)*	*(4)*	*(5)*	*(6)*	*(7)*	*(8)*	*(9)*
Vehicles turning left								
Vehicles proceeding straight ahead								
Vehicles turning right								

(*ii*) **Electric contact.** A pair of steel strips are contained in a rubber pad which is burried beneath the surface. On being pressed by the weight of a moving axle the steel strips come into contact with each other and cause the electric current to flow. Electric contact detectors, while retaining the advantages and disadvantages of pneumatic tube detectors, have the ability to detect vehicles in individual lanes.

(*iii*) **Co-axial cable.** A co-axial cable is clamped across the road surface, with the capability of generating signals with the passage of axles. These signals actuate a transistorised counter. The advantages associated with this type of detectors are their better reliability and performance on inferior type of surfaces and their lesser susceptibility to damage.

(*iv*) **Photo-electric.** On one end of the road is a source of light which emits a beam across the road. At the other end is a photo-cell which can distinguish between the light beam and absence of the light beam. The passage of a vehicle in the path of the light beam obstructs the beam and causes detection by the photo-cell. The difficulty with this otherwise simple technique is that obstructions can be caused by pedestrians and that more than one vehicle in the different traffic lanes, but positioned in the line of the beam simultaneously, will register only one vehicle.

(*v*) **Radar.** *Doppler Effect* is a well-known phenomenon in Physics which enables detection of vehicles moving at a speed. When a moving object approaches or recedes from the source of a signals, the frequency of the signal received back from the moving object will be different from the frequency of the signal emitted by the source. This difference in the two frequencies causes detection of a moving object. The initial cost of this device is no doubt high, but its accuracy and reliability, apart from its non-susceptibility to damage by traffic, have much to appeal.

(*vi*) **Infra-red and ultrasonic.** Infra-red sensors can detect the heat radiated from a vehicle or can react to the reflection from the vehicle of infra-red radiation emitted by the sensors.

An ultrasonic sensor is also used for vehicle detection.

Both these types have the same advantages as a Radar, though both are equally expensive as a Radar.

(*vii*) **Magnetic.** The process employs the disturbance caused in a magnetic field by a passing vehicle as the basis of sensing. The magnetic field itself is provided by a wire coil buried beneath the road surface.

4.6.3. Recording mechanism

The second part of an automatic traffic counting device is the recording mechanism. Many types of mechanisms are in use, such as :

(*i*) **Counting register.** This is simply an accumulating counter indicating directly the number of vehicles on a meter. The readings must be taken before and after the counting period.

(*ii*) **Printed output.** This device prints the accumulated totals at regular intervals of time on a roll of paper, resetting the counting register to zero at the end of each time interval. The time of the day and the number of vehicles are printed side by side, thus giving directly the traffic count between the previous time reading and the time reading against the printed count. The paper rolls out at the speed of the clock. The mechanism is synchronised such that the printed time corresponds to the local time by the clock. The time interval usually selected for printing is one hour, though a 15 minute or 30 minute interval is also used.

(*iii*) **Electronic system.** Modern systems use electronic registering of data on floppies or discs.

4.6.4. Maintenance of automatic devices

For accuracy and reliability, proper maintenance and unkeep of the mechanical devices is a necessity. The accuracy of clock-work mechanism needs to be checked periodically. The recording devices usually run on 6 volt batteries, which need to be charged at regular intervals. Battery charging facilities may often be a necessary adjunct to a counting system embracing a number of counters. The pneumatic tubes, magnetic devices, radar, infra-red and ultra sonic equipment and underground cables also should receive frequent attention.

4.6.5. Video Photography

Video photography gives a permanent record of volume counts. Its analysis can be done conveniently in the office by replaying the casette on a TV monitor.

4.7. Planning and Programming Traffic Counts

4.7.1. General considerations

Since the purpose and scope of traffic counts is vast and varied, meticulous care is needed in planning and programming the counts. The traffic engineer has to understand clearly what sort of data would be appropriate to his needs and what financial constraints have been prescribed for the cost of data accumulation. The accuracy demanded will greatly influence the cost. In most of the developing countries, financial resources are necessarily scarce and considerable accuracy has to be sacrificed.

In most of the countries, some system of road classification already exists. Accurate information is also available on the length of the roads of different classes. Given these two, the programming of traffic counts lies in choosing a suitable number of census points so that the flows that would be recorded are fairly accurately representative of the flows in the entire road system. Sampling techniques find useful application here. The principle is that if the census points are chosen randomly on each class of roads, the recorded average flow at the census points can be expected to give an estimate of the average flow over that class of the road. The average flow of that class of road multiplied by the length of all the roads in that class then gives an estimate of the amount of travel in vehicle-kilometres. The number of randomly located points can be chosen on the criterion that, with a given probability, the average traffic at the count points should not differ from the true average by more than a certain proportion of the overall sample average. Preliminary counts should first of all be made at a number of randomly chosen points within each class of the road, and the means and standard deviations determined. Suppose the following values are obtained as shown in Table 4.7.

Table 4.7

	Road classification	*Preliminary Counts*		*Road length*	*Proportion of total road length (P)*
		Mean	*Standard Deviation (σ)*		
1.	National highways	$\bar{x}_1$	σ_1	L_1	$p_1 = \frac{L_1}{L}$
2.	State highways	$\bar{x}_2$	σ_2	L_2	$p_2 = \frac{L_2}{L}$
3.	District roads	$\bar{x}_3$	σ_3	L_3	$p_3 = \frac{L_3}{L}$
4.	Village roads	$\bar{x}_4$	σ_4	L_4	$p_4 = \frac{L_4}{L}$

Given that the overall sample average is not to differ from the true average by more than a proportion K of the sample average with a probability of $P = 0.95$, the standard error of the overall average,

$$\sigma \bar{x} = \sqrt{\frac{(\Sigma p\sigma)^2}{N}} = \frac{K\bar{x}}{2}$$, 2 being the value of *za* for $P = 0.95$

(see Chapter 8)

where N = Number of samples (*i.e.*, in this case the number of census stations).
$\bar{x}$ = Overall average.

The above equation determines the number of census stations, N, which can then be allocated to the various classes of road in proportion to the values ($L.\sigma$) for each class.

As regards whether the counting programme should be continuous or only for a short duration, here again the prime consideration is cost. Whereas some countries like U.K. and U.S.A. have a continuous system of traffic counts on selected permanent stations, this will be considered a luxury in developing countries. At least a 7-day count once or twice a year, would be all that one could think of in the developing countries.

The practice in some of the developing countries is reviewed by Howe (Ref. 4) who concludes that manual counts (simple random counts or replicated random counts) cannot provide estimates of sufficient accuracy.

If the counts are for a 7-day period, once or more during a year, the exact timing of the count will be very important. Obviously abnormal traffic conditions like a seasonal fair should not be selected. The practice in India is to conduct the survey once during the peak season of harvesting and again during the lean season.

4.7.2. Indian practice

The practice recommended by the Indian Road Congress for the traffic census on non-urban roads is outlined in IRC : 9–1972 (Ref. 3). A repetitive 7-day count taken twice every year, once during the peak season of harvesting and marketing and the other during the lean season, is prescribed at preselected stations. The operations are generally intended to embrace only the important trunk routes like National Highways, State Highways and Major District Roads. No definite criteria are laid down for the selection of the number of census stations. The census figures on the National Highways system are transmitted to the Roads Wing of the Ministry of Surface Transport.

Recently, under an important research scheme sponsored by the Ministry of Surface Transport (Roads Wing), Government of India, a continuous count of traffic at a number of stations in India was carried out for (Ref. 9–13) a year. The objective of the project was to get an accurate idea of the hourly, daily, weekly monthly and seasonal variation of traffic. If the count stations are continued on more or less permanent basis, the data can also give an idea of annual rate of growth of traffic. Further extension of the scheme has been taken up.

An account of the permanent traffic counting station at Warangal, using the Strester Amet Automatic Printer Counter is given by Raghavachari (Ref. 14).

Some of the findings from the Indian continuous counts (Ref. 9–13) are summarised below :

1. The peak hour traffic is generally in the range of 6–10 percent of the AADT. This shows that there is not a predominant peak-hour traffic on Indian rural highways.

2. The variation in daily traffic is marginal. Saturdays and Sundays generally attract higher volumes.

3. Monsoons and winter season have less traffic.

4. The composition of traffic varies from site to site. The percentage of fast traffic is generally higher than that of slow traffic. Roughly two-thirds of the total traffic is fast and the balance one-third is slow.

5. Of the fast traffic, trucks account for a big share, 50–70 percent, buses account for 10–20 percent, cars account for 5–20 percent, and the balance consists of motorised two-wheelers, tractors, three-wheelers etc.

6. Cycles constitute a big share of slow traffic, about 70–90 per cent. The balance represents animal-drawn traffic.

The above traffic pattern has undergone some changes during the past 25 years. On a typical National Highway, data gathered in 2005 (Ref. 17) has revealed the following :

(*i*) The percentage of slow-moving traffic has come down, and is currently about 10 percent. Animal-drawn vehicles are almost vanishing on the highways.

(*ii*) A typical composition of traffic is as under :

Cars and light vehicles	15%
Two-wheelers	6%
Buses	15%
Trucks	54%
Slow-moving	10%

(*iii*) The peak-hourly traffic is 6 per cent, showing that the traffic is spread over a long period of the day, without significant peaking.

(*iv*) All the seven days of the week have similar volume of traffic.

Vehicle flow counts on urban streets have so far been isolated and far and few between. No organised effort to programme and schedule regular or periodic counts seems to be in evidence. This points to the great scope for future action.

4.7.3. U.K. practice

4.7.3.1. For rural road planning and design, the practice in U.K. is to measure the 7-day average flow for 16 hours (6 am–10 pm) in the month of August (Ref. 5), in which month average travel demands have been found consistently to be at their highest. For observations taken during any other months, average adjustment factors are applied to convert them to their August equivalent.

4.7.3.2. 50-Point traffic census in U.K.

One of the examples of organised planning and programming for regular traffic census in any country can be had in the 50-Point Traffic Census in U.K. launched in 1956 (Ref. 6).

The road system is classified into five classes in U.K. : trunk, class I, class II, class III and unclassified. Even though the unclassified roads formed about half the total length of roads, they carried less than one-fifth of the total amount of motor traffic, and this consideration led to the elimination of this class of roads from the census. This helped to simplify the selection of points and to save effort on maintaining continuous counts on roads of little importance. Both urban roads and rural roads were considered, so that, in effect, 8 classes of roads were included. 16 points were selected on the urban road system and 34 on the rural road system.

The actual selection of the location of the points was done on the basis of random numbers, compatible with the desirability of locating appropriate number of points in the various geographical and administrative divisions of the country.

At each of these locations, the automatic counters have been in operation ever since, supplemented by information about the various classes of vehicle by manual counts. The results have yielded valuable information on the trends from year to year, seasonal variations, hourly patterns and day-to-day variations.

4.7.4. U.S.A. practice

4.7.4.1. Rural roads

For rural roads, the U.S. Bureau of Public Roads (B.P.R.) recommends the following procedure (Ref. 7) :

1. *Permanent continuous count stations.* An average of 30 to 40 permanent continuous count stations are selected throughout the States to obtain data representative of the different classes to roads.

2. *Control count stations.* Repeat counts are made on a sufficient number of road sections to obtain seasonal and daily traffic characteristics and factors required to expand single counts to average annual figures. Two types of control counts are recommended.

(*a*) **Major control counts.** These are to be repeated each month of the year, with each count covering three week days, one Saturday and one Sunday. Hourly recording type portable recorder should be used. Roughly one major control station should be needed for each 90 coverage stations.

(*b*) **Minor control counts.** These are to be repeated four to six times a year, in alternate months, with each count covering a minimum of 48 hours on week days. Roughly two minor control stations should be needed for each 90 coverage stations.

3. *Coverage count stations.* A sufficient number of coverage counts are needed to provide a reasonably accurate determination of ADT on each road section. Roads having an ADT of less than 25 need not be covered by these counts. Counts are made only once during the year for a minimum of 48 hours on week days, with cumulative type portable recorders. The number of coverage count stations required may be obtained by dividing the total mileage by 2, then dividing the quotient by the average length of each road section (which may be considered as 2 miles). Thus in a State having 72,000 miles, there will be 18,000 coverage count stations, 400 minor control stations and 200 major control stations.

4. *Classification counts.* Classification counts are made at selected control stations.

4.7.4.2. Urban roads

For urban roads, the recommended practice of the National Committee on Urban Transportation (Ref. 8) is as below :

(1) **Street Classification.** The streets are classified into various types.

(2) **Selecting control stations.** The results of volume counts at these stations provide the basis for the determination of control factors necessary to record volume counts on a common basis. Major control stations and minor control stations are selected to represent the flow on the major streets and the minor streets respectively.

(3) Coverage counts are made on the major and minor streets.

(4) Central business district cordon count is made to measure the transportation activity developed by the CBD.

(5) Screen-line counts are taken to detect long-ranges changes in volume and direction of traffic.

4.8. Vehicle Occupancy Surveys

Vehicle occupancy surveys are easy to conduct for cars and motorised two-wheelers. The observers can note the number of occupants of every one of such vehicles by standing by the roadside. An Indian study (Ref. 15) has reported the average occupancy of cars as 2.7 for sub-urban conditions and 4.5 for rural situations (including the drivers).

For buses, the occupancy can be determined by examining the ticket sales. An Indian Study, covering 640 buses for nearly two years, has concluded that the average occupancy of buses is 43 passengers, in addition to the driver and the conductor (Ref. 16).

A more recent Indian study (Ref. 18) has yielded the average occupancy of vehicles as under :

Car	:	4.8
Two-wheeler	:	1.5
Bus	:	43

I. MULTIPLE CHOICE QUESTIONS

1. Incorrect steering axis inclination (S.A.I.) causes
   (*a*) tendency to assume toe-out orientation
   (*b*) poor recovery of the steering wheel after making a turn
   (*c*) the vehicle to pull to the side of lesser inclination
   (*d*) generation of a braking effect at tight corners
2. A traction control system (TCS) in automobiles controls the
   (*a*) vibrations on the steering wheel
   (*b*) torque that is transmitted by the tyres to the road surface
   (*c*) stopping distance in case of emergency
   (*d*) engine power during acceleration
3. When brakes are applied on a moving vehicle; the kinetic energy is converted to
   (*a*) mechanical energy (*b*) heat energy
   (*c*) electrical energy (*d*) potential energy

4. **The force required to stop a vehicle is dependent on**

(*a*) the weight of vehicle **(*b*) the deceleration rate**

(*c*) both (*a*) and (*b*) **(*d*) none of the above**

5. **Select the correct statement.**

(*a*) Nagpur road plan formula take into account the towns with very large population.

(*b*) Nagpur road plan has a target road length of 32 km per 100 square km.

(*c*) second 20 year plan has provided 1600 km of expressways out of the proposed National highway.

(*d*) second 20 year plan allowed deduction of length of railway track in the area while calculating the length of roads.

6. **The sequence of four stages of survey in a highway alignment is**

(*a*) reconnaissance, map study, preliminary survey detailed survey map study,

(*b*) preliminary survey, reconnaissance and detailed survey map study,

(*c*) reconnaissance, preliminary survey and detailed survey

(*d*) preliminary survey, map study, reconnaissance and detailed survey.

7. **The shape of the camber, best suited for cement concrete pavements, is**

(*a*) straight line **(*b*) parabolic**

(*c*) elliptical

(*d*) combination of straight and parabolic

8. **For water bound macadam roads in localities of heavy rainfall, the recommended value of camber is**

(*a*) 1 in 30 **(*b*) 1 in 36** **(*c*) 1 in 48** **(*d*) 1 in 60**

Answers

1. (*b*)	**2.** (*b*)	**3.** (*b*)	**4.** (*a*)	**5.** (*c*)
6. (*c*)	**7.** (*a*)	**8.** (*b*)		

II. SOLVED QUESTIONS AND ANSWERS

Q. 1. What is the need of traffic volume studies?

Ans. Traffic volume studies are conducted to determine the number, movements, and classifications of roadway vehicles at a given location. These data can help identify critical flow time periods, determine the influence of large vehicles or pedestrians on vehicular traffic flow, or document traffic volume trends. The length of the sampling period depends on the type of count being taken and the intended use of the data recorded.

For example, an intersection count may be conducted during the peak flow period. If so, manual count with 15-minute intervals could be used to obtain the traffic volume data.

Q. 2. What are the methods to conduct traffic volume counts?

Ans. Two methods are available for conducting traffic volume counts: (1) manual and (2) automatic.

Manual counts are typically used to gather data for determination of vehicle classification, turning movements, direction of travel, pedestrian movements, or vehicle occupancy.

Automatic counts are typically used to gather data for determination of vehicle hourly patterns, daily or seasonal variations and growth trends, or annual traffic estimates.

Q. 3. What are manual count recording methods?

Ans. Manual counts are recorded using one of three methods: tally sheets, mechanical counting boards, or electronic counting boards.

Tally sheets: Recording data onto tally sheets is the simplest means of conducting manual counts. The data can be recorded with a tick mark on a pre-prepared field form. A watch or stopwatch is necessary to measure the desired count interval. A blank traffic volume count intersection tally sheet is provided.

Mechanical counting boards: Mechanical count boards consist of counters mounted on a board that record each direction of travel. Common counts include pedestrian, bicycle, vehicle classification, and traffic volume counts. Typical counters are push button devices with three to five registers. Each button represents a different stratification of type of vehicle or pedestrian being counted. The limited number of buttons on the counter can restrict the number of classifications that can be counted on a given board. A watch or a stopwatch is also necessary with this method to measure the desired count interval.

Electronic counting boards: Electronic counting boards are battery-operated, hand-held devices used in collecting traffic count data. They are similar to mechanical counting boards, but with some important differences. Electronic counting boards are lighter, more compact, and easier to handle. They have an internal clock that automatically separates the data by time interval. Special functions include automatic data reduction and summary. The data can also be downloaded to a computer, which saves time.

Q. 4. What are the key steps to a manual count method?

Ans. Key steps to a manual count study. A manual count study includes three key steps:

***(i)* Perform necessary office preparations:** Office preparations start with a review of the purpose of the manual count. This type of information will help determine the type of equipment to use, the field procedures to follow, and the number of observers required. For example, an intersection with multiple approach lanes may require electronic counting boards and multiple observers.

***(ii)* Select proper observer location:** Observers must be positioned where they have a clear view of the traffic. Observers should be positioned away from the edge of the roadway. If observers are positioned above ground level and clear of obstructions they usually have the best vantage point. Visual contact must be maintained if there are multiple observers at a site. If views are unobstructed, observers may count from inside a vehicle.

***(iii)* Label data forms and record observations:** Manual counts may produce a large number of data forms; therefore, the data forms should be carefully labeled and organized. On each tally sheet (a blank tally sheet is provided in Appendix B), the observer should record the location, time and date of observation, and weather conditions. Follow the data recording methods discussed earlier.

Q. 5. What is automatic count method?

Ans. The automatic count method provides a means for gathering large amounts of traffic data. Automatic counts are usually taken in 1 hour intervals for each 24 hour period. The counts may extend for a week, month, or year. When the counts are recorded for each 24 hour time period, the peak flow period can be identified.

Q. 6. Explain automatic count recording methods.

Ans. Automatic counts are recorded using one of three methods: portable counters, permanent counters, and videotape.

Portable counters: Portable counting is a form of manual observation. Portable counters serve the same purpose as manual counts but with automatic counting equipment. The period of data collection using this method is usually longer than when using manual counts. The portable counter method is mainly used for 24-hour counts. Pneumatic road tubes are used to conduct this method of automatic counts. Specific information pertaining to pneumatic road tubes can be found in the users' manual.

Permanent counters: Permanent counters are used when long-term counts are to be conducted. The counts could be performed everyday for a year or more. The data collected may be used to monitor and evaluate traffic volumes and trends over a long period of time. Permanent counters are not a cost-effective option in most situations. Few jurisdictions have access to this equipment.

Videotape: Observers can record count data by videotaping traffic. Traffic volumes can be counted by viewing videotapes recorded with a camera at a collection site. A digital clock in the video image can prove useful in noting time intervals. Videotaping is not a cost-effective option in most situations. Few small jurisdictions have access to this equipment.

Q. 7. What are intersection counts?

Ans. Intersection counts are used for timing traffic signals, designing channelization, planning turn prohibitions, computing capacity, analyzing high crash intersections, and evaluating congestion. The manual count method is usually used to conduct an intersection count. A single observer can complete an intersection count only in very light traffic conditions.

The intersection count classification scheme must be understood by all observers before the count can begin. Each intersection has 12 possible movements. The intersection movements are through, left turn, and right turn. The observer records the intersection movement for each vehicle that enters the intersection.

Q. 8. What is meant by pedestrian counts?

Ans. Pedestrian count data are used frequently in planning applications. Pedestrian counts are used to evaluate sidewalk and crosswalk needs, to justify pedestrian signals, and to time traffic signals. Pedestrian counts may be taken at intersection crosswalks, midblock crossings, or along sidewalks. When pedestrians are tallied, those 12 years or older are customarily classified as adults. Persons of grade school age or younger are classified as children. The observer records the direction of each pedestrian crossing the roadway.

Q. 9. Explain vehicle classification counts.

Ans. Vehicle classification counts are used in establishing structural and geometric design criteria, computing expected highway user revenue, and computing capacity. If a high percentage of heavy trucks exists or if the vehicle mix at the crash site is suspected as contributing to the crash problem, then classification counts should be conducted.

Typically cars, station wagons, pickup and panel trucks, and motorcycles are classified as passenger cars. Other trucks and buses are classified as trucks.

School buses and farm equipment may be recorded separately. The observer records the classification of the vehicles and the vehicles' direction of travel at the intersection.

Q. 10. What is average daily traffic and annual average daily traffic counts?

Ans. Average daily traffic (ADT) counts represent a 24 hour count at any specified location. These counts are obtained by placing an automatic counter at the analysis location for a 24 hour period. Accuracy of the ADT data depends on the count being performed during typical roadway, weather, and traffic demand conditions. Local levels of government will typically conduct this type of count.

Annual average daily traffic (AADT) counts represent the average 24 hour traffic volume at a given location averaged over a full 365 day year.

Q. 11. What are uses of AADT volume counts?

Ans. AADT have the following uses:

***(i)* Information gathering:** Before a jurisdiction contacts an engineering consulting firm to perform a traffic volume count study, a variety of information may need to be collected. Any information may aid the consulting firm in adequately completing the study. The following is a list of possible information that an engineering consulting firm may request:

- issue at hand
- historic volume counts
- existing zoning
- proposed future land use changes
- traffic impact statements if available
- citizen input
- location map
- appropriate contact persons
- any other relevant information

***(ii)* Videotape:** Observers can record count data by videotaping traffic. Traffic volumes can be counted by viewing videotapes recorded with a camera at a collection site. A digital clock in the video image can prove useful in noting time intervals. Videotaping is not a cost-effective option in most situations. Few small jurisdictions have access to this equipment.

Q. 12. What are the types of counts?

Ans. Various types of traffic counts are carried out, depending on the anticipated use of the data to be collected. They include:

(i) Cordon count: These are made at the perimeter of an enclosed area (CBD, shopping center etc.). Vehicles or persons entering and leaving the area during a specified time period are counted.

(ii) Screen line count: These are classified counts taken at all streets intersecting an imaginary line (screen line) bisecting the area. These counts are used to determine trends, expand urban travel data, traffic assignment etc.

(iii) Pedestrian count: These are used in evaluating sidewalk and crosswalk needs, justifying pedestrian signals, traffic signal timings etc.

(iv) Intersection count: These are measured at the intersections and are used in planning turn prohibitions, designing channelization, computing capacity, analyzing high accident intersections etc.

III. REVIEW QUESTIONS

1. What is the difference between short-term and long term count?
2. Discuss the types of count.
3. What do you mean by sensors? What are the principles sensors operate?
4. Discuss the recording mechanism.
5. How will you conduct vehicle occupancy surveys?

REFERENCES

1. Roads Wing, Ministry of Shipping and Transport, Government of India, Report of the Technical Group set up by the Government of India regarding norms for maintenance of National Highways, New Delhi.
2. United Nations, Manual on Traffic Surveys, New Delhi.
3. Indian Road Congress, Traffic Census on Non-urban Roads, IRC. 9-1972, New Delhi, 1972.
4. J.D.G.F. Howe, A Review of Rural Traffic Counting Methods in Developing Countries, R.R.L. Reports LR 427, 1972.
5. Traffic Prediction for Rural Roads, Ministry of Transport (U.K.) Her Majesty's Stationery Office, London, 1968.
6. Tanner J.C. and J.R. Scott, 50-point Traffic Census : the First 5 years, Department of Scientific Industrial Research, Road Research Technical Paper No. 63, London, 1962.
7. Traffic Counting Classification and Weighing, U.S. Bureau of Public Roads, Washington, 1957.
8. Measuring Traffic Volumes, National Committee on Urban Transportation, U.S.A.
9. Data on Growth of Highway Traffic, Technical Report, Maharashtra Engineering Research Institute, Nasik, 1980.
10. Report of Research Project R-3, Gujarat Engineering Research Institute, Vadodara, 1980.
11. Report on Research Project R-3 P.W.D. (B & R) Rajasthan, Jaipur, 1979.
12. Report on Research Project R-3, First Year Results of Continuous Traffic Count Station on NH 45 in Tamil Nadu, Highway Research Station, Madras, 1980.
13. Report on Research Project R-3 U.P. P.W.D. Research Institute, Lucknow, 1980.
14. Raghavachari, S. Location of Permanent Traffic Counting Stations in Andhra Pradesh, Regional Engineering College, Warangal, 1979.
15. Kadiyali, L.R., *et al*, Average occupancy of Indian Cars, RUCS Technical Paper No. 65. Thirteen Quarterly Report, Central Road Research Institute, New Delhi, 1981.
16. Road User Cost Study, Final Report, Vol. V. Special Studies on Accidents and Value of Time, Central Road Research Institute, New Delhi, 1982.
17. Four/Six Laning of Bangaluru-Kolar-Mulbagal Section of NH 4, Final Feasibility Report prepared for National Highways Authority of India, L.R. Kadiyali & Associates, New Delhi, 2006.
18. Manual on Economic Evaluation of Highway Projects in India, Special Publication IRC : SP : 30 Indian Roads Congress, New Delhi, 2009.

5. Origin-Destination Survey

5.1. Need for O.D. Survey

5.1.1. In a transportation study, it is often necessary to know the exact origin and destination of the trips. It is not only necessary to know how many trips are made, but also group these trips with reference to the zones of their origin and destination. Other information yielded by the O-D survey includes land-use of the zones of origin and destination, household characteristics of the trip-making family, time of the day when the journeys are made, trip purpose and mode of travel. Origin is defined as the place where the trip begins and destination is defined as the place where the trip ends.

5.1.2. The specific uses to which O-D survey data can be put are :

(*i*) To determine the amount of by-passable traffic that enters a town, and thus establish the need for a bypass.

(*ii*) To develop trip generation and trip distribution models in transport planning process.

(*iii*) To determine the extent to which the present highway system is adequate and to plan for new facilities.

(*iv*) To assess the adequacy of parking facilities and to plan for future.

5.2. Preliminaries

The purpose for which an O-D survey is required determines the extent of preliminary preparations for organising the survey. For instance, if the survey is needed as part of an urban transportation planning process for a large town, it will be necessary to define the study area and sub-divide it into zones. The origins and destinations can then be related to these zones. In the study of transportation problems of towns, big or small, it is usually necessary to define external cordon lines, which are imaginary lines at the boundary of the study area. In big towns, it may also be necessary to select some internal cordon lines, which may be concentric rings. For checking the accuracy of survey data, it may be necessary to have screen-lines, which are imaginary lines dividing the area into parts. More details of the selection of zones, cordon lines and screen lines are given in Chapter 30.

5.3. Survey Methods

5.3.1. The following are some of the methods available for conducting an O-D survey :

(*a*) Home interview survey

(*i*) Full interview technique

(*ii*) Home-questionnaire technique

(*b*) Road-side interview survey

(*c*) Post-card questionnaire survey

(*d*) Registration number plate survey

(*e*) Tags on vehicles.

A full description of the procedures involved in these surveys and the comparative merits of each are given in Chapter 30.

5.3.2. The National Committee on Urban Transportation, (U.S.A.) (Ref. 1) has brought out Procedure Manuals, laying down the methodology for conducting O-D surveys. By a combination of some of the methods mentioned in para 5.3.1 above, it is possible to economically collect O-D data for towns of different sizes. Usually it is adequate to carry out only an external cordon O-D survey in towns with populations less than 5,000. This is because in such small towns the traffic from outside is the one which influences the traffic pattern. For towns with population in the range of 5,000–75,000 the external-internal cordon surveys are most appropriate. The external cordon is selected at the edge of urban development and the internal is located at the limits of the central business district. Road-side interviews at the intersection of the roads with these cordons yield the O-D data. For large cities, with population more than 75,000, the most effective method is to combine the external cordon survey with a home-interview survey.

5.4. Checking the Accuracy of O-D Survey Data

The data obtained from the home-interview survey is usually cross-checked with the data obtained from screen-line and cordon counts.

5.5. Presentation of Results

O-D surveys yield a vast amount of data. To understand them it is necessary to present them in convenient tabular or pictorial form.

The most convenient form is an O-D matrix, in which the origin zones and destination zones are represented. The horizontal axis of the matrix represents the destination zones and the vertical axis of the matrix represents the origin zones. The zones may be further classified into internal and external zones if the survey covers both the internal and external zones. The number of trips are entered in the cells of the matrix. The matrix is represented in Fig. 5.1 below :

In the above matrix t_{2-3} represents the number of trips originating in zone 2 and terminating in zone 3.

The most popular pictorial representation is by means of a desire line chart. In this chart, the trips between any pair of zones are represented by a straight line connecting the centroids of the two zones and having a band width drawn to a suitable scale to represent the actual volume of trips. A typical desire line chart is given in Fig. 5.2.

			Destination zones						
			Internal				External		
			1	2	3		72	73	74
Origin zones	Internal	1							
		2							
		3							
	External	72							
		73							
		74							

Fig. 5.1. O-D matrix.

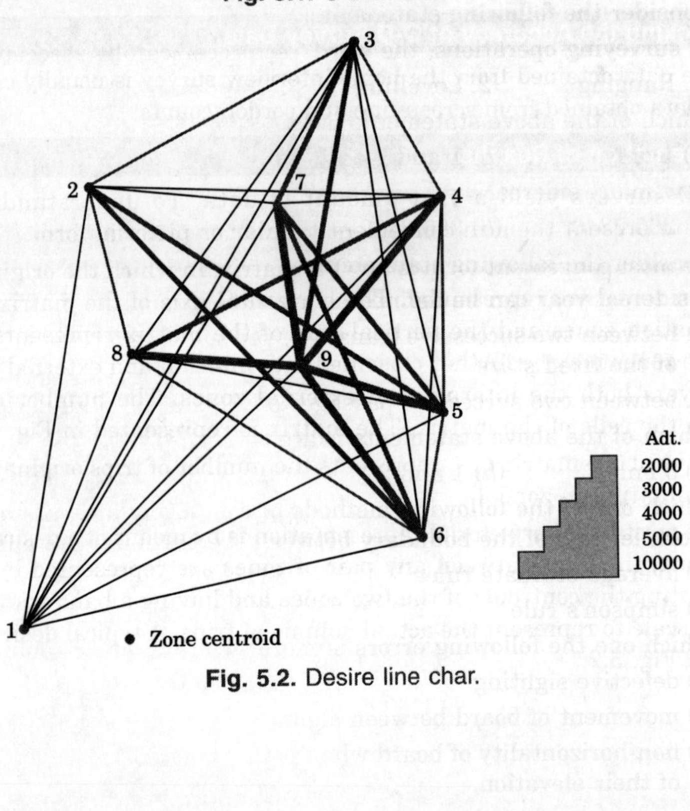

Fig. 5.2. Desire line char.

I. MULTIPLE CHOICE QUESTIONS

1. **Which one of the following tests is performed in the laboratory to determine the extent of weathering of aggregates for road works?**

(*a*) soundness test **(*b*) crushing test**

(*c*) impact test **(*d*) abrasion test**

2. **Which one of the following geometric features requires the magnitudes of weaving angle and weaving distance for its design?**

(*a*) rotary design **(*b*) right-angle intersection**

(*c*) roundabout **(*d*) grade-separated junction**

3. **Which one of the following methods is used in the design of rigid pavements?**

(*a*) CBR method **(*b*) group index method**

(*c*) Westergaard's method **(*d*) McLeod's method**

4. **In which one of the following yards, are reception, sorting and dispatch of railway wagons done?**

(*a*) loco yard **(*b*) marashalling yard**

(*c*) goods yard **(*d*) passenger yard**

5. **Which one of the following is not a desirable property of the sub-grade soil as a highway material?**

(*a*) stability **(*b*) ease of compaction**

(*c*) good drainage **(*d*) bitumen adhesion**

6. **Consider the following statements:**

In surveying operations, the word 'reciprocal' can be associated with

1. Ranging **2. Levelling** **3. Contouring**

Which of the above statements is/are correct?

(*a*) 1 only **(*b*) 1 and 2 only** **(*c*) 2 and 3 only** **(*d*) 1, 2 and 3**

7. **How many sidereal days are there in a solar year?**

(*a*) 365.2840 **(*b*) 366.2422** **(*c*) 360.2500** **(*d*) 365.0000**

8. **Consider the following statements:**

A sidereal year can be defined as the time interval:

1. between two successive transits of the sun through the meridian of any of the fixed stars

2. between two successive passages of the sun through perigee

Which of the above statements is/are correct?

(*a*) 3 only **(*b*) 1 and 2** **(*c*) 2 and 3** **(*d*) 1 only**

9. **Which one of the following methods of computing area assumes that the short, lengths of the boundary between the ordinates are parabolic ares?**

(*a*) average ordinate rule **(*b*) middle ordinate rule**

(*c*) simpson's rule **(*d*) trapezoidal rule**

10. **Which one the following errors is more severe in plane-table surveying?**

(*a*) defective sighting **(*b*) defective orientation**

(*c*) movement of board between sights

(*d*) non-horizontality of board when points sighted are at large differences of their elevation.

Answers

1. (*a*)	**2.** (*a*)	**3.** (*c*)	**4.** (*b*)	**5.** (*d*)
6. (*b*)	**7.** (*b*)	**8.** (*b*)	**9.** (*c*)	**10.** (*b*)

II. QUESTIONS AND ANSWERS

Q. 1. What is origin?

Ans. Origin is the location where a trip begins. It is described in terms of a street address, a monument (*e.g.*, a well-known location, an office building, a school, etc.), a street intersection or a district. All survey origins have been geo-coded for precision.

Q. 2. What do you mean by destination?

Ans. Destination is the location where a trip ends. It has similar attributes as the trip origin.

Q. 3. What is OD?

Ans. Origin-destination describes both 'ends' of a single trip. This term is commonly abbreviated to 'O-D.

Q. 4. What is trip, or person trip?

Ans. Trip, or person trip, is a single or one-directional movement of one person from one point (origin) to a second point (destination), for a single purpose. For example, the commute from home to work represents one trip.

Q. 5. What do modes describe?

Ans. Modes describe the types of transportation services that are used by residents in the region. For the purposes of the survey, they include: auto driver, auto passenger (as distinct from the driver), public transit (bus or rail), Para transit, school bus, motorcycle, taxi, ferry, walking and cycling.

Q. 6. What is modal share?

Ans. Modal share is the proportion of trips by any given mode out of the total trips by all modes, for a given time period

Q. 7. Define origin destination. What does it provide?

Ans. Origin-destination (O-D) surveys provide a detailed picture of the trip patterns and travel choices of a city's or region's residents. These surveys collect valuable data related to households, individuals and trips. This information allows stakeholders to understand: Travel patterns and characteristics measure trends provide input to travel demand model development forecasting, and planning for area-wide transportation needs and services progress in implementing transportation policies.

Q. 8. What is the purpose of origin destination survey?

Ans. In a transportation study, it is often necessary to know the exact origin and destination of the trips. It is not only necessary to know how many trips are made, but also group these trips with reference to the zones of their origin and destination. Other information yielded by the O-D survey includes: Land-use of the zones of origin and destination. Household characteristics of the trip-making family. Time of the day when journeys are made.

Q. 9. What are the specific objectives of OD survey?

Ans. The specific objectives to which O-D survey data can be put are:

- To determine the amount of by-passable traffic that enters a town, thus establish the need for a bypass.

- To develop trip generation and trip distribution models in the transportation planning process.
- To determine the extent to which the present highway system is adequate and to plan new facilities. To assess the adequacy of parking facilities and to plan for future.

Q. 10. What is home interview survey? Explain.

Ans. It is one of the most reliable type of surveys for collection of origin and destination data. It mainly focuses on travel pattern of the residents of the household and general characteristics of the household influencing trip-making. Travel pattern includes number of trips made their origin and destination purpose of trip travel mode time of departure and arrival home school play ground.

Q. 11. Discuss road-side interview survey.

Ans. Road-side interview survey is one of the methods of carrying out a screen-line or cordon survey. The road-side interview survey can be done either directly interviewing drivers of vehicles at selected survey points or by issuing prepaid post cards containing questionnaire to all or sample of the drivers. The survey points are selected along the junction of the cordon-line or screen-line with the roads. Less than 5000 population – External cordon is sufficient. For dual carriageways or roads with very little traffic, the traffic in both the directions is dealt with simultaneously. In other cases the traffic in 2 directions will be interviewed at different times. If the survey covers most of the day it may be sufficient to interview traffic in one direction only and to assume that the journeys in the opposite direction are the same as the direction interviewed.

Q. 12. What do you mean by post-card questionnaire survey? Explain.

Ans. In this method, reply-paid questionnaires are handed over to each of the drivers or a sample of them at the survey points and requesting them to complete the information and return by post. The method avoids delay caused to the drivers by direct road-side interview method but suffer from disadvantage that the response may not be good.

Q. 13. What is registration number plate survey?

Ans. Registration number plate survey consists of nothing the registration numbers of vehicles entering or leaving an area at survey points located on the cordon line. By matching the registration numbers of vehicles at the points of entry and exit from the area, one is enabled to identify 2 points on the paths of the vehicle.

Q. 14. Briefly explain the method of selecting the survey sample for origin-destination (O-D) survey.

Ans. However, after operating the stations for a short time, it became apparent that a much larger sample could easily be obtained without jeopardizing traffic safety. As soon as the vehicles previously selected for interviews had left the survey station and merged back into traffic on road, the traffic constables and interviewers would direct the next five vehicles over to the roadside (rather than waiting for 25 vehicles to pass by before starting to select again). While these vehicles were being interviewed, all other traffic was permitted to pass by the survey station. When it was time to direct more vehicles over to the survey station, the traffic constables and interviewers would normally choose the next five vehicles.

III. REVIEW QUESTIONS

1. What is the need for OD Surveys?
2. Explain the preliminaries of OD survey.
3. How will you check the accuracy of OD survey data?
4. Discuss the survey methods.

REFERENCES

1. National Committee on Urban Transportation, Procedure Manuals 2 A (Origin-Destination and Land Use) and 2 B (Conducting a Home Interview Origin-Destination Survey), Public Administration Service, Chicago.

6. Parking Surveys

6.1. Need for Parking Surveys

Parking is one of the serious problems that confront the urban planner and the traffic engineer. Before any measures for the betterment of the conditions can be formulated basic data pertaining to the availability of parking space, extent of its usage and parking demand are essential. If it is proposed to implement a system of parking charges it will also be necessary to know how much to charge and what will be the effect of the pricing policy on parking. Parking surveys are intended to supply all this kind of information.

6.2. Definition of Common Terms

First of all, it is desirable to understand clearly the meaning of certain terms associated with parking. These terms and their meanings are given below.

Parking Accumulation. The total number of vehicles parked in an area at a specified moment. The curve of parking accumulation for a typical day is given in Fig. 6.1.

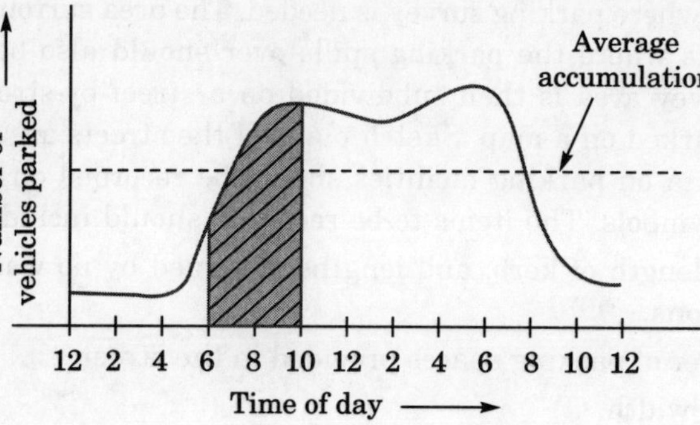

Fig. 6.1. Parking accumulation diagram.

Parking volume. The number of vehicles parking in a particular area over a given period of time. It is usually measured in vehicles per day.

Parking load. The area under the parking accumulation curve during a specified period. For example, in Fig. 6.1, the hatched area represents the parking load in vehicle-hour for a period of 4 hours from 6 AM to 10 AM.

Parking duration. The length of time spent in a parking space.

Parking index. Percentage of parking bays actually occupied by parked vehicles as compared to the theoretical number available.

$$\text{Parking index} = \frac{\text{Number of bays occupied}}{\text{Theoretical number of bays available}} \times 100$$

Parking turn-over. Rate of the usage of the available parking space. Thus if there were 10 parking spaces used by 100 vehicles in a period of, say 12 hours, then the parking turnover would be :

= 100/10 vehicles per space in a period of 12 hours.

6.3. Types of Parking Surveys

The type of parking survey to be conducted for formulating a comprehensive parking plan for an area can be vary detailed in scope. The data collected and the degree of sophistication employed depend upon the funds available. Detailed manuals for conducting such a survey give general guidance (Ref. 1–3).

The following are the types of parking surveys usually conducted :

(*i*) Parking space inventory.

(*ii*) Parking usage survey by patrol.

(*iii*) Questionnaire type parking usage survey.

(*iv*) Cordon count.

These will be discussed in detail.

6.4. Parking Space Inventory

The first step in a parking survey is to collect data on the amount, type and location of space actually or potentially available for parking in an area. The area to be surveyed should first be delineated. The central business district is usually the area where parking survey is needed. The area surrounding the central business districts where the parking spills over should also be included in the survey. The survey area is then subdivided on a street-by-street basis and the sub-divisions marked on a map. Sketch plans of the streets are then prepared in advance. The data on parking facilities should be recorded on the sketch plans using suitable symbols. The items to be recorded should include the following :

(*a*) Total length of kerb, and lengths governed by no waiting and limited waiting restrictions.

(*b*) Number of parking spaces provided in the street.

(*c*) Street width.

(*d*) Location of bus stops, bus bays, pedestrian crossings, fire hydrants, loading zones, taxi stands, driveways and other features that are likely to affect the use of the street for parking.

(*e*) Traffic management measures in force, such as prohibited turns, one way streets, exclusive bus lanes etc.

(*f*) Number and type of traffic signs for regulation of parking.

(*g*) Private streets, service and rear-access alleys.

(*h*) Vacant or unused land suitable for temporary or permanent parking space.

A typical sketch plan for a street with the above information recorded is given in Fig. 6.2.

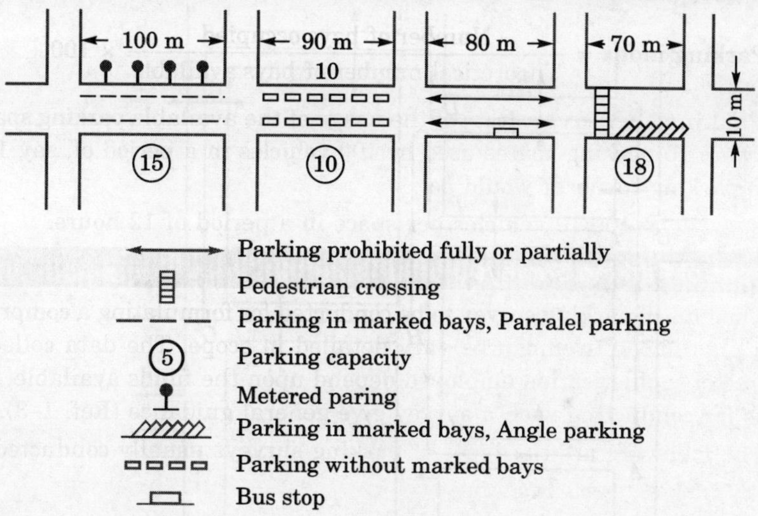

Fig. 6.2

Off-street parking facilities should be recorded separately, indicating the type, capacity, parking charges, parking pattern, entrances and exits.

6.5. Parking Usage Survey by Patrol

6.5.1. Purpose

The purpose of parking usage survey is to obtain data on the extent of usage of parking spaces. The survey will include counts of parked vehicles at regular intervals through a period, covering both the morning and evening peak period, and the parking accumulation and turnover.

The survey can be for on-street and off-street parking. The general methodology for both the surveys is similar, though minor details can be different. The method consists of making periodic observations of parked vehicles on each patrol. For off-street observations, the entire parking space can be patrolled or, alternatively, the entrances and exits may be observed continuously.

6.5.2. Mapping the street system

The first step is to prepare a map of the street system that will be covered by the patrol, showing therein its sub-division into sections. Street junctions make convenient points for determining the sections. A typical map is given in Fig. 6.3.

The recording can be for both sides of the road or separately for each side. The map and the forms should clearly show the direction of travel by the patrol-man and the side or sides where observations are to be recorded (by arrows). The length of streets to be covered by a patrol is limited by :

(*i*) the speed of walking while noting the registration numbers and (*ii*) the frequency of patrol.

A speed of 900 m in half an hour is useful guide, which should be altered to suit the density of parking.

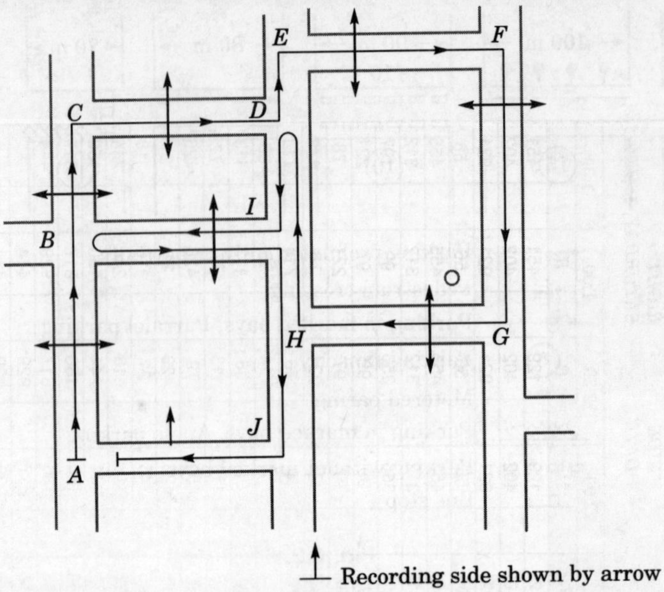

Fig. 6.3. Patrol map.

6.5.3. Frequency of patrol

Though more frequent patrols result in more accurate data the field work and subsequent analysis becomes more tedious. A frequency of ½ an hour is considered to be satisfactory for on-street parking, while a frequency of 1 hour could be used for off-street parking. A frequency of ½ an hour is likely to miss short-term parkers (up to 29 minutes duration) and this makes it necessary to have more frequent patrols in selected areas where short-term parking may be significant, *e.g.*, near banks, post offices etc.

6.5.4. Method of observation

Usually patrols are by foot, but where vehicles are not parked too close to one another a moving car may also be used. As an aid, a tape recorder may be used to record the registration numbers of vehicles.

6.5.5. Timing of survey

The survey should be done on a typical week day, free from factors likely to result in non-representative characteristics. The period of the survey is usually 10 to 12 hours, so as to cover the arrival and departure of commuters and shoppers.

6.5.6. Equipment and form of recording

Each observer will be equipped with a watch, a pencil, a supply of forms, a map of the street and a board. A specimen form is shown in Fig. 6.4.

6.5.7. Analysis

The analysis of the data is simplified if it is done in a methodical manner (Ref. 3) as indicated in Forms 1, 2, 3, 4 and 5. In these forms the data recorded in Fig. 6.4 have been used.

PARKING SURVEY
ON STREET : DURATION

PATROL : *2*
STREET : *KING ST.*
SECTION : *EF*
SIDE : *RIGHT*
DATE : *SATURDAY, 15.4.2017*

A.M. P.M.

8.00		8.30		9.00		9.30		10.00		10.30		11.00		11.30		12.00		12.30		1.00		1.30	
L.	*R.*	*L.*	*R.*	*L.*	*R.*	*L.*	*R.*	*L.*	*R.*	*L.*	*R.*	*L.*	*R.*	*L.*	*R.*	*L.*	*R.*	*L.*	*R.*	*L.*	*R.*	*L.*	*R.*
	421		421		402		336		814		836		129		129		129		129		451		451
	545		356		421		402		336		402		402		402		402		402		402		402
	817		27		353		421		402		921		921		921		921		921		921		921
			98		350		333		921		421		56		56		56		56		56		56
			113		114		350		421		455		455		455		455		455		455		455
			545		27		114		333		333		312		312		312		312		312		312
			120		714		114		356		356		356		356		356		356		99		197
			817		98		27		114		114		99		99		99		99		97		221
					113		97		27		97		97		97		97		97		221		76
					656		714		97		221		221		221		221		221		76		172
					545		98		221		65		76		76		76		76		714		714
					212		113		714		714		714		714		714		714		98		98
					149		450		98		95		98		98		98		98		113		113
					120		656		113		113		113		113		113		113		77		77
					88		545		450		450		450		450		450		79		426		426
					817		212		428		426		426		426		426		426		21		21
					44		149		656		656		345		345		345		21		118		118
							150		118		118		118		118		118		118		244		244
							120		212		212		212		212		244		244		150		150
							817		149		150		150		150		150		150		817		817
							44		250		149		136		136		136		817		929		444
							929		817		817		817		817		817		929		444		
									929		929		929		929		929		929		444		
											444		444		444		444						

Fig. 6.4. Specimen form of recording data on parking useage by patrol survey.

PARKING ANALYSIS : Form 1 (NUMBER OF TIMES SEEN) ACCEPTABLE NUMBER OF PARKING SPACES : 24 PARKING RESTRICTIONS IN FORCE : None																PATROL : 2 STREET : KING ST. SECTION : EF SIDE : RIGHT DATE : SATURDAY, 15.4.2017					
Veh.	NUMBER OF TIMES SEEN																				
	8.00	8.30	9.00	9.30	10.00	10.30	11.00	11.30	12.00	12.30	1.00	1.30	2.00	2.30	3.00	3.30	4.00	4.30	5.00	5.30	Total
421	6	6	6	6	6	6															6
545	1																				1
817	1																				1
356		9	9	9	9	9	9	9	9	9											9
27		4	4	4	4																4
98		11	11	11	11	11	11	11	11	11	11	11									11
113		11	11	11	11	11	11	11	11	11	11	11									11
etc.																					
etc.																					
Total	8	8	17	29	21	24	24	24	24	23	22	20	22	24	24	24	24	23	21	16	411
Arrival	3	7	9	7	2	6	7	1	1	2	1	2	6	5	4	6	5	5	4	5	89
Departure		2		1	4	2	7	1	1	3	2	4	4	3	4	6	5	6	6	1	62

Fig. 6.5. Parking Analysis, Form : 1.

PARKING ANALYSIS : FORM 2 (DURATION)																PATROL : 2 STREET : KING ST. SECTION : EF SIDE : RIGHT DATE : SATURDAY, 15.4.2017						
No. of Times Seen	Avg. Duration (Hrs.)	Number of vehicles parked																				Total
		8.00	8.30	9.00	9.30	10.00	10.30	11.00	11.30	12.00	12.30	1.00	1.30	2.00	2.30	3.00	3.30	4.00	4.30	5.00	5.30	
1	1/2	2		1	2		1	1					2				1	2		1	1	14
2	1			1	1							1	1			2			2			4
3	1.5		1	1	2	1	1	2	3	3	3	2	2			3	3	3				10
4	2		2	6	6	6	4			1	1	1	1			2	6	6	6	4		13
etc.																						
etc.																						
																						89
Total		3	8	17	23	21	24	24	24	24	23	22	20	22	24	24	24	24	23	21	16	411

Fig. 6.6. Parking analysis : Form 2.

Note. (1) The total number of times each vehicle was seen over the whole survey period should be recorded in each half-hourly column to facilitate subsequent analysis.

(2) An arrival is recorded for the period of the first recording.

(3) A departure is recorded for the period of the first omission of a recording.

PARKING ANALYSIS : *FORM 3* AREAS : 1, 2 & 3
OCCUPANCY : *DERIVED FROM FORM 1* DATE : SATURDAY, 15.4.2017

Area No.	Capacity of Area		8.00	8.30	9.00	9.30	10.00	10.30	11.00	11.30	12.00	12.30	1.00	1.30	2.00	2.30	3.00	3.30	4.00	4.30	5.00	5.30
1.	206	No. of Veh.	44	78	141	192	210	212	228	210	212	210	210	204	212	212	212	212	210	212	172	102
		%	21	38	69	95	102	103	101	102	103	102	102	109	103	103	103	103	102	103	84	50
2.	340	No. of Veh.	50	165	272	251	345	346	344	345	344	356	339	340	342	346	350	351	342	330	280	100
		%	15	49	80	98	102	102	101	102	101	99	100	100	101	102	103	103	102	97	82	29
3.	128	No. of Veh.	28	39	80	91	120	110	112	114	108	109	102	100	104	118	124	126	127	116	84	52
		%	22	30	63	71	94	86	88	89	84	85	80	78	81	92	97	98	99	91	66	41

Fig. 6.7. Parking Analysis : Form 3.

Note. **The extreme right column records the total number of vehicles parked for a given period during the day, *i.e.*, the number of vehicles parked in each half-hourly period divided by the number of time each vehicle was seen.**

PARKING ANALYSIS : FORM 4 PARKING THROUGHOUT THE SURVEY PERIOD (DERIVED FROM FORM 2)				AREA NO. 1 DATE : SATURDAY, 15.4.2017				
NO. OF TIMES SEEN	**1**	**2**	**3-4**	**5-8**	**9-12**	**13-20**	**21-22**	**TOTAL**
Average Duration in Hours	½	1	1½-2	2½-4	4½-6	6½-10	Over 10	
Total Parked	712	414	398	342	162	84	20	2132
%	39	19	19	17	7	4	1	100
Vehicle-Hours of occupation	356	414	695	1110	850	694	215	4334
%	8	10	16	25	20	16	5	100

Fig. 6.8. Parking Analysis : Form 4.

PARKING ANALYSIS : FORM 5 PARKING DURATION THE PEAK PERIOD (DERIVED FROM FORM 2)				AREA NO. 1 DATE : SATURDAY, 15.4.2017				
NO. OF TIMES SEEN	**1**	**2**	**3-4**	**5-8**	**9-12**	**13-20**	**21-22**	**TOTAL**
Average Duration in Hours	½	1	1½-2	2½-4	4½-6	6½-10	Over 10	
Total Parked	23	40	72	81	74	62	10	362
%	6	11	20	22	21	17	3	100
Vehicle-Hours of occupation	12	40	126	264	388	512	108	1450
%	1	3	9	18	27	35	7	100

Fig. 6.9. Parking Analysis : Form 5.

6.6. Questionnaire Type Parking Useage Survey

6.6.1. The questionnaire type parking useage survey involves interviews with the drivers who use the parking facilities. As a result, it is possible to collect information on the extent to which the existing facilities are being used, the parking requirements at the prices existing at the time of the survey, the parking demand at different prices, the distribution of demand over area and time and the journey purposes of car parkers (Ref. 5).

6.6.2. The survey can either be made by making enquiries among the car owners living in the vicinity of the survey area, or by making enquiries among the drivers of cars seen to park in the area at the time of the survey. In the former method, a sample of car owners is selected from the list of car owners and the questionnaires are delivered to them. Non-respondents can be followed up and the requirements of the sample are grossed up to the requirements of the whole population. The second method suffers from a disadvantage that it misses information about potential parkers who never bring their cars to the survey area for want of space, but this can be overcome by a supplementary sample.

6.6.3. In the interview of actual parkers, the information collected should include :

(*i*) address of origin of the trip

(*ii*) address of destination of the trip

(*iii*) trip purpose

(*iv*) time of arrival at the parking place

(*v*) time of departure from the parking place

(*vi*) type of parking space used

(*vii*) type of vehicle.

Normally one interviewer is required to cover about fifteen spaces. All the parkers in 8 or 10 hours period are interviewed. The duration of the survey may be a single day (working day) or may be spread over a number of days.

6.7. Cordon Count

In this method, the area to be surveyed is demarcated by a cordon line which is crossed by the roads emanating from the area. Counting stations are established at these crossing points and a count is made of all the vehicles entering and leaving the area. The difference between the two traffic gives the number of vehicles parked or in motion in the area. The counting can be manual or automatic.

6.8. Photographic Methods

Photographic methods are finding useful application in parking surveys. These are dealt with in Chapter 7.

I. MULTIPLE CHOICE QUESTIONS

1. Which of the following factors are not strictly related to design of traffic rotary intersections?

   1. Radius of central island 2. Weaving length

   3. Ramps and interchanges 4. Acceleration lanes

   Select the correct answer using the codes given below:

   (*a*) 1 and 2 (*b*) 1 and 4

   (*c*) 2 and 3 (*d*) 3 and 4 (*IES, 2008*)

2. Which one of the following geometric features requires the magnitudes of weaving angle and weaving distance or its design?

   (*a*) rotary design (*b*) right-angle intersection

   (*c*) groundabout (*d*) grade-separated junction

   (*IES, 2009*)

3. Which set of traffic studies is needed for functional design as well as for 'highway capacity' design?

   (*a*) origin and destination studies (*b*) parking and accident studies

   (*c*) speed and volume studies (*d*) axle load studies (*IES, 2009*)

4. Ratio of the width of the car parking area required at kerb for 30o parking relative to 60″ Parking is approximately.

   (*a*) 0.5 (*b*) 0.7

   (*c*) 0.8 (*d*) 2.0 (*IES, 1996*)

5. Match List-I (Traffic survey) with List-II (Method) and select the correct answer using the codes given below the lists:

List-I	List-II
A. Spot speed	1. By video tape
B. Traffic volume	2. By road side interview
C. O-D survey	3. By Doppler radar
D. Parking survey	4. By pneumatic tube

Codes:

	A	B	C	D
(*a*)	3	1	2	4
(*b*)	2	4	3	1
(*c*)	3	4	2	1
(*d*)	2	1	3	4

(*IES, 2005*)

6. Ratio of the width of the car parking area required at kerb for 30° parking relative to 60° parking is approximately

(*a*) 0.72 (*b*) 0.57 (*c*) 0.83 (*d*) 2.0

Answers

1. (*d*) **2.** (*a*) **3.** (*c*) **4.** (*c*) **5.** (*c*)
6. (*d*)

II. SOLVED QUESTIONS AND ANSWERS

Q. 1. What is parking?

Ans. Parking is one of the major problems that is created by the increasing road traffic. It is an impact of transport development. The availability of less space in urban areas has increased the demand for parking space especially in areas like central business district. This affects the mode choice also. This has a great economical impact.

Q. 2. What is meant by street parking?

Ans. On street parking means the vehicles are parked on the sides of the street itself. This will be usually controlled by government agencies itself. Common types of on-street parking are as listed below. This classification is based on the angle in which the vehicles are parked with respect to the road alignment. As per IRC the standard dimensions of a car is taken as 5 × 2.5 meters and that for a truck is 3.75 × 7.5 meters.

Q. 3. What is parallel parking? Illustrate with example. Illustrate with figure.

Ans. The vehicles are parked along the length of the road. Here there is no backward movement involved while parking or unparking the vehicle. Hence, it is the most safest parking from the accident perspective. However, it consumes the maximum curb length and therefore only a minimum number of vehicles can be parked for a given kerb length. This method of parking produces least obstruction to the on-going traffic on the road since least road width is used. Parallel parking of cars is shown in Fig. 1.

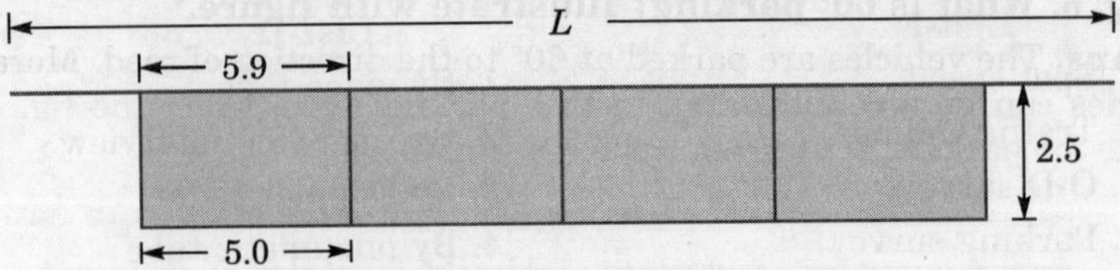

Fig. 1. Illustration of parallel parking

The length available to park N number of vehicles, $L = N/5.9$

Q. 4. What is 30° parking? Illustrate with figure.

Ans. In 30° parking, the vehicles are parked at 30° with respect to the road alignment. In this case, more vehicles can be parked compared to parallel parking. Also there is better maneuverability. Delay caused to the traffic is also minimum in this type of parking. An example is shown in Fig. 2. From the Fig. 1.

AB	$= OB \sin 30°$	$= 1.25$
BC	$= OP \cos 30°$	$= 4.33$
BD	$= DQ \cos 60°$	$= 5$
CD	$= BD - BC = 5 - 4.33$	$= 0.67$
$AB + BC$	$= 1.25 + 4.33$	$= 5.58$

For N vehicles, $L = AC + (N - 1)CE = 5.58 + (N - 1)5 = 0.58 + 5N$

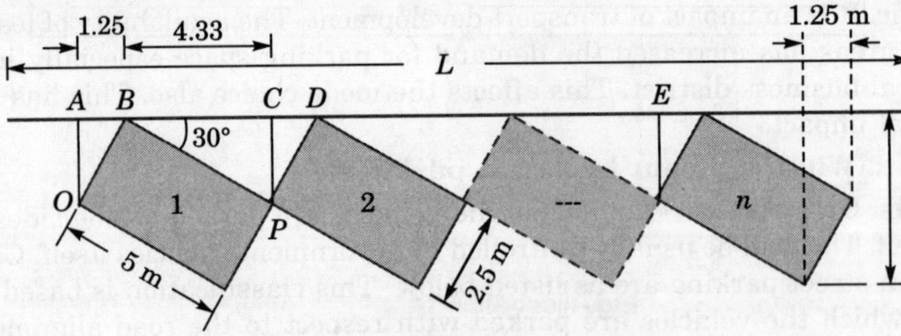

Fig. 2. Illustration of 30° parking

Q. 5. Explain about 45° parking. Illustrate with fig.

Ans. As the angle of parking increases, more number of vehicles can be parked. Hence compared to parallel parking and 30° parking, more number of vehicles can be accommodated in this type of parking. From fig, length of parking space available for parking N number of vehicles in a given kerb is $= 3.54N + 1.77$

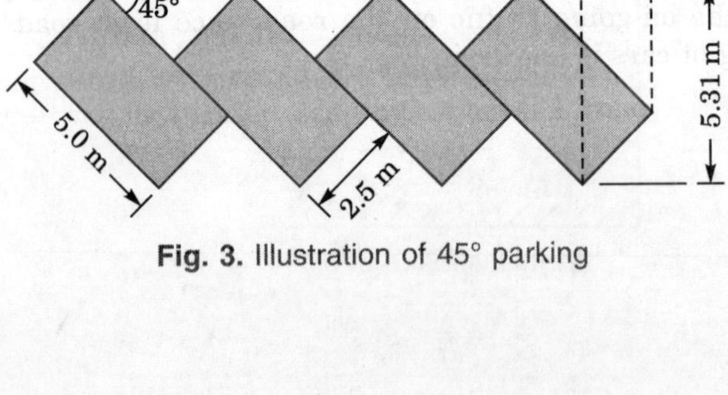

Fig. 3. Illustration of 45° parking

Q. 6. What is 60° parking? Illustrate with figure.

Ans. The vehicles are parked at 60° to the direction of road. More number of vehicles can be accommodated in this parking type. From the fig. 4, length available for parking N vehicles = 2.89 N + 2.16.

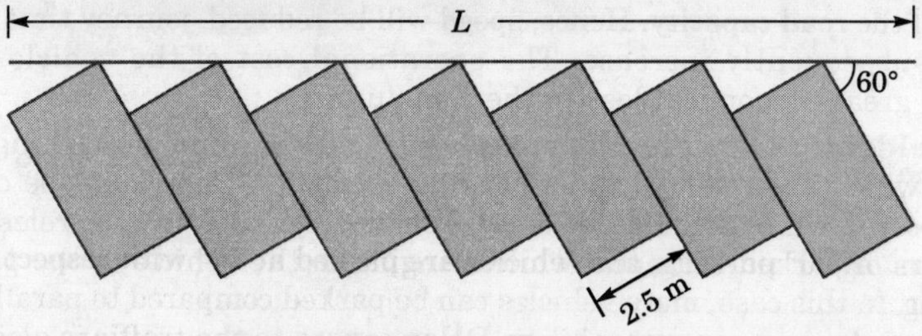

Fig. 4. Illustration of 60° parking

Q. 7. What is right angle parking? Illustrate with figure.

Ans. In right angle parking or 90° parking, the vehicles are parked perpendicular to the direction of the road. Although it consumes maximum width kerb length required is very little. In this type of parking, the vehicles need complex maneuvering and this may cause severe accidents. This arrangement causes obstruction to the road traffic particularly if the road width is less. However, it can accommodate maximum number of vehicles for a given kerb length. An example is shown in fig. Length available for parking N number of vehicles is $L = 2.5\ N$.

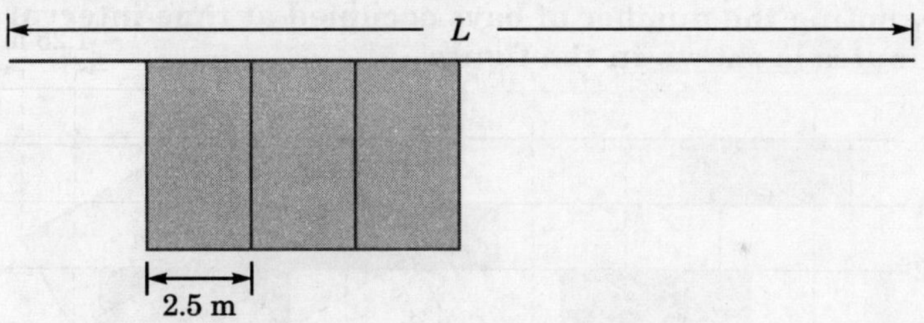

Fig. Illustration of 90° parking

Q. 8. What is off street parking?

Ans. In many urban centers, some areas are exclusively allotted for parking which will be at some distance away from the main stream of traffic. Such a parking is referred to as off-street parking. They may be operated by either public agencies or private firms. A typical layout of an off-street parking is shown in fig.

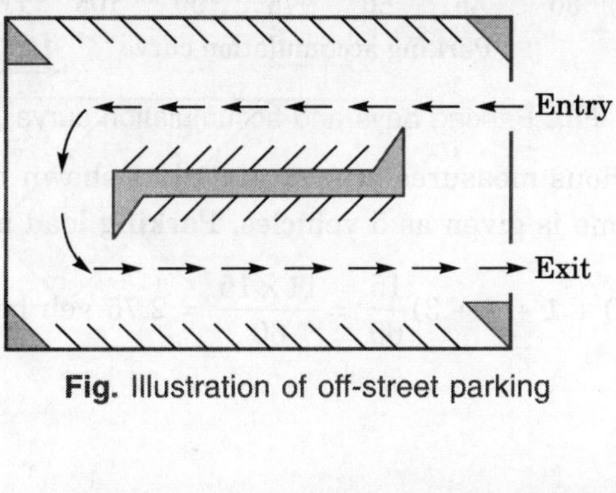

Fig. Illustration of off-street parking

Q. 9. Discuss the effects of parking.

Ans. Parking has some ill-effects like congestion, accidents, pollution, obstruction to fire-fighting operations etc.

Congestion: Parking takes considerable street space leading to the lowering of the road capacity. Hence, speed will be reduced, journey time and delay will also subsequently increase. The operational cost of the vehicle increases leading to great economical loss to the community.

Accidents: Careless maneuvering of parking and unparking leads to accidents which are referred to as parking accidents. Common type of parking accidents occur while driving out a car from the parking area, careless opening of the doors of parked cars, and while bringing in the vehicle to the parking lot for parking.

Environmental pollution: They also cause pollution to the environment because stopping and starting of vehicles while parking and unparking results in noise and fumes. They also affect the aesthetic beauty of the buildings because cars parked at every available space creates a feeling that building rises from a plinth of cars.

Obstruction to fire fighting operations: Parked vehicles may obstruct the movement of firefighting vehicles. Sometimes they block access to hydrants and access to buildings.

Q. 10. To illustrate the various measures, consider a small example in figure, which shows the duration for which each of the bays are occupied(shaded portion). Now the accumulation graph can be plotted by simply noting the number of bays occupied at time interval of 15, 30, 45 etc. minutes is shown in the figure.

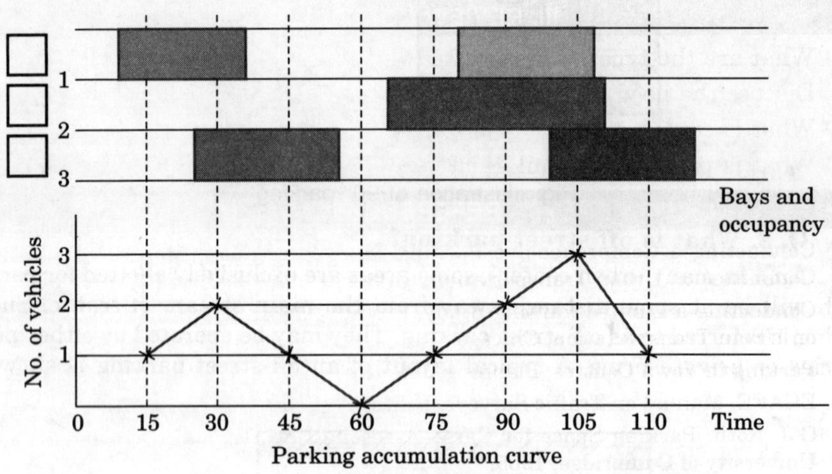

Fig. Parking bays and accumulation curve

Ans. The various measures are calculated as shown below:

Parking volume is given as 5 vehicles. Parking load is given as

$$(1 + 2 + 1 + 0 + 1 + 2 + 3)\frac{15}{60} = \frac{11 \times 15}{60} = 2.75 \text{ veh hour}$$

Average parking duration is computed as $\frac{2.75 \text{ veh hour}}{5 \text{ veh}} = 33$ minutes.

Parking turnover is obtained as $\frac{5 \text{ veh/2 hour}}{5 \text{ veh}} = 0.83$ veh/hr/bay. Parking index is calculated as $\frac{2.75 \text{ veh hour} \times 100}{3 \times 2 \text{ veh hours}} = 45.38\%$.

Q. 11. Why parking surveys are conducted?

Ans. Parking surveys are conducted to collect the above said parking statistics. The most common parking surveys conducted are in-out survey, fixed period sampling and license plate method of survey.

In this survey, the occupancy count in the selected parking lot is taken at the beginning. Then the number of vehicles that enter the parking lot for a particular time interval is counted. The number of vehicles that leave the parking lot is also taken. The final occupancy in the parking lot is also taken. Here the labor required is very less. Only one person may be enough. But we wont get any data regarding the time duration for which a particular vehicle used that parking lot. Parking duration and turn over is not obtained. Hence we cannot estimate the parking fare from this survey. For quick survey purposes, a **fixed period sampling** can also be done. This is almost similar to in-out survey. All vehicles are counted at the beginning of the survey. Then after a fixed time interval that may vary between 15 minutes to 1 hour, the count is again taken. Here there are chances of missing the number of vehicles that were parked for a short duration.

III. REVIEW QUESTIONS

1. What are the types of surveys?
2. Discuss the need and purpose of parking surveys.
3. What is cordon count?
4. What is parking accumulation?

REFERENCES

1. Conducting a Comprehensive Parking Study, Procedure Manual 3*D,* National Committee on Urban Transportation, Chicago.
2. Conducting a Limited Parking Study, Procedure Manual 3C National Committee on Urban Transportation, Chicago.
3. Parking in Town Centres, Planning Bulletin 7, H.M.S.O., London 1965.
4. ECAFE, Manual on Traffic Surveys, United Nations, New York, 1971.
5. G.J. Roth, Parking Space for Cars—Assessing the Demand, Occasional Papers, University of Cambridge, 1965.

7. Use of Photographic Techniques in Traffic Survey

7.1. Historical

Though photography itself was invented some two hundred years ago, it was the First World War which gave an impetus to rapid developments in the application of aerial photography in producing maps. It was at this time that traffic on roads had just started growing. The congestion caused by the ever increasing number of motor vehicles on the road soon gave birth to the new science of traffic engineering. Engineers began to study the behaviour of different aspects of road traffic such as speed, volume, concentration, accidents and geometric design. It soon became evident that traffic studies constituted major part of traffic engineering and elaborate instrumentation was needed for this purpose. Data collection and analysis by manual methods was proving increasingly cumbersome and inadequate to pursue the complex interaction between man, machine, environment, time and space. The sophistication in instrumentation has, by now, reached a high degree, sided by the developments in electronics and computers. Photographic techniques fall into this category of useful instruments.

The first traffic study utilising aerial surveys reported in literature dates back to 1927, when the technique was used for a study of traffic between Baltimore and Washington (Ref. 1). Greenshields, an early pioneer in traffic engineering, was a proponent of photographic methods in studying traffic behaviour (Ref. 2). In 1947, he dealt with the potential use of photographs in traffic analysis (Ref. 3) and the experiments he reported used a helicopter for taking aerial photographs. The following years saw a remarkable development in the field. Techniques such as stereo—continuous strip photography and time lapse photography have been found to be extremely useful tools in the hands of the traffic engineer.

7.2. Different Methods of Photography in Traffic Studies

The different methods involving the principle of photography in traffic studies are :

(1) Continuous stereoscopic strip photography from moving aeroplanes or helicopters.

(2) Stereo-camera photos from a stationary object on the ground.

(3) Conventional air photography, from aeroplanes, or helicopters.

(4) Time-lapse photography from aeroplanes, balloons or helicopters.

(5) Time-lapse photography from fixed position of the ground.

(6) Video system.

7.3. Appraisal of Photographic Methods for Traffic Studies

Photographic methods for traffic studies offer a number of advantages. They present more accurate data, use limited manpower and provide a permanent record of traffic conditions on which a variety of analyses can be performed without returning to the field. The driver behaviour is unaffected by the study, unlike manual methods, in which the presence of personnel on the roadside might seriously affect driver behaviour. The photographic methods obviate much of the tedium of manual count and data collection. They are ideal for a comparison of the *before* and *after* conditions. They afford opportunity for easily determining the inter-relationship between various traffic parameters, such as volume, speed and concentration, as against conventional techniques wherein each study tends to be limited to a particular object.

On the other hand, aerial photographic surveys can be expensive, inspite of the fact that they require less manpower. The greatest difficulty is in analysis of the film which is complicated, time consuming and at times expensive. Some of these disadvantages, however, do not apply to time-lapse photography which uses simple instruments.

To sum up, it may be stated that photographic techniques can touch many important aspects of traffic engineering and can offer much benefits. Its potentialities are virtually unlimited and it is a powerful tool both for research and for operational studies (Ref. 7,8, 9).

7.4. Conventional Photography

Conventional photographs can be taken from a moving plane or from a stationary platform, such as a helicopter. Ordinary air-survey camera with a 230 mm × 230 mm or 70 mm × 70 mm format is commonly used. A 60 per cent before and after overlap and 25 per cent lateral overlap is most common, though overlaps upto 90 per cent have also been used.

Speed can be determined by measuring the distance travelled between consecutive prints, the time interval between the prints being known. Acceleration and deceleration can be measured by the change in vehicle speeds where there are three or more images of the vehicle on consecutive prints. Distance headways can simply be obtained by measuring the distance between consecutive vehicles on any single photograph. Time headways can be obtained by dividing the distance headways by the speed of the following vehicle. Traffic density (*i.e.* concentration) can be measured directly by counting the vehicles on the photographs. In the case of traffic volume determinations, some difficulty arises because the camera and the vehicles both move and the entire survey area is not photographed instantaneously by the camera. The difficulty is overcome by introducing a correction on the basis of average speeds and equal distance headways. A study of car parking by conventional photographs is very convenient. If the photographs are taken at say, 15 min. intervals, relatively few short-stay vehicle are missed when carrying out a parking duration survey.

7.5. Continuous Strip Photography (Ref. 4, 5, 6)

The continuous strip photography method utilises stereoscopic photography, but is unlike conventional photography in that two continuously overlapping strips of photography are produced. Usually a double lens system is used. A narrow slit

(¼ – ½ mm wide) lies across the width of the film near the focal plane of a pair of lenses. In order to provide stereoscopic effect, the lenses are inclined, one forward and the other aft. The photograph taken through the forward lens will record a ground object before the aeroplane passes over it and that taken through the aft lens will record the same object after the aeroplane passes over it. Thus a stereo-pair of photographs result. It is possible to determine parameters such as speed, volume, lateral spacing, headway, parking characteristics, vehicle classification etc.

7.6. Time-Lapse Photography

7.6.1. Principle behind time-lapse photography technique

Photographs taken at continuous intervals of time lapse provide a record of the continuous changes or development taking place in a system. For example, if a series of photographs at short intervals are taken of a concrete cube under a compression test, the gradual behaviour of the cube with time can be permanently recorded. The stages in which the initial hair cracks appear, how these hair cracks grow into bigger ones and how finally the cube fails can be gradually recorded in the same sequence in which they occurred. If these photographs are projected at any selected speed, the entire process which has been photographed can be reproduced for the viewer or any of the individual frames containing the photographs can be studied.

Road traffic, which consists of a sequence of events spread over time, with vehicle and pedestrian movements taking place continuously, is ideally suited for study by the time lapse technique. Traffic is a complicated interaction of a number of factors and ordinary manual methods of collection of data cannot fully cope up with the demands of a comprehensive study embracing different parameters such as volume, concentration, delay, speed etc. A photographic film can, on the other hand, permanently record the entire sequence of events, so that the analyst can at a time study any particular aspect he wishes to.

A continuous film at high speed (l/16th of a second per exposure) for movies will be wasteful of film and will be expensive for traffic study. Films taken at a lapse of every second are generally sufficient for the purpose. Each individual film taken at this interval of one second is called a *frame.* Each *frame* then represents the true traffic condition at the particular instant it was exposed. In particular, the position of different vehicles in the traffic stream is pictorially represented in the frame. Successive frames will contain different positions of the same vehicles, and a comparison of these will yield the traffic data desired, knowing the time interval between the frames being compared.

7.6.2. Equipment needed for the study

The three basic items of equipment needed for the study are :

(*i*) A cine camera capable of exposing films at the intervals desired.

(*ii*) A timing device for accurately effecting the exposure at the intervals desired.

(*iii*) A cine projector for analysing the films.

7.6.3. Camera

7.6.3.1. The camera for a time-lapse study has to be of a special type, and in particular should meet the following basic requirements (Ref. 10).

(*i*) The camera should be able to run for periods upto one and half hours so as to cover the peak traffic flow period. This implies that at the rate of one frame per second, a length of 60 m of film would be needed and the camera should be able to contain this length of film.

(*ii*) The camera must be able to expose one frame on receipt of an impulse from an electrical or mechanical timing device.

(*iii*) The camera must be portable and self-contained.

7.6.3.2. A number of cameras are described in different studies (Ref. 10, 12, 13, 14). A 35 mm photography has been described by Forbes and Reiss in 1952 (Ref. 11).

A typical camera is powered from a small 12 volt battery. It is capable of operating at cine speeds of 8, 16 or 24 frames/second and, in addition, can also operate for time-lapse photography. In the latter case, the shutters operate on receipt of electric impulses at predetermined intervals. The camera can hold a 60 m length of 16 mm film. Shorter films of 15 m or 30 m can be loaded. It can be mounted on a tripod or can be strapped to a lighting standard. The exposure times range from 1/250 second to 1/20 second. By using a different circuit it can be arranged for one pulse to open the shutter and a second pulse to close it thus completing the exposure, which can be of any desired duration.

The timing device is an equally important piece of equipment. Its accuracy controls the correctness of the entire data acquired and it is, therefore, highly necessary that the timing device should be simple, reliable and at the same time extremely accurate. A typical unit consists of a D.C. motor run on a 9-volt battery, and one revolution of final output corresponds to a period of one second. A counter, which can be set to zero at the start of the work, advances by one unit each time a pulse is sent to the exposure. An extremely good accuracy is generally possible.

7.6.4. Cine projector for analysing the film

A number of cine projectors are available for analysing the film. It is generally possible to project at the slow motion speed of 2 frames per second, in addition to the normal projection speed of 16 frames per second. A still picture can be shown for any duration of time and the film can be progressed frame by frame by operating a push-button switch on the projector base, or by remote control. There is provision for hand winding the film backwards, enabling a sequence to repeated without rethreading. A resettable frame counter enables a single frame or sequence to be identified. This also permits the duration of an operation to be determined. The projector can be placed conveniently on a small desk and the film projected on a small screen.

7.6.5. Analysis

7.6.5.1. The analysis is facilitated by a perspective grid which should be drawn on a sheet of white paper which will serve as the screen.

7.6.5.2. Volumes

If a total of V vehicles pass a given point between frames m and n, the frames being at t sec intervals,

$$\text{The traffic volume} = \frac{V}{(n-m)t} \times 3600 \text{ Vehicle per hour}$$

Usually the mid-point of the section is chosen for counting the volumes.

7.6.5.3. Concentration

The concentration of vehicles can be measured by counting the number of vehicles in a chosen length, say 100 m or 150 m. The number of vehicles in this length is counted on each series of frames and the mean value of concentration can be found by averaging out the concentration in a number of frames.

7.6.5.4. Speed

The speed can be computed by measuring the time interval required for a vehicle to travel a known distance. If a vehicle passes the end points of a section in frames m and n the time taken for the vehicle to travel the length of the section is $(m-n)\,t$ sec. If the length of the section is d m, the speed of the vehicle is given by :

$$\frac{(m-n)t}{d} \text{ m/sec.}$$

7.6.5.5. Gaps and headway

The time gaps between successive vehicles can be found by the difference in frame numbers when successive vehicles pass through a given point. Thus if m and n are the frame numbers when two successive vehicles pass a given point, then :

$$\text{Time gap} = (n - m)\,t \text{ sec.}$$

7.6.6. Appraisal of the time-lapse photographic technique

Time-lapse photographic technique renders it possible to obtain a large number of observations of several traffic parameters with only one observer. This it is very economical of man-power. The photographic film is a permanent record of the traffic behaviour and can be analysed in detail in the office. The analyst can devote any length of time at a particular picture to study the peculiarities of traffic. The ability to stop the film, reverse and repeat the sequence is a very versatile device which can be used to great advantage. It is easy to study *before* and *after* behaviour of traffic when any improvements in road geometry or junction layout are contemplated. Pedestrian behaviour can be studied accurately. Vehicle paths can be traced exactly, thus leading to exact geometrical layouts of roads and junctions.

The major disadvantage is the tedium involved in manual processing of the data. One hour's film may take 50 hours or more of time of an operator. The labour can be minimised if help is taken of computers for analysis.

7.7. Video System

The latest development in the field of instrumentation for traffic studies is the video system (Ref. 15, 16, 18). The Central Road Research Institute in New Delhi acquired this instrumentation system, developed by the Australian Road Research Board. The system can be mounted on a car, at a stationary point by the side of a road or assembled in a laboratory. The instrumentation consists of the following components :

1. Video camera to take photographs of traffic, from a moving vehicle or a stationary position.

2. Video recorder and player to record the images taken from the camera on to video casettes and to-play back the recorded casettes at any time later.

3. Monitors (TV Screens).

The versatility of the system is of an extremely high order. By suitable electronic circuitry, information such as the time from a video clock, speed from radar speed-meters, distance travelled etc., can be displayed by numerical images on the monitor. An arrangement for analysis frame-by-frame can also be built in. By another arrangement the two images from two separate cameras can be combined and brought out simultaneously on the screen, the top half representing the view from one camera and the bottom half from another.

The system can be used for speed-volume, speed, fuel consumption, overtaking and crossing studies. The simulation model for predicting speed and fuel consumption under Indian conditions developed in India utilises this instrumentation system (Ref. 17).

I. MULTIPLE CHOICE QUESTIONS

1. Match List-I (Method of traffic volume counts) with List-II (Equipment used) and select the correct answer using the codes given below the lists:

List-I	List-II
A. Manual count	1. Video-recorder
B. Combination of manual and mechanical methods.	2. Pneumatic tube
C. Automatic devices	3. Watch
D. Photographic method	4. Multiple pen recorder

Codes:

	A	B	C	D
(*a*)	3	4	1	2
(*b*)	4	3	1	2
(*c*)	3	4	2	1
(*d*)	4	3	2	1

(*IES 2003*)

2. All survey data is .... through suitable computer software.

(*a*) coded (*b*) punched
(*c*) analysed (*d*) all of these

Answers

1. (*b*) **2.** (*d*)

II. SOLVED QUESTIONS AND ANSWERS

Q. 1. Why do we need photographic techniques in traffic survey?

Ans. There is a real need among traffic engineers, not only for more data, but for data of a more complex nature, in order to improve their understanding of traffic trends and the effects of changes in the existing infrastructure on these trends. While traditional ground survey methods seem incapable of keeping abreast of these expanding data needs, photographic techniques have the potential to do so, particularly when combined with modern co-ordinate reader and computer technologies. Although traffic data has been collected by photographic means for more than 50 years, the techniques, in spite of their advantages, have never become generally accepted. The major problem has always been the difficulties associated with extracting and analysing the vast quantities of data available from photographs.

Q. 2. What is meant by traffic survey?

Ans. Traffic surveys are the means of obtaining information about traffic. This is a systematic way of collecting data to be used for various traffic engineering purposes.

Q. 3. What are the main purposes of traffic survey?

Ans. The main purposes of traffic survey are: traffic monitoring, traffic control and management, traffic enforcement, traffic forecasting, model calibration and validating etc.

Q. 4. Discuss the parts of traffic studies.

Ans. Traffic studies include:

- Inventory of road traffic physical features
- Traffic stream characteristics-volume, speed, density, occupancy studies etc.
- Capacity studies of streets and intersections
- System usage studies- Travel time and delay, O-D survey
- Travel demand home interview survey
- Road users cost.

Q. 5. What is photo survey?

Ans. It follows from the above discussion that a photographic survey may have several advantages apart from the facto that a distress history will be in permanent record for future reference. Several photographic techniques have been used in the condition surveying of concrete pavement at the Center for Highway Research. Movies have been used in a survey of CRCP in different states and have been proven to be a way of recording the condition of a section in the field and that then allows a review of its condition later in the office. Because of the slow speed, 0 to 5 miles per hour, this method is not considered to be practical in urban area, as mentioned before.

Q. 6. What is photographic equipment?

Ans. Photographic surveys are possible during daylight hours and at night, with the assistance of a strobe. The latter method however poses two important restrictions. The first problem is that photo coverage only of one traffic lane is possible, which requires "manual" surveying of the other lanes unless the extra cost of photo surveying the other lanes is warranted. Manual surveying implies the counting of structural failures in the form of punchouts and repairs which are easily detected at 30 miles per hour (48 km/h). Thus, daylight surveying is required. The second restriction is that in order to photograph a continuous strip of pavement at 30 miles per hour (48 krn/h) for determination of crack spacing, a camera speed of at least 4 frames per second is required.

Thus, a strobe with an output of 4 or more flashes per second for about 8 seconds is required, which will require an extensive layout.

Q. 7. What is conventional photography?

Ans. Conventional photographs can taken from a moving plane or from a stationary platform, such as a helicopter. Ordinary survey camera with a 230 mm × 230 mm or 70 mm × 70 mm format is commonly used. A 60% n for and after overlap and 25% lateral overlap is most common, though overlaps up to 90% have also been used.

Q. 8. What is continuous strip photography?

Ans. The continuous strip photography method utilizes stereoscopic photography, but is unlike conventional photography in that two continuously overlapping strips of photography in that two continuously overlapping strips of photography are produced.

Usually a double lens system is used. A narrow slit lies across the width of the film near the focal plane of a pair of lenses. In order to provide stereoscopic effect, the lenses are inclined, one forward and the other aft. The photograph taken through the forward lens will record a ground object before the aeroplane passes over it.

III. REVIEW QUESTIONS

1. What is the photographic technique?
2. Discuss the appraisal of photographic methods.
3. What is continuous strip photography?
4. What is time-lapse photographic technique?

REFERENCES

1. Johnson, A.N., Maryland Aerial Traffic Density Survey, Proceedings, Highway Research Board, 7 (1). Washington, 1927.
2. Greenshields, B.D., The Photographic method of Studying Traffic Behaviour, Proceedings, Highway Research Board, Vol. 13. Washington, 1933.
3. Greenshields, B.D., The Potential Use of Air Photographs in Traffic Analysis, Proceedings, Highway Research Board, Vol. 27, Washington, 1947.
4. Wohl. M., Vehicle Speeds and Volume Using Sonne Stereo Strip Continuous photography, Traffic Engineering, 1959.
5. Mc Carland, W.T., Comparison of Two Techniques of Aerial Photography for Application in Freeway Traffic Operation Studies, Highway Research Record No. 65, Highway Research Board Washington.
6. Wohl. M. and S.M. Sickle, Continuous Strip Photography and Approach to Traffic Studies, Photogrammetic Engineering, Vol. XXV, Washington, 1959.
7. Wagner, F.A. and A.D. May, Use of Aerial Photography in Freeway Traffic Operations Studies, HRR No. 19, Highway Research Board, Washington, 1963.
8. Jordon, T.D. Development of the Sky Count Technique for Traffic Analysis, HRR No. 19 Highway Research Board, Washington, 1963.
9. Rice, R.F. Adoption of Aerial Survey Methods for Traffic Operations, HRR No. 19, Highway Research Board, Washington, 1963.
10. Constantine, T., Time Lapse Cinematography for Traffic Studies, Traffic Engineering and Control, March, 1974.
11. Forbes T.W. and R.J. Reiss, 35 mm Airphotos for the Study of Driver Behaviour, Bulletin 60, Highway Research Board, 1952.
12. White, R., Application of time Lapse Cinematography to Urban Traffic Engineering Problems, Traffic Engineering and Control, 7 (9), 1966.
13. White, R., the Application of Time-lapse Cinematography to Urban Traffic Problems, Lecture given to the East Midlands Traffic Engineering Group of the Institution of Civil Engineers (U.K.).
14. Worral, R.D., Time-lapse Cinematography Aids Traffic Studies, Traffic Engineering and Control, December, 1962.
15. Kadiyali, L.R., *et al* Background Paper on Traffic Simulation Experiments and Instrumentation Procedures, RUCS Technical Paper-No. 54 Twelfth Quarterly Report, Central Road Research Institute, New Delhi, 1980.
16. Troutbeck, R.J. and R. Ranga Rao, Some notes on Video equipment for overtaking studies, RUCS Technical Paper No. 81, Fifteenth Quarterly Report, Central Road Research Institute, New Delhi, 1981.
17. Traffic Simulation Modelling Study, Central Road Research Institute, New Delhi, 1985.
18. Viswanathan, E, *et al*, Video system for overtaking and crossing manoeuvres study, Vol. 50-2, Journal of the Indian Roads Congress, New Delhi, 1989.

SECTION 3

ANALYSIS AND INTERPRETATIONS OF TRAFFIC STUDIES

8. Statistical Methods for Traffic Engineering

8.1. Need

In ordinary sense, statistics deals with the collection and compilation of data. The subject of statistics has outgrown this simple definition and is now more and more concerned about drawing conclusions, and inferences from the observed data. Statistical methods find important applications in many fields of science and engineering. Traffic engineering is a subject which deals extensively with quantitative data in the planning, design and operation of transportation facilities. Statistical methods are a powerful tool to analyse and interpret these data. In this Chapter, the elementary principles of statistics are outlined, with particular reference to their application to traffic engineering studies.

8.2. Elementary Concepts of Probability

8.2.1. Definition of probability

In statistical methods, probability theories are useful in understanding and estimating the properties of a population and in testing some hypothesis about the population. Many statistical problems are couched in the language of probability and an elementary knowledge of the same and some fundamental concepts of probability will be essential.

When an experiment is repeated a large number of times under identical conditions and each time an event A does or does not occur, then the probability of occurrence of A is the proportion of times in which A occurs. Thus :

Probability of occurrence of A

$$= \frac{\text{Number of experiments in which } A \text{ occurs}}{\text{Total number of experiments}} \quad \text{...(8.1)}$$

The probability is usually expressed as a fraction and can assume values from 0 to 1. It is expressed symbolically as :

$$P(A) = \text{Probability of occurrence of event } A.$$

For example, consider the situation where approximately half the population of India are males. If the *experiment* consists of 'choosing a person living in India and the event A is that the person so chosen is a male, then the probability of occurrence of event A,

$$P(A) = 0.5.$$

8.2.2. Some simple laws of probability

(*a*) The probability of an event has a minimum possible value of 0 and a maximum possible value of 1.

Thus : $0 << P(A) << 1$

(*b*) If $P(A)$ is the probability of occurrence of an event A, then the probability of non-occurrence of the event, symbolically expressed as $P(\sim A)$, is $1 - P(A)$.

Thus : $P(\sim A) = 1 - P(A)$ ...(8.2)

As an example, if the probability of the arrival of a vehicle in a particular time interval is 0.6, the probability that no vehicle arrives in that time interval is 1 – 0.6 = 0.4.

(*c*) If an event A is a composite event, which is a collection of simple events $e_1, e_2, e_3, \ldots e_n,$ then the probability that the composite events A will occur is the sum of the probabilities of the simple events of which it is composed.

Thus : $P(A) = P(e_1) + P(e_2) + P(e_3) + \ldots + P(e_n)$ ...(8.3)

As an illustration, consider a simple count of vehicle arrivals at a point in fixed time intervals of 10 seconds. Let the probabilities of 0, 1, 2, 3, 4, 5, 6 and 7 and more vehicles arriving in 10 seconds interval be as follows :

$$P(0) = 0.13$$
$$P(1) = 0.27$$
$$P(2) = 0.27$$
$$P(3) = 0.18$$
$$P(4) = 0.09$$
$$P(5) = 0.03$$
$$P(6) = 0.01$$
$$P(7) = \underline{0.02}$$

1.00 (Total)

Then the probability of 3 and more vehicles arriving in the interval is,

$$P(r >> 3) = P(3) + P(4) + P(5) + P(6) + P(7)$$
$$= 0.18 + 0.09 + 0.03 + 0.01 + 0.02 = 0.33$$

(*d*) Total probability of 2 mutually exclusive events.

If two events, A and B, are mutually exclusive (*i.e.*, if A occurs, B cannot occur and *vice versa*), the total probability that one of these events occurs is :

$$P(A \text{ or } B) = P(A) + P(B) \quad \text{...(8.4)}$$

In the above example, the probability that one or two vehicles arrive during a time interval of 10 sec is :

$$P(1 \text{ or } 2) = P(1) + P(2) = 0.27 + 0.27 = 0.54$$

The above rule can be extended to more than two mutually exclusive events by applying the formula as many times as required. For example, if the probability of one, two or three vehicles arriving in a 10 seconds time interval is needed from the data above,

$$P(1 \text{ or } 2 \text{ or } 3) = P(1) + (2) + P(3)$$
$$= 0.27 + 0.27 + 0.18 = 0.72$$

(*e*) **Joint probability.** If two events *A* and *B*, are independent (*i.e.*, the occurrence of *A* has no influence on the occurrence of *B* and *vice versa*), the probability that both will occur together is :

$$P(A \text{ and } B) = P(AB) = P(A) \times P(B) \quad ...(8.5)$$

As an illustration, consider the simple problem of picking two balls from a box of 10, half of which are white and half black. After the first ball is picked it is returned to the box before the next ball is picked.

The probability of picking a white ball at the first attempt is 5/10.

The probability of picking a white ball at the second attempt is also 5/10.

The probability of picking 2 white balls consecutively is then given by

Pr (2 white balls picked) = *Pr* (first ball picked is white) × *Pr* (second ball picked is white)

$$= \frac{5}{10} \times \frac{5}{10} = 1/4.$$

If, in the above problem, the first ball is not returned to the box, and if it happens to be white, then there are only 9 balls now left out of which 4 are white. The probability of picking a white ball at the second attempt is then 4/9. In this case, the probability of picking 2 white balls consecutively is given by

Pr (2 balls picked are white) = *Pr* (first ball picked is white) × *Pr* (second ball picked is white, subject to the assumption that the first ball picked is also white) $= \frac{5}{10} \times \frac{4}{9} = \frac{2}{9}$.

The slight variation in the problem above introduces what is known as *conditional probability*. It is symbolically written as *P*(*B*/*A*) and is read as probability that *B* occurs when it is known that *A* has occurred. The joint probability rule can be written down as follows for conditional probability :

$$P(A \text{ and } B) = P(A) \,.\, P(B/A) \quad ...(8.6)$$

8.2.3. Probability distribution

Consider the example in the previous section. The probability of arrival of 0, 1, 2, 3, 4, 5, 6 and 7 and more vehicles in a given time interval are plotted, as shown in Fig. 8.1, with the *x*-axis indicating the number of arrivals, *k*, and the *y*-axis indicating the probability of occurrence of the event, *P*(*k*). The ordinate at each point gives the probability that the vehicle arrivals are as indicated by the point on the *x*-axis.

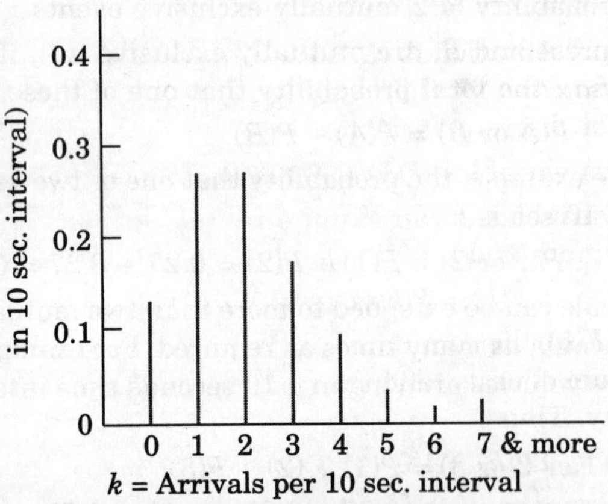

Fig. 8.1. Probability of vehicle arrivals per unit time. (Probability mass function).

In the other words, the above figure gives the possible values of a random variable (number of arrivals in this case) with the associated probabilities. These random variables can be continuous (height of persons, headway in seconds between vehicles etc.) or can be discrete (number of arrivals of vehicles in a stated time interval, number of vacant parking spaces at any instant, number of accidents etc.). If the random variable is continuous, the points giving the values of random variable and their associated probabilities can be joined by curve. If the variable is discrete, the graph is represented by vertical lines as in the above figure.

The above representation is termed as probability mass function for the variable and is denoted as $f(x)$, $f(k)$, $P(x)$ or $P(k)$. The letter x is used for continuous variables and the letter k for discrete variables.

The above data can be represented in a slightly different form. Suppose instead of the individual probability of occurrence of as event, if we plot the cumulative probability upto a particular value of the random variable, we get the cumulative probabilities for each value of the variable. Fig. 8.2 has been constructed such that the ordinate at each point represents the probability that there will be less than the stated number of arrivals.

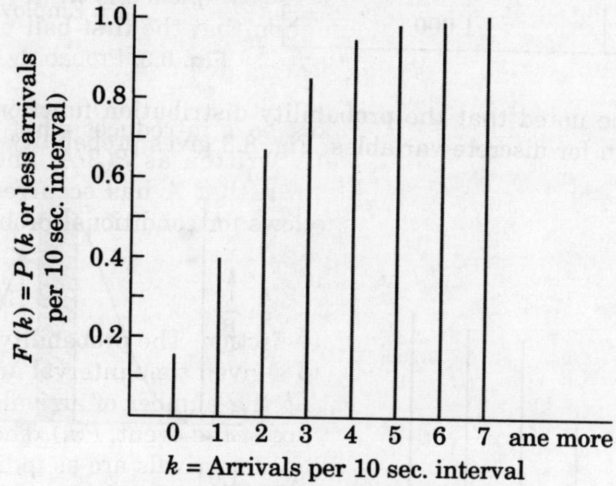

Fig. 8.2. Cumulative distribution function.

The above representation is called the *cumulative distribution function* or simply *distribution function for* the variables. It is denoted as $F(x)$ for continuous variables and $F(k)$ for discrete variables.

For continuous variables, $F(x) = P(x \text{ or less})$.

For discrete variables, k can assume values such as 1, 2, 3, 4, ... etc. upto the admissible maximum and

$$F(k) = P(k \text{ or less}).$$

The values of $F(k)$, the cumulative distribution function, will be a minimum of zero and a maximum of unity, and the sum of all values of the probability mass function will be unity. Thus

$$\Sigma P(k) = 1.0$$

This is demonstrated in the problem below.

Problem 8.1. *The probability that a particular household of 4 members has 0, 1, 2, 3, and 4 employed residents is given below :*

$$P(0) = 0.316, P(1) = 0.422$$

$$P(2) = 0.211, P(3) = 0.047, P(4) = 0.004$$

Plot the probability mass function and the cumulative distribution function.

Solution. The household has 4 members and this gives all the possible values of K as 0, 1, 2, 3, and 4. The sum of all the values of

$$P(k) = 0.316 + 0.422 + 0.211 + 0.047 + 0.004 = 1.000$$

The values of the cumulative distribution function $F(k)$ are tabulated below :

Values of K	*Values of Cumulative Distribution Function F(k)*
0	0.316
1	0.738
2	0.949
3	0.976
4	1.000

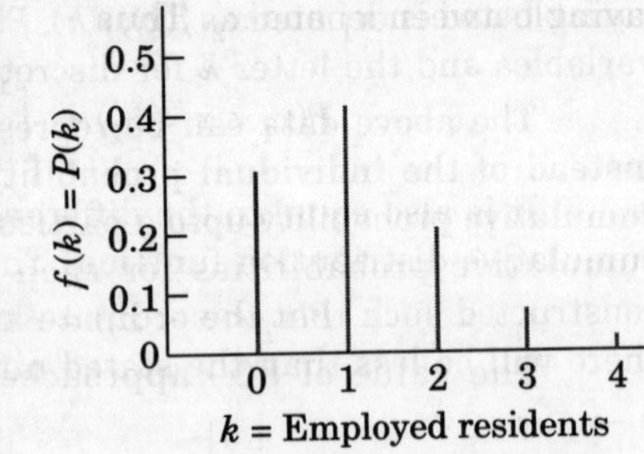

Fig. 8.3. Probability mass function.

It may be noted that the probability distribution function Fig. 8.4, is a stepped function for discrete variables. Fig. 8.3 gives probability mass function.

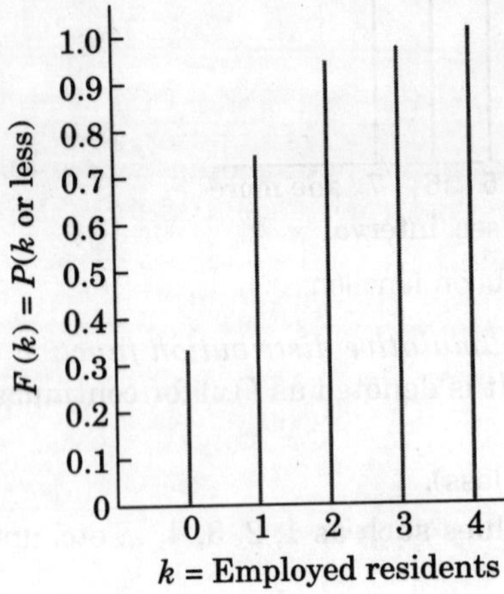

Fig. 8.4. Probability distribution function.

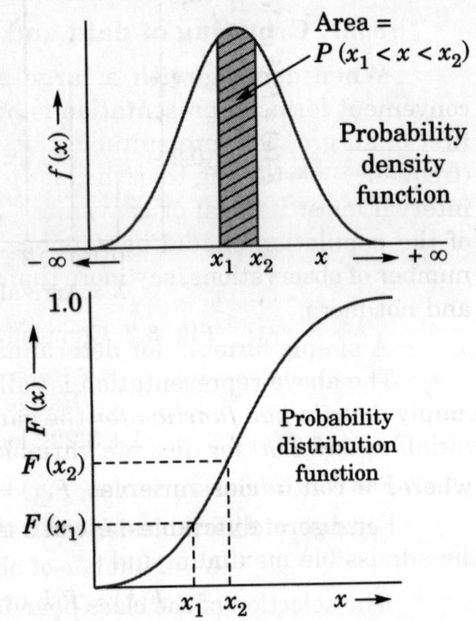

Fig. 8.5. Probability density and distribution functions for a continuous variable.

For continuous variables, the cumulative distribution function

$$F(x) = P(x \text{ or less})$$

$$\int_{-\infty}^{x} f(x)\, dx,$$

where, $f(x)$ is called the probability density function and is defined by :

$$f(x) = \frac{dF(x)}{dx}$$

The probability distribution and probability density function for a continuous variable are given in Fig. 8.5. It may be noted that the area under the probability density function is unity. Also, the area under the probability density function between two points x_1 and x_2 gives the probability of the variable laying between x_1 and x_2. Thus

$$P(x_1 < x < x_2) = \int_{x_1}^{x_2} f(x)\, dx \qquad ...(8.7)$$

It is also equal to the difference in $F(X)$ values at these two points on the cumulative distribution function. Thus

$$P(x_1 < x < x_2) = F(x_2) - F(x_1) \qquad ...(8.8)$$

The value of $F(x)$ approaches unity as $x \to +\infty$ and approaches 0 as $x \to -\infty$.

The above figures represent the condition when x varies between any two numbers x_1 and x_2 and the above relations still hold good in principle except that the limits for x will X_1 and X_2 instead of $-\infty$ and $+\infty$.

8.3. Mean, Standard Deviation and Variance

8.3.1. Grouping of data and presentation

When dealing with a large amount of data on variables, one of the convenient forms of presentation is by means of a frequency table. The data are first of all grouped into suitable class intervals and the number of observations (frequency) pertaining to each class is recorded. In selecting the size of the class interval, most natural or convenient classes should be selected and the picture of the population should be as smooth as possible. When dealing with a large number of observations, say more than 200, one should have about 10 to 20 classes and not more.

A simple formula for determining the size of the class interval is :

$$i = \frac{\text{Range}}{1 + 3.222 \log_{10} \text{N}}$$

where, i = class interval

Range = difference between the largest and smallest item,

N = Number of observations.

The selection of the class boundaries should be carefully done so that there is no doubt about placing a particular value in any particular class interval. The process will be clear from the following example.

Problem 8.2. *Table 8.1, below indicates the values of spot speeds observed at a certain location. Classify the data into a frequency table.*

Table 8.1

Obs. No.	*Speed (K.P.H.)*	*Obs. No.*	*Speed (K.P.H.)*
1	40.8	25	55.0
2	52.8	26	56.2
3	51.6	27	49.9
4	46.2	28	57.3
5	55.1	29	35.2
6	48.8	30	54.5
7	41.2	31	49.2
8	51.6	32	56.2
9	47.8	33	66.6
10	52.6	34	46.2
11	32.2	35	54.7
12	57.5	36	37.8
13	43.4	37	51.8
14	58.2	38	58.2
15	47.8	39	48.4
16	53.7	40	50.9
17	42.6	41	74.2
18	58.1	42	47.5
19	62.2	43	51.7
20	53.2	44	36.4
21	63.3	45	50.6
22	43.2	46	59.8
23	64.5	47	61.2
24	64.1	48	42.9

As examination of the 48 observations indicates that the least value is 32.2 K.P.H. and the highest values is 74.2 K.P.H. A selection of classes with an interval of 5 K.P.H. appears logical, covering the speeds from 30 K.P.H. to 75 K.P.H.

The speeds are recorded correct to one decimal place. In order to avoid ambiguity of recording a value such as 35.0, the class limits are made equal to 29.95–34.95, 34.95–39.95 and so on. With this arrangement, a reading of 34.9 will fall in the class 29.95–34.95, and the nest reading of 35.0 will fall in the next class of 34.95–39.95.

Table 8.2, illustrates the frequency table of data in Table 8.1. The mid-point of the class limit is entered in column 2. Column 3 indicates a convenient way of counting the observations. The relative frequency entered in column 5 is the actual frequency divided but the total number of observations. Column 6 gives the percentage frequency and column 7 the cumulative percentage frequency.

The graphical presentation of the above data by a histogram is possible where the horizontal axis represents the class-limits and the vertical axis the percentage frequency. The histogram for the data in table 8.2 is shown in Fig. 8.6.

Table 8.2. Frequency Table of Data in Table 8.2

Class limits	*Mid point*	*Tabulation*	*Frequency*	*Relative frequency*	*Percentage frequency*	*Cumulative percentage frequency*
(1)	*(2)*	*(3)*	*(4)*	*(5)*	*(6)*	*(7)*
29.95–34.95	32.45	𝍷	1	0.02	2.0	2.0
34.95–39.95	37.45	𝍷𝍷𝍷	3	0.06	6.0	8.0
39.95–44.95	42.45	𝍸 𝍷	6	0.13	13.0	21.0
44.95–49.95	47.95	𝍸 𝍷𝍷𝍷𝍷	9	0.19	19.0	40.0
49.95–54.95	52.45	𝍸 𝍸 𝍷𝍷	12	0.25	25.0	65.0
54.95–59.95	57.45	𝍸 𝍸	10	0.21	21.0	86.0
59.95–64.95	62.45	𝍸	5	0.10	10.0	96.0
64.95–69.25	67.45	𝍷	1	0.02	2.0	98.0
69.95–74.95	72.45	𝍷	1	0.02	2.0	100.0
		Total	48	1.00	100.0	

The cumulative percentage frequency diagram or ogive curve is also shown in Fig. 8.6. The vertical axis of this diagram indicates the percentage frequency of speeds less than a given speed. For this purpose, the upper-class limits of the class-intervals are considered when plotting the various points on the curve.

8.3.2. Parameters of distribution

The frequency table, histogram and the cumulative frequency curve give only rough idea of the distribution and their inherent characteristics. An accurate idea of the distribution can be had from the parameters of distribution, such as a measure of the central tendency and a measure of the dispersion. The most common measure of the central tendency is the arithmetic mean, or mean. Suppose there are n observations of a variable x (*i.e.*, sample size = n and these are denoted by $x_1, x_2, x_3, \ldots\ldots, x_n$), then the mean $\bar{x}$ is given by the formula :

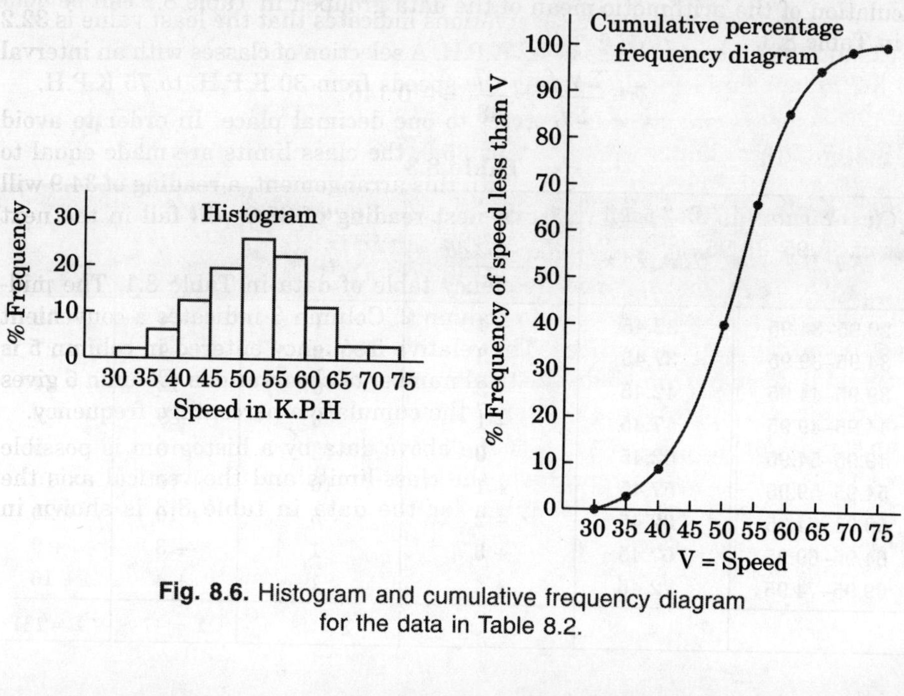

Fig. 8.6. Histogram and cumulative frequency diagram for the data in Table 8.2.

$$\bar{x} = \frac{x_1 + x_2 + x_3 \ldots\ldots x_n}{n}$$

$$= \frac{\sum_{i=1}^{n} x_i}{n} \qquad \ldots(8.9)$$

The above equation applies to individual observations of the variable x. When the observations are grouped into different classes, as was done in Table 8.2, the calculation of the mean is done with reference to the mid-points of the classes and the respective frequencies of the classes. Let $x_1, x_2, x_3, \ldots\ldots, x_m$ represent the mid-points of the m classes, $f_1, f_2, f_3, \ldots\ldots, f_m$ representing the frequencies of observations. The mean $\bar{x}$ is then given by the following formula :

$$\bar{x} = \frac{\sum_{i=1}^{m} x_i f_i}{\sum_{i=1}^{m} f_i} \qquad \ldots(8.10)$$

The computations can be at times simplified if a *false origin* is used. The variables x_1 can be transformed to a variable y_i by using false origin constants as below :

$$y_i = \frac{x_i - k}{c} \qquad \ldots(8.11)$$

and $$\bar{y} = \frac{\bar{x} - k}{c} \qquad \ldots(8.12)$$

where, k and c are constants.

The class-interval can be conveniently used as the constant c. The calculation of the arithmetic mean of the data grouped in Table 8.2 can be done as in Table 8.3.

$$\bar{y} = \frac{\Sigma y_i f_i}{\Sigma f_i} = \frac{-7}{48} = -0.146.$$

Table 8.3

Class Limits (K.P.H.)	*Mid-point (K.P.H.)* x_i	$y_i = \frac{x_i - 52.45}{5}$	*Frequency* f_i	$y_i f_i$	$y_i^2 f_i$
29.95–34.95	32.45	– 4	1	– 4	+ 16
34.95–39.95	37.45	– 3	3	– 9	+ 27
39.95–44.95	42.45	– 2	6	– 12	+ 24
44.95–49.95	47.45	– 1	9	– 9	+ 9
49.95–54.95	52.45	0	12	0	0
54.95–59.95	57.45	+ 1	10	+ 10	+ 10
59.95–64.95	62.45	+ 2	5	+ 10	+ 20
64.95–69.95	67.45	+ 3	1	+ 3	+ 9
69.95–74.95	72.45	+ 4	1	+ 4	+ 16
			Σ = 48	Σ = –7	Σ = 131

But $\bar{y} = \frac{\bar{x} - k}{c}$ from Eq. (8.12)

$$\therefore \quad -0.146 = \frac{\bar{x} - 52.45}{5}$$

$$\therefore \quad \bar{x} = 0.73 + 52.45 = 51.72, \text{ say } 51.7 \text{ K.P.H.}$$

A simple measure of dispersion is the range which is the difference between the largest and the smallest measurement. A better measure is the variance, which is defined below.

$$\text{Variance} = s^2 = \frac{\sum_{i=1}^{n} (x_i - \bar{x})^2}{n-1} \qquad \text{...(8.13)}$$

The standard deviation, s, is the square root of the variance and is given by :

$$\text{Standard deviation} = s = \sqrt{\frac{\sum_{i=1}^{n} (x_i - \bar{x})^2}{n-1}} \qquad \text{...(8.14)}$$

The following formula is something more convenient ;

$$s = \sqrt{\frac{\sum_{i=1}^{n} \left(x_i^2 - n\bar{x}^2\right)}{n-1}} \qquad \text{...(8.15)}$$

If a false origin constant k is used to simplify the calculations, the following formula can be used :

$$s = \sqrt{\frac{\sum_{i=1}^{n} (x_i - k)^2 - n(\bar{x} - k)^2}{n-1}} \qquad \text{...(8.16)}$$

For grouped data consisting of m classes the following formula can be used:

$$s = \sqrt{\frac{\sum_{i=1}^{m} (x_i - \bar{x})^2 f_i}{\sum_{i=1}^{m} f_i - 1}} \qquad \text{...(8.17)}$$

$$= \sqrt{\frac{\sum_{i=1}^{m} f_i x_i - \left(\sum_{i=1}^{m} f_i\right) \bar{x}^2}{\sum_{i=1}^{m} f_i - 1}} \qquad \text{...(8.18)}$$

For grouped data with false origin k,

$$s = \sqrt{\frac{\sum_{i=1}^{m} (\bar{x}_i - k)^2 f_i + \left(\sum_{i=1}^{m} f_i\right)(\bar{x} - k)^2}{\sum_{i=1}^{m} f_i - 1}} \quad ...(8.19)$$

Sometimes it is necessary to transform the variable x_i to another variable y_i by using two constants as below :

$$y_i = \frac{x_i - k}{c} \quad ...(8.20)$$

Then the standard deviation of y is first found from Eq. (8.21)

$$s_y = \sqrt{\frac{\sum_{i=1}^{m} (y_i)^2 f_i - \left(\sum_{i=1}^{m} (f_i)\right)(\bar{y})^2}{\sum_{i=1}^{m} f_i - 1}} \quad ...(8.21)$$

The standard deviation of x is then found from :

$$s_x^2 = s_y^2 c^2 \quad ...(8.22)$$

For the data contained in Tables 8.1 and 8.2 the last column in Table 8.3 gives $y_i^2 f_i$. The standard deviation of y is found from Eq. (8.21) thus :

$$s_y = \sqrt{\frac{131 - 48 \times \left(-\frac{7}{48}\right)\left(-\frac{7}{48}\right)}{48 - 1}} = \sqrt{\frac{131 - 1.02}{47}} = 1.66$$

$\therefore$ $s_x^2 = s_y^2 \times c^2$ from Eq. (8.22) $= (1.66)^2 \times 5 \times 5$

$\therefore$ $s_x = 1.66 \times 5 = 8.30$ K.P.H.

8.4. Poisson and Binomial Distributions

8.4.1. One of the relationship a traffic engineer finds extremely useful in dealing with the random properties of traffic is the Poisson distribution, named after Poisson, a French mathematician.

The Poisson distribution, is the limiting case of the more generalised Binomial distribution which describes the probability of occurrence of a particular event in a specified number of trials.

8.4.2. The Binomial distribution is based on the following assumptions

(*i*) Each trial has only two possible outcomes, which are mutually exclusive. For example, success or failure, defective or non-defective, head or tail, good or bad etc. are mutually exclusive.

(*ii*) The population from which the samples are drawn or trials are made is infinite, such that the withdrawal of a sample or conducting a trial does not alter the relative proportion of samples with either of the two characteristics. If the population is finite, yet the same condition as above may be achieved provided the item drawn is returned to the population after a single trial, but before the next item is drawn.

Table 8.4. Binomial Coefficient ^nC_r or $\binom{n}{r}$ or $\dfrac{n!}{r!(n-r)!}$

n \ r	0	1	2	3	4	5	6	7	8	9	10
1	1	1									
2	1	2	1								
3	1	3	3	1							
4	1	4	6	4	1						
5	1	5	10	10	5	1					
6	1	6	15	20	15	6	1				
7	1	7	21	35	35	21	7	1			
8	1	8	28	56	70	56	28	8	1		
9	1	9	36	84	126	126	84	36	9	1	
10	1	10	45	120	210	252	210	120	45	10	1
11	1	11	55	165	330	462	462	330	165	55	11
12	1	12	66	220	495	792	924	792	495	220	66
13	1	13	78	286	715	1287	1716	1716	1287	715	286
14	1	14	91	364	1001	2002	3003	3432	3003	2002	1001
15	1	15	105	455	1365	3003	5005	6435	6435	5005	3003
16	1	16	120	560	1820	4368	8008	11440	12870	11440	8008
17	1	17	136	680	2380	6188	12376	19448	24310	24310	19448
18	1	18	153	816	3060	8568	18564	31824	43758	48620	43758
19	1	19	171	969	3876	11628	27132	50388	75582	92378	92378
20	1	20	190	1140	4875	15504	38760	77520	125970	167960	184756

(*iii*) The probability of occurrence of a certain event at each trial is a constant (say, p) so that probability of the non-occurrence of the event is also a constant (say q).

By the law of total probability, then

$$p + q = 1$$

$$\therefore \quad q = 1 - p.$$

(*iv*) The distribution is valid for only discrete values of the number of times an event can occur in a given number of trials.

The Binomial distribution gives :

$$P(r) = \frac{n!}{r!(n-r)!} p^r q^{n-r} \qquad \text{...(8.23)}$$

where, $P(r)$ = Probability of occurrence of an event r times in n trials, r and n being discrete numbers.

p = Probability of occurrence of the event in a single trial.

q = Probability of non-occurrence of the event in a single trial which is $1 - p$ as indicated earlier.

The Binomial Coefficient $\dfrac{n!}{r!(n-r)!}$ is also sometimes designated as ^nC_r or (r^n) and its value can be found from Table 8.4.

The following example illustrates the use of the Binomial distribution :

Problem 8.3. *In analysing the employment structure of house-holds in an urban area, in connection with a trip generation study it has been found that there are 2500 households of 4 members in each. Find the probability that a particular household of this size has 0, 1, 2, 3 and 4 employed residents.*

Solution. In this example, p, the probability of a resident being employed

$$= \frac{2500}{10{,}000} = \frac{1}{4}$$

$\therefore$ $q = 1 - p$ = Probability of a resident being unemployed = ¾

n = the number of residents in a household = 4

$\therefore$ Probability that a particular household has r employed residents out of 4,

$$P(r) = \frac{n!}{r!(n-r)!} p^r q^{n-r} \qquad \text{...(8.23)}$$

$$P(r = 0) = \frac{4!}{0!4!}\left(\frac{1}{4}\right)^0\left(\frac{3}{4}\right)^4 = \left(\frac{3}{4}\right)^4 = 0.316.$$

[Note. 0 ! = 1, (¼)⁰ = 1]

$$P(r = 1) = \frac{4}{1!3!}\left(\frac{1}{4}\right)^1\left(\frac{3}{4}\right)^3 = 0.422$$

$$P(r = 2) = \frac{4!}{2!2!}\left(\frac{1}{4}\right)^2\left(\frac{3}{4}\right)^2 = 0.211$$

$$P(r = 3) = \frac{4!}{3!1!}\left(\frac{1}{4}\right)^3\left(\frac{3}{4}\right)^1 = 0.047$$

$$P(r = 4) = \frac{4}{4!0!}\left(\frac{1}{4}\right)^4\left(\frac{3}{4}\right)^0 = 0.004$$

Note that total of all the above probabilities is

= 0.316 + 0.422 + 0.211 + 0.047 + 0.004 = 1.00.

The above data is plotted graphically in Fig. 8.7. Note that the graph is valid only for discrete values of r, the number of employed residents and the probability distribution is represented by erecting an ordinate of height $P(r)$ over each value of r.

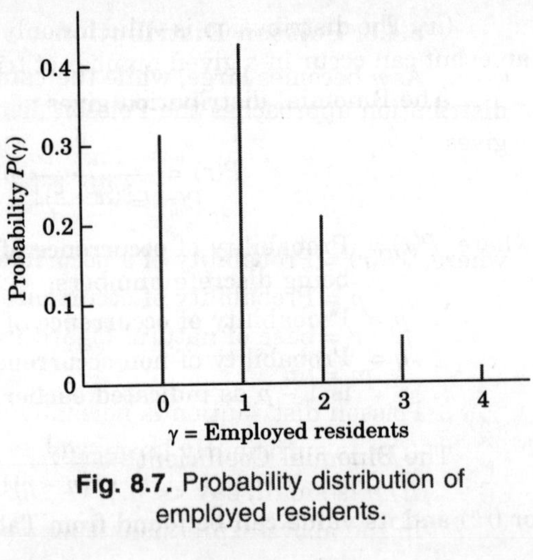

Fig. 8.7. Probability distribution of employed residents.

Problem 8.4. *At an uncon-trolled T junction, past experience indicates that the probability of a vehicle arriving on the side road during a 15 second interval and turning right into the main road is 1/5. Find the probability that in a period of 1 minute, there will be 0, 1, 2, 3 or 4 vehicles arriving and turning right.*

Solution. Here, the possibility p of a vehicle arriving on the side road and turning right, is one of the two mutually exclusive events, the other being the probability of a vehicle arriving and not turning right.

Hence, $p = \frac{1}{5}$

and $q = 1 - p = \frac{4}{5}$

n = Number of 15 second intervals in one minute = 4.

$$P(r) = \frac{n!}{r!(n-r)!} p^{(r)} q^{(n-r)} \qquad \text{...(8.23)}$$

$$P(r = 0) = \frac{4!}{0!4!}\left(\frac{1}{5}\right)^0 \left(\frac{4}{5}\right)^4 = \frac{256}{625}$$

$$P(r = 1) = \frac{4!}{1!3!}\left(\frac{1}{5}\right)^1 \left(\frac{4}{5}\right)^3 = \frac{4 \times 1}{5} \times \frac{64}{125} = \frac{256}{625}$$

$$P(r = 2) = \frac{4!}{2!2!}\left(\frac{1}{5}\right)^2 \left(\frac{4}{5}\right)^2 = \frac{4 \times 3}{2} \times \frac{16}{625} = \frac{96}{625}$$

$$P(r = 3) = \frac{4!}{3!1!}\left(\frac{1}{5}\right)^3 \left(\frac{4}{5}\right)^1 = \frac{4 \times 4}{625} = \frac{16}{625}$$

$$P(r = 4) = \frac{4!}{4!0!}\left(\frac{1}{5}\right)^4 \left(\frac{4}{5}\right)^0 = \frac{1}{625}$$

$$\text{Total probability} = \frac{256}{625} + \frac{256}{625} + \frac{96}{625} + \frac{16}{625} + \frac{1}{625} = \frac{625}{625} = 1.$$

8.4.3. Poisson Distribution

As n becomes large, while the value of np is a finite constant, the Binomial distribution approaches the Poisson distribution as a limit. Poisson distribution gives :

$$P(r) = \frac{np^r e^{-np}}{r!} = \frac{m^r e^{-m}}{r!} \qquad \text{...(8.24)}$$

where, $P(r)$ = Probability of a occurrence of an event r times in n trials.

p = Probability of occurrence of the event in a single trial.

e = base of natural logarithms.

$m = np$

Poisson distribution is normally valid when :

(*i*) n is sufficiently large, say >> 50, and

(*ii*) p is small, say << 0.1.

In the above expression, it may be noted that

$np = m$ = average number of times the event occurs.

Poisson distribution has found useful application in many aspects of traffic engineering. As early as 1936, Adams (Ref. 1) discussed certain aspects of the application. The most widely used situation is the arrival pattern of vehicles, which is a random phenomenon.

The expression $P(r) = \dfrac{np^r e^{-m}}{r!} = \dfrac{m^r \varepsilon^{-m}}{r!}$

can be conveniently modified to a more suitable form when dealing with the arrival pattern of vehicles, m in the above expression denotes the average number of occurrences, which can be determined from observed data or in the absence of observed data can be assumed from past experience. The observations can be instantaneous, or counting events during a stated time interval or counting units in a definite area etc. The general definition for m then is :

$$m = \frac{\text{Total number of events observed}}{\text{Total number of trials or time intervals}}$$

In the case of arrival of traffic, m becomes the average number of vehicles per any stated time interval. The time interval can be 10 seconds, 15 seconds, 20 seconds or any value suitable to a particular study, and can be designated as t seconds for the generalised case. If λ is the average rate of arrival per second, then $m = \lambda.t$

and the Poisson probability expression can be written as :

Probability of arrival of r vehicles in any time interval t seconds.

$$P(r) = \frac{(\lambda t)^r e^{-\lambda t}}{r!} \qquad \text{...(8.25)}$$

If V is the number of vehicles per hour, the value of λ is simply :

$$= \frac{V}{3600}$$

It should be noted that λ and t in the above expression should be in the same units, *viz.*, seconds, minutes or hour. It is convenient to work with seconds in most situations.

As an example of the use the Poisson distribution the following data pertaining to the arrival pattern of vehicles at a toll booth will be considered.

Problem 8.5. *On a motorway, the number of vehicles arriving from one direction in successive 10 seconds intervals was counted and recorded in Table 8.5.*

Table 8.5

Vehicles arriving in 10 seconds intervals	*Frequency*
0	*11*
1	*28*
2	*30*
3	*18*
4	*8*
5	*4*
6	*1*
7 and over	*0*

Find out the mean rate of arrival and with the help of Poisson distribution and compare the observed frequency with the theoretical frequency. Does the data suggest that the arrival pattern can be considered as random ?

Solution. The first step is to calculate the mean arrival rate, *i.e.*, the mean number of vehicles arriving per unit time.

The working is facilitated by a tabular form as shown in Table 8.6.

Table 8.6

No. of vehicles arriving in 10 sec. interval	*Observed frequency (f_i)*	*Total No. of vehicles col (1) × col (2)*	*Total time col (2) × 10 sec.*	*Theoretical probability of stated no. of arrivals in 10 sec.*	*Theoretical probability col (5) × Σf_i*
(1)	*(2)*	*(3)*	*(4)*	*(5)*	*(6)*
0	11	0	110	0.1353	13.53
1	28	28	280	0.2706	27.06
2	30	60	300	0.2706	27.06
3	18	54	180	0.1804	18.04
4	8	32	80	0.0932	9.02
5	4	20	40	0.0361	3.61
6	1	6	10	0.0120	1.20
7 and over	0	0	0	0.0048	0.48
	Σ = 100	Σ = 200	Σ = 1000	Σ = 1.0000	Σ = 100.00

Dividing total of column 3 by total of column 4,

λ = arrival per unit time (one second)

$$= \frac{\text{Total no. of vehicles arriving during the observation}}{\text{Total time of observation}}$$

$$= \frac{200}{1000} = 0.2 \text{ vehicles per second.}$$

In this example, t = time interval selected for observation

= 10 seconds

$$P(r) = \frac{(\lambda t)^r e^{\lambda t}}{r!}$$

$$= \frac{(0.2 \times 10)^r e^{-(0.2 \times 10)}}{r!} = \frac{2^r e^{-2}}{r!}$$

Calculation of $P(r)$ are facilitated by the following procedure :

$$P(r = 0) = \frac{(\lambda t)^0 e^{-\lambda t}}{0!}$$

$$= e^{-2} = 0.1353$$

$$P(1) = \frac{(\lambda t)^1 e^{-\lambda t}}{1!} = P(0) \times \frac{\lambda t}{1} = 0.2706.$$

$$P(2) = \frac{(\lambda t)^2 e^{-\lambda t}}{2!} = P(1) \times \frac{\lambda t}{2} = 0.2706.$$

$$P(3) = \frac{(\lambda t)^3 e^{-\lambda t}}{3!} = P(2) \times \frac{\lambda t}{3} = 0.1804.$$

$$P(4) = \frac{(\lambda t)^4 e^{-\lambda t}}{4!} = P(3) \times \frac{\lambda t}{4} = 0.0902.$$

$$P(5) = \frac{(\lambda t)^5 e^{-\lambda t}}{5!} = P(4) \times \frac{\lambda t}{5} = 0.0361.$$

$$P(6) = \frac{(\lambda t)^6 e^{-\lambda t}}{6!} = P(5) \times \frac{\lambda t}{6} = 0.0120.$$

$$P(7) = \frac{(\lambda t)^7 e^{-\lambda t}}{7!} = P(6) \times \frac{\lambda t}{7} = 0.0035.$$

and $$P(r \geq 7) = 1 - 0.9952 = 0.0048.$$

In the above procedure, the only computation involving $e^{-\lambda t}$ was performed for calculating $P(0)$ and the rest of the computations were simple.

Table 8.7. Some typical values of e^{-x} (e = 2.71828)

x	e^{-x}	x	e^{-x}	x	e^{-x}
0	1.0000	2.90	0.0550	6.70	0.0012
0.01	0.9900	3.00	0.0498	6.80	0.0011
0.02	0.9802	3.10	0.0450	6.90	0.0010
0.03	0.9704	3.20	0.0408	7.00	0.0009
0.04	0.9608	3.30	0.0369	7.10	0.0008
0.05	0.9512	3.40	0.0334	7.20	0.0007
0.06	0.9418	3.50	0.0302	7.30	0.0007
0.07	0.9324	3.60	0.0273	7.40	0.0006
0.08	0.9231	3.70	0.0247	7.50	0.0006
0.09	0.9139	3.80	0.0224	7.60	0.0005
0.10	0.9048	3.90	0.0202	7.70	0.0005
0.20	0.8187	4.00	0.0183	7.80	0.0004
0.30	0.7408	4.10	0.0166	7.90	0.0004
0.40	0.6703	4.20	0.0150	8.00	0.0003
0.50	0.6065	4.30	0.0136	8.10	0.0003
0.60	0.5488	4.40	0.0123	8.20	0.0003
0.70	0.4966	4.50	0.0111	8.30	0.0002
0.80	0.4493	4.60	0.0101	8.40	0.0002
0.90	0.4066	4.70	0.0091	8.50	0.0002
1.00	0.3629	4.80	0.0082	8.60	0.0002
1.10	0.3329	4.90	0.0074	8.70	0.0002
1.20	0.3012	5.00	0.0067	8.80	0.0002
1.30	0.2725	5.10	0.0061	9.00	0.0001
1.40	0.2466	5.20	0.0055	9.10	0.0001
1.50	0.2231	5.30	0.0050	9.20	0.0001
1.60	0.2019	5.40	0.0045	9.30	0.0001
1.70	0.1827	5.50	0.0041	9.40	0.0001
1.80	0.1653	5.60	0.0037	9.50	0.0001
1.90	0.1496	5.70	0.0033	9.60	0.0001
2.00	0.1353	5.80	0.0030	9.70	0.0001
2.10	0.1225	5.90	0.0027	9.80	0.0001

2.20	0.1108	6.00	0.0025	9.90	0.0001
2.30	0.1003	6.10	0.0022	10.00	0.0000
2.40	0.0907	6.20	0.0020		
2.50	0.0821	6.30	0.0018		
2.60	0.0743	6.40	0.0017		
2.70	0.0672	6.50	0.0015		
2.80	0.0608	6.60	0.0014		

Table 8.8. Poisson Probabilities, $P(r) = \dfrac{m^r e^{-m}}{r\,!}$

r	m									
	0.005	.01	.02	.03	.04	.05	.06	.07	.08	.09
0	.9950	.9900	.9802	.9704	.9608	.9512	.9418	.9324	.9231	.9139
1	.0050	.0099	.0192	.0291	.0384	.0476	.0565	.0653	.0738	.0823
2	.0000	.0000	.0002	.0004	.0008	.0012	.0017	.0023	.0030	.0037
3	.0000	.0000	.0000	.0000	.0000	.0000	.0000	.0001	.0000	.0001
	0.1	0.2	0.3	0.4	0.5	0.6	0.7	0.8	0.9	1.0
0	.9048	.8187	.7408	.6703	.6065	.5488	.4966	.4493	.4066	.3679
1	.0905	.1637	.2222	.2681	.3033	.3293	.3476	.3595	.3659	.3679
2	.0045	.0164	.0333	.0536	.0758	.0988	.1217	.1438	.1647	.1839
3	.0002	.0011	.0033	.0072	.0126	.0198	.0284	.0383	.0494	.0613
4	.0000	.0001	.0002	.0007	.0016	.0030	.0050	.0077	.0111	.0153
5	.0000	.0000	.0000	.0001	.0002	.0004	.0007	.0012	.0020	.0031
6	.0000	.0000	.0000	.0000	.0000	.0000	.0001	.0002	.0003	.0005
7	.0000	.0000	.0000	.0000	.0000	.0000	.0000	.0000	.0000	.0001
	1.1	1.2	1.3	1.4	1.5	1.6	1.7	1.8	1.9	2.0
0	.3329	.3012	.2725	.2466	.2231	.2019	.1827	.1653	.1496	.1353
1	.3662	.3614	.3543	.3452	.3347	.3230	.3106	.2975	.2842	.2707
2	.2014	.2169	.2303	.2417	.2510	.2584	.2640	.2678	.2700	.2707
3	.0738	.0867	.0998	.1128	.1255	.1378	.1496	.1607	.1710	.1804
4	.0203	.0260	.0324	.0395	.0471	.0551	.0636	.0723	.0812	.0902
5	.0045	.0062	.0084	.0111	.0141	.0176	.0216	.0260	.0309	.0361
6	.0008	.0012	.0018	.0026	.0035	.0047	.0061	.0078	.0098	.0122
7	.0001	.0002	.0003	.0005	.0008	.0011	.0015	.0020	.0027	.0034
8	.0000	.0000	.0001	.0001	.0001	.0002	.0003	.0005	.0006	.0009
9	.0000	.0000	.0000	.0000	.0000	.0000	.0001	.0001	.0001	.0002
	2.1	2.2	2.3	2.4	2.5	2.6	2.7	2.8	2.9	3.0
0	.1225	.1108	.1003	.0907	.0821	.0743	.0672	.0608	.0550	.0498
1	.2572	.2438	.2306	.2177	.2052	.1931	.1815	.1703	.1596	.1494
2	.2700	.2681	.2652	.2613	.2565	.2510	.2450	.2384	.2314	.2240
3	.1890	.1966	.2033	.2090	.2138	.2176	.2205	.2225	.2237	.2240
4	.0992	.1082	.1169	.1254	.1336	.1414	.1488	.1557	.1622	.1680
5	.0417	.0476	.0538	.0602	.0668	.0735	.0804	.0872	.0940	.1008
6	.0146	.0174	.0206	.0241	.0278	.0319	.0362	.0407	.0455	.0504
7	.0044	.0055	.0068	.0083	.0099	.0118	.0139	.0163	.0188	.0216
8	.0011	.0015	.0019	.0025	.0031	.0038	.0047	.0057	.0068	.0081
9	.0003	.0004	.0005	.0007	.0009	.0011	.0014	.0018	.0022	.0027

r	m									
	2.1	2.2	2.3	2.4	2.5	2.6	2.7	2.8	2.9	3.0
10	.0001	.0001	.0001	.0002	.0002	.0003	.0004	.0005	.0006	.0008
11	.0000	.0000	.0000	.0000	.0000	.0001	.0001	.0001	.0002	.0002
12	.0000	.0000	.0000	.0000	.0000	.0000	.0000	.0000	.0000	.0001
	3.1	3.2	3.3	3.4	3.5	3.6	3.7	3.8	3.9	4.0
0	.0450	.0408	.0369	.0334	.0302	.0273	.0247	.0224	.0202	.0183
1	.1397	.1304	.1217	.1135	.1057	.0984	.0915	.0850	.0789	.0733
2	.2165	.2087	.2008	.1929	.1850	.1771	.1692	.1615	.1539	.1465
3	.2237	.2226	.2209	.2188	.2158	.2125	.2087	.2046	.2001	.1954
4	.1734	.1781	.1823	.1858	.1888	.1912	.1931	.1944	.1951	.1954
5	.1075	.1140	.1203	.1264	.1322	.1377	.1429	.1477	.1522	.1563
6	.0555	.0608	.0662	.0716	.0771	.0826	.0881	.0936	.0989	.1042
7	.0246	.0278	.0312	.0348	.0385	.0425	.0466	.0508	.0551	.0595
8	.0095	.0111	.0129	.0148	.0169	.0191	.0215	.0241	.0269	.0298
9	.0033	.0040	.0047	.0056	.0066	.0076	.0089	.0102	.0116	.0132
10	.0010	.0013	.0016	.0019	.0023	.0028	.0033	.0039	.0045	.0053
11	.0003	.0004	.0005	.0006	.0007	.0009	.0011	.0013	.0016	.0019
12	.0001	.0001	.0001	.0002	.0002	.0003	.0003	.0004	.0005	.0006
13	.0000	.0000	.0000	.0000	.0001	.0001	.0001	.0001	.0002	.0002
14	.0000	.0000	.0000	.0000	.0000	.0000	.0000	.0000	.0000	.0001
	4.1	4.2	4.3	4.4	4.5	4.6	4.7	4.8	4.9	5.0
0	.0166	.0150	.0136	.0123	.0111	.0101	.0091	.0082	.0074	.0067
1	.0679	.0630	.0583	.0540	.0500	.0462	.0427	.0395	.0365	.0337
2	.1393	.1323	.1254	.1188	.1125	.1063	.1005	.0948	.0894	.0842
3	.1904	.1852	.1798	.1743	.1687	.1631	.1574	.1517	.1460	.1404
4	.1951	.1944	.1933	.1917	.1898	.1875	.1849	.1820	.1789	.1755
5	.1600	.1633	.1662	.1687	.1708	.1725	.1738	.1747	.1753	.1755
6	.1093	.1143	.1191	.1237	.1281	.1323	.1362	.1398	.1432	.1462
7	.0640	.0686	.0732	.0778	.0824	.0869	.0914	.0959	.0002	.1044
8	.0328	.0360	.0393	.0428	.0463	.0500	.0537	.0575	.0614	.0653
9	.0150	.0168	.0188	.0209	.0232	.0255	.0280	.0307	.0334	.0363
10	.0061	.0071	.0081	.0092	.0104	.0118	.0132	.0147	.0164	.0181
11	.0023	.0027	.0032	.0037	.0043	.0049	.0056	.0064	.0073	.0082
12	.0008	.0009	.0011	.0014	.0016	.0019	.0022	.0026	.0030	.0034
13	.0002	.0003	.0004	.0005	.0006	.0007	.0008	.0009	.0011	.0013
14	.0001	.0001	.0001	.0001	.0002	.0002	.0003	.0003	.0004	.0005
15	.0000	.0000	.0000	.0000	.0001	.0001	.0001	.0001	.0001	.0002
	5.1	5.2	5.3	5.4	5.5	5.6	5.7	5.8	5.9	6.0
0	.0061	.0055	.0050	.0045	.0041	.0037	.0033	.0030	.0027	.0025
1	.0311	.0287	.0265	.0244	.0225	.0207	.0191	.0176	.0162	.0149
2	.0793	.0746	.0701	.0659	.0618	.0580	.0544	.0509	.0477	.0446
3	.1348	.1293	.1239	.1185	.1133	.1082	.1033	.0985	.0938	.0892
4	.1719	.1681	.1641	.1600	.1558	.1515	.1472	.1428	.1383	.1339
5	.1753	.1748	.1740	.1728	.1714	.1697	.1678	.1656	.1632	.1606
6	.1490	.1515	.1537	.1555	.1571	.1584	.1594	.1601	.1605	.1606
7	.1086	.1125	.1163	.1200	.1234	.1267	.1298	.1326	.1353	.1377
8	.0692	.0731	.0771	.0810	.0849	.0887	.0925	.0962	.0998	.1033
9	.0392	.0423	.0454	.0486	.0519	.0552	.0586	.0620	.0654	.0668
10	.0200	.0220	.0241	.0262	.0285	.0309	.0334	.0359	.0386	.0413

r	m									
	5.1	5.2	5.3	5.4	5.5	5.6	5.7	5.8	5.9	6.0
11	.0093	.0104	.0116	.0129	.0143	.0157	.0173	.0190	.0207	.0225
12	.0039	.0045	.0051	.0058	.0065	.0073	.0082	.0092	.0102	.0113
13	.0015	.0018	.0021	.0024	.0028	.0032	.0036	.0041	.0046	.0052
14	.0000	.0007	.0008	.0009	.0011	.0013	.0015	.0017	.0019	.0022
15	.0002	.0002	.0000	.0003	.0004	.0005	.0006	.0007	.0008	.0009
16	.0001	.0001	.0001	.0001	.0001	.0002	.0002	.0002	.0003	.0003
17	.0000	.0000	.0000	.0000	.0000	.0001	.0001	.0001	.0001	.0001
	6.1	6.2	6.3	6.4	6.5	6.6	6.7	6.8	6.9	7.0
0	.0022	.0020	.0018	.0017	.0015	.0014	.0012	.0011	.0010	.0009
1	.0137	.0126	.0116	.0106	.0098	.0090	.0082	.0076	.0070	.0064
2	.0417	.0390	.0364	.0340	.0318	.0296	.0276	.0258	.0240	.0223
3	.0848	.0806	.0765	.0726	.0688	.0652	.0617	.0584	.0552	.0521
4	.1294	.1249	.1205	.1162	.1118	.1076	.1076	.0992	.0952	.0912
5	.1579	.1549	.1519	.1487	.1454	.1420	.1385	.1349	.1314	.1277
6	.1605	.1601	.1595	.1586	.1575	.1562	.1546	.1529	.1511	.1490
7	.1399	.1418	.1435	.1450	.1462	.1472	.1480	.1486	.1489	.1490
8	.1066	.1099	.1130	.1160	.1188	.1215	.1240	.1263	.1284	.1304
9	.0723	.0758	.0791	.0825	.0858	.0891	.0923	.0954	.0985	.1014
10	.0441	.0469	.0498	.0528	.0558	.0588	.0618	.0649	.0679	.0710
11	.0245	.0265	.0285	.0307	.0330	.0353	.0377	.0401	.0426	.0452
12	.0124	.0137	.0150	.0164	.0179	.0194	.0210	.0227	.0245	.0264
13	.0058	.0065	.0073	.0081	.0089	.0098	.0108	.0119	.0130	.0142
14	.0025	.0029	.0033	.0037	.0041	.0046	.0052	.0058	.0064	.0071
15	.0010	.0012	.0014	.0016	.0018	.0020	.0023	.0026	.0029	.0033
16	.0004	.0005	.0005	.0006	.0007	.0008	.0010	.0011	.0013	.0014
17	.0001	.0002	.0002	.0002	.0003	.0003	.0004	.0004	.0005	.0006
18	.0000	.0001	.0001	.0001	.0001	.0001	.0001	.0002	.0002	.0002
19	.0000	.0000	.0000	.0000	.0000	.0000	.0000	.0001	.0001	.0001
	7.1	7.2	7.3	7.4	7.5	7.6	7.7	7.8	7.9	8.0
0	.0008	.0007	.0007	.0006	.0006	.0005	.0005	.0004	.0004	.0003
1	.0009	.0054	.0049	.0045	.0041	.0038	.0035	.0032	.0029	.0023
2	.0208	.0194	.0180	.0167	.0156	.0145	.0134	.0125	.0116	.0107
3	.0492	.0464	.0438	.0413	.0389	.0366	.0345	.0324	.0305	.0286
4	.0874	.0836	.0799	.0764	.0729	.0696	.0663	.0632	.0602	.0573
5	.1241	.1204	.1167	1130	1094	.1057	.1021	.0986	.0951	0916
6	.1468	.1445	.1420	.1394	.1367	.1339	.1311	.1282	.1252	.1221
7	.1489	.1486	.1481	.1474	.1465	.1454	.1442	.1428	.1413	.1396
8	.1321	.1337	.1351	.1363	.1373	.1382	.1388	.1392	.1395	.1396
9	.1042	.1070	.1096	.1121	.1144	.1167	.1187	.1207	.1224	.1241
10	.0740	.0770	.0800	.0829	.0858	.0887	.0914	.0941	.0967	.0993
11	.0478	.0504	.0531	.0558	.0585	.0613	.0640	.0667	.0695	.0722
12	.0283	.0303	.0323	.0344	.0366	.0388	.0411	.0434	.0457	.0481
13	.0154	.0168	.0181	.0196	.0211	.0227	.0243	.0260	.0278	.0296
14	.0078	.0086	.0095	.0104	.0113	.0123	.0134	.0145	.0157	.0169
15	.0037	.0041	.0046	.0051	.0057	.0062	.0069	.0075	.0083	.0090

r	m									
	7.1	7.2	7.3	7.4	7.5	7.6	7.7	7.8	7.9	8.0
16	.0016	.0019	.0021	.0024	.0026	.0030	.0033	.0037	.0041	.0045
17	.0007	.0008	.0009	.0010	.0012	.0013	.0015	.0017	.0019	.0021
18	.0003	.0003	.0004	.0004	.0005	.0006	.0006	.0007	.0008	.0009
19	.0001	.0001	.0001	.0002	.0002	.0002	.0003	.0003	.0003	.0004
20	.0000	.0000	.0001	.0001	.0001	.0001	.0001	.0001	.0001	.0002
21	.0000	.0000	.0000	.0000	.0000	.0000	.0000	.0000	.0000	.0001
	8.1	8.2	8.3	8.4	8.5	8.6	8.7	8.8	8.9	9.0
0	.0003	.0003	.0002	.0002	.0002	.0002	.0002	.0002	.0001	.0001
1	.0025	.0023	.0021	.0019	.0017	.0016	.0014	.0013	.0012	.0011
2	.0100	.0092	.0086	.0079	.0074	.0068	.0063	.0058	.0054	.0050
3	.0269	.0252	.0237	.0222	.0208	.0195	.0183	.0171	.0160	.0150
4	.0544	.0517	.0491	.0466	.0443	.0420	.0398	.0377	.0357	.0337
5	.0882	.0849	.0816	.0784	.0752	.0722	.0692	.0663	.0635	.0607
6	.1191	.1160	.1128	.1097	.1066	.1034	.1003	.0972	.0941	.0911
7	.1378	.1358	.1338	.1317	.1294	.1271	.1247	.1222	.1197	.1171
8	.1395	.1392	.1388	.1382	.1375	.1366	.1356	.1344	.1332	.1318
9	.1256	.1269	.1280	.1290	.1299	.1306	.1311	.1315	.1317	.1318
10	.1017	.1040	.1063	.1084	.1104	.1123	.1140	.1157	.1172	.1186
11	.0749	.0776	.0802	.0828	.0853	.0878	.0902	.0925	.0348	.0970
12	.0505	.0530	.0555	.0579	.0604	.0629	.0654	.0679	.0703	.0728
13	.0315	.0334	.0354	.0374	.0395	.0416	.0438	.0459	.0481	.0504
14	.0182	.0196	.0210	.0225	.0240	.0258	.0272	.0289	.0306	.0324
15	.0098	.0107	.0116	.0126	.0136	.0147	.0158	.0169	.0182	.0194
16	.0050	.0055	.0060	.0066	.0072	.0079	.0086	.0093	.0101	.0109
17	.0024	.0026	.0029	.0033	.0036	.0040	.0044	.0048	.0053	.0058
18	.0011	.0012	.0014	.0015	.0017	.0019	.0021	.0024	.0026	.0029
19	.0005	.0005	.0006	.0007	.0008	.0009	.0010	.0011	.0012	.0014
20	.0002	.0002	.0002	.0003	.0003	.0004	.0004	.0005	.0005	.0006
21	.0001	.0001	.0001	.0001	.0001	.0002	.0002	.0002	.0002	.0003
22	.0000	.0000	.0000	.0000	.0001	.0001	.0001	.0001	.0001	.0001
	9.1	9.2	9.3	9.4	9.5	9.6	9.7	9.8	9.9	10.0
0	.0001	.0001	.0001	.0001	.0001	.0001	.0001	.0001	.0001	.0000
1	.0010	.0009	.0009	.0008	.0007	.0007	.0006	.0005	.0005	.0005
2	.0046	.0043	.0040	.0037	.0034	.0031	.0029	.0027	.0025	.0023
3	.0140	.0131	.0123	.0115	.0107	.0100	.0093	.0087	.0081	.0076
4	.0319	.0302	.0285	.0269	.0254	.0240	.0226	.0213	.0201	.0189
5	.0581	.0555	.0530	.0506	.0483	.0460	.0439	.0418	.0398	.0378
6	.0881	.0851	.0822	.0793	.0764	.0736	.0709	.0682	.0656	.0631
7	.1145	.1118	.1091	.1064	.1037	.1010	.0982	.0955	.0928	.0901
8	.1302	.1286	.1269	.1251	.1232	.1212	.1191	.1170	.1148	.1126
9	.1317	.1315	.1311	.1306	.1300	.1293	.1284	.1274	.1263	.1251
10	.1198	.1210	.1219	.1228	.1235	.1241	.1245	.1249	.1250	.1251
11	.0991	.1012	.1031	.1049	.1067	.1083	.1098	.1112	.1125	.1137
12	.0752	.0776	.0799	.0822	.0844	.0866	.0888	.0908	.0928	.0948
13	.0526	.0549	.0572	.0594	.0617	.0640	.0662	.0685	.0707	.0729
14	.0342	.0361	.0380	.0399	.0419	.0439	.0459	.0479	.0500	.0521

r	m									
	9.1	9.2	9.3	9.4	9.5	9.6	9.7	9.8	9.9	10.0
15	.0208	.0221	.0235	.0250	.0265	.0281	.0297	.0313	.0330	.0347
16	.0118	.0127	.0137	.0147	.0157	.0168	.0180	.0192	.0204	.0217
17	.0063	.0069	.0075	.0081	.0088	.0095	.0103	.0111	.0119	.0128
18	.0032	.0035	.0039	.0042	.0046	.0051	.0055	.0060	.0065	.0071
19	.0015	.0017	.0019	.0021	.0023	.0026	.0028	.0031	.0034	.0037
20	.0007	.0008	.0009	.0010	.0011	.0012	.0014	.0015	.0017	.0019
21	.0003	.0003	.0004	.0004	.0005	.0006	.0096	.0007	.0008	.0009
22	.0001	.0001	.0002	.0002	.0002	.0002	.0003	.0003	.0004	.0004
23	.0000	.0001	.0001	.0001	.0001	.0001	.0001	.0001	.0002	.0002
24	.0000	.0000	.0000	.0000	.0000	.0000	.0000	.0001	.0001	.0001

Comparing columns 2 and 6, it is seen that there is good agreement between the observed data and the frequency to be expected under Poissonian distribution. This is, however, an approximate approach for judging the apparent agreement between the two. The statistical test for the goodness of fit can be done by the χ^2 (chi square) test, which is reserved for treatment later on.

In computations involving the use of Poisson distribution, one comes across computations involving an exponential of e, the base of natural logarithms. Table 8.7 gives some of the values of e^{-x}. It is also possible to get the Poisson probabilities directly from Table 8.8.

Another application of the use of Poisson distribution is in analysis of parking. The following example illustrates this.

Problem 8.8. *The data of Table 8.9 shows the occupancy of parking spaces in a parking lot consisting of 50 spaces. The count was taken as 15 minute intervals during the 4 hours duration between 11-00 A.M. and 3-00 P.M. on 6 weak (days Monday through Saturday). Find by inspection whether the number of vacant spaces during any count follows a Poisson distribution.*

Table 8.9

Occupancy of parking Spaces	*Frequency*
50	*6*
49	*15*
48	*21*
47	*20*
46	*15*
45	*10*
44	*5*
43	*2*
42	*1*
41	*1*
40 and less	*0*

Solution. The calculations are done in a tabular form vide Table 8.10.

$$m = \frac{288}{96} = 3 \text{ vacant spaces per count.}$$

$$P(r) = \frac{m^r e^{-m}}{r!}$$

$$P(0) = e^{-m} = e^{-3} = 0.0498$$
$$P(1) = P(0) \times 3/1 = 0.1494$$
$$P(2) = P(1) \times 3/2 = 0.2241$$
$$P(3) = P(2) \times 3/3 = 3.2241$$
$$P(4) = P(3) \times 3/4 = 0.1681$$
$$P(5) = P(4) \times 3/5 = 0.1009$$
$$P(6) = P(5) \times 3/6 = 0.0505$$
$$P(7) = P(6) \times 3/7 = 0.0216$$
$$P(8) = P(7) \times 3/8 = 0.0081$$
$$P(9) = P(8) \times 3/9 = 0.0027$$
$$\Sigma = 0.9993$$
$$P\ (r \geq 10) = 1 - 0.9993 = 0.0007.$$

Inspection of columns (2) and (5) indicates a fair agreement between observed and theoretical frequencies. A correct assessment of the statistical goodness of fit can be made by the χ^2 test, described later.

Table 8.10

No. of vacant spaces out of 50 total spaces available = 50 – spaces occupied	*Observed frequency (f_i)*	*Total no. of vacant spaces Col. (1) × col (2)*	*Theoretical probability of stated no. of vacant space*	*Theoretical frequency of stated no. of vacant spaces*
(1)	(2)	(3)	(4)	(5)
0	8	0	0.0498	4.8
1	15	15	0.1494	14.3
2	21	42	0.2241	21.5
3	20	60	0.2241	21.5
4	15	60	0.1681	16.5
5	10	50	0.1009	9.6
6	5	30	0.0505	4.8
7	2	14	0.0216	2.1
8	1	8	0.0081	0.8
9	1	9	0.0027	0.3
10 or more	0	0	0.0007	0.1
	Σ = 96	Σ = 288	Σ = 1.0000	Σ = 96.0

8.5. Normal Distribution

8.5.1. The most important distribution of a continuous variable is the normal distribution. The normal distribution has a number of useful applications in traffic engineering. One of the illustrative examples of a normal distribution is the speed of vehicles (Chapter 9). Normal distribution is very useful in dealing with sampling, since it is found that irrespective of the distribution of the population, means of random samples taken from the population tend to assume a normal distribution. Normal distribution can also be used for approximating other types of distribution.

8.5.2. The normal density function, also called the Gaussian function, is given by the following equation :

$$f(x) = \frac{1}{\sigma\sqrt{2\pi}} e^{-\frac{1}{2}\left(\frac{x-\mu}{\sigma}\right)^2} \qquad ...(8.26)$$

for $\qquad - \infty < x + \infty$

A typical graph of this function is given in Fig. 8.8.

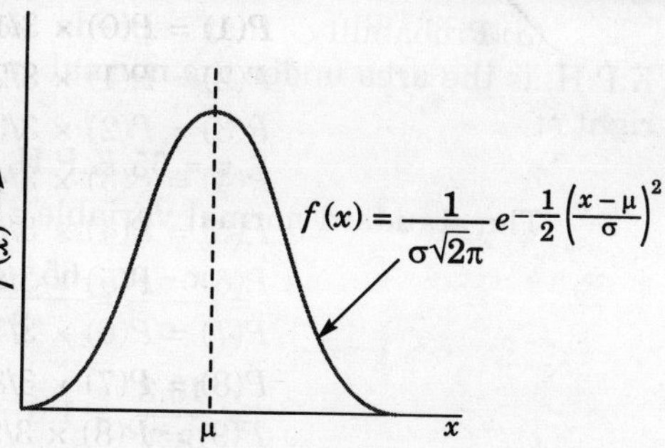

Fig. 8.8. Typical normal density curve.

It will be seen that the curve is bell-shaped and is symmetrical with respect to μ, which is the population mean. A knowledge of μ will fix the location of the curve along the x-axis. The other parameter that determines the distribution is the standard deviation, σ. If $\mu = 0$ and $\sigma = 1$, we get the standard normal density curve as in Fig. 8.9.

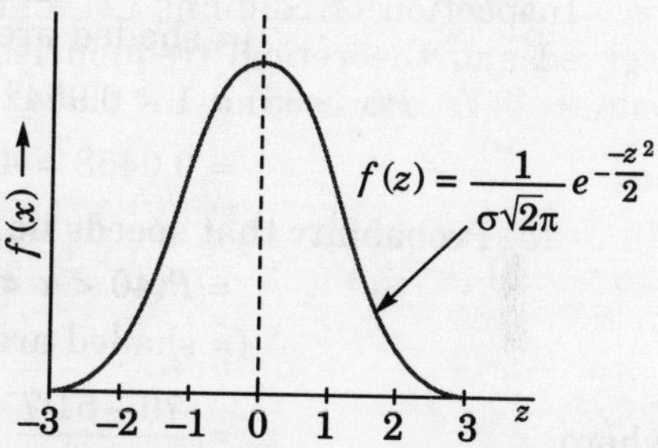

Fig. 8.9. Standard normal density curve.

The equation of the above function, in terms of the standard normal variate z would be,

$$f(z) = \frac{1}{\sqrt{2\pi}} e^{-\frac{z^2}{2}} \qquad ...(8.27)$$

for $\qquad - \infty < x + \infty$

The relationship between the variable x and the standard normal variate z is given by

$$z = \frac{x-\mu}{\sigma} \qquad ...(8.28)$$

Standard tables are available giving the value of the area under any part of the normal curve for variable z, that is, for the normal curve having $\mu = 0$ and $\sigma = 1$. Table 8.11 gives these values. In this table, the cumulative unit normal distribution, or $F(z)$, also denoted as $\phi(z)$, is plotted for values of z above 0. By symmetry, for values of z less than 0, $\phi(-z) = 1 - \phi(z)$. The use of the table is illustrated by the following examples.

Problem 8.7. *The spot speeds at a particular location are normally distributed with a mean of 51.7 K.P.H. and a standard deviation of 8.3 K.P.H. What is the probability that*

(a) the speeds exceed 65 K.P.H. ?

(b) and the speeds lie between 40 K.P.H. and 70.00 K.P.H. ?

(c) What is the 85th percentile speed ?

Solution. $\qquad \mu = 51.7$

$\qquad \sigma = 8.3$

(*a*) Probability that speeds exceed 65 K.P.H. is the area under the normal curve to the right of

$$x = 65 \text{ K.P.H.}$$

The standard normal variable z

$$= \frac{x - \mu}{\sigma} = \frac{65 - 51.7}{8.3}$$

$$= \frac{13.3}{8.3} = 1.6$$

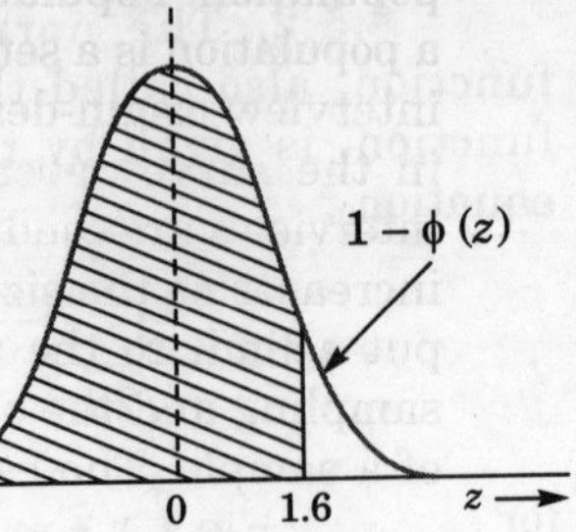

Fig. 8.10

From Table 8.11,

$$\phi\,(1.6) = 0.9542$$

(= shaded area in Fig. 8.10)

$$P(x \geq 65) = 1 - 0.9542$$

i.e., $= 0.0458 = 4.58$ percent.

(*b*) Probability that speeds lie between 40 and 70 K.F.H.

$$= P(40 < x < 70)$$

(= shaded area in Fig. 8.11) $= \phi(z_2) - \phi(z_1)$

where, $z_2 = \dfrac{70 + 51.7}{8.3} = 2.21$

and $z_1 = \dfrac{40 \;\; 51.7}{8.3} = -1.41$

$$\phi(z_2) = \phi(2.21) = 0.9864$$

$$\phi(z_1) = \phi(-1.41) = 1 - \phi\,(1.41)$$

$$= 1 - 0.9207 = 0.0793$$

$\therefore$ $P(40 < x < 70) = 0.9864 - 0.0793$

i.e., $= 0.9071 = 90.71$ percent.

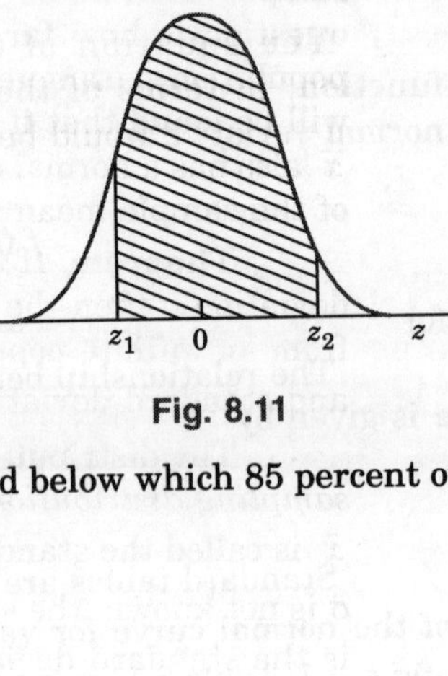

Fig. 8.11

(*c*) 85th percentile speed is defined as the speed below which 85 percent of all the speeds lie. (See Chapter 9)

$\therefore$ $\phi(z) = 0.85$

$\therefore$ $z = 1.04$

$$z = \frac{x - \mu}{\sigma} = \frac{x - 51.7}{8.3} = 1.04$$

$\therefore$ $x = 1.04 \times 8.3 + 51.7 = 8.6 + 51.7$

$= 60.3$ K.P.H. (85th percentile speed).

8.6. Sampling Theory and Significance Testing

8.6.1. Need for sampling

The traffic engineer has to collect and analyse data on a variety of subjects such as traffic volume, origins and destinations, travel modes and pattern, speed and delay, and parking. It is impossible and impractical to collect full data and one normally takes recourse to sampling. The idea is to obtain information about a population by studying the characteristics of a sample drawn from the population. Population is a class or set of objects, finite or infinite. A sample from a population is a set of objects, drawn from the population. For example, in a home-interview origin-destination study, the population is the set of all dwelling units in the survey area, whereas the sample is the set of dwelling units where interviews are conducted. The accuracy of the information obtained from sampling increases as the size of the sample increases. But considerations of cost and time put a limit to the size of the sample. This section deals with the principles of sampling and the estimation of the characteristic of a population from the study of a sample. The estimation is done utilizing the concept of probability.

8.6.2. Types of samples

Samples can be of different types, such as random samples, systematic samples, stratified samples, sequential samples and so on. Most common is the random sampling, where the sampling is done in such a way that each member of the population has the same chance of being chosen. The other types of sampling methods have definite applications in the work of the traffic engineer. For the purposes of the limited treatment given here, it will be assumed that the samples are selected at random.

8.6.3. Distribution of mean when sampling a normal population

By taking a random sample from a population, it is possible to obtain information about the population distribution. The quantities such as $\bar{x}$, the sample mean and s^2, the sample variance give a description of the sample. The question is how far reliable are $\bar{x}$ and s^2 as estimates of the corresponding population parameters μ and σ^2. If repeated sampling experiments are done, it will be found that if the population has a normal distribution, the sample mean $\bar{x}$ also has a normal distribution. The following theorem expresses the distribution of the sample mean $\bar{x}$.

Theorem. If x possesses a normal distribution with mean μ and standard deviation σ then the sample mean $\bar{x}$ based on a random sample of size n drawn from an infinite population will also possess a normal distribution with mean μ and standard deviation $\sigma/\sqrt{n}$.

The distribution of the mean $\bar{x}$ given by the above theorem is called the *sampling distribution* of the mean. The standard deviation of the distribution of $\bar{x}$ is called the standard error of the mean, and is denoted by $\sigma_{\bar{x}}$. If the value of σ is not known, the standard error can be assumed to be given by $s/\sqrt{n}$, where 's' is the standard deviation of the sample. The above theorem means that as the simple size increases, the distribution of the sample mean becomes more and more concentrated about the population mean, and we become more and more certain that the sample mean is a good estimate of the population mean.

The following problem will illustrate the above principles :

Problem 8.8. *The spot speeds at a particular location on a highway are known to be normally distributed with a mean of 80 K.P.H. and a standard deviation of 15 K.P.H. What is the probability that if a sample of 100 vehicles are tested that the mean speed observed will exceed 75 K.P.H. ?*

Solution. $\mu = 80, \sigma = 15$

$$\sigma_{\bar{x}} = \frac{\sigma}{\sqrt{n}} = \frac{15}{\sqrt{100}} = 1.5$$

$P\{\bar{x} < 75\}$ = Shaded area in Fig. 8.12

$$= 1 - \phi\left(\frac{75-80}{1.5}\right)$$

$$1 - \phi(-3.33) = 1 - [1 - \phi\ (3.33)]$$
$$= 1 - 1 + \phi\ (3.33)$$
$$= 0.9996, \text{ using Table 8.11.}$$

The probability is 0.9996 that the mean speed of sample of size of 100 vehicles will exceed 75 K.P.H.

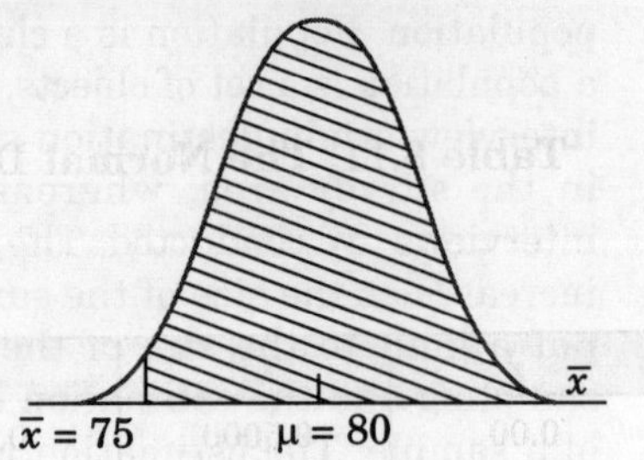

Fig. 8.12

8.6.4. Distribution of mean where sampling a non-normal population–Central limit theorem

If the variable x does not possess a normal distribution, what can be predicted about the distribution of $\bar{x}$, the sample mean ? It has been found that if the sample size is sufficiently large, say larger than 25, the distribution of $\bar{x}$ will be normal, even though the distribution of x may be non-normal. This principle is known as the "Central limit theorem" and can be stated as follows :

Theorem. If x possess a distribution with mean μ and standard deviation σ, then the sample mean $\bar{x}$, based on a sample of size n, will have a distribution that approaches the distribution of a normal variable with mean μ and standard deviation $\sigma/\sqrt{n}$, as n becomes infinite.

8.6.5. Point estimate and interval estimate

A population can be studied if the parameters of the distribution can be estimated. For example, for a normal distribution, the parameters μ and σ determine the properties of a population, and it will be important to be able to estimate these parameters.

In connection with the estimation of parameters, two types of estimates are in common use : point estimate and interval estimate. A point estimate is a particular value which may be used to estimate the unknown value of a parameter. For example, the sample mean $\bar{x}$ is a point estimate of the population mean μ. A point estimate has its own limitations, because even the best point estimate may be considerably away from the parameter and a single value may not be deemed satisfactory. Thus, it is more usual to estimate a parameter within an interval, determined by two numbers obtained to contain the value of the parameter within the bounds of the two values. Thus, if t_1 and t_2 are two values of a random variable t and it is desired to estimate the parameter θ, then

$$P[t_1 \leq \theta \leq t_2] = \gamma \qquad ...(8.29)$$

where γ is some specified probability. In other words, it can be said that the probability of θ lying in the interval bounded by two values

Table 8.11. The Normal Distribution Function

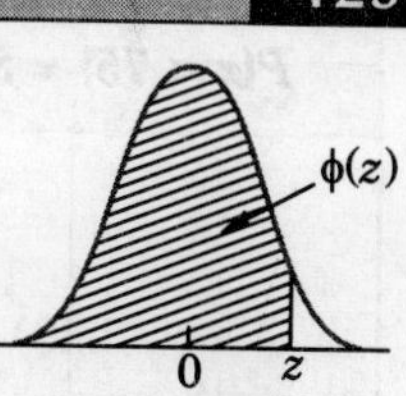

z	$\phi(z)$	z	$\phi(z)$	z	$\phi(z)$
0.00	0.5000	0.35	0.6368	0.70	0.7580
.01	.5040	.36	.6406	.71	.7611
.02	.5080	.37	.6443	.72	.7642
.03	.5120	.38	.6480	.73	.7673
.04	.5160	.39	.6517	.74	.7704
0.05	0.5199	0.40	0.6554	0.75	0.7734
.06	.5239	.41	.6591	.76	.7764
.07	.5279	.42	.6628	.77	.7794
.08	.5319	.43	.6664	.78	.7823
.09	.5359	.44	.6700	.79	.7852
0.10	0.5398	0.45	0.6736	0.80	.7881
.11	.5438	.46	.6772	.81	.7910
.12	.5478	.47	.6808	.82	.7939
.13	.5517	.48	.6844	.83	.7967
.14	.5557	.49	.6879	.84	.7995
0.15	0.5596	0.50	0.6915	0.85	0.8023
.16	.5636	.51	.6950	.86	.8051
.17	.5675	.52	.6985	.87	.8078
.18	.5714	.53	.7019	.88	.8103
.19	.5753	.54	.7054	.89	.8133
0.20	0.5793	0.55	0.7088	0.90	0.8159
.21	.5832	.56	.7123	.91	.8186
.22	.5871	.57	.7157	.92	.8212
.23	.5910	.58	.7190	.93	.8238
.24	.5948	.59	.7224	.94	.8264
0.25	0.5987	0.60	0.7257	0.95	0.8289
.26	.6026	.61	.7291	.96	.8315
.27	.6064	.62	.7324	.97	.8340
.28	.6103	.63	.7357	.98	.8365
.29	.6141	.64	.7389	.99	.8389
0.30	0.6179	0.65	0.7422	1.00	0.8413
.31	.6217	.66	.7454	.01	.8438
.32	.6255	.67	.7486	.02	.8461
.33	.6293	.68	.7517	.03	.8485
.34	.6331	.69	.7549	.04	.8508
1.05	0.8531	1.55	0.9394	2.05	0.97982
.06	.8554	.56	.9406	.06	.98030
.07	.8577	.57	.9418	.07	.98077
.08	.8599	.58	.9429	.08	.98124
.09	.8621	.59	.9441	.09	.98169
1.10	.8643	1.60	0.9452	2.10	0.98214
.11	.8665	.61	.9463	.11	.98257
.12	.8686	.62	.9474	.12	.98300
.13	.8708	.63	.9484	.13	.98341
.14	.8729	.64	.9495	.14	.98382

z	φ(z)	z	φ(z)	z	φ(z)
1.15	0.8749	1.65	0.9505	2.15	0.98422
.16	.8770	.66	.9515	.16	.98461
.17	.8790	.67	.9525	.17	.98500
.18	.8810	.68	.9535	.18	.98537
.19	.8830	.69	.9545	.19	.98574
1.20	0.8849	1.70	0.9554	2.20	0.98610
.21	.8869	.71	.9564	.21	.98645
.22	.8888	.72	.9573	.22	.98679
.23	.8907	.73	.9582	.23	.98723
.24	.8925	.74	.9591	.24	.98745
1.25	0.8944	1.75	0.9599	2.25	0.98778
.26	.8962	.76	.9608	.26	.98809
.27	.8980	.77	.9616	.27	.98840
.28	.8997	.78	.9625	.28	.98870
.29	.9015	.79	.9633	.29	.98899
1.30	0.9032	1.80	0.9641	2.30	0.98928
.31	.9049	.81	.9649	.31	.98956
.32	.9066	.82	.9656	.32	.98983
.33	.9082	.83	.9664	.33	.99010
.34	.9099	.84	.9671	.34	.99036
1.35	0.9115	1.85	0.9678	2.35	0.99061
.36	.9131	.86	.9686	.36	.99086
.37	.9147	.87	.9693	.37	.99111
.38	.9162	.88	.9699	.38	.99134
.39	.9177	.89	.9706	.39	.99158
1.40	0.9192	1.90	0.9713	2.40	0.99180
.41	.9207	.91	.9719	.41	.99202
.42	.9222	.92	.9726	.42	.99224
.43	.9236	.93	.9732	.43	.99245
.44	.9251	.94	.9738	.44	.99266
1.45	0.9265	1.95	0.9744	2.45	0.99286
.46	.9279	.96	.9750	.46	.99305
.47	.9292	.97	.9756	.47	.99324
.48	.9306	.98	.9761	.48	.99343
.49	.9319	.99	.9767	.49	.99361
1.50	0.9332	2.00	0.97725	2.50	0.99379
.51	.9345	.01	.97778	.51	.99396
.52	.9357	.02	.97831	.52	.99413
.53	.9370	.03	.97882	.53	.99430
.54	.9382	.04	.97932	.54	.99446
2.55	0.99461	2.75	0.99702	2.95	9.99841
.56	0.99477	.76	.99711	.96	.99846
.57	.99492	.77	.99720	.97	.99851
.58	.99506	.78	.99728	.98	.99856
.59	.99520	.79	.99736	.99	.99861
2.60	0.99534	2.80	0.99744	3.00	0.99865
.61	.99547	.81	.99752	.10	.99903
.62	.99560	.82	.99760	.20	.99931
.63	.99573	.83	.99767	.30	.99952
.64	.99585	.84	.99774	.40	.99966

z	$\phi(z)$	z	$\phi(z)$	z	$\phi(z)$
2.65	0.99598	2.85	0.99781	3.50	0.99977
.66	.99609	.86	.99788	.60	.99984
.67	.99621	.87	.99795	.70	.99989
.68	.99632	.88	.99801	.80	.99993
.69	.99643	.89	.99807	.90	.99995
2.70	0.09653	2.90	0.99813	4.00	0.99997
.71	.99664	.91	.99819		
.72	.99674	.92	.99825		
.73	.99683	.93	.99831		
.74	.99693	.94	.99836		

t_1 and t_2 is γ. The set of values between t_1 and t_2 inclusive is called the *confidence interval of* θ. The values of t_1 and t_2 are called the confidence limits. The probability measure γ in the probability equation (8.29) is called the *confidence level*. The confidence interval is also referred to as 100 γ percent confidence interval.

The confidence level of an interval estimate for a parameter is the proportion of the samples for which the interval includes the true value.

From the normal distribution table (Table 8.11), the values of the standard normal variable z_a associated with various confidence levels γ are summarised in Table 8.12. It may be noted that

$$a = \frac{1+\gamma}{2}.$$

Table 8.12. Standard Normal Variables z Associated with Various Confidence Levels γ

Confidence level, γ	$a = \frac{1+\gamma}{2}$	z_a
0.80	0.90	1.282
0.90	0.95	1.645
0.95	0.975	1.960
0.98	0.99	2.326
0.99	0.995	2.576
0.999	0.9995	3.291
0.9999	0.99995	3.891
0.99999	0.999995	4.417

The following problems illustrate the above principles :

Problem 8.9. *A series of observations of spot speeds of vehicles was taken and the data is presented in Table 9.1 (Chapter 9). The mean and standard deviation of the sample have been worked out as 39.7 K.P.H. and 6.8 K.P.H. respectively. Give a 95 per cent confidence interval for the population mean speed.*

Solution. In this problem, n = 363

$\bar{x}$ = 39.7 K.P.H. ; s = 6.8 K.P.H.

The distribution of $\bar{x}$ follows a normal distribution with mean as μ and a standard deviation of $\sigma/\sqrt{n}$ (known as standard error of mean).

Since σ can be estimated by *s*, the standard deviation of the sample.

$$\therefore \qquad \sigma_{\bar{x}} = \frac{s}{\sqrt{n}} = \frac{6.8}{\sqrt{363}} = 0.36$$

$\bar{x}$ has a normal distribution with mean as μ and standard deviation of 0.36.

For a 95 per cent confidence level, γ =0.95 and hence the normal variable *z* has a value of 1.96 (See Table 8.12).

With 95 per cent confidence, the population mean has a value between $\bar{x} + 1.96\,\sigma_{\bar{x}}$ and $x - 1.96\,\sigma_{\bar{x}}$.

i.e., between 39.7 + 1.96 × 0.36 and 39.7 – 1.96 × 0.36, *i.e.*, between 40.4 and 39.0 K.P.H.

In other words, the probability is 95 per cent that the population mean speed lies between 40.4 and 39.0 K.P.H.

Problem 8.10. *In connection with a survey for determining the spot speed at a given location of a highway, it is desired to obtain the average speed within 2 K.P.H. with a probability of 0.95. Previous studies have indicated that the standard deviation of the speeds is 8 K.P.H. What size of sample should be selected ?*

Solution. The population mean can be estimated from the sample mean as per the formula

$$\mu = \bar{x} + z_a\,\sigma_{\bar{x}} \qquad \text{...(8.30)}$$

where, μ = population mean

$\bar{x}$ = sample mean

$\sigma_{\bar{x}}$ = standard error of mean

z_a = normal variable associated with a probability γ

$$\text{where,} \qquad a = \frac{1+\gamma}{2}, \quad \mu - \bar{x} = z_a\,\sigma_{\bar{x}}$$

If $\mu - \bar{x}$ is denoted by *e*, then

e = difference between population mean and sample mean = $z_a\,\sigma_{\bar{x}}$

$$e = z_a \frac{\sigma}{\sqrt{n}} \qquad \text{...(8.31)}$$

$$\therefore \qquad n = \frac{z_a^2\,\sigma^2}{e^2} \qquad \text{...(8.32)}$$

In this case, σ = 8 K.P.H.

e = 2 K.P.H.

z_a = 1.96 (From Table 8.12)

$$\therefore \qquad n = \frac{(1.96)^2 \times (8)^2}{2^2} = 61.5, \text{ say about 60 observations.}$$

8.6.6. Significance testing

8.6.6.1. An important problem of statistics is the testing of hypothesis about population. A statistical hypothesis is an assumption about the distribution of one or more random variable. A test of a statistical hypothesis, called *significance or hypothesis test*, is a procedure for deciding, on the basis of a sample and on the grounds of probability, whether to "accept" or "reject" the hypothesis. Typical examples of the application of significance testing for traffic engineering problems are :

(*i*) Has the average spot speed of vehicles at a particular location improved significantly after modifications to the geometric standards of the highway ?

(*ii*) Has the installation of traffic signals at an intersection resulted in a significant reduction in the accident situation ?

(*iii*) Does the rate of arrival of vehicles at a particular location follow a Poisson distribution ?

8.6.6.2. Hypothesis testing starts with defining the hypothesis. The null hypothesis, designated H_0, is the statistical hypothesis under test, and this usually takes the form of "no effect", "no difference between samples", "the process used at present is as good as the new one" and so on. The alternative hypothesis, designed H_1 is the hypothesis alternative to the null hypothesis. For example, when testing whether any significant reduction in the accident rate has been caused by the installation of traffic signals, the null hypothesis would be that there is no reduction. The alternative hypothesis would be that the installation of signals has resulted in a significant reduction in the accident rate. The next step would be to select the significance level or the size of the test, denoted by α. The significance level is the probability of making a wrong decision when the null hypothesis is true, that is probability of wrongly deciding to reject H_0. A significance level of 5 per cent or 1 per cent is common for many problems. The procedure then requires a rule for "accepting" or "rejecting" the null hypothesis. For this, estimator of the parameter of the hypothesis is selected and the set of all possible values of an estimator is divided into two sets—one being in the region of rejection, called the *critical region*, and the other region of acceptance. If the value of the estimator falls within the critical region the null hypothesis is rejected. It may be noted that the probability of the estimator falling within the critical region is the significance level of the test. On the basis of this procedure, a decision whether to accept or reject the null hypothesis can be made. There are two types of errors in reaching this decision :

(*i*) Type I error, when we reject the null hypothesis when in fact it is true.

(*ii*) Type II error, when we accept the null hypothesis when it is false.

The probability of making a Type I error is the same as the significance level of the test, α. The probability of making a Type II error is denoted by β. The following table summarises the types of errors associated with hypothesis testing.

Table 8.13. Types of Error in Hypothesis Testing

	H_0 *true*	H_1 *true*
Accept H_0	Probability = $1 - \alpha$	Probability = β Type II error
Accept H_1	Probability = a Type I error	Probability = $1 - \beta$

The location of the critical region can lead to two types of tests: one-sided test or two-sided test. One-sided test occurs when the critical region is located only at one end of the scale of measurements of the test. Two-sided test occurs when the critical region is made up of two parts located equally at both ends of the scale of measurements of the test.

8.6.6.3. The above principles will be illustrated by the following problems.

Problem 8.11. *The spot speeds at a particular location on an expressway are known to be normally distributed with a mean of 80 K.P.H. A new radar speedometer was bought by Traffic Department and a set of 100 observations of speeds was taken. The mean speed observed was 77.3 K.P.H. and the standard deviation was 15 K.P.H. Is there any evidence to prove that*

(i) the new speedometer might have been faulty and

(ii) the new speedometer is showing lesser speeds than actual ?

Solution. $\mu = 80,\ \bar{x} = 77.3,\ s = 15,\ n = 100$

$$\sigma_{\bar{x}} = \frac{\sigma}{\sqrt{100}} = \frac{15}{\sqrt{100}} = 1.5, \text{ since } \sigma = s$$

(*i*) The null hypothesis is that the new speedometer is not faulty, such that $H_0 : \mu = 80$.

The alternative hypothesis is that the speedometer is faulty and the speeds indicated by it vary significantly from the true speeds.

$$H_1 : \mu \neq 80$$

The hypothesis will be tested for a significance level of $\alpha = 0.05$. In the present case, we are interested in knowing whether the observed value is significantly different from the population mean, and hence we are interested in both the tails of the normal distribution curve. A 2-sided test is, therefore, appropriate, with the area of the critical region being

$$\frac{\alpha}{2} = 0.025$$

at each side. The limits of the critical region are given by :

$$\bar{x}_1 = \mu + z\sigma_x \ ; \ \bar{x}_2 = \mu - z\,\sigma_{\bar{x}}$$

The value of z corresponding to an area of 0.025 at either tail of the normal curve is 1.96 (Table 8.12).

$\therefore$ $\bar{x}_1 = 80 + 1.96 \times 1.5 = 80 + 2.94 = 82.94$

and $\bar{x}_2 = 80 - 1.96 \times 1.5 = 80 - 2.94 = 77.06$

The observed value of $\bar{x}$ is 77.3 and it does not fall in the critical region (Fig. 8.13).

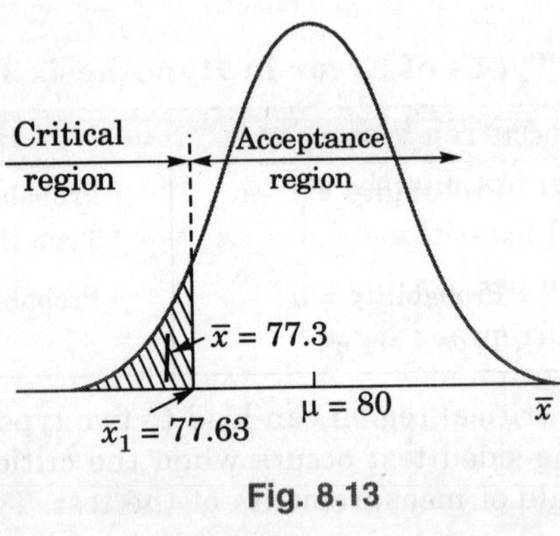

Fig. 8.13

Hence accept the null hypothesis H_0 and assert that there is no significant difference between the speed readings recorded by the new radar speedometer and true mean speed at 5 per cent significance level.

(*ii*) In this part of the question, we are interested to know whether the speedometer is showing lesser speeds than actual. Hence we are interested in only the left-side tail of the curve. Thus :

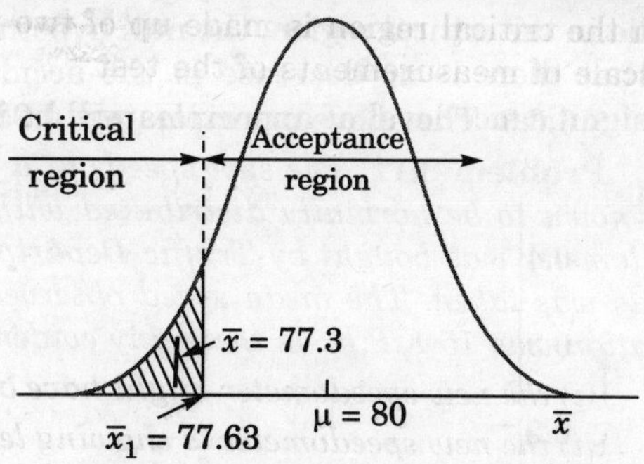

Fig. 8.14

$$H_0 : \mu = 80 \; ; \; H_1 : \mu < 80$$

The area of the critical region at the left-side tail is $\alpha = 0.05$, all the limit of $\bar{x}$, for $\alpha = 0.05$ is given by : $\bar{x}_1 = \mu - z\,\sigma_{\bar{x}}$

where z, the normal variable, corresponding to an area of 0.05 is 1.645 (Table 8.12).

$$\therefore \qquad \bar{x}_1 = 80 - 1.645 \times 1.5 = 80 - 2.47 = 77.53$$

The value of $\bar{x}$ is 77.53 and it falls in the critical region (Fig. 8.14).

Hence reject the null hypothesis H_0 and accept the alternative hypothesis H_1. This leads to the conclusion that the speeds indicated by the radar speedometer are significantly less than the true mean at 5 per cent significant level.

Problem 8.12. *It has been observed from the records over a number of years that the number of accidents caused at a curve with poor visibility is 9 per year. After improving the visibility at the curve, the accident rate observed in the next year was 5. Is there evidence to prove that the improvement has resulted in a decrease in the accident rate ? Assume a Poisson distribution for the number of accident in a year.*

Solution. To null hypothesis is that there is no decrease in the accident rate, so that : $H_0 : \mu = 8$.

The alternative hypothesis is that $H_1 : < 8$, and we have to use one side test.

According to Poisson distribution,

$$P(r) = \frac{m^{-r} e^{-m}}{r!} \qquad \text{[Equation 8.24]}$$

where, $P(r)$ = Probability of occurrence of an event r times in a given period.

m = average number of times the event occurs in the same given period.

e = base of natural logarithms.

In this example, m = 9 per year.

The probability of occurrence of events can be directly read from Table 8.8 and has been plotted in the graph at Fig. 8.15. Since the Poisson distribution is discrete, the critical region for a significance level of $\alpha = 0.05$ cannot be precisely 0.05. It will be seen that $P(r \le 3) = 0.0212$, whereas $P(r \le 4) = 0.0549$. Hence the critical level can be deemed to be upto $r = 3$. Since the observed result 5 does not

lie in the critical region, the null hypothesis can be accepted. Hence we can assert that there is no decrease in the accident situation after the improvement, at significance level of approximately 0.05.

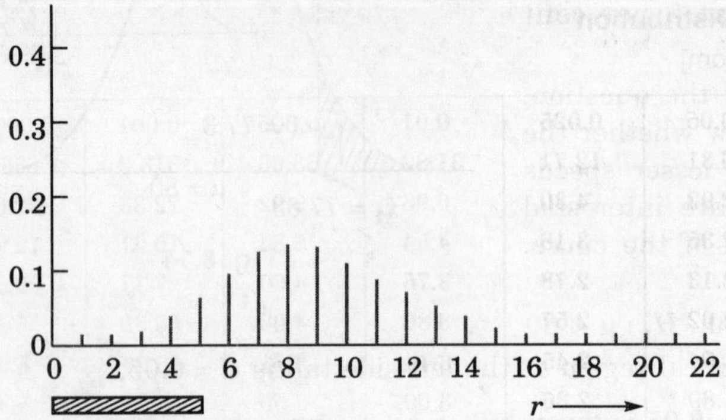

Critical region (α = 0.05)

r	P(r)	r	P(r)
0	0.0001	11	0.0970
1	0.0011	12	0.0728
2	0.0050	13	0.0504
3	0.0150	14	0.0324
4	0.0337	15	0.0194
5	0.0607	16	0.0109
6	0.0911	17	0.0058
7	0.1191	18	0.0029
8	0.1318	19	0.0014
9	0.1318	20	0.0006
10	0.1186	21	0.0003
		22	0.0001

Fig. 8.15. Probability of accident occurrence (Problem 8.12).

8.6.7. Small sample test for a mean

In the problem solved in the previous section (Problem 8.11), it was assumed that σ can be replaced by *s*. This assumption gives accurate results only for large size of samples, say *n* > 25. For sample sizes smaller than this, a new variable, called the Student's variable, is introduced, with its value given by :

$$t = \frac{\bar{x} - \mu}{s}\sqrt{n} \qquad ...(8.33)$$

where, $\bar{x}$ = sample mean

s = standard deviation of the sample

n = sample size

μ = population mean.

The Students' *t* distribution does not require a knowledge of σ, the population standard deviation. It can be used when only the sample standard deviation is known. The *t* distribution can, however, be used only when the basic variable *x* possesses a normal distribution.

Table 8.14, gives the values of the variable *t* for various "degree of freedom". The degree of freedom, *v*, is given by : $v = n-1$

The following problem illustrates the use of the *t* distribution for finding out the confidence interval.

Problem 8.13. ***The mean spot speed of 20 vehicles passing a given location was 68.3 K.P.H. and the standard deviation was 8.6 K.P.H. Give a 95 per cent confidence interval for the true mean speed of vehicles at the location.***

Solution. **In this example,**

$$n = 20$$

$$\bar{x} = 68.3 \text{ K.P.H.}$$

$$s = 8.6$$

For a 95 percent confidence interval, the area of the two tails will be 0.025 each.

Degrees of freedom

$$v = n - 1 = 19$$

Table 8.14. Student's 't' Distribution

[v = Degrees of freedom]

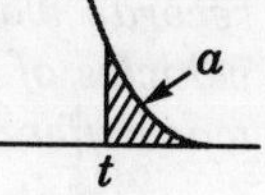

α	0.125	0.05	0.025	0.01	0.005	0.001	0.0005
v = 1	2.41	6.31	12.71	31.82	63.66	318.3	636.6
2	1.60	2.92	4.30	6.96	9.92	22.33	31.60
3	1.42	2.35	3.18	4.54	5.84	10.21	12.92
4	1.34	2.13	2.78	3.75	4.60	7.17	8.61
5	1.30	2.02	2.57	3.36	4.03	5.89	6.87
6	1.27	1.94	2.45	3.14	3.71	5.21	5.96
7	1.25	1.89	2.36	3.00	3.50	4.79	5.41
8	1.24	1.86	2.31	2.90	3.36	4.50	5.04
9	1.23	1.83	2.26	2.82	3.25	4.30	4.78
10	1.22	1.81	2.23	2.76	3.17	4.14	4.59
12	1.21	1.78	2.18	2.68	3.05	3.93	4.32
15	1.20	1.75	2.13	2.60	2.95	3.73	4.07
20	1.18	1.72	2.07	2.53	2.85	3.55	3.85
25	1.18	1.71	2.06	2.49	2.80	3.47	3.75
30	1.17	1.70	2.04	2.46	2.75	3.39	3.65
40	1.17	1.68	2.02	2.42	2.70	3.31	3.55
60	1.16	1.67	2.00	2.39	2.66	3.23	3.46
120	1.16	1.66	1.98	2.36	2.62	3.16	3.37
∞	1.15	1.64	1.96	2.33	2.58	3.09	3.29

From Table 8.14, for v = 19

and area of tail = 0.025

$$t_c = 2.093.$$

Confidence interval is given by the two values μ_1 and μ_2 as follows :

$$\mu_1 = \bar{x} + t_c \frac{s}{\sqrt{n}} = 68.3 + 2.093 \times \frac{8.6}{\sqrt{20}}$$

$$= 68.3 + 4.0 = 72.3$$

$$\mu_2 = \bar{x} - t_c \frac{s}{\sqrt{n}}$$

$$= 68.3 - 2.093 \times \frac{8.6}{\sqrt{20}}$$

$$= 68.3 - 4.0 = 64.3$$

With 95 percent confidence, one can say that the true mean spot speed of the population lies between 72.3 K.P.H. and 64.3 K.P.H.

The following problem illustrates the use of t test for significance testing with a small sample.

Problem 8.14. *The mean spot speeds of 15 vehicles observed on a Sunday at a particular location of a highway was 81.2 K.P.H. and the standard deviation was 10.2 K.P.H. The mean speed of all vehicles at this location as per previous records was 75.5 K.P.H. Is there sufficient evidence to show that the speed of vehicles of the particular Sunday (i) was higher than the average speed and (ii) was different from the average speed ?*

Solution. $\mu = 75.5$

$\bar{x} = 81.2$

$n = 15$

$s = 10.2$

(*i*) The null hypothesis is that there is no difference between that

$$H_0 : \mu = 75.5$$

The alternative hypothesis is that there is an increase in the speeds, such that

$$H_1 : \mu > 75.5$$

Since the number of samples is only 15, a small sample *t* test is appropriate.

The significance level chosen will be 0.05 and the critical area will be the right hand side tail of the curve. From Table 8.14, the value of t_c for one-sided tail area of 0.05 and degrees of freedom 14 (*i.e.*, 15–1) is 1.76.

The variable *t* is given by Equation (8.33)

$$t = \frac{\bar{x} - \mu}{s}\sqrt{n}$$

Substituting the critical value of *t* for a significance level of 0.05.

$$\bar{x} = \mu + t_c \frac{s}{\sqrt{n}} = 75.5 + 1.76 \times \frac{10.2}{\sqrt{15}}$$

$$= 75.5 + 4.64 = 80.14 \text{ K.P.H.}$$

The observed value of $\bar{x} = 81.2$ falls within the critical region, and hence the null hypothesis is rejected (Fig. 8.16). We, therefore, conclude that the increase in speed observed on the particular. Sunday is not merely due to chance alone and there is a significant difference (at 5 percent significant level) between this value and the true population mean.

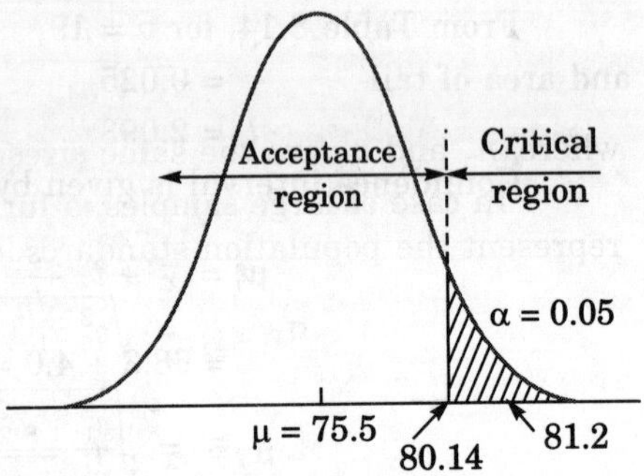

Fig. 8.16

(*ii*) In part (*ii*) the problem, the null hypothesis

is $H_0 : \mu = 75.5$, whereas

the alternative hypothesis

is $H_1 : \mu \neq 75.5$.

Here we are not particularly interested, whether the mean is above or below the true value, but only whether the mean differs from the true mean. Thus a two-sided test is needed. With a significance level of 0.05, the critical area at each tail will be 0.025. The value of t_c corresponding to an area of 0.025 at one tail is 2.145 (degrees of freedom = 14).

$$\bar{x} = \mu \pm t_c \frac{s}{\sqrt{n}}$$

$$= 75.5 \pm 2.145 \times \frac{10.2}{\sqrt{15}}$$

$$= 75.5 \pm 5.65$$

i.e., 81.15 and 69.85 K.P.H.

The observed value of 81.2 is just within the critical region and is in fact a marginal case. On the basis of the present sample (Fig. 8.17), it can be said that there is a significant difference between the observed value and the true mean, but because of the result being just marginal, it would be desirable to test a larger sample before concluding either way.

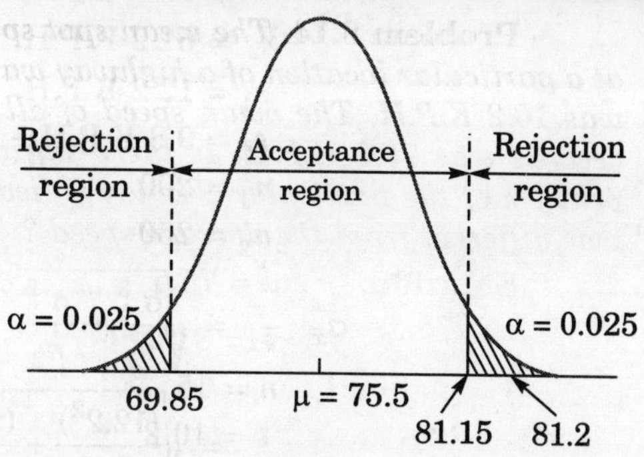

Fig. 8.17

8.6.8. Testing difference between two means

8.6.8.1. Consider the case of sampling from two populations with one large sample from each ($n > 30$). Let $\bar{x}_1$ be the mean and s_1 the standard deviation of the sample from one population. Let μ_1 and σ_1 denote the mean and standard deviation respectively of the first population. Similarly, let $\bar{x}_2$ and s_2 be the mean and standard deviation of the sample from the second population, and let μ_2 and σ_2 denote the mean and standard deviation respectively of the second population. If the variable x_1 and x_2 are known to possess independent normal distribution, then the variable $\bar{x}_1 - \bar{x}_2$ (the difference of the sample means) will also possess a normal distribution with mean $\mu_1 - \mu_2$ and standard deviation given by the formula :

$$\sigma_{\bar{x}_1 - \bar{x}_2} = \sqrt{\sigma^2_{\bar{x}_1} + \sigma^2_{\bar{x}_2}}$$

$$= \sqrt{\frac{\sigma_1^2}{n_1} + \frac{\sigma_2^2}{n_2}} \quad \text{...(8.35)}$$

where, n_1 and n_2 are the same sizes.

In case of large samples, a further assumption can be made that s_1 and s_2 represent the population standard deviations, σ_1 and σ_2, so that :

$$\sigma_{\bar{x}_1 - \bar{x}_2} = \sqrt{s^2 \bar{x}_1 + s^2 \bar{x}_2}$$

$$= \sqrt{\frac{s_1^2}{n_1} + \frac{s_2^2}{n^2}} \quad \text{...(8.36)}$$

Problem 8.15. *At a highway location with poor geometrics, the mean spot speed observed with a sample of 200 vehicles was 58.3 K.P.H. with a standard deviation of 12.2 K.P.H. After effecting improvements to the geometrics, the mean speed observed with a sample of 250 vehicles was 61.2 K.P.H. with a standard deviation of 9.8 K.P.H. Has there been a significant increase in the speed after the improvements ?*

Solution. In this problem,

$$\bar{x}_1 = 58.3 \text{ K.P.H.}$$

$$\bar{x}_2 = 61.2 \text{ K.P.H.}$$
$$s_1 = 12.2 \text{ K.P.H.}$$
$$s_2 = 9.8 \text{ K.P.H.}$$
$$n_1 = 200$$
$$n_2 = 250$$

$$\sigma_{\bar{x}_1 - \bar{x}_2} = \sqrt{\frac{\sigma_1^2}{n_1} + \frac{\sigma_2^2}{n_2}} \approx \sqrt{\frac{s_1^2}{n_1} + \frac{s_2^2}{n_2}}$$

$$= \sqrt{\frac{(12.2^2)}{200} + \frac{(9.8^2)}{250}} = 1.06$$

The null hypothesis is that there has not been any increase in the speeds, such that $H_0 : \mu_1 - \mu_2 = 0$.

The alternative hypothesis is that there has been an increase in the speeds, such that $H_1 : \mu_2 > \mu_1$.

The difference between the means $\bar{x}_1 - \bar{x}_2$ has a normal distribution with $\mu_1 - \mu_2 = 0$ as the mean and a standard error = 1.06.

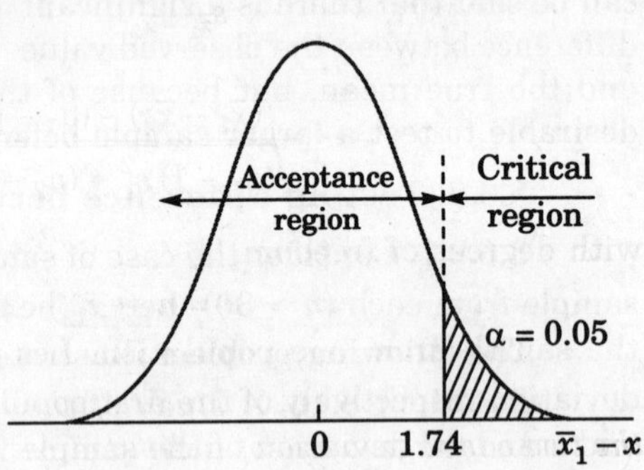

Fig. 8.18

A significance level of $\alpha = 0.05$ will be used, and since we are only interested in the right hand side of the normal curve, $z_c = 1.645$ (Table 8.12).

The critical region, therefore, lies in that parts of the horizontal axis to the right of a value of $\bar{x}_1 - \bar{x}_2$ given by :

$$\sigma_{(\bar{x}_1 - \bar{x}_2)_{z_c}} = 1.645 \times 1.06 = 1.74 \text{ K.P.H.}$$

The observed difference in the means is :

$$61.2 - 58.3 = 2.9 \text{ K.P.H.}$$

and lies in the critical region (Fig. 8.18).

The null hypothesis is, therefore, rejected. We conclude that there is a significant increase (at 5 per cent significance level) in the spot speeds after the improvements to geometrics.

8.6.8.2. When the samples are small in number, the Student's t test is used. This requires an assumption to be made that the two basic variables x_1 and x_2 possess independent normal distributions with equal variances. A joint estimate of the common variance is made with the following formula,

$$s^2 = \frac{(n_1 - 1)\, s_1^2 + (n_2 - 1)\, s_2^2}{n_1 + n_2 - 2} \qquad \text{...(8.37)}$$

where, n_1 = sample size of variable x_1

n_2 = sample size of variable x_2

s_1 = standard deviation of sample of variable x_1.

s_2 = standard deviation of sample of variable x_2.

The difference of the mean then follows that Student's t distribution, with a standard error given by,

$$s_{\bar{x}_2 - \bar{x}_2} = s\sqrt{\frac{1}{n_1} + \frac{1}{n_2}} \qquad \text{...(8.38)}$$

$$= \sqrt{\frac{(n_1 - 1)\, s_1^2 + (n_2 - 1)\, s_2^2}{(n_1 + n_2 - 2)}} \sqrt{\frac{(n_1 + n_2)}{n_1 n_2}} \qquad \text{...(8.39)}$$

The value of the variable t is given by,

$$t = \frac{(\bar{x}_1 - \bar{x}_2) - (\mu_1 - \mu_2)}{s_{\bar{x}_1 - \bar{x}_2}}$$

$$= -\frac{(\bar{x}_1 - \bar{x}_2) - (\mu_1 - \mu_2)}{\sqrt{(n_1 - 1)\, s_1^2 + (n_2 - 1)\, s_2^2}} \times \sqrt{\frac{n_1 n_2 (n_1 + n_2 - 2)}{n_1 + n_2}} \qquad \text{...(8.40)}$$

with degrees of freedom,

$$v = n_1 + n_2 - 2 \qquad \text{...(8.41)}$$

The following problem illustrates the use of the above principle.

Problem 8.16. *A study of the reaction time of drivers was conducted on the two carriage–ways of a divided carriageway facility. The mean reaction time, sample size and standard deviation obtained are tabulated below :*

East carriageway :

Mean $\bar{x}_1 = 1.70$ *seconds*

Standard deviation, $s_1 = 1.32$ *seconds*

Sample size $n_1 = 20$

West carriageway :

Mean $x_2 = 1.79$ *seconds*

Standard deviation, $s_2 = 1.52$ *seconds*

Sample size $n_2 = 22.$

Test whether the population means are equal, assuming that the populations are normal with equal variance.

Solution. The null hypothesis is that the populations means are equal, such that $H_0 : \mu_1 - \mu_2 = 0$. The alternative hypothesis is that the population means are not equal, such that $H_1 : \mu_1 \neq \mu_2$.

The difference of the mean of the samples has a t distribution, with the value of t given by Eq. (8.40). Substituting,

$$t = \frac{-0.09 - 0}{\sqrt{(20-1)(1.32)^2 + (22-1)(1.52)^2}} \sqrt{\frac{20 \times 22\,(20 + 22 - 2)}{(20 + 22)}}$$

$$= \frac{-0.09}{\sqrt{33.1 + 48.5}} \sqrt{\frac{20 \times 22 \times 40}{42}} = -0.0204.$$

For a significance level of $\alpha = 0.05$, the critical area of the curve at either end will be 0.025. This corresponds to a critical value of t of –2.02 and + 2.02, with (20 + 22 – 2) = 40° of freedom.

The observed value of –0.024 falls within the acceptance region (Fig. 8.19). Hence the null hypothesis is accepted. We conclude that there is no difference between the reaction time of drivers in the two carriageways.

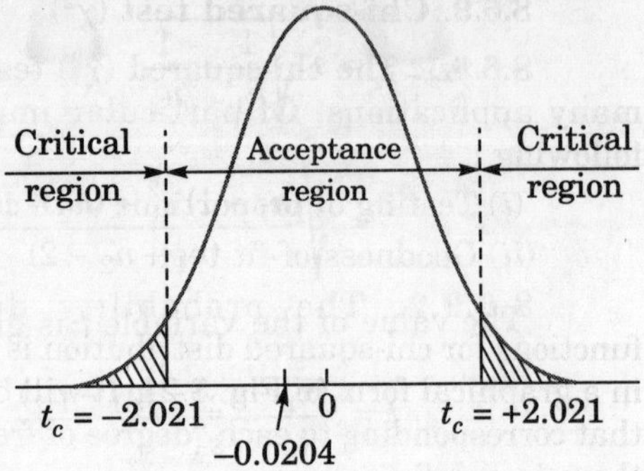

Fig. 8.19

Problem 8.17. *To test whether the sex of a driver had any effect on the speed of a vehicle, a spot speed study was conducted. A sample of 15 male drivers yielded a mean speed of 70.2 KPH with a variance of 98.2 and a sample of 17 female drivers yielded a mean speed of 65.5 KPH with a variance of 84.2. Test whether the mean speed of female drivers is significantly below the mean speed of male drivers. The samples come from normal populations with the same variance.*

Solution. In this problem, let suffix 1 denote male drivers and 2 denote female drivers.

Thus mean speed $x_1 = 70.2$, $x_2 = 65.5$

Variance $s_1^2 = 98.2$, $s_2^2 = 84.2$

$n_1 = 15$, $n_2 = 17$.

Let μ_1 and μ_2 be the population mean speeds. The null hypothesis is that the mean speed of female drivers is the same as that of the male drivers and the observed difference is merely due to chance. Thus, $H_0 : \mu_2 - \mu_2 = 0$.

The alternative hypothesis is that the mean speed of male drivers is higher than that of female drivers, such that, $H_1 : \mu_1 > \mu_2$. A significance level of $\alpha = 0.05$ is chosen, and we are interested in only one tail of the t distribution, *i.e.*, the right hand side tail. The values of t_c corresponding to the critical area 0.05 is 1.70, for a degree of freedom 15 + 17 – 2 = 30 (Table 8.14).

The standard error of the difference of mean is given by Eq. (8.39). Substituting

$$s_{\bar{x}_1 - \bar{x}_2} = \sqrt{\frac{(15-1)\,98.2 + (17-1)\,84.2}{30}}\sqrt{\frac{15+17}{15 \times 17}}$$

$$= \sqrt{\frac{2722 \times 32}{30 \times 15 \times 17}} = \sqrt{11.4} = 3.38$$

Critical value of $\bar{x}_1 - \bar{x}_2$ is given by

$$(\bar{x}_1 - \bar{x}_2)_c = t_c \times s_{\bar{x}_1 - \bar{x}_2}$$

$$= 1.70 \times 3.38 = 5.74$$

The actual difference $\bar{x}_1 - \bar{x}_2$ observed is 70.2 – 65.5 = 4.7 K.P.H. which does not fall in the critical region (Fig. 8.20). Hence the null hypothesis is accepted and we conclude that there is no significant difference between the mean speeds of male and female drivers (at 0.05 significance level). The observed difference is merely due to chance.

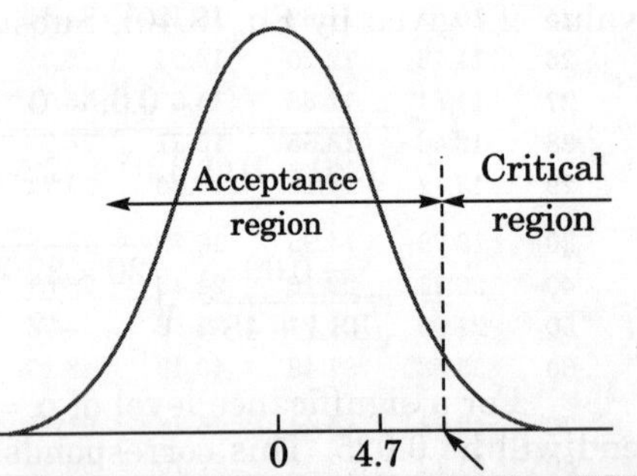

Fig. 8.20

8.6.9. Chi-squared test (χ^2)

8.6.9.1. The chi-squared (χ^2) test is a very useful statistical tool and has many applications. Of particular importance to the traffic engineer are the following :

(*i*) Testing of proportions with contingency tables.

(*ii*) Goodness-of-fit-test.

8.6.9.2. The probability density functions for chi-squared distribution is shown in a graphical form in Fig. 8.21. It will be seen that corresponding to each "degree of freedom" there is a definite curve.

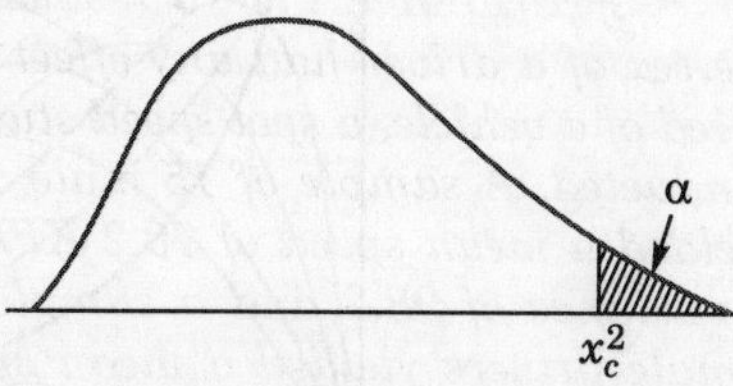

Table 8.15. Critical χ_c^2 values for various degrees of freedom (v)

α	99.5	99	97.5	95	10	5	2.5	1	0.5	0.1
v = 1	0.04393	0.03157	0.03982	0.00393	2.71	3.84	5.02	6.63	7.88	10.83
2	0.0100	0.0201	0.0506	0.103	4.61	5.99	7.38	9.21	10.64	13.81
3	0.0717	0.115	0.216	0.352	6.25	7.81	9.35	11.34	12.80	16.27
4	0.207	0.297	0.484	0.711	7.78	9.49	11.14	13.28	14.86	18.47
5	0.412	0.554	0.831	1.15	9.24	11.07	12.83	15.09	16.75	20.52
6	0.676	0.872	1.24	1.64	10.64	12.59	14.45	16.81	18.55	22.46
7	0.989	1.24	1.69	2.17	12.02	14.07	16.01	18.48	20.28	24.32
8	1.34	1.65	2.18	2.73	13.36	15.51	17.53	20.09	21.95	26.12
9	1.73	2.09	2.70	3.33	14.68	16.92	19.02	21.67	23.59	27.88
10	2.16	2.56	3.25	3.94	15.99	18.31	20.48	23.21	25.19	29.59
11	2.60	3.05	3.82	4.58	17.28	19.68	21.92	24.73	26.76	31.26
12	3.07	3.57	4.40	5.23	18.55	21.03	23.34	26.76	28.30	32.91
13	3.57	4.11	5.01	5.89	19.81	22.36	24.74	27.69	29.82	34.53
14	4.07	4.66	5.63	6.57	21.05	23.68	26.12	29.14	31.32	36.12
15	4.60	5.23	6.26	7.26	22.31	25.00	27.49	30.58	32.80	37.70
16	5.14	5.81	6.91	7.96	23.54	26.30	28.85	32.00	34.27	39.25
17	5.70	6.41	7.56	8.67	24.77	27.59	30.19	33.41	35.72	40.79
18	6.26	7.01	8.23	9.39	25.99	28.87	31.53	34.81	37.16	42.31
19	6.84	7.63	8.91	10.12	27.20	30.14	32.85	36.19	38.58	43.82
20	7.43	8.26	9.59	10.86	28.41	31.41	34.17	37.57	40.00	45.30
21	8.03	8.90	10.28	11.59	29.68	32.67	35.48	38.93	41.40	46.81
22	8.64	9.54	10.98	12.34	30.81	33.92	36.78	40.29	42.80	48.27
23	9.26	10.20	11.69	13.09	32.01	35.17	38.08	41.64	44.18	49.73
24	9.89	10.86	12.40	13.85	33.20	36.42	39.36	42.98	45.56	51.18
25	10.52	11.62	13.12	14.61	34.38	37.65	40.65	44.31	46.93	52.62
26	11.16	12.20	13.84	15.38	35.56	38.89	41.92	45.64	48.29	54.05
27	11.81	12.88	14.57	16.15	36.74	40.11	43.19	46.96	49.64	55.48
28	12.46	13.56	15.31	16.93	37.92	41.34	44.46	48.28	50.99	56.89
29	13.12	14.26	16.05	17.71	39.09	42.56	45.72	49.59	52.34	58.30
30	13.79	14.95	16.79	18.49	40.26	43.77	46.98	50.89	53.67	59.70
40	20.71	22.16	24.43	26.51	51.81	55.76	59.34	63.69	66.77	73.40
50	27.99	29.71	32.36	34.76	63.17	67.50	71.42	76.15	79.49	86.65
60	35.53	37.48	40.48	43.19	74.40	79.08	83.30	88.38	91.95	99.61
70	43.28	45.44	48.76	51.74	85.53	90.53	95.02	100.4	104.2	112.3
80	51.17	53.54	57.15	60.39	96.58	101.9	106.6	112.3	116.3	124.8
90	59.20	61.75	65.65	69.13	107.6	113.1	118.1	124.1	128.3	137.2
100	67.33	70.06	74.22	77.93	118.5	124.3	129.6	135.8	140.2	149.4

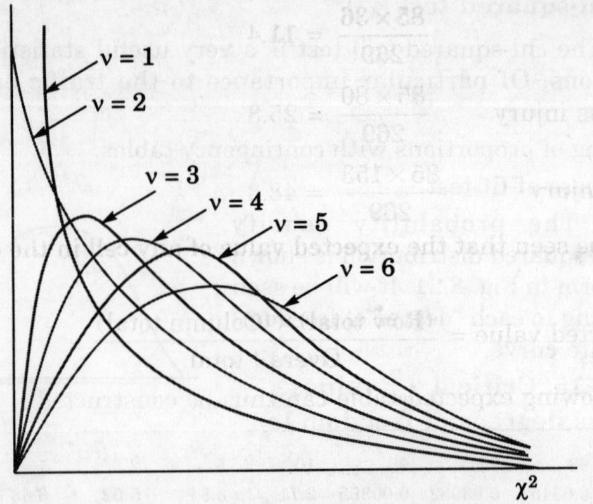

Fig. 8.21. Distribution of χ^2 for various degrees of freedom.

Table 8.15 gives the value of χ^2 for various degrees of freedom and significance levels.

8.6.9.3 A convenient application of the χ^2 test is in testing the comparability of observed and expected values in two way tables, known as contingency tables. The following problem illustrates the application.

Problem 8.18. *In order to assess whether there is any relationship between the accident severity and the age of drivers, an analysis of previous data yielded the following :*

		Age group of drivers			*Total*
		18—30	*30—50*	*Above 50*	*Accidents*
Accident severity	*Fatal*	12	8	16	36
	Grievous injury	25	14	41	80
	Minor injury	48	35	70	153
	Total Accident	85	57	127	269

Test whether there is any significant relationship between the accident severity and the age of drivers.

Solution. The null hypothesis is that the two classification, *viz.*, the accident severity and the age structure are independent. The alternative hypothesis is that the two classifications are interrelated, or correlated.

If the accident severity is independent of the age classification the distribution of the accident numbers by severity should be the same for each age group. Thus, the ratio of fatal, grievous and minor injury accidents should be 36 : 80 : 153 for each age group. The total accidents for age group 18—30 being 85, the expected distribution of this number by accident severity would be,

Fatal $\frac{85 \times 36}{269} = 11.4$

Grievous injury $\frac{85 \times 80}{269} = 25.3$

Minor injury $\frac{85 \times 153}{269} = 48.3$

It will be seen that the expected value of any cell in the contingency table is given by,

$$\text{Expected value} = \frac{(\text{Row total}) \times (\text{Column total})}{\text{Overall total}} \quad ...(8.42)$$

The following expected table can thus be constructed :

		Age group of drivers			*Total*
		18—30	*30—50*	*Above 50*	*Accidents*
Accident Severity	Fatal	11.4	7.6	17.0	36
	Grievous injury	25.3	17.0	37.7	30
	Minor injury	48.3	32.4	72.3	153
	Total	85	57	127	269

The χ^2 value observed from a comparison of the actual and expected tables is found as follows :

$$\chi^2\,(\text{obs}) = \sum_{j=1}^{c} \frac{(O_j - E_j)^2}{E_j} \quad ...(8.43)$$

where, O_j = Observed value in the jth cell

E_j = Expected value in the same cell

c = Total number of cells.

In the present case,

$$\chi^2\,(\text{obs}) = \frac{(12-11.4)^2}{11.4} + \frac{(8-7.6)^2}{7.6} + \frac{(16-17)^2}{17}$$

$$+ \frac{(25-25.3)^2}{25.3} + \frac{(14-17.0)^2}{17.0}$$

$$+ \frac{(41-37.7)^2}{37.7} + \frac{(48-48.3)^2}{48.3}$$

$$+ \frac{(35-32.4)^2}{32.4} + \frac{(70-72.3)^2}{72.3}$$

$$= 0.032 + 0.021 + 0.050 + 0.004 + 0.53 + 0.289$$

$$+ 0.002 + 0.208 + 0.074 = 1.219$$

The degree of freedom is given by :

$$v = (m-1)\,(n-1) \quad ...(8.44)$$

where, v = Degree of freedom
m = Number of columns
n = Number of rows.

In the present case, $v = (3 - 1)(3 - 1) = 1$

For a significance level of 0.05 and $v = 4$, we find from Table 8.15.

$$\chi^2 \text{ critical} = 9.49$$

χ^2 (obs) < χ^2 (critical), and the observed value does not fall in the critical region. Hence accept the null hypothesis that the severity and the age group are not interrelated.

8.6.9.4. Another useful application of the chi-squared distribution is in the goodness-of-fit test. Under this test, the measure of the discrepancy between a set of observed data and the values that are to be expected if the results follow a hypothesis distribution is evaluated. For example, the analyst will have an answer to the following types of questions :

(*i*) It is observed from past experience that the spot speeds follow a normal distribution. Are the data obtained from a particular study also in line with the previous experience ?

(*ii*) The arrival pattern of vehicles is known to be generally random and follows the Poisson distribution. Do the data obtained from a particular study also indicate a Poissonian arrival pattern ?

The χ^2 value is estimated from the following equation :

$$\chi^2 \text{ (observed)} = \sum_{j=1}^{c} \frac{(O_j - E_j)^2}{E_j} \quad ...(8.45)$$

where, O_j = Observed frequency of the jth class or event
E_j = Expected frequency of the jth class or event as per hypothesised distribution
c = Number of classes or events.

The degree of freedom, v, for a goodness-of-fit test depends upon the particular distribution being tested. It will be given by the number of classes or events (c) minus the number of parameters that were utilised from the observed data to calculate the theoretical values. The following table gives the degrees of freedom for various distributions :

Table 8.16. Degrees of freedom for various distribution for use in χ^2 goodness-of-fit test

Distribution	*Degrees of freedom*
Binomial	$c-2$
Normal	$c-3$
Poisson	$c-2$
Exponential	$c-2$

When the individual values of the expected frequencies are small, it is usual to re-group some of these classes so that the theoretical frequency is not less than about 5.

8.7. Linear Regression and Correlation

8.7.1. The traffic engineer is commonly faced with the problem of predicting whether any relationship can be established between two or more variables. The examples of such situations are :

(*i*) The total number of trips produced in an area is seen to depend upon some variables such as family income, family size, employment structure, etc. A large data have been collected giving the number of trips on one hand and the values of family income, family size, employment structure etc. on the other. Can a relationship be determined between these ?

(*ii*) Plot of a number of observations of speed and concentration of vehicles on a long crowded street indicates a scatter. The general tendency observed from the scatter is that as the concentration increases the speed tends to fall. Can a linear relationship be fitted to the scatter of points in the best possible manner ?

The technique of predicting the values of one variable (called the dependent variable) from measurements of the other (called the independent variable) is called regression analysis. If the relationship between the independent and the dependent variables is linear, the analysis is known as linear regression. In the elementary treatment that is possible here, only linear regression will be dealt with. If the independent variables are two or more in number, the analysis is known as multiple linear regression. Thus example (*i*) above deals with a multiple linear regression analysis, where the number of trips is the dependent variable and the factors such as family income, family size and employment structure are the independent variables.

8.7.2. If the dependent variable is denoted by *Y* and the independent variable by *X*, the linear relationship between the two can be of the following form :

$$Y = a + bX + \varepsilon \qquad ...(8.46)$$

where *a* and *b* constants and ε it is a term denoting the random error. The random error term ε will be small if the two variables are closely related. Thus if the observed values of *X* and *Y* are :

$(x_1, y_1), (x_2, y_2), ... (x_n, y_n)$, then

$$y_i = a + bx_i + \varepsilon_i \qquad ...(8.47)$$

where, the subscript *i* refers to the *i*th point (Fig. 8.22).

The problem is to estimate values of *a* and *b* such that the sum of the ε_i's (*i.e.*, $\varepsilon_1 + \varepsilon_2 + \varepsilon_3 ... + \varepsilon_n$ is as small as possible, and thus minimise the likely error in prediction. This is done by the method of least squares thus :

a and *b* are to be chosen to make :

$$R = \sum_{i=1}^{n} \varepsilon_i^2$$

a minimum, where *n* is the total number of points to observed.

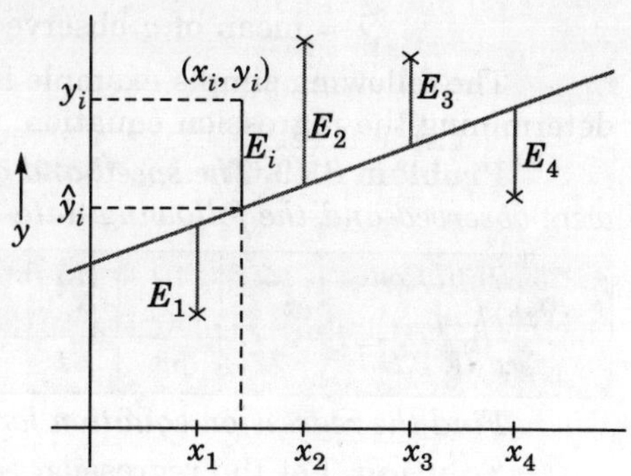

Fig. 8.22

This is done by differentiating *R* with respect to *a* and *b* and equating $\frac{\partial R}{\partial a}$ and $\frac{\partial R}{\partial b}$ to zero.

From Eq. (8.47), $\varepsilon_i = y_i - a - bx_i$

$$\therefore \quad R = \sum_{i=1}^{n} (y_i - a - bx_i)^2$$

$$\therefore \quad \frac{\partial R}{\partial a} = \frac{\partial(na^2)}{\partial a} \frac{\partial}{\partial a}(2a\sum y_i) + \frac{\partial}{\partial a}(2ab\sum x_i) + 0$$

(**Note : 0 from terms not involving *a***)

$$= 2\,na - 2\,\Sigma y_i + 2\,b\,\Sigma x_i = 0$$

$$\therefore \quad a = \bar{y} - b\,\bar{x} \qquad ...(8.48)$$

$$\frac{\partial R}{\partial b} = \frac{\partial}{\partial b}\left(b^2 \sum x_i^2 + 2ab\sum x_i - 2b\sum x_i y_i\right) = 0$$

(**Note : 0 from terms not involving *b***)

$$= 2b\sum x_i^2 + 2a\sum x_i - 2\sum x_i y_i = 0$$

$$\therefore \quad b\sum x_i^2 + a\sum x_i = \sum x_i y_i$$

Substituting the value of $a = \bar{y} - b\bar{x}$ from Eq. (8.48) above,

$$b\sum x_i^2 + \bar{y}\sum x_i - b\bar{x}\sum x_i = \sum x_i y_i$$

$$\therefore \quad b\left(\sum x_i^2 - n\bar{x}\right)^2 = \sum x_i\, y_i - n\bar{x}\,\bar{y}$$

$$\therefore \quad b = \frac{\sum x_i y_i - n\bar{x}\;\bar{y}}{\sum x_i - n\bar{x}^2} = \frac{\sum (x_i - \bar{x})\,(y_i - \bar{y})}{(\sum x_i - x)^2} \qquad ...(8.49)$$

The values of *a* and *b* obtained from Eqs. (8.48) and (8.49) are estimates *a* and *b* of the true regression coefficients so that :

$$\hat{a} = \bar{y} - \hat{b}\bar{x}$$

and
$$\hat{b} = \frac{\sum x_i\, y_i - n\bar{x}\;\bar{y}}{\sum x_i^2 - n\bar{x}^2} = \frac{\sum (x_i - \bar{x})\,(y_i - \bar{y})}{\sum (x_i - \bar{x})^2} \qquad ...(8.50)$$

It may be noted that in the above equation,

$\bar{x}$ = mean of *n* observed values of x_i

$\bar{y}$ = mean of *n* observed values of y_i

The following simple example illustrates the use of the above formulae in determining the regression equation.

Problem 8.19. *The speed and concentration of vehicles in a traffic stream were observed and the following data were obtained :*

Concentration (Veh/km)	5	10	15	20	25	30	35	40	45	50
Speed (K.P.H.)	72	68	61	52	47	39	32	27	20	13

Find the regression equation for determining the speed from concentration.

Solution. Let the regression equation be of the form :

$$y = a + bx$$

where, y = speed in K.P.H.

x = concentration in veh/km.

a and *b* are constants.

x_i	y_i	$x_i y_i$	x_i^2
5	72	360	25
10	68	680	100
15	61	915	225
20	52	1040	400
25	47	1175	625
30	39	1170	900
35	33	1155	1125
40	27	1080	1600
45	20	900	2025
50	13	650	2500
Σ = 275	Σ = 432	Σ = 9125	Σ = 9625

$$\bar{x} = \frac{275}{10} = 27.5, \quad \bar{y} = \frac{432}{10} = 43.2$$

Using Eq. (8.51), $\hat{b} = \dfrac{\Sigma x_i\, y_i - n\, \bar{x}\, \bar{y}}{\Sigma x_i^2 - n\, \bar{x}^2}$

$$= \frac{9125 - 10 \times 27.5 \times 43.2}{9625 - 10 \times (27.5)^2} = -\frac{2775}{2075} = -1.34$$

Using Eq. (8.20),

$$\hat{a} = \bar{y} - \hat{b}\, x = 43.2 - (-1.34)\; 27.5$$
$$= 43.2 + 36.8 = 80.0$$

The regression equation is :

$$y = 80 - 1.34x$$

The graph in Fig. 8.23 indicates the scatter of points as per data recorded and the regression line. It can be seen that the line $y = 80 - 1.34\,x$ is very closely fitting the observed data.

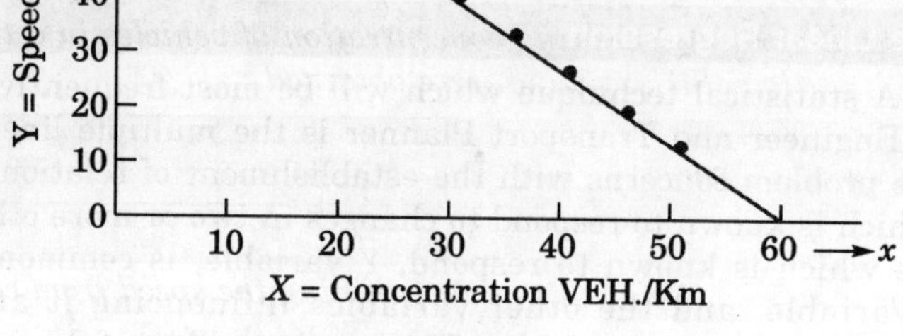

Fig. 8.23

8.7.3. A measure of how good is the relation between the two variables can be had by the correlation coefficient, r, which is defined as :

$$r = \frac{\sum_{i=1}^{n} (x_i - \bar{x})(y_i - \bar{y})}{\sqrt{\sum_{i=1}^{n} (x_i - \bar{x})^2 \sum (y_i - \bar{y})^2}} \quad ...(8.51)$$

The value of the correlation coefficient varies between – 1 and + 1. It is zero if x and y are completely independent of each other, and ± 1 if they fully linearly dependent.

The square of r, r^2, is known as the coefficient of determination. This measure provides an estimate of the fraction of the variability in Y_i, values explained by the estimated regression equation. A high r^2 value, nearing 1, signifies that a major proportion of the variability is explained by the regression equation.

Other statistical measures as the confidence intervals for b and y require a knowledge of the analysis of variance approach and are beyond the range of the elementary treatment. Nevertheless, they are frequently needed by the traffic engineer in his work and the reader is advised to refer standard books on statistics for this purpose.

8.7.4. It often becomes convenient to transform a non-linear relationship to a linear one and then carry out the regression analysis. A classic example of this procedure is fitting a growth model to a time-series data. Often, such data conform to a compound interest rate of growth. For example, if the population of cars over a period of time is analysed, the following equation can be fitted :

$$C_n = C_o (1 + r)^n$$

where, C_n = population of cars in the nth year

C_o = population of cars in the base year ($n = o$)

r = rate of growth per annum expressed as a fraction.

Taking logs on both sides,

$$\text{Log } C_n = \text{Log } C_o + n \text{ Log } (1 + r)$$

This is of the standard linear form : $Y = a + b X$

where, $Y = \text{Log } C_n$

$a = \text{Log } C_o$

$X = n$

$b = \text{Log } (1 + r)$

8.8. Multiple Linear Regression

8.8.1. A statistical technique which will be most frequently encountered by a Traffic Engineer and Transport Planner is the multiple linear regression analysis. The problem concerns with the establishment of relationship between a variable which is known to respond to changes in two or more other variables. The variable which is known to respond, Y variable, is commonly called the dependent variable, and the other variables influencing it are called the independent variables, *i.e.*, X variables. The function will be of the following form :

$$Y = a_0 + a_1x_1 + a_2x_2 + a_3x_3 + \ldots\ldots a_mx_m$$

where, Y = true estimate of the dependent variable, y

$x_1, x_2, x_3 \ldots\ldots x_m$ = m independent variables

a_0 = regression constant

$a_1, a_2, a_3 \ldots\ldots a_m$ = regression coefficients of the respective m independent variables.

8.8.2. Assumptions in the multiple linear regression analysis : The following are some of the conditions which must be satisfied if a multiple linear regression analysis is to be used :

1. All the independent variables must be independent of each other and there should be no correlation between them.

2. All the variables are normally distributed.

3. All the variables are continuous.

4. A linear relationship exists between the dependent variable and the independent variables.

5. The influence of the independent variable is additive, *i.e.*, the inclusion of each variable in the equation contributes a distinct portion in the estimation of the dependent variable and all the variables together contribute additively in the estimation.

8.8.3. The regression coefficients are determined from a given set (n) of observed values of y and $x_1, x_2, \ldots\ldots x_m$, by the method of least squares.

The method of least squares requires the residual error to be minimised.

The sum of residual errors from n set of observations

$$= R = \sum_{i=1}^{n} (y_i - Y_i)^2$$

where, y_i = individual observation and Y_i the estimated value from the regression equation.

By differentiating the residual sum of errors and equating to zero, a number of equations are generated, the solution of which yields the regression coefficients and the regression constant.

The analysis of variance approach is used to test the predictor equation. The total sum of squares is split into the regression sum of squares and the error sum of squares.

Total sum of squares = Regression sum of squares + Error sum of squares.

$$S^2, \text{ variance of error} = \frac{\text{Error sum of squares}}{\text{Degree of freedom}}$$

$$= \frac{\text{Error sum of squares}}{n - m - 1}$$

where, n = number of sets of observations

and m = number of independent variables

Standard error of estimate

$$= \sqrt{S^2}$$

Standard error of estimate expressed as percentage of response mean

$$= \frac{\sqrt{S^2}}{\text{Response mean}} \times 100$$

The smaller the value of the above percentage, then better is the equation.

The co-efficient of determination, R^2, is another indicator of the strength of the relationship. Its value is given by

$$R^2 = \frac{\text{Regression sum of squares}}{\text{Total sum of squares}}$$

The R^2 value lies between 0 and 1. The closer it is to 1, the better is the equation. The R^2 value is the proportion of the variability in the dependent variable explained by the different independent variables.

The F ratio gives another indication of the adequacy of the equation.

$$= \frac{\text{Mean sum of regression squares}}{\text{Mean sum of residual squares}}$$

$$= \frac{\dfrac{\text{Sum of regression squares}}{\text{Degrees of freedom}}}{\dfrac{\text{Sum of residual squares}}{\text{Degrees of freedom}}}$$

$$= \frac{\dfrac{\text{Sum of regression squares}}{m}}{\dfrac{\text{Sum of residual squares}}{n-m-1}}$$

Reference to standard F tables will indicate whether the null hypothesis H_0, is true or not at the selected level of significance.

The significance of the regression co-efficients is determined by the t statistic. The t value is given by the ratio of the regression co-efficient and its standard error. If the 't' value of any regression co-efficient is less than the t value from standard tables at the degree of freedom, it means that the particular independent variable does not contribute significantly to the explanation in the dependent variable.

The solution to multiple linear regression analysis by manual methods is extremely tedious. Computers facilitate the quick analysis. Standard statistical packages are used for the purpose. Step-wise regression is a technique in the computer programmes whereby the computer selects different combinations of the variables in an orderly manner and presents the results of each step. The step-up programme starts by selecting one independent variable which has the highest correlation with the dependent variable.

Thereafter, the programme builds up by adding further variables in the order of their correlation and terminates the programme by including all variables which are significant at a selected level of significance.

For a more rigorous treatment on regression analysis, a reference to standard textbooks may be made (For example, Ref. 6).

I. MULTIPLE CHOICE QUESTIONS

1. The chi square (χ^2) test is a very useful statistical tool and has many applications. Of particular importance to the traffic engineering are the following:

   (*a*) testing of proportions with contingency tables

   (*b*) goodness of fit test

   (*c*) none of these

   (*d*) both (*a*) and (*b*)

2. **The ruling minimum radius of horizontal curve of a national highway in plain terrain for a ruling design speed of 100 km/hour with e = 0.07 and f = 0.15 is close to**

(*a*) 250 m (*b*) 360 m (*c*) 36 m (*d*) 300 m

3. **A car is moving at a speed of 72 km/hr on a road having 2% upward gradient. The driver applies brakes when he sees an obstruction. If his reaction time is 1.5 seconds, assuming that the co-efficient of friction between the pavement and tyre as 0.15, calculate the distance traversed before the car finally stops.**

(*a*) 24 m (*b*) 150 m (*c*) 1056 m (*d*) 324 m

4. **Ruling minimum radius of horizontal curve of a national highway in plain terrain for a ruling design speed of 100 kmph with e = 0.07 and f = 0.15 is done**

(*a*) 250 m (*b*) 360 m (*c*) 36 m (*d*) 300 m

5. **Co-efficient of correlation values lies between**

(*a*) – 1 and + 1 (*b*) 0 and 1 (*c*) – 1 and 0 (*d*) None of these

6. **If more than one items is assigned the same rank .... adjustment is made.**

(*a*) $\frac{1}{12}^{(m^3 - m)}$ (*b*) $\frac{1 - r^2}{\sqrt{N}}$

(*c*) r^2 (*d*) r + P.E. to r – P.E.

7. **Probable error for co-efficient of correlation can be found by the formula ....**

(*a*) $\frac{1 - r^2}{\sqrt{N}}$ (*b*) r^2 (*c*) $\frac{1}{12}^{(m^3 - m)}$

(*d*) r + P.E. to r – P.E.

Answers

1. (*d*)	**2.** (*b*)	**3.** (*b*)	**4.** (*b*)	**5.** (*a*)
6. (*d*)	**7.** (*d*)			

II. SOLVED QUESTIONS AND ANSWERS

Q. 1. What is speed?

Ans. **Speed is one of the basic parameters of traffic flow and time mean speed and space mean speed are the two representations of speed. Time mean speed and space mean speed and the relationship between them will be discussed in detail in this chapter. The relationship between the fundamental parameters of traffic flow will also be derived. In addition, this relationship can be represented in graphical form resulting in the fundamental diagrams of traffic flow. The space mean speed also averages the spot speed, but spatial weightage is given instead of temporal.**

Q. 2. Explain the traffic engineering.

Ans. It is the art and science of estimating traffic demand, highway capacity and measuring and determining the relationships between traffic variables and the application of this knowledge to planning, designing, operating and administering highway facilities in order to achieve the safe and efficient movement of persons and goods.

Be very useful to traffic engineer to evaluate the results of their action through the statistical treatment of the situations before and after a change surgery and to identify trends and forecast parameters related to road traffic.

Q. 3. What is multiple linear regression?

Ans. Multiple linear regression is the most common form of linear regression analysis. As a predictive analysis, the multiple linear regression is used to explain the relationship between one continuous dependent variable from two or more independent variables.

Q. 4. What is correlation?

Ans. The term correlation indicates the relationship between two variables in which with changes in the value of one variable, the values of the other variable also change. Correlation has been defined by various eminent statisticians, mathematicians and economists.

Q. 5. What is histogram?

Ans. Histograms are best used for large sets of data, especially when the data has been grouped into classes. They look a little similar to bar charts or frequency diagrams. In histograms, the frequency of the data is shown by the area of the bars and not just the height. Histograms are most commonly used for continuous data. Histograms often have bars of varying width, *i.e.,* unequal class intervals. This is not a problem as we are dealing with area, not just the height.

The vertical axis of a histogram is labelled **frequency density** and is calculated by the following formula:

$$\text{Frequency density} = \frac{\text{Frequency}}{\text{Class width}}$$

Q. 6. What is cumulative frequency?

Ans. Cumulative frequency is kind of like a running total. We add each frequency to the ones before to get an 'at least' total.

These cumulative frequencies ('at least' totals) are plotted against the upper class boundaries to give us a cumulative frequency curve. The cumulative frequency column is the column you will be expected to add for yourself.

Q. 7. What are the important properties of the correlation co-efficient *r*?

Ans. 1. The co-efficient of correlation lies between – 1 and + 1. Symbolically, or.

2. The co-efficient of correlation is independent of change of scale and origin of the variables X and Y.

3. The co-efficient of correlation is the geometric mean of two regression co-efficients.

III. REVIEW QUESTIONS

1. Give examples of linear regression.
2. What is the mean spot speed?
3. The mean spot-speeds of 15 vehicles observed on a Sunday at a particular location of a highway was 81.2 K.P.H and the standard deviation was 10.2 K.P.H. The mean speed of all vehicles at this location as per previous records was 75.5 K.P.H. Is their sufficient evidence to show that the speed of vehicles of the particular Sunday (*i*) was higher than the average speed and (*ii*) was different from the average speed?

REFERENCES

1. Adams, W.P., Road Traffic Considered as a Random Series, Journal of the Institution of Civil Engineers, Vol. 4, Nov. 1936, London.
2. Votaw, D.F., and H.S. Levinson, Elementary Sampling for Traffic Engineers, ENO Foundation for Highway Traffic Control, Saugatuck, Connecticut, 1962.
3. Hoel, P.G., Elementary Statistics, John Wiley and Sons, Inc., New York, 1971.
4. Lowell Wing, R., Statistics for Scientists and Engineers, Prentice-Hall of India (Private) Ltd., New Delhi, 1966.
5. Moroney, M.J., Facts from Figures, Penguin Books, Inc., Baltimore, 1957.
6. Chatterjee S. and B. Price, Regression Analysis by example, John Wiley and Sons, New York, 1977.

9. Speed Studies

9.1. Introduction

The spot speeds measured at any particular location will depend upon a number of factors such as the geometric layout of a road, the volume of traffic, the composition of traffic, the condition of the road, environmental influences (rain, fog, etc), the human element associated with individual drivers and the characteristics of vehicles. As a result, the speeds measured show a considerable amount of scatter. The mathematical analysis of the observed data makes it possible to understand the speed characteristics. Statistical tools, already discussed earlier, become handy for this purpose. The present chapter is concerned with the analysis of speed data by statistical methods and the results of speed studies.

9.2. Grouping of Spot Speed Data and Tabular Presentation

To facilitate easy computations, the first step in the analysis of speed data is to group them into speed class intervals and a frequency distribution table is constructed. A sample is given in Table 9.1, in which a total of 363 observations are grouped into class-intervals of 4 K.P.H.

Table 9.1

Speed class limits (K.P.H.) (1)	*Mid-point speed (K.P.H.)* (x_1) (2)	*No. of vehicles* (f_1) (3)	*Frequency* % (4)	*Cumulative frequency* % (5)
26–29.9	28	9	2.5	2.5
30–33.9	32	74	20.4	22.9
34–37.9	36	79	21.7	44.6
38–41.9	40	75	20.7	65.3
42–45.9	44	66	18.2	83.5
46–49.9	48	33	9.1	92.6
50–53.9	52	17	4.7	97.3
54–57.9	56	6	1.6	98.9
58–61.9	60	1	0.3	99.2
62–65.9	64	3	0.8	100.0

9.3. Histogram and Cumulative Frequency Curves

The data from Table 9.1, can be used to construct the histogram, where the horizontal axis represents the speed-class limits and the vertical axis the percentage frequency. The histogram is shown in Fig. 9.1. This diagram shows that speeds of vehicles tend to cluster about the mean value and the frequency drops as the speeds depart from the mean.

The cumulative percentage frequency in column 5 of the Table 9.1 can be plotted against the upper limit of the speed-class in column 1 to give the cumulative frequency distribution curve, Fig. 9.1. This is a very commonly used graphical representation of the speed distribution and is of great use in determining the percentile speed. The percentile speed is the speed below which a specified percentage of vehicles are travelling. The percentile speeds of interest to a traffic engineer are :

(*i*) 98th percentile speed, the speed below which 98 per cent of all the vehicle travel, is used as a design speed in geometric design. Sometimes 95th percentile speed is also used for this purpose.

(*ii*) 85th percentile speed, the speed below which 85 per cent of all the vehicle travel, is used for determining the speed limits for traffic regulation.

(*iii*) 50th percentile speed, or the median speed, is the speed at which there are as many vehicles going faster as there are going slower.

(*iv*) 15th percentile speed, the speed below which 15 percent of all the vehicles travel, is used to determine the lower speed limit on major highway facilities such as Expressways. Vehicles travelling below this speed cause interference with the traffic stream and may cause hazards.

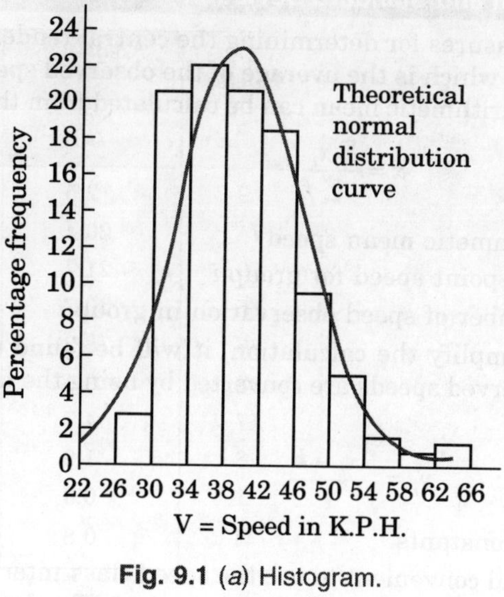

Fig. 9.1 (*a*) Histogram.

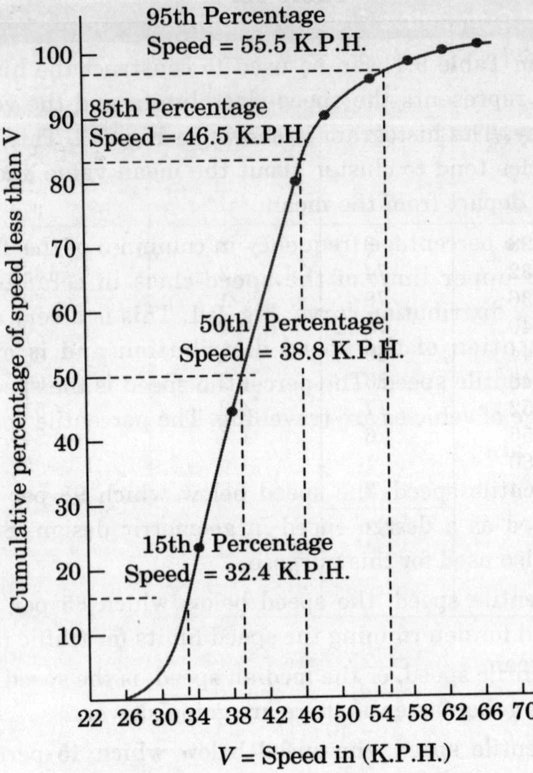

Fig. 9.1 (*b*) Histogram and cumulative frequency curve.

The above values are plotted in Fig. 9.1.

9.4. Arithmetic Mean and Standard Deviation

One of the measures for determining the central tendency of speed data is the arithmetic mean, which is the average of the observed speed. Since the speed data is grouped, the arithmetic mean can be calculated from the following formula

$$\bar{x} = \frac{\Sigma x_i f_i}{\Sigma f_i} \qquad ...(9.1)$$

where, $\bar{x}$ = arithmetic mean speed

x_i = mid-point speed for group i

f_i = number of speed observation in group i

In order to simplify the calculation, it will be found convenient to use *a false origin*. The observed speeds are converted by using the false origin'constants as below

$$y_i = \frac{x_i - k}{c} \qquad ...(9.2)$$

where, k and c are constants.

It will be found convenient to use the speed class-interval as the value of c. The calculations in Table 9.2 illustrates the method. The data used are the same as in Table 9.1.

Table 9.2

Speed class limits (K.P.H.) (1)	*Mid-point (K.P.H.) (x_i)* (2)	*No. of vehicles (f_i)* (3)	$y_i = \frac{x_i - k}{c}$ *where k = 40 and c = 4* (4)	$y_i f_i = \left(\frac{x_i - 40}{4}\right) f_i$ (5)	$y_i^2 f_i$ *Col.(5) × Col. (4)* (6)
26–29.9	28	9	– 3	– 27	+ 81
30–33.9	32	74	– 2	– 148	+ 296
34–37.9	36	79	– 1	– 79	+ 79
38–41.9	40	75	0	0	0
42–45.9	44	66	+ 1	+ 66	+ 66
46–49.9	48	33	+ 2	+ 66	+ 132
50–53.9	52	17	+ 3	+ 51	+ 153
54–57.9	56	6	+ 4	+ 24	+ 96
58–61.9	60	1	+ 5	+ 5	+ 25
62–65.9	64	3	+ 6	+ 18	+ 108
Total		363		– 254 + 230	+ 1036
				Σ = – 24	

Arithmetic mean, $\overline{y} = \frac{\Sigma y_i f_i}{\Sigma f_i} = -\frac{24}{363} = -0.066$

$$y_i = \frac{\overline{x}_i - 40}{4} \quad \text{and} \quad \overline{y} = \frac{\overline{x} - 40}{4}$$

$$\therefore \quad \overline{x} = 4\overline{y} + 40 = 4 \times (-0.066) + 40$$

$$= 40 - 0.264 = 39.736, \text{ say } 39.7.$$

The measure of dispersion of the speed data is the standard deviation, which is calculated from the following formulae :

$$s^2 = \frac{\Sigma (x_i - \overline{x})^2}{n - 1} \text{ for ungrouped data} \qquad ...(9.3)$$

$$s^2 = \frac{\Sigma (x_i - k)^2 - n(\overline{x} - k)^2}{n - 1} \qquad ...(9.4)$$

for ungrouped data with a false origin k.

$$s^2 = \frac{\Sigma (x_i - \overline{x})^2 f_i}{\Sigma f_i - 1} \text{ for grouped data} \qquad ...(9.5)$$

$$s^2 = \frac{\Sigma (x_i - k)^2 - (\Sigma f_i)(\overline{x} - k)^2}{\Sigma f_i - 1} \qquad ...(9.6)$$

for grouped data with false origin k.

$$s^2 (yi) = \frac{\Sigma (y_i)^2 f_i - (\Sigma f_i)(\overline{y})^2}{\Sigma f_i - 1} \qquad ...(9.7)$$

for grouped data with false origin where $y_i = \frac{x_i - k}{c}$

$$s_x^2 = s_y^2 c^2 \qquad ...(9.8)$$

In the above formulae, s = Standard deviation,

s = Variance

x_i = Individual speed measurement for ungrouped data and individual speed measurement in group i for grouped data

f_i = Number of speed observation in group

$\bar{x}$ = Arithmetic mean speed.

$$y_i = \frac{x_i - k}{c}$$

where, k and c are constants (false origin)

$\bar{y}$ = Arithmetic mean of y_i

s_y = Standard deviation of the variable y.

In the example given in Table 9.2, using formula (9.7)

$$s_y^2 = \frac{1036 - 363 \times \frac{24}{363} \times \frac{24}{363}}{363 - 1}$$

$$= \frac{1036 - 1.6}{362} = 2.86$$

$$s^2 = 2.86 \times (4)^2$$

$$s = \sqrt{2.86 \times 4} = 6.76, \text{ say } 6.8 \text{ (K.P.H.)}$$

9.5. Fitting a Normal Distribution Curve to Observed Speed Data

The spot speeds can be represented by a *normal distribution*, which has already been described in Chapter 8. The parameters that describe such a distribution are μ, the true mean of the population and σ, the true standard deviation of the population. The values $\bar{v}$ and s calculated from the observations are unbiased estimates of μ and σ respectively. By using the Statistical Tables of area under a normal distribution curve, it is possible to calculate the theoretical frequency of the grouped speed observations. The calculations are illustrated in Table 9.3, for the same data as given in Tables 9.1 and 9.2.

Column (1) gives the upper speed class limits (v_i). Column (2) gives the difference between the upper speed class limit and the arithmetic mean of observed speeds, *i.e.* $v_i - \bar{v}$. Column (3) gives the value of $v_i - \bar{v}/s$, where s is the observed standard deviation, 6.8 in the present case. Column (4) gives the area under the normal curve upto the upper speed class-limits under consideration and is obtained from Statistical Tables directly (Table 8.11). Column (5) gives the probability of speeds lying between successive values of upper speed class limits, and is obtained by the difference between successive values of column 4. The theoretical frequency is calculated by multiplying column (5) by the total number of observations Σq_i, which is 363 in the present case. Column (7) gives the percentage theoretical frequency.

The values in column (7) have been plotted in Fig. 9.1 (*a*), super imposed on the histogram. It may be noted that the values are plotted against the mid point speed values are not against the upper class speed values. Fbr plotting the cumulative frequency diagram, Fig. 9.1 (*b*), the upper class speed values are selected.

For checking the *goodness-of-fit* between the theoretical distribution, obtained as above and the observed distribution, *a* χ^2 test is usually performed. In Table 9.3, column (8) gives the observed frequency (*O*), whereas column (6) gives the Expected frequency (*E*). Summing up the values of $\frac{(O-E)^2}{E}$, we get

Table 9.3

Upper speed class limits x_i	$x_i - \bar{x}$	$\frac{x_i - \bar{x}}{s}$	$\phi \frac{x_i - \bar{x}}{s}$ *(area under Normal Curve from Statistical Tables)*	*Probability of speed lying between successive values of* x_i *= Difference between successive values of Col. (4)*	*Theoretical Frequency Col. (5)* $\times \Sigma f_t = E$	*Theoretical % frequency*	*Observed frequency = O*	$\frac{(O-E)^2}{E}$
(1)	(2)	(3)	(4)	(5)	(6)	(7)	(8)	(9)
26	– 13.7	– 2.02	0.0217	0.0217	7.8	2.17	–	7.80
30	– 9.7	– 1.43	0.0764	0.0547	19.9	5.47	9	5.96
34	– 5.7	– 0.84	0.2005	0.1241	45.0	12.41	74	18.70
38	– 1.7	– 0.25	0.4013	0.2008	73.0	20.08	79	0.49
42	+ 2.3	+ 0.34	0.6331	0.2318	84.0	23.18	75	0.95
46	+ 6.3	+ 0.93	0.8238	0.1907	69.0	19.07	66	0.16
50	+ 10.3	+ 1.52	0.9357	0.1119	40.6	11.19	33	1.42
54	+ 14.3	+ 2.10	0.9821	0.0464	16.8	4.64	17	0.00
58	+ 18.3	+ 2.69	0.9964	0.0143	5.2	1.43	6	2.02
62	+ 22.3	+ 3.28	0.9995	0.0031	1.1 (6.4)	0.31	1 (10)	
66	+ 26.3	+ 3.87	0.9999	0.0004	0.1	0.04	3	
			Total	0.9999	362.7	99.99	363	37.50

$$\chi^2 \text{ obs} = \frac{\Sigma (O-E)^2}{\Sigma E} = 37.50$$

It may be noted that where the values of *E* are less than 5, the observations are combined so that the minimum value of *E* is 5. In this example, the values of *E* for v_i = 62 and 66 are less than 5, and hence the values of *E* for v_i = 58, 62 and 66 are obtained to give a value greater than 5.

The number of degrees of freedom for a normal distribution curve is *c*–3, where *c* = 5, number of rows in column (9), and 3 representing the number of constraints adopted in the calculations (*i.e.*, mean speed, standard deviation and the total number of speed observations) *vide* Table 8.16.

In this case, degrees of freedom = 9 – 3 = 6.

For a 5 percent level of significance, we get the value of χ^2 (critical) from Statistical Table ($d.f = 6$) [Table 8.15], as 12.59. Since χ^2 (observed) is greater than χ^2 (critical), we conclude that there is a significant difference between the theoretical values and the observed value of speed frequency (at 5 percent significance level). In this particular case, the observed data do not show a good fit with the theoretically expected distribution.

9.6. Accuracy of Sampling

It is not always possible and practical to determine the spot speed of each and every vehicle in a busy stream of traffic. Sampling is, therefore, a necessity. The speed characteristics determined from sample can be assumed to represent the characteristics of the speeds of all the vehicles in the traffic stream provided the sampling has been done in an unbiased manner. Statistical techniques are available to determine the accuracy of sampling and these have been discussed in an earlier Chapter 8.

9.7. General Trends in Speed Data

9.7.1. Speed as related to level of service and volume

Speed is governed by a number of factors some of which also determine the level of service under which a highway operates (see Chapter 21 for definition of level of service).

Fig. 9.2, illustrates the cumulative frequency curves (Ref. 1) for different traffic volumes and levels of service on a two-lane highway.

9.7.2. Speed trends

The general trend over the years has been an increase in the speed of all types of vehicles, when drivers are free to choose their own speeds. Thus, in U.S.A., the national average for the speed of passenger car was 92 K.P.H. in 1963, whereas it was only 74 K.P.H. in 1946 (Ref. 1). The advancement in the vehicle design over the years is reflected in these figures. At the same time, congestion has been increasing in the city streets and the speeds in the urban areas have been generally decreasing (Ref. 2).

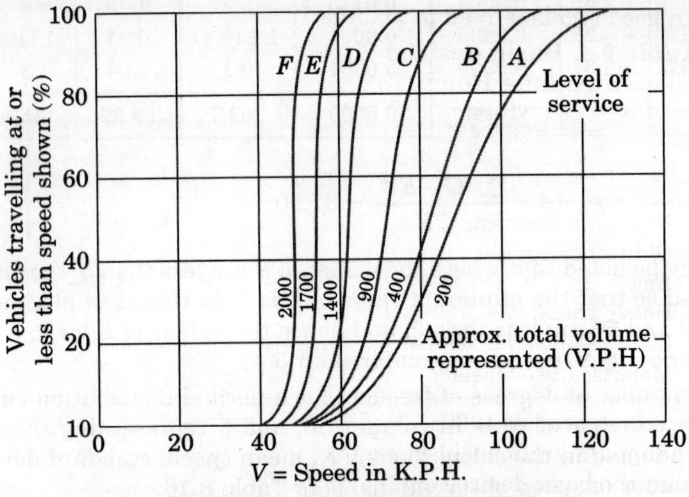

Fig. 9.2. Distribution of passenger car speeds on a 2 lane highway. (Source : Ref. 1).

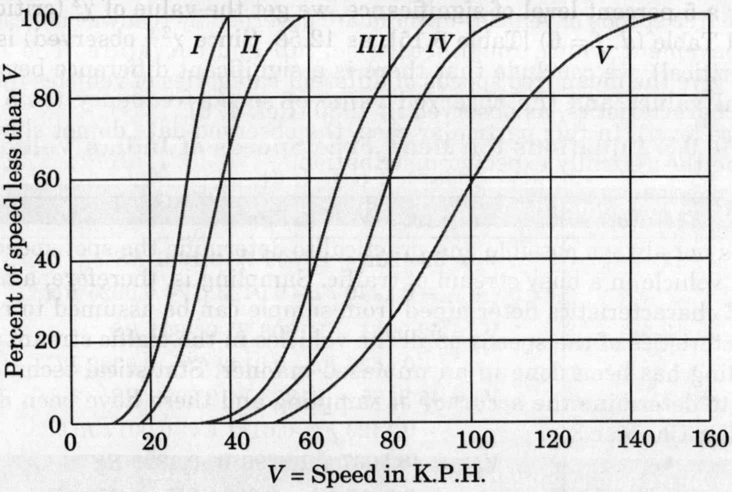

I – Urban Street
II – Inner Suburban Road
III – Outer Suburban Road
IV – Rural Road
V – Motorway

Fig. 9.3. Distribution of speeds on five different roads. (*Source* : Ref. 2)

9.7.3. Speeds on different classes of roads

On superior type of facilities such as the motorway, drivers are free to choose their own speeds. One can thus expect that the range of speeds observed on these facilities would be very wide. On the other hand, the range of speeds observed on an urban street will be very small. The cumulative frequency curves for different classes of roads from U.K. data are given in Fig. 9.3 (Ref. 2).

9.7.4. Indian speed studies

A large amount of research into speeds of vehicles under Indian conditions has been carried out recently (Ref. 4, 5, 6, 7, 8).

Some of the findings are given below. The basic desired speeds of Indian vehicles on a good two-lane road in plain terrain are as below :

Table 9.4. Basic desired Speed of Vehicles in India on Two-lane Roads in Plain Terrain (Ref. 7, 8)

S.No.	*Vehicles type*	*Basic desired mean speed (K M P H)*	*Standard deviation (K M P H)*	*Coefficient of variation*
1.	New technology car	91.76	12.66	0.14
2.	Old technology car	68.03	10.39	0.15
3.	Bus	75.34	9.59	0.13
4.	Heavy truck	63.28	8.34	0.13
5.	Light commercial vehicle	69.56	8.69	0.12
6.	Motorised two-wheeler	56.50	7.77	0.14

More recent studies (2001) (Ref. 9) indicate that the basic desired speeds have not changed significantly between 1992 and 2001. The 2001 study also concluded that the basic desired speeds on four-lane divided carriageways are not appreciably different from values found from two-lane bi-directional carriageways with paved shoulders.

The free speeds of vehicles in India are governed by roadway characteristics such as pavement width, roughness, rise, fall and curvature. The following equations give the mean free speeds of different categories of vehicles in terms of roadway characteristics, as observed in 1980 (Ref. 4, 6).

Table 9.5. Equations for Mean Free Speeds of Indian Vehicles
(Source : Ref. 4, 6)

Eqn. No.	*Eqn*
1.	$V_c = 52.4202 + 1.7609\ W - 0.1621\ RS - 0.1240\ FL - 0.0123\ CV - 0.0025\ RG$
2.	$V_B = 52.9951 + 1.1608\ W - 0.2601\ RS - 0.1626\ FL - 0.0142\ CV - 0.0022\ RG$
3.	$V_T = 45.9185 + 1.1883\ W - 0.2099\ RS - 0.1420\ FL - 0.0131\ CV - 0.0017\ RG$
4.	$V_{2-w} = 46.5067 + 1.0296\ W - 0.1329\ RS - 0.0817\ FL - 0.0054\ CV - 0.0020\ RG$

Note. (1) V_C, V_B, V_T and V_{2-w} are the mean free speeds of cars, buses, trucks and motorised two-wheelers respectively in KMPH.

(2) W = Pavement width in m.

(3) RS = Rise in m/km.

(4) FL = fall in m/km.

(5) CV = Curvature (Bendiness), degrees/km.

(6) RG = Roughness, measured by British Towed Fifth Wheel Bump Integrator, in mm/km.

The free speeds of vehicles on Indian roads are somewhat lower than in the developed countries abroad. Various possible reasons are :

1. The drivers are conscious of the danger from pedestrians and animals on the roads.

2. The drivers are conscious of the high fuel consumption if they drive at high speeds. This factor might have become important after the fuel crisis.

3. The poor condition of the shoulders forces the drivers to be careful.

4. The generally lower standards of driving discipline prompts the drivers to adopt safe speeds.

5. The lower horsepower of the passenger cars and commercial vehicles in India do not permit very high speeds.

6. Heavy overloading of trucks and buses results in lower speeds in these categories of vehicles.

7. The high occupancy rate of cars and buses results in lower speeds.

8. The low horse-power of scooters brings down the speed of two-wheelers.

The free speeds of different types of vehicles follow a normal distribution with a coefficient of variation in the range of 0.15 to 0.20.

The free speeds at night are significantly lower than free speeds under daylight condition.

The free speeds of vehicles on single lane roads (3.75 m) with brick shoulders are nearly equal to those on an intermediate lane road (5.5 m).

The free speeds on wet surfaces are significantly lower than those on dry surfaces. Raining conditions reduce the free speeds further.

9.8. Time Mean Speed and Space Mean Speed

The mean speed of vehicles over a period of time at a point in space is called the time mean speed. The mean speed over a space at a given instant is called the *space mean speed.*

The relation between the time-mean speed and space-mean speed has been given by Wardrop (Ref. 3). The relation can be derived as follows :

Consider a stream of traffic with a total flow of Q, consisting of subsidiary streams with flows $q_1, q_2, q_3 \ldots\ldots\ldots, q_c$ and speeds $v_1, v_2, v_3 \ldots\ldots\ldots, v_c$.

$$Q = q_1 + q_2 + q_3 + \ldots\ldots\ldots + q_c = \sum_{i=1}^{c} q_i$$

Consider the subsidiary stream with flow q_1 and speed v_1. The average time head-way between vehicles in this stream is $1/q_1$ and the distance travelled in this time is v_1/q_1.

The density of this stream in space, *i.e.* the number of vehicles per unit length at any instant (the concentration) is given by :

$$k_i = \frac{q_i}{v_i}, \; t = 1, 2, 3, \ldots\ldots, c \qquad \ldots(9.9)$$

The total concentration is given by $K = \sum_{i=1}^{c} k_i$

Putting $f_i' = \dfrac{k_i}{k}$ gives the frequency f_i' of v_i in space.

The time mean speed is defined as

$$\bar{v}_t = \sum_{i=1}^{c} \frac{q_i \, v_i}{Q} \qquad \ldots(9.10)$$

and the space mean speed is defined as

$$\bar{v}_s = \sum_{i=1}^{c} \frac{k_i \, v_i}{K} \qquad \ldots(9.11)$$

But from Eq. (9.9)

$$k_i \, v_i = q_i \qquad \ldots(9.12)$$

$$\text{Therefore, } \bar{v}_s = \sum_{i=1}^{c} \frac{q_i}{K} = \frac{Q}{K} \qquad \ldots(9.13)$$

$$\text{From Eq. (9.13), } \bar{v}_s = \frac{Q}{K} = \frac{Q}{\Sigma \, k_i} = \frac{Q}{\sum \frac{q_i}{v_i}}$$

From Eqs. (9.10) and (9.12)

$$v_t = \sum_{l=1}^{c} \frac{k_i \, v_i^2}{Q}$$

$$= K \frac{\sum_{i=1}^{c} f_i' \, v_i^2}{Q} = \frac{\sum_{i=1}^{c} f_i' \, v_i^2}{\bar{v}_s} \qquad \text{from (9.13)}$$

$$= \frac{\sum_{i=1}^{c} f_i' \{\overline{v}_s + (\overline{v}_i - \overline{v}_s)\}^2}{\overline{v}_s}$$

$$= \frac{\left\{\left(\sum_{i=1}^{c} f_i'\right)(\overline{v}_s)^2 + \sum_{i=1}^{c} f_i' (v_i - \overline{v}_s)^2\right\}}{\overline{v}_s}$$

since $\sum_{i=1}^{c} f_i' (v_i - \overline{v}_s) = 0$

$$\therefore \quad \overline{v}_t = \overline{v}_s + \frac{\sigma_s^{\,2}}{\overline{v}_s} \qquad \text{...(9.14)}$$

where, $\sigma_s = \sqrt{\left\{\sum_{i=1}^{c} f_i' (v_i - \overline{v}_s)^2\right\}}$

= standard deviation of the space mean speed.

The equation $\overline{v}_t = \overline{v}_s + \frac{\sigma_s^2}{\overline{v}_s}$ shows that the time mean speed is never less than the space mean speed.

If v_i is the individual speed in time, then

$$\overline{v}_t = \frac{\sum v_t}{n} \qquad \text{...(9.15)}$$

where, n is the number of observations.

The space mean speed is obtained by :

$$\overline{v}_s = \frac{n}{\sum \frac{1}{\overline{v}_i}} \qquad \text{...(9.16)}$$

Problem 9.1. *Speed observations from a radar speed meter have been taken, giving the speeds of the subsidiary streams composing the flow along with the volume of traffic of each subsidiary stream. The readings are as under :*

Speed (km/hr)	*Volume of subsidiary stream (Vehicle hour)*
2–5	*1*
6–9	*4*
10–13	*0*
14–17	*7*
18–21	*20*
22–25	*44*
26–29	*80*
30–33	*82*
34–37	*79*
38–41	*49*
42–45	*36*
46–49	*26*
50–53	*9*
54–57	*10*
58–61	*3*

Calculate : (i) Time-mean speed (ii) Space-mean speed
(iii) Variance about space mean speed.

Solution : The mid-point of speed class (vi) and q_i are tabulated below, along with $q_i\ v_i$ and $k_i = q_i/v_i$

Speed Range	*Volume* (q_i)	*Speed mid-point* (v_i)	$q_i v_i$	$k_i = q_i/v_i$
2–5	1	3.5	3.5	0.29
6–9	4	7.5	30.0	0.54
10–13	0	11.5	0	0
14 –17	7	15.5	108.5	0.45
18–21	20	19.5	390.0	1.02
22–25	44	23.5	1034.0	1.87
26–29	80	27.5	2200.0	2.91
30–33	82	31.5	2583.5	2.60
34–37	79	35.5	2804.5	2.23
38–41	49	39.5	1935.5	1.24
42–45	36	43.5	1566.0	0.83
46–49	26	47.5	1235.0	0.55
50–53	9	51.5	493.5	0.17
54–57	10	55.5	555.0	0.18
58–61	3	58.5	178.5	0.05
	450		15087.0	14.93

$$\Sigma q_i = Q = 450 \text{ veh/hour}$$

$$\Sigma v_i q_i = 15087.0$$

$$\Sigma k_i = k = 14.93$$

$$\bar{v}_t = \frac{\Sigma q_i v_i}{Q} = \frac{15087}{450} = 33.53 \text{ km/hr.}$$

$$\bar{v}_s = \frac{Q}{k} = \frac{450}{14.93} = 30.1 \text{ km/hr.}$$

$$\bar{v}_t = \bar{v}_s + \frac{\sigma_s^2}{\bar{v}_s}$$

$$\therefore \quad \sigma_s^2 = (33.53–30.1)\ 30.1 = 103.5.$$

Problem 9.2. *Twenty five spot speed observations were taken and were as under :*

50, 40, 60, 54, 45, 31, 72, 58, 43, 52, 46, 56, 43, 65, 33, 69, 34, 51, 47, 41, 62, 43, 55, 40, 49.

Calculate :

(i) Time-mean speed *(ii) space-mean speed*

and verify the relation between the two.

Solution:

$$\Sigma v_i = 50 + 40 + 60 + 54 + 45 + 31 + 72 + 58 + 43 + 52 + 46 + 56 + 43 + 65 + 33 + 69 + 34 + 51 + 47 + 41 + 62 + 43 + 55 + 40 + 49 = 1239$$

$$\bar{v}_t = \frac{1239}{25} = 49.56,\ \Sigma\,(v_i - 49.56)^2 = 2972.41$$

$$\Sigma 1/v_i = 1/50 + 1/40 + 1/60 + 1/54 + 1/45 + 1/31 + 1/72 + 1/58 + 1/43 + 1/52 + 1/46 + 1/56 + 1/43 + 1/65 + 1/33 + 1/69 + 1/34 + 1/51 + 1/47 + 1/41 + 1/62 + 1/62 + 1/43 + 1/55 + 1/40 + 1/49 = 0.529.$$

$$\therefore \quad \overline{v}_s = \frac{25}{0.529} = 47.26$$

$$s^2 = \frac{2972.41}{25-1} = 123.85$$

The relationship between $\overline{v}_t$ and $\overline{v}_s$ is

$$\overline{v}_t = \overline{v}_s + \frac{s^2}{\overline{v}_s} = 47.26 + \frac{123.85}{47.26}$$

$$= 47.26 + 2.62 = 49.88$$

This may be compared with the value of 49.56 obtained earlier.

I. MULTIPLE CHOICE QUESTIONS

1. It was noted that on a section of road the free speed was B0 kmph and the jam density was 70 vpkm' The maximum flow in vph that could be expected on this road is
(*a*) 800 (*b*) 1400 (*c*) 2800 (*d*) 5600
(IES, 1998)

2. If the normal flows on two approach roads at an intersection are respectively 500 pcu per hr and 300 pcu per hr. the saturation flows are 1600 pcu per hr on each road and the total lost time per signal cycle is 16 s, then the optimum cycle time by Webster's method is
(*a*) 72.5 s (*b*) 58 s (*c*) 48 s (*d*) 19.3 s
(IES, 1999)

3. For a road with camber of 3% and the design speed of 80 km/hr, the minimum radius of the curve, beyond which NO super elevation is needed, is
(*a*) 1680 m (*b*) 948 m (*c*) 406 m (*d*) 280 m

4. The length of Summit Curve on a two lane two way highway depends upon.
(*a*) allowable rate of change of centrifugal acceleration
(*b*) co-efficient of lateral friction
(*c*) required stopping sight distance
(*d*) required overtaking sight distance

5. What is the minimum length of overtaking zone for a design speed of 96 kmph assuming acceleration as 0.69 m/s^2, reaction time as 2 sec and traffic road as one way
(*a*) 342 m (*b*) 684 m (*c*) 1026 m (*d*) 1710 m

6. For detecting whether a bitumen is cracked (over heated) or not the test/ performed is/are
(*a*) spot test (*b*) float test (*c*) ductility test (*d*) all of these

Answers

1. (*b*) 2. (*b*) 3. (*d*) 4. (*c*) 5. (*c*)
6. (*c*)

II. SOLVED QUESTIONS AND ANSWERS

Q. 1. What is speed?

Ans. Speed is an important transportation consideration because it relates to safety, time, comfort, convenience, and economics. Spot speed studies are used to determine the speed distribution of a traffic stream at a specific location. The data gathered in spot speed studies are used to determine vehicle speed percentiles, which are useful in making many speed-related decisions. Spot speed data have a number of safety applications, including the following:

Q. 2. How will you determine the existing traffic operations and evaluation of traffic control devices?

Ans. We determine in following ways:

- Evaluating and determining proper speed limits
- Determining the 50th and 85th speed percentiles (explained below)
- Evaluating and determining proper advisory speeds
- Establishing the limits of no-passing zones
- Determining the proper placements of traffic control signs and markings
- Setting appropriate traffic signal timing

Q. 3. Explain the ways of establishing roadway design elements.

Ans. The ways of establishing roadways:

- Evaluating and determining proper intersection sight distance
- Evaluating and determining proper passing sight distance
- Evaluating and determining proper stopping sight distance

Q. 4. How will assess the roadway safety questions?

Ans. The ways to assess the roadway safety are:

- Evaluating and verifying speeding problems
- Assessing speed as a contributor to vehicle crashes
- Investigating input from the public or other officials
- Monitoring traffic speed trends by systematic ongoing speed studies
- Measuring effectiveness of traffic control devices or traffic programs, including signs and markings, traffic operational changes, and speed enforcement programs.

Q. 5. What are methods to gather spot speed?

Ans. Spot speed data are gathered using one of three methods:

(1) Stopwatch method, (2) Radar meter method, or (3) Pneumatic road tube method.

These methods are described in this chapter in order from least expensive to most expensive. The stopwatch method is the least expensive and least accurate of the methods.

Q. 6. Explain the radar meter method.

Ans. A radar meter is a commonly used device for directly measuring speeds in spot speed studies.

This device may be hand-held, mounted in a vehicle, or mounted on a tripod. The effective measuring distance for radar meters ranges from 200 feet up to 2 miles (Parma 2001). A radar meter requires line-of-sight to accurately measure speed and is easily operated by one person. If traffic is heavy or the sampling strategy is complex, two radar units may be needed.

Different sized vehicles and the detection of the observation vehicle may affect radar readings (Currin 2001). Large vehicles such as trucks and buses send the strongest return signal to the radar meters and as a result smaller vehicles may not be detected. If there is a presence of large vehicles, the observer may need to record the speeds of vehicles that are alone. Also, some vehicles are equipped with radar detectors to warn them that a radar unit is operating in their vicinity. Drivers will slow down when warned by a detector. It is not unusual for other drivers to slow down also. This slowing will affect the study results. The radar unit may be turned off while not in use so radar detectors cannot detect it.

Q. 7. What are the key steps to a radar meter pot speed?

Ans. A radar meter spot speed study includes four key steps:

- Select proper location and placement of radar meter.
- Determine an appropriate selection strategy.
- Record observations on radar meter spot speed study data form.
- Generate frequency distribution table and determine speed percentiles.

Q. 8. What is pneumatic road tube method?

Ans. The pneumatic road tube method is normally used for longer data collection time periods than those of either the stopwatch or radar meter method. Using this method, pneumatic tubes are placed in the travel lanes and are connected to recorders located at the side of the road.

Q. 9. Explain the pneumatic road tube method.

Ans. The pneumatic road tube method is normally used for longer data collection time periods than those of either the stopwatch or radar meter method. Using this method, pneumatic tubes are placed in the travel lanes and are connected to recorders located at the side of the road.

Q. 10. What is information gathering?

Ans. Information Gathering. Before a jurisdiction contacts an engineering consulting firm to perform a spot speed study, a variety of information may need to be collected. Any available information may aid the consulting firm in adequately completing the study. The following is a list of possible information that an engineering consulting firm may request:

- written description of the issue at hand
- map of posted speed limits in the area
- preliminary speed studies
- proposed future land use changes
- documented citizen input
- location map
- appropriate contact persons
- any other relevant information

Q. 11. If the spot speeds are 50, 40, 60, 54 and 45, then find the time mean speed and space mean speed.

Ans. Time mean speed v_t is the average of spot speed. Therefore,

$$v_t = \frac{\Sigma v_t}{n} = \frac{50 + 40 + 60 + 54 + 45}{5} = \frac{249}{5} = 49.8.$$

Space mean speed is the harmonic mean of spot speed. Therefore,

$$v_s = \frac{n}{\Sigma \frac{1}{v_i}} = \frac{5}{\frac{1}{50} + \frac{1}{40} + \frac{1}{60} + \frac{1}{54} + \frac{1}{45}} = \frac{5}{012} = 48.82.$$

Q. 12. The results of a speed study is given in the form of a frequency distribution table. Find the time mean speed and space mean speed.

Speed range	*Frequency*
2-5	1
6-9	4
10-13	0
14-17	7

Ans. The time mean speed and space mean speed can be found out from the frequency table given below. First, the average speed is computed, which is the mean of the speed range.

For example, for the first speed range, average speed, $v_i = \frac{2+5}{2} = 3.5$ seconds. The volume of flow q_i for that speed range is same as the frequency. The terms $v_i.q_i$ and q_i/v_i are also tabulated, and their summations given in the last row. Time mean speed can be computed as $v_t = \frac{\Sigma\, q_i v_i}{\Sigma\, q_i} = \frac{142}{12}$. Similarly, space mean speed can be computed as, $v_s = \frac{\Sigma\, q_i}{\Sigma \frac{q_i}{v_i}} = \frac{12}{3.28} = 3.65.$

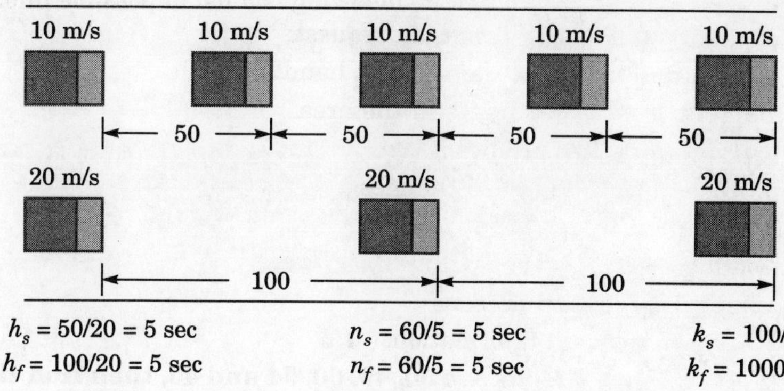

Fig. Illustration of relation between time mean speed and space mean speed

No.	*Speed range*	*Average speed* (v_i)	*Volume of flow* (q_i)	$v_i\ q_i$	q_i/v_i
1	2-5	3.5	1	3.5	2.29
2	6-9	7.5	4	30.0	0.54
3	10-13	11.5	0	0	0
4	14-17	15.5	7	108.5	0.45
	Total		12	142	3.28

REFERENCES

1. Highway Capacity Manual, Special Report 97, Highway Research Board, Washington, 1965.
2. Research on Road Traffic, Road Research Laboratory (U.K.), H.M.S.O., London, 1965.
3. J.G. Wardrop, Some Theoretical Aspects of Road Traffic Research, Proceedings of the Institution of Civil Engineers, 2 (Part 1), London, 1952.
4. Kadiyali, L.R., *et al*, Free Speeds of vehicles on Indian roads, Paper No. 343, Journal of the Indian Roads Congress, Vol. 42-3, New Delhi, 1981.
5. Kadiyali, L.R. *et al*, Studies on basis desired speed of vehicles, RUCS Technical Paper No. 31, Tenth Quarterly Report, Central Road Research Institute, New Delhi, 1980.
6. Road User Cost Study, Final Report, Vol III, Speed-Flow Studies (Analytical), Central Roads Research Institute, New Delhi, 1982.
7. Study for Updating Road User Cost Data in India, Final Report, New Delhi, 1992.
8. Kadiyali, L.R., *et al*, Speed-Flow characteristics on Indian Highways, Vol. 52-2, Indian Road Congress, New Delhi.
9. Reddy, T.S., *et al*, Updation of Road User Cost and Revised Software for Evaluation of Highway Projects, Paper No. 493, Journal, Indian Roads Congress, New Delhi, 2003.

10. Traffic Forecasting

10.1. Need for Traffic Forecasting

Investments in the transport sector constitute a significant part of the total investment. This is especially true in the case of developing nations, where transport is the catalyst for all round development and is one of the basic infrastructures. When the capital available is scarce and has competing demands, the investments in a transport project have to be planned carefully, keeping in view not only the present demand but also the requirements for a reasonable period in future. This underlines the need for estimating the future traffic accurately, whether the plan be for the construction of a new facility or the improvement of existing facilities. To a great extent, the accurate estimate of future traffic will influence the engineering design of the facility and the economic decision whether to take up the project or not.

10.2. Limitations of Traffic Forecasting

Traffic forecasting, in the present state of knowledge, can at best be approximate. Traffic is generated as a result of the interplay of a number of contributory factors. Forecasts of traffic have, therefore, to be dependent on the forecasts of factors such as population, gross domestic product, vehicle ownership, agricultural output, fuel consumption and so on. Future pattern of change in these factors can be estimated with only a limited degree of accuracy and hence traffic forecasting cannot be done more precisely than this. Inspite of this inherent drawback, traffic forecasting is a very important job a transport planner is after called upon to do.

10.3. Types of Traffic

It is necessary to distinguish between the following types of traffic met with when dealing with traffic forecasting :

10.3.1. Current Traffic—existing and attracted (or diverted).

This represents the volume of traffic that would use an improved highway if it were open to traffic. It consists of the existing traffic plus or minus the existing traffic attracted to it/or diverted traffic that is lost to it from/to other facilities when the improvements are completed.

10.3.2. Traffic Increase

(*i*) Normal growth of traffic, representing the increase in traffic on the existing facility if no improvement is made. This is due to general increase in the number and usage motor vehicles.

(*ii*) Diverted traffic, representing the traffic diverted on to, or away from, the route or mode being studied.

(*iii*) Induced traffic, representing the new traffic because of new travellers making use of the improved or new facility.

Items (*ii*) and (*iii*) above are also sometimes known as generated traffic.

(*iv*) Development traffic, representing the increase in traffic due to improvements on adjacent land over and above the development which would have taken place had not the new or improved highway been constructed.

10.4. Forecasts Based on Past Trends and Extrapolation

The simplest method of forecasting is to analyse the past data for a number of years and to extrapolate the past trends assuming that the conditions will continue to change in the future at the same rates as in the past. Obviously, such a simplification suffers from many disadvantages, although it is relatively easy and cheap. It is good enough in a stable environment, which is beyond the influence of any major change in production, population and so on (Ref. 1). The analyst has to carefully study the past data and look out for any indicators that are likely to influence the future pattern.

The equation derived from the time series data on Bus-passenger-km travelled in a recent study is (Ref. 9) :

$$BUPK_n = 29.31\ (1.100)^n$$

where, $BUPK_n$ = Bus-passenger-km in billions at the end of the nth period.

n = Year – 1951.

In another method, data on certain known contributors to traffic are analysed and then the traffic is forecast based on any relationship between traffic and these contributors. Population, gross domestic product vehicle ownership, agricultural and industrial output, fuel consumption etc. are some of these.

Population data are available in all the countries for the past number of years. Elaborate methods to predict future population are available with the demographers, and the transport planner should consult the experts in field for help. The past data on *Gross Domestic Product* (GDP) or *Gross National Product* (GNP) are easily available. Traffic is highly related to GDP or GNP.

Forecasts of this parameter may not be very reliable, but a reasonable accuracy should be acceptable for purposes of traffic forecasting.

In developing these models, it is often convenient to transform the variables to the log form as under (Ref. 10) :

$$\log_e T = A_0 + A_1 \log_e GNP$$

where, T = Transport demand, GNP = Gross national product

A_0 and A_1 are co-efficients.

In the above equation, the co-efficient A_1 directly gives the elasticity of Transport Demand with respect to *GNP*. In other words, one per cent increase in the *GNP* will result in A_1 per cent increase in Transport Demand.

A typical equation relating the freight movement by road in terms of *GNP* in India is as under (Ref. 9).

$$\log_e T = -\ 20.4677 + 2.3563 \log_e GNP$$

where, T = Total freight moved by road in billion tonne-km.

GNP = Gross National Income in crores of rupees at 1970-71 prices.

Econometric models have been developed to determine car ownership rates. For example, Khan and Willumsen (Ref. 11) give the following model :

$$\log C_{1000} = -362 + 70.5 \log_e GNPH - 0.373 \log_e PURTAX$$
$$- 2.58 \log_e OWNTAX - 0.682 \log_e IMPDUTY$$
$$- 29.4 \log_e FUELPR - 2.04 \log_e POPDEN$$
$$R^2 = 0.86$$

where, C_{1000} = Car ownership per 1000 persons

$GNPH$ = Gross National Product per capita

$PURTAX$ = Purchase tax associated with cars

$OWNTAX$ = Ownership tax (license fee)

$IMPDUTY$ = Import duty for cars

$FUELPR$ = Fuel price per litre of fuel

$POPDEN$ = Population density

Vehicle ownership is an important determinant of traffic. If the vehicle population is known, total amount of travel per year expressed in vehicle-kilometres, can be found by multiplying the vehicle population with the average run of a vehicle in a year. The vehicle ownership is commonly found to be following a logistic curve (Ref. 2), Fig. 10.1, assuming no catastrophic events and no major substitution in the transport modes.

The equation for the curve contains only one unknown, the saturation level *a*, and is uniquely determined if *a* is known. The curve tends towards this saturation level with passage of time. The percentage rate of growth decreases steadily as the upper limit is approached. In U.K., a saturation level 0.55 motor vehicles (all vehicles) per person has been determined (Ref. 2).

Equation : $$x > \frac{aX}{X + (a - x)e^{-art/(a-x)}}$$

Data on fuel consumption in the past can be a good indicator of the amount of travel. Difficulties arise because of the use of petrol and diesel for purposes other than the motor vehicle, and the varying fuel consumption pattern by different vehicles under diverse conditions.

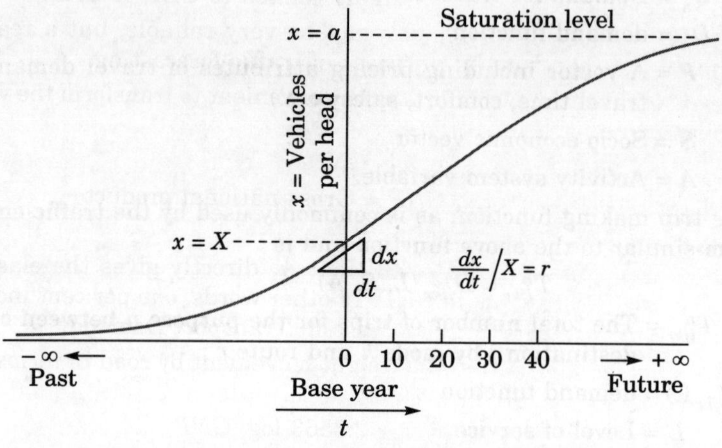

Fig. 10.1. Logistic growth curve for vehicle ownership.

Data on agricultural, industrial and mining output and consumption, including exports and imports, can be useful in discerning the past trends and arriving at a future forecast.

10.5. Forecasts and Mathematical Models

10.5.1. Forecasts of travel demand (trip making) based on mathematical modelling are popular in Transportation Planning. The prediction of future traffic based on such models is an accurate method, provided the models are calibrated with care after the necessary data are collected. The enormous work involved in modelling can be lessened to a great extent by the use of computers.

10.5.2. Travel demand functions

Demand for travel is an economic activity. As such it can be understood by reference to well-known economic principles dealing with consumer choice behaviour in relation to common goods.

Purchase of $Q_1, Q_2, \ldots\ldots, Q_n$ quantities of n items of goods priced at $P_1, P_2, \ldots\ldots, P_n$ involves a certain level of utility, which can be represented by :

$$U(Q_1, Q_2, \ldots\ldots, Q_n)$$

Every consumer would like to maximise the utility derived by purchasing a combination of items of goods, each of a certain quantity, subject to his budgetary constraint. He would thus like to maximise $U(Q_1, Q_2, \ldots\ldots, Q_n)$ subject to

$$\sum_{i=1}^{n} P_i Q_i \leq B$$

where, P_i is the price of the ith item, Q_i the quantity and B is the total budget.

The quantity of ith item consumed, Q_i, can be considered as a demand function, relating the quantity to the prices of goods and the income level.

$$Q_i = D_i (P_1, P_2, \ldots\ldots, P_n)$$

The above demand function can be simplified to deal with travel demand as follows :

$$Q_n = D^n (P, S, A)$$

where, Q_n = Demand for travel for purpose n

D_n = demand function

P = A vector including pricing attributes of travel demand (price, travel time, comfort, safety etc.)

S = Socio-economic vector

A = Activity system variable.

The trip making function, as is commonly used by the traffic engineer is of the form similar to the above function and is :

$$T^n_{ijkr} = D^n_{ijkr} (L, S, A)$$

where, T^n_{ijkr} = The total number of trips for the purpose n between origin i to destination j by mode k and route r

D^n_{ijkr} (.) = demand function

L = Level of service

S = Socio-economic factor

A = Activity system, variable.

10.5.3. Sequential and simultaneous models

The trip making behaviour consists of four discriptors :

1. Trip frequency
2. Destination
3. Mode
4. Route and other choice.

In the sequential models, the trip-making behaviour is built-up in a sequential order, and each travel decision made independently of others. In the simultaneous models, all attributes of travel choice are considered together.

10.5.4. Aggregate and disaggregate models

Aggregate models deal with the estimation of travel of a group of travellers. Disaggregate models deal with the smallest decision making unit, the individual traveller.

10.6. Period for Forecasting

The question as to what period should be selected for traffic forecasting is difficult to answer. Since traffic forecasting is needed for transport plans, the design period selected for the transport plans should be sufficient for traffic forecasting. Long-range transport plans are not particularly useful (Ref. 3), since such plans do not exist in other sectors. In general, transport plans are for a period of about ten years-five years in detail and an additional five years in less detail. In U.K., it is customary to forecast traffic for a design period of 15 years (Ref. 4) when dealing with rural roads. In India, National Highways are designed for 15 years after completion of the work.

I. MULTIPLE CHOICE QUESTIONS

1. Operations generated forecasts often not to do with
   (*a*) inventory requirements (*b*) resource needs
   (*c*) time requirements (*d*) sales
2. Which of the following is not true for forecasting?
   (*a*) forecasts are rarely perfect
   (*b*) the underlying casual system will remain same in the future
   (*c*) forecast for group of items is accurate than individual item
   (*d*) short range forecasts are less accurate than long range forecasts
3. Which of the following is not a forecasting technique?
   (*a*) judgemental (*b*) time series (*c*) time horizon (*d*) associative
4. In which of the following forecasting technique, subjective inputs obtained from various sources are analyzed?
   (*a*) judgmental forecast (*b*) time series forecast
   (*c*) associative model (*d*) all of the above
5. In which of the following forecasting technique, data obtained from past experience is analyzed?
   (*a*) judgmental forecast (*b*) time series forecast
   (*c*) associative model (*d*) all of the above
6. Delphi method is used for
   (*a*) judgmental forecast (*b*) time series forecast
   (*c*) associative model (*d*) all of the above

7. **Short term regular variations related to the calendar or time of day is known as**
   (*a*) trend (*b*) seasonality
   (*c*) cycles (*d*) random variations
8. **The demand for period *t*-2 and *t*-1 is 10 and 12 cases respectively. As per naïve method, the demand for next period '*t*' is**
   (*a*) 10 (*b*) 11 (*c*) 12 (*d*) 14

Answers

1. (*d*)	**2.** (*d*)	**3.** (*c*)	**4.** (*a*)	**5.** (*b*)
6. (*a*)	**7.** (*b*)	**8.** (*d*)		

II. SOLVED ANSWER AND QUESTIONS

Q. 1. What is traffic survey? What is its adjectives?

Ans. Traffic survey is one of the important inputs required for a highways project. This chapter deals with various traffic studies carried out and the analysis of the data obtained from these studies. The main objectives of the traffic survey are:

To determine the traffic movement characteristics:

- Capacity assessment and determination of future requirements based on traffic projections;
- To determine turning movements for geometric design of intersections;
- To determine travel pattern, type and weight of commodities;
- To determine axle loads and vehicle damage factor for pavement design;
- To identify need for bypass and service roads;
- To identify accident-prone areas and suggest methods for improvement;
- To study parking characteristics and identify parking locations;
- To determine speed-delay characteristics.

Q. 2. What is traffic flow characteristics?

Ans. Traffic flow characteristics for the capacity analysis is based on the traffic volume count survey data (km 30.000, km 55.300 and km 87.200). The traffic at km 3.000 is influenced by city traffic and does not represent the NH traffic on entire stretch. The traffic at km 55.300 is on lower side compared to other locations and cannot be adopted for design of entire stretch. On the basis of traffic characteristics, project reach has been split up into two sections, one from km 0/000 to 46/300 and other from km 46/300 to 92/000. The traffic on the basis of volume count carried out at km 30.000 has been taken as the representative traffic for section-1 and volume count at km 87.200 for section-Z.

Q. 3. What is trun movement count survey?

Ans. Turning movement survey has been conducted at important junctions to understand the nature and quantum of turning movements. The study is helpful in understanding the percentage of traffic getting diverted as well as to help the design of intersections. The traffic survey was conducted for 12 hours (8.00 AM to 8.00 PM).

Q. 4. What are types of roadway traffic forecasts?

Ans. Different types of roadway traffic forecasts are utilized in different situations. Forecasts are used for anything from determining future traffic volumes and turning movements at intersections (in order for traffic lights to have proper timing, for example) to providing traffic volumes and truck percentages so a roadway can be designed correctly for its capacity and material composition. Different types of forecasts are used to determine whether a future roadway should be planned for, designed, and built at a given location in order to accommodate future traffic volumes. Per WisDOT guidelines, several types of roadway forecasts exist, including but not limited to:

- Roadway mainline (daily and peak forecasts)
- Roadway turning movement (daily and peak forecasts)
- Roadway mainline and turning movement forecasts for special land use analysis
- Traffic impact analysis (TIA)
- Trip origin-destination or expansion analysis
- Trip generation studies
- Alternatives analysis.

Q. 5. What are the reasons for forecasts and travel analysis?

Ans. Transportation forecasting is the process of estimating the number of people or vehicles that will use a specific transportation facility in the future. It can be utilized in a variety of different situations and with different modes of transport, from estimating traffic volumes on a specific segment of road or highway to estimating ships in a port or passenger volumes on a city's buses.

Forecasts explain what the needs of the future might be and provide benchmarks for proper design and efficient transportation system operation. Transportation forecasts are fundamentally important inputs in developing infrastructure – from developing overall transportation policy, to planning studies, to the engineering design of specific projects. Example applications of forecast information include:

- Development of infrastructure capacity and design calculations (*e.g.*, the operations of an existing
- Or proposed roadway or bridge, or the thickness or type of roadway pavements)
- Estimation of the financial and/or social viability of projects (*e.g.*, developing benefit-cost analyses and/or social impact assessments)
- Calculation of environmental impacts, such as air and noise pollution

III. REVIEW QUESTIONS

1. Why is there a need for traffic forecasting?
2. What are the limitations of traffic forecasting?
3. Discuss the types of traffic.
4. How do you present forecasts and mathematical models? Give examples.

REFERENCES

1. ECAFE Manual on Traffic Surveys, United Nations, New York, 1971.
2. Research of Road, Road Traffic, Road Research Laboratory, (U.K.). H.M.S.O., 1965.
3. Adler, H.A., Sector and Project Planning in Transportation, World Bank Staff Occasional Papers No. 4, Distributed by the John Hopkins Press, 1967.
4. Traffic Prediction for Rural Roads, Ministry of Transport (U.K.), H.M.S.O., London, 1968.
5. Goldberger, A.S. Econometric Theory, Wiley, 1964.

6. Stowers, J.R. and Kanwit, E.L., The Use of Behavioural Surveys in Forecasting Transportation Requirements, Highway Research Record, No. 106, Washington, 1966.
7. Urban Travel Demand Forecasting, H.R.B. Special Report, 143, Washington, 1972.
8. Raghavachari, S., Lectures on Urban Transportation Planning Techniques, I.S.T.E. Summer School, Regional College of Engineering, Warangal, 1978.
9. Kadiyali, L.R., Road Transport Demand Forecast for 2000 AD, Vol. 48-3, Journal of the Indian Roads Congress, New Delhi, 1987.
10. Guidelines for Traffic Prediction on Rural Highways, IRC : 108—1996, The Indian Roads Congress, New Delhi, 1996.
11. Khan, A. and Willumsen, L.G., Modelling car ownership and use in developing countries, Traffic Engineering and Control, 27(11), 1986.

SECTION 4
GEOMETRIC DESIGN

11. Geometric Design

11.1. Introduction

11.1.1. Geometric design is an aspect of highway design dealing with the visible dimensions of a roadway. It is dictated, within economic limitations, by the requirements of traffic and includes the design elements of horizontal and vertical alignment, sight distance, cross-section components, lateral and vertical clearances, intersection treatment, control of access, etc.

11.1.2. The safe, efficient and economic operation of a highway is governed to a large extent by the care with which the geometric design has been worked out. Safety or the lack of it is an immediate corollary of the various features of the highway. Efficient and comfortable operation of traffic is possible only if the design elements have been meticulously considered. A well designed highway has to be consistent with economy. Too liberal standards may not fit in with the available resources, whereas if the standards are too low the cost operation may mount up.

11.1.3. The design engineer has to consider the following points when selecting the design standards for a highway :

(*i*) Adequate geometric design in planning a highway facility ensures that the facility will not become obsolescent in the foreseeable future. Hence the volume and composition of traffic in the design year should be the basis of design.

(*ii*) Faulty geometrics are costly to rectify at a later date and so due consideration should be given to geometric design at the initial stage itself.

(*iii*) The design should be consistent and the standards proposed for the different elements should be compatible with one another. Abrupt changes in the design should be avoided.

(*iv*) The design should embrace all aspects of geometrics of the road, including signs, markings, proper lighting, intersections, access control, etc.

(*v*) The highway should be considered as an element of the total environment and its location and design should enhance rather than degrade the environment. The highway should be aesthetically satisfying. The design elements should strive to control pollution.

11.1.4. Standards in this country and abroad

Geometric design standards in this country are prescribed by the Indian Roads Congress (Ref. 1, 2). The AASHTO Standards (Ref. 3, 4) represents the

American practice, whereas the Department of the Environment (U.K.) standards (Ref. 5, 6), give the current British practice. These standards will be profusely quoted in the various sections to follow.

11.2. Highway Classification

11.2.1. The highway system is generally classified into various distinct groups. For a road designer, this will enable him to relate the geometric and structural design standards to roads under the different classes. For a planner, this provides a basis for long-term planning where different priorities could be assigned to different classes. The system of classification also serves various administrative needs. The highways are basically of two types, *viz.* urban and rural (non-urban). The design standards and criteria for these two types are different.

11.2.2. The classification system for urban highways in this country (Ref. 1, 7), along with a brief description of each class is given below :

(*i*) **Express ways.** These are defined as divided arterial highways for motor traffic, with full or partial control of access and provided generally with grade separation at intersections. Their main function is to provide for movement of heavy volumes of motor traffic at high speeds under free-flow conditions. They connect major points of traffic generation and are intended to serve trips of medium and long lengths between residential areas, industrial or commercial concentrations and the central business district. They are generally provided with grade separations at intersections. Parking, loading and unloading of goods and pedestrian traffic are not permitted on these highways.

(*ii*) **Arterial streets.** The term *arterial street* is a general term denoting a street primarily meant for through traffic usually on a continuous route. These, along with expressways (where they exist), serve as the principal network for through traffic flow. Significant intra-urban travel such as between central business district and outlying residential area or between major suburban centres is served by this facility. These streets should be coordinated with existing and proposed expressway systems to provide for distribution and collection of through traffic to and from sub-arterial and collector street system. Continuity is the prime essential of arterial streets to ensure efficient movement of through traffic. A properly developed and designated arterial street system would help to identify residential neighbourhoods, industrial sites and commercial areas. These streets are generally spaced at less than 1.5 km in highly developed central business area and at 8 km or more in sparsely developed urban fringes. The arterial streets are generally divided highways with full or partially controlled access. Parking, loading and unloading activities are usually restricted and regulated. Pedestrians are allowed to cross only at intersections.

(*iii*) **Sub-arterial streets.** These are streets of somewhat lower level of travel mobility than the arterial streets. The emphasis on access to adjoining areas is more in case of these streets than in the case of arterial streets. Parking, loading and unloading are usually restricted and regulated. Pedestrians are allowed to cross only at intersections. Their spacing varies from about 0.5 km in the central business district to 3 to 5 km in the sub-urban fringe.

(*iv*) **Collector streets.** A collector street is one intended for collecting and distributing the traffic to and from local streets and also for providing access to arterial streets. These may be located in residential neighbourhoods, business areas and industrial areas. Normally, full access is allowed on these streets from abutting properties. There are few parking restrictions except during peak hours.

(*v*) **Local streets.** A local street is one primarily intended for access to residence, business or other abutting property. Such a street normally does not carry large volume of traffic. The traffic carried either originates or terminates along its length. A local street may be residential, commercial or industrial, depending upon the prominent use of the adjoining land. A local street allows unrestricted parking and pedestrian movements.

11.2.3. As regards rural roads, the classification in India is based on the Nagpur Plan (Ref. 8). The classes and their description are given below :

(1) **National Highways.** National Highways are the main highways running through the length and breadth of the Indian Union, connecting ports, foreign highways and capitals of States and including roads of strategic and military value. They constitute the frame on which the entire road communication system of the country is based. They must give uninterrupted road communication throughout India and be of fairly high grade construction. All National Highways vest in the Union of India, as per the National Highway Act, 1956, and it is the responsibility of the Central Government to develop and maintain in proper repair all National Highways. The National Highways have to be declared as such by the Central Government by notification in the official Gazette.

(2) **State Highways.** State highways are the other main trunk or arterial roads of a State, connecting up with the National headquarters and important cities within the State. The State Highways are the main arteries of traffic within a State. They are to be of the same standard as national highways.

(3) **District Roads.** District roads are roads traversing each district, serving area of production and markets and connecting these with each other or with National and State Highways or Railways or important navigational routes. They should be capable of taking road traffic into the heart of the rural areas throughout the year with only minor interruptions. District Roads are divided into two classes on the basis of traffic :

(*i*) Major District Roads (M.D.R.) for the higher order of traffic.

(*ii*) Other District Roads (O.D.R.) for the lower order of traffic.

(4) **Village Roads.** Village roads are roads connecting villages of groups of villages with each other and to the nearest district roads, National or State Highway or railway or navigational routes. They are in essence roads from the villages to a market (*mandi*) or to a main route. In India, the O.D.Rs. and Village Roads are considered together as Rural Roads.

In the Road Development Plan (1981-2001), a new class of roads known as Expressways has been included in the suggested classification (Ref. 40). These are superior facilities on arterial routes restricted for use by any motor vehicles, to which frontal access is prohibited and on which grade separation is provided at all intersections with or without provision for an interchange. Expressways are known by the term "Freeways" in U.S.A. and "Motorways" in U.K. In U.S.A., the Interstate routes are access controlled and high-speed facilities. In Germany, they are known by the term "Autobahns".

The main features of Expressways are :

(*i*) Exclusion of pedestrians, animals, animal-drawn vehicles, cycles, agricultural tractors and motorised two-wheelers having an engine capacity of less than 50 c.c.

(*ii*) Complete separation of opposing traffic streams by a median.

(*iii*) Elimination of all crossing at the road level.

(*iv*) Provision of specially designed inter-changes at reasonably long intervals and the elimination of other accesses.

(*v*) Moderate gradients and gentle curves with adequate sight distances permitting high speeds.

(*vi*) Hard shoulders on the outer edge of each carriageway to facilitate parking of vehicles of carriageway.

(*vii*) Telephones at convenient intervals for use during emergency.

(*viii*) Servicing and refreshment facilities at convenient intervals.

(*ix*) Police and Traffic aid posts at convenient intervals.

(*x*) Highway Maintenance Depots at convenient intervals.

11.2.4. Classification of Terrain

The classification of terrain is normally done by means of the cross-slope of the country, *viz.*, the slope approximately perpendicular to the centre-line of the highway location. The following classification is generally followed (Ref. 2, 6, 10).

Table 11.1. Terrain Classification

S. No.	*Terrain Classification*	*Percent cross slope of country*
1.	Level	0 to 10
2.	Rolling	10 to 25
3.	Mountainous	25 to 60
4.	Steep	Greater than 60

An extensive survey of 42,000 km of roads in India as part of the Road User Cost Study (Ref. 39) has enabled the development of the following system of terrain classification :

Table 11.2. Terrain Classification Suggested in Road User Cost Study

S. No.	*Terrain Classification*	*Rise and fall (m/km)*	*Curvature (deg/km)*
1.	Plain (*a*) Low curvature	0–15	0–50
	(*b*) High curvature	0–15	Above 51
2.	Rolling (*a*) Low curvature	16–30	0–100
	(*b*) High curvature	16–30	Above 101
3.	Hilly (*a*) Low curvature	Over 31	0–200
	(*b*) High curvature	Over 31	Above 201

The quantification of curvature and rise and fall has been done on the basis of Fig. 11.1.

Plan

Average curvature of Section AB, CV = $\dfrac{\phi_1 + \phi_2 + \phi_3 + \ldots\ldots + \phi_n}{\text{DISTANCE AB (Km)}}$ Expressed in Degrees/km

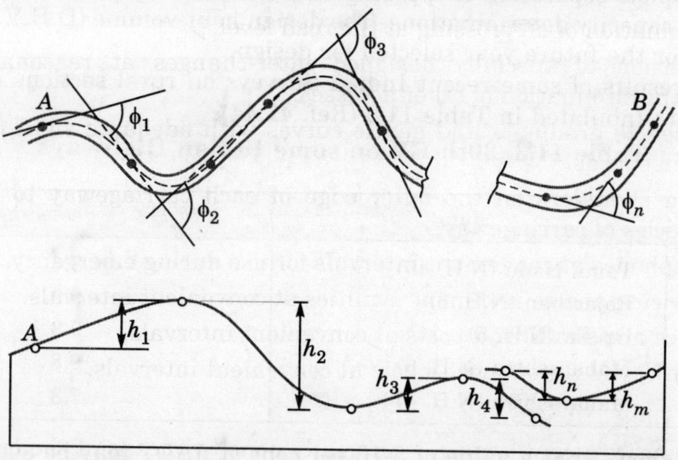

Fig. 11.1. Quantification of rise, fall and curvature.

Longitudinal Profile

Average Rise of Section AB, RS $= \dfrac{h_1 + h_3 + ... + h_m \text{ (metres)}}{\text{DISTANCE AB (Km)}}$, **Expressed in m/km**

Average Fall of Section AB, FL $= \dfrac{h_2 + h_4 + ... + h_n \text{ (metres)}}{\text{DISTANCE AB (Km)}}$, **Expressed in m/km**

Average Rise and Fall of section AB, RF

$= \dfrac{h_1 + h_2 + h_3 + h_4 + ... + h_n + h_m \text{ (metres)}}{\text{DISTANCE AB (Km)}}$, **Expressed in m/km**

11.3. Traffic

11.3.1. Importance of Traffic Data in Geometric Design

Of crucial importance in highway design is the traffic data-both current and future estimates. Traffic volume indicates the service for which the highway is being planned and directly affects the geometric features such as width, alignment, grades etc. Without traffic data, it is futile to design any highway.

11.3.2. Design Hour Volume

The general unit for measuring traffic on a highway is the annual average daily traffic volume, abbreviated as AADT. It is equal to the total annual volume of the traffic divided by the number of days in the year (Ref. 11). AADT can be easily estimated where continuous traffic counts are maintained. Where only periodic counts are maintained, it can be established by applying correcting factors. A knowledge of traffic in terms of AADT is not of much use in geometric design, since it does not represent the variations in traffic during various months of the year, days of the week and hours of the day. A commonly used unit for geometric design is the 30th highest hourly volume, abbreviated as 30 HV. It is defined as the hourly volume that is exceeded by 29 hourly volumes during a designated year. For average American conditions, the 30 HV is about 15 per cent of the AADT in the case of Rural highways (Ref. 3) and is about 8-12 percent of the AADT in the case of Urban Streets (Ref. 4). It may be noted that a design hour volume of 8-10 percent of the AADT has been suggested for Indian conditions

(Ref. 12) for capacity determinations. The design hour volume (D.H.V.) should be the 30 HV for the future year selected for design.

The results of some recent Indian Surveys on rural sections of National Highways are tabulated in Table 11.3 (Ref. 41.44).

Table 11.3. 30th HV on some Indian Highways

S. No.	*State*	*30th HV as percentage AADT*
1.	Tamil Nadu (N.H. 45)	8.4
2.	Rajasthan (N.H. 8)	9.9
3.	Orissa (N.H. 5)	9.3
4.	Maharashtra (N.H. 6)	9.8
5.	Maharashtra (N.H. 4)	7.3

This shows that a value of 8-10 per cent of AADT may be assumed to be the 30th HV under Indian conditions. From the recent traffic data available from the National Highway Development Programme (NHDP) in India, it is seen that the peak hour traffic on rural sections is in the range of 6-8 percent.

11.3.3. Directional Distribution of Traffic

For 2-lane highways, the design hour volume is the total traffic in both directions of travel. For highways with more than 2-lanes, it is desirable to know the directional distribution of traffic. Though this distribution has to be found from traffic surveys and is found to be in the range of 55-70 percent (Ref. 4) a rough approximation can be to assume 67 percent of total traffic to travel in one direction under the design conditions. The design has to take into account both the morning and evening peak situations.

11.3.4. Traffic Compositions

Traffic composition has a vital effect on capacity and other design considerations. In this country, the traffic is heterogeneous in character, consisting of fast driven cars, trucks, buses and slow animal drawn vehicles. It is customary in this country to express the traffic volume in terms of the passenger car units. The values indicated in Table 21.2 have been suggested for Indian conditions.

11.3.5. Future Traffic Estimates

Obviously, the design of the geometric elements has to be prepared for the traffic likely to use the road in the design year. The design period used generally varies from 15 to 25 years. A period of 20 years is widely used as basis for design. The future traffic estimates should be calculated to include the following components :

(1) *Current Traffic*—Existing and attracted (or diverted)

(2) *Future Traffic* :

(*i*) normal traffic growth

(*ii*) generated traffic — Diverted traffic / Induced traffic

(*iii*) development traffic.

These have been discussed in detail under Chapter 10, where the methods available for traffic forecast have also been outlined.

11.4. Design Vehicle Dimensions

11.4.1. What is a 'Design Vehicle' ?

A *Design Vehicle* is a selected motor vehicle, the weight, dimensions and operating characteristics of which are used to establish highway design controls to accommodate vehicles of a designated type. The dimensions and operating characteristics of a vehicle profoundly influence the geometric design aspects such as radii, width of pavements, clearances, parking geometries etc. The weight of axles and the weight of the vehicles affect the structural design of the pavement and structures, as also the operating characteristics of vehicles on grades. Because of its crucial importance the standardisation of the dimensions and weights of design vehicles is the first step in formulating geometric design standards. This has been done in most countries. In India, the current standard (Ref. 13) of the Indian Road Congress dates to 1954 and has been now revised (Ref. 14). The AASHTO standards (Ref. 3, 4) represent American practice. In U.K., the Ministry of Transport Regulations (Ref. 15) govern the features of the design vehicle.

11.4.2. Types of design vehicles

The AASHTO Standard (Ref. 4) classifies the follow-in types of design vehicles :

(1) Passenger car — P
(2) Single unit truck — SU
(3) Single unit bus — BUS
(4) Semi-trailer combination, intermediate — WB-12
(5) Semi-trailer combination, large — WB-15
(6) Semi-trailer-full trailer combination — WB-18

The IRC Standard (Ref. 14) contains only commercial vehicles of the following types :

(1) Single unit truck (2) Semi-trailer
(3) Truck-trailer combination.

11.4.3. Summary of Standards for Dimensions of Vehicles

Table 11.4 gives a summary of standards of dimensions for design vehicles for various authorities :

Table 11.4. Standards of Dimensions for Design Vehicles by Various Authorities (in metres)

Authority Country	*Maximum Width*	*Maximum Height*	*Maximum Overall Length*				
			Passenger Car	*Single Unit Truck*	*Semi-Trailer*	*Truck Trailer*	*Single Unit Bus*
AASHTO	2.6	4.1	5.8	9.1	15.2-16.7	19.9	18.3
U.K.	2.5	4.57 (Bus)	5.5	11.0	13.0	18.0	
IRC (1984)	2.5	3.8–4.2 (Truck) 4.75 (Double Decker Bus)	–	11.0	16.0	18.0	12.0

The turning radii for the various AASHTO design vehicles are given in Table 11.5.

Table 11.5. Turning Radii of AASHTO Vehicles

	Design Vehicle Type	*Passenger Car (P)*	*Single Unit Truck (SU)*	*Semi-trailer (large) (W-15)*	*Semi-trailer full-trailer (WB-18)*
1.	Minimum Turning Radius (*m*)	7.3	12.8	13.7	13.7
2.	Minimum Turning Radius (Inside) (*m*)	4.2	8.5	5.8	6.8

The turning radii for some of the vehicles in common use are given in Fig. 11.2 (Ref. 4).

11.4.4. Axle loads and weights of vehicles

The maximum axle loads of vehicles as per standards in some of the countries are given in Table 11.6.

Table 11.6. Maximum Axle Loads in Different Countries (tonnes)

	Single Axle	*Tandem Axle*
AASHTO	9.0	14.5
U.K.	10.0	–
IRC (1983)	10.2	18.0

The maximum weight of vehicles depends upon the number configuration and spacing of axles.

11.4.5. Selection of the design vehicle

The selection of the design vehicle for the design of a highway is governed by the type and volume of traffic that is expected to use the highway. For instance, the design of superior facility such as a motorway or an expressway should be based on the largest design vehicle. The design of streets and junctions primarily in residential areas can be done by using the passenger car design vehicle.

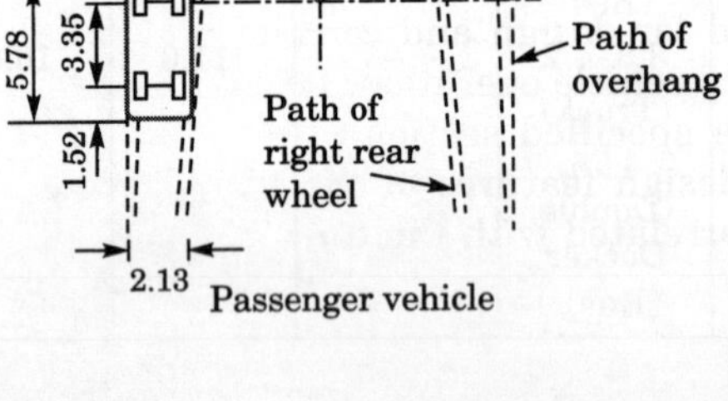

Passenger vehicle

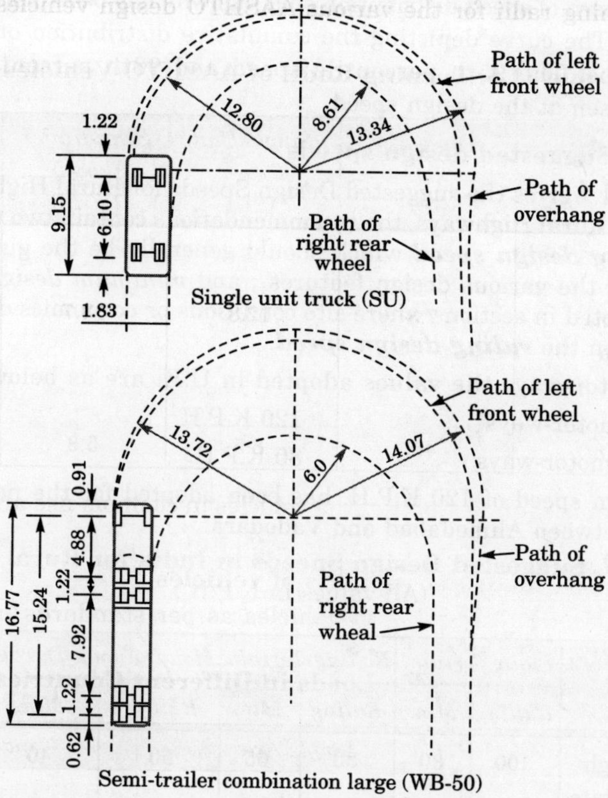

Fig. 11.2. Turning Radii of some U.K. Design Vehicles. (*Source* : Ref. 6)

11.5. Design Speed

11.5.1. Speed as a design factor

The value of a highway is largely indicated by the speed, safety and convenience afforded by the facility for travel. Speed is important for economic operation and has a great bearing on the safety of the highway. It plays a vital role in determining the geometrical design.

11.5.2. Design speed

There is considerable variation in the speed adopted by different drivers and by different types of vehicles. This raises the question of what value should be adopted for design. This value selected should accommodate nearly all demands with reasonable adequacy and yet the design should not fail completely under the severe or extreme load. The speed adopted should satisfy nearly all drivers, excepting, of course, those few who drive at extremely high speed. *Design speed* is a speed determined for design and correlation of the physical features of a highway that influence vehicle operation. It is the maximum safe speed that can be maintained over a specified section of a highway when conditions are so favourable that the design features of the highway govern. The design speed obviously has to be correlated with the terrain conditions and the classification of the highway.

The pattern of speed distribution on a highway has been discussed in detail in Chapter 9. The curve depicting the cumulative distribution of speeds has a typical S-shape. The 95th percentile speed and 98th percentile speed are frequently chosen at the design speed.

11.5.3. Suggested design speeds

Table 11.7 gives the suggested Design Speeds for Rural Highways in India (Ref. 2, 9). For Rural Highways, the recommendations contain two types of design speeds : ***ruling design speed*** which should generally be the guiding criterion for correlating the various design features ; and *minimum design speed* which should be adopted in sections where site conditions or economics do not permit a design based on the ***ruling design speed***.

For motor-ways the values adopted in U.K. are as below (Ref. 5, 6) :

Rural motor-ways : 120 K.P.H.

Urban motor-ways : 80 K.P.H.

A design speed of 120 K.P.H. has been adopted for the new expressway constructed between Ahmedabad and Vadodara.

Table 11.7. Suggested Design Speeds in India for Rural Highways
(All values in K.P.H.)

Classification of Roads	*Plain Terrain*		*Rolling Terrain*		*Mountainous Terrain*		*Steep Terrian*	
	Rulling	*Min.*	*Ruling*	*Min.*	*Ruling*	*Min.*	*Ruling*	*Min.*
1. National highways and state highways	100	80	80	65	50	40	40	30
2. Major district roads	80	65	65	50	40	30	30	20
3. Other district roads	65	50	50	40	30	25	25	20
4. Village roads	50	40	40	35	25	20	25	20

For urban streets, the U.K. design speeds as below (Ref. 6) :

Primary distributor (Urban motorway) : 80 KPH

Primary distributor (All-purpose) : 65 KPH

District-distributor, local distributor, important access road : 50 KPH

The design speeds suggested by IRC for urban streets are as under (Ref. 1) :

Arterial : 80 K.P.H.

Sub-arterial : 60 K.P.H.

Collector : 50 K.P.H.

Local : 30 K.P.H.

11.6. Capacity

The subject of highway capacity has been discussed in detail in Chapter 21. The number of lanes to be provided in a highway facility is directly governed by the traffic volume and the capacity. The design service volumes that can be catered to in different highways at different levels of service have been indicated in that Chapter.

11.7. Horizontal Alignment

11.7.1. Basic formula for movement of vehicles on curves

When a vehicle is moving on a curved path, it is subjected to an outward force, commonly known as the centrifugal force. In order to resist this force, it is the usual practice to superelevate the roadway cross-section. Fig. 11.3 shows the forces acting on the vehicle at a superelevated section.

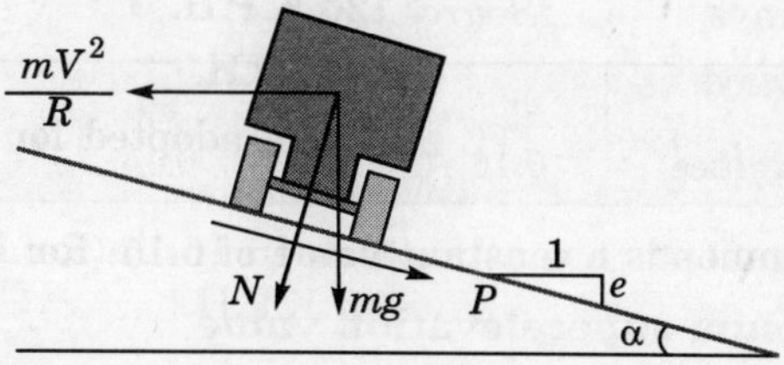

Fig. 11.3. Forces acting on a vehicle at a super-elevated section.

Let, m = mass of the vehicle,

v = speed of the vehicle, m/second

R = radius of the curve, m

g = acceleration due to gravity = 9.81 m/second^2

P = side friction force resisting the centrifugal force = $N\mu$

N = normal force

μ = co-efficient of lateral friction

α = angle of super elevation

e = super elevation, $\tan \alpha$

The centrifugal force = $\frac{mv^2}{R}$

For equilibrium,

$$\frac{mv^2}{R} \cos \alpha = m\,g \sin \alpha + P = m\,g \sin \alpha + N\mu$$

$$= mg \sin \alpha + \mu \left(m\,g \cos \alpha + \frac{m\,gv^2}{R} \sin \alpha \right)$$

Therefore, $\frac{v^2}{gR} = \tan \alpha + \mu + \frac{\mu v^2}{gR} \tan \alpha$

The term $\mu v^2/gR \tan \alpha$ is small and can be neglected. Thus,

$$\frac{v^2}{gR} = e + \mu$$

Expressing speed as V in K.P.H., and putting $g = 9.81$,

$$\frac{V^2}{127R} = e + \mu \qquad \text{...(11.1)}$$

Equation (11.1) above is the basic equation relating the speed of vehicles, the radius of the curve, the superelevation and the co-efficient of lateral friction. The equation forms the basis of design horizontal curves.

11.7.2. Value of co-efficient of lateral friction

The value of the co-efficient of lateral friction depends upon a number of factors, chief among them being the vehicle speed, type and condition of roadway surfaces, and type and condition of the tyre. AASHTO (Ref. 4) recommends the value given in Table 11.8.

Table 11.8. Co-efficient of Lateral Friction as Recommended by AASHTO

(*Source* : Ref. 4)

Design speed (K.P.H.)	50	65	80	100	110	120
Max. lateral friction values	0.16	0.15	0.14	0.12	0.10	0.08

The IRC recommends a constant value of 0.15 (Ref. 9, 16, 17)

11.7.3. Maximum superelevation value

If equation (11.1) is to be used for design, it is desirable to know the maximum superelevation that can be permitted. Practice in this regard varies from country to country. The AASHTO practice limits it to 0.12, whereas the U.K. practice limits it to 0.07 (Ref. 5, 6). In India the IRC recommendations for hill roads are a maximum value of 0.07 for snowbound areas and 0.10 for area not affected by snow (Ref. 9). In other cases, a value of 0.07 is considered maximum. The presence of bullockcart traffic on Indian roads has necessitated the adoption of this value.

11.7.4. Minimum radii of curves

Equation (11.1) can be rewritten as :

$$R = \frac{V^2}{127\,(e + \mu)} \qquad \text{...(11.2)}$$

Adopting the value of $e = 0.07$ or 0.10 and $\mu = 0.15$ as per current Indian practice the formula becomes :

$R = 0.0357\ V^2$ (for plain and rolling terrain) ...(11.3)

$R = 0.0315\ V^2$ (for hill roads which are not snow bound) ...(11.4)

$R = 0.0357\ V^2$ (for hill roads which are snow bound) ...(11.5)

Since the current Indian Standards give two values for Design Speed (*i.e.*, Ruling and Minimum), the above formulae yield two values for radius, the ruling radius and the absolute minimum radius.

These values are indicated in Table 11.9.

11.7.5. Superelevation rates

Superelevating on curves is intended to counteract a part of the centrifugal force, the remaining part being resisted by the lateral friction. Also, super-elevating results in economies in maintenance. This is because skidding and unequal pressure on the wheels of vehicles, which result from high value of sideway force between the tyres and the roadway surface, give rise to pot-holes and necessitate frequent attention to the surface.

Table 11.9

Minimum Radii of Horizontal Curves for Different Terrain Conditions

(*Source* : Ref. 2)

Classification Road	*Plain terrain*		*Rolling terrain*		*Mountainous terrain*				*Steep terrain*			
					Areas not affected by snow		*Snow bound areas*		*Areas not affected by snow*		*Snow bound areas*	
	Ruling Minimum	*Absolute Minimum*	*Ruling Minimum*	*Absolute Minimum*	*Ruling Minimum*	*Absolute Minimum*	*Ruling Minimum*	*Absolute Minimum*	*Ruling Minimum*	*Absolute Minimum*	*Ruling Minimum*	*Absolute Minimum*
1. National highways and state highways	360	230	230	155	80	50	90	60	50	30	60	33
2. Major district Roads	230	155	155	90	50	30	60	33	30	14	33	15
3. Other district roads	155	90	90	60	30	20	33	23	20	14	23	15
4. Village roads	90	60	60	45	20	14	23	15	20	14	23	15

Superelevation can be provided either fully to counteract the centrifugal force or to counteract a fixed proportion of the centrifugal force. In the former case, the superelevation needed would be more than 1 in 15 on sharp curves causing inconvenience to slow moving vehicles. Since superelevation is limited to 0.07 or 0.10 as per Indian practice, maximum friction would have to be relied upon when the sharpest possible curve is traversed. When a vehicle negotiates a flat curve, friction would not be developed to the maximum. This is not a balanced design. It is desirable that the superelevation should be such that a moderate amount of friction is developed while rounding flat curves and friction not exceeding the maximum allowable be developed at sharp curves. Therefore, designing the superelevation to fully counteract the centrifugal force developed at a fraction of the design speed will provide the necessary balance.

The above is achieved as per U.K. practice by providing full superelevation for a speed of 63.2 percent of the design speed, such that 40 per cent of the centrifugal force is counteracted by superelevation (Ref. 5). Thus,

$$e = \frac{(0.632\,V^2)}{127R} = \frac{V^2}{314R} \qquad \text{...(11.6)}$$

As per Indian practice, super-elevation is calculated on the assumption that it should counteract the centrifugal force developed at three-fourth the design speed. Thus :

$$e = \frac{(0.75\,V^2)}{127R} = \frac{V^2}{225R} \qquad \text{...(11.7)}$$

The superelevation calculated from Equation (11.7) is restricted to 0.07 and 0.10 indicated in 11.7.3.

11.7.6. Radii curves for which no superelevation is required

The normal cambered section of a highway can itself be continued on a curve where the superelevation calculated is less than the camber. From formula (11.7),

$$R = \frac{V^2}{225e} \qquad \text{...(11.8)}$$

Substituting the value of the camber for e in Equation (11.8), the minimum radius beyond which no superelevation is required is obtained. Table 11.10 gives the values for various design speeds and cambers.

Table 11.10. Radii Beyond which Superelevation is not Essential for Different Cambers and Speeds (Indian Practice)

(*Source* : Ref. 2)

Design Speed (K.P.H.)	*Radius in metres beyond which no superelevation is needed for values of camber of*				
	4%	*3%*	*2.5%*	*2%*	*1.7%*
20	50	60	70	90	100
25	70	90	110	140	150
30	100	130	160	200	240
35	140	180	220	270	320
40	180	240	280	350	420
50	280	370	450	550	650
65	470	620	750	950	1100
80	700	950	1100	1400	1700
100	1100	1500	1800	2200	2600

In such cases where the radii is greater than those given in Table 11.10, it is however, desirable to remove the adverse crown in the outer half of the carriageway and superelevate at the normal crown slope.

11.7.7. Method of attainment of superelevation

The normal cambered surface on a straight reach of road is changed into a superelevated surface in two stages. In the first stage, outer half of the camber is gradually raised until it is level, *vide*. Fig. 11.4 (*b*).

It is desirable to accomplish the raising of the outer-half till it is level before the starting point of the transition curve. The raising of the outer edge should be done in a slope not exceeding 1 in 150 for plain and rolling terrain and 1 in 60 for hilly terrain.

In the second stage, three methods may be adopted to attain the full superelevation :

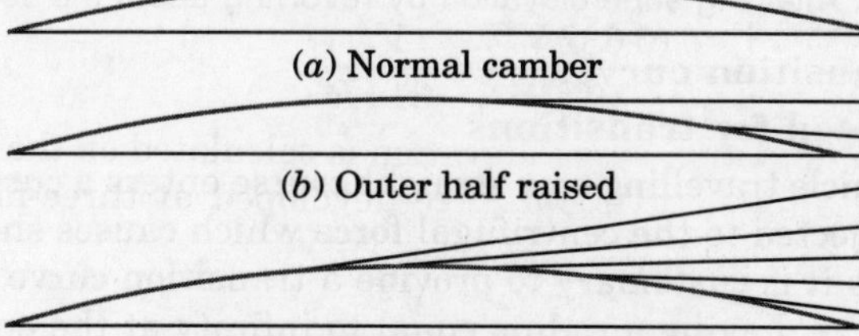

(*a*) Normal camber

(*b*) Outer half raised

(*c*) Superelevation equal to camber

Fig. 11.4. First stage in attaining super elevation by bringing superelevation equal to camber.

(*i*) The surface of the road is rotated about the centre-line of the carriageway, gradually lowering the inner edge and raising the upper edge, keeping the level of the centre line constant [Fig. 11.5 (*a*)].

(*ii*) The surface of the road is rotated about the inner edge, raising the centre and the outer edge [Fig. 11.5 (*b*)].

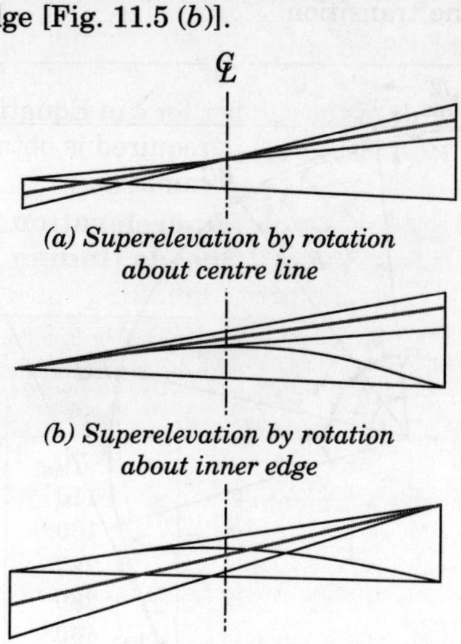

(a) Superelevation by rotation about centre line

(b) Superelevation by rotation about inner edge

(c) Superelevation by rotation about outer edge

Fig. 11.5. Second stage in attaining superelevation.

(*iii*) The surface of the road is rotated about the outer edge, depressing the centre and the inner edge [Fig. 11.5 (*c*)].

Method (*i*) is most generally used as it results in the least distortion of the pavement. Fig. 11.6 shows the method of attaining superelevation as per method (*i*).

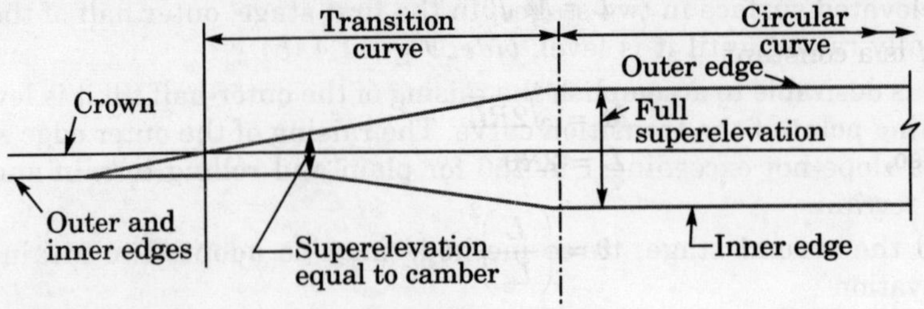

Fig. 11.6. Attaining superelevation by revolving about the centre line.

11.7.8. Transition curves

11.7.8.1. Need for transitions

When a vehicle travelling on a straight course enters a curve of finite radius it is suddenly subjected to the centrifugal force which causes shock and sway. In order to avoid this it is customary to provide a transition curve at the beginning of the circular curve, having a radius equal to infinity at the end of the straight and gradually reducing the radius to the radius of the circular curve where the curve begins. Incidentally, the transition portion is also used for the gradual application of the superelevation and the curve widening.

11.7.8.2. The spiral

Various forms of curves are suitable for highway transitions, but the one most popular and recommended for use in this country is the Spiral (Ref. 17). It is easy, to set out in the field and the rate of change of acceleration is uniform through the length of the transition.

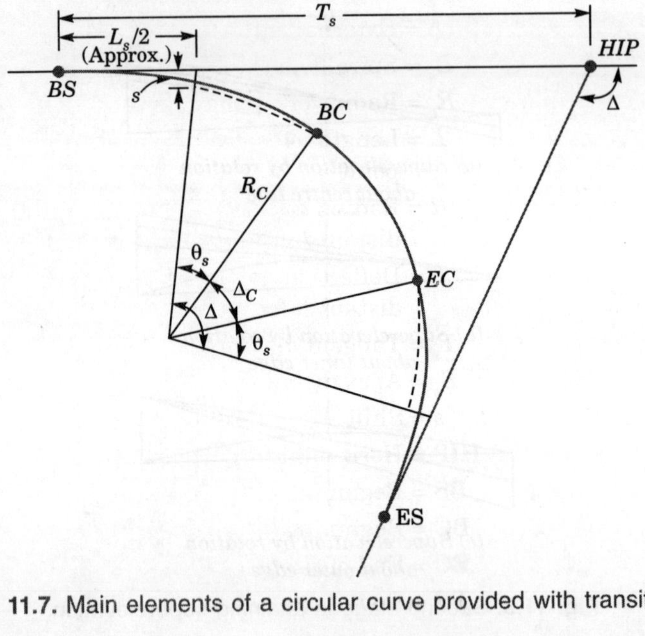

Fig. 11.7. Main elements of a circular curve provided with transitions.

Fig. 11.7 indicates the main elements of a circular curve provided with spiral for transitions at its two ends.

Some of the important properties of the spirals are given below :

(*i*) $$L_s R_c = LR = \text{constant} \quad ...(11.9)$$

(*ii*) $$L = M\sqrt{\theta} \quad ...(11.10)$$

where, M is a constant

$$M = \sqrt{2RL} \quad ...(11.11)$$

Also, $$L = 2R\theta \quad ...(11.12)$$

(*iii*) $$\theta = \left(\frac{L}{L_s}\right)^2 \theta_s \quad ...(11.13)$$

(*iv*) $$\theta_s = \frac{L}{2R_c} \text{ radians}$$

$$= \frac{28.65\, L_s}{R_c} \text{ degrees} \quad ...(11.14)$$

(*v*) $$T_s = \frac{L_s}{2} + (R_c + s)\tan\frac{\Delta}{2} \quad ...(11.15)$$

(*vi*) $$S = \frac{L_s^2}{24\, R_c} \quad ...(11.16)$$

(*vii*) $$E_s = (R_c + s)\sec\frac{\Delta}{2} - R_c \quad ...(11.17)$$

The following nomenclature may be noted :

θ_s = Spiral angle

Δ_c = Angle of the circular curve

Δ = External angle

L_s = Spiral length

R_c = Radius of circular curve

L = Length of spiral from starting point to any point

R = Radius of curvature of the spiral at the point distant L from starting point

θ = Deflection angle at any point of the spiral distant L from the starting point

T_s = Tangent distance

E_s = Apex distance

s = Shift

HIP = Horizontal intersection point

BS = Beginning of spiral

BC = Beginning of circular curve

EC = End of circular curve

ES = End of spiral.

Table 11.11. Minimum Transition Lengths for Different Speeds and Curve Radii

(*Source* : Ref. 2)

Plain and rolling terrain							*Mountainous and steep terrain*					
Curve radius R (metres)	*Design speed (KPH)*						*Curve radius (metres)*	*Design s speed (KPH)*				
	100	*80*	*65*	*50*	*40*	*35*		*50*	*40*	*30*	*25*	*20*
	Transotopm length-meters							*Transition length-meters*				
45					NA	70	14				NA	30
60				NA	75	55	20				35	20
90				75	50	40	25			NA	25	20
100			NA	70	45	35	30			30	25	15
150			80	45	30	25	40		NA	25	20	15
170			70	40	25	20	50		40	20	15	15
200		NA	60	35	25	20	55		40	20	15	15
240		90	50	30	20	NR	70	NR	30	15	15	15
300	NA	75	40	25	NR		80	55	25	15	15	NR
360	130	60	35	20			90	45	25	15	15	
400	115	55	30	20			100	45	20	15	15	
500	95	45	25	NR			125	35	15	15	NR	
600	80	35	20				150	30	15	15		
700	70	35	20				170	25	15	NR		
800	60	30	NR				200	20	15			
900	55	30					250	15	15			
1000	50	30					300	15	NR			
1200	40	NR					400	15				
1500	35						500	NR				
1800	30											
2000	NR											

NA–Not applicable.
NR–Transition not required.

11.7.8.3. Length of transition

The length of the transition should be determined from the following two considerations :

(*i*) The rate of change of centrifugal acceleration adopted in the design should not cause discomfort to the drivers.

If C is the rate of change of acceleration,

$$C = \frac{v^2}{R_c} \div t$$

$$= \frac{v^2}{R_c} \div \frac{L_s}{v} = \frac{v^3}{R_c L_s} \quad ...(11.18)$$

where, R_c and L_s have the same meaning as in 11.7.8.2,

t = time in second to ravel a distance of L_s

v = speed in m/second.

Thus, $$L_s = \frac{V^3}{(3.6)^3 CR_c} = \frac{0.0215\, V^3}{R_c} \quad ...(11.19)$$

where, **V = speed in K.P.H.**

The value of C is given by the following equation (Ref. 17) :

$$C = \frac{80}{75 + V} \qquad \text{...(11.20)}$$

subject to a maximum of 0.8 and minimum of 0.5.

(ii) The rate of change of superelevation should be such as not to cause higher gradients and unsightly appearances. This could be kept 1 in 150 for roads in plain and rolling terrain and 1 in 60 in hilly terrain (Ref. 2).

The higher of the values given by considerations *(i)* or *(ii)* above may be used. Suggested values are given in Table 11.11.

11.7.9. Widening on curves

Widening of pavements is needed on curves for the following reasons :

(i) On curves the vehicles occupy a greater width because the rear wheels track inside the front wheels (Fig. 11.8).

(ii) On curves drivers have difficulty in steering their vehicles to keep to the centre line of the line.

From Fig. 11.8, l^2 considering the right-angled triangle OBC,

$$l^2 = (R - m)^2 = R^2$$

$$\therefore \quad l^2 = 2\,Rm - m^2$$

Neglecting m^2, $\quad l = 2\,Rm$

$$m = \frac{l^2}{2R} \qquad \text{...(11.21)}$$

Assuming a wheel base of 6 m for a vehicle corresponding to AASHTO Single Unit (SU).

Widening in metres,

$$m = \frac{18}{R} \qquad \text{...(11.22)}$$

where, $\quad R$ = radius in metres.

The widening due to psychological reasons is function of speed and can be assumed to be given by the empirical formula :

$$\frac{0.1V}{\sqrt{R}} \qquad \text{...(11.23)}$$

where, $\quad V$ = speed in K.P.H.

R = radius in metres.

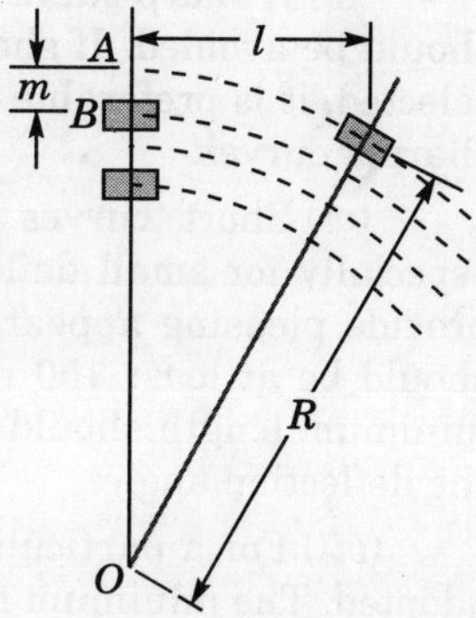

Fig. 11.8. Widening on curves.

The total widening for pavements whose width is in excess of two lanes is given by the formula :

$$W_e \text{ (in m)} = n\,\frac{18}{R_c} + 0.1\,\frac{V}{\sqrt{R_c}} \qquad \text{...(11.24)}$$

where, $\quad n$ = number of lanes.

Table 11.12 gives the widening necessary for single lane and 2 lane pavements (Ref. 2, 17).

Table 11.12. Extra Width of Pavement at Horizontal Curve

Radius of curve (m)	Upto 20	21 to 40	41 to 60	61 to 100	101 to 300	Above 300
Extra width (m)						
Two-lane	1.5	1.5	1.2	0.9	0.6	Nil
Single-lane	0.9	0.6	0.6	Nil	Nil	Nil

11.7.10. General controls for horizontal alignment

The following general controls for horizontal alignment should be kept in view in a sound design practice.

(*i*) Alignment should be as directional as possible.

(*ii*) The alignment should be consistent with the topography and should conform generally to the natural contours.

A line cutting across the contours involves high fills and large cuts, mars the landscape and is difficult for maintenance.

(*iii*) The number of curves should, in general, be kept to a minimum.

(*iv*) The alignment should be fluent and avoid abrupt turns. Winding alignment consisting of short curves should be avoided, since it is the cause of erratic operation.

(*v*) A sharp curve at the end of a long tangent is extremely hazardous and should be avoided. If sharp curvature is unavoidable over a portion of the route selected, it is preferable that this portion of the road be preceded by successfully sharper curves.

(*vi*) Short curves giving the appearance of kinks should be avoided, especially for small deflection angles. The curves should be sufficiently long to provide pleasing appearance and smooth driving on important highways. They should be at least 150 metres long for a deflection angle of 5 degrees, and the minimum length should be increased by 30 metres for each 1 degree decrease in the deflection angle.

(*vii*) For a particular design speed, as large a radius as possible should be adopted. The minimum radii should be reserved only for the critical locations.

(*viii*) The use of sharp curves should be avoided on high fills. In the absence of cut slopes, shrubs, trees, etc., above the roadway, the drivers may have difficulty in estimating the extent of curvature and fail to adjust to the conditions.

(*ix*) While abrupt reversal in curvature is to be avoided, the use of reverse curves becomes unavoidable in hilly terrain. When they are provided, adequately long transitional curves should be inserted for superelevation run off.

(*x*) Curves in the same direction separated by short tangents, say 300-500 metres long are called *broken-back* curves.

They should be avoided as they are not pleasing in appearance and are hazardous.

(*xi*) Compound curves may be used in difficult topography in preference to a *broken-back* arrangement, but they should be used only if it is impossible to fit in a single circular curve. To ensure safe and smooth transition from curve to curve, the radius of the flatter curve should not be disproportional to the radius of the sharper curve. A ratio of 2 : 1 or preferably 1.5 : 1 should be adopted.

(xii) The horizontal alignment should blend with the vertical harmoniously. General controls for the combination of horizontal and vertical alignments are indicated in section 11.6.

11.8. Vertical Alignment

11.8.1. Gradients

11.8.1.1. Maximum Grades

One of the important considerations in designing a highway is the gradient. The cost of operation of vehicles and the capacity of a highway are profoundly affected by the grades provided. On mountainous and steep terrain, the grades are not only influenced by the ability of vehicles to negotiate them, but also by the altitude of a road above the sea level since rarefaction of air causes loss of engine power. A peculiar consideration for Indian conditions is the presence of animal drawn traffic. The grades should be selected so as to encourage uniform operation as far as possible for a given terrain and the vertical profile should be as smooth as economically possible. Roads once constructed to a given vertical profile are very difficult to be upgraded to flatter gradients at a subsequent stage without great loss to the initial investment.

Table 11.13 gives the Indian practice as regards maximum gradients (Ref. 2, 9).

In Table 11.13 gradients up to the *ruling* may be used as a matter of course in design. The *limiting* gradients may be used where the topography of the area compels this or where the adoption of gentler gradients would enormously add to the cost. *Exceptional gradients* are meant to be used only in very exceptional situations and for short lengths prescribed in the Table 11.13.

Table 11.13. Gradients for Different Terrain Types-Indian Practice

S. No.	*Terrain type*	*Ruling gradient %*	*Limiting gradient %*	*Exceptional gradient %*
1.	Plain	3.3 (1 in 30)	5 (1 in 20)	6.7 (1 in 15) (for short distances not exceeding 100 m at a stretch)
2.	Rolling	3.3 (1 in 30)	5 (1 in 20)	6.7 (1 in 15) (for short distances not exceeding 100 m at a stretch)
3.	Mountainous	5 (1 in 20)	5 (1 in 16.7)	7 (1 in 14.3) (for a distance not more than 400 m)

4.	**Steep**			
(*i*)	**Upto 3000 m height above mean sea level**	**6 (1 in 16.7)**	**7 (1 in 14.3)**	**8 (1 in 12.5) (for short distances not exceeding 100 m)**
(*ii*)	**Above 3000 m height above mean sea level**	**5 (1 in 20)**	**6 (1 in 16.7)**	**7 (1 in 14.3) (for short distances not exceeding 100 m) 8 (1 in 12) (for distance not exceeding 100 m)**

11.8.1.2. Minimum gradients for drainage

Where the drainage of a road is to be effected by means of longitudinal ditches, as in cut sections or hill roads, or along the curbed edge of pavements, a certain minimum gradient is necessary for efficient drainage. The values given in Table 11.14 are commonly adopted.

Table 11.14. Minimum Gradients for Drainage

(*Source : Ref. 2*)

S. No.	*Site condition*	*Minimum gradients (%) (Desirable)*
1.	Longitudinal drainage along curbed edge of pavements, (lined).	0.5
2.	Longitudinal drainage along road side ditches, (unlined).	1.0

11.8.2. Vertical Curves

11.8.2.1. General considerations

Vertical curves should be provided at points of change of grade so as to achieve gradual transition. If the point of vertical intersection (termed PVI) is above the road surface, the curve is a summit or crest curve and if it is below the road surface, the curve is a sag curve. If the total change in gradient from one tangent to another does not exceed 0.5 per cent, vertical curves can be dispensed with. The parabola is a vertical curve most frequently used because of the ease with which it can be laid out.

11.8.2.2. Length of summit curves

For summit curves, the most important consideration in determining the length of the curve is the sight distance requirement. The following basic formulae can be used to determine the length :

Case I. When the sight distance (S) is less than the length of the curve (L) :

$$L = \frac{AS^2}{\left(\sqrt{2h_1} + \sqrt{2h_2}\right)^2} \quad ...(11.25)$$

Case II. When the sight distance (S) is greater than the length of the curve (L).

$$L = 2S - \frac{2\left(\sqrt{h_1} + \sqrt{h_2}\right)^2}{A} \qquad ...(11.26)$$

In the above formulae,

S = sight distance in metres

L = length of vertical curve, in metres

h_1 = height of eye above the roadway surface

h_2 = height of object above the roadway surface.

A = algebraic difference in slopes expressed in absolute values in a decimal form

Adopting $h = 1.2$ m

and $h_2 = 0.15$ m

as per current Indian practice (Ref. 18, 45) Eqs. (11.25) and (11.26) transform as under for safe stopping sight distance :

When S is less than L,

$$L = \frac{AS^2}{4.4} \qquad ...(11.27)$$

When S is greater than L,

$$L = 2S - \frac{4.4}{A} \qquad ...(11.28)$$

If the gradient is expressed in per cent, the above formulae become :

$$L = \frac{NS^2}{440}$$

$$L = 2S - \frac{440}{N}$$

For intermediate or overtaking sight distance, $h_1 = h_2 = 1.2$ m.

This gives the following relationships :

When L is greater than S,

$$L = \frac{AS^2}{9.6}$$

When L is less than S,

$$L = 2S - \frac{9.6}{A}$$

11.8.2.3. Sag curves

For sag curves, one of the criteria for determining the length is rider comfort. The other criteria are vehicle headlight, sight distance, drainage control and general appearance. As per rider comfort as the basis of design of sag curves (Ref. 19), α, the vertical radial acceleration, is given by,

$$\alpha = \frac{v^2}{R}$$

where α is in m/sec^2.

$$v = \text{speed in m/sec.} = \frac{V^2}{13R}$$

where, V = speed in K.P.H. ...(11.29)

Since the radial acceleration changes from zero to α in a length of $L/2$, where L = length of the sag curve, the rate of change of radial acceleration C is given by :

$$C = \frac{V^2}{13R} + \frac{L \times 3.6}{2V} = \frac{V^3}{23.4\,LR} \text{ m/sec}^3 \quad ...(11.30)$$

The recommended value of C for design is 0.6 m/sec (Ref. 3,19). Substituting this in (11.30), we get

$$L = \frac{V^2}{14R} \quad ...(11.31)$$

$$\therefore \quad LR = \frac{V^3}{14} \quad ...(11.32)$$

For a transition curve of the form of spiral or a cubic parabola, it can be shown that

$$L = (2RLA)^{1/2} \quad ...(11.33)$$

where, A = algebraic difference in grades of the straights, expressed as a decimal.

Substituting Eq. (11.32), in (11.33),

$$L = \left(2\frac{V^3}{14}A\right)^{1/2} = \frac{V^{3/2}\;A^{1/2}}{2.65} \quad ...(11.34)$$

Another criterion for design of valley curves is that for night travel, the design must ensure that the roadway ahead is illuminated by vehicle headlights for a sufficient length, which enables the vehicle to brake to a stop, if necessary. This is known as the headlight sight distance and is equal to the safe stopping distance (Table 11.15).

For designing valley curves for the above needs, the following points should be kept in view (Ref. 18) :

(*i*) height of headlight above the road surface is 0.75 m,

(*ii*) the useful beam of headlight is one degree upwards from the grade of the road ; and

(*iii*) the height of the object is nil.

11.8.3. General Controls for Vertical Alignment

The following general controls for vertical alignment should be kept in view in designing the vertical profile of a highway :

(*i*) The grade line selected should be smooth with gradual changes, consistent with the class of highway and terrain. Numerous brakes and short lengths of grades should be avoided.

(*ii*) The *roller-coaster* or *hidden type* of profile should be avoided as it is hazardous and aesthetically unpleasant.

(*iii*) Undulating grade lines, involving substantial lengths of momentum grades, should be appraised for their effect upon traffic operation. Such profiles permit heavy trucks to operate at higher overall speeds than when an upgrade is not preceded by a down-grade but may encourage excessive speeds of trucks with consequent hazard to traffic.

(*iv*) A *broken-back* grade line (two vertical curves in the same direction separated by short section of tangent grade) should generally be avoided.

(*v*) On long continuous grades, it may be preferable to place the steepest grades at the bottom and flatten the grades near the top. Alternatively, the long grades may be broken by short intervals of flatter grades.

(*vi*) Intersections on grades should be avoided as far as possible. Where unavoidable, the approach gradients and the gradient through the intersections should be flattened to the maximum possible extent.

11.9. Combination of Horizontal and Vertical Alignment

The following general controls should be kept in view in designing a highway for combination of horizontal and vertical alignment :

(*i*) The overall appearance of highway is enhanced with a judicious selection of horizontal and vertical elements. Best results are obtained when there is a proper balance between curvature and grades. Straight alignment or flat horizontal curves imposed on a road with steep or long grades or excessive curvature imposed on a road with a flat grade are both poor designs. A compromise between the two offers the best in safety, capacity, ease and uniformity of operation.

(*ii*) Vertical curvature superimposed upon horizontal curvature or *vice-versa*, generally results in a pleasing effect.

(*iii*) From the safety point of view, sharp horizontal curvature should not be introduced at or near the top of a pronounced summit vertical curve. In such situations, the horizontal curve is made longer than the vertical curve.

(*iv*) Sharp horizontal curves should be avoided at or near the low point of a pronounced sag curve.

(*v*) On two-lane highways, the need for safe passing sections at frequent intervals and for an appreciable percentage of the length of the highway often superceedes the general desirability for combination of horizontal and vertical alignment. In these situations, it is desirable to have long straight sections to secure, sufficient passing sight distance in design.

(*vi*) Horizontal curvature and vertical profile should be made as flat at feasible at highway intersections when sight distance is important and vehicles may have to slow down or stop.

(*vii*) On divided highways, variation in the width of the median and the use of separate profiles and horizontal alignments should be considered to derive design and operational advantages of one-way roadways.

11.10. Sight Distance

11.10.1. General

The design of a highway with adequate sight ahead of a travelling vehicle results in safe operation. The distance along the road surface at which a driver has visibility of objects stationary or moving at a specified height above the carriageway is known as the sight distance. A knowledge of the slight distance requirement is needed in designing vertical curves. It also governs the set-back of buildings, slopes, fences and other obstructions adjacent to a carriageway on a horizontal curve. Sight distance requirement can be considered under two heads : (*i*) Stopping sight distance and (*ii*) overtaking (passing) sight distance.

11.10.2. Stopping sight distance

Stopping sight distance is the distance required by a driver of a vehicle travelling at a given speed to bring his vehicle to a stop after an object on the roadway becomes visible. Because of its importance to safety, all highways must be designed for the minimum stopping sight distance. It is made up of two components.

(*i*) the distance travelled during perception and brake reaction time ; and

(*ii*) the distance travelled during the time the brakes are under application till the vehicle comes to a stop.

When sensations received through the eyes, ears or body are strong enough to be recognised and interpreted, they becomes perceptions. In the case of a motorist, it is the time which elapses between the instant the driver perceives the object on the carriageway and the instant that the realizes that braking of the vehicle is needed. The time lag or the brief interval between the perception of danger and the effective application of the brakes is called the brake reaction time. The perception time and the brake reaction time depend upon a variety of factors, *viz.*, age, sex, alertness and visual acuity of the driver, visibility, vehicle design, the size and type of the object etc.

As per current Indian Practice (Ref. 18), a value of 2.5 second is considered reasonable for the perception time and brake reaction time taken together. The distance travelled during this interval is given by

$$d_1 = vt$$

$$d_1 = 0.278\ Vt = 0.278\ V \times (2.5)$$

$$= 0.695\ V \qquad \text{...(11.35)}$$

where, d_1 = Distance travelled in metres

V = Speed in K.P.H.

t = Perception and brake reaction time

= 2.5 seconds.

The braking distance is the distance in which a moving vehicle comes to a stop after the application of brakes. On a level road, the distance is given by :

$$d_2 = \frac{V^2}{254\ f} \qquad \text{...(11.36)}$$

where, d_2 = Braking distance is meters

V = Speed in K.P.H.

f = Coefficient of longitudinal friction between the type and the payment.

The coefficient of friction is assumed to vary from 0.40 at 20 K.P.H. to 0.35 at 100 K.P.H. (Ref. 18).

The above considerations yield the values given in Table 11.15 recommended by IRC (Ref. 2, 18).

Table 11.15. Stopping Sight Distance as per Indian Practice

Design speed K.P.H.	*Safe stopping sight distance (metres)*
20	20
25	25
30	30
40	45
50	60
60	80
65	90
80	120
100	180

Some slight adjustments are needed in the values of the braking sight distance to take into account for the effect of grades. The following amended formula may be used to calculate d_2 :

$$d_2 = \frac{V^2}{254\,(f \pm 0.01G)} \quad \text{...(11.37)}$$

where, G = longitudinal grade in per cent.

The positive sign may be used inside the bracket if the gradient is upgrade and the negative sign may be used if the gradient is downward. Correction for grade should not be applied on undivided roads with two-way traffic, but must invariably be considered for divided highway which have independently designed profiles.

The safe stopping sight distance is measured between two points, one 1.2 m above the carriageway representing the driver's eye and the other 0.15 m height representing on object (Ref. 18).

11.10.3. Overtaking sight distance

Overtaking becomes necessary because some vehicles travel fast while some travel slow. The highway should be designed so that the vehicles travelling at the design speed are afforded chance to overtake the slower vehicles. Overtaking manoeuvre is possible only if the drivers have sufficient sight distance available to them so that the whole operation can be accomplished without danger. On undivided roads, overtaking involves the occupation of the road space normally used by opposing traffic. The overtaking sight distance is that distance which should be available to enable the driver to overtake another vehicles safely and comfortably without interfering with the speed of an oncoming vehicle travelling at the design speed should it come into view after the overtaking manoeuvre is started.

For computing the minimum overtaking sight distance on two-lane two-way highways, the following assumptions are made (Ref. 18) :

(i) The overtaken vehicle travels at a uniform speed which is 16 K.P.H. less than the design speed of the road.

(ii) The overtaking vehicle has reduced its speed and trails the overtaken vehicle as it enters the overtaking section.

(iii) The overtaking vehicle follows the vehicle ahead for a short while to perceive the clear road ahead before beginning the overtaking movement.

(*iv*) The overtaking is done by acceleration to the design speed and hurriedly returning to its own side of the road.

(*v*) When the overtaking vehicle returns to its lane, it arrives alongside an oncoming vehicle in the opposite lane.

[Some authorities such as AASHTO (Ref. 3) lay down that a certain clear distance should be available between the two].

Fig. 11.9 below indicates the elements that go to make up the overtaking sight distance.

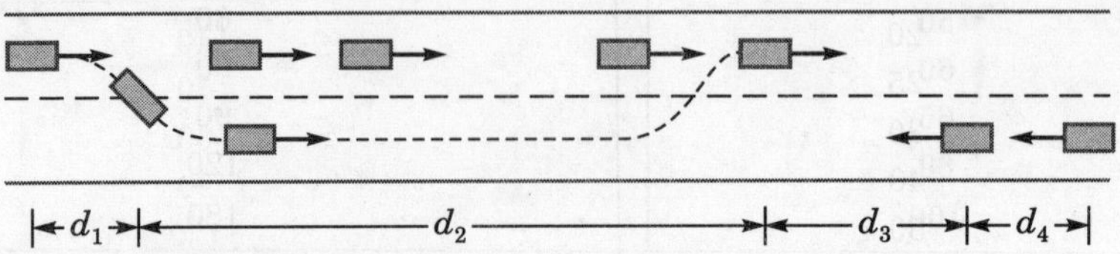

Fig. 11.9. Elements of overtaking sight distance.

The dimension d_1 represents the distance travelled during the perception and reaction time and during the initial acceleration to the point of encroachment on the right lane. AASHTO practice (Ref. 3) gives a value of 3.6 to 4.5 seconds for this initial manoeuvre. The dimension d_2 represents the actual distance covered by the overtaking vehicle during the overtaking manoeuvre. The overtaking vehicles are assumed in AASHTO practice to take 9.3 to 10.4 seconds for this purpose. The dimension d_3 is the distance between the overtaking vehicle at the end of its manoeuvre and the opposing vehicle and is known as the clearance length. It is assumed to vary from 34 to 90 m as per AASHTO practice (Ref. 3). The distance d_4 represents the distance travelled by an opposing vehicle at the design speed while the overtaking manoeuvre is taking place. Strictly speaking at should be $d_1 + d_2$, but in practice it is taken as 2/3 d_2, on the reasoning that the overtaking vehicle can always return to its lane if the driver sees an opposing vehicle.

As per latest Indian practice (Ref. 18), the calculation have been simplified by assuming that it takes about 9 to 14 seconds for the overtaking manoeuvre. About 2/3rds of this value is accounted for the distance travelled by the opposing vehicle. The final values as per Indian practice are given in Table 11.16.

Table 11.16. Overtaking Sight Distance (*Source* : Ref. 18)

Speed (K.P.H.)	*Safe overtaking sight distance (m)*
40	165
50	235
60	300
65	340
80	470
100	640

The overtaking distance is measured between two points both 1.2 m above the carriageway, representing the position of the eye of the driver and the object.

As per Indian practice (Ref. 18), for sections of roads where the overtaking sight distance cannot be provided, as a second preference, and intermediate value between the safe stopping sight distance and the safe overtaking sight distance is recommended. An intermediate sight distance is considered to afford reasonable opportunities for overtaking. Recommended values are given in Table 11.17.

Table 11.17. Intermediate Sight Distance for Two-lane Highways

(*Source* : Ref. 11)

Speed (*K.P.H.*)	*Intermediate sight distance* (*m*)
20	40
25	50
30	60
40	90
50	120
60	160
65	180
80	240
100	360

11.10.4. Sight distance for multi-lane roads

Divided highways with 4 or more lanes need only be designed for a safe stopping sight distance. Undivided highways with 4 lanes have enough opportunities for overtaking within one-half of the carriageway. Such roads, therefore, need only be designed for safe stopping sight distance.

11.10.5. Set-back distance at obstructions of horizontal curves

On horizontal curves with obstructions on the inside an important consideration is the lateral clearance so as to obtain the sight distance. Fig. 11.10 indicates the elements of design.

The following formulae give the value of the set-back distances :

(*a*) $$M = \frac{S^2}{8R} \qquad \text{...(11.38)}$$

(for cases where the required sight distance is wholly within the curved road)

(*b*) $$M = \frac{E(2L - S)}{8R} \qquad \text{...(11.39)}$$

(for cases where the required sight distance is greater than the available length of curve).

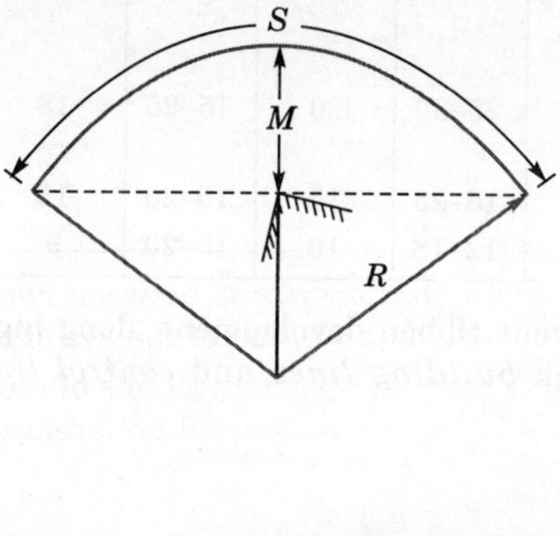

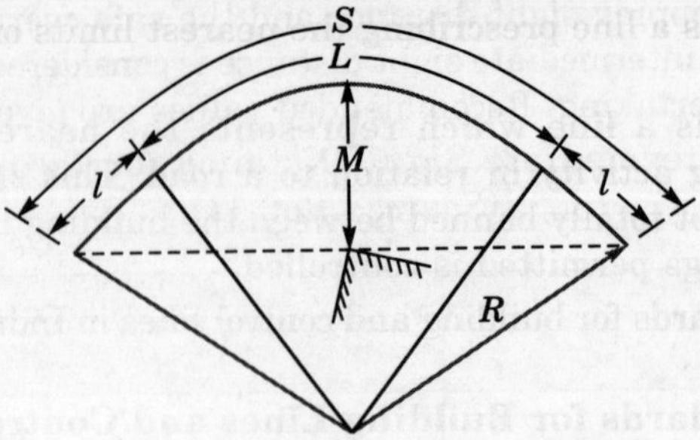

Fig. 11.10. Elements of set-back distance at obstructions of horizontal curves.

11.11. Cross-sectional Elements

11.11.1. General

The cross-sectional elements in the highway design pertain to those features which deal with its width. They embrace aspects such as right-of-way width, roadway width, pavement width, central reservations (median), shoulder, camber, side slopes, horizontal and vertical clearances etc.

11.11.2. Right-of-way

The road land width or the right-of-way width is the width of land secured and preserved to the public for road purposes. The right-of-way should be adequate to accommodate all the elements that make up the cross-section of the highway and may reasonably provide for future development. Indian practice (Ref. 20) is summarised in Table 11.18.

Table 11.18. Right-of-way Width for Different Classes of Roads in India

(*Source* : Ref. 2, 9)

Class of Road	*Right-of-way width in metres*							
	Plain and rolling terrain				*Mountainous and steep terrain*			
	Rural Areas		*Urban Areas*		*Rural Areas*		*Urban Areas*	
	Normal	*Range*	*Normal*	*Range*	*Normal*	*Exceptional*	*Normal*	*Exceptional*
1. National and state highways	45	30–60	30	30–60	24	18	20	18
2. Major district roads	25	25–30	20	15–25	18	15	15	12
3. Other district roads	15	15–25	15	15–20	15	12	12	9
4. Village roads	12	12–18	10	15–20	9	9	9	9

In order to prevent ribbon development along highways, it is sometimes necessary to establish *building lines* and *control lines* with the following definitions :

Building line is a line prescribing the nearest limits of future buildings in relation to a road.

Control line is a line which represents the nearest limits of future uncontrolled building activity in relation to a road. This signifies that though building activity is not totally banned between the building line and control line, the nature of buildings permitted is controlled.

Current standards for building and control lines in India (Ref. 20) are given in Table 11.19.

Table 11.19. Standards for Building Lines and Control Lines in India
(*Source* : Ref. 20)

Class of Road	*Plain rolling terrain*			*Mountainous and steep terrain*			
	Rural area		*Urban and industrial area*	*Distance between building line and boundary (metre)*			
	Width between building lines (over all width) (metres)	*Width between control lines (over all width) (metres)*	*Distance between building line and road boundary (metres)*	*Rural areas*		*Urban areas*	
				Normal	*Exceptional*	*Normal*	*Exceptional*
1. National and State Highways	80	150	3–6	5	3	5	3
2. Major District Roads	50	100	3	5	3	5	3
3. Other District Roads	25/30*	35	–	5	3	5	3
4. Village Roads	25	30	–	5	3	5	3

* If the land width is equal to the width between building lines indicated in this column, the building lines shall be set 2.5 m from the road boundary lines.

The terms right-of-way, building line and control lines are indicated in Fig. 11.11.

11.11.3. Space standards for urban roads

In urban situations the space available is generally restricted and costly. Table 11.20 below represents current Indian practice. (Ref. 1, 7).

Table 11.20. Space Standards for Urban Roads
(*Source* : Ref. 1 and 7)

S. No.	*Category of road*	*Minimum space in metres*
1.	Expressway (6 lane divided)	60
2.	Arterial street (4 lane divided with cycle tracks).	50–60
3.	Sub-arterial street (4 lane divided)	30–40
4.	Collector street (4 lane)	20–30
5.	Local street such as main residential street, shopping street in residential area, minor residential street and culde-sac.	10–20

11.11.4. Roadway width

Current Indian Practice as regards roadway width in summarised in Table 11.21.

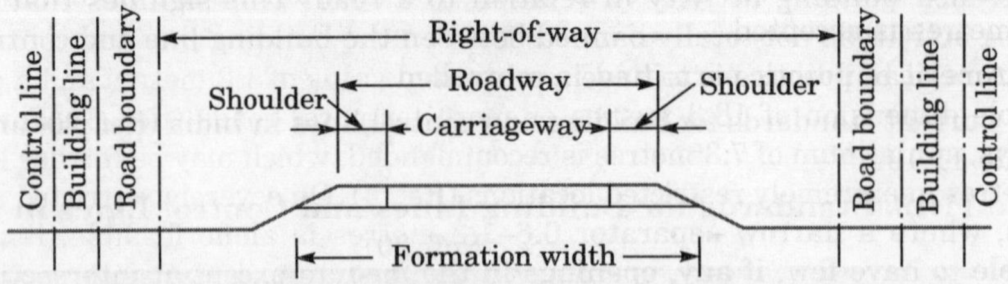

Fig. 11.11. Right of way, building line and control line.

Table 11.21. Standards for Roadway Width-Indian Practice

Highway classification	*Width of Roadway (in metres)*	
	Plain and rolling terrain	*Mountainous and Steep terrain*
1. National and state highways		
(*a*) Single lane		6.25
(*b*) Two lane	12.0	8.8
2. Major district roads	9.0	
3. Other district roads	9.0	4.75
(*i*) Two lanes		
(*ii*) Single lane	7.5	4.75
4. Village roads	7.5	4.0

For roads with more than two lanes, the roadway width has to be worked out to suit the number of lanes, medians (if any) and shoulders.

11.11.5. Carriageway width

The width of a traffic lane governs the safety and convenience of traffic and has a profound influence on the capacity of a road. Generally, a width of 3–4 metres is adopted for each traffic lane. In India, single lane pavements are generally 3.75 metres wide, whereas two lane pavements (without raised curbs) are 7.0 metres wide (Ref. 2, 21). When raised curbs are provided, two lane pavements are made 7.5 metres in width. An intermediate width of 5.5 metres is commonly used in India for less important roads with two lanes. For multi-lane pavements. Indian pratice is to provide 3.5 metre per lane. Single lane pavements in India are often provided with paved shoulder 0.9 metre wide on either side to facilitate crossing and overtaking manoeuvres.

For village roads, a carriageway width of 3.0 metres is generally deemed sufficient.

11.11.6. Central reservations (Median)

A central reservation or a median is the longitudinal space separating dual carriageways. The functions of the central reservation are :

(*i*) To separate the opposing streams of traffic

(*ii*) To minimise head light glare

(*iii*) To include space for safe operation of crossing and turning vehicles at intersections at grade, *e.g.* right turning pockets.

(*iv*) To provide a stopping area in case of emergencies.

Indian practice is to provide a minimum width of 5.0 metres for rural areas, which is relaxed to 3.0 metres in restricted conditions (Ref. 2). For urban situations, a width of 5.0 metres is considered desirable, but a minimum width of 1.2 metres is accepted.

American practice is to have a minimum value of 1.2 metres and a greater desirable dimension of 18.3 metres or more for rural roads (Ref. 3). For urban freeways, a minimum of 7.3 metres is recommended, which may be further lowered to 5 metres in extremely restricted locations (Ref. 4). On severely restricted arterial streets, where a narrow separator 0.6–1.2 metres is alone feasible, it may be desirable to have few, if any, openings in the median except at intersections.

In India, it is recommended (Ref. 2) that the minimum width of medians on rural highways should be 5 metres, but this could be reduced to 3 metres where land is restricted. On long bridges and viaducts, the width of median may be reduced to 1.5 metres, but in any case it should not be less than 1.2 m.

As far as possible, the median should be of uniform width in a particular section of the highway. However, where changes are unavoidable, a transition of 1 in 15 to 1 in 20 must be provided.

11.11.7. Shoulders

The shoulder is that portion of the roadway contiguous with the travelled way and is intended for accommodation of stopped vehicles, emergency use and lateral support of base and the surface courses.

The width of the shoulder should be adequate for giving working space around a stopped vehicle. American practice (Ref. 3) recommends a metre width for high type facilities and a width 1.2 to 2.4 metres for low type facilities. U.K. practice for rural roads recommends widths ranging from 1.2 to 3.65 metres depending upon the type of road and the nature of the kerb side treatment. Current Indian practice for 2-lane roads suggests a shoulder width of 2.5 metres for roads in rural areas (Ref. 2).

11.11.8. Curbs (kerbs)

A curb is a vertical or sloping member along the edge of a pavement or shoulder forming part of a gutter, strengthening or protecting the edge, and clearly defining the edge to vehicle operators. Its functions are :

(*i*) To facilitate and control drainage

(*ii*) To strengthen and protect the pavement edge

(*iii*) To delineate the pavement edge

(*iv*) To present a more finished appearance

(*v*) To assist in the orderly development of the roadside.

Curbs are generally needed on urban roads.

Curbs are classified as *barrier* and *mountable*. Barrier curbs are designed to discourage vehicles from leaving the pavement. The face may be vertical or sloping and the height may vary from 15 to 25 cm. A typical design for a nearly vertical barrier curb is given in Fig. 11.12.

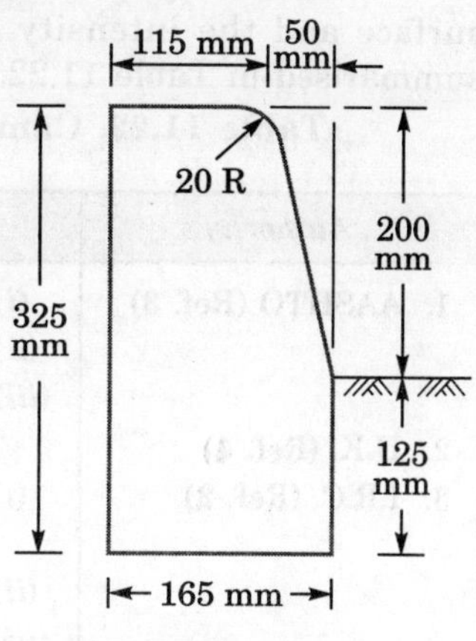

Fig. 11.12. Nearly vertical barrier curb.

In Fig. 11.12 a smaller height (say 15 cm) may be adopted for pedestrian refuge is lands. A larger height (say 22.5 cm) is adopted for bridges (Ref. 22).

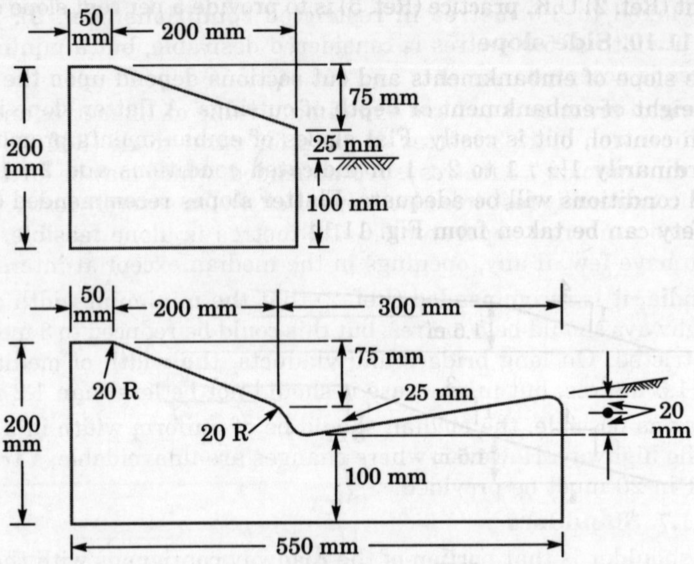

Fig. 11.13. Typical designs of mountable curbs.

Mountable curbs are those which can be easily crossed by vehicles, if required. They are used at medians and channelizing islands. Two typical designs are given in Fig. 11.13.

11.11.9. Camber

Camber, also known as cross-slope, facilitates drainage of the pavement laterally. The pavement can have a crown or a high point in the middle with slopes downwards toward both the edges. This type is favoured on two-lane roads and wider undivided roads. On divided roads, the individual carriageways may be centrally crowned separately or a unidirectional slope may be provided across the entire carriageway width.

The amount of camber to be provided depends upon the smoothness of the surface and the intensity of rainfall. The practice by various authorities is summarised in Table 11.22.

Table 11.22. Camber Prescribed by Various Authorities

Authority	*Type of surface*	*Camber (%)*
1. AASHTO (Ref. 3)	(*i*) High	1–2
	(*ii*) Intermediate	1.5–2
	(*iii*) Low	2–4
2. U.K. (Ref. 4)		2.5
3. I.R.C. (Ref. 2)	(*i*) Cement concrete or high type bituminous	1.7–2.0
	(*ii*) Thin bituminous surfacing	2.0–2.5
	(*iii*) Water-bound macadam	2.5–3.0
	(*iv*) Earth	3.0–4.0

The cross-slope for the shoulders should be generally steeper than for the pavement by about 0.5 percent to facilitate quick drainage subject to a minimum of 3 percent (Ref. 2) U.K. practice (Ref. 5) is to provide 5 per cent slope on shoulder.

11.11.10. Side slope

The slope of embankments and cut sections depend upon the type of soil and the height of embankment or depth of cuttings. A flatter slope is conducive for erosion control, but is costly. Flat slopes of embankments promote safety of traffic. Ordinarily 1½ : 1 to 2 : 1 in indicated conditions and 2 : 1 to 3 : 1 in inundated conditions will be adequate. Flatter slopes recommended (Ref. 23) for traffic safety can be taken from Fig. 11.14.

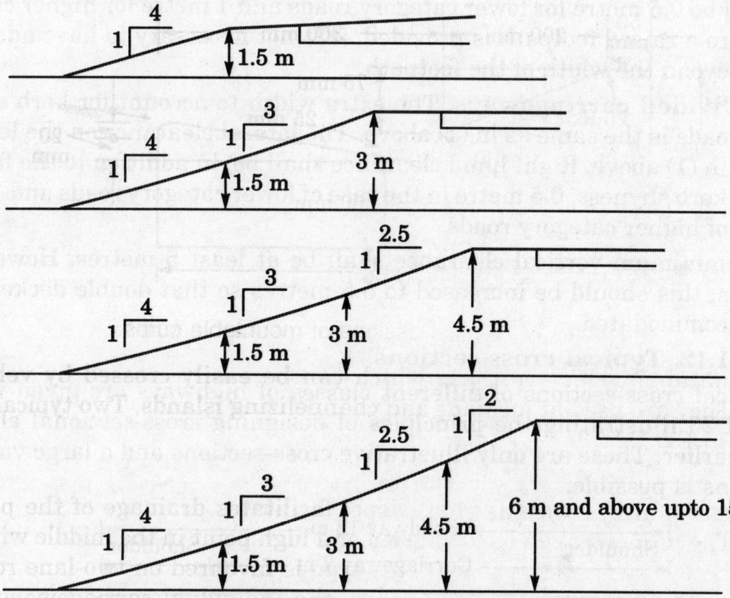

Fig. 11.14. Side slopes for road embankments (*Source* : Ref. 23)

11.11.11. Lateral and vertical clearances

Lateral clearance is the distance between the extreme edge of the carriageway to the face of a structure. Vertical clearance is the height above the highest point of the travelled way to the lowest point of the overhead structure.

Current Indian standards (Ref. 24) for lateral clearance are summarised below :

Type of roads		*Minimum lateral clearances*
A. Rural roads		
1. Single carriageway		
(*i*) National and state highways	:	Normal 2.5 metres, exceptional 2.0 metres
(*ii*) Major district roads and other district roads	:	Normal 2.0 metres, exceptional 1.5 metres
(*iii*) Village Roads	:	Normal 1.5 metres, exceptional 1.0 metre

If footpaths are provided, the lateral clearance should be the width of the footpaths plus one metre.

2. Divided carriageway

(*i*) Left hand clearance as in 1 above

(*ii*) Right and side clearance : 2.0 metres desirable
1.5 metres minimum

B. Urban roads.

1. **Single carriageways.** For kerbed roads, the carriageway width is widened on both sides under structures (underpass etc.) by 0.25 metre in the case of lower category urban roads and by 0.5 metre in the case of higher category urban roads to account for kerb shyness. If a footpath does not form the cross-section of the urban road, the minimum lateral clearance in addition to the above values shall be 0.5 metre for lower category roads and 1 metre for higher category roads. Where a raised footpath is provided, it is not necessary to have additional clearance beyond the width of the footpath.

2. **Divided carriageways.** The extra width to account for kerb shyness on kerbed roads is the same as in (1) above. The lateral clearance on the left shall be same as in (1) above. Right hand clearance shall be, in addition to the factor to account for kerb shyness, 0.5 metre in the case of lower category roads and 1 metre in the case of higher category roads.

The minimum vertical clearance shall be at least 5 metres. However, in urban areas, this should be increased to 5.5 metres so that double decker buses could be accommodated.

11.11.12. Typical cross-sections

Typical cross-sections of different classes of highways are given in Figs. 11.15 to 11.23 illustrating the principles of designing cross-sectional elements discussed earlier. These are only illustrative cross-sections and a large variety of combinations is possible.

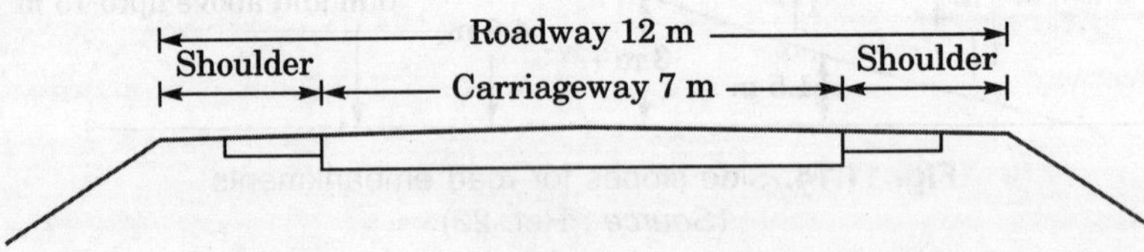

Fig. 11.15. Typical 2 lane National or State highway. (Rural Section)

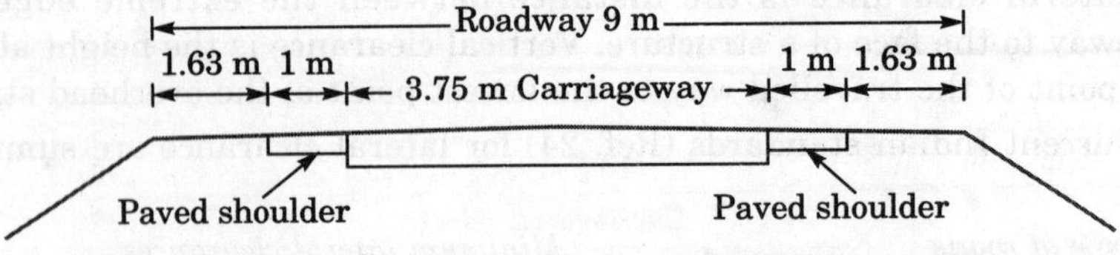

Fig. 11.16. Typical single lane road with paved shoulders. (Major District Road or other District Road)

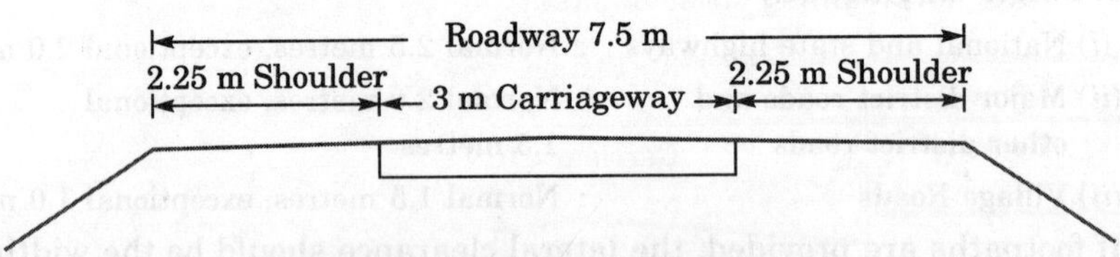

Fig. 11.17. Typical village road.

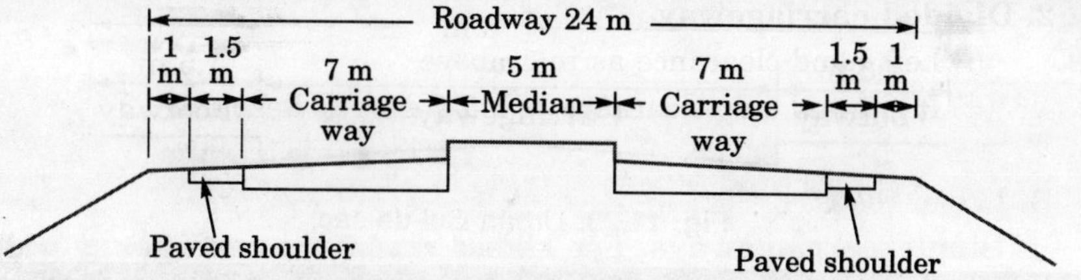

Fig. 11.18. Typical dual carriageway (2-lane each).

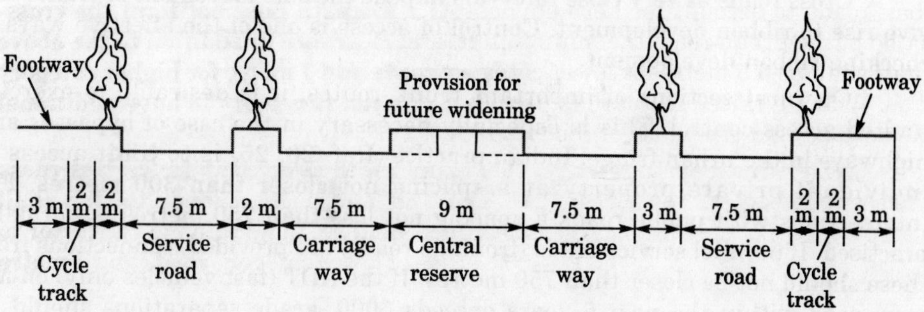

Fig. 11.19. Urban arterial street.

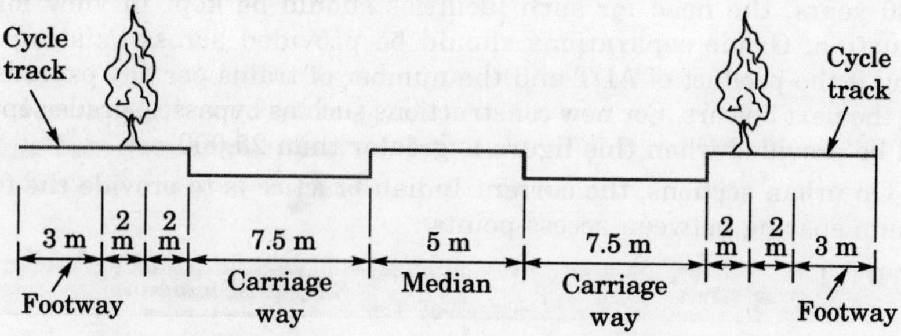

Fig. 11.20. Urban sub-arterial street through open area.

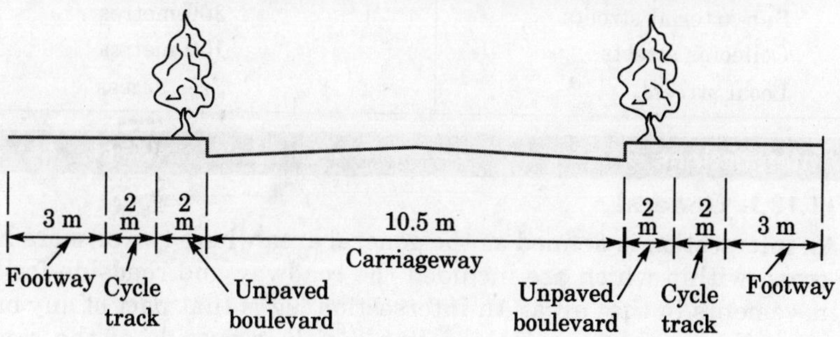

Fig. 11.21. Urban collector street through residential area.

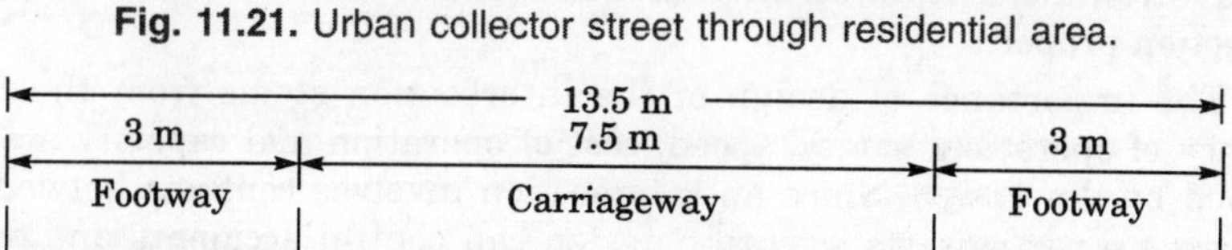

Fig. 11.22. Urban residential street.

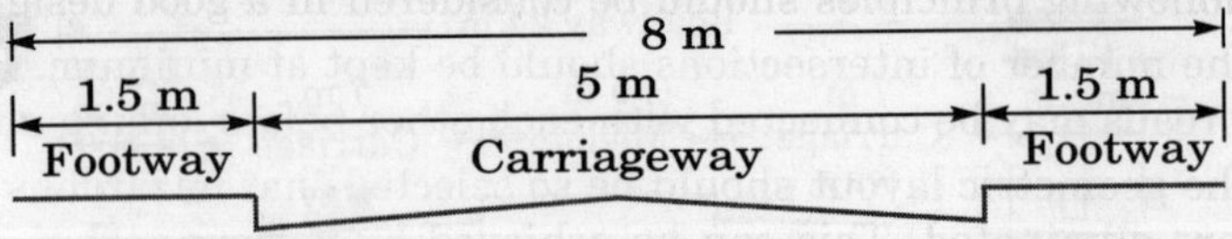

Fig. 11.23. Urban Cul-de-sac.

11.12. Control of Access

Cross roads at very close intervals impede the safe movement of traffic and give rise to ribbon development. Control of access is one of the effective ways of checking ribbon development.

On rural sections of important trunk routes, it is desirable to exercise limited access control. This is especially necessary in the case of bypasses and highways in the urban fringe. Indian practice (Ref. 20, 25) is to limit access to individual private property at a spacing not closer than 300 metres. For intersections with public roads a spacing not less than 750 metres is normally practised. If parallel service roads (frontage roads) are provided, connections from these should not be closer than 750 metres. If the ADT (fast vehicles only) on the cross-road within the next 5 years exceeds 5000, grade separations should be provided at the intersections. Where this traffic figure will be reached within the next 20 years, the need for such facilities should be kept in view for future construction. Grade separations should be provided across existing railway crossing if the product of ADT and the number of trains per day exceeds 50,000 within the next 5 years. For new constructions such as bypasses, grade separations should be provided when this figure is greater than 25,000.

On urban sections, the current Indian practice is to provide the following minimum spacing between access points :

Type of street	*Suggested minimum spacing*
Expressways	1000 metres
Arterial streets	500 metres
Sub-arterial streets	300 metres
Collector streets	150 metres
Local streets	Free access

11.13. Intersections

11.13.1. General

An intersection is defined as the general area where two or more highways join or cross, within which are included the roadway and roadside facilities for traffic movements in that area. An intersection leg is that part of any one of the roadways radiating from an intersection which is outside of the area of the intersection proper.

The importance of design of the intersection stems from the fact that efficiency of operation, safety, speed, cost of operation and capacity are directly governed by the design. Since an intersection involves conflicts between traffic in different directions, its scientific design can control accidents and delay and can lead to orderly movement of traffic. Intersections represent potentially dangerous locations from point of view of traffic safety. It is believed that well over half the fatal and serious road accidents in built-up areas occur at junctions.

The following principles should be considered in a good design :

(*i*) The number of intersections should be kept at minimum. If necessary, some minor roads may be connected with each other before joining a major road.

(*ii*) The geometric layout should be so selected that hazardous movements by drivers are eliminated. This can be achieved by various techniques such as channelizing and staggering.

(*iii*) The design should permit the driver to discern quickly either from the layout or from traffic signs the path he should follow and the actions of merging and diverging. This can be achieved by a good layout, traffic islands, signs and carriageway markings. Good visibility improves safety.

(*iv*) The layout should follow the natural vehicle paths. Smoothness, in contrast to abrupt and sharp corners, should guide minor streams into stopping or slowing down positions.

(*v*) The number of conflict points should be minimised by separating some of the many cutting, merging or diverging movements.

(*vi*) Vehicles that are forced to wait in order to cross a traffic stream should be provided with adequate space at the junctions.

11.13.2. At-grade and grade separated junctions

An intersection where all roadways join or cross at the same level is known as an at-grade intersection. An intersection layout which permits crossing manoeuvres at different levels is known as grade separated intersection.

The choice between an at-grade and grade separated junctions at a particular site depends upon various factors such as traffic, economy, safety, aesthetics, delay etc. Grade separated junctions generally are more expensive initially and are justified in certain situations. These are :

(*i*) On high type facilities such as expressways, freeways and motorways.

(*ii*) Certain at-grade intersections which have reached the maximum capacity and where it is not possible to improve the capacity further by retaining the at-grade crossing.

(*iii*) At certain locations which have a proven record of bad accident history when functioning as at-grade junctions.

(*iv*) At junctions where the traffic volume is heavy and the delays and loss caused justify economically the provision of grade-separation.

(*v*) At certain specific topographical situations where it is logical to provide a grade-separated structure rather than an at-grade intersection, which may involve considerable earthwork or acquisition of land.

11.13.3. Basic forms of at-grade intersections

Fig. 11.24 gives the basic forms of at-grade intersections.

11.13.4. Alignment and vertical profile

In view of the conflict and hazard involved at intersections, the alignment and vertical profile of the intersecting roads should permit the drivers to discern the danger and make the manoeuvres necessary for safety. This requires that the alignment be as straight and gradients as low as practicable.

It is desirable that the intersection roads meet at or nearly at right angles. For this purpose, it may be necessary to modify the alignment of the roads as shown in Fig. 11.25.

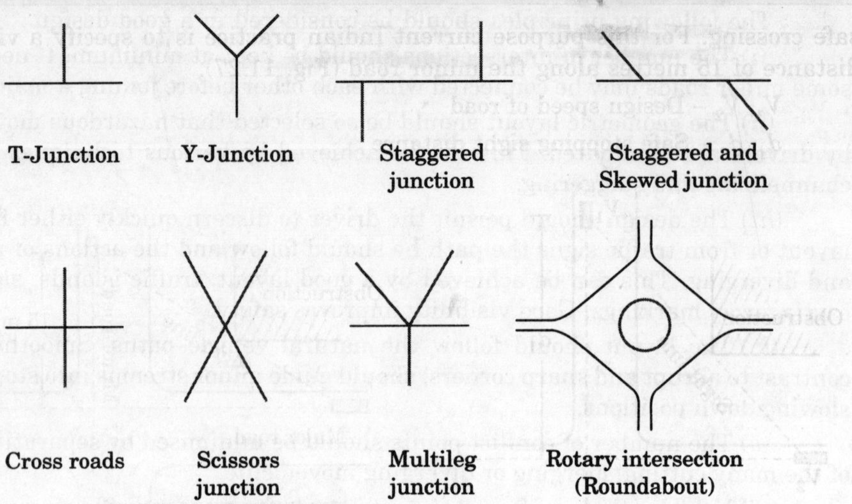

Fig. 11.24. Basic forms of at-grade intersections.

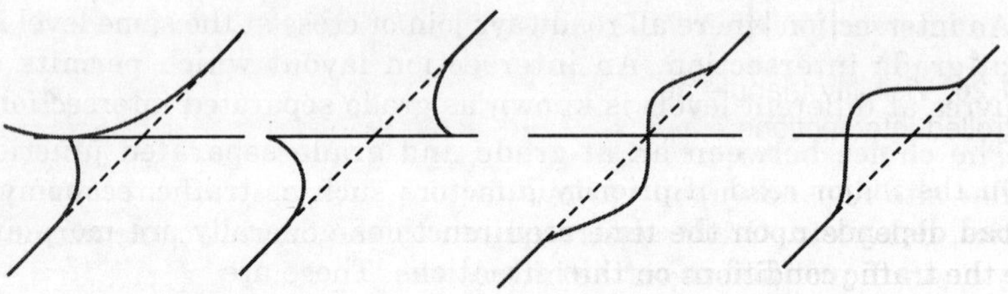

Fig. 11.25. Modification in alignment of intersecting roads to give a nearly right-angle crossing.

11.13.5. Visibility at intersections

11.13.5.1. The safety of traffic can be ensured only if the visibility is full and unimpeded along both roads. Any obstructions should be clear of the minimum visibility triangle for a height of 1.2 metres above the roadway.

11.13.5.2. Rural road intersections

At intersection where the intersecting roads are of more or less equal importance and there is no established priority (generally known as uncontrolled intersections) visibility should be provided on the principle that a driver approaching the intersection on either highway must be able to perceive a hazard and halt the vehicle, if that becomes essential, before reaching the intersection. The stopping distances, d_1 and d_2 appropriate for the design speeds of V_1 and V_2 respectively in the two intersecting highways, may be taken from Table 11.15, which gives the current Indian practice (Ref. 18). Fig. 11.26 explains the requirement of the visibility triangle.

On intersections involving a major road and a minor road, it is customary to control traffic on the minor road by STOP or GIVE WAY sign or road markings. Such intersections are known as *Priority Intersections* (Ref. 6). On priority intersections, the visibility provided should be such that drivers approaching from the minor road are able to see the vehicles on the major road in good time and judge whether the required gap is available in the main road traffic stream for a safe crossing. For this purpose current Indian practice is to specify a visibility distance of 15 metres along the minor road (Fig. 11.27).

V_1, V_2 – Design speed of road
d_1, d_2 – Safe stopping sight distance

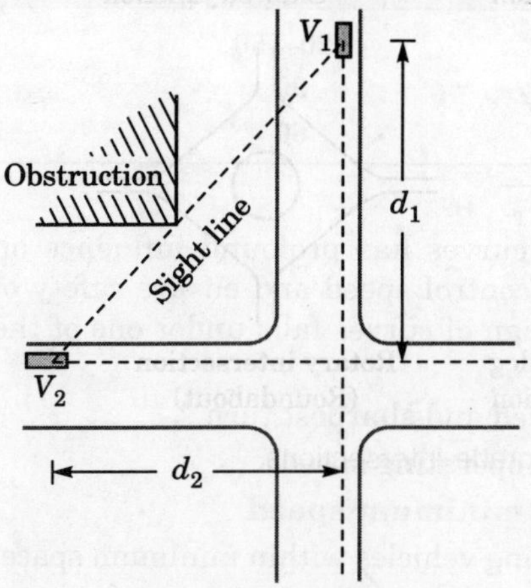

Fig. 11.26. Visibility triangle at uncontrolled intersections.

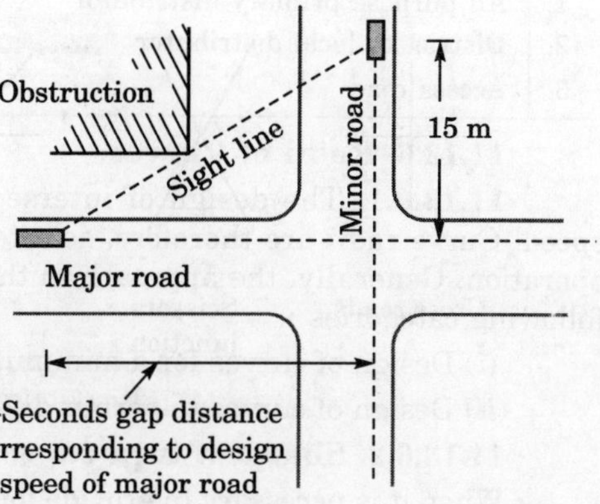

Fig. 11.27. Visibility triangle at priority intersections.

On the major road of priority junctions, the visibility needed along the major road depends upon the time required by the driver on the minor road to perceive the traffic conditions on the intersection, evaluate the gaps in the vehicle stream, take a decision about the actual crossing and finally accelerate the vehicle to complete the manoeuvre. A time of 8 seconds may be considered appropriate for this purpose. On this basis, Table 11.23 gives the visibility distance along the major road (Ref. 48).

Table 11.23. Minimum Visibility Distance along Major Roads at Priority Intersections on Rural Roads

(*Source* : Ref. 48)

Design speed of major road in K.P.H.	*Minimum visibility distance along a major road (metres)*
100	270
80	180
65	145
50	110

11.13.5.3. Urban road intersections

On urban intersections, the visibility distance on a liberal scale as on rural intersections is not possible and is perhaps not necessary due to reduced speed of operation. Current U.K. practice is to limit the visibility distance along the minor road in a priority intersection to about 4.5–9.0 metres. On the major road, the visibility distance is provided as shown in Table 11.24.

Table 11.24. Visibility Distance Along Major Roads on Priority Junction (*Adapted from Reference 6*)

	Type of road	*Visibility distance (in metres) along a major road (metres)*
1.	All purpose primary distributor	120–150
2.	District or local distributor	90
3.	Access road	60

11.13.6. Radii of Curves

11.13.6.1. The design of intersection curves has profound influence on speed. Curve radii are therefore, selected to control speed and ensure safety of operation. Generally, the approach to the design of curves falls under one of the following categories :

(*i*) Design of curves for a minimum speed and sharpest turn.

(*ii*) Design of curves to permit different operating speeds.

11.13.6.2. Sharpest turn curves for minimum speed

When it is necessary to provide for turning vehicles within minimum space, the minimum turning paths of design vehicles apply.

Table 11.25. Minimum Edge of Pavement Designs for Turns at Intersections (*Adapted from Reference 3*)

AASHTO design vehicle	*Angle of Turn (degrees)*	*Radius of Simple Curve (m)*	*3-centred compound curve symmetric*		*3-centred compound curve symmetric*	
			Radii, (m)	*Offset, (m)*	*Radii, (m)*	*Offset, (m)*
(1)	*(2)*	*(3)*	*(4)*	*(5)*	*(6)*	*(7)*
P	30	18	–	–	–	–
SU		30	–	–	–	–
WB–40		45	–	–	–	–
WB–50		60	–	–	–	–
P	45	15	–	–	–	–
SU		23	–	–	–	–
WB–40		36	–	–	–	–
WB–50		52	60–30–60	0.9	–	–
P	60	12	–	–	–	–
SU		18	–	–	–	–
WB–40		27	–	–	–	–
WB–50		–	60–23–60	1.7	60–23–84	0.6–1.8
P	75	10	30–7.5–30	0.6	–	–
SU		17	36–14–36	0.6	–	–
WB–40		26	36–14–36	1.5	36–14–60	0.6–2.0
WB–50		–	45–15–45	1.8	45–15–70	0.6–3.0

AASHTO design vehicle (1)	Angle of Turn (degrees) (2)	Radius of Simple Curve (m) (3)	3-centred compound curve symmetric		3-centred compound curve symmetric	
			Radii, (m) (4)	Offset, (m) (5)	Radii, (m) (6)	Offset, (m) (7)
P	90	9	30–6–30	0.75	–	–
SU		15	36–12–36	0.9	–	–
WB–40		–	36–12–36	1.5	36–12–60	0.6–1.8
WB–50		–	55–18–55	1.8	36–12–60	0.6–3.0
P	105	–	30–6–30	0–75	–	–
SU		–	30–10.5–30	0.9	–	–
WB–40		–	30–10.5–30	1.5	30–10.5–60	0.6–2.5
WB–50		–	55–14–55	2.5	45–12–64	0.6–3.0
P	120		30–6–30	0.6		–
SU		–	30–9–30	0.9		–
WB–40		–	36–9–36	1.8	30–9–55	0.6–2.7
WB–50		–	55–12–45	2.6	47–10.5–67	0.6–3.7
P	135	–	30–6–30	0.45	–	–
SU		–	30–9–30	1.2	–	–
WB–40		–	36–9–36	2.0	30–7.5–55	0.9–4.0
WB–50		–	50–10.5–50	2.7	40–9–57	0.9–4.3
P	150	–	23–5.5–23	0.6	–	–
SU		–	30–9–30	1.2	–	–
WB–40		–	30–9–30	1.8	27–7.5–50	0.9–3.4
WB–50		–	50–10.5–50	2.1	36–9–55	0.9–4.3
P	180	–	15–4.5–55	0.15	–	–
SU		–	30–9–30	0.45	–	–
WB–40		–	30–6–30	2.9	26–6–45	1.8–4.0
WB–50		–	40–7.5–40	2.9	30–7.5–55	1.8–4.0

The turning radii for AASHTO design vehicles have been given earlier in Table 11.5. Considering these, the minimum edge of pavement designs for sharpest turns for different angles of turns for AASHTO vehicles are given in Table 11.25 (Ref. 3).

Compound curves are generally three-centred, as in Fig. 11.28.

The sharpest turns are usually provided for unchannelized intersections and for curbs of urban streets at intersection. In residential streets, a curb radius of 6 metres is regarded as the minimum, though radii of 4.5–7.5 metres are satisfactory. For junctions using commercial vehicles, a radius of 9–15 metres will be necessary.

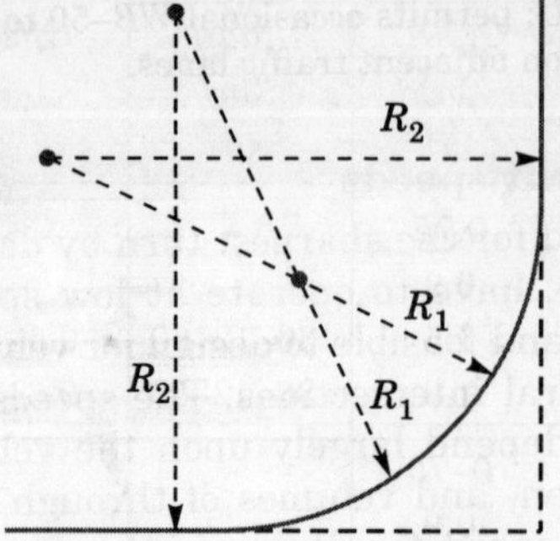

Fig. 11.28. Compound curve for corner design.

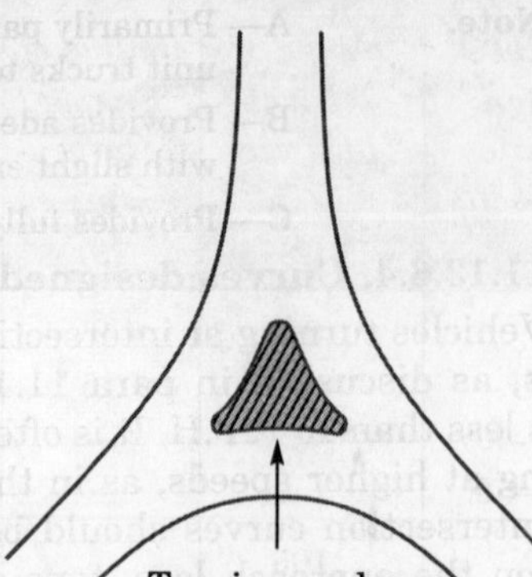

Fig. 11.29. Turning roadway.

11.13.6.3. Minimum design for turning roadways

Where the inner edges of pavement for left turns at intersections are designed to accommodate semi-trailer combinations, or where the design permits passenger vehicles to turn at speeds 25 K.P.H. or, more, the pavement area at the intersections may become excessively large for proper control of traffic. To avoid this a corner island is provided to form a separate turning roadway. A turning roadway is a connecting roadway for traffic turning between two intersection legs, Fig. 11.29.

AASHTO standards for turning roadways are summarised in Table 11.26 (Ref. 3).

Table 11.26. Minimum Design for Turning Roadways

(Adapted from Ref. 3)

Angle of turn (degrees)	*Design Designation*	*3-centred compound curve Radii (m)*	*Offset (m)*	*Width of lane (m)*	*Approximate island size (sq. m)*
75	A	45–22.5–45	1.5	4.25	5.5
	B	45–22.5–45	1.5	5.5	4.5
	C	55–27.5–55	1.1	6.0	4.5
90	A	45–15–45	0.9	4.25	4.5
	B	45–15–45	1.5	5.5	7.5
	C	55–20–55	1.8	6.0	11.5
105	A	36–12–36	0.6	4.5	6.5
	B	30–11–30	1.5	6.5	4.5
	C	55–14–45	2.5	9.0	5.5
120	A	30–9–30	0.75	5.0	11.0
	B	30–9–30	1.5	7.5	8.5
	C	55–12–55	2.6	10.5	20.0
135	A	30–9–30	0.75	5.0	43.0
	B	30–9–30	1.5	8.0	34.0
	C	50–11–50	2.75	11.0	60.0
150	A	30–9–30	0.75	5.0	130.0
	B	30–9–30	1.8	9.0	110.0
	C	50–11–50	2.2	11.5	160.0

Note. A— Primarily passenger vehicles ; permits occasional design single unit trucks to turn with restricted clearances.

B— Provides adequately for *SU*; permits occasional *WB*–50 to turn with slight encroachment on adjacent traffic lanes.

C— Provides fully for *WB*–50.

11.13.6.4. Curves designed for different speeds

Vehicles turning at intersections designed for the sharpest turn by design vehicles, as discussed in para 11.13.6.2 above, have to operate at low speed, perhaps less than 15 K.P.H. It is often desirable and feasible to design for vehicles operating at higher speeds, as in the case of rural intersections. The speeds for which intersection curves should be designed depend largely upon the vehicle speeds on the approach legs, type of intersection, and volumes of through and turning traffic. Generally, a desirable turning speed for design is the average running speed, which may be roughly taken as 0.7 of the design speed. Design at such speeds offers little hindrance to smooth flow of traffic and may be justified on some intersections for certain movements involving little or no conflict with pedestrians or other vehicular traffic.

Table 11-27 gives the AASHTO practice (Ref. 3) for the design of such curves.

Table 11.27. Minimum Radii for Intersection Curves for Various Turning Speeds *(Adapted from Ref. 3)*

Design turning speed V(K.P.H.)	25	30	40	50	60	65
Side friction factor, f	0.32	0.27	0.23	0.20	0.18	0.16
Assumed minimum superelevation, e	0.00	0.02	0.04	0.06	0.08	0.09
Total $e + f$	0.32	0.29	0.27	0.26	0.26	0.25
Calculated min. R (m) $R = \dfrac{V^2}{127\,(e+f)}$	14	28	47	70	96	129
Suggested minimum R (m)	15	27	45	70	95	130

11.13.7. Widths of carriageways in junctions

The carriageway width needs to be increased in curves as already explained in 11.4.9.

The following values of carriageway widths are as per IRC practice (Ref. 48):

Table 11.28. Widths of Carriageways at Intersections *(Adapted from Ref. 48)*

Inner Radius (m)	*Design speed (K.P.H.)*	*Single lane width (m)*	*Two lane width for one or two way traffic (m)*
10.5	18	5.5	11.5
15	23	5.5	10.5
20	27	5.0	10.0
30	32	4.5	9.0
40	37	4.5	9.0
50	41	4.5	8.0
75	50	4.5	8.0

100	57	4.5	8.0
125	62	4.5	8.0
150	64	4.5	8.0
Straight		4.5	7.0

11.13.8. Speed change lanes

Drivers entering into or leaving an intersection have to necessarily change their speed. In the former case, they will reduce the speed to a safe value at which the intersection can be negotiated. In the latter case, they will increase the speed until the desired open highway speed is reached. When such deceleration or acceleration is accomplished on the highway travelled way, traffic is disrupted and hazard may be caused. To avoid this, speed change lanes are normally provided on superior facilities such as expressways and are also frequently provided on other main highways. IRC practice is to provide acceleration lanes where the future traffic on the acceleration lane is expected to be more than 1000 PCUs per day (Ref. 48). These lanes also increase the capacity of the junction.

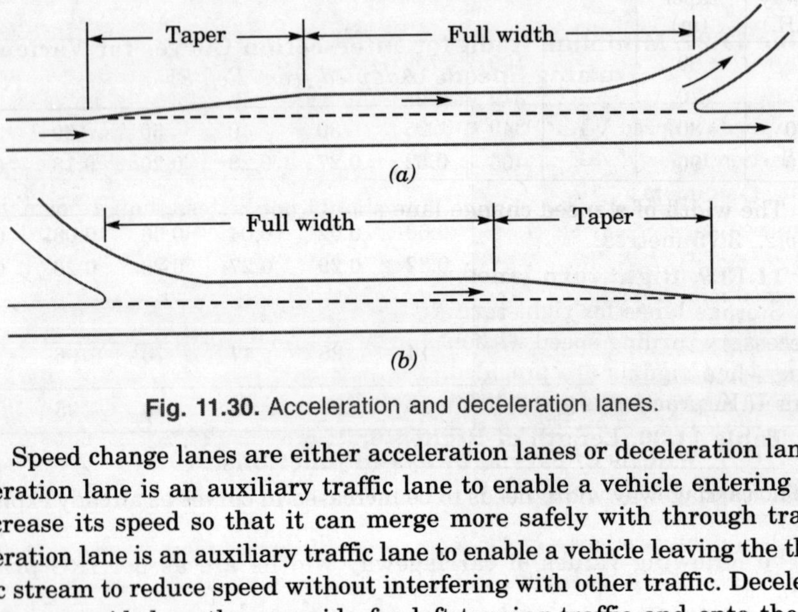

Fig. 11.30. Acceleration and deceleration lanes.

Speed change lanes are either acceleration lanes or deceleration lanes. An acceleration lane is an auxiliary traffic lane to enable a vehicle entering a road to increase its speed so that it can merge more safely with through traffic. A deceleration lane is an auxiliary traffic lane to enable a vehicle leaving the through traffic stream to reduce speed without interfering with other traffic. Deceleration lanes are provided on the nearside for left-turning traffic and onto the right-turning lane where provision is made for right turning traffic.

Acceleration and deceleration lane are usually provided with a taper, vide Fig. 11.30.

The length of a deceleration lane depends upon the manoeuvring speed and the deceleration characteristics. The length of an acceleration lane depends upon the speed at which the drivers merge with through traffic and the acceleration characteristics. Tables 11.29 and 11.30 give the lengths of deceleration and acceleration lanes as per IRC practice (Ref. 48).

Table 11.29. Design Length of Deceleration Lanes (Grade: 2 Per cent or Less) (*Adapted from Ref. 48*)

Design speed of turning roadways curve, K.P.H.		Stop condition	25	30	40	50	60	65	75	80
Minimum curve radius (m)		–	15	27	45	70	95	130	170	210
Design speed of highway (K.P.H.)	Length of taper (m)	Total length of deceleration lane, including taper (m)								
65	60	100	90	85	75	60	–	–	–	–
80	70	130	120	115	110	100	90	–	–	–
100	80	150	150	145	140	130	120	100	90	–
110	90	165	165	160	150	145	140	115	100	–
Design speed of highway (K.P.H.)	Length of taper (m)	Total length of acceleration, lane, including taper (m)								
65	60	–	100	75	70	–	–	–	–	–
80	70	–	210	190	180	150	120	–	–	–
100	80	–	340	320	300	270	240	180	120	–
110	90	–	465	450	420	400	370	300	250	175

The width of a speed change lane should not be less than a normal traffic lane, *viz*., 3.75 metres.

11.13.9. Right turn lanes

Storage lanes for right-turning traffic enable the vehicle to slow down to the necessary turning speed without impeding the flow of through traffic and for waiting when signals are provided. The length of such right turn lanes as per current U.K. practice is given below (Ref. 6).

Table 11.30. Length of Right-turn Lanes (*Adapted from Ref. 6*)

Design speed of major road (K.P.H.)	80	65	50
Length of right-turn lane including 30 metres taper (m)	120	100	80

11.13.10. Channelization

11.13.10.1. The direction of traffic flow at intersections to definite paths, by means of traffic markings, islands or other means is known as channelization. A channelised intersection is one in which traffic is directed into definite paths by islands and markings. An unchannelised intersection, on the other hand, is one without islands for directing traffic into definite paths. An unchannelised intersection is the most simple type but is the most dangerous and inefficient. All important junctions should, therefore, provide for channelizations.

11.13.10.2. Channelisation serves the following purposes :

(*i*) **Separation of conflicts.** To diminish the number of possible vehicle conflicts, to reduce the possible area of conflicts in the carriageway and to present drivers with only one decision at a time.

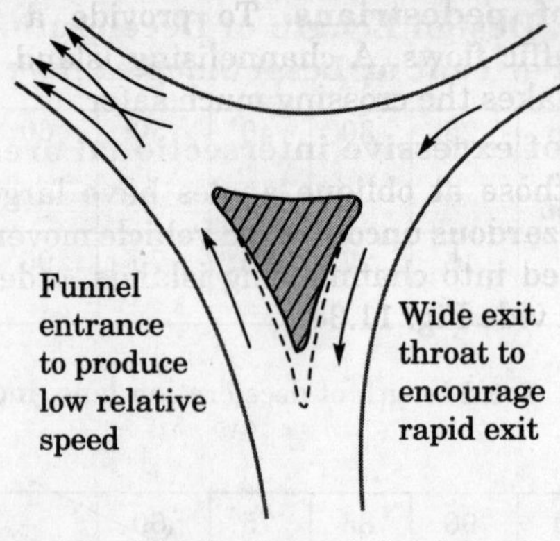

Fig. 11.31. Channelising island.

(*ii*) **Control of angle of conflict.** Small angles of crossing cause severe accidents if they occur. Severity is reduced if the angle of conflict is controlled.

(*iii*) **Control of speed.** To reduce the speed of traffic entering the intersection and increase the speed of traffic leaving the intersection, bending or funnelling by suitable channelization techniques is resorted to, vide Figs. 11.31 and 11.32.

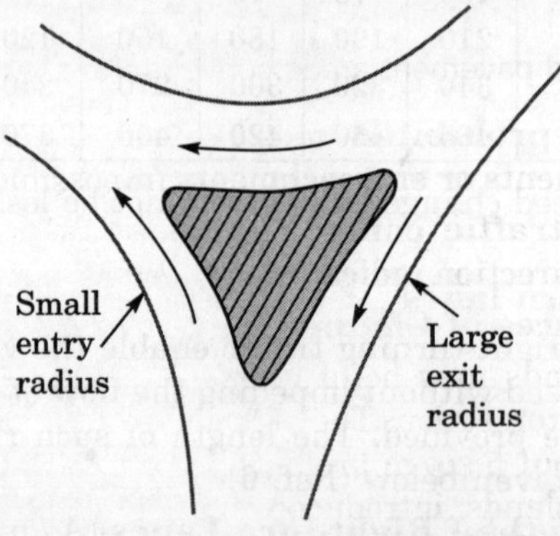

Fig. 11.32. Control of radius of entry and exit for control of speed.

(*iv*) **Protection of traffic for vehicles leaving or crossing the main traffic stream.** This is exemplified by the separate storage pockets for right turning traffic at an intersection and the adjacent island, vide Fig. 11.33.

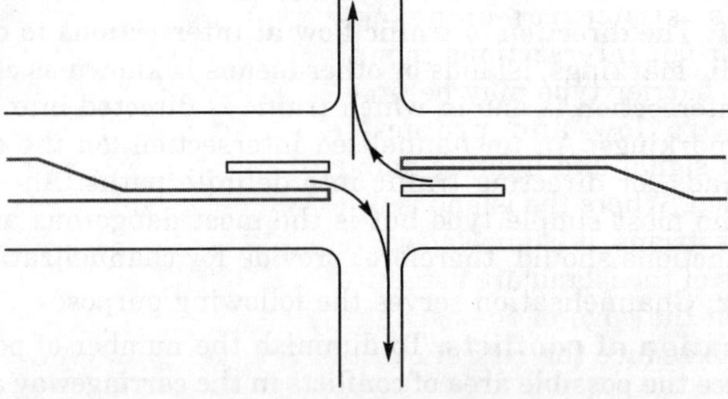

Fig. 11.33. Channelising to protect right turning traffic.

(*v*) **Protection of pedestrians.** To provide a haven or refuge for pedestrians between traffic flows. A channelising island such as in Fig. 11.33 serves as a refuge and makes the crossing much safer.

(*vi*) **Elimination of excessive intersectional areas.** Intersections with large corner radii and those at oblique angles have large paved areas, which permit and encourage hazardous uncontrolled vehicle movements. If these unused paved areas are converted into channelising islands, orderly movement results and hazards are reduced vide Fig. 11.34.

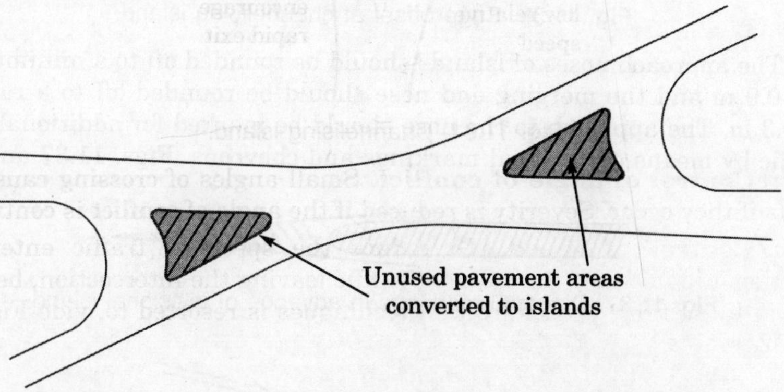

Fig. 11.34. Unused pavement areas converted to channelising islands.

(*vii*) **Blockage of prohibited movements.** To support regulations by making improper movements or encroachments impossible or inconvenient.

(*viii*) **Location of traffic control devices.** To provide space for traffic control devices such as direction indicators, reflectors, signs, etc.

11.13.10.3. Features of Channelising Islands

Channelising islands may be of many different shapes and sizes guided by the geometry of the intersection. They should be sufficiently large to command attention. An area of about 5 sq. m preferably 7 sq. m., should be the minimum. Elongated or divisional islands, introduced on undivided highways to alert drivers and regulate traffic through the intersections, should be at least 1.2 m wide and 3.5 to 6 m long. An example of a divisional island, which is in continuation of a triangular channelising island at a rotary intersection is given in Fig. 11.35.

In rural intersections, island curbs should always be of a mountable type, except where there is a need for a barrier, as at preferable for most situations, structures or pedestrian crossings. For urban intersections mountable curbs are but a barrier type may be provided, where speeds are low and protection of pedestrians with signal and lighting standards and signs is needed. Where the island is used for protection of pedestrians, it is known as a refuge island. The edges of the island are usually offset 0.3 to 0.6 m from the edge of the carriageway, vide Fig. 11.36, to reduce the risk of overriding the islands.

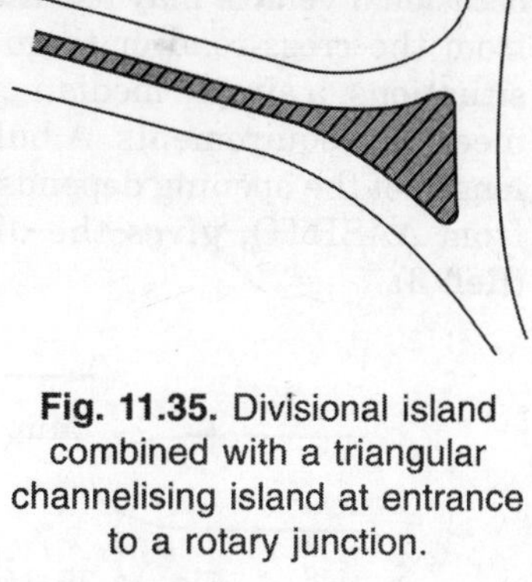

Fig. 11.35. Divisional island combined with a triangular channelising island at entrance to a rotary junction.

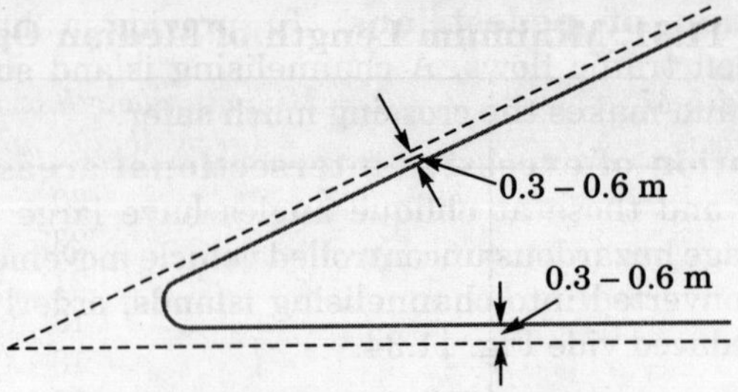

Fig. 11.36. Edge offset at channelising island.

The approach noses of islands should be rounded off to a minimum radius of 0.6–0.9 m and the merging end nose should be rounded off to a radius of at least 0.3 m. The approach to the nose should be marked for additional guidance of traffic by means of diagonal markings and chevrons. Figs. 11.37 and 11.38.

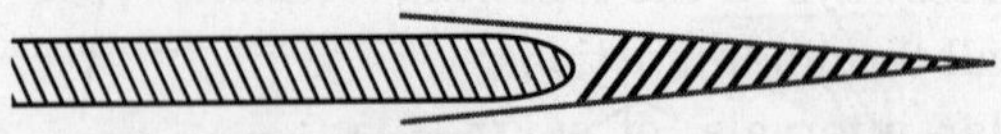

Fig. 11.37. Diagonal markings in advance of divisional island.

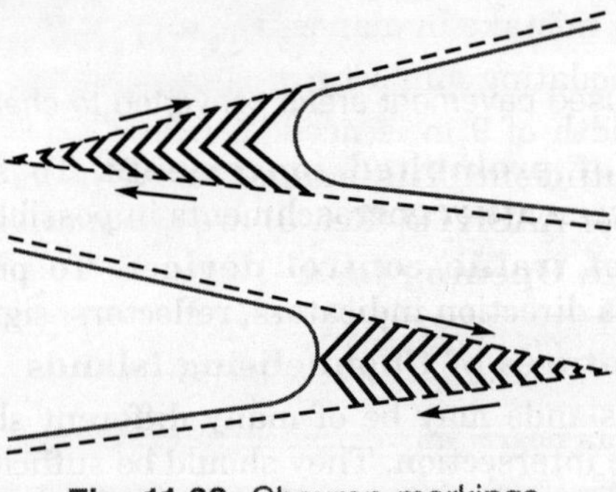

Fig. 11.38. Chevron markings.

11.13.11. Median openings

11.13.11.1. Minimum design

Where the traffic pattern at the intersection is such that nearly all the traffic travels on the divided highway and the volume is below capacity, the occasional vehicle may use the median opening either to come to the through road from the cross-road or to go to the cross-road from the through road. In such situations, a simple median opening designed for the minimum turning path will meet the requirements. A bullet nosed median end is preferable, Fig. 11.39. The length of the opening depends upon the width of the median. Table 11.31, adapted from AASHTO, gives the dimensions of the opening for SU design vehicles (Ref. 3).

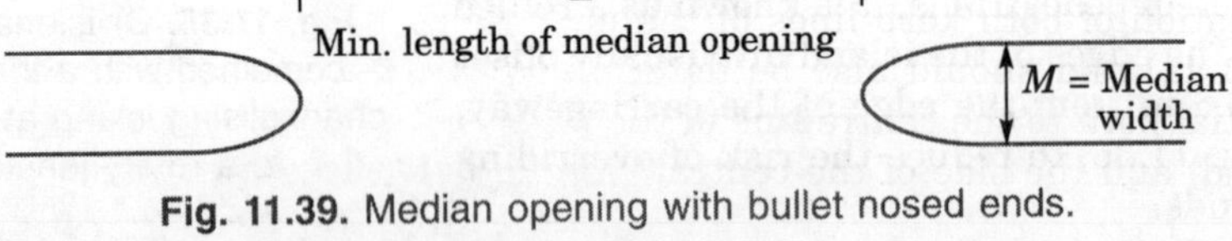

Fig. 11.39. Median opening with bullet nosed ends.

Table 11.31. Minimum Length of Median Opening

Median width M (in metres)	*L = Minimum length of median opening (in m) (bullet nose ends)*
1.2	30
1.8	23
2.5	21
3.0	19
3.6	18
4.3	16
5.0	14
6.0	13
7.2 and above	12 minimum

11.13.11.2. Design for U-turns

Median openings designed only to accommodate vehicles making U-turns are needed on expressways in the following locations :

(*i*) Near to all terminal junctions.

(*ii*) At regular intervals, of say 1.5 km, between junctions to facilitate maintenance operation, policing, repair of stalled vehicles etc.

(*iii*) Beyond important intersections to enable drivers unfamiliar with the locality, to correct any mistake in manoeuvring.

(*iv*) For accommodating minor turning movements.

A minimum width of 9 m is needed, except for passenger vehicles alone when only 6 m is sufficient. The accommodations for U-turns provided by minimum design as per AASHTO (Ref. 3) are summarised in Table 11.32.

Table 11.32. Median Opening for U-Turns Design (*Adapted from Ref. 3*)

Median width (m)	*Type of manoeuvre accommodated on 4 lane divided highway*	*AASHTO design vehicle provided for*
18	Permits nearly all vehicles to turn from inner to inner lane	All vehicles
12	Permits all passenger vehicles to U-turn from inner to inner lane ; permits some trucks to turn from outer to outer lane, with large trucks partially encroaching on shoulder.	P and SU
10	Permits passenger vehicles to U-turn from inner to inner lane, permits all trucks to U-turn with encroachment on shoulders.	P and SU
6	Permits passenger vehicles to U-turn from inner to outer lane, U-turn impossible for trucks without backing.	P only

11.13.11.3. Gap in median at junctions

To ensure that large vehicles can turn right without difficulty to or from a major road, the gap in the median should normally extend 3 m beyond the continuation of both kerb lines of the minor road to the edge of the major road (Fig. 11.41) and should also be determined by 12–15 m radius control circles tangential both to the centre line of the minor road (or to the sides of any refuge or island) and the side of the central reserve away from the minor road.

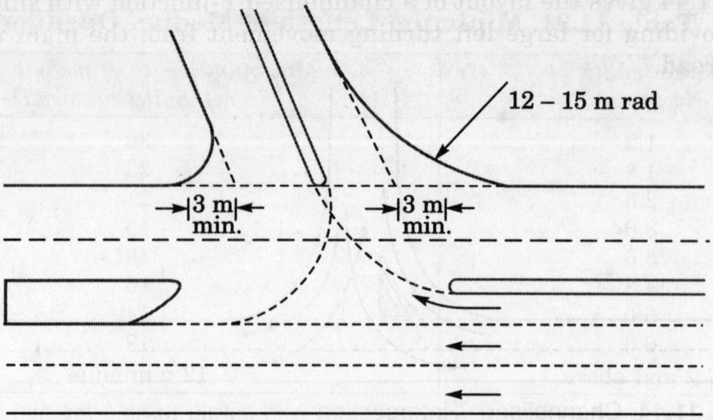

Fig. 11.40. Gap in median at junction.

11.13.12. T-Intersection

Fig. 11.41 shows an unchannelised *T*-junction for very light traffic and for minor and local roads. Obviously it is hazardous to use this in locations where speeds are high and turning movements are large.

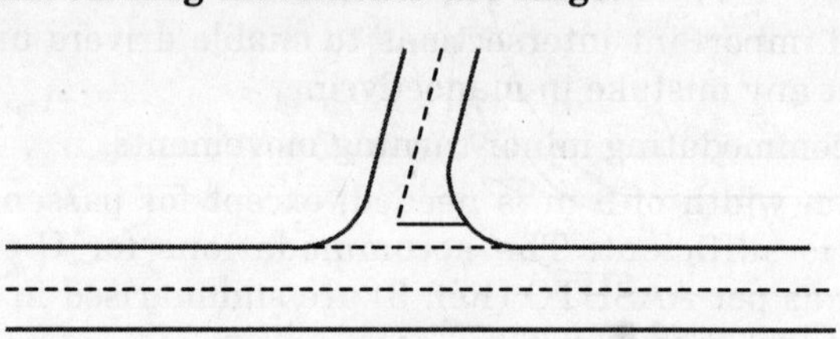

Fig. 11.41. Unchannelised *T*-junction.

Fig. 11.42 is an improvement since the main road is flared to provide an additional lane to facilitate left turns.

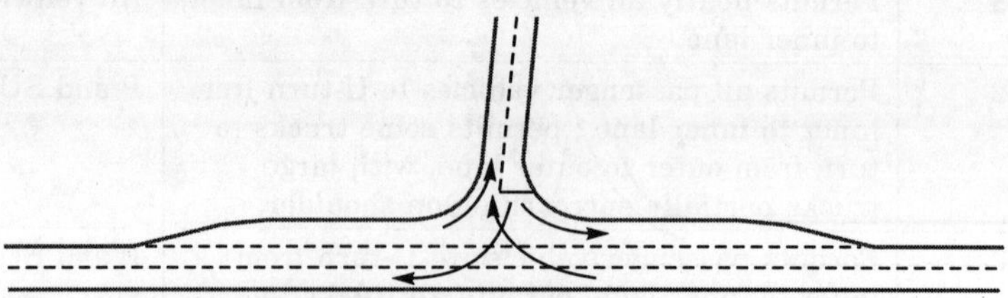

Fig. 11.42. Flared Unchannelised *T*-intersection to facilitate left turning movements.

If there is heavy right turning movement, the flaring can be done opposite the intercepted road, as in Fig. 11.43.

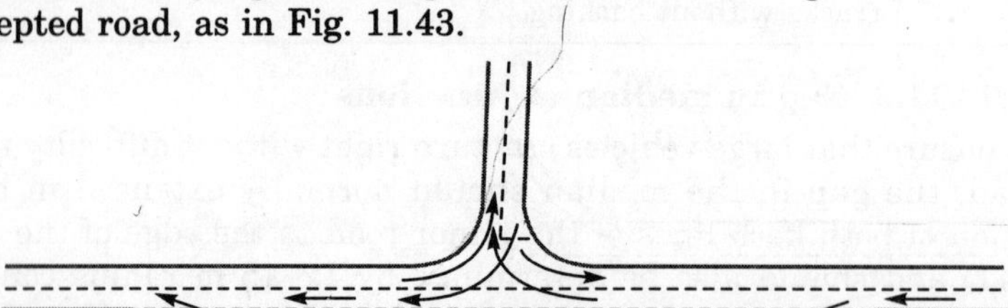

Fig. 11.43. Flared unchannelised intersection to facilitate right turning movements.

Fig. 11.44 gives the layout of a channelised *T*-junction with single turning roadway, providing for large left turning movement from the main road to the intercepted road.

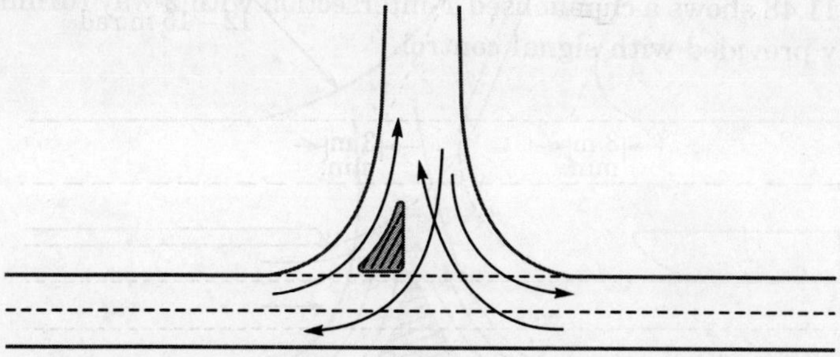

Fig. 11.44. Channelised *T*-intersection with single turning roadways.

Fig. 11.45 gives the layout of a channelised *T*-intersection with a pair of turning roadways.

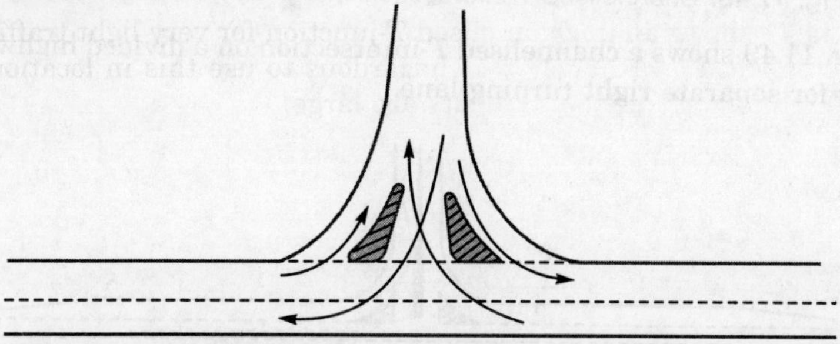

Fig. 11.45. Channelised *T*-intersection with a pair of turning roadways.

Fig. 11.46 shows a channelised *T*-intersection with a divisional island.

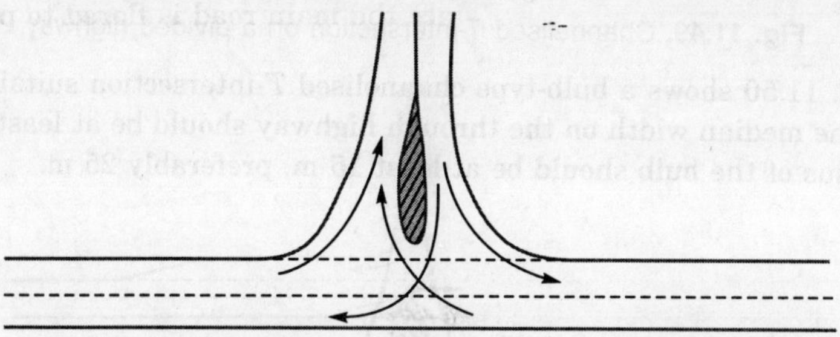

Fig. 11.4. Channelised *T*-intersection with a divisional island.

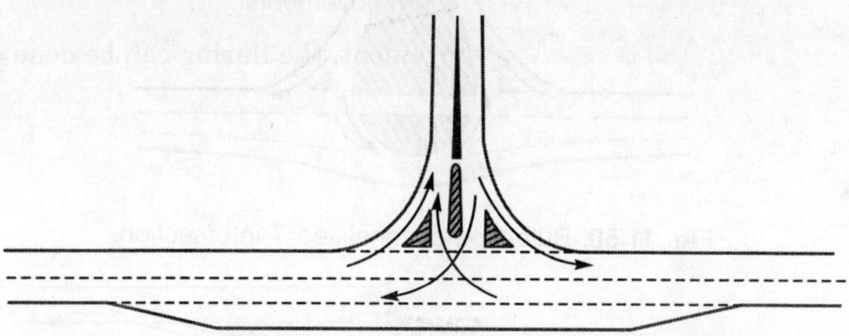

Fig. 11.47. Channelised *T*-intersection with a divisional island and two turning roadways.

Fig. 11.47 shows a channelised *T*-intersection with a divisional island and two turning roadways.

Fig. 11.48 shows a channelised *T*-intersection with 2-way turning roadways. It is usually provided with signal control.

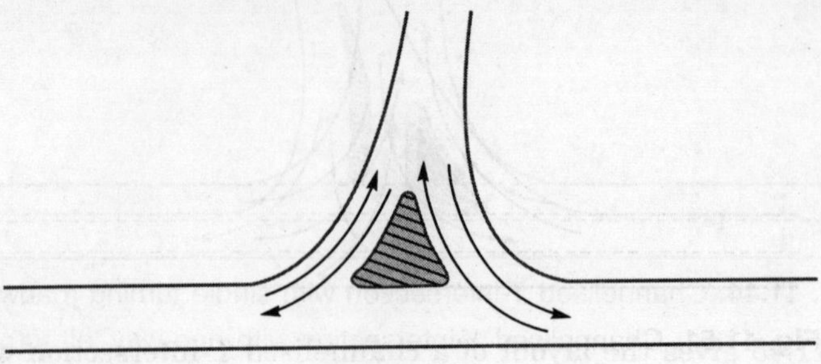

Fig. 11.48. Channelised *T*-intersection with 2-way turning roadways.

Fig. 11.49 shows a channelised *T*-intersection on a divided highway facility providing for separate right turning lane.

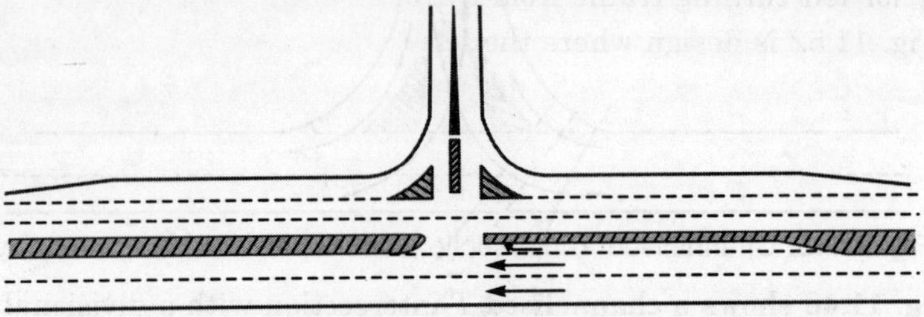

Fig. 11.49. Channelised *T*-intersection on a divided highway.

Fig. 11.50 shows a bulb-type channelised *T*-intersection suitable for high speeds. The median width on the through highway should be at least 12 m. The inner radius of the bulb should be at least 15 m, preferably 25 m.

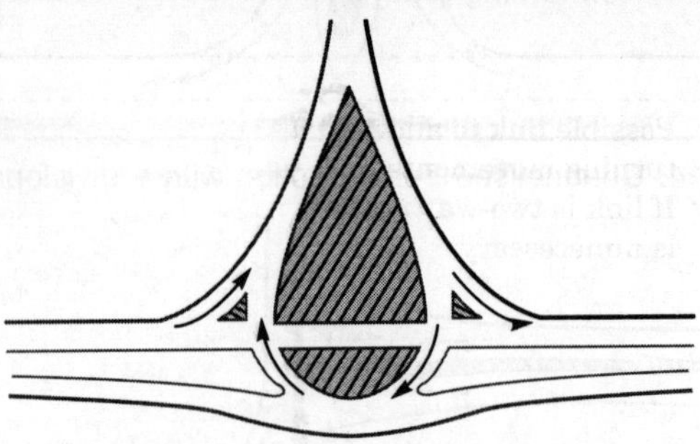

Fig. 11.50. Bulb-type channelised *T*-intersection.

11.13.13. Y-intersection

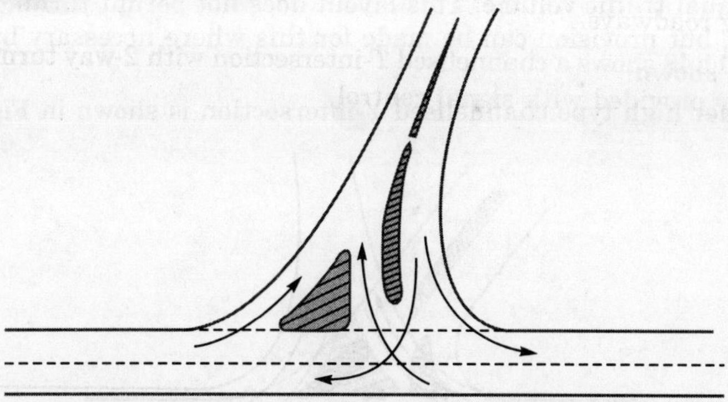

Fig. 11.51. Channelised *Y*-intersection with one-way left turning roadway from the main road.

For most of *Y*-intersections, the design can suitably follow the same features as for *T*-intersections, except where the intersection angle is very acute.

Fig. 11.51 indicates a channelised *Y*-intersection with a one-way turning roadway for left turning traffic from the main road.

Fig. 11.52 is design where the left turning roadway is 2-way.

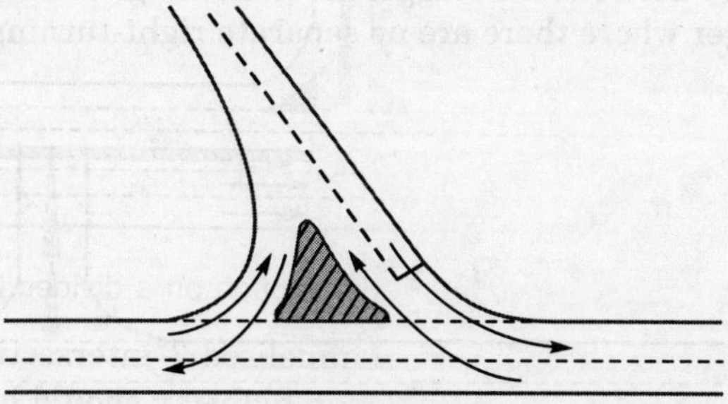

Fig. 11.52. Channelised *Y*-intersection with 2-way left turning roadway from the main road.

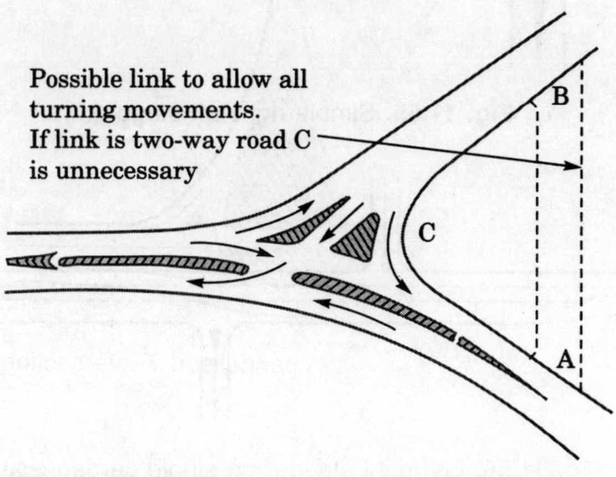

Fig. 11.53. High type channelised *Y*-intersection.

Fig. 11.53 is a high type channelised Y-intersection, where the branches of Y carry equal traffic volume. This layout does not permit turning movements from A to B, but provision can be made for this where necessary by means of a link road as shown.

Another high type channelised Y-intersection is shown in Fig. 11.54.

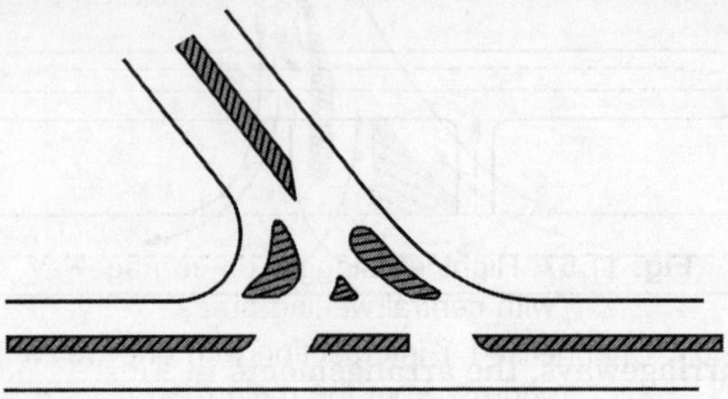

Fig. 11.54. High type channelised intersection on the *Y*-type.

11.13.14. Staggered intersection

Many cross-roads can be conveniently converted into a properly designed staggered intersection. This will increase the safety. A simple right/left stagger suitable for an unimportant junction is shown in Fig. 11.55. A minimum distance of 35 m is necessary between the staggered roads. A right/left stagger is preferable to left/ right stagger where there are no separate right-turning lanes in the main road.

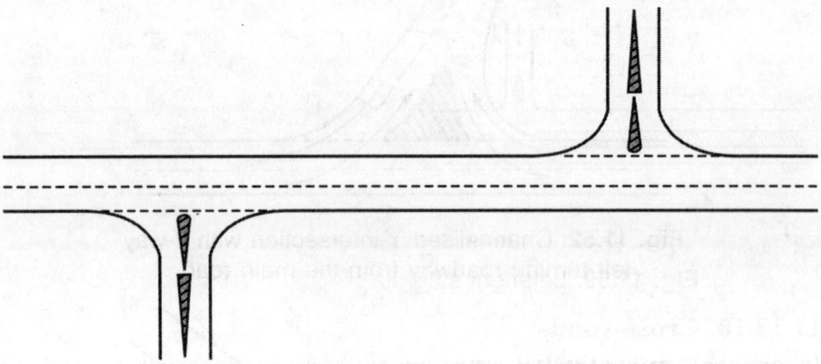

Fig. 11.55. Simple right/left stagger.

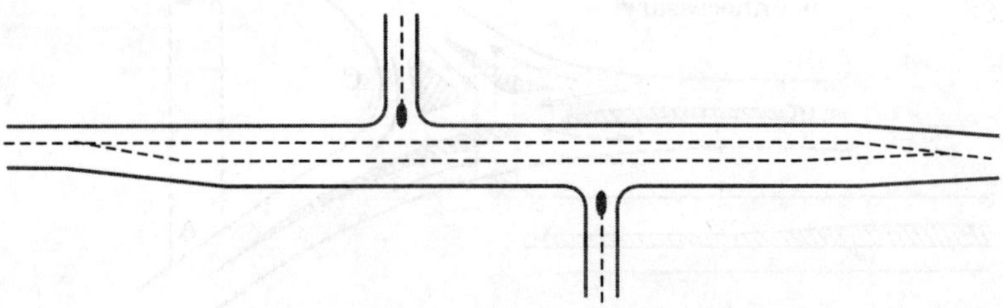

Fig. 11.56. Left/right stagger on single carriageway with central waiting areas.

Figs. 11.56 and 11.57 indicate suitable layout for staggered junctions on single carriageways, where a central lane is reserved for the use of right turning traffic.

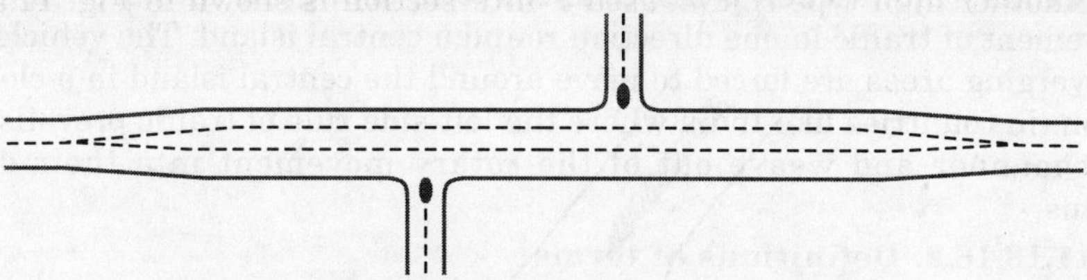

Fig. 11.57. Right/left stagger on carriageway with central waiting areas.

For dual carriageways, the arrangements in Figs. 11.58 and 11.59 can be adopted.

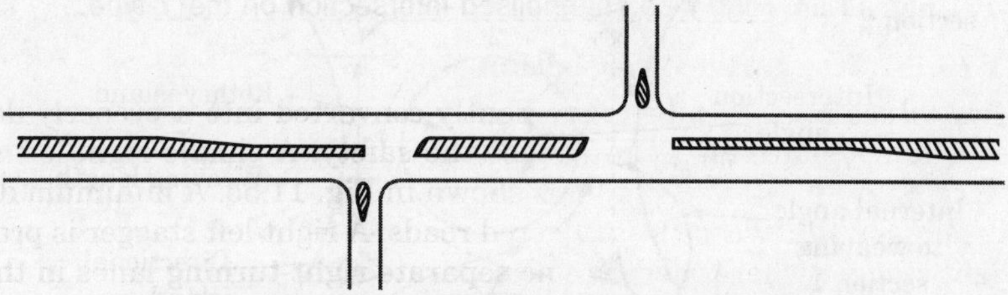

Fig. 11.58. Right/left stagger on dual carriageways.

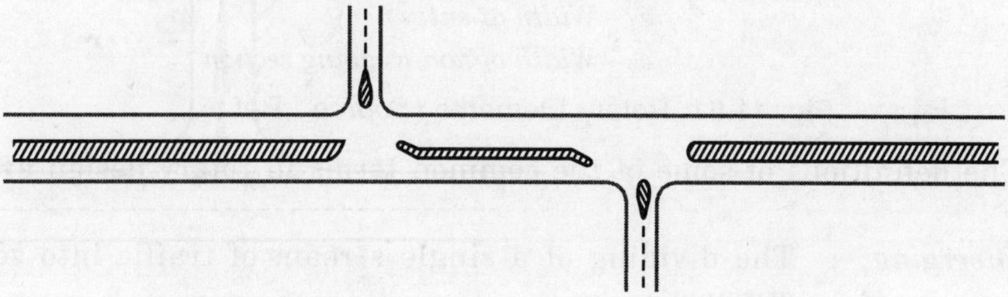

Fig. 11.59. Left/right stagger on dual carriageways.

11.13.15. Cross-roads

In general, uncontrolled cross-roads are unsatisfactory due to the high accident rate. If possible, they should be converted to staggered intersections. A layout suitable for a divided carriageway crossing on undivided road is given in Fig. 11.60.

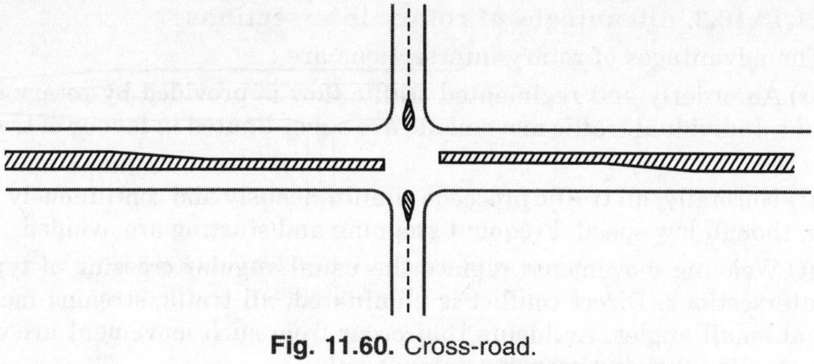

Fig. 11.60. Cross-road.

11.13.16. Rotary intersection (round about)

11.13.16.1. General

A rotary intersection is a specialised form of *at grate* intersection laid out for movement of traffic in one direction round a central island. The vehicles from the converging areas are forced to move around the central island in a clockwise direction (in countries like India where the left-side rule of traffic prevails) in an orderly manner and weave out of the rotary movement into their desired directions.

11.13.16.2. Definitions of terms

A sketch of the rotary alongwith some of its elements is given in Fig. 11.61.

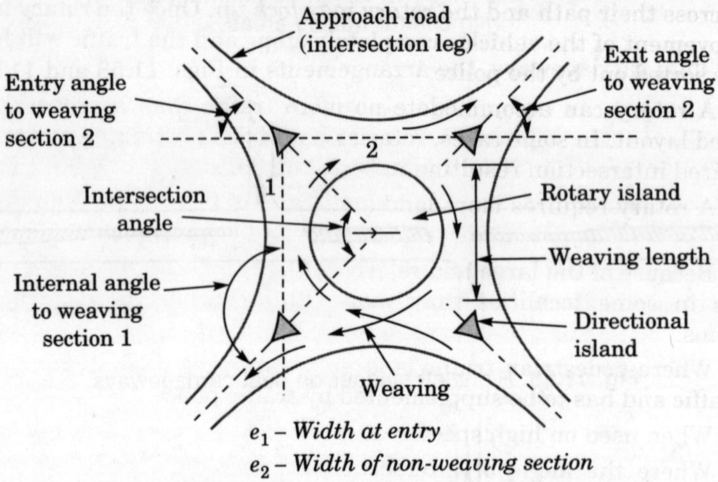

Fig. 11.61. Rotary Elements (*Source* : Ref. 5).

The definitions of some of the common terms in rotary design are given below:

Diverging	:	The dividing of a single stream of traffic into separate streams.
Merging	:	The converging of separate streams of traffic into a single stream.
Weaving	:	The combined movement of merging and diverging of traffic streams moving in the same general direction.
Weaving length	:	The length of a section of a rotary in which weaving occurs.

11.13.16.3. Advantages of rotary intersections

The advantages of rotary intersections are :

(*i*) An orderly and regimented traffic flow is provided by rotary one-way movement. Individual traffic movements are subordinated in favour of traffic as a whole.

(*ii*) Normally, all traffic proceeds simultaneously and continuously at fairly uniform, though low speed. Frequent stopping and starting are avoided.

(*iii*) Weaving movements replace the usual angular crossing of typical at-grade intersections. Direct conflict is eliminated, all traffic streams merging of leaving at small angles. Accidents that occur from such movement are of minor nature, usually involving property damage only.

(*iv*) All turns can be made with ease, although little extra travel distance is required for all movements except left turns.

(*v*) A rotary is especially suited for intersections with five or more intersection legs, and/or where there are heavy right-turning movements.

(*vi*) For moderate traffic, rotaries are self-governing and need no control by police or traffic signals.

11.13.16.4. Disadvantages of rotary intersections

(*i*) As the flow increases and reaches the capacity, *weaving* generally gives way to a *stop and go* motion as vehicles force their way into the rotary, being followed by vehicle waiting in the queue behind them. Under such conditions, vehicles, once having got into the rotary, may not be able to get out of it, because of vehicles across their path and the rotary may *lock up*. Once the rotary has *locked up*, the movement of the vehicles completely stops and the traffic will have to be ultimately sorted out by the police.

(*ii*) A rotary can accommodate no more traffic than a properly designed channelized layout. In some cases, rotaries have been eliminated and replaced by a channelized intersection resulting in better operations.

(*iii*) A rotary requires more land and may not be feasible in many built-up locations.

(*iv*) Because of the large and relatively flat land area required, topographic conditions in some localities may make it impracticable to develop rotary intersections.

(*v*) Where pedestrian traffic is large, a rotary by itself is not sufficient to control traffic and has to be supplemented by traffic police.

(*vi*) When used on high speed roads, rotaries require extremely large size.

(*vii*) Where the angle of intersection between two roads is too acute, it becomes difficult to provide adequate weaving length.

(*viii*) When provided at close intervals they make travel troublesome.

(*ix*) Traffic turning right has to travel a little extra distance.

(*x*) A rotary requires many warning and directional signs for safety. The central island and entrances and exits must be well lighted at night. These tend to make it costly.

(*xi*) The rotary is not readily adaptable to stage development. Attempts at stage development generally result in some over-design when viewed from immediate traffic needs.

11.13.16.5. Guidelines for selecting a rotary type of intersection

Considering some of the above advantages and disadvantages, the following general guidelines may be kept in view when selecting a rotary type of intersection :

(*i*) Rotaries are most apt where volumes entering from different intersection legs are approximately equal.

(*ii*) A total volume of 3000 vehicles per hour entering from all the intersection legs appears to be the maximum practical capacity of high type rotaries.

(*iii*) A rotary design is most appropriate when the proportion of turning traffic is very high. It is considered (Ref. 6) that at a four-leg intersection a rotary may need less land than traffic signals if right-turning traffic exceeds about 30 per cent of all approaching traffic.

(*iv*) A rotary is a good choice when there are more than four approaches to the junction.

(*v*) When the approach widths are restricted and it is impossible to provide separate lanes for through and turning traffic, a rotary may be desirable.

(*vi*) Rotaries are not generally warranted for intersections carrying very light traffic. Normally, the lowest traffic volume for which a rotary design should be considered is about 500 vehicles per hour (Ref. 26).

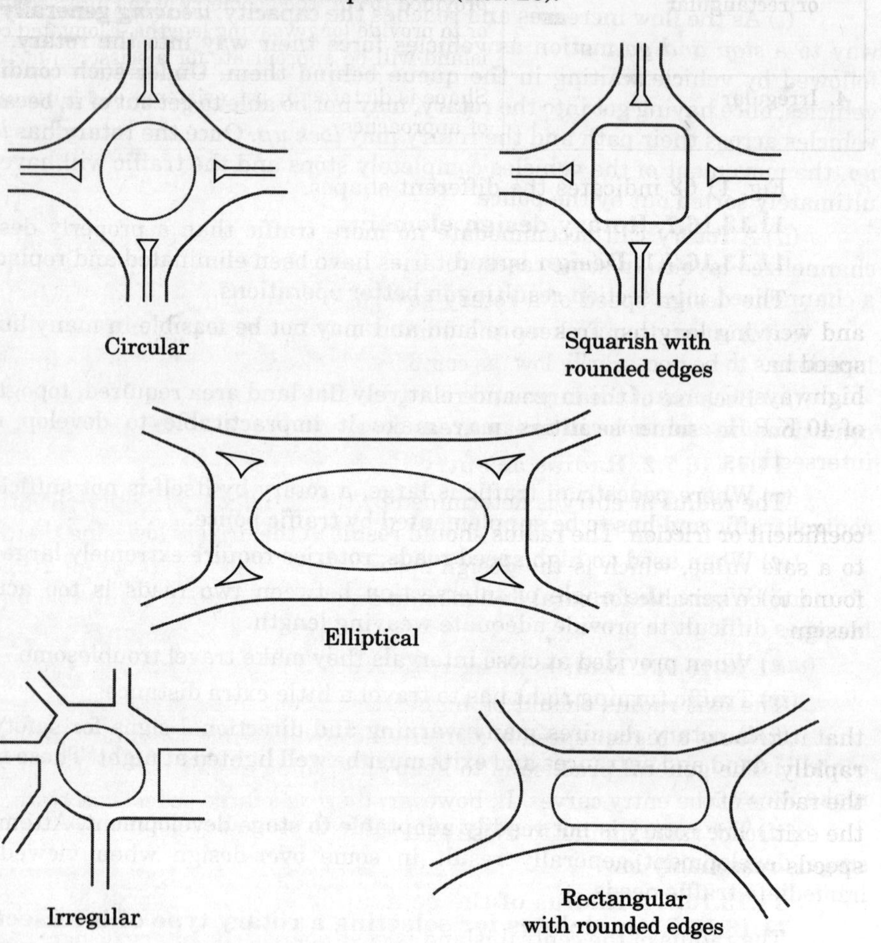

Fig. 11.62. Shape of Rotary Island.

(*vii*) Rotaries should be a good choice for moderately busy intersections in urban and suburban areas, where otherwise the alternative may be to go in for a complicated channelised layout or traffic signals.

11.13.16.6. Shape of rotary island

The shape and layout of the rotary island depends upon the layout of the intersecting roads and the traffic flow pattern. The following brief remarks about the different types in common use are intended to be general guidelines.

Type	*Remarks*
1. Circular	Suited where roads of equal importance carrying approximately equal volumes intersect at nearly equal angles.
2. Squarish with rounded edges	Suited for predominatly straight ahead flows.
3. Elliptical, elongated, oval or rectangular	Intended to favour through traffic. They may be provided to suit the geometry of the intersecting legs or to provide long weaving lengths. Elongated central island will be appropriate for a scissors junction.
4. Irregular	Shape is dictated by the existence of a large number of approaches.

Fig. 11.62 indicates the different shapes.

11.13.16.7. Rotary design elements

11.13.16.7.1. Design speed

The design speed of a rotary governs the various elements such as radii and weaving lengths. To keep the dimensions within practical limits the design speed has to be necessarily low, as compared to the design speed of the intersecting highways. Current Indian practice is to design rotaries in rural areas for a speed of 40 K.P.H. and those in urban areas to a speed of 30 K.P.H.

11.13.16.7.2. Radius at entry

The radius at entry is determined by the design speed, super-elevation and coefficient or friction. The radius should result in the drivers lowering their speeds to a safe value, which is the design speed of the rotary. A range of 20–35 m is found to be suitable for rural designs and a range of 15–20 m is suitable for urban design.

11.13.16.7.3. Radius at exit

The exit radius should be higher than the radius of the rotary island so that it favours a higher speed by drivers. This will enable the rotary to be cleared rapidly. The general practice is to keep the radius of exit curves 1½ to 2 times the radius of the entry curves. If, however, there is a large pedestrian traffic across the exit road, radii similar to those at entrances may be provided to keep the exit speeds reasonably low.

11.13.16.7.4. Radius of the central island

The radius of the central island is governed by the rotary design speed and theoretically it should be equal to the radius at entry. In practice, however, (Ref. 26) the radius of the central island may be kept slightly larger than that of the curve at entry, this being an attempt to give a slight preference to the traffic already on the rotary and to slow the approaching traffic. A value of 1.33 times the radius of entry curve is probably adequate for this purpose.

11.13.16.7.5. Weaving lengths

The weaving length determines the ease with which the traffic can merge and diverge. The weaving length is decided on the basis of factors such as the width of the weaving section, the average width of entry, total traffic and the proportion of weaving traffic in it. As a general rule, it is desirable to prevent direct traffic cuts and this can be achieved by making the ratio of the weaving length to the weaving width large enough. A ratio of 4 : 1 is regarded as a minimum (Ref. 5). Current Indian practice (Ref. 26) as regards weaving length is summarised in Table 11.33.

Table 11.33. Minimum Length of Weaving Section

Design speed (K.P.H.)	*Minimum weaving length (m)*
40	45
30	30

It is also necessary to put a limit to the maximum weaving length to discourage speeding. Twice the values given above can serve as the upper limits.

11.13.16.7.6. Width of carriageway at entry and exit

The carriageway width of the intersection legs is governed by the design year traffic entering and leaving the intersection. A minimum width of carriageway of 5 m, with provision for extra widening due to curvature, may be provided for the entrance and exit. Table 11.34 summarises current Indian practice (Ref. 26).

Table 11.34. Width of Carriageway at Entrance and Exit

(*Source* : Ref. 26)

Carriageway width of approach road	*Radius at entry (m)*	*Width of carriageway at entry and exit (m)*
7 m (2 lanes)		6.5
10.5 m (3 lanes)	25–35	7.0
14 m (4 lanes)		8.0
21 m (6 lanes)		13.0
7 m (2 lanes)		7.0
10.5 m (3 lanes)		7.5
14 m (4 lanes)	15–25	10.0
21 m (6 lanes)		15.0

11.13.16.7.7. Width of rotary carriageway

Fig. 11.63 indicates the design elements.

The width of the non-weaving section should be equal to be widest single entry into the rotary, and should generally be less than the width of the weaving section. The width of the weaving section (w) should be one traffic lane (3.5 m) wider than the mean entry width thereto.

Thus is : $w = \dfrac{e_1 + e_2}{2} + 3.5$, with reference to Fig. 11.63.

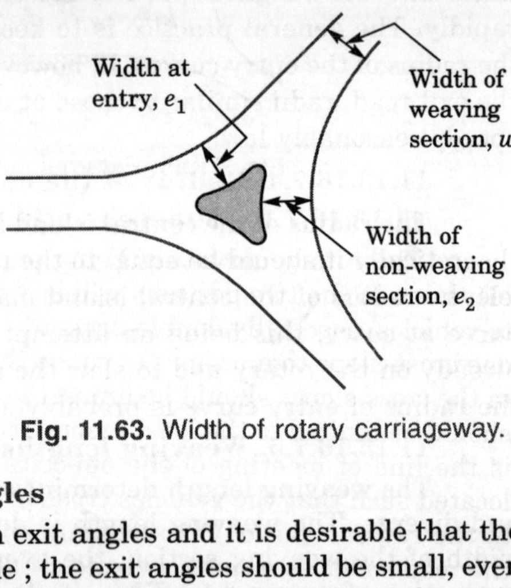

Fig. 11.63. Width of rotary carriageway.

11.13.16.7.8. Entry and exit angles

Entry angles should be larger than exit angles and it is desirable that the entry angles should be about 60° if possible ; the exit angles should be small, even tangential. A design with an entry angle of 60° and exit angle of 30° is shown in Fig. 11.64.

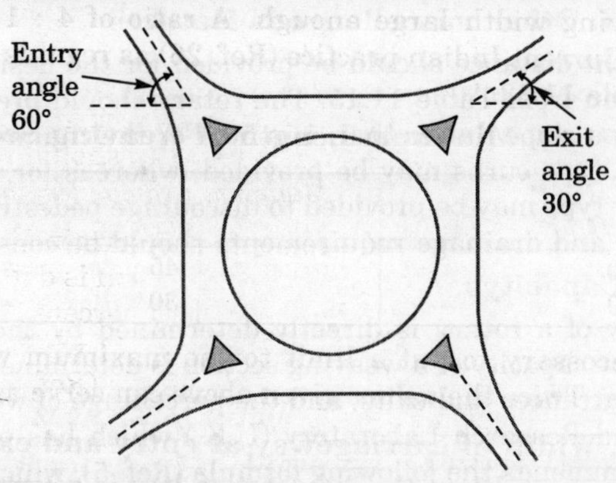

Fig. 11.64. An ideal layout with 60° entry angle and 30° exit angle.

11.13.16.7.9. External curb line

The external curb line of the weaving sections should not be normally re-entrant, but consist of a straight or large radius curve of the same sense as the entry and exit curves. Such an arrangement eliminates waste of area, which is unlikely to be used by traffic, vide Fig. 11.65.

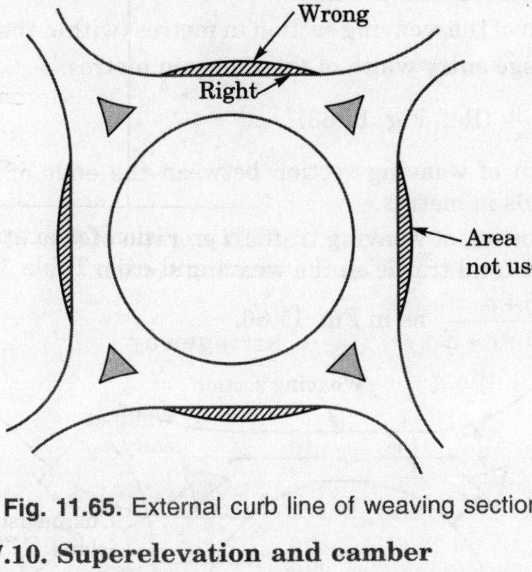

Fig. 11.65. External curb line of weaving sections.

11.13.16.7.10. Superelevation and camber

The rotary curvature being opposite to that of entry and exit, the superelevation in the two portions will be opposite to each other. This causes difficulty to vehicles, especially double-decker buses and top heavy trucks, in changing from one cross-slope to another. To cause the least discomfort, the algebraic difference in the cross-slopes should be limited to about 0.07. The individual superelevations should be limited to the least amount consistent with speed. The crown-line, which is the line of meeting of the opposite cross-slopes should as far as possible, be located such that the vehicles cross it while travelling along the common tangent to the reverse curve. Channelising islands should be situated on the peak with the road surfaces sloping away to all the sides. The cross-slope at entrance should be carried around on the outer edge of the rotary to the adjacent exit making slight adjustments as may be needed from the point of view of curvature and speed.

11.13.16.7.11. Other miscellaneous features

Stopping sight distance should be provided for the design speed adopted. Values may be taken from Table 11.15. The rotary should preferably be located on a ground having a slope flatter than 1 in 50. For the channelising and central islands, mountable type curbs may be provided, whereas for the outer edges of the rotary a barrier type may be provided to discourage pedestrians from crossing over. Land-scaping and drainage requirements should be considered.

11.13.16.8. Capacity

The capacity of a rotary is directly determined by the capacity of each weaving section. The capacity of a weaving section is determined by the geometric layout, including entrances and exits, and the percentage of weaving traffic. The Transport and Road Research Laboratory (U.K.) which has pioneered research on this aspect, recommends the following formula (Ref. 5), which is a modification of the well-known Wardrop formula (Ref. 27):

$$Q_p = \frac{280w\left(1+\frac{e}{w}\right)\left(1+\frac{p}{3}\right)}{1+\frac{w}{l}} \qquad \text{...(11.40)}$$

where, Q_p = practical capacity of the weaving section of the rotary in passenger car units, PCUs per hour.

w = width of the weaving section in metres (within the range of 6–18 m)

e = average entry width of the rotary in metres.

$= \frac{e_1 + e_2}{2}$ (Ref. Fig. 11.63)

l = length of weaving section between the ends of the channelising islands in metres.

p = proportion of weaving traffic, *i.e.*, ratio of sum of crossing streams to the total traffic on the weaving section.

$= \frac{b+c}{a+b+c+d}$ as in Fig. 11.66.

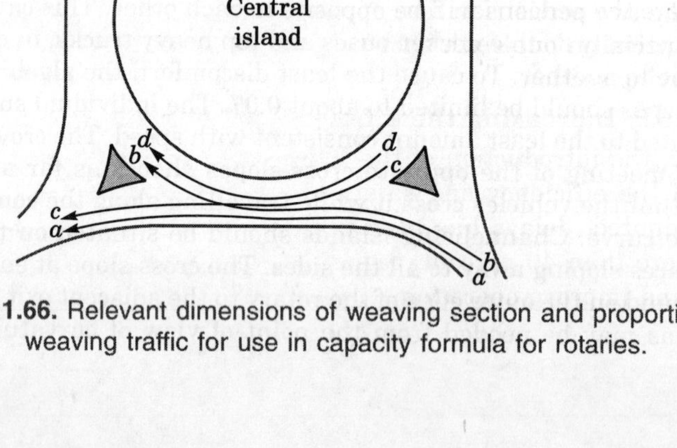

Fig. 11.66. Relevant dimensions of weaving section and proportion of weaving traffic for use in capacity formula for rotaries.

In the above formula, the equivalency factors recommended by TRRL (Ref. 5, 6) and IRC (Ref. 26) are given in Table 11.35.

Table 11.35. Equivalency Factors for Use in Rotary Design
(Adapted from Ref. 5, 6, 26)

	Vehicle type	*PCU*
1.	Cars and light commercial vehicles, including 3-wheelers	1.0
2.	Buses and medium heavy commercial vehicles	2.8
3.	Motorcycles and scooters	0.75
4.	Pedal cycles	0.5
5.	Animal drawn vehicles	4 to 6

The above formula is valid under the following conditions :

(*i*) There are no standing vehicles on the approaches to the rotary.

(*ii*) The site of the rotary is level and approach gradients do not exceed 1 in 25.

(*iii*) $\frac{e}{w}$ should be between 0.4 and 1.0.

(*iv*) $\frac{w}{l}$ should be between 0.12 and 0.40.

(*v*) p should be between 0.4 and 1.0.

(*vi*) l should be between 18 and 90 m.

In the above formula, certain adjustments need to be made to take into account the geometric layout. These are :

(*i*) Where the entry angle is between 0° and 15°, deduct 5 percent from the capacity of the weaving section.

(*ii*) Where the entry angle is between 15° and 30°, deduct 2½ percent from the capacity of the section.

(*iii*) Where the exit angle is between 60° and 75°, deduct 5 percent from the capacity of the weaving section.

(*iv*) Where the exit angle is greater than 75°, deduct 5 percent from the capacity of the weaving section.

(*v*) Where the internal angle is greater than 95°, deduct 5 percent from the capacity of the weaving section.

(*vi*) Where the pedestrian flow across an exit is more than 300 per hour, an arbitrary reduction on one-sixth should be made in the practical capacity of the preceding weaving section.

11.13.16.9. Increasing the efficiency of rotaries

In U.K., the introduction of the "Give way to traffic from the right" rule has increased the efficiency and safety of rotaries (Ref. 6). Locking has been virtually eliminated, delays has been reduced and the accident rate has been reduced by more than 40 percent. Signals on the approaches can also increase the capacity and improve operation.

11.13.16.10. An example of the design of a rotary intersection

Problem 11.1. *Traffic flow in an urban sections at the intersection of two highways in the design year are given below :*

Approach	*Left turning*			*Straight ahead*			*Right turning*		
	Cars	*Commer-cial*	*Scoo-ters*	*Cars*	*Commer-cial*	*Scoo-ters*	*Cars*	*Commer-cial*	*Scoo-ters*
N	*200*	*50*	*100*	*250*	*100*	*150*	*150*	*50*	*80*
E	*180*	*60*	*80*	*220*	*50*	*120*	*200*	*40*	*120*
S	*250*	*80*	*100*	*150*	*50*	*90*	*160*	*70*	*90*
W	*220*	*50*	*120*	*180*	*60*	*100*	*250*	*60*	*100*

The highways at present intersect at right angles and have a carriageway width of 15 m. Design a rotary intersection making suitable assumptions.

Solution. The rotary is located in an urban section and hence a design speed of 30 K.P.H. is appropriate. Since the intersection legs carry almost equal traffic, a round-shaped central island will be adopted. The entrance and exist angles will be 45° each. A radius of 20 m at entry, 40 m at exit and 80 m for the central island will be adopted. These values conform to current Indian standards discussed earlier.

The traffic in terms of PCUs from each leg is depicted in Fig. 11.67.

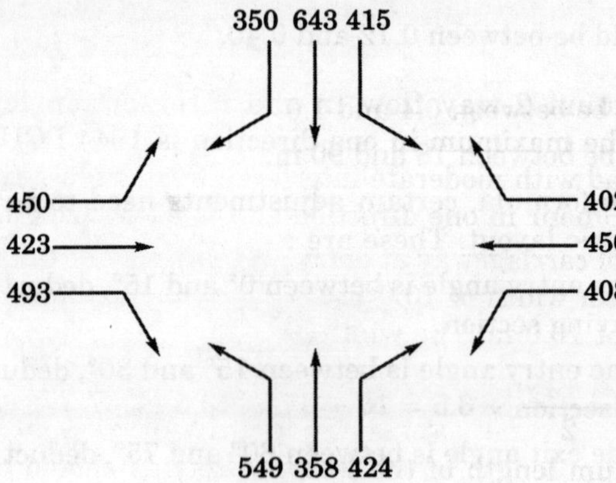

Fig. 11.67

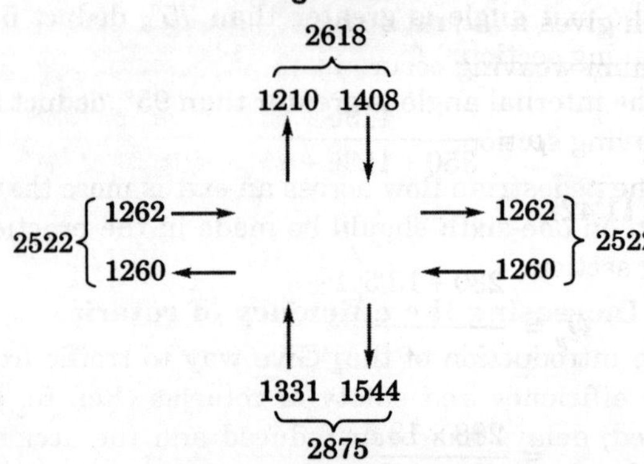

Fig. 11.68

The traffic in terms of PCUs on each leg is depicted in Fig. 11.68.

The traffic is assigned to the network in Fig. 11.69.

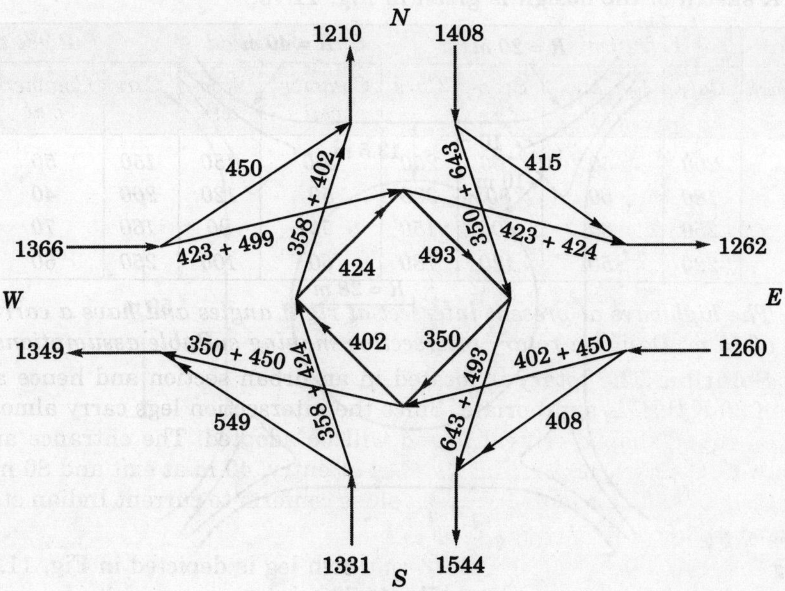

Fig. 11.69

The maximum 2-way flow in the intersection leg (South) is 2875 PCUs/hour, and the maximum in one direction is 1544 PCUs/hour. From Table 11.13, a 4 lane road with moderate interference from cross-traffic has a capacity of 1400–1800 PCU/hour in one direction and is good enough.

The width of carriageway at entry and exit may be taken from Table 11.31. For a 4-lane road, a width of 10.0 m is in order. The width of the non-weaving section is also kept 10.0 m. The width of the weaving section is :

$$\frac{e_1 + e_2}{2} + 3.5 = 10 + 3.5 = 13.5 \text{ m.}$$

The minimum length of the weaving section should be 30 m from Table 11.33, but this fails the requirement that the ratio l/w should be at least 4. So adopt 55 m, which gives a l/w ratio slightly greater than 4.

The maximum weaving occurs in the E-S section.

$$p = \frac{1136 + 852}{350 + 1136 + 852 + 408} = \frac{1988}{2746} = 0.72$$

From Eq. (11.42),

$$Q_p = \frac{280 + 13.5\left(1 + \frac{10}{13.5}\right)\left(1 - \frac{0.72}{3}\right)}{1 + \frac{13.5}{55}}$$

$$= \frac{288 \times 13.5 \times 1.74 \times 0.76}{1.245} = 4000 \text{ PCU/hour.}$$

This is very much higher than the traffic flow of 2746 PCU/ hour, and the design is acceptable.

A sketch of the design is given in Fig. 11.70.

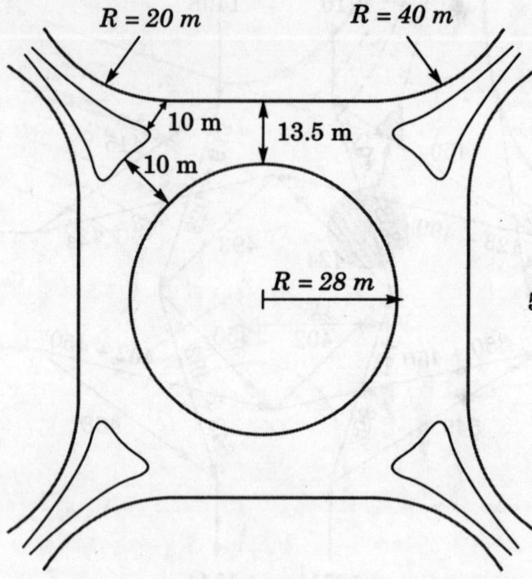

Fig. 11.70

11.13.17. Mini Round abouts

11.13.17.1. A serious drawback of the conventional (large diameter island) roundabout is that very often the traffic locks itself. A locked roundabout causes large traffic delays and unlocking can only be achieved by means of police help. The conventional round-about is thus not truely *self-regulatory* at all times.

11.13.17.2. In the United Kingdom, which is one of the countries where roundabouts still retain their popularity, a major change in the approach to roundabout design was the off-shoot of the introduction of the *priority* ***from-the-right*** rule in 1966. This rule was intended to get over the drawback of locked roundabouts, but other advantages soon began to offer themselves. The absence of *weaving* under the new rule made possible a drastic reduction in the size of the central island and an availability of greater carriageway space with the existing layout. The width of approach at the entry to the junction could be increased considerably thus rendering multi-vehicle flow from the Give-Way line. The ***mini-roundabout*** has made its appearance and was soon to become very popular in many cities in the U.K. Experiments were initiated at the T.R.R.L. (Ref. 28) and these showed that the use of small diameter islands could bring in higher capacities than large diameter roundabouts. This has formed the basis of current U.K. practice (Ref. 29).

The basic principles of design of a mini-round about illustrated in Fig. 11.71 are :

(*i*) Provision of small diameter island of a diameter of about one-third that of a hypothetical circle inscribed within the outer carriageway boundaries, but normally not less than 8 m. For new layouts, space for a larger diameter (15 to 25 m) may be reserved if such a provision becomes necessary in future.

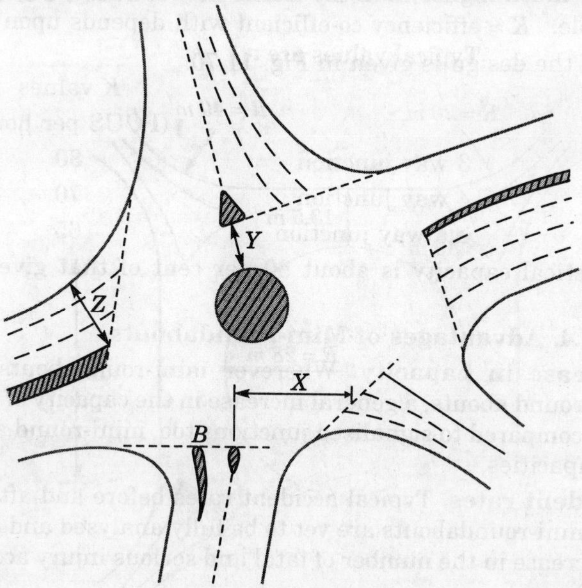

Fig. 11.71. Typical mini-roundabout as per U.K. practice.

(*ii*) An increase in the number of lanes at the Give-Way line.

(*iii*) A minimum stopping distance of 26 m between the Give-Way line and the point of conflict with a vehicle from the left. This is the dimension *X* in Fig. 11.71.

(*iv*) A width between traffic islands and the roundabout (shown as dimension *Y* in Fig. 11.71) which is not less than the total lane width at the entry preceding it, *i.e.*, dimension shown as *Z* in Fig. 11.71.

(*v*) An entry taper (about 1 in 6) that is about twice sharp as the exit taper (about 1 in 12).

(*vi*) A deflection island (shown as *B* in Fig. 11.71) intended to ensure that straight-through movements do not occur.

(*vii*) To maintain junction capacity, a single lane approach should be tapered to give at least three lanes at the junction and a 2-lane approach tapered to at least four lanes at the junction. At the exits, the merge should normally be from four lanes to two lanes and two lanes to one lane.

11.13.17.3. Blackmore (Ref. 26) gives the capacity of a mini-round about as:

$$q = K(\Sigma w + a^{1/2}) \qquad \text{...(11.43)}$$

where, q = total entry volume in PCU/hour.

Σw = sum of the basic road width (not half widths) used by traffic in both direction to and from the intersection, in metres.

a = area of the junction widening, *i.e*,. the area within the intersection outline (including islands) which lies outside the area of the basic cross-roads, in sq. m.

K = efficiency co-efficient with depends upon site conditions. Typical values are :

	K values (PCUS per hour)
3 way junction	**80**
4 way junction	**70**
≥ 5 way junction	**65**

The practical capacity is about 80 per cent of that given by the above equation.

11.13.17.4. Advantages of Mini-roundabouts

(*i*) Increase in capacity. Wherever mini-roundabouts have replaced large diameter round abouts, a general increase in the capacity of the junction has been noted. As compared to signalised junctions too, mini-round abouts can bring about higher capacities.

(*ii*) Accident rates. Typical accident rates before and after conversion of junctions into mini-roundabouts are yet to be fully analysed and evaluated, but a trend in the decrease in the number of fatal and serious injury accidents has been noticed (Ref. 30).

11.13.17.5. Limitation for use

As indicated earlier, mini-round abouts are the result of the introduction of the *priority-from-the-right* rule. They can be introduced only in situations where such rules are in force and are strictly followed. Non-observance of this rule can only lead to chaos and serious accidents.

The mini-roundabout requires careful considerations on the design speed and ways of reducing speeds at junctions to this value. A very high standard of pavement markings and road signing, a great care in provision of guide islands, alignment of the kerbs and location of the central island are necessary for the satisfactory functioning of this type of junctions.

Adequate visibility to the approaching drivers is of prime importance. This alone will enable him to give way to the traffic on his right and to check on the traffic on his left for giving way to him.

11.14. Grade Separated Intersections

11.14.1. Criteria for provision

The general site conditions favouring a grade-separated inter-section have been enumerated earlier in Section 11.13. Current Indian practice (Ref. 25) requires grade separation across streets and highways to be provided as per the following criteria :

Urban streets. Grade separation to be provided if the estimated traffic volumes within the next 5 years are in excess of the capacity of the intersection; when traffic projections show that volumes within the next 20 years will exceed the capacity of an at-grade intersection, the need for a grade separated facility should be kept in view for future construction.

Rural highways. Grade separation should be provided at intersections of divided rural highways if the ADT (fast vehicles only) on the cross-road within the next 5 years exceeds 5000. Where this figure will be reached within the next 20 years, the need for such facilities should be kept in view.

Grade-separation is of great importance across railway lines. Current Indian practice (Ref. 25) for rural highways requires grade separation to be provided across existing railway lines if the product of ADT (fast vehicles only) and the number of trains per day exceeds 50,000 within the next 5 years. For new constructions such as byepasses, grade separations should be provided when this figure is greater than 25,000.

11.14.2. Types of grade-separated intersections

Basically two types are met with :

(*i*) Grade-separated intersections without interchange.

(*ii*) Grade-separated intersections with interchange.

Interchange is a system whereby facility is provided for movement of traffic between two or more roadways at different levels in the grade separated junction. A structure without interchange is an over-bridge or underpass or flyover, whereby the traffic at different levels moves separately without a provision for an interchange between them.

The different forms of a grade-separated junction can be considered under the number of legs the intersection serves. Thus the interchanges can be classified as three-leg, four-leg and multileg, and these in turn can be subdivided into various types as below :

A. Three-leg interchange.

1. *T* interchange
2. *Y* interchange
3. A partial rotary interchange.

B. Four-leg interchange.

1. Diamond interchange
2. Half clover leaf interchange
3. Clover leaf interchange
4. Rotary interchange
5. Directional interchange

C. Multi-leg interchange.

1. Rotary interchange.

These types will be discussed in detail.

11.14.3. Three-leg interchange

If one of the intersection legs meets a highway at approximately a right angle but does not cross it, and if the junction is provided with interchanging facility, the junction is called at *T* interchange. If the intersection angle is acute or obtuse, the junction is called a *Y* interchange. A widely used form of a *T* or *V* interchange is a ***trumpet***, shown in Fig. 11.72 (*a*) and (*b*). (*a*) is the usual form but (*b*) can also be adopted under particular site conditions, but this has the drawback that vehicles leaving the major road have to turn through a small

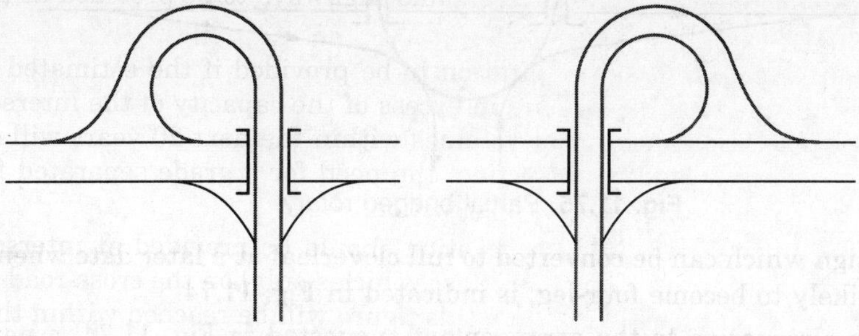

(*a*) (*b*)

Fig. 11.72. Trumpet interchange.

radius. These designs can be adopted for junctions of major streets with expressway.

Designs shown in Fig. 11.73 (*a*), (*b*) and (*c*) are *Y*-shaped and these permit limited movements.

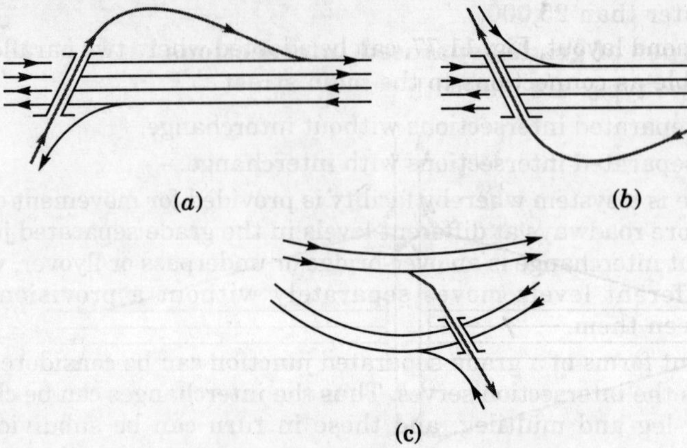

Fig. 11.73. *Y*-shaped interchanges.

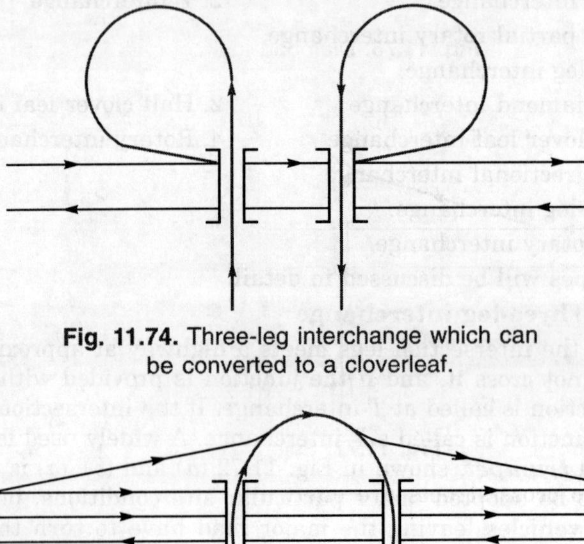

Fig. 11.74. Three-leg interchange which can be converted to a cloverleaf.

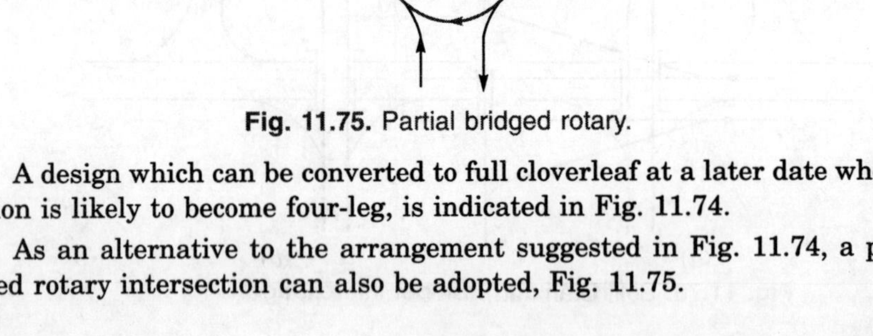

Fig. 11.75. Partial bridged rotary.

A design which can be converted to full cloverleaf at a later date when the junction is likely to become four-leg, is indicated in Fig. 11.74.

As an alternative to the arrangement suggested in Fig. 11.74, a partial bridged rotary intersection can also be adopted, Fig. 11.75.

11.14.4. Diamond interchange

A popular form of interchange in urban locations involving a major-minor crossing is a diamond interchange. It can be designed for a relatively narrow right-of-way of the major road. A simple form of diamond interchange is given in Fig. 11.76.

A split-diamond layout, Fig. 11.77, can be adopted where two parallel cross-streets are available as connections to the main street.

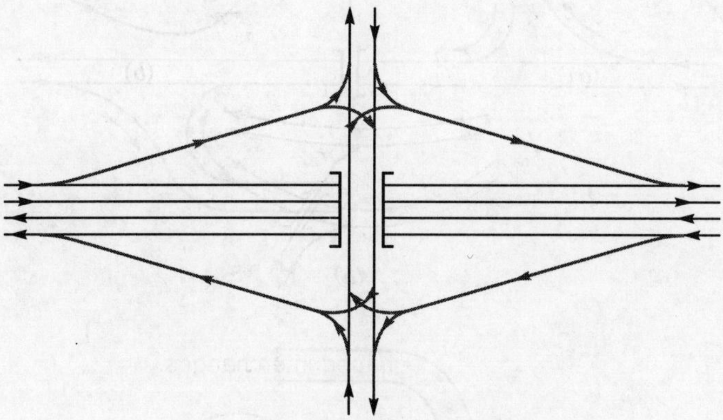

Fig. 11.76. Diamond Interchange.

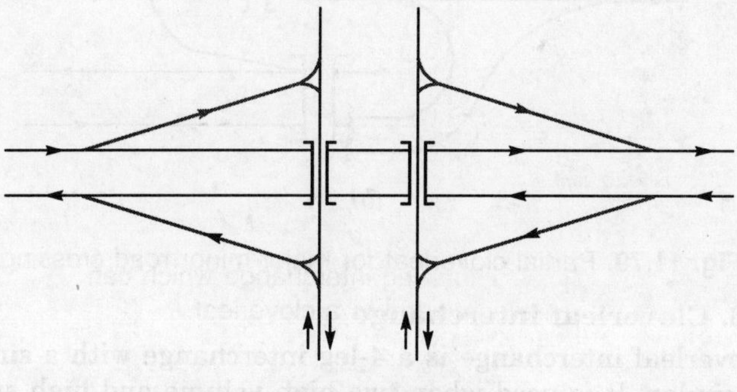

Fig. 11.77. Split Diamond.

If the two cross-streets are one-way, the layout in Fig. 11.78 may be adopted.

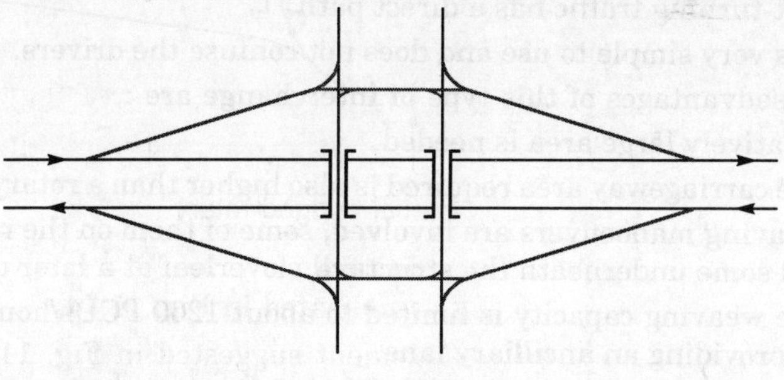

Fig. 11.78. Split Diamond Interchange with one-way streets.

11.14.5. Half cloverleaf interchange

When a major road crosses a minor road (not more than 3 lanes), it is possible to provide a partial cloverleaf. Two such designs are given in Fig. 11.79 (*a*) and (*b*).

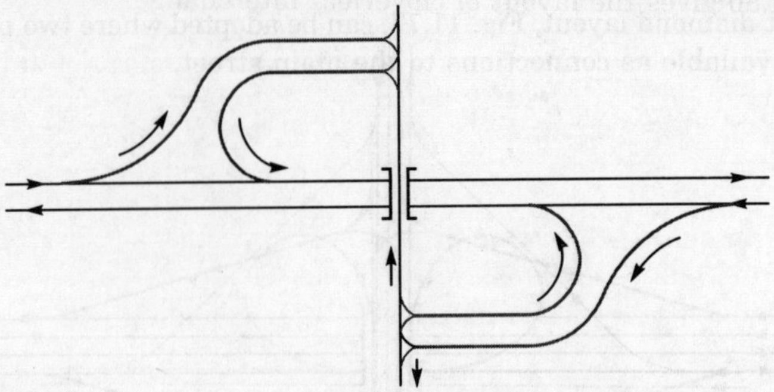

(*a*)

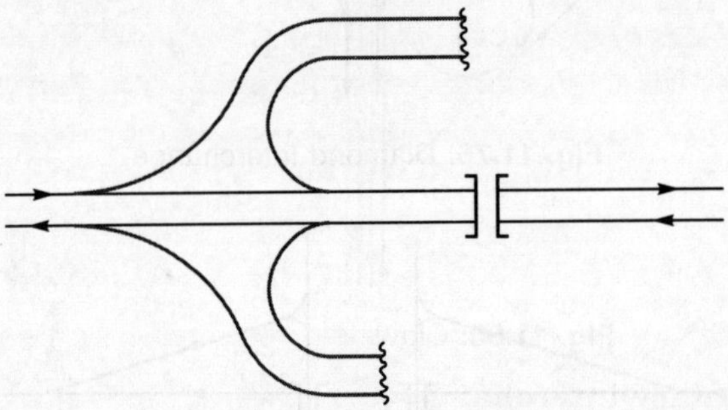

(*b*)

Fig. 11.79. Partial cloverleaf for major-minor road crossing.

11.14.6. Cloverleaf interchange

The cloverleaf interchange is a 4-leg interchange with a single structure and is very popular. It is used when two high volume and high speed facilities intersect each other. Some of its advantages are :

(*i*) Through traffic on both roads is unimpeded.

(*ii*) Only one structure is required.

(*iii*) Left-turning traffic has a direct path.

(*iv*) It is very simple to use and does not confuse the drivers.

The disadvantages of this type of interchange are :

(*i*) Relatively large area is needed.

(*ii*) The carriageway area required is also higher than a rotary interchange.

(*iii*) Weaving manoeuvers are involved, some of them on the roadway of the structure and some underneath the structure.

(*iv*) The weaving capacity is limited to about 1200 PCUs/hour, but may be increased by providing an ancilliary lane.

(*v*) The U-turns are long and operationally difficult.

(*vi*) Loop design speeds have to be low and speeds above 50 KPH are likely to increase the cost.

(*vii*) Right-turning traffic has to travel extra distance.

(*viii*) The capacity of the loop is also restricted. A capacity of 800–1200 vehicles per hour is almost the limit.

Fig. 11.80 gives the layout of cloverleaf interchange.

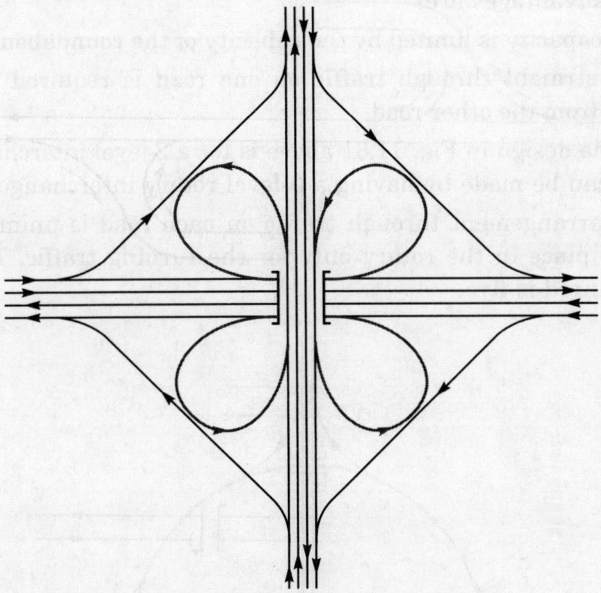

Fig. 11.80. Cloverleaf Interchange.

11.14.7. Rotary interchange

A rotary interchange is another form of 4-leg interchange popular in some countries, especially U.K. It can also be used for multi-leg interchange. Fig. 11.81 gives the layout.

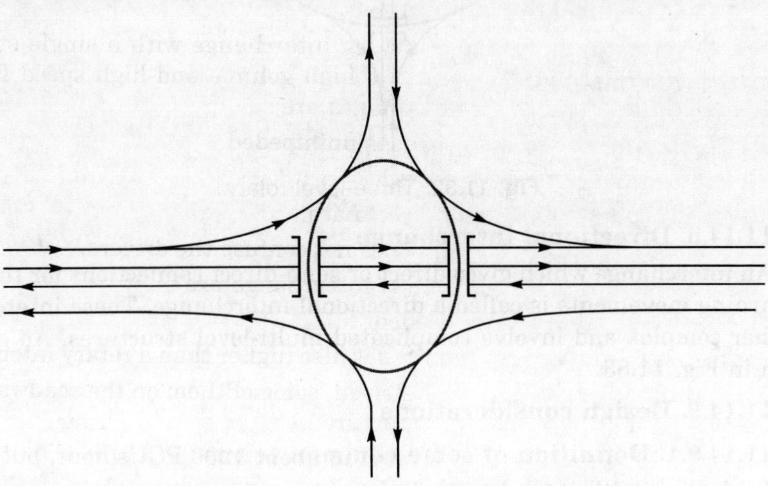

Fig. 11.81. Grade-separated rotary interchange.

The advantages claimed for this type of interchange are :

(*i*) It occupies relatively less land area.

(*ii*) The carriageway area is also less than many other types.

(*iii*) U-turns are easy.

(*iv*) Suitable for most sites.

The disadvantages are:

(*i*) The capacity is limited by the capacity of the roundabout itself.

(*ii*) The straight through traffic on one road is required to weave with turning traffic from the other road.

While the design in Fig. 11.81 above is for a 2-level interchange, a further improvement can be made by having a 3-level rotary interchange (Fig. 11.82).

In this arrangement through traffic on each road is unimpeded and the weaving takes place in the rotary only for the turning traffic. The number of structures required is five.

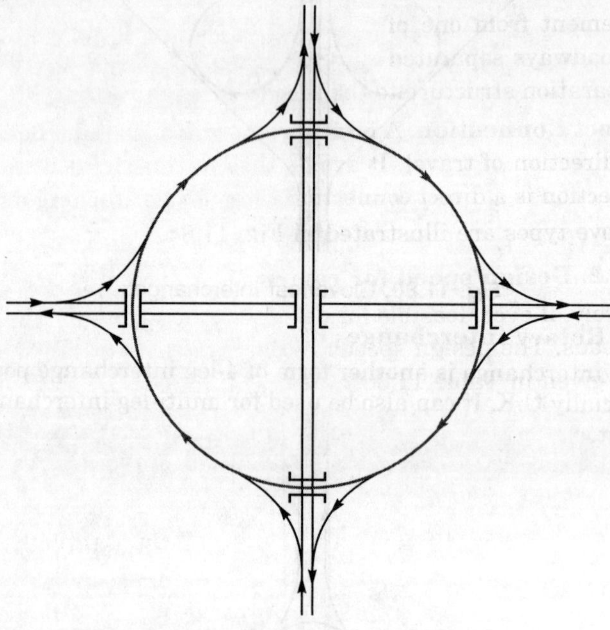

Fig. 11.82. Three-level rotary.

11.14.8. Directional interchange

An interchange which gives direct or semi-direct connections for the major right-turning movements is called a directional interchange. These interchanges are rather complex and involve complicated multi-level structures. An example is given in Fig. 11.83.

11.14.9. Design considerations

11.14.9.1. Definition of some common terms

In dealing with interchange design, some common terms are met with, which are defined below :

1. **Interchange Ramp.** (Access Ramp or Slip Road). An interconnecting roadway of a traffic interchange or any connection between highway facilities at different levels, on which vehicles may enter or leave a designated roadway. It may be in the form of a loop, outer connection or direct connection.

2. **Loop.** A one-way turning roadway that curves about 270° to the left to accommodate a right-turning movement.

3. **Outer Connection.** A ramp used by traffic destined for left-turn movement from one of the through roadways separated by a grade separation structure to the second through roadway.

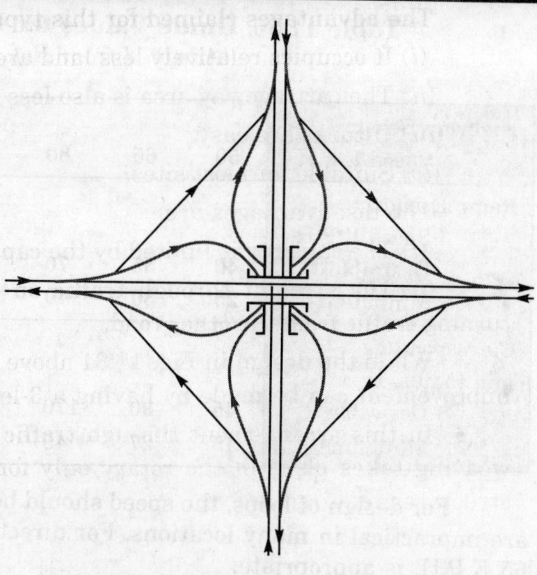

Fig. 11.83. Directional interchange.

4. **Direct Connection.** A form of ramp which does not deviate much from the intended direction of travel. It avoids the loop for right-turning movements. An outer connection is a direct connection ramp for left-turning movements.

The above types are illustrated in Fig. 11.84.

11.14.9.2. Design speed for ramps

It is rarely, if ever, feasible for designing the ramps for the same speed as the through roads. The design speeds recommended for ramps as per AASHTO (Ref. 3, 4) are given in Table 11.36.

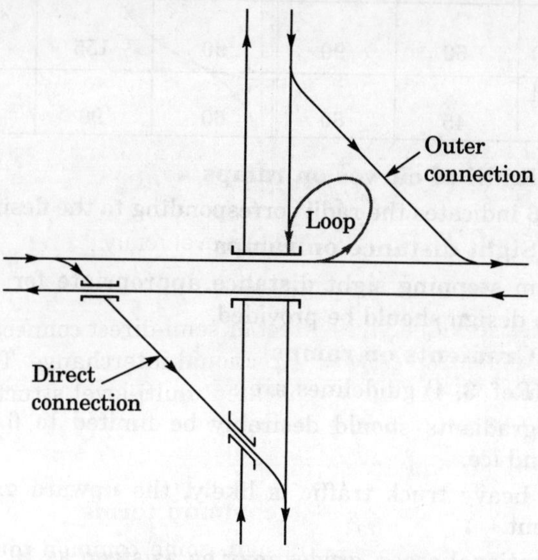

Fig. 11.84. Types of interchange.

Table 11.36. Guide Values for Ramp Design Speeds
(Adapted from Ref. 3 and 4)

Highway Design Speed K.P.H.	50	65	80	100	105	115	120	135
Ramp Design Speed K.P.H.								
Desirable	40	55	70	80	90	100	100	105
Minimum	25	30	40	50	50	50	55	65
Corresponding min radius, m								
Desirable	45	90	170	210	260	320	320	380
Minimum	15	27	45	70	70	70	90	130

For design of loops, the speed should be about 40–45 K.P.H. Higher values are impractical in many locations. For direct connections, a design speed of 60–65 K.P.H. is appropriate.

The standards prescribed for the ramp design elements for interchanges in urban areas in India (Ref. 46) are reproduced in Table 11.37.

Table 11.37. Standards for Ramp Design in India for Urban Locations

	Particulars	*Design values for major highway*				*For loop ramps*	
		80 K.P.H.		*100 K.P.H.*			
		Minimum	*Desirable*	*Minimum*	*Desirable*	*Minimum*	*Desirable*
1.	Ramp Design Speed (K.P.H.)	40	50	50	65	30	40
2.	Radius of Curvature (m)	60	90	90	155	30	60
3.	Stopping sight distance (m)	45	60	60	90	25	45

11.14.9.3. Radii of curves on ramps

Table 11.36 indicates the radii corresponding to the design speed selected.

11.14.9.4. Sight distance on ramps

A minimum stopping sight distance appropriate for the design speed selected for ramp design should be provided.

11.14.9.5. Gradients on ramps

AASHTO (Ref. 3, 4) guidelines are :

(*i*) Ramp gradients should desirably be limited to 6 percent in areas subject to snow and ice.

(*ii*) Where heavy truck traffic is likely, the upward gradient should be limited to 4 percent.

(*iii*) In exceptional cases, grades may be as steep as 10 percent, on minor ramps and low volume ramps.

11.14.9.6. Width of ramps

Ramps should preferably carry one-way traffic only. The width of pavements should be selected as appropriate for different radii.

11.14.9.7. Capacity

The practical capacity of through lanes can be taken as 1500 PCUs per hour (Ref. 6) per lane for interchange design. For ramps, a capacity of 1200 PCUs per hour per lane is appropriate. The capacity of weaving sections may be calculated from the principles discussed in the Highway Capacity Manual, 1965 (Ref. 31) and briefly stated in Chapter 21.

11.15. Bus Facilities

11.15.1. Bus stops

The following guidelines for the location of bus stops may be kept in view :

(*i*) Since bus traffic near intersections interferes with the traffic flow if not properly designed for, it is desirable that the bus stops should be located at least 75 m away from the intersections (Ref. 32).

(*ii*) It is preferable to locate bus stops on the exit side of the intersection (Ref. 32).

(*iii*) Bus stops on opposite sides of two-way undivided carriageways should be staggered, preferably so that buses stop tail to tail and move off from each other. A suggested stagger is 60-90 m (Ref. 6).

(*iv*) Bus stops should be as far as possible so located as to disembark the passengers at safe places such as curbs or islands.

(*v*) For buses intending to turn right at an intersection, if it becomes necessary to locate the bus stop before the intersection, the stop should be sufficiently away from the intersection so that the bus can be manoeuvred from the curb to the extreme right lane before the intersection.

11.15.2. Bus bays

It is desirable to recess the curb sufficiently for accommodating the bus-bays. Typical designs are given in Fig. 11.85.

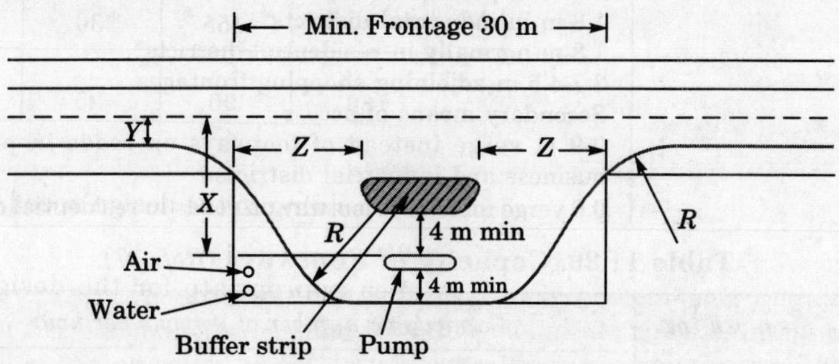

Minimum distance to intersection : 60 m for main roads and 45 m for others.

X Not less than 13 m where no cycle tracks are required now or in future not less than 20 m where cycle tracks exist or may be required in future.

Y Half the roadway width but not less than 6 m for national and state highways and 5 m for others.

Z Not less than 9 m.

R Radius not less than 11 m.

Fig. 11.85. Layout of a motor-fuel filling station.

11.16. Driveways

Driveways are roads providing access to abutting property owners and utilities. An important example is the access road to a petrol-filling facility. The geometrics of the driveways can have a profound influence on the traffic flow and safety of travel on the through road. The entrance and the exit should be wide with easy curves. A layout suggested by the I.R.C., (Ref. 33) for motor-fuel filling stations is given in Fig. 11.85.

11.17. Design for Pedestrian Facilities

11.17.1. Footways

For safety of pedestrians, separate footways should be provided alongside the carriageway of urban streets. They should be provided on either side of the road and should preferably be raised above the general carriageway level. To attract full use by pedestrians, they should be properly surfaced or paved. They should be sloped adequately to drain away the rain water. Crossfalls within the range of 1 in 40 to 1 in 30 are satisfactory.

The widths of footways as per current U.K. practice (Ref. 6) are given in Table 11.38.

The capacity of footways recommended by I.R.C. are given in Table 11.39 (Ref. 47).

Table 11.38. Recommended Footway Widths
(Adopted from Ref. 6)

Type of Road	*Recommended minimum footways widths*
Primary distributor	
Urban motorway	No footways.
All-purpose road	2.7 m*
District distributor	2.7 m in principal business and industrial districts*. 2.4 m in residential districts.
Local distributor	2.7 m in principal business and industrial districts* 1.8 m in residential districts*.
Access road	Principal means of access : 2.7 m in principal business districts*. 1.8 m in industrial districts*. 1.8 m normally in residential districts*. 3.7-4.5 m adjoining shopping frontages. Secondary means of access : 0.9 m verge instead of footways on roads in principal business and industrial districts. 0.6 verge instead of footway on roads in residential districts.

Table 11.39. Capacity of Footways *(Ref. 47)*

Width of Footpath (m)	*Capacity in number of persons per hour*	
	All in one direction	*In both directions*
1.50	1,200	800
2.00	2,400	1,600
2.50	3,600	2,400
3.00	4,800	3,200
4.00	6,000	4,000

*If no footway is required, provide at least 0.9 m verge.

The capacity of pedestrian underpasses or overbridges is 50 persons per minute per metre width on level or upto 1 in 20 ramp, and 35 persons per minute per metre width on steps or ramps steeper than 1 in 20.

11.17.2. Pedestrian bridges and subways

The flows should not normally exceed 20 persons per 300 mm width per minute on level or upto 1 in 20 gradient and 14 persons per 300 mm width per minute on stairs or ramps steeper than 1 in 20 (Ref. 34). A dead width of 0.75 m is usually allowed adjoining any display windows in sub-ways (Ref. 6).

The gradients of continuous ramps should not be steeper than 1 in 10. If a landing is provided at mid-height, a maximum slope of 1 m 7 can perhaps be allowed (Ref. 35). A minimum height of 2.30 m and a width of 2.45 m should be provided for subways upto about 23 m length, and for longer subways the dimensions need to be increased to 2.60 m height × 2.75 m width.

11.18. Design Criteria for Separate Cycle Tracks

Segregation of cycle tracks promotes safety. In veiw of the large number of cyclists on the streets of Indian towns and cities, it is desirable that separate cycle tracks be provided wherever feasible.

The design criteria for cycle tracks are laid down by the Indian Roads Congress (Ref. 36). The design criteria commonly adopted are :

	Design Element	*Guideline*		
1.	Justification for separate cycle tracks	Where the peak hour cycle traffic is 400 or more on routes with a vehicular traffic of 100–200 veh/hr. Where the motor vehicle traffic is more than 200 veh/hour, separate cycle tracks are justified even if the cycle traffic is only 100 per hour.		
2.	Capacity (per day)		*One-way traffic*	*Two-way traffic*
		2-lanes	2000–5000	500–2000
		3-lanes	> 5000	2000–5000
		4-lanes	–	> 5000
3.	Horizontal curves (Radius)	(*i*) Minimum 10 m (*ii*) Where the gradient is steeper than 1 in 40, minimum radius shall be 15 m.		
4.	Vertical curves	Minimum Radius (*i*) Summit curves : 200 m (*ii*) Valley curves : 100 m		
5.	Gradients	*Gradient*	*Maximum length* (*m*)	
		1 in 30	90	
		1 in 35	125	
		1 in 40	160	
		1 in 45	200	
		1 in 50	250	
		1 in 55	300	
		1 in 60	360	
		1 in 65	425	
		1 in 70	500	
		(Gradients of 1 in 20 and 1 in 30 may be allowed for short lengths of 20 m and 50 m respectively).		

	Design Element	*Guideline*
6.	Sight distance	Clear view $\nless$ 25 m. In gradients of 1 in 40 or steeper, clear view shall not be less than 60 m.
7.	Lane width	1 metre per lane.
8.	Width of pavement	Minimum : 2 m (for 2 lanes)
9.	Clearance	(*i*) Vertical : 2.25 m minimum 2.50 desirable (*ii*) Horizontal : 0.25 m.

Other general considerations for the design of separate cycle tracks are :

(*i*) Preferably cycle tracks should be provided on both sides of the road.

(*ii*) They should be separated from the main carriageway be a verge or a berm, which should be preferably 0.5-1.0 m wide.

(*iii*) Cycle tracks should be constructed and maintained with care and should have good riding qualities. Black-topped and concrete surfaces provide smooth riding qualities, whereas water bound macadam and earthen tracks are not inviting for the cyclists to ride upon.

(*iv*) The tracks should be clear of obstructions such as hedges' ditches tree roots, kerbs etc. by at least 0.5 m.

Design speeds for cycles is about 32 km/hr (Ref. 38).

11.19. Geometric Design Standards for Expressways

The following extracts from the draft geometric design standards under consideration by the Indian Roads Congress (Ref. 49).

1. **Right of Way**

Plain and rolling terrain : 90 m
Mountain terain :
Same level : 60 m
Split level : 30 m

2. **Design Speeds KPH**

Terrain	*Ruling Speed*	*Minimum Speed*
Plain	120	100
Rolling	100	80
Mountainous	80	60

3. **Carriageway width**

Lane : 3.75 m for plain and rolling terrain
3.5 m for mountainous terrain

Carriageway should be divided

4. **Shoulder width**

Paved shoulder on median side : 0.75 m
Paved shoulder on outer (left) side : 3.0 m for plain and rolling terrain
2.5 m for mountainous terrain
Earthen shoulder (left or outer side) : 1.5 m

5. **Median Width**

   Without barrier : 12 m

   With barrier : 4.5 m

6. **Median Openings**

   To be provided with detachable barriers at every 2 to 5 km spacing for traffic management and maintenance works.

7. **Cross-fall**

   With rainfall 100 cm or more : 2.5%

   With rainfall less than 100 cm : 2.0%

8. **Stopping Sight Distance**

Design speed (KPH)	*Stopping Sight Distance (m)*
120	250
100	180
80	120
60	80

9. **Radius of horizontal curves**

Design speed (KPH)	*Radius (m)*
120	670 (absolute min.) 1000 (desirable)
100	440 (absolute min.) 700 (desirable)
80	260 (absolute min.) 400 (desirable)
60	140 (absolute min.) 200 (desirable)

10. **Gradient**

Terrain	*Ruling Grade (%)*
Plain	3
Rolling	4
Mountainous	5

In addition to the above, the following standards for interchange elements may be noted:

Interchange design speed	:	Desirable 80 KPH Minimum 60 KPH
Interchange ramp gradient	:	Desirable 4% Maximum 6%
Interchange carriageway width	:	Desirable 7 m Minimum 5.5 m

A typical cross-section of an Expressway as adopted for the Taj Expressway (Ref. 50) is given in Fig. 11.86.

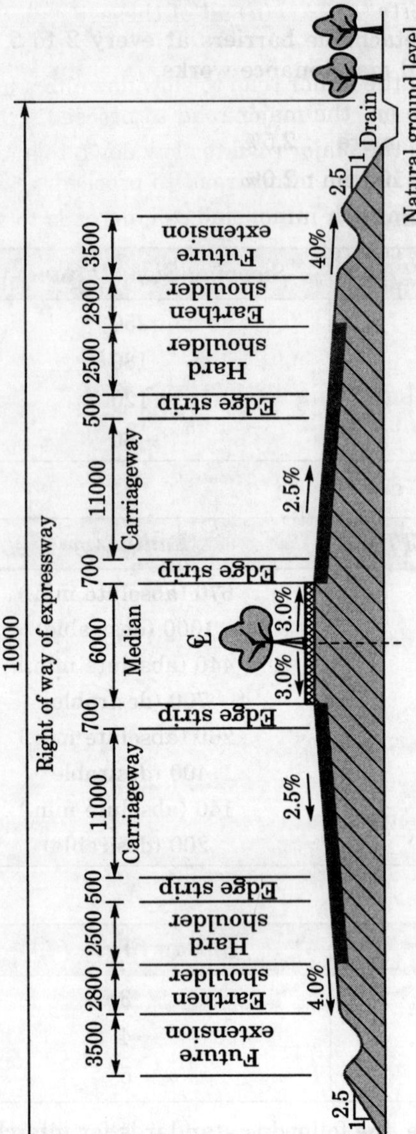

Fig. 11.86. Typical crosssection of 2 × 3 lane expressway. (Taj Expressway)
(All dimensions are in mm).

I. MULTIPLE CHOICE QUESTIONS

1. A local streets is one primarily intended for access to:
(a) residence (b) business
(c) other abutting property (d) all of these

2. The GIVE WAY control
(a) requires the driver in the minor road to slow down to a minimum speed and allow the vehicle on the major road to proceed.
(b) requires the driver in the major road to slow down to a minimum speed and allow the vehicle on the minor road to proceed.
(c) requires the drivers on both minor and major roads to stop.
(d) is similar to one way control.

3. Traffic signal is an example of
(a) passive control (b) no control
(c) active control (d) none of these

4. When the path travelled along the road surface is more than the circumferential movement of the wheels due to rotation, then it results in:
(a) slipping (b) skidding
(c) turning (d) revolving

5. Co-efficient of friction is less when the pavement surface is
(a) rough (b) dry
(c) smooth and dry (d) smooth and wet

6. The stopping sight distance depends upon the
(a) reaction time (b) speed of vehicle
(c) efficiency of brakes (d) all of the above.

7. Highway facilities are designed for
(a) annual average hourly volume (b) annual average daily traffic
(c) thirtieth highest hourly volume (d) peak hourly volume of the year

8. For highway geometric design purposes the speed used is
(a) 15th percentile (b) 50th percentile
(c) 85th percentile (d) 98th percentile

9. Which of the following is known as design capacity?
(a) basic capacity (b) theoretical capacity
(c) possible capacity (d) practical capacity

10. If the average centre to centre spacing of vehicles is 20 metres, then the basic capacity of a traffic lane at a speed of 50 kmph is
(a) 2500 vehicles per day (b) 2000 vehicles per hour
(c) 2500 vehicles per hour (d) 1000 vehicles per hour

11. Consider the following statements in the context of geometric design of roads
I. A simple parabolic curve is an acceptable shape for summit curves
II. Comfort to passengers is an important consideration in the design of summit curves

The correct option evaluating the above statements and their relationship is (*a*) I is true, II is false

(*b*) I is true, II is true, and II is the correct reason for I

(*c*) I is true, II is true, and II is NOT the correct reason for I

(*d*) I is false, II is true (*GATE 2010*)

Answers

1. (*d*)	**2.** (*a*)	**3.** (*c*)	**4.** (*b*)	**5.** (*d*)
6. (*d*)	**7.** (*c*)	**8.** (*d*)	**9.** (*a*)	**10.** (*c*)
11. (*a*)				

III. SHORT ANSWERS AND QUESTIONS

Q. 1. What is geometrical design?

Ans. This is a branch of computational geometry. It deals with the construction and representation of free-form curves, surfaces, or volumes and is closely related to geometric modelling. Core problems are curve and surface modelling and representation. It studies especially the construction and manipulation of curves and surfaces given by a set of points using polynomial, rational, piecewise polynomial, or piecewise rational methods. The most important instruments here are parametric curves and parametric surfaces, such as Bézier curves, spline curves and surfaces.

Q. 2. What are the points to be considered by selecting the standard design?

Ans. The following points are to be considered:

- Highway classification
- Arterial streets
- Requirements of safety barriers
- Traffic safety barriers
- Object markers
- Express ways
- Local streets
- Delineators
- Roadway indicators

Q. 3. What are the types of roads?

Ans. There are various types of roads:

- Local roads
- Collector and distributor roads

  Sub-arterial and arterial roads: These are the major highways, motorways and freeways. Also, while not technically a road, bikeways provide the general community with an alternative means of travel.

Q. 4. What is the importance of traffic data in geometric design?

Ans. The geometric design of roads is the branch of highway engineering concerned with the positioning of the physical elements of the roadway according to standards and constraints. The basic objectives in geometric design are to optimize efficiency and safety while minimizing cost and environmental damage. Geometric design also affects an emerging fifth objective called "livability," which is defined as designing roads to foster broader community goals, including providing access to employment, schools, businesses and residences, accommodate a range of travel modes such as walking, bicycling, transit, and automobiles, and minimizing fuel use, emissions and environmental damage.

Q. 5. What is directional distribution?

Ans. Directional distribution is the split of traffic during the design hour commonly expressed as percent in the peak and off-peak flow directions.

Q. 6. What is design vehicle?

Ans. Design vehicle is defined as the vehicle that must regularly be accommodated on a roadway without encroachment into other travel lanes. It is a selected motor vehicle, the weight, dimension and operating characteristic of which are used to establish highway design controls to accommodate vehicles. The design vehicle's dimensions and movements can play a large role in the physical characteristics of a roadway, such as the appropriate lane width and the radii of curves at intersections and driveway corners.

Q. 7. What are the characteristics which influence design vehicle selection for residential neighborhoods?

Ans. The main characteristics are:

- Large truck traffic is uncommon
- Transit vehicles are often present
- Significant pedestrian presence is likely

Q. 8. What do you mean by vertical curves?

Ans. Vertical curves are the second of the two important transition elements in geometric design for highways, the first being horizontal curves. A vertical curve provides a transition between two sloped roadways, allowing a vehicle to negotiate the elevation rate change at a gradual rate rather than a sharp cut.

Q. 9. When does a vertical curves applied?

Ans. It is applied at an intersection of two slopes on a highway or a roadway to provide a safe and comfort ride for vehicles on a roadway.

Q. 10. What are the types of vertical curves?

Ans. Two kinds of vertical curve

- Crest Vertical Curves + Type I and Type II
- Sag Vertical Curves + Type III and Type IV.

Q. 11. What are intersections?

Ans. It is defined as the general area where two or more highways in or cross, within which are included the roadway and roadside facilities for traffic movements in the area. An intersection leg is that part of any one of the roadways radiating from an intersection which is outside of the area of the intersection proper.

Q. 12. What is channelization?

Ans. It is the separation or regulation of conflicting traffic movements into definite paths of travel by traffic islands or pavement marking to facilitate the safe and orderly movements of both vehicles and pedestrians.

Q. 13. What is conflict?

Ans. It is defined as the demand for the same highway space by two or more users of the highway. Conflicts are classified into mainly three types:

(*a*) Crossing conflicts (*b*) Diverging conflicts (*c*) Merging conflicts

Q. 14. What is angle of intersection?

Ans. The angle of intersection is that formed by the centerlines of the intersecting streets. Where the angle of intersection departs significantly from right angles, the intersection is referred to as a skewed intersection.

Q. 15. What is refuge areas?

Ans. The area which is used to give refuge to the pedestrians crossing a street (the open area between two medians) is known as a refuge area.

Q. 16. What are the main objectives of channelization?

Ans. The main objectives of channelization can be summarized as follows:

(*i*) Separation of maneuver areas: The drivers should be presented with only one decision at a time to reduce confusion and the influence of operations caused due to the overlapping of maneuver areas.

(*ii*) Reduce excessively large paved areas: The spread of the paved area can be considerably reduced by the construction of raised islands and medians where these are considered safe and necessary.

(*iii*) Control of maneuver angle: The intensity of accidents can be reduced to a large extent by providing small angles for merging, diverging and weaving (at low relative speeds) and approximately right angles for crossing (at high relative speeds). The maneuver angle can be easily controlled by constructing islands of appropriate shapes and sizes.

(*iv*) Favor predominant turning movements: Channelization is also directed for giving preference to turning movements at an intersection where the proportion of such traffic is high.

(*v*) Control of speed: Channelization is also used for supporting stop or speed regulations by removing differentials in speed for merging, diverging, weaving and crossing by using the bending and funneling techniques.

(*vi*) Protection and storage of turning and crossing vehicles: To shadow slow or stopped vehicles from other traffic flows.

(*vii*) Blockage of prohibited movements: Proper channelization also helps maintain traffic regulations by making prohibited movements impossible or inconvenient.

(*viii*) Provide space for traffic control devices: To provide space for traffic control devices when the ideal location for the same is within the intersection area.

(*ix*) Segregation of non-homogeneous flows: Channelization provides separate channels for turning and through, fast and slow, and opposite direction traffic.

Q. 17. What is traffic islands?

Ans. An island is a defined area between traffic lanes for control of vehicle movements. Within an intersection area, a median or an outer separation is considered to be an island. It may range from an area delineated by barrier curbs to a pavement area marked by paint.

Traffic islands usually serve more than one function, but may be generally classified in three separate types:

(*i*) Channelizing islands. These are designed to control and direct traffic movement, usually turning.

***(ii)* Refuge islands.** Pedestrian islands are provided to serve as safety zones for the aid and protection of persons on foot. If a divisional island is located in an urban area where pedestrians are present, portions of each island can be considered a refuge island.

Q. 18. What are auxiliary lanes?

Ans. Auxiliary lanes are used under conditions of relatively high traffic volumes in the intersections. In these cases, traffic congestion problems can be significantly alleviated with auxiliary lanes to handle turning movements. The median lane should be 12 feet (3.6 m), but not less than 10 feet (3.0 m) wide and should be clearly marked for this purpose. Auxiliary lanes can also be introduced to provide for both left turns and right turns at intersections. The need for such lanes is determined by capacity analysis and the acceptable level of service designated for the facility.

Q. 19. What is grade separation?

Ans. Grade separation is the name given to a method of aligning a junction of two or more surface transport axes at different heights (grades) so that they will not disrupt the traffic flow on other transit routes when they cross each other.

Q. 20. What is interchange?

Ans. It is a system where traffic between two or more roadways flows at different levels in the *grade separated* junctions.

Q. 21. What does a common types of interchange?

Ans. A common types of interchange include trumpet interchange, diamond interchange, and cloverleaf interchange.

Q. 22. What is trumpet interchange?

Ans. Trumpet interchange is a popular form of three leg interchange.

Q. 23. What is DDI?

Ans. It is a diverging **diamond interchange** (DDI), also called a double crossover **diamond interchange** (DCD), is a type of **diamond interchange** in which the two directions of traffic on the non-freeway road cross to the opposite side on both sides of the bridge at the freeway.

Q. 24. What is half cloverleaf interchange?

Ans. A **cloverleaf interchange** is a two-level interchange in which left turns (*reverse directions in left-driving regions*), are handled by ramp roads (US: ramps, UK: slip roads). To go left (in right-hand traffic), vehicles first continue as one road passes over or under the other, then exit right onto a one-way three-fourths loop ramp (270°) and merge onto the intersecting road. The objective of a cloverleaf is to allow two highways to cross without the need for any traffic to be stopped by red lights, even for left and right turns. The limiting factor in the capacity of a cloverleaf interchange is traffic weaving.

Q. 25. Briefly explain intersection.

Ans. An intersection is the area shared by the joining or crossing of two or more roads. Since the main function of an intersection is to enable the road user to make a route choice, it is a point of decision. Hence the problems that are encountered by the motorist while passing through an intersection must be recognized and the design should be in such a way that the driving task is as simple as possible.

Intersection is also a point of large number of major conflicts, besides a point of decision. These conflicts may be due to the crossing maneuvers of vehicles moving in different directions. Good intersection design results from a minimization of the magnitude and characteristics of the conflicts and a simplification of driver route selection process.

Q. 26. What is grade separated intersection?

Ans. It is a bridge that eliminates crossing conflicts at intersections by vertical separation of roadways in space. Grade separated intersection are otherwise known as Interchanges. Grade separated intersections cause less hazard and delay than grade intersections. Route transfer at grade separations is accommodated by interchange facilities consisting of ramps. Interchange ramps are classified as Direct, Semi-Direct and Indirect. Interchanges are described by the patterns of the various turning roadways or ramps. The interchange configurations are designed in such a way to accommodate economically the traffic requirements of flow, operation on the crossing facilities, physical requirements of the topography, adjoining land use, type of controls, right-of-way and direction of movements.

Q. 27. What is underpass?

Ans. An underpass or a tunnel is an underground passageway, completely enclosed except for openings for ingress and egress, commonly at each end. A tunnel may be for foot or vehicular road traffic, for rail traffic. If an underpass is constructed for pedestrians and/or cyclists beneath a road or railway, allowing them to reach the other side in safety, then such a construction is termed as a Subway. These are constructed when it is necessary for pedestrians to cross a railroad or a limited-access highway. Subways may also be constructed for the benefit of wildlife.

Q. 28. What is overpass?

Ans. An overpass also known as a flyover, is a bridge, road, railway or similar structure that crosses over another road or railway. A pedestrian overpass allows pedestrians safe crossing over busy roads without impacting traffic. And Railway overpasses are used to replace at-grade crossing as a safer alternative. Overpasses allows for unobstructed rail traffic flow from mixing with vehicular and pedestrian traffic. Stack interchanges are made up of many overpasses.

Q. 29. What is trumpet interchange?

Ans. Trumpet interchanges have been used where one highway terminates at another highway. These involve at least one loop ramp connecting traffic either entering or leaving the terminating expressway with the far lanes of the continuous highway. These interchanges are useful for highways as well as toll roads, as they concentrate all entering and exiting traffic into a single stretch of roadway, where toll booths can be installed. Trumpets are suitable at the locations where the side road exists on only one side of the freeway, and traffic is relatively low.

Q. 30. Explain diamond interchange.

Ans. The diamond interchange is the simplest form of grade separated intersection between two roadways. The conflicts between through and crossing traffic are eliminated by a bridge structure. This particular intersection has four one way ramps which are essentially parallel to the major artery. The left turn crossing movement conflicts are considerably reduced by eliminating the conflict with the traffic in opposite direction. All the remaining left turn conflicts, merging and diverging maneuver conflicts take place at the terminal point of each ramp. Limitation in application of this design depends on the operations of these terminals. So, it is suitable for locations where the volume of left turn traffic is relatively low.

Q. 31. Define cloverleaf interchange.

Ans. The full clover interchange eliminates all crossing movement conflicts by the use of weaving sections. This weaving section is a critical element of cloverleaf design. It replaces a crossing conflict with a merging, followed some distance farther by a diverging conflict. There are two points of entry and exit on each through roadway. The first exit is provided before the cross road structure allows right turn movements. The second exit, immediately after the cross road structure, allows for left turn movements. A weaving section is created between the exit and entry points near the structure. Sufficient length and capacity is to be provided to allow for a smooth merging and diverging operation.

Q. 32. What do you mean by partial cloverleaf interchange?

Ans. This is another variation of the cloverleaf configuration. Partial clover leaf or parclo is a modification that combines some elements of a diamond interchange with one or more loops of a cloverleaf to eliminate only the more critical turning conflicts. This is the most popular freeway -to- arterial interchange. Parclo is usually employed when crossing roads on the secondary road will not produce objectionable amounts of hazard and delay. It provides more acceleration and deceleration space on the freeway.

Q. 33. What is directional interchange?

Ans. A directional interchange provides direct paths for left turns. These interchanges contain ramps for one or more direct or semi-direct left turning movements. Interchanges of two freeways or interchanges with one or more very heavy turning movements usually warrant direct ramps, which have higher speeds of operation and higher capacities, compared to loop ramps. Some designers do not favor entrance of merging traffic in the left lane, which is a characteristic of most direct-connection bridges. The principal limitations of this type of interchange is higher cost of construction and requirement relatively large amount of land when compared to the diamond interchanges and in some cases than cloverleaf interchange. Various combinations of directional, semi-directional and loop ramps may be appropriate for certain conditions. They are the basic patterns that use the least space, have the fewest or least complex structures, minimize internal weaving and appropriate for the common terrain and traffic conditions.

Q. 34. Explain acceleration lane.

Ans. An acceleration lane is defined as extra pavement, of constant or variable width, placed parallel or nearly so, to a merging maneuver area to encourage merging at low relative speed. The major difference in opinion concerning acceleration design stems from lack of information on driver performance. Field observations have indicated that drivers desire to follow the direct path even though extra width or tapered section is provided.

Q. 35. What are deceleration lanes?

Ans. Deceleration lanes are defined as extra pavement of constant or variable width, placed parallel or nearly so, to a diverging maneuver area to encourage diverging at low relative speed. The lengths of deceleration lanes are based on the difference in the speed of traffic of the combined flow (in advance of the collision area) and the speed at which drivers negotiate the critical diverging channel curve, as well as the deceleration practices of drivers.

III. REVIEW QUESTIONS

1. What is horizontal alignment?
2. Define vertical alignment.
3. Illustrate the combination of horizontal and vertical alignment.
4. What is trumpet interchange?
5. Explain diamond interchange.
6. What is cloverleaf interchange?
7. What is partial cloverleaf interchange?
8. What is directional interchange?
9. What is bridged rotar?

REFERENCES

1. Geometric Design of Urban Roads, I.R.C : 86-1983, Indian Roads Congress, New Delhi, 1983.
2. Geometric Design Standards for Rural (Non-urban) Roads, Indian Roads Congress, New Delhi, 1980.
3. A Policy on Geometric Design of Highways and Streets, American Association of State Highway and Transportation Officials, Washington, 1994.
4. A Policy on Design of Highways and Streets, American Association of State Highway Officials, Washington, 1973.
5. Layout of Road in Rural Areas, Development of Environment (U.K.) H.M.S.O., London, 1968.
6. Roads in Urban Areas, Department of Environment (U.K.), H.M.S.O., London, 1966.
7. Space standards in Urban Areas, Indian Roads Congress, New Delhi, 1977.
8. Post-war Road Development in India, Proceedings of the conference of Chief Engineers, (Nagpur Plan), Published by the Indian Roads Congress, New Delhi.
9. Recommendations about the Alignment Survey and Geometric Design of Hill Roads, I.R.C : 52–1973, Indian Roads Congress, New Delhi, 1973.
10. ECAFE, Highway Classification and Design Standards for Asian Highway, Journal of the Indian Roads Congress, Vol. XXXIII-1, New Delhi, 1970.
11. Highway Capacity Manual (1965), Highway Research Board Special Report 87, Washington, 1965.
12. Guidelines on Capacity of Roads in Rural Areas, I.R.C : 64–1990, Indian Roads Congress, New Delhi, 1990.
13. Dimensions and Weights of Road Design Vehicles, I.R.C : 3–1954, Indian Roads Congress, New Delhi, 1954.
14. Dimensions and Weights of Road Design Vehicles, I.R.C : 3–1984 Indian Roads Congress, New Delhi, 1983.
15. Statutory Instruments 1963, No. 1646. The Motor Vehicles (Construction and Use) Regulations 1963, H.M.S.O., London, 1963.
16. Horizontal and Transition Curves for Highways, Paper No. 119. Journal of the Indian Roads Congress, Vol. XI-3, New Delhi, 1947.
17. Guidelines for Design of Horizontal Curves for Highways and Design Tables, I.R.C : 38–1988, Indian Roads Congress, New Delhi, 1988.
18. Recommended Practice for Sight Distances on Rural Highways, Indian Roads Congress, I.R.C : 66–1976 New Delhi, 1976.

19. Vertical Curves for Highways, Paper No. 156, Journal of the Indian Roads Congress, Vol. XIV-1, New Delhi, 1951.
20. Ribbon Development Along Highways and its Prevention, Special Report 15, Indian Roads Congress, New Delhi, 1974.
21. Geometrics of Roads, Transport and Communications Monthly Review, May 1966, Indian Roads Congress, New Delhi.
22. Standard Specifications and Code of Practice for Road Bridges, Section I, I.R.C : 5–1970. Indian Roads Congress, 1970.
23. Recommended Practice for the Construction of Earth Embankment for Road Works, I.R.C : 36–1970, Indian Roads Congress, New Delhi, 1970.
24. Lateral and Vertical Clearances at Underpasses for Vehicular Traffic, I.R.C : 54-1974, Indian Roads Congress, New Delhi, 1974.
25. Guidelines for Control of Access on Highways, I.R.C : 62-1976, Indian Roads Congress, New Delhi, 1976.
26. Recommended Practice for Traffic Rotaries, I.R.C : 65-1976, Indian Roads Congress, New Delhi, 1976.
27. Wardrop, J.G., The Traffic Capacity of Weaving Sections of Roundabouts, Proceedings of the First International Conference on Operational Research, English Universities Press, London, 1957.
28. Blackmore, F.C., Capacity of Single Level Intersections, Ministry of Transport, R.R.L. Report No. LR 356.
29. Junction Design (Interim), Department of Environment, (U.K.) Tech, Memo No. H 7/71, June, 1971.
30. Country Surveyors Society, Mini-round abouts in England and Wales, Essex Country Council, Old Court, Chelmsford, Essex, 1972.
31. Highway Capacity Manual, Highway Research Board Special Pub-lication 87, Washington, 1965.
32. Guidelines for the Regulation and Control of Mixed Traffic in Urban Areas, Traffic Engineering Committee of the Indian Roads Congress, Vol. 36-3, Journal of IRC, New Delhi, 1975.
33. Recommended Practice for Location and Layout of Roadside Motor Fuel Filling Stations, I.R.C : 13-1967 Indian Roads Congress, New Delhi, 1967.
34. Pedestrian Safety, Department of the Environment, H.M.S.O., London, 1973.
35. A Manual for Planning Pedestrian Facilities, U.S. Dept. of Transportation, Washington, 1974.
36. Recommended Practice for the Design and Layout of Cycle Tracks, Indian Roads Congress, New Delhi, 1971.
37. Manual on Route Location, Design, Construction and Maintenance of Rural Roads (Other District Roads and Village Roads), Indian Roads Congress, New Delhi, 1979.
38. Guide for Development of New Bicycle Facilities, American Association of State Highway and Transportation Officials, Washington, 1981.
39. Road User Cost Study in India, Final Report, Central Road Research Institute, New Delhi, 1982.
40. Road Development Plan for India (1981-2001), Indian Roads Congress, New Delhi, 1985.
41. Data on Growth of Highway Traffic, Maharashtra Engineering Research Institute, Nasik, 1980.
42. Data on Growth of Highway Traffic on N.H. 5 in Orissa, National Highway Division, Cuttack, 1980.
43. Report on Research Project, R-3 Scheme, P.W.D. (B and R), Rajasthan, Jaipur, 1979.

44. First Year Results of Continuous Traffic Count Station on N.H. 45 in Tamil Nadu, Highway Research Station, Chennai, 1980.

45. Vertical Curves for Highways, I.R.C. Special Publication 23, Indian Roads Congress, New Delhi, 1983.

46. Guidelines for the design of interchanges in urban areas, I.R.C : 92–1985. The Indian Roads Congress, New Delhi, 1985.

47. Guidelines for Pedestrian Facilities, I.R.C : 103-1988, Indian Roads Congress, New Delhi, 1989.

48. Guidelines for the Design of At-grade Intersections in Rural and Urban Areas, I.R.C SP 41, Indian Roads Congress, New Delhi, 1994.

49. Geometric Design Standards for Expressway, Draft under consideration by the H-1 Committee of the Indian Roads Congress, New Delhi, 2008.

50. Taj Expressway, Detailed Project Report, Jaiprakash Industries Ltd., New Delhi, 2006.

SECTION 5
PARKING

12. Parking

12.1. Traffic and Parking Problems

One of the problems created by road traffic is parking. Not only do vehicles require street space to move about, but also do they require space to park where the occupants can be loaded and unloaded. It is roughly estimated that out of 8,760 hours in a year, the car runs on an average for only 400 hours, leaving 8,360 hours when it is parked (Ref. 1). Every car owner would wish to park the car as closely as possible to his destination so as to minimise his walking. This results in a great demand for parking space in the CBD and other areas where the activities are concentrated. With the growing population of motor vehicles, the problem of parking has assumed serious proportions. A systematic study of the parking characteristics and demand and regulatory measures that are possible for controlling parking is of great help to a traffic engineer as well as a town planner.

12.2. Ill-effects of Parking

12.2.1. Congestion

One of the serious ill-effects of parking is the loss of street space and the attendant traffic congestion. The capacity of the streets is reduced, the journey speed drops down and the journey time and delay increase. The operational cost of vehicles is thereby increased, causing serious economic loss to the community.

12.2.2. Accidents

The manoeuvres associated with parking and unparking are known to cause road accidents. Careless opening of the doors of parked vehicles, moving out of a parked position and bringing a car to the parking location from the mainstream of traffic are some of the common causes of parking accidents.

12.2.3. Obstruction to fire-fighting operations

Parked cars obstruct the movement of fire-fighting vehicles and greatly impede their operations. They block access to hydrants and access to buildings.

12.2.4. Environment

Parked vehicles degrade the environment of the town centre. Stopping and starting of vehicles result in noise and fumes. Cars parked into every little available space debase the visual aesthetics and "buildings seem to rise from a plinth of cars" (Ref. 2).

12.2.5. Inspite of the above ill-effects, parking has to be allowed on the streets or off the streets close to the users' destination. The well-being of the community and the town centre demands that great thought should be bestowed on the parking needs and how best to fulfil them.

12.3. Zoning and Parking Space Requirement Standards

While on-street parking and its regulation will be an important aspect of the overall parking policy of a town, one way of ensuring that future development of land-use and building activity adequately takes care of parking needs is to promulgate zoning ordinances whereby all new or remodelled buildings will be required to have within their curtilage a prescribed parking space. The advantages of zoning and land-use controls for fostering safe and efficient traffic have been well recognised (Ref. 3). Parking demand is a function of the land-use, though variations can exist within the same land-use. Notwithstanding the difficulties that arise in prescribing a uniform parking space requirement standard for all communities, it is possible to prescribe minimum standards based on extensive experience. The values given in Table 12.1 are the I.R.C. Standards (Ref. 4) and have been drawn up based on local experience.

Table 12.1. Desirable Parking Space Standards for Different Land Use

(*Source* : Ref. 4)

S. No.	*Land use*	*Parking space standard*
1.	Residential	
	(*i*) Detached, semi detached row houses :	
	Plot area upto 100 sq. m.	No private or community parking space is required.
	Plot area from 101 to 200 sq. m.	Only community parking space is required.
	Plot area from 201 to 300 sq. m.	Only community parking space is required.
	Plot area from 301 to 500 sq. m.	Minimum one-third of the open area should be earmarked for parking.
	Plot area from 501 to 1000 sq. m.	Minimum one-fourth of the open area should be earmarked for parking.
	Plot area 1001 sq. m. and above	Minimum one-sixth of the open area should be earmarked for parking.
	(*ii*) Flats.	One space for every two flats of 50 to 99 sq. m. or more of floor area.
	(*iii*) Special, costly developed area	One space for every flat of 50 to 100 sq. m. of floor area. One and a half spaces for every flat of 100 to 150 sq. m. of floor area. Two spaces for every flat of above 150 sq. m. of floor area.
	(*iv*) Multi-storeyed group	One space for every four dwell-housing scheming except in cities like Kolkata and Mumbai where the demand may be more.
2.	Offices	One space for every 70 sq. m. of floor area.

(*Cont...*)

S. No.	Land use	Parking space standard
3.	Industrial premises	One space for upto 200 sq. m. of initial floor area. Additional spaces at the rate of one for every subsequent 200 sq. m. of fraction thereof.
4.	Shops and markets	One space for every 80 sq. m. of floor area.
5.	Restaurants	One space for every 10 seats.
6.	Theatres and cinemas	One space for every 20 seats.
7.	Hotels and motels	
	(*i*) Five and four-star hotels	One space for every 4 guest rooms.
	(*ii*) Three-star hotels	One space for every 8 guest rooms.
	(*iii*) Two-star hotels	One space for every 10 guest rooms.
	(*iv*) Motels	One space for each guest room.
8.	Hospitals	One space for every 10 beds.

When applying the above standards, the parking space requirement of a car can be considered as 3 m × 6 m when individual parking space is required and 2.5 m × 5 m when community parking is envisaged. For two-wheelers, a stall 0.8 m × 2.5 m is desirable. (Ref. 11).

For loading and unloading of commercial vehicles, the requirements given in Table 12.2 can be recommended for Indian conditions (Ref. 4).

Table 12.2. Commercial Vehicle Loading and Unloading Space Standards

(*Source* : Ref. 4)

S. No.	Description	Loading and unloading space standard
1.	All kinds of developments excepting warehouses and godowns.	One berth for initial 500–1500 sq. m. of floor area. Additional berths at the rate of one for every subsequent 1000 sq. m. or part thereof.
2.	For warehouses and godowns.	Two berths for initial 500 to 1500 sq. m. of floor area. Additional berths at the rate of one for every subsequent 500 sq. m. or part thereof.

The space requirement for one commercial vehicle is 3.75 m × 7.5 m (Ref. 4).

The parking area required per bicycle is 1.4–1.8 sq. m.

12.4. Design Standards for On-street Parking Facilities

12.4.1. Common methods of on-street parking

These are :

(i) Parallel parking (*ii*) 30° angle parking (*iii*) 45° angle parking (*iv*) 60° angle parking (*v*) Right angle parking.

Parallel parking consumes the maximum curb length which decreases as the angle of parking increases. The minimum curb length is consumed by right angle parking, which accommodates nearly 2 times the number of vehicles as parallel parking.

On the other hand, parallel parking makes the least use of the width of the street, and this is an important consideration in narrow streets. As the parking angle increases the width of street used also increases.

From the point of view of manoeuvrability, angle parking seems to be better than parallel parking which usually involves a backing motion. Delay to traffic is minimum with angle parking.

As regards safety, it has been noticed that angle parking is more hazardous than parallel parking (Ref. 1, 5).

Considering the above, it is recommended that in general parallel parking should be favoured on streets. On exceptionally wide (wider than 20 m) and low volume streets, consideration might be given for angle parking.

12.4.2. Fig. 12.1 gives some of the typical layouts and dimensions for on-street parking stalls.

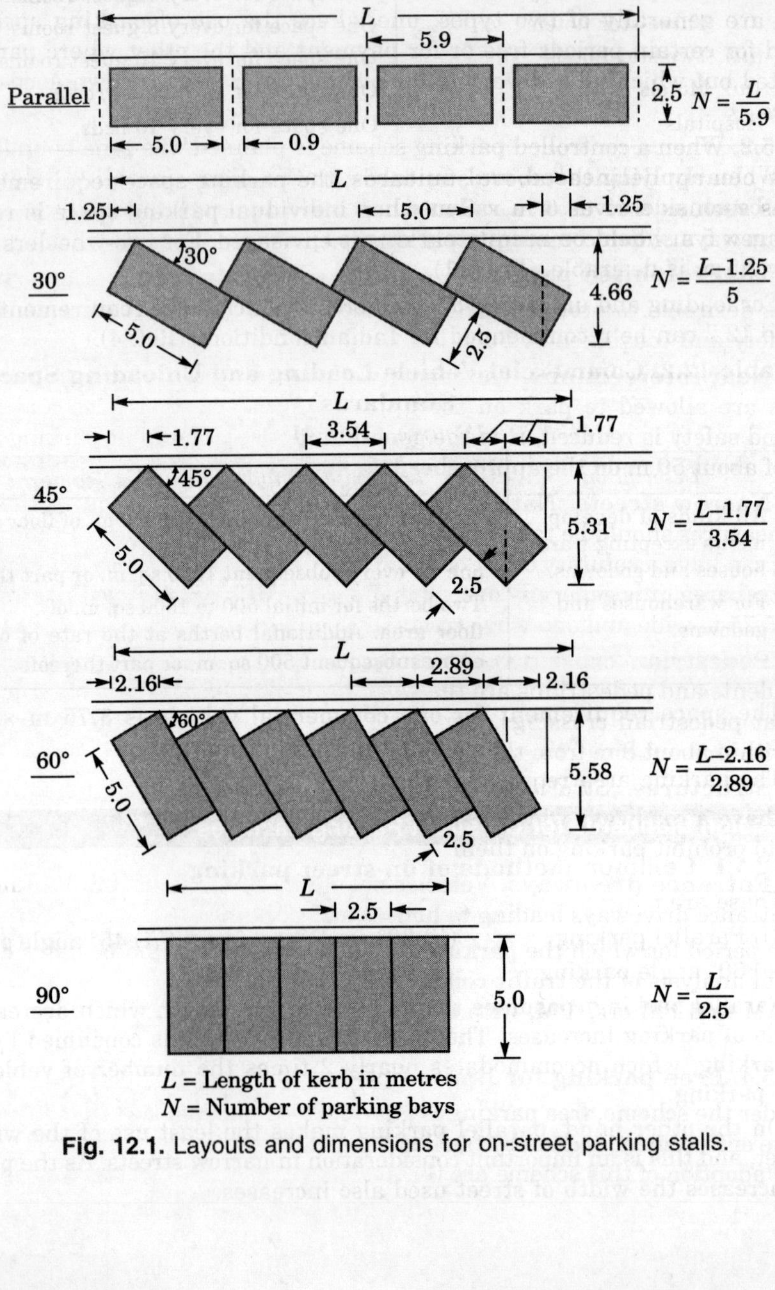

Fig. 12.1. Layouts and dimensions for on-street parking stalls.

12.5. Traffic Regulatory Measures for On-street Parking

12.5.1. Though on-street parking is an extravagant use of the precious street space, it cannot entirely be prohibited. A judicious application of appropriate traffic management measures will help mitigate some of the ill-effects of on-street parking. These measures should be a part of the comprehensive parking policy for the town, formulated after a careful survey of the situation. The regulatory measures range from waiting restrictions in a street to the comprehensive control over an extended area, laying down where vehicles may or may not park, what should be the waiting time and what charges should be levied for parking. The measures should be periodically reviewed and altered, if necessary. The regulatory measures are generally of two types, one where the use of parking space is authorised for certain periods free or for payment and the other where parking is prohibited but which allows waiting for picking up or setting down goods and passengers.

12.5.2. When a controlled parking scheme is planned, the zone boundaries should be clearly delineated and suitable traffic signs installed. Natural boundaries such as a river or a railway line could be advantageously adopted. The carriageways should be adequately marked.

12.5.3. Prohibited parking

It is desirable to prohibit parking at certain locations and for a period to ensure safety and convenience. Such locations are :

(*i*) **Near intersections.** The capacity of an intersection is greatly reduced if vehicles are allowed to park on the approaches. Visibility is also adversely affected and safety is reduced. It is the general practice to prohibit parking for a distance of about 50 m on the approaches to a major intersection.

(*ii*) **Narrow streets.** Narrow streets with heavy traffic require that all possible measures should be taken to remove obstacles to traffic flow. Prohibition of parking can have a salutory effect on traffic flow and congestion. In busy streets of the central area, it is generally desirable to prohibit parking on two-way streets less than 5.75 m wide and one-way streets less than 4.0 m wide.

(*iii*) **Pedestrian crossings.** As already mentioned, parking manoeuvres cause accidents and pedestrians are the worst sufferers. Parked cars obstruct the visibility at pedestrian crossings. For these reasons, it is desirable to prohibit parking within about 8 m from the pedestrian crossing.

(*iv*) **Structures.** Structures such as bridges, tunnels and underpasses generally have a roadways width less than the highway and for this reason it is desirable to prohibit parking on them.

(*v*) **Entrance driveways.** Vehicles should be prohibited from parking in front of entrance driveways leading to houses and buildings.

The period for which the parking is to be prohibited should be determined by a careful analysis of the traffic conditions. Normally the period can be from 8 AM or 9 AM to 6 PM or 7 PM, with relaxations on Sunday or such other closed days.

12.5.4. Free parking for limited period

Under the scheme, free parking is allowed in designated stalls for a limited period. The enforcement is done by traffic police or wardens. The conditions which favour the adoption of this scheme are the availability of sufficient parking spaces to meet the demand and the knowledge that the parkers usually leave their vehicles for a time which is less than the posted time limits.

12.5.5. Parking meters

Under the parking meter scheme a controlled zone is established within which, during specified days and hours, parking may only take place at a metered parking bay, but within which, unless specially prohibited by signs or markings, stopping for the purpose of picking up or setting down passengers or collecting or delivering goods is permitted anywhere without charge, including at metered parking bays for the time being unoccupied (Ref. 6). The controlled zone is distinctly marked by traffic signs to alert the drivers.

The meters are of two general types, the manual meter and the automatic meter. The manual meter is operated by inserting the appropriate coin(s) and working a handle or lever. This activates the clockwork mechanism of the meter. The automatic meter is wound up periodically (say once a week) by an attendant. Upon the insertion of the coin(s) a needle shows in the time scale the time bought and it will start moving across the time scale on the face of the meter until the period of time bought has expired. Thereupon a flag or other indication will appear on the meter and the excess charge period comes into operation. When the excess charge period has also expired, another flag, usually red, appears on the meter face, denoting that the motorist is liable to prosecution and the vehicle may be towed away at the owner's expense.

The meters can be of two general types, *i.e.* the "cancelling" or "non-cumulative" type and the "cumulative" type. In the former, any unexpired time remaining on the meter is cancelled out by the insertion of the next coin. This results in the next motorist getting no more time than he actually pays for. With the latter type, the unexpired time left is added to the time paid for by the next motorist, the total time being limited to the maximum which the scheme allows to be purchased.

Since the metering plan is intended to favour short-term parking and to increase the parking turnover, the time limit has to be judiciously selected. One-hour is a satisfactory limit for the central area of a city, whereas a two-hour limit can be more appropriate to the areas fringing the central area. Near post-offices, banks and public utility offices, a duration of 15–30 min should suffice for the motorists to finish their work.

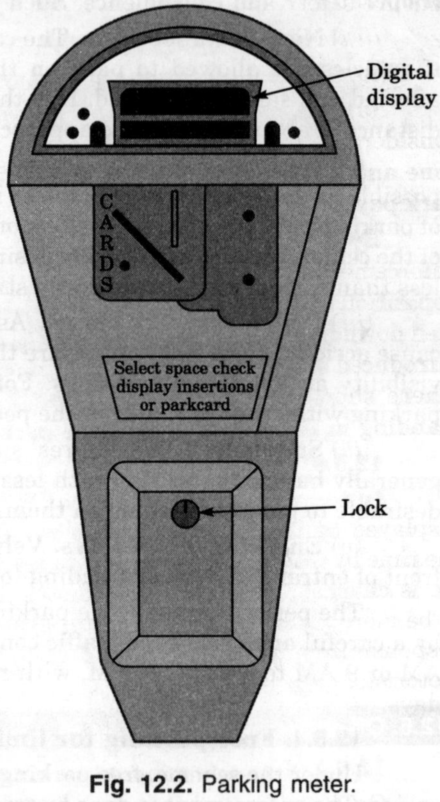

Fig. 12.2. Parking meter.

The meter scheme cannot be successful unless strong enforcement measures are concurrently taken to punish the violators. To guide the motorists the traffic signs and carriageway marking should be provided carefully.

A sketch of a parking meter commonly in use is given in Fig. 12.2 (Ref. 11).

The advantages of a parking meter scheme are :

(*i*) The meter puts a stop to all-day parking and favours the short-term parker. As many as 4 or 5 cars can use a single parking space during the course of a day.

(*ii*) The time check is very accurate and it is easily possible to detect the violators.

(*iii*) Since the parking bays are precisely marked, close and dangerous parking is avoided.

(*iv*) It yields revenue which can be used for improvement of traffic conditions.

(*v*) Parking in front of important buildings and near intersections is avoided.

The following are some of the disadvantages of the metered parking scheme :

(*i*) Strict enforcement is necessary for its success. Laxity can breed disrespect for its proper use.

(*ii*) Since the bays are to be uniformly marked with reference to the largest car operating in the country, it is wasteful of space.

(*iii*) Metered scheme is a solution to only a part of the parking problem.

(*iv*) The meter columns are not aesthetically desirable.

(*v*) The motorist wanting to park the car for a longer period is put to considerable inconvenience. He has to either park his car outside the controlled zone and walk a long distance or he has to park his car in a multi-storeyed car park paying a high price.

By and large, it can be said that wherever it has been introduced and strict enforcement measures applied, metered parking has been a success. The initial opposition of shopkeepers has given way to appreciation. Their business has not died down but in fact has prospered. All the same, the scheme should be carefully introduced after studying the parking habits. It should be favoured in situations where short-term parking is greatly in demand, but long-term parkers are standing in the way of short-term parkers.

12.5.6. Parking discs

In some of the smaller towns in U.K., a system is in use whereby a disc is displayed by the motorist on the wind-screen, showing the time of arrival and the time by which the vehicle has to leave the parking space. Obviously no parking fee is charged. The scheme promotes short-term parking. The enforcement has to be more vigilant than the meter system since abuse can be more common, the responsibility for setting and displaying the disc correctly being that of the motorists.

12.6. Off-street Parking Facilities

12.6.1. Considerations in locating off-street parking facilities

On-street parking can only solve a part of the parking problem of major cities. For a satisfactory answer, various types of off-street facilities become unavoidable. The types of off-street facilities commonly considered are :

(*i*) Surface car parks
(*ii*) Multi-storey car parks
(*iii*) Roof parks
(*iv*) Mechanical car parks
(*v*) Underground car parks.

There are certain basic considerations which govern the location of these facilities. Since these facilities are costly to provide and maintain, a comprehensive study should be done before hand to help determine the location, types and size of these facilities.

Obviously, the prime consideration should be the nearness to the place of useage by the prospective customers. The proposed facility should be close to major parking generators. Examples of such location are major airport terminals, railways terminals, shopping district, sports stadium, major industrial establishment, office buildings and commercial establishments. If a car park is intended to serve the carowners coming from a particular part of the city, it would be preferable to locate the facility on the side of the central area towards the part of the city from where they come. The facility should be served by good access streets. It is desirable to locate the facility such that no exit or entrance is within a distance of at least 50 m from an intersection.

12.6.2. Surface car parks

Surface car parks, properly located and developed on a piece of vacant land or surrounding an office complex or super market, are very popular with the motorists. Great care is needed in their design and operation. The overall aesthetics of the area should receive due attention.

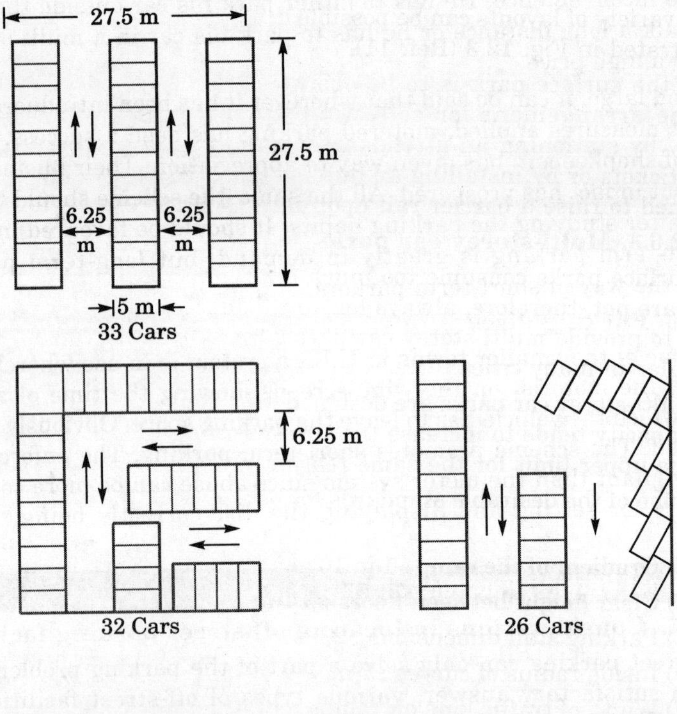

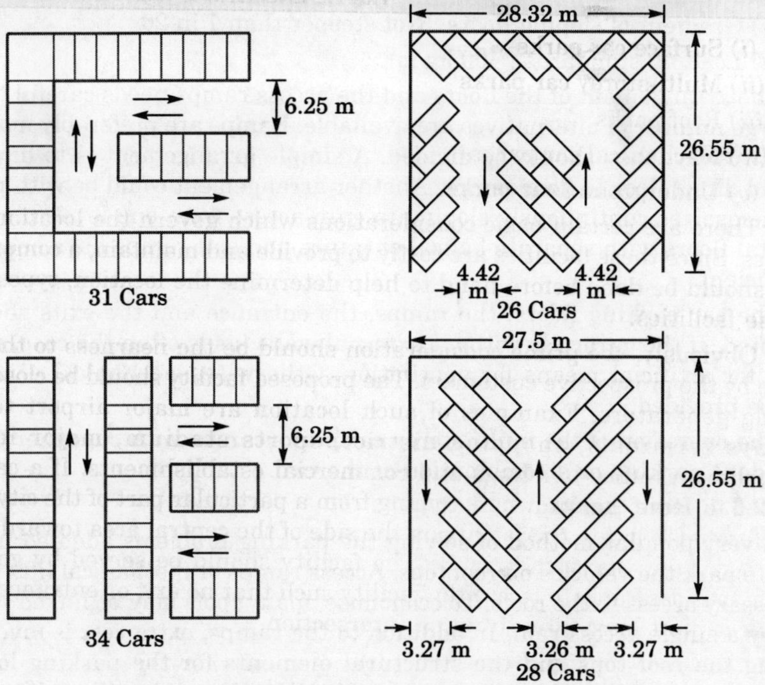

Fig. 12.3. Typical layouts of parking areas.

A stall size of 2.5 m × 5 m is probably adequate for Indian conditions (Ref. 4), predominated by small-size cars, whereas for American conditions 2.6 m × 5.5 m is recommended (Ref. 7).

A variety of layouts can be possible depending upon the area. A few of them are illustrated in Fig. 12.3 (Ref. 11).

If the surface park is to be operated with a fee-charging system, there should be arrangement for collecting the money. This can be done either manually by stationing an attendant in a kiosk at the entrance who sells the parking tickets or by installing an automatic ticket vending machine which can be designed to raise a barrier rail upon insertion of a coin.

12.6.3. Multi-storey car parks

Surface parks consume too much of the precious land in the heart of the city and are not, therefore, always feasible. One of the alternatives when land is costly is to provide multi-storey car parks. Such facilities have become common and popular in many cities.

Multi-storey car parks are designed for a capacity of about 400 to 500 cars. Larger capacity tends to increase the time for unparking a car. About five floors is also the upper limit for the same reason.

Some of the desirable standards for designing of the multi-storey car parks are :

(*i*) Gradient of the ramp : 1 in 10 generally and 1 in 8 for very short ramps.

(*ii*) Clear height between floors : 2.1 m

(*iii*) Parking stall dimensions : 2.5 m × 5 m

(*iv*) Inside radius of curves : 7 m

(*v*) Width of traffic lane on ramps and entrances : 3.75 m

(*vi*) Gradient of sloping floors : Not steeper than 1 in 20

(*vii*) Loading standards : 400 kg/m^2.

The arrangement of the floors and the access ramps needs careful thought and a large number of alternatives are available. Ramps are preferably made one-way. If two-way, they should be divided. A simple arrangement is to have level floors and direct floor to floor ramps. Another arrangement could be with parking floors themselves continuously sloping to gain access from one level to another. Horizontal floors with separate helical entrance and exit ramps have been found to be efficient.

The car parking floors, the ramps, the entrance and the exits should be well lighted. If the garage is without external walls, as is often the case, there is no need for artificial means for ventilation. Otherwise, mechanical ventilators should be provided.

The operation of the multi-storey car parks can be with customer parking or attendant parking or a combination of the two.

12.6.4. Roof parks

A very popular method of solving the parking problems adopted in many cities is to park the vehicles on roof tops. Access ramps or mechanical lifts provide the necessary access to the roofs. To economise, many roofs may be linked together served by a single access ram. In addition to the ramps, extra cost is involved in designing the roof tops and the structural elements for the parking load. An extensive system of linked car parks at roof level, integrated with multistoreyed parking garages is in use at Coventry, U.K. (Ref. 8).

12.6.5. Mechanical car parks

Mechanical car parks provide for lifting of the cars from floor to floor by means of a lift and transfer of cars to and from the parking stall by means of wheeling or mechanically operated transfer dollies or cradles. Since the ramps and aisles are eliminated in this system, it is more economical in space as compared to the ramped system, multistorey garages. The disadvantages are the higher maintenance costs and the possibility of breakdown due to mechanical or power failure. The latest innovation in mechanical parking is *Robotic parking,* where the parking is done by robots replacing man (Ref. 12), thus saving in costs.

12.6.6. Underground car parks

The great advantage of underground car parks is the least intrusion they cause to the aesthetics of a place. An excellent example is the underground parking space below Hyde Park in London and in Connaught Place, New Delhi. These parks can be built in the basement of any multi storeyed building or below open spaces. Since the work involves large quantities of excavation, construction of retaining walls, ventilation and lighting, such car parks tend to be very costly. Underground car parks can be single storeyed or multi storeyed though the latter tend to be very costly.

12.7. Peripheral Parking Schemes

Since the centre of the town is the worst hit by the parking problem, it is natural to think in terms of providing parking facilities at the periphery of the town and induce the motorists to park there and travel to the busy town centre by some other mode. There are a number of schemes which have been devised to secure the desired objectives. They are :

(*i*) Park and walk (*ii*) Park and ride (*iii*) Kiss and ride.

12.7.1. Park and walk

Under this scheme the motorists are induced to park at the outskirts of the town and walk down to the town centre. The inducement is in the form of lower parking charges at the periphery than at the town centre or no parking charges at all. The inducement should be really attractive if the scheme has to succeed.

12.7.2. Park and ride

This scheme provides for peripheral parking facilities and public transport side to the destinations in the town centre. This is an attractive scheme and has been tried with success in many towns. Since the motorist voluntarily parks his car at the fringe under this schemes, the scheme should be really appealing to him. For success, the total travel time including parking time, waiting time at the bus stop and travel time by bus should not be excessively high to make him look at it with disfavour and decide to bring the car to the town centre inspite of the high parking charges there. Also, the cost of peripheral parking plus the charges for to and fro journey by bus should be less than the cost of travel by his car plus the charges for parking at the town centre.

12.7.3. Kiss and ride

This scheme derives its name from the goodbye kiss of a wife dropping her husband in the car in the morning near a bus stop from where the husband goes to work in a bus. The reverse operation takes place in the evening. Adequate space for parking of cars near the bus stops where the husbands can be dropped off or their arrivals awaited is an essential prerequisite for the success of the scheme.

12.8. Loading and Unloading Facilities

12.8.1. Bus-bays

Properly designed bus-bays recessed into the curb facilitate loading and unloading of passengers without the vehicles blocking the stream of traffic on the carriageway. The following are the guidelines for the location of the bus-bays :

(*i*) The bus stops should not be located too close to the intersections. A minimum distance of 75 m from the intersections is desirable for urban situation and for rural locations a distance of 300 m is desirable. The bus stops should be located preferably on the farther side of the intersection.

(*ii*) Bus stops should be as far as possible so located as to disembark the passengers at safe places such as curbs or islands.

(*iii*) For buses intending to turn right at an intersection, the stop should be sufficiently away from the intersection so that the bus can be manoeuvered from the curb to the extreme right lane well before the intersection.

The length of the recess should be about 12–15 m for a single bus with an additional length of 12–15 m for every additional bus. The taper on either side should be about 8 : 1, the maximum value being 6 : 1.

A simple design of the bus-bay directly abutting the footway is given in Fig. 12.4 (Ref. 11).

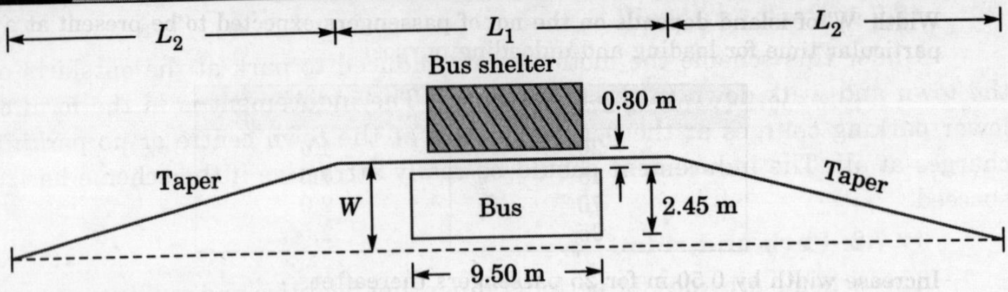

Notes :

1. L = 12 m minimum for congested locations for single bus, add 12 m for every additional bus
   = 15 m desirable for single bus, add 15 m for every additional bus
2. W = 3 m minimum for congested areas = 3.5 m desirable
3. L = 6 W minimum for congested areas = 8 W desirable

Fig. 12.4. Layout of bus-bay abutting footway.

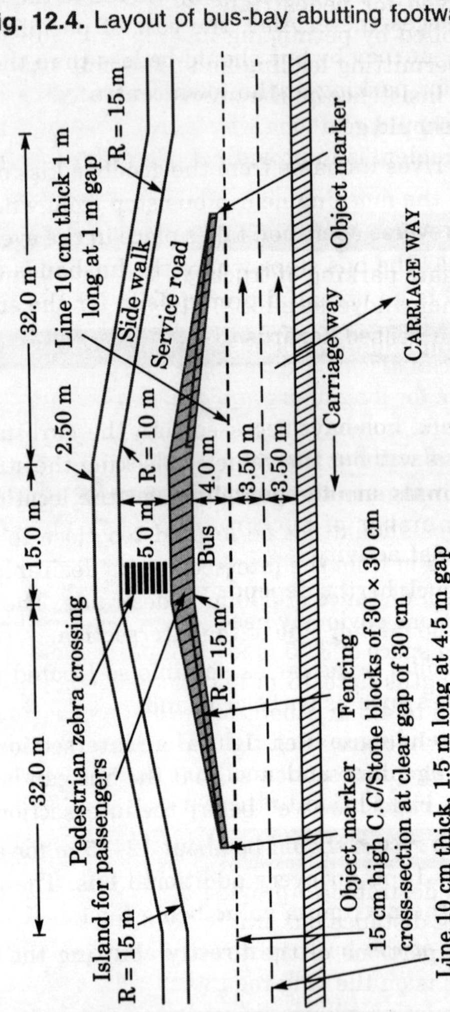

Notes :

Width 'W' of island depends on the no. of passengers expected to be present at a particular time for loading and unloading purposes.

No. of Passengers	*Width*
20	1.0
45	1.5
70	2.0
95	2.5

Increase width by 0.50 m for 25 passengers thereafter.

Fig. 12.5. Layout of bus-bay with a separate lane for slow moving traffic.

A layout of the bus-bay when a separate lane is provided for slow moving vehicles is given in Fig. 12.5 (Ref. 11).

12.8.2. Commercial traffic

Loading and unloading of goods by trucks on the roadside makes demands upon the space reserved for pedestrians as well as for moving vehicles. The problem can be controlled by permitting loading and unloading only between 6 PM and 8 PM, or by permitting loading and unloading only at specified location. For new development, insistence on the provision of loading and unloading spaces indicated in Sec. 12.3 should go a long way in overcoming the difficulty. The long-term solution of the problem is in providing adequately designed truck terminals outside the cities.

12.9. Truck Terminals

The indiscriminate parking of trucks in the streets and upon spaces of the towns and cities has many adverse effects. The foremost is the degradation of the environment and the increased hazards to traffic. Security of the goods contained in the parked vehicles can also be a serious problem. The needs of the truck drivers who have often to drive for long hours and consequently need rest, toilet facilities and food deserve careful consideration. The right approach is to provide well designed truck terminals at the fringes of the cities.

The truck terminals should be located on important arterial routes and at the confluence of a number of such routes. They should be close to major generators of commercial activity.

The length of truck berths depends upon the type of trucks to be handled. Truck-trailer combinations obviously need longer lengths than single unit trucks. For single unit trucks a space of 3.75 m × 7.5 m per vehicle is adequate. For truck-tailer combinations lengths upto 15 m may be needed. The width of the loading platforms should be 3.5 to 4.5 m. About 600–750 trucks can be accommodated in a parking area of one hectare (Ref. 10). The area required for the entrance forecourt, hotel, resting place and ancillary facilities (weigh bridge, fuelling facilities, workshop) is in addition to the parking area requirement.

12.10. Long Distance Bus Terminals

In a town, it is desirable to provide a properly designed bus terminal to handle exclusively long distance bus traffic. Such terminals should preferably be outside the congested portions of the town. If a town has ring road, the ideal location in many cases is on the ring road itself. The terminal should be planned such that one-way circulation of buses is achieved. The entrances and exits should be located with this in view. Pedestrian movements will be heavy in a terminal and should, as far as possible, not conflict with vehicular movements. Provision for facilities such as cafetaria, book-stalls, toilets etc. should be made. The platforms where pedestrians wait should be raised. Parking facilities for cars, taxis, cycles, scooters, etc. should be liberally provided. The terminals should be planned for the anticipated future traffic in the design year.

The bus width is generally 2.5 m and it is desirable to have 3.3–3.75 m wide lanes. The length of a bus about 10 m. The vertical clearance should be at least 3.75 m. The passenger platforms should have a minimum width of 2.5 m.

I. MULTIPLE CHOICE QUESTIONS

1. When the path travelled along the road surface is more than the circumferential movement of the wheels due to rotation, then it results in
   (*a*) slipping (*b*) skidding (*c*) turning (*d*) revolving
2. What is the closest distance you may park on the approach side of a children's crossing?
   (*a*) 25 metres (*b*) 20 metres (*c*) 10 metres (*d*) 9 metres
3. It is early morning and there is no other traffic about. You approach an intersection when the traffic lights change from green to yellow. What should you do?
   (*a*) keep going (*b*) drive through and sound your horn
   (*c*) stop
4. In surveying operations, the word 'reciprocal' can be associated with
   1. ranging 2. levelling 3. contouring
   Which of the above statements is/are correct?
   (*a*) 1 only (*b*) 1 and 2 only (*c*) 2 and 3 only (*d*) 1, 2 and 3
5. Which set of traffic functional design as well as for 'highway capacity' design?
   (*a*) Origin and destination studies (*b*) Parking and accident studies
   (*c*) Speed and volume studies (*d*) Axle load studies
6. Which one of the following traffic survey schemes is most relevant when deciding on locating major 'routes' in a city?
   (*a*) Traffic volume survey (*b*) Origin and destination survey
   (*c*) Speed survey (*d*) Traffic capacity survey

Answers

1. (*b*) **2.** (*b*) **3.** (*c*) **4.** (*b*) **5.** (*a*)
6. (*b*)

II. SOLVED QUESTIONS AND ANSWERS

Q. 1. What is parking?

Ans. Parking is one of the major problems that is created by the increasing road traffic. It is an impact of transport development. The availability of less space in urban areas has increased the demand for parking space especially in areas like central business district. This affects the mode choice also. This has a great economical impact.

Q. 2. What is parking load?

Ans. Parking load gives the area under the accumulation curve. It can also be obtained by simply multiplying the number of vehicles occupying the parking area at each time interval with the time interval. It is expressed as vehicle hours.

Q. 3. What is parking volume?

Ans. It is the total number of vehicles parked to a given duration of time. This does not account for repetition of vehicles. The actual volume of vehicles entered in the area is recorded.

Q. 4. What do you mean by parking accumulation?

Ans. Parking accumulation is the number of vehicles parked at a given instant of time. Normally this is expressed by accumulation curve. Accumulation curve is the graph obtained by plotting the number of bays occupied with respect to time.

Q. 5. What is traffic congestion?

Ans. Traffic congestion is a condition on transport networks that occurs as use increases, and is characterized by slower speeds, longer trip times, and increased vehicular queueing. The most common example is the physical use of roads by vehicles. When traffic demand is great enough that the interaction between vehicles slows the speed of the traffic stream, this results in some congestion.

Q. 6. What is road rage?

Ans. Road rage is aggressive or angry behaviour by a driver of an automobile or other motor vehicle. Such behaviour might include rude gestures, verbal insults, deliberately driving in an unsafe or threatening manner, or making threats. Road rage can lead to altercations, assaults, and collisions which result in injuries and even deaths. It can be thought of as an extreme case of aggressive driving.

Q. 7. When does traffic congestion take place?

Ans. Traffic congestion occurs wherever demand exceeds the capacity of the transportation system. This lecture gives an overview of how congestion is generated, how it can be measured or quantified; and also the various countermeasures to be taken in order to counteract congestion. Adequate performance measures are needed in order to quantify congestion in a transportation system. Quality of service measures indicates the degree of traveller satisfaction with system performance and this is covered under traveller perception. Several measures have been taken in order to counteract congestion. They are basically classified into supply and demand measures. An overview of all these aspects of congestion is dealt with in this lecture.

Congestion has a large number of ill effects on drivers, environment, health and the economy in the following ways.

Drivers who encounter unexpected traffic may be late for work and other appointments causing a loss in productivity and their valuable time.

Q. 8. What is roof parks?

Ans. In recent years, roof parks have become more popular. Space within cities is limited and buildings often replace expensive, green parts. At the same time, office owners and property managers are looking for ways to provide a better and greener work environment for their employees. By creating a roof park, you create extra space and the possibility to relax, have meetings and have lunch while getting some fresh air. Hotels and restaurants are also increasingly making use of the unique possibilities roofs have to offer, like growing their own vegetables and herbs. Our team of designers, architects and landscapers can transform your roof into an ideal and accessible outside space from A to Z.

Q. 9. Why a roof park?

Ans. Green has a strong and soothing influence on the mood of your tenants, employees or guests. The addition of a green outside space to your building will increase sustainability and create multifunctional spaces. This way the roof park can even be an attraction for guests, relations and potential clients. Furthermore the roof park can offer an attractive spot for butterflies and bees, storage for rainwater and cooling down your building.

Q. 10. What is multi-storey car park?

Ans. A multi-storey car park (also called a parking garage, parking structure, parking ramp, parkade, parking building, parking deck or indoor parking) is a building designed for car parking and where there are a number of floors or levels on which parking takes place. It is essentially a stacked car park.

Q. 11. What do you mean by kiss and ride?

Ans. Many railway stations and airports feature a "kiss-and-ride" or "kiss-and-fly" area in which cars can stop briefly to discharge or, less commonly, pick up passengers. The term first appeared in a 20 January 1956 report in the Los Angeles Times. It refers to the nominal scenario whereby a passenger is driven to the station by spouse or partner; they kiss each other goodbye before the passenger catches the train.

Some high-speed railway stations in Taiwan have signs outside stations reading "Kiss and Ride" in English, with Chinese characters above the words that read "temporary pick-up and drop-off zone". Most people in Taiwan have no idea what the colloquialism in English means white or yellow writing and symbols on a background of blue green brown black.

III. REVIEW QUESTIONS

1. Discuss the traffic and parking problems in India.
2. What are the ill-effects of parking?
3. Explain zoning and parking space requirement standards.
4. Explain off street parking facilities.
5. What is meant by long distance bus terminals?

REFERENCES

1. J. Brierly, Parking of Motor Vehicles, C.R. Books Ltd., London 1962.
2. Traffic in Towns, The Buchanan Report, H.M.S.O., London, 1963.
3. Witheford D.K. and C.E. Kannan, Zoning, Parking and Traffic, ENO Foundation for Transportation, Saugatuck, 1972.
4. Tentative Recommendations on the Provision of Parking Spaces for Urban Areas, Special Publication 12, 1.R.C., New Delhi, 1973.
5. Evans, H.K., Editor, Traffic Engineering Handbook, Institute of Traffic Engineers, New Haven, 1950.
6. Hogg, A.W., Control of Street Parking, Institution of Municipal Engineers, London, 1960.
7. Burrage, R.H. and E.G. Mogren, Parking, ENO Foundation for Highway Traffic Control, Saugatuck, 1957.

8. Berry, G., Roof-top Parking and Mechanical Parking, Institution of Municipal Engineers, London, 1960.
9. Guidelines on regulation and control mixed traffic in Urban areas, IRC : 70-1977 Indian Roads Congress, New Delhi, 1977, Road Congress
10. Department of the Environment, Lorry Parking, H.M.S.O., London.
11. Development of Aesthetic Designs for Roadside Furniture, State-of-Art Report, Ministry of Surface Transport, prepared by Dr. L.R. Kadiyali and Associates, New Delhi, 1996.
12. Gupta, O.P., Vijay Gupta and Chandan Gupta, Robotic parking garages are the need of Metropolitan cities in India, paper presented in the National Seminar on Innovative Materials and Mechanised Techniques and Technology in Building Construction, New Delhi, 2009.

SECTION 6
TRAFFIC CONTROL

13. Traffic Signs

13.1. Importance of Traffic Signs

Traffic signs perform a number of functions :

(*i*) They give timely warning of hazardous situations when they are not self evident.

(*ii*) They are of great help in regulating traffic by imparting messages to the drivers about the need to stop, give way and limit their speeds.

(*iii*) They give information as to highway routes, directions and points of interest.

In view of the above, traffic signs occupy a significant place in traffic engineering.

13.2. Need for International Standardization

In order to facilitate people travelling from one country by road in another country where the language may be different, standardization of traffic signs assumes considerable importance. Realizing the need for such standardization, the United Nations convened a conference of Road and Motor Transport in Geneva in 1949. As a result, a Protocol on Road Signs and Signals was signed by a number of countries and was ratified subsequently by the various countries (Ref. 1, 2). For the ECAFE Region, the U.N. has evolved a code of system for signs, signals and markings (Ref. 3). Even though the adoption of a uniform system of traffic signs by various countries is beset with a number of problems, one of the principles commonly agreed to by all the signatory countries is the adoption of graphic symbols to convey messages, rather than lettered messages.

13.3. The Situation in India

According to the Motor Vehicles Act, 1988 (Ref. 4), the State Governments are required to erect traffic signs which have been prescribed in the Act. The Traffic Engineering Committee of the Indian Road Congress have prepared a Standard on Road Signs (Ref. 5) and it has been incorporated in a recent revision of the relevant Section of the Motor Vehicles Act. India was one of the countries which have signed the United Nations Protocol in 1949 and it was, therefore, felt necessary that the traffic signs be modified on the lines of international practice. The I.R.C. Standards have been evolved keeping in view various international conventions.

13.4. General Principles of Traffic Signing

The objectives in pursuance of which traffic signs are installed can best be achieved if some general principles are observed in their installation. These are :

(*i*) Traffic signs should be installed only by the authority of law. Unofficial and nonessential signs should not be permitted.

(*ii*) For imparting a sense of respect towards signs, proper enforcement measures should be undertaken.

(*iii*) Signs should be put up only after traffic engineering studies and sound engineering judgment.

(*iv*) Excessive use of signs should not be resorted to. A conservative use of regulatory and warning signs is recommended as these signs, if used in excess, tend to lose their effectiveness.

(*v*) The signs should be designed for the foreseeable traffic conditions and speeds on the roads on which they are proposed to be used.

(*vi*) The basic requirements of a sign are that it be legible to those using it and that it be understood in time to produce a proper response. This means :

(*a*) high visibility, both during night and day.

(*b*) lettering or symbols of adequate size for being read from far away by a speeding driver without diverting the gaze through too great an angle.

(*c*) location at a conspicuous position to be able to be seen by drivers, without being obscured by vehicles and other objects.

(*d*) Simplicity and uniformity in design, position and applications.

(*vii*) It is desirable that there should be two sizes for each type of sign, namely (*a*) standard size and (*b*) a reduced size where conditions do not permit or the safety of road users does not require the erection of the standard size.

13.5. Types of Traffic Signs

According to the U.N. Protocol (Ref. 2), the international system of traffic signs comprises of the following classes :

(*a*) Danger signs, also known as warning signs or cautionery signs.

(*b*) Signs giving definite instruction, also known as regulatory signs in some countries (Ref. 5). These are further subdivided into :

(*i*) Prohibitory signs. (*ii*) Mandatory signs.

(*c*) Information signs, further sub-divided into :

(*i*) Indication signs.

(*ii*) Advance direction signs and direction signs.

(*iii*) Place and route identification signs.

The I.R.C. Standard (Ref. 4) generally conforms to the above classification.

13.6. Danger Signs (Warning Signs or Cautionery Signs)

Danger signs are used when it is deemed necessary to warn traffic of existing or potentially hazardous conditions on or adjacent to a highway or street. Warning signs are of great help in ensuring safety of traffic. The use of warning signs should be kept to a minimum, however, because their unnecessary use tends to breed disrespect for all signs.

The U.N. Protocol recommends an equilateral triangle with one point upwards. The side of the triangle, as per the I.R.C. Standard, is 900 mm for a standard size and 600 mm for a reduced size. These signs have a red border and the symbols indicated therein are in black colour against a white background.

The warning signs recommended by the I.R.C. Standard are illustrated in Fig. 13.1, along with the message they intend to convey.

13.7. Prohibitory Signs

These signs are part of the regulatory signs, which are intended to inform the highway users of traffic laws or regulations. The prohibitory signs give definite negative instructions prohibiting the motorist from making particular manoeuvres. They may be of the following types :

(*i*) Movement prohibition, such as prohibition of certain turns, prohibition of entry, prohibition of overtaking, one-way traffic, exclusion of certain types of vehicles or all vehicles etc.

(*ii*) Waiting restriction signs, such as prohibited waiting etc.

(*iii*) Restrictions on dimensions, weight or speed of vehicle.

According to the I.R.C. Standards, the prohibitory signs are of a circular shape, with a diameter of 0.60 m for signs of the standard size and 0.40 m for the reduced size. The signs have a red border. The colour of the background is white for speed control, blue for waiting and parking restrictions, and blue for direction control and other signs. The symbols are in black colour for prohibitory signs and white in colour for direction control signs. These signs are illustrated in Fig. 13.2 along with the messages intended to be conveyed by them.

13.8. Mandatory Signs

Mandatory signs are part of regulatory signs and are intended to convey definite positive instructions when it is desired that motorists take some positive action. The two most important mandatory signs are the (*i*) STOP sign and the (*ii*) YIELD or GIVE WAY sign.

The STOP sign requires all vehicles to come to a halt before the stop line. Because it causes a substantial inconvenience to motorists, it should not be indiscriminately used. It is generally used at an intersection where the following conditions exist :

(*i*) Intersection of a less important road with a main road where the application of the normal right-of-way rule is unduly hazardous.

(*ii*) Street entering a through highway or street.

(*iii*) Unsignalised intersection in a signalised area.

(*iv*) Other intersections where a combination of high speed, restricted view and serious accident record indicates a need for control by the STOP sign.

The STOP sign should not be used :

(*i*) on through roadways or expressways.

(*ii*) for speed control.

(*iii*) at signalised intersections.

The various designs used for the STOP sign are :

American practice. Octagon with written message STOP in white on a red background, with white border.

European practice. Octagon with written message STOP and border in white or yellow on a red background.

U.N. practice. Red triangle with point downwards inscribed within a red circle, the word STOP may be written within the triangle.

I.R.C. Standards. Octagon with a white border and red background, the side of the octagon being 900 mm for the standard sized sign, and 600 mm for the smaller sized sign. It shall be used in combination with a definition plate carrying the message STOP.

The YIELD or GIVE WAY sign is used to assign right-of-way to traffic on certain approaches to an intersection. Vehicles controlled by a YIELD sign need stop only when necessary to avoid interference with other traffic that is given the right-of-way. It is used under the following conditions :

(*i*) On a minor road at the entrance to an intersection where it is necessary to assign right-of-way to the major road, but where a stop is not necessary at all times.

(*ii*) On the entrance ramp to an expressway when an acceleration lane is not provided.

(*iii*) Within an intersection with a divided highway where a STOP sign is present at the entrance to the first roadway and further control is necessary at the entrance to the second roadway and where the median width between the two roadways exceeds 9 m.

(*iv*) Where there is a separate or channelised left-turn lane, without an adequate acceleration lane.

(*v*) At any intersection where a special problem exists and where an engineering study indicates the problem to be susceptible of correction by the use of the YIELD sign.

THE YIELD sign should not be used

(*i*) to control the major flow of traffic at an intersection.

(*ii*) on the approaches of more than one of the intersection streets or highways or at any intersection where there are STOP signs on one or more of the approaches, except, under special circumstances, to provide minor movement control within complex intersections.

(*iii*) on the through roadways or expressways.

The following are the designs used for the YIELD or GIVE WAY signs by various authorities :

American practice. Downward pointing equilateral triangle having a red border band, a white interior, with the word YIELD in red.

U.K. practice. Downward pointing equilateral triangle having a red border.

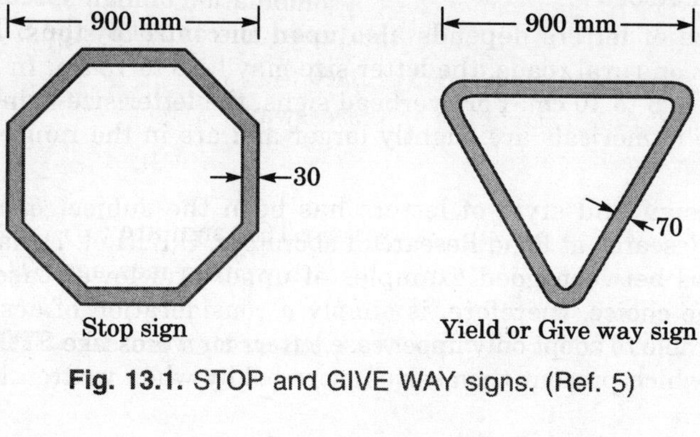

Fig. 13.1. STOP and GIVE WAY signs. (Ref. 5)

I.R.C. Standards. Downward pointing equilateral triangle having a red border and a white background. The side of equilateral triangle is 900 mm long in the standard sized sign and 600 mm long in the smaller sized sign. It shall be used in combination with a definition plate carrying the message "GIVE WAY"

Fig. 13.1 gives the STOP and GIVE WAY signs recommended in the standard of the I.R.C. (Ref. 5).

13.9. Informatory Signs

Informatory signs are intended to guide the motorist along streets and highways, to inform him of intersecting routes, to direct him to cities, towns, villages or other important destinations, to identify nearby rivers and streams, parks, forests and historical sites, and generally to give him such information as will help him along his way in the most simple, direct manner possible.

Informatory signs do not lose their effectiveness by over-use and it is desirable to erect them as frequently as is necessary, and in any case, at locations where the motorist would be in doubt.

A considerable of importance in dealing with the provision of informatory signs is the size of lettering. The letters should be of such a size as can be easily read when the vehicles are moving at speed. The American practice (Ref. 6) is to relate the size of letter to the highway type rather than to variable speeds on any class of highway. Work at the Road Research Laboratory, U.K. has resulted in the following formulae for height of letters.

$$x = \frac{S}{10} + \frac{V}{100}(N + 6) \text{ inches}$$

$$H = \left[\frac{S}{10} + \frac{V}{100}(N + 6)\right] \text{ inches}$$

where, x = height of small lower case letters such as x and z

H = height of upper case letters such as H and I

S = Distance of the sign from the vehicle path in ft.

V = Velocity of the vehicle in miles/hour.

N = Number of names on the sign.

If N is assumed to be four, the above formulae further simplify to :

$$x = \frac{S + V}{10} \text{ 9}, \quad H = \frac{4}{3}\left(\frac{S + V}{10}\right)$$

In the above formulae, the speed limit in force may be adopted as the value of V wherever such a limit is in force, and in other cases the design speed may be used as the value of V.

The size of letters depends also upon the type of signs. For advance direction signs on rural roads, the letter size may be 8 to 15 cm. In urban areas, the letter size is 8 to 10 cm. For overhead signs, the letter size is in the range of 20–25 cm; the numericals are slightly larger and are in the range of 25–30 cm (Ref. 6).

The design and style of letters has been the subject of a number of experiments. Research at Road Research Laboratory, U.K. (Ref. 7), has shown that the differences between good examples of upper and lower case scripts are negligible. The choice, therefore, is simply a consideration of aesthetics. It is perhaps preferable to adopt only upper case letters for words like STOP, DANGER, POLICE etc. which present themselves more boldly when written in upper case letters. The American practice is to use uppercase letters generally, except that destination names may be in lower case lettering, with initial upper case. In India, the present practice is to adopt only upper case letters (Ref. 8).

13.10. Indication Signs

One of the sub-classes of Informatory signs are Indication signs. They generally provide information of facilities such as hospital, filling station, telephone, eating place, rest house, first-aid post etc. They are generally rectangular in shape. The I.R.C. standard provides for a size of 600 mm × 450 mm with a black symbol, against a white rectangle and blue background. The indication signs recommended in this code are given in Fig. 13.2.

Information regarding parking facilities are frequently needed. Some of the signs recommended in the I.R.C. Standard are given in Fig. 13.2.

13.11. Direction Signs, Advance Direction Signs and Place Identification Signs

Direction signs indicate the name of the place and are rectangular in shape, terminating in the form of an arrow head. A typical form is given in Fig. 13.2.

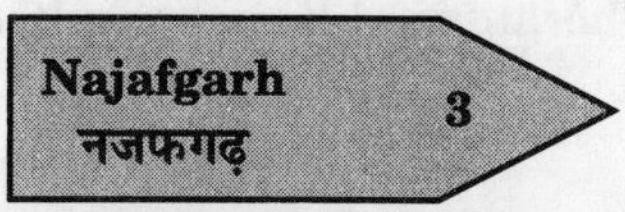

Fig. 13.2. Direction sign.

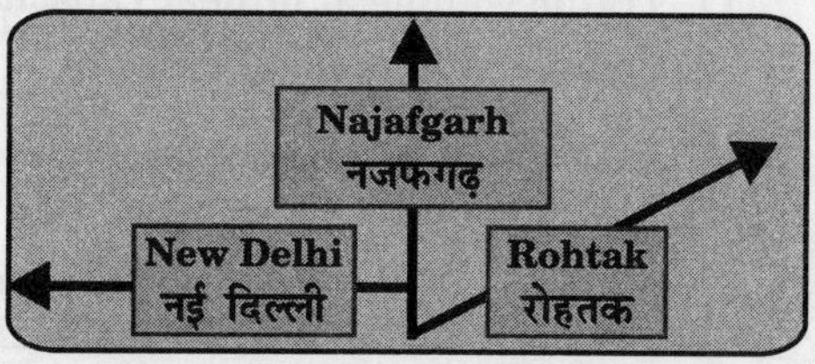

Fig. 13.3. Advance direction sign.

Advance direction signs are necessary at intersections of roads. They are generally rectangular in shape. An example of an advance direction recommended by the I.R.C. is given in Fig. 13.3.

A destination sign indicates the names of places and the distances thereto. Fig. 13.4 gives such a sign.

Fig. 13.5 gives a Place Identification Sign.

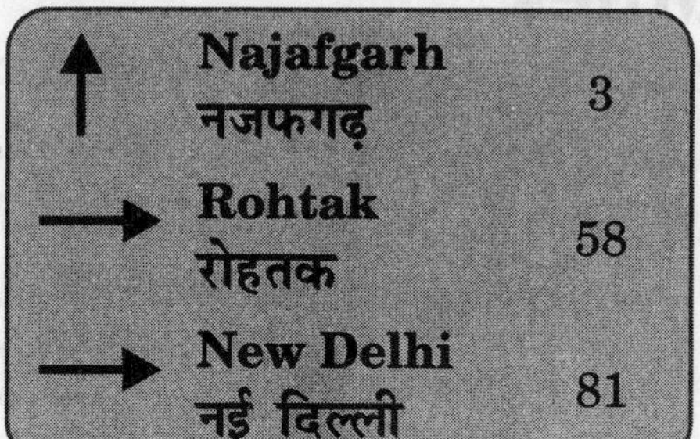

Fig. 13.4. Destination sign.

Fig. 13.5. Place Identification Sign.

A Reassurance Sign reassures the traveller about the places ahead and the distances thereto Fig. 13.6 gives such a sign.

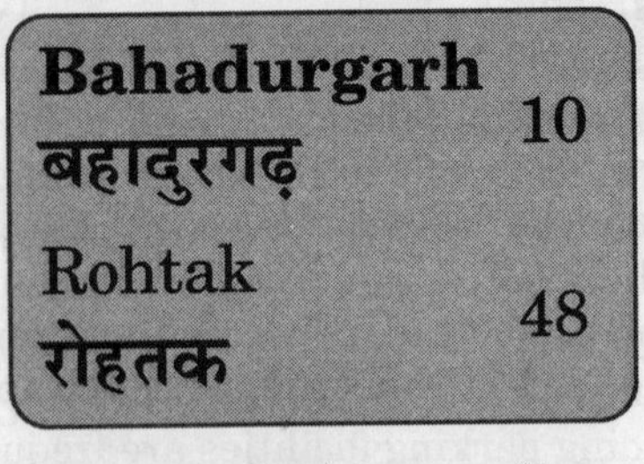

Fig. 13.6. Reassurance Sign.

13.12. Overhead Signs

Overhead signs are provided at locations where the following considerations exist (Ref. 6) :-

1. Traffic volume at or near capacity
2. Complex interchange

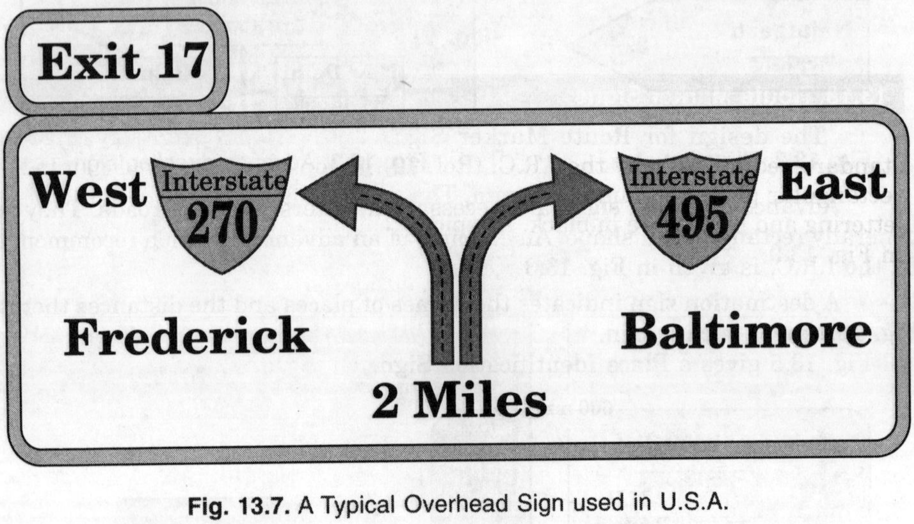

Fig. 13.7. A Typical Overhead Sign used in U.S.A.

3. Three or more lanes in each direction
4. Restricted sight distance
5. Closely spaced interchanges
6. Multi-lane exits
7. Large percentage of trucks
8. Street lighting background
9. High speed traffic
10. Consistency of sign message location through a series of interchanges
11. Insufficient space for ground mounted signs
12. Junction of an interstate route with another freeway
13. Right exit ramps.

The existence of any or more of the conditions listed does not automatically justify the use of overhead signs. Some of the elements listed above can be made less critical by close coordination between design and operation.

A typical overhead sign used in U.S.A. is given in Fig. 13.7.

The height of the panel ranges from 125 cm to 305 cm, depending upon the number of lines and messages. The letter size is in the range of 25-30 cm. The maximum panel length is about 7 m. The vertical clearance needed in India is 5.5 m (Ref. 11).

The support system can be any one of the following :

1. Cantilever with one post
2. Butterfly with one post
3. Two post unit.

These are illustrated in Fig. 13.8.

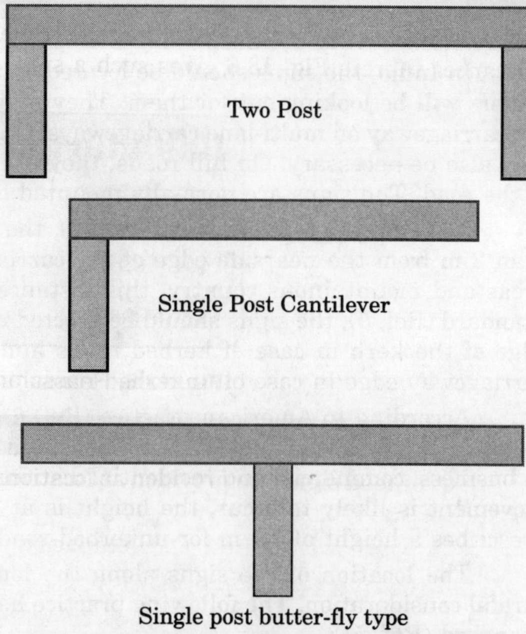

Fig. 13.8. Support system for overhead signs.

13.13. Route Marker Signs

The design for Route Marker Signs for National Highways has been standardised in India by the I.R.C. (Ref. 10). It consists of a shield painted on a rectangular plate 450 mm × 600 mm. The sign has a yellow background and the lettering and border are in black. A typical assembly of route marker sign is given in Fig. 13.9.

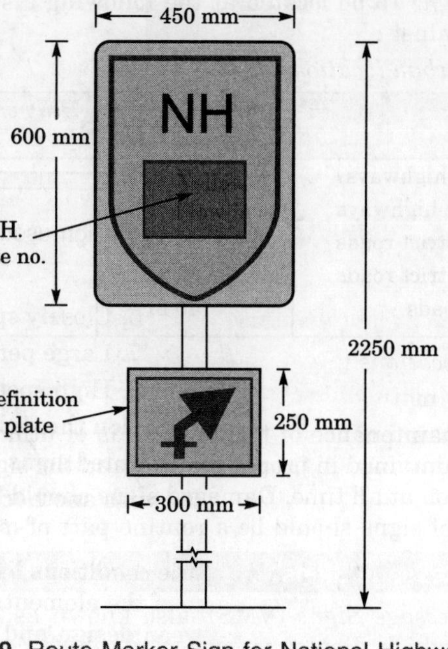

Fig. 13.9. Route Marker Sign for National Highway.

13.14. Location, Height and Maintenance of Traffic Signs

In India, the signs should be located on the left side of the road where the drivers will be looking out for them. They may be repeated on the other side of the carriageway on multi-lane carriageways. On wide expressways, overhead signs may also be necessary. On hill roads, they are normally fixed on the valley side of the road. The signs are normally mounted on posts.

According to U.N. Protocol (Ref. 2), the signs should be erected not more than 2 m from the nearside edge of the carriageway in open areas. In built-up areas and mountainous country, this distance can be 0.5 m. As per the I.R.C. Standard (Ref. 5), the signs should be erected not less than 60 cm away from the edge of the kerb in case of kerbed roads and at a distance of 2.3 m from the carriageway edge in case of unkerbed roads.

According to American practice (Ref. 6), the signs in rural areas shall be mounted at a height of a least 1.5 m (measured from the bottom to the pavement). In business, commercial and residential districts where parking and/or pedestrian movement is likely to occur, the height is at least 2.1 m. The I.R.C. Standard prescribes a height of 1.5 m for unkerbed roads and 2 m for kerbed roads.

The location of the signs along the longitudinal axis of the road needs careful consideration. The following practice has been recommended in the I.R.C. Standard (Ref. 5) :

STOP Sign. To be located at the point where the vehicle is required to stop or as near thereto as possible, say 1.5 to 3 metres. Where there is a marked pedestrian crossing, the stop sign shall be erected 1.2 m in advance of the pedestrian crossing stop line.

Give Way Sign. To be located as near to the point where the vehicle is to stop, say at a distance of 1.5 to 3 metres. Where there is a marked Give Way line, it should be erected 1.2 metres in advance of the marking.

Warning Sign. To be located at the following distances in advance of the hazard warned against :

(*a*) *In non-urban locations* :

	Plain and rolling terrain	*Hilly or mountainous terrain*
(*i*) National highways and State highways	120 m	60 m
(*ii*) Major district roads	90 m	50 m
(*iii*) Other district roads	60 m	40 m
(*iv*) Village roads	40 m	30 m

(*b*) *Urban locations* :

50 m

Adequate maintenance of traffic signs is of utmost importance. The sign posts should be maintained in proper position and the signs should be maintained in a legible condition at all time. Damaged signs should be replaced immediately. Periodic painting of signs should be a routine part of maintenance.

13.15. Variable Message Signs

Variable Message Signs (VMS), also known as *Dynamic Message Signs* (DMS) are signs which display information which can be changed whenever needed by means of remote control at a centralised location. (Ref. 12), They give traveller information such as severe weather conditions, incident notification (accident, road closure), congestion, travel time between destinations. VMS display board uses high intensity LEDs, generally yellow in colour. The size of the letters is generally 380-400 mm. An example of a VMS is given in Fig. 13.10.

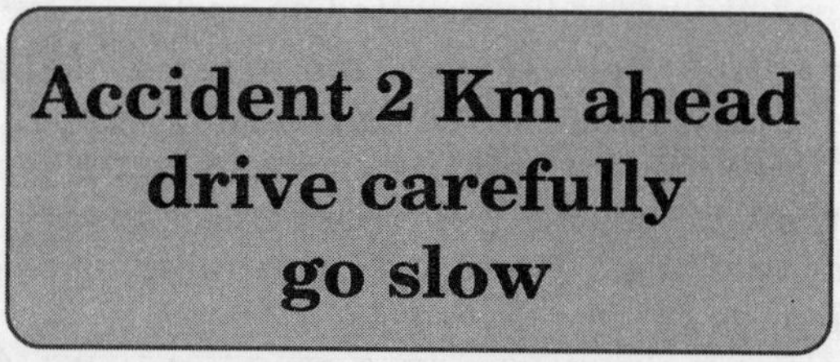

Fig. 13.10. A VMS displaying occurrence of an accident ahead.

I. MULTIPLE CHOICE QUESTIONS

1. Stop sign comes under
   (*a*) regulatory signs (*b*) cautionary signs
   (*c*) informative signs (*d*) none of these
2. The triangular sign in the picture means
   (*a*) a warning sign
   (*b*) a prohibitive sign
   (*c*) an indicative sign
   (*d*) a directional sign

3. The sign in the picture means
   (*a*) downhill road
   (*b*) uphill road
   (*c*) steep downhill road
   (*d*) steep uphill road

4. The sign in the picture means
   (*a*) steep downhill road
   (*b*) sliding
   (*c*) downhill road
   (*d*) steep uphill road

5. The sign in the picture means
   (*a*) carriageway narrow
   (*b*) road narrows on the left side
   (*c*) road narrows on the right side
   (*d*) reduce speed

6. The sign in the picture means
(*a*) watch for falling rocks
(*b*) dangerous hillside road
(*c*) mind downhill sliding
(*d*) steep downhill road

7. The sign in the picture means
(*a*) dangerous hillside road
(*b*) watch for pedestrians
(*c*) watch for children
(*d*) men at work

8. The sign in the picture means
(*a*) hump road
(*b*) house
(*c*) tunnel
(*d*) village

9. The sign in the picture means
(*a*) accident ahead
(*b*) no passing
(*c*) stop
(*d*) danger

Answers

1. (*a*)	**2.** (*a*)	**3.** (*d*)	**4.** (*a*)	**5.** (*a*)
6. (*a*)	**7.** (*d*)	**8.** (*c*)	**9.** (*d*)	

II. SOLVED QUESTIONS AND ANSWERS

Q. 1. What are traffic signs?

Ans. Traffic signs are the silent speakers on the road. Be it the person behind the wheel or a pedestrian, having a sound knowledge about road safety is absolutely necessary for all before hitting the roads.

Traffic signs give information about the road conditions ahead, provide instructions to be followed at the major crossroads or junctions, warn or guide drivers, and ensure proper functioning of road traffic. Being unaware of road signs is akin to throwing caution to the wind. It can lead to loss of life and property. A person is supposed to be familiar (get through a written or oral test) with the traffic signs and symbols before acquiring a driving license in India.

Q. 2. What are the types of road safety signs?

Ans. Road safety signs are primarily of three types:

(*i*) **Mandatory signs:** These signs are used to ensure free movement of traffic and make the road users cognisant of certain laws and regulations, restrictions and prohibitions. Violation of these signs is an offence, as per law.

(*ii*) **Cautionary signs:** These signs make the road users conscious of hazardous conditions on the road beforehand. The drivers, accordingly, take necessary actions to handle the situation

(*iii*) **Informatory signs:** These signs guide the road users about destinations, distance, alternative routes, and prominent locations like food joints, public toilets, nearby hospitals, etc.

Q. 3. What is meant by route marker signs?

Ans. Route marker signs give you directions and information for the route you are:

- approaching
- travelling on
- exiting from.

Q. 4. Define route numbering.

Ans. Route numbers identify roads as part of major routes connecting towns and cities. They are there to reassure you that you are continuing in the right direction for your journey.

Route numbers can appear as a stand-alone sign along the road or with a directional sign.

Q. 5. What are route letters?

Ans.

- **A routes:** An A route is a major arterial route in built-up areas that connects routes between cities and towns in rural areas.
- **B routes:** B routes are alternative routes that connect cities and towns. They are normally a longer journey than an A route.
- **C routes:** C route roads also connect road users to cities and towns, but they are not major routes like A or B routes.
- **M routes:** An M route may replace an A route on freeways and motorways that extend out from the city.

Q. 6. What is meant by national highway route markers?

Ans. A national highway route marker identifies a highway as a national route. The number on the sign shows which national route you are travelling on.

National highway route signs have yellow writing or symbols on a green background.

Q. 7. What do you mean by national route markers?

Ans. A national route marker indicates the road you are on has national significance. The number on the sign shows which national route you are driving on.

National route marker signs have black writing or symbols on a white background.

Q. 8. Explain state route markers.

Ans. A state route marker identifies a road as a state route. The number on the sign shows what state route you are driving on. State route marker signs have white writing or symbols on a blue background.

Q. 9. Discuss tourist drive signs.

Ans. A tourist drive is a scenic drive or route that connects a number of tourist attractions. Tourist drive signs are brown with white writing or symbols.

Q. 10. What are kilometre plates?

Ans. Kilometre plates show the distance to towns and cities. You will see these signs on arterial routes in built-up and rural areas.

- The letter is the first letter of the town or city. For example, W may stand for Warwick.
- Kilometre plates have white writing or symbols on a green background.
- The national sign (yellow writing on a green background) is a supplementary sign that identifies national routes.

Q. 11. Explain alphanumeric route numbers.

Ans.

- A yellow letter and number without a border or shield shows which numbered route the marked road is.
- These numbers may appear on a standalone sign or as part of a larger direction sign.

Q. 12. Route number start and end signs.

Ans. Route number start and end signs show the start and end of a numbered route.

They have a green background and the:

- words 'start' and 'end' in white writing
- route number in yellow writing.

Q. 13. Discuss route number change signs.

Ans. Route number change signs show where an M route becomes an A route, or *vice versa*. While the letter on the route number might change, the number will not.

For example, the Warrego Highway heading west through the Ipswich region is an M2 route and changes to an A2 route near Brassall.

They have a green background and the:

- words 'end' and 'start' in white writing
- route numbers in yellow writing.

Q. 14. What is guidance marker signs?

Ans. A guidance marker sign shows that you are continuing on your route through built-up areas. The sign will show the direction you should go to continue your route. They have a green background and:

- a white arrow
- the route number in yellow writing.

Q. 15. Explain reassurance marker signs.

Ans. A reassurance marker sign shows that you are still travelling on the specified route.

Reassurance marker signs are placed:

- outside a town or city facing departing traffic
- after an intersection with another numbered route

- after an intersection where the numbered route turns.
- They have the route number in yellow writing on a green background.

III. REVIEW QUESTIONS

1. What is the importance of traffic signs?
2. Discuss the need for international standardisation.
3. Discuss the general principles of traffic signing.
4. Write a brief note on:

   (*a*) prohibitory signs (*b*) mandatory signs

   (*c*) informatory signs (*d*) indication signs

REFERENCE

1. Protocol on Road Signs and Signals, Command 3454, H.M.S.O. London.
2. United Nations Conference on Road and Motor Transport, United Nations Publication, Lake Success, New York, 1950.
3. Code on a Uniform System of Road Signs and Signals, Pavement Markings and Signs for Road Works in the ECAFE Region, United Nations Economic Commission for Asia and the Far East, Saigon, 1964.
4. The Motor Vehicles Act, 1988, New Delhi, 1988.
5. Code of Practice for Road Signs, I.R.C : 67–2010 Indian Roads Congress, New Delhi, 2010.
6. Manual or Uniform Traffic Control Devices, U.S. Deptt. of Transportation, Federal Highway Administration, Washington, 1978.
7. Road Research Laboratory, Research on Road Traffic, H.M.S.O., London, 1965.
8. Standard Letters and Numerals of Different Heights for use on. Highway signs, I.R.C. : 30–1968, Indian Roads Congress, New Delhi, 1988.
9. Traffic Signs, Journal of the Indian Roads Congress, Vol. XXV-3. New Delhi, 1960.
10. Route Marker Signs for National Highways, I.R.C : 2–1968, Indian Roads Congress, 1968.
11. Specifications for Road and Bridge Works, Ministry of Surface Transport, published by the Indian Roads Congress, New Delhi, 2013.
12. Guidelines for Variable Message Signs, I.R.C : SP : 85–2010, Indian Roads Congress, New Delhi, 2010.

14. Road Markings

14.1. Function

Road markings are used as a means of controlling and guiding traffic. They are highly important on roads and intersections as they promote road safety and bring about smooth and harmonious flow of traffic along guided paths of travel. They also serve to supplement the messages conveyed by road signals and signs. In some cases, they are used alone to convey certain regulation, information or warning that cannot otherwise be effectively made known to the road users.

Despite certain limitations inherent in them, such as obliteration by snow, poor visibility when wet and the need for frequent renewals when painted on surfaces exposed to traffic, road markings have a useful role to play in traffic engineering. One advantage with road markings is that they convey the required information to the driver without distracting his attention from the carriageway.

14.2. Types of Road Markings

Road markings are basically of two types : carriageway markings and object markings. As the name implies, the former type of markings are those that are applied to the carriageway itself. The latter type covers markings on objects such as abutment, piers, kerbs, traffic islands, culvert headwalls, etc.

Carriageway markings are of the following categories :

1. Centre line.
2. Traffic lane lines.
3. No-overtaking zone markings.
4. Pavement edge lines.
5. Carriageway width reduction transition markings.
6. Obstruction approach markings.
7. Stop lines
8. Pedestrian crossings.
9. Cyclist crossings.
10. Route direction arrows etc.
11. Word messages.
12. Markings at approaches to intersections.
13. Parking space limits.
14. Bus stops.z

Object markings are of the following categories :

1. Objects within the carriageway.
2. Kerb marking for visibility.
3. Kerb marking for parking restriction.
4. Objects adjacent to the carriageway.

14.3. General Principles of Longitudinal Pavement Markings

The practice of pavement markings in different countries varies. However, some general principles of longitudinal pavement markings are common in many countries. Some of these are (Ref. 1, 2, 3, 4) :

1. Solid lines are restrictive in nature and it is an offence to cross or straddle such a line.

2. Broken lines are restrictive in character, and vehicles can cross such a line provided this can be done with safety.

3. When a combination of a solid and broken lines is used, in countries where the traffic moves to the left (right), a vehicle should not cross the continuous line adjacent to and to the left (right) of a broken line on the right (left) of the lane in which it is moving.

4. Double lines indicate maximum restrictions.

14.4. Material and Colour

The material commonly used for pavement, curb and object markings is paint. Ordinary paints for road markings are covered by Indian Standards IS : 164–1981 (First Revision Reaffirmed in 1986) (Ref. 7). Those paints do not have a good degree of brightness and hence fail to catch instantaneously the drivers eyes. They also have short life. The better alternative is hot applied thermoplastic paints, which are now-a-days specified for roads (Ref. 8). Improved night visibility is obtained by the use of minute glass beads incorporated in the markings to produce a "retro-reflective" surface.

Other materials that may also be used include pre-fabricated sheet materials, glue-down plastic stripes, metal and plastic inserts— and road studs.

The commonly used colours for road markings are white and yellow. The useage of these colours in India is summarised in Table 14.1.

Table 14.1. Colour of Roads Markings as per Indian Practice

Colour	*Uses*
White	All carriageway markings except those intended for parking restrictions.
Yellow	(*i*) Markings intended for parking restrictions. (*ii*) Continuous centre and barrier line markings.
Alternate bands of white and black.	Kerb and object markings.

14.5. Centre Lines

A centre line marking is provided to demarcate the centre of a carriageway and to separate traffic in opposite directions.

The pattern of centre line markings depends upon whether the road is in an urban section or rural section. On rural sections, the segments and gaps are double in length than for an urban location.

The length of gap is shorter near approaches to intersection and on curves than on straight reaches. The gap is half the value on straight sections.

Fig. 14.1 gives typical centre line markings and lane markings.

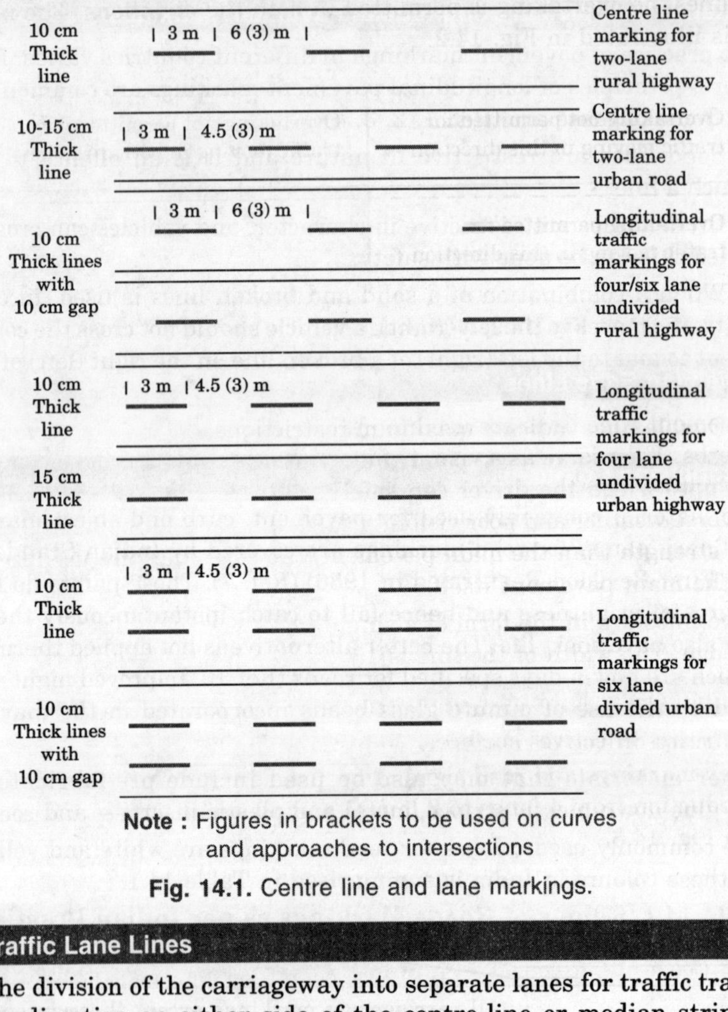

Note : Figures in brackets to be used on curves and approaches to intersections

Fig. 14.1. Centre line and lane markings.

14.6. Traffic Lane Lines

The division of the carriageway into separate lanes for traffic travelling in the same direction on either side of the centre-line or median strip helps to promote travel in proper lanes and curb the meandering tendency of the drivers, thus promoting safety and ensuring maximum capacity.

Traffic lane lines are broken lines which permit lane changing with caution.

Fig. 14.1 above also indicates the traffic lane lines.

14.7. No Overtaking Zone Markings

No overtaking zone markings are provided on summit curves, horizontal curves and elsewhere in two and three lane highways where overtaking and passing manoeuvres must be prohibited because of non-availability of safe overtaking sight distances or other hazardous conditions.

The marking for a "no overtaking" zone consists of a combination line along the centre line. The combination line consists of a double line, the left hand element of which shall be a solid barrier line. The right hand element will be either a normal broken centre line or solid barrier line governing the traffic from the opposite direction. Where a solid barrier line is to the right of the broken line, the overtaking restriction shall apply only to the opposing traffic. If both the lines are solid lines, no overtaking is permitted in both the directions. The use of this marking is illustrated in Fig. 14.2.

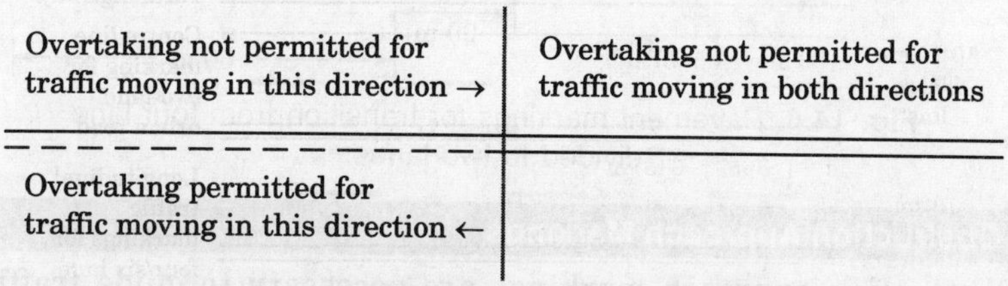

Fig. 14.2. No overtaking zone markings.

14.8. Pavement Edge Lines

Pavement edge lines are used to indicate the edges of carriageways which have no kerbs. They serve as a visual guidance for the drivers, indicating to them the limits upto which the driver can safely venture. They especially are useful during adverse weather and poor visibility. Where the paved shoulder is of a lesser structural strength than the main pavement, the edge lines are used to promote travel on the main pavement itself.

Edge lines are in the form of a single continuous line placed about 15 cm from the edge. The width of the line is 15–20 cm.

14.9. Carriageway Width Reduction Transition Markings

Reduction in the carriageway width is made known by a combination of carriageway markings and road signs. A typical design for carriageway marking showing reduction from 4 lanes to 2 lanes as per current Indian practice (Ref. 4) is given in Fig. 14.3.

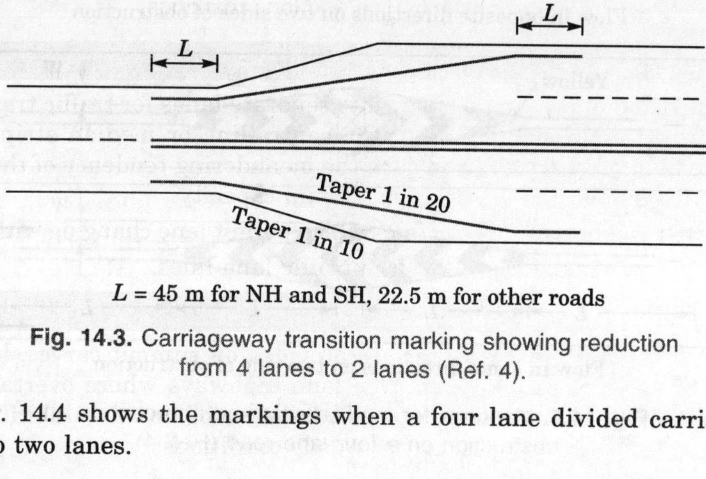

L = 45 m for NH and SH, 22.5 m for other roads

Fig. 14.3. Carriageway transition marking showing reduction from 4 lanes to 2 lanes (Ref. 4).

Fig. 14.4 shows the markings when a four lane divided carriageway is reduced to two lanes.

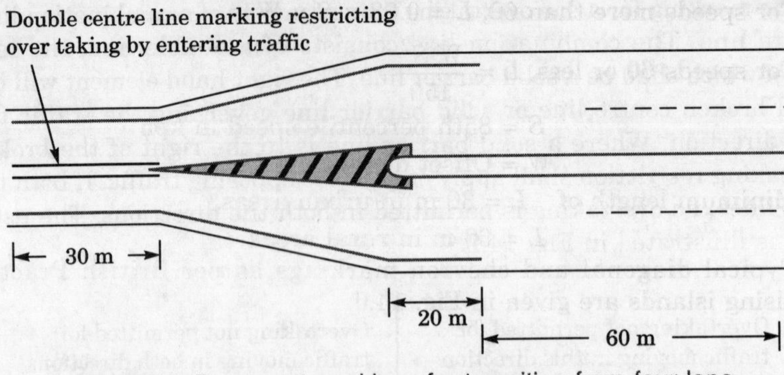

Fig. 14.4. Pavement markings for transition from four-lane divided to two lanes.

14.10. Obstruction Approach Markings

Obstruction approach markings are necessary to guide traffic on the approach to fixed obstructions within the carriageway. The markings must be designed to guide the traffic away from the obstruction. Typical markings for median strip approach and approach to an obstruction on a four lane road as per current Indian practice are given in Fig. 14.5.

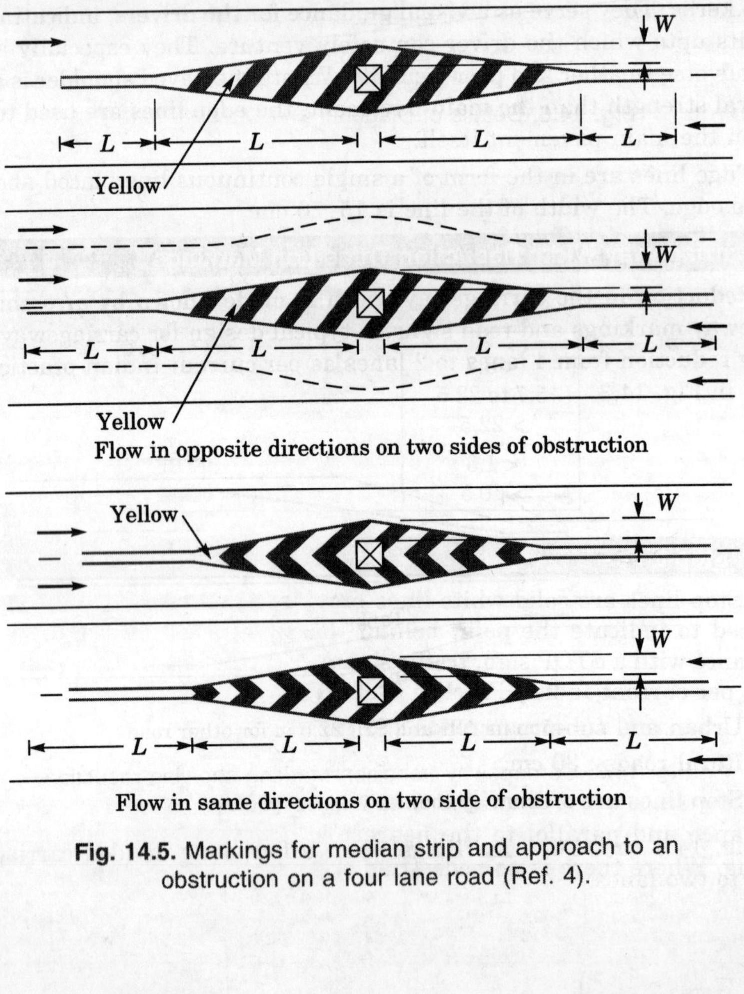

Fig. 14.5. Markings for median strip and approach to an obstruction on a four lane road (Ref. 4).

For speeds more than 60, $L = 0.63 \times S \times W$

For speeds 60 or less, $L = \dfrac{W S^2}{150}$

S = 85th percentile speed in kph

W = Offset distance in M

Minimum length of L = 30 m in urban areas

L = 60 m in rural areas

Typical diagonal and chevron markings as per British Practice near channelising islands are given in Fig. 14.6.

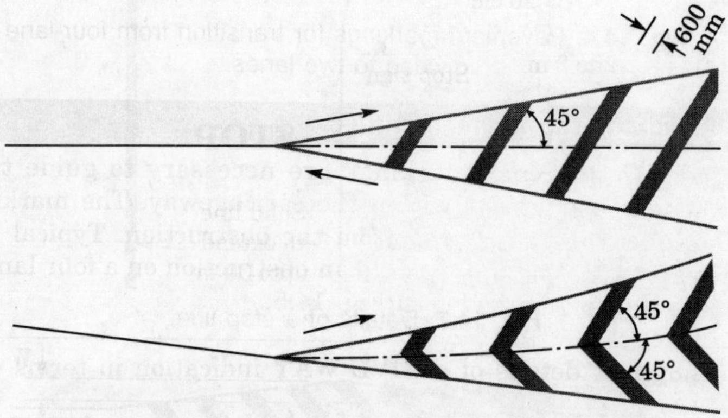

Fig. 14.6. Details of Diagonal and Chevron Markings.

Total length of marking (m)	*Spacing between bars or Chevrons (mm)*	
	Low Speed (< 75 km/h)	*High Speed (> 75 km/h)*
< 5.7	2100	. . .
5.7 to 22.5	3500	. . .
> 22.5	5000	. . .
< 1.5	. . .	4000
> 10.5	. . .	8000

14.11. Stop Lines

Stop lines are solid white lines provided transversely to the carriageway and used to indicate the point behind which vehicles are required to stop in compliance with a STOP sign, traffic signal or traffic police. The width of the stop line as per current Indian Practice (Ref. 4) is :

Urban and sub-urban roads : 20 cm.

Rural roads : 30 cm.

Stop lines are ordinarily located not less than 2 m and not more than 3 m in advance and parallel to the nearest boundary of the pedestrian crossing marking. Where there is no pedestrian crossing, the stop line is placed not less than 1.25 m and not more than 9 m from the nearest carriageway edge of the intersecting road.

Fig. 14.7 gives details of a stop line

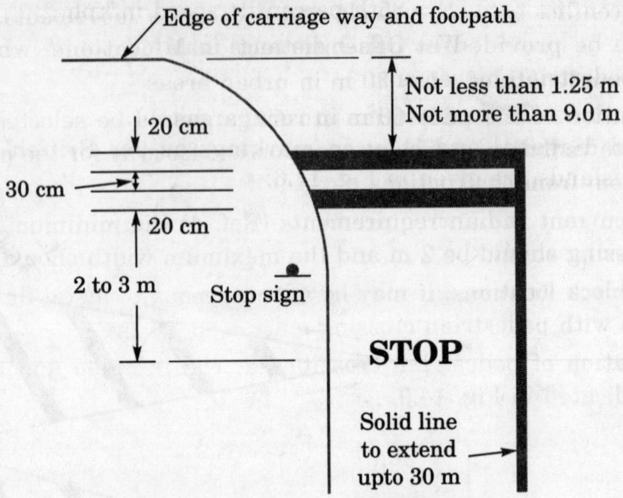

Fig. 14.7. Details of a stop line.

Fig. 14.8 gives details of a GIVE WAY indication in terms of pavement markings.

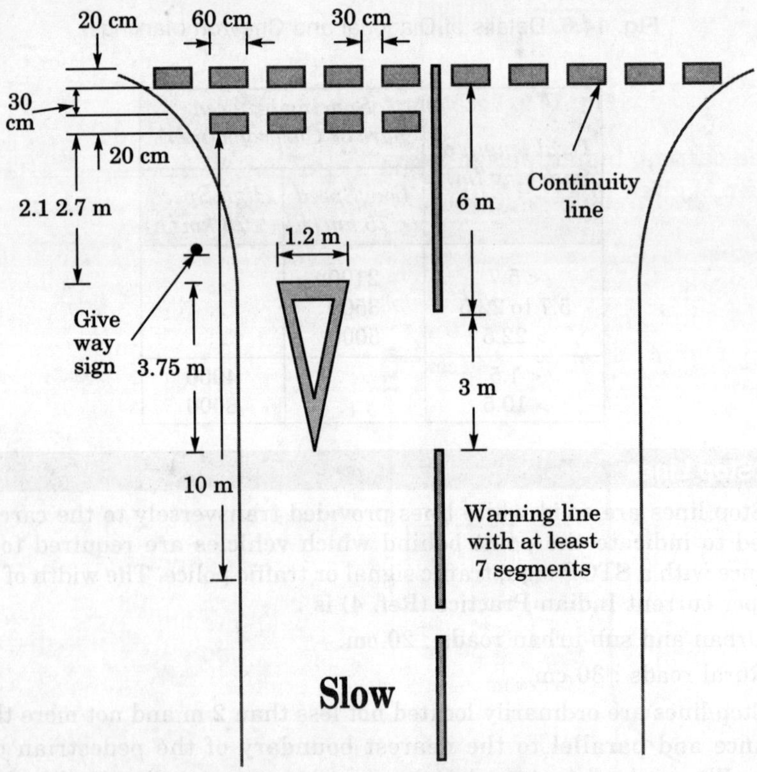

Fig. 14.8. Markings for GIVE WAY indication.

14.12. Pedestrian Crossings

Pedestrian crossings are very important for pedestrian safety as they guide pedestrians in the proper paths. They are marked at all intersections where there is substantial conflict between vehicle and pedestrian movements. In addition, they may also be provided at non-intersectional locations, where there is concentrated pedestrian movement.

The location of the pedestrian crossings should be selected properly to ensure adequate visibility, sufficient space on the footway for the pedestrians to wait and freedom from obstructions.

As per current Indian requirements (Ref. 4) the minimum width of the pedestrian crossing should be 2 m and the maximum width should be 4 m.

At mid-block locations, it may be advantageous to instal flashing signals in combination with pedestrian crossing markings.

The location of pedestrian crossings at right-angled and acute angled crossings is indicated in Fig. 14.9.

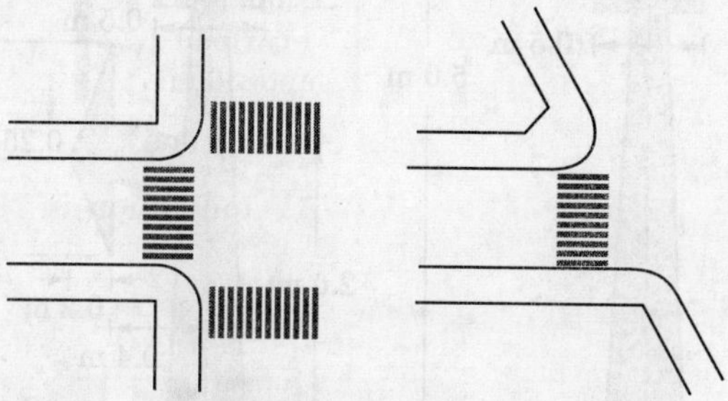

Fig. 14.9. Location of pedestrian crossing.

The current Indian design for pedestrian crossings at a nonsignalised intersection is given in Fig. 14.10.

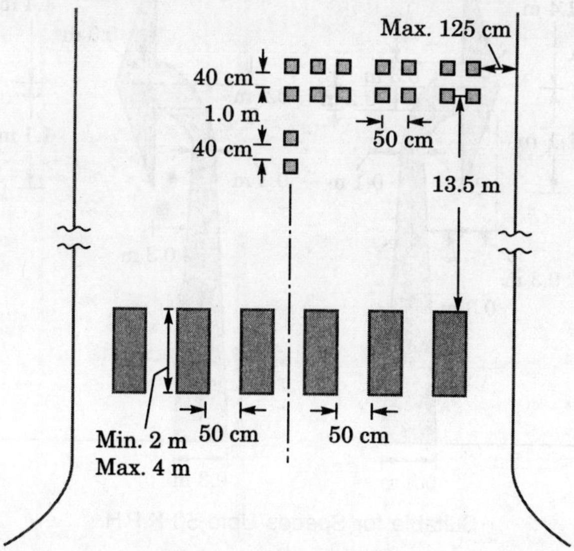

Fig. 14.10. Indian practice for pedestrian markings.

14.13. Cyclist Crossings

Cyclist crossings should have the same width as the cycle tracks. They are marked by means of squares 50 cm × 50 cm at a centre-to-centre spacing of 100 cm.

14.14. Route Direction Arrows

Route direction arrows are used to guide effectively the traffic in the correct direction and lanes. Because of low angle at which such markings are viewed by the drivers, these markings must be elongated in the direction of traffic to be properly legible. Recommended designs for such arrows as per current Indian practice are given in Fig. 14.11.

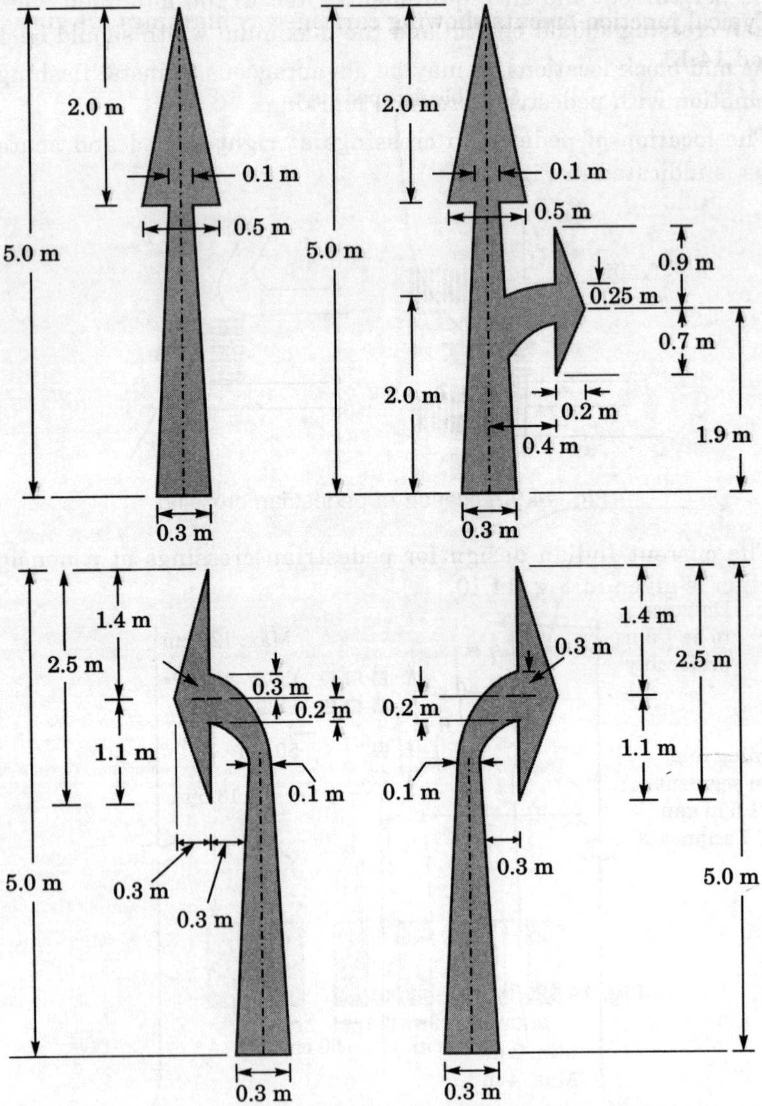

Suitable for Speeds Upto 50 K.P.H.

Fig. 14.11. Recommended designs for route direction arrows. (Ref. 4)

14.15. Word Messages

Word messages are used to convey information to guide, warn or regulate traffic. Some of the common messages used are : STOP, SLOW, SPEED 25 etc. Standard designs for letterings and numerals are available for different speeds (Ref. 4).

14.16. Markings at Approaches to Intersections

For orderly and guided traffic movement at intersections, markings can be profitably employed. Markings at intersections can be a combination of centre-lines, turn markings, lane markings, stop lines, route direction arrows etc.

Typical junction layouts showing carriageway markings are given in Figs. 14.12 and 14.13.

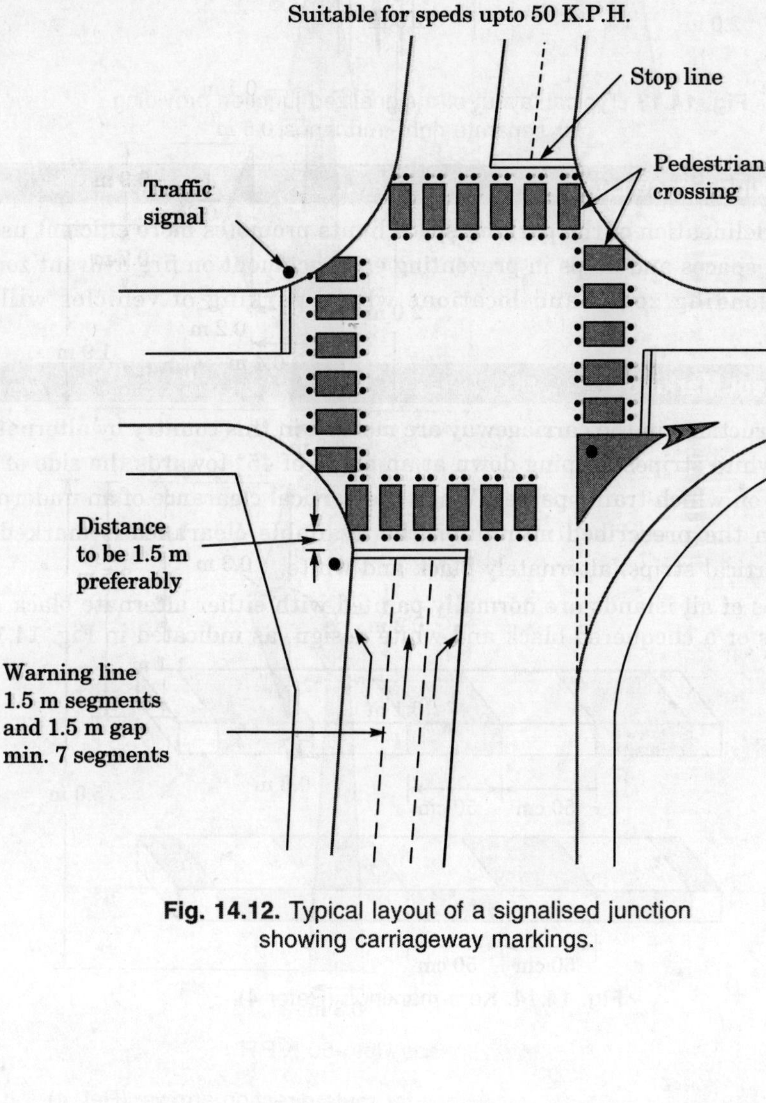

Fig. 14.12. Typical layout of a signalised junction showing carriageway markings.

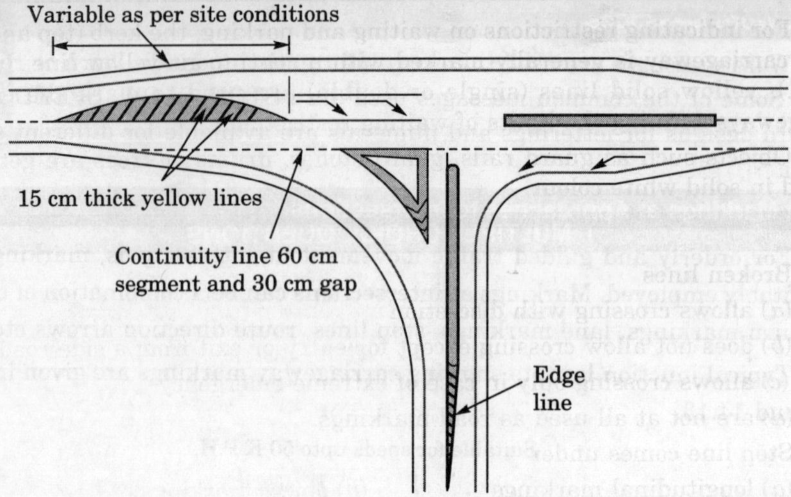

Fig. 14.13. Typical layout of a signalized junction providing for separate right-trun lanes.

14.17. Parking Space Limits

The delineation of the parking space limits promotes more efficient use of the parking spaces and helps in preventing encroachment on fire hydrant zones, bus stops loading zones and locations where parking of vehicles will be undesirable.

14.18. Object Markings

Obstructions in the carriageway are marked in this country by alternating black and white stripes, sloping down at an angle of 45° towards the side of the obstruction on which traffic passes. When the vertical clearance of an underpass is less than the prescribed minimum, the available clearance is marked by painting vertical strips, alternately black and white.

Kerbs of all islands are normally painted with either alternate black and white strips or a chequered black and white design, as indicated in Fig. 14.14.

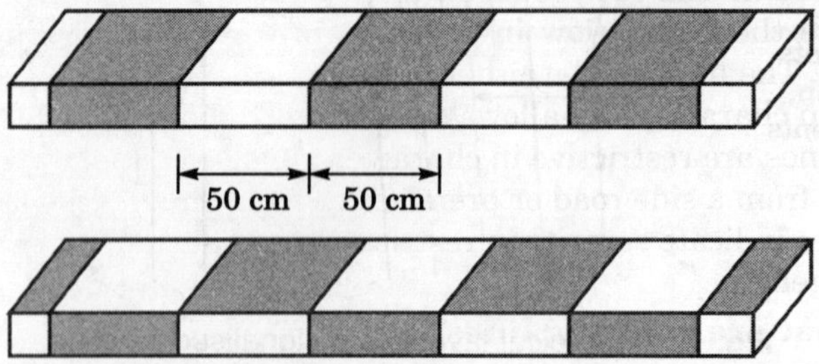

Fig. 14.14. Kerb markings (Refer 4).

For indicating restrictions on waiting and parking, the kerb (top and face) or the carriageway is generally marked with a continuous yellow line. In U.K. (Ref. 1), yellow solid lines (single or double) are used to mark the edge of carriageways defining the limits of waiting restrictions.

Objects such as guard rails, guard stones, drums or trees are generally marked in solid white colour.

I. MULTIPLE CHOICE QUESTIONS

1. Broken lines
   (*a*) allows crossing with discretion
   (*b*) does not allow crossing except for entry or exit from a side road
   (*c*) allows crossing only in case of extreme emergency
   (*d*) are not at all used as road markings.
2. Stop line comes under
   (*a*) longitudinal markings (*b*) object markings
   (*c*) transverse markings (*d*) none of these

Answers

1. (*a*) **2.** (*c*)

II. SOLVED QUESTIONS AND ANSWERS

Q. 1. What is meant by road markings?

Ans. The road markings are defined as lines, patterns, words or other devices, except signs, set into applied or attached to the carriageway or kerbs or to objects within or adjacent to the carriageway, for controlling, warning, guiding and informing the users. The road markings are classified as longitudinal markings, transverse markings, object markings, word messages, marking for parkings, marking at hazardous locations etc.

Q. 2. What are longitudinal marking?

Ans. Longitudinal markings are placed along the direction of traffic on the roadway surface, for the purpose of indicating to the driver, his proper position on the roadway. Some of the guiding principles in longitudinal markings are also discussed below. Longitudinal markings are provided for separating traffic flow in the same direction and the predominant colour used is white. Yellow colour is used to separate the traffic flow in opposite direction and also to separate the pavement edges. The lines can be either broken, solid or double solid. Broken lines are permissive in character and allows crossing with discretion, if traffic situation permits. Solid lines are restrictive in character and does not allow crossing except for entry or exit from a side road or premises or to avoid a stationary obstruction. Double solid lines indicate severity in restrictions and should not be crossed except in case of emergency.

Q. 3. What are warning lines?

Ans. Warning lines warn the drivers about the obstruction approaches. They are marked on horizontal and vertical curves where the visibility is greater than prohibitory criteria specified for no overtaking zones. They are broken lines with 6 m length and 3 m gap.

Q. 4. Explain edge lines.

Ans. Edge lines indicate edges of rural roads which have no kerbs to delineate the limits upto which the driver can safely venture. They should be at least 150 mm from the actual edge of the pavement. They are painted in yellow or white.

Q. 5. What are transverse markings?

Ans. Transverse markings are marked across the direction of traffic. They are marked at intersections etc. The site conditions play a very important role. The type of road marking for a particular intersection depends on several variables such as speed characteristics of traffic, availability of space etc.

Q. 6. What is meant by stop line markings?

Ans. Stop line markings, markings for pedestrian crossing, direction arrows, etc. are some of the markings on approaches to intersections. Stop line indicates the position beyond which the vehicles should not proceed when required to stop by control devices like signals or by traffic police. They should be placed either parallel to the intersecting roadway or at right angles to the direction of approaching vehicles.

Q. 7. Discuss pedestrian crossings.

Ans. Pedestrian crossings are provided at places where the conflict between vehicular and pedestrian traffic is severe. The site should be selected that there is less inconvenience to the pedestrians and also the vehicles are not interrupted too much. At intersections, the pedestrian crossings should be preceded by a stop line at a distance of 2 to 3 m for unsignalized intersections and at a distance of one metre for signalized intersections.

Q. 8. What are directional arrows?

Ans. In addition to the warning lines on approaching lanes, directional arrows should be used to guide the drivers in advance over the correct lane to be taken while approaching busy intersections. Because of the low angle at which the markings are viewed by the drivers, the arrows should be elongated in the direction of traffic for adequate visibility.

Q.9. What do you mean by word messages? Explain.

Ans. Information to guide, regulate, or warn the road user may also be conveyed by inscription of word message on road surface. Characters for word messages are usually capital letters. The legends should be as brief as possible and shall not consist of more than three words for any message. Word messages require more and important time to read and comprehend than other road markings. Therefore, only few and important ones are usually adopted.

Some of the examples of word messages are

STOP, SLOW, SCHOOL, RIGHT TUN ONLY etc.

The character of a road message is also elongated so that driver looking at the road surface at a low angle can also read them easily.

III. REVIEW QUESTIONS

1. What are the types of road markings?
2. Discuss the general principles of longitudinal pavement.
3. Explain the pavement edge lines.
4. What is cyclist crossings?
5. Explain route direction arrows.

REFERENCES

1. Ministry of Transport, Traffic Signs, Manual, Her Majesty's Stationery Office, London, 1965.
2. Manual or Uniform Traffic Control Devices, U.S. Deptt. of Transportation, Federal Highway Administration, Washington, 1978.
3. Code on a Uniform System of Road Signs and Signals, Pavement Markings and Signs for Road Works in the ECAFE Region, United Nations Economic, Commission for Asia and the Far East, Saigon, 1964.
4. Code of Practice for Road Markings (with paints), I.R.C. : 35–1970, First Revision the I.R.C., New Delhi, 1990.
5. N.S. Srinivasan and Sharfudadin, Durability of Road Marking Materials, Journal of the I.R.C., Vol., XXIX 4, New Delhi, 1966.
6. Roads in Urban Areas, Department of the Environment, Her Majesty's Stationery Office, London, 1966.
7. Specifications for ready mixed paints for road markings, IS : 164— 1981 (First Revision Reaffirmed in 1986), Bureau of Indian Standards, New Delhi, 1986.
8. Specification for Road and Bridge works, Ministry of Surface Transport, published by the Indian Roads Congress, New Delhi, 2013.

15. Traffic Signals

15.1. Introduction

The use of traffic signals for control of conflicting streams of vehicular and pedestrian traffic is extensive in most of the towns and cities. *The first traffic signal* is reported to have been used in *London* as early as in 1868 and was of the *semaphore-arm* type with red and green lamps for night use (Ref. 1). During the hundred years since then traffic signals have been developed to a high degree of sophistication. In this Chapter it is proposed to deal with types, design, location, indications and maintenance of the signals. A brief review will also be given of the latest trends in Area Control System whereby linked-signal control can be exercised over a large area under the control of a master computer.

15.2. Advantages and Disadvantages of Traffic Signals

Traffic signals, when properly designed, located and operated, have one or more of the following advantages :

(*i*) They can provide for an orderly movement of traffic.

(*ii*) When proper geometric layouts and control measures are employed, they can increase the traffic handling capacity of the intersection.

(*iii*) They can reduce the frequency of certain types of accidents, especially the right-angle type and pedestrian accidents.

(*iv*) Under favourable conditions, they can be co-ordinated to provide for continuous or nearly continuous movement of traffic at a definite speed along a given route.

(*v*) They can be used to interrupt heavy traffic at intervals to permit other traffic–vehicular or pedestrian to cross.

(*vi*) Traffic signals dispense with police control and can thus be economical.

(*vii*) If properly designed and set, they can assign right of way impartially to traffic, unlike manual controls which can stop and interrupt traffic streams at the personal whim of the traffic controller.

As regards their disadvantages, the following may be mentioned, especially if the signals are installed improperly :

(*i*) Excessive delay to vehicles may be caused, particularly during off-peak hours.

(*ii*) Unwarranted signal installations tend to encourage the disobedience of the signal indications.

(*iii*) Drivers may be induced to use less adequate and less safe routes to avoid delays at signals.

(*iv*) Accident frequency, especially of the rear-end type, may increase.

(*v*) When the installations break down, due to any fault in the system, total and widespread confusion and difficulties can result.

15.3. Signal Indications

15.3.1. The types, meaning and sequence of signal indications vary in different countries.

British practice. In British practice, the signal sequence is red, red/amber shown together, green and amber. The common practice is to use circular red, amber and green signals, though in certain circumstances green-arrow signals are also used. When the red signal is displayed, the right of way is denied to the traffic from entering the intersection. The red/amber signifies an alert to the drivers that the signal aspect is about to change to green so that they can be in readiness to go. The green signal aspect denotes that the right of way is given to the drivers to enter the intersection. The amber signal alerts the drivers that the red signal aspect is about to commence shortly and the green aspect is about to be terminated. A green arrow aspect permits the drivers to enter the intersection to make the movement indicated by the arrow.

American practice. In American practice, the signal sequence is Red, Green and Yellow. Red indication prohibits entry into the section, whereas the green permits entry. Yellow indication warns the traffic that the related green movement is being terminated and the red indication is about to commence. Thus, while allowing entry into the intersection, the yellow aspect requires the traffic to clear the intersection before the yellow expires. In addition to circular red, green and yellow, American practice permits red arrow, green arrow and yellow arrow indications to control traffic in certain direction (Ref. 2).

Indian practice. The Indian practice (Ref. 15) is to have an amber period of 2 seconds as a transition interval between termination of related green movement and exhibition of a red indication or between termination of a red indication and commencement of related green movement.

15.3.2. Pedestrian Signal Indications

In U.K. practice, the *don't cross* indication is given by a red standing man. The *cross* indication is a green walking man (Ref. 3), whereas a flashing green signifies *don't start to cross*.

The Indian Standard on Traffic Signals (Ref. 4) prescribes the following symbols for pedestrians (Fig. 15.1). The red standing man represents the *don't cross* indication and the green walking man represents *cross* indication.

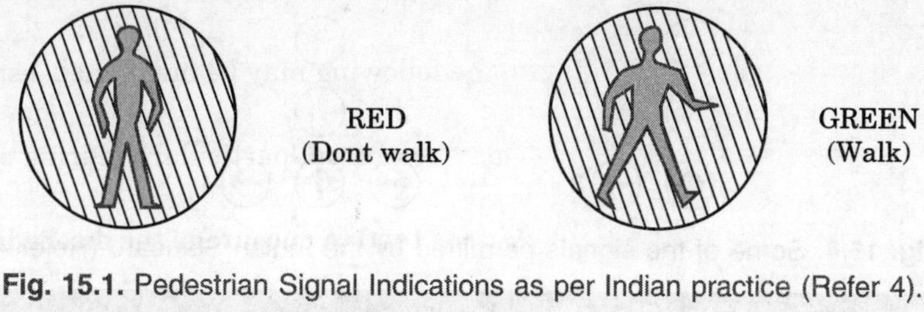

Fig. 15.1. Pedestrian Signal Indications as per Indian practice (Refer 4).

Flashing amber. A flashing amber signal is a hazard identification beacon is normally used to warn of obstruction and intersections to supplement regulatory signs and to warn of midblock cross-walks.

15.4. Signal Face

The minimum number of lenses in a signal face is three-red, amber and green, and the maximum number in American practice is five. The lenses in a signal face can be arranged in a vertical or horizontal straight line. The relative positions (top to bottom or left to right) are : red, amber and green. A simple signal face with three lenses in a vertical line is indicated in Fig. 15.2.

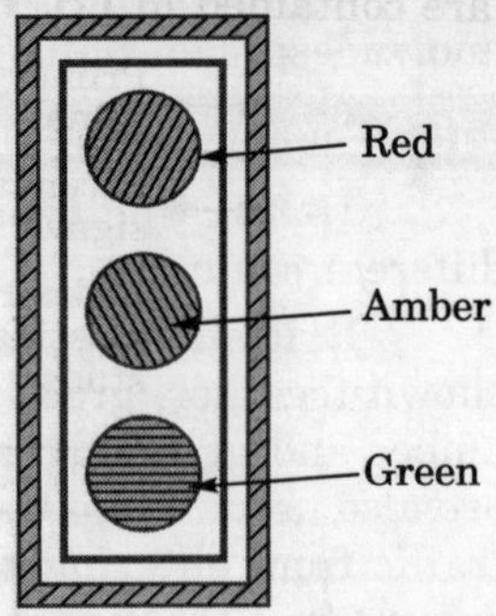

Fig. 15.2. Signal face.

The lenses are normally of two sizes, *viz.*, 200 mm and 300 mm diameter. The larger size is used where the 85th percentile approach speeds exceed 65 K.P.H., for special problem locations, for all arrow indications, for intersections where signalization may be unexpected and for intersections where drivers may view both traffic control and lane direction control signs simultaneously. The Indian Standard (Ref. 4) and the British Standard (Ref. 6) recommended a size of 200 mm for light signals intended for drivers, 300 mm for green arrow signals and 300 mm for signals intended for pedestrians.

The arrows are pointed vertically upward to indicate a straight through movement and in a horizontal direction to indicate a turn at approximately right angles. When the angle of the turn is substantially different from a right angle, the arrow can be positioned on an upward angle approximately equal to that of the turn.

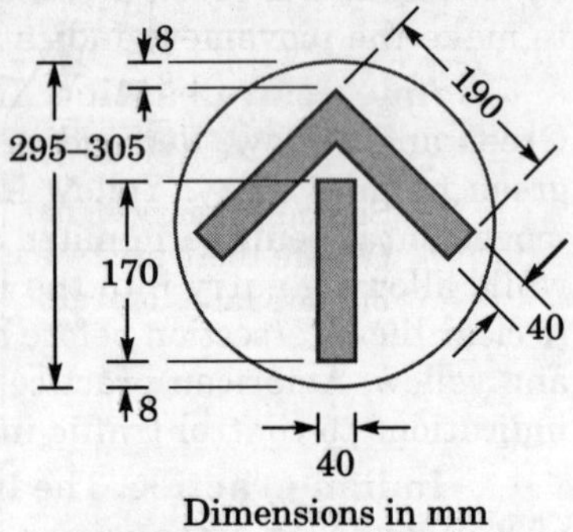

Fig. 15.3. Standard arrow recommended by Indian Standard (Refer 4).

The arrow specified in the Indian Standard is given in Fig. 15.4.

Some of the positions of signals recommended by the Indian Standard are given in Fig. 15.4. A suggested layout of the signal post is given in Fig. 15.5.

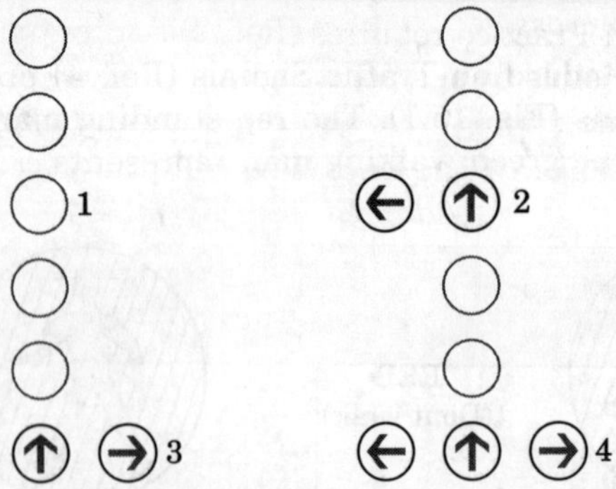

Fig. 15.4. Some of the signals permitted by the Indian Standard (Refer 4).

15.5. Illumination of the Signals

The American Practice requires that the signals should be so illuminated as to be visible for a distance of at least 0.4 km under normal atmospheric conditions. Detailed specifications for the illumination of the lamps for signals are contained in I.S. 7537–1974 (Ref. 4).

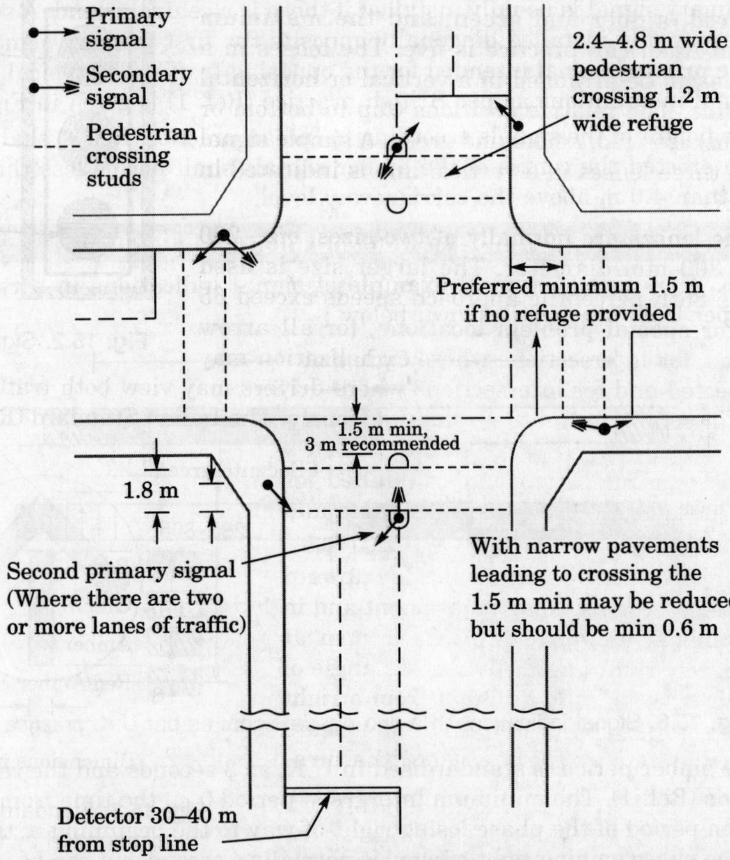

Fig. 15.5. Typical layout of traffic signal installations in U.K. (From Refer 1).

15.6. Number and Location of Signal Faces

The American Practice requires (Ref. 2) a minimum of two signal faces to be provided and be visible from a point at least the following distances in advance of and to the stop line :

Table 15.1. Location of Signal Faces

(Adapted from Refer 2)

85 percentile speed (K.P.H.)	*Minimum visibility (m)*
30	30
40	55
50	75
60	100
65	120
75	145
80	170
90	190
100	210

A signal face is, however, permissible for control of an exclusive turn lane.

Normally one primary signal is installed at 0.9 m from the stop line and a second primary signal is usually installed if there is a central island. A secondary signal is commonly installed diagonally opposite the first primary signal on the back of the primary signal intended for the opposing traffic. The typical layout of traffic signal installations as per British practice (Ref. 1) is given in Fig. 15.5.

The height of the signals according to Indian practice (Ref. 4) shall be such that when erected the centre of the amber signals shall not be less than 2.4 m nor more than 4.0 m above the carriageway level.

15.7. Amber Period. Red/Amber Period and Intergreen Period

U.K. Practice. A typical example of signal indications in a two-phase signal as per U.K. practice is shown below :

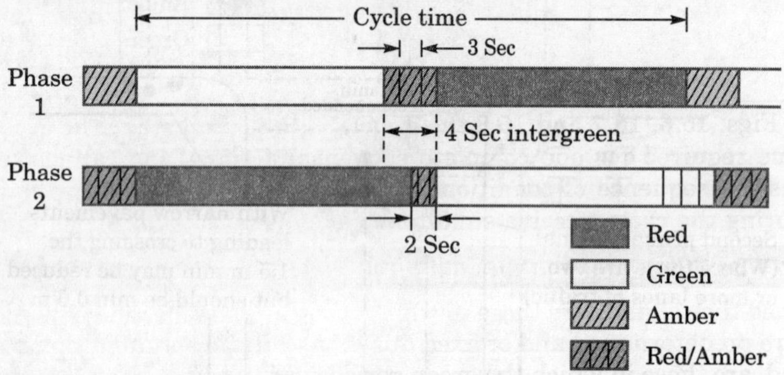

Fig. 15.6. Signal indications in a two phase signal as per U.K. practice.

The amber period is standardised in U.K. as 3 seconds and the red/amber at 2 seconds (Ref. 1). The minimum intergreen period (*i.e.*, the time from the end of the green period of the phase losing right-of-way to the beginning of the green period of the phase gaining right-of-way) is normally 4 second, but can be increased to suit particular needs such as pedestrian crossing requirement.

American Practice. In American practice, a typical example of signal indications in a two phase signal is shown in Fig. 15.7.

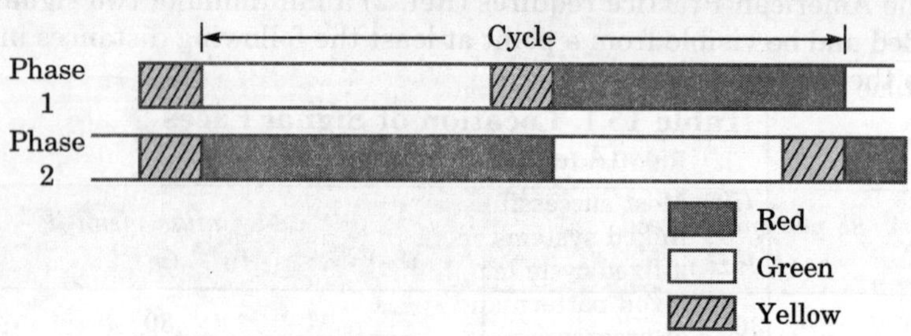

Fig. 15.7. Signal indications in a two phase signal as per American practice.

The Manual on Uniform Traffic Control Devices recommends a yellow interval of 3 to 6 seconds, longer intervals being appropriate to higher approach speeds. Sometimes a short all red clearance interval may be provided immediately after the yellow period to permit the intersection to clear before cross traffic is released.

Indian Practice. The Indian practice is Red, Green and Amber (Ref. 15). The amber interval is a transition interval between termination of related green movement and exhibition of a red indication or between termination of a red indication and commencement of related green movement. In the first case it is called "Clearance Amber" and in the second case, it is called "Initial Amber". The amber period is generally 2 seconds.

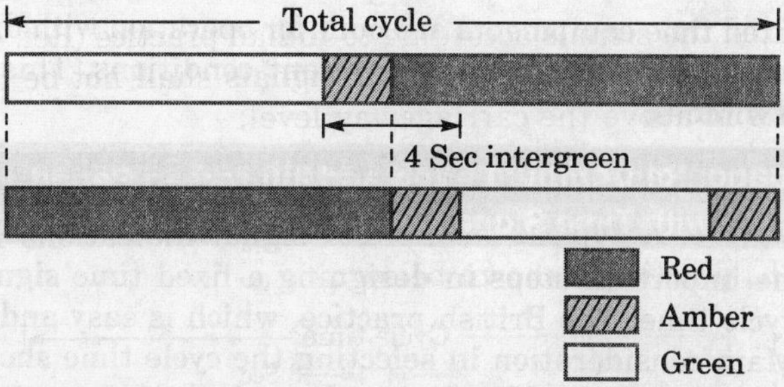

Fig. 15.8. Indian Practice of signal indications.

In Figs. 15.6, 15.7 and 15.8 the cycle length is indicated. The cycle length is the time required for one complete sequence of signal indications. Phase is defined as the sequence of conditions applied to one or more streams of traffic which, during the cycle, receive simultaneous identical signal indications.

15.8. Fixed-Time Signals and Vehicle-Actuated Signals

Fixed time signals are those in which the green periods, and hence the cycle lengths are predetermined and of fixed duration. Vehicle actuated signals, on the other hand, are those in which the green periods vary and are related to the actual demands made by traffic. This is made possible by installing detectors on all the approaches. An intermediate type, semi-vehicle actuated signals, is also available, in which the right of way normally rests with the main road and detectors are located only on the side roads.

Vehicle actuated signals are very popular in U.K., whereas in the U.S.A. fixed time signals are far more numerous than vehicle actuated types.

The advantages and disadvantages of the three types are briefly given below :

Type	*Advantages*	*Disadvantages*
1. Fixed time	(*i*) Simple in construction. (*ii*) Relatively inexpensive. (*iii*) Most successfully used in linked systems requiring a fixed cycle length for a given pattern and speed of progression.	(*i*) Inflexible and hence may cause avoidable delay. (*ii*) Require careful setting.
2. Vehicle-actuated.	(*i*) They are flexible and are able to adjust to changing traffic conditions automatically. (*ii*) Delay is held to a minimum and maximum capacity is achieved.	(*i*) Require costly equipment such as detectors and sophisticated controllers. (*ii*) Cannot provide signal co-ordination.

Type	*Advantages*	*Disadvantages*
3. Semi-vehicle-actuated	Useful for junction of a side street having low traffic volume with a main street having heavy flow.	They are believed to cause high accident rates at times of light traffic.

Modern fixed time equipments are built for operation with different settings at certain periods of the day, to cover different conditions. This is achieved by providing time switches.

15.9. Determination of Optimum Cycle Length and Signal Settings for an Intersection with Time Signals

One of the important steps in designing a fixed time signal system is to determine the cycle time. The British practice, which is easy and simply, will be outlined here. Main consideration in selecting the cycle time should be that the least delay is caused to the traffic passing through the intersection. In selecting a cycle time, another guiding factor is the proportion of the time lost (in the intergreen period and by starting delays) to the cycle time. If the cycle time is small, the proportion of the time lost to the cycle time will be high, resulting in an inefficient signal operation and lengthy delays. On the other hand, if the cycle is large, the proportion of the time lost to the cycle time will be small, and the signal operation will be more efficient. At the same time, if the cycle time is too large, there is a danger that a good portion of the green time will be used by unsaturated flow of traffic, which again leads to inefficiency. For each traffic flow volume there is an optimum cycle time which results in the minimum delay to the vehicles.

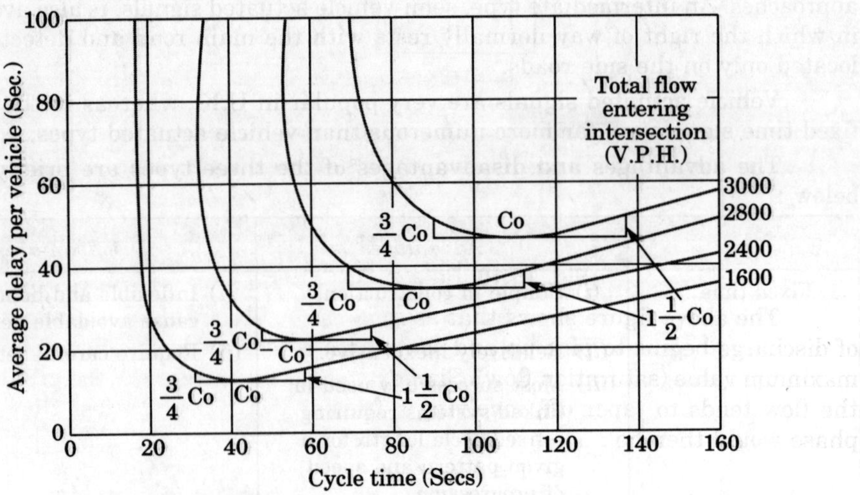

2 phase 4 arm intersection equal flows on all arms equal saturation flows : 1800 V.P.H. Equal Green Times
Total Lost Time Per Cycle : 10 Sec.

Fig. 15.9. Effect of variation of cycle time on delay. (From Refer 1)

The above subject was studied in the Road Research Laboratory (U.K.) by means of computer simulation of flow at the traffic signals. Some examples of the variation of delay with cycle time are shown in Fig. 15.9. It will be seen therein that for cycle times within the range of three quarters to one-and-a-half times the optimum value the delay is never more than 10–20 percent above that given by the optimum cycle time. The result could thus be used in determining the compromise cycle time that would suit variations in flow during the day.

By differentiating the equation for the total delay for the intersection with respect to cycle time, the following equation for the optimum cycle time has been obtained :

$$C_0 = \frac{1.5\,L + 5}{1 - Y} \text{ seconds} \qquad ...(15.1)$$

where C_0 = optimum cycle time

L = total lost time per cycle (in seconds)

$Y = y_1 + y_2 + ... + y_n$

and $y_1, y_2 ...., y_n$ are the maximum ratios of flow to saturation flow for phases 1, 2, ...*n* (*i.e.*, *q*/*s* where *q* is the flow and *s* is the saturation flow).

The lost time *L* in the above formula can be understood with reference to Fig. 15.10, indicating the rate of flow against time.

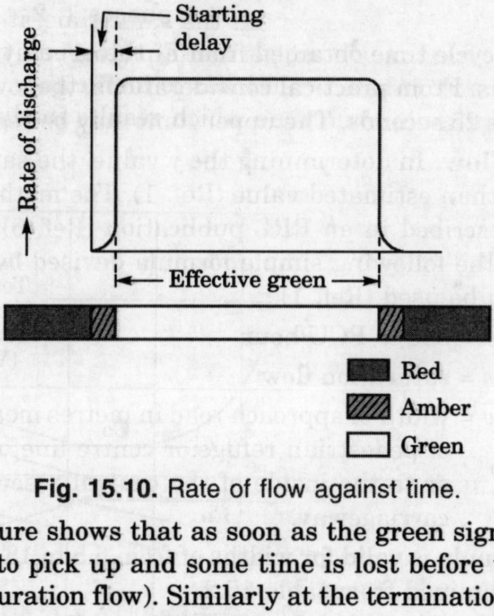

Fig. 15.10. Rate of flow against time.

The above figure shows that as soon as the green signal is given, the rate of discharge begins to pick up and some time is lost before the flow reaches the maximum value (saturation flow). Similarly at the termination of the green phase, the flow tends to taper off, involving a further lost time. The lost time for the phase would then be :

$$l = k + a - g \qquad ...(15.2)$$

where, l = lost time for the phase

k = green time for the phase

a = amber time for the phase

g = effective green time = b/s

b = number of vehicles discharged on the average during a saturation flow.

s = saturation flow.

The total lost time due to starting delays per cycle will be nl if there are n phases in the cycle. In addition to this lost time, the time R during each cycle when all signals display red (including red/ amber) simultaneously is also lost to the traffic. Thus the total lost time L can be expressed as :

$$L = nl + R \qquad ...(15.3)$$

The value of Y is the sum of y values for each phase. Each phase will handle one or more intersection, each approach having its own traffic flow and saturation flow. But for the purposes of the above formula, it has been shown by Webster (Ref. 5) that the y value for the phases may be taken as the highest ratio of traffic flow to saturation flow.

The effective green time available in a cycle can be apportioned to the different phases by the following rule, so as to give the least overall delay to the traffic using intersection (Ref. 5) :

$$g_1 : g_2 ...... g_n = y_1 : y_2 : ...... : y_n \qquad ...(15.4)$$

where $g_1 : g_2, ... g_n$ = effective green times allotted to phases 1, 2, ..., n respectively.

$y_1 : y_2, ... : y_n$ = maximum of y values

$$\left(= \frac{\text{Flow}}{\text{Saturation flow}} = \frac{q}{s} \right)$$

for phases 1, 2, ..., n respectively.

The optimum cycle time obtained from Eq. (15.1) may be very short under light traffic conditions. From practical considerations, the lower limit of the cycle time may be taken as 25 seconds. The upper limit may be regarded as 2 minutes.

Saturation Flow. In determining the y value, the saturation flow should be measured rather than estimated value (Ref. 1). The method of measuring the saturation flow is described in an RRL publication (Ref. 6). For designing new signal installations, the following simple formula devised by the Road Research Laboratory, U.K. can be used (Ref. 1) :

$$s = 525\, w \text{ PCU/hour} \qquad ...(15.5)$$

where s = saturation flow

w = width of approach road in metres measured kerb to inside of pedestrian refuge or centre line, whichever is nearer, or to the inside of the central reserve in case of a dual carriageway.

The above formula is valid for widths of from 5.5 to 18 m. For lesser widths the values may be obtained from table 15.2.

Table 15.2. Saturation Flow for Widths 3 to 5.5 m

(Adopted from Refer 1)

Width w in metres	3.0	3.5	4.0	4.5	5.0	5.5
s (PCU/hour)	1850	1890	1950	2250	2250	2900

When the approaches are in a gradient, the saturation flow needs some adjustment. Approximately this can be done by decreasing the saturation flow by 3 per cent for each 1 per cent uphill gradient and increasing the saturation flow by 3 per cent for each 1 per cent of downhill gradient.

The effect of composition of vehicles can be accounted for in measuring the flow and saturation flow by converting into PCU equivalents as per values given in Table 15.3.

Table 15.3. PCU Equivalents for Traffic Signal Computations

Types of Vehicle	*PCU Equivalent*
Heavy or medium goods vehicles	1.75
Light goods vehicle	1.00
Bus	2.25
Motor cycle, moped or scooter	0.33
Pedal cycle	0.20

The effect of right-turning traffic on the saturation flow can be accounted for in the following manner :

(i) No opposing flow, no exclusive right turning lanes. An overall figure of saturation flow for the approach, irrespective of turning movements, can be obtained using the rules given above.

(ii) No opposing flow, exclusive right turning lanes. The saturation flow of right turning stream through a right angle should be obtained separately by the following formula :

$$s = \frac{1800}{1 + 1.52 / r} \text{ PCU/hour for single file streams} \quad ...(15.6)$$

and $$s = \frac{3000}{1 + 1.52/r} \text{ PCU/hour for double file streams} \quad ...(15.7)$$

where r is the radius of curvature (in metres) of the right turning stream through a right angle.

(iii) Opposing flow, no exclusive right turning lanes. Three effects are possible under these circumstances. Firstly, because of the opposing traffic, the right turners are themselves delayed and consequently delay other non-right-turning vehicles in the same stream. Secondly, their presence tends to inhibit the use of the off side lane by straight ahead vehicle. These two effects can be allowed by assuming that on the average each right turning vehicle is equivalent to 1.75 straight ahead vehicles. The third effect pertains to the discharge of right turners through suitable gaps in the opposing stream. The following equation gives the maximum number of right turning vehicles per cycle (n_r) that can take advantage of gaps in the opposing stream :

$$n_r = s_r \times \left(\frac{gs - qc}{s - q} \right) \quad ...(15.8)$$

where s_r = right turning saturation flow

g = green time

c = cycle time

q = flow in opposing arm

s = saturation flow for opposing arm.

(The derivation of this equation is given in Appendix 15.1).

If g and c are in seconds in the above equation, s should be flow per second.

If the average number of right turners per cycle is more than n_r, the difference between the two (n_w) will have to wait at the intersection at the termination of the green time. For allowing all these n_w vehicles to clear the intersection, the intergreen time can be made equal to $2\frac{1}{2}\, n_w$ seconds, assuming each vehicle takes 2½ seconds to clear.

(iv) Opposing flow, exclusive right turning lanes. There will be no delay to the straight ahead traffic using the same approach as the right turners, but there will be an effect on the cross-phase and this should be calculated as outlined in *(iii)* above.

The effect of left turners on the saturation flow can be disregarded if the percentage of left turners is less than 10. If more, a correction is made for the excess over 10 percent by assuming each left turner is equivalent to 1.25 straight ahead vehicles.

The effect of site characteristics can be considered by applying the factors given in Table 15.4.

Table 15.4. Effect of Site Characteristics on Saturation Flow (*Ref. 1*)

Site Designation	*Description*	*Percentage of standard saturation flow*
Good	Dual carriageway. No interference from pedestrians, parked vehicles, right-turning traffic, (either owing to their absence or because special provision is made for them). Good visibility and adequate turning radii.	120
Average	Average sites. Some characteristics of 'Good' and 'Poor'.	100
Poor	Average speed low. Some interference from standing vehicles, pedestrians, right turning traffic. Poor visibility and/or poor alignment of intersection. Busy shopping street.	85

The following examples illustrate the use of the approach described so far in determining the settings of fixed time signals.

Problem 15.1. *A fixed time 2 phase signal is to be provided at an intersection having a North-South and an East-West road where only straight-ahead traffic is permitted. The design hour flows from the various arms and the satuartion flows for these arms are given in the following table :*

	North	*South*	*East*	*West*
Design hour flow (q) in PCU s/hour	*800*	*400*	*750*	*1000*
Saturation flow (s) in PCU s/hour	*2400*	*2000*	*3000*	*3000*

Calculate the optimum cycle time and green times for the minimum overall delay. The intergreen time should be the minimum necessary for efficient operation. The time lost per phase due to starting delays can be assumed to be 2 seconds. The value of the amber period is 2 seconds. Sketch the timing diagram for each phase.

Solution :

	N	S	E	W
q	800	400	750	600
s	2400	2000	3000	3000
$y = q/s$	0.33	0.20	0.25	0.20
y (max) values	0.33		0.25	

Minimum intergreen period $(I) = 4$ seconds
Amber period $(a) = 3$ seconds
Red/amber period $= 2$ seconds
} as per British practice

Time lost due to starting delays = 2 seconds per phase.

$$L = \Sigma(I - a) + \Sigma l$$

$$= (4-2) + (4-2) + 2 + 2, \text{ since there are 2 phases.}$$

$$= 8 \text{ seconds.}$$

$$\text{Optimum cycle time} = C_0 = \frac{1.5\,L + 5}{1 - Y} \qquad \text{from (Eq. 15.1)}$$

$$= \frac{1.5\,(8) + 5}{1 - 0.33 - 0.25} = \frac{17}{0.42} = 40 \text{ seconds}$$

Effective green time per cycle

$$= C_0 - L = 40 - 8 = 32 \text{ seconds.}$$

The effective green times for the phases can be calculated as follows :

$$g\,N\,S = \frac{y\,N\,S}{Y}\,(C_0 - L) = \frac{0.33}{0.58} \times 32 = 18 \text{ seconds}$$

$$g\,E\,W = \frac{y\,E\,W}{Y}\,(C_0 - L)$$

$$= \frac{0.25}{0.58} \times 32 = 14 \text{ seconds.}$$

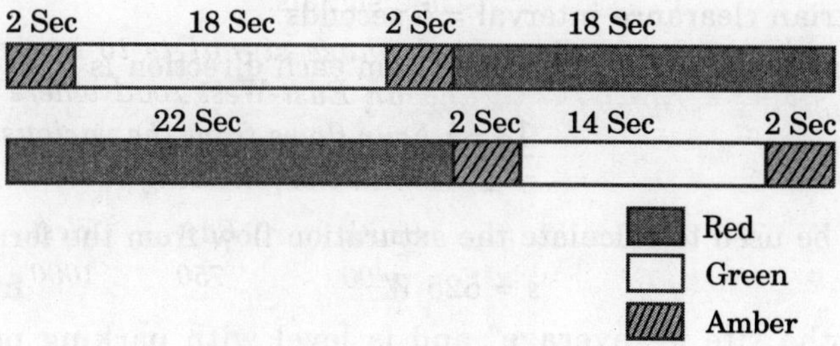

Fig. 15.11. Timing Diagram.

The timing diagram is indicated in Fig. 15.11.

Problem 15.2. *A three-phase traffic signal is to be installed at a right angled crossing of two city streets. The site is "average" and the approaches are 12 metres wide between kerbs. The approaches are straight and level and parking is prohibited on them. One of the phases is to be a "pedestrian only" phase occurring at the end of each cycle. Starting delay may be taken as 2 seconds. An "all-red"* *period of 4 seconds is to be provided after each vehicle phase to allow clearance of right turning vehicles left over in the crossing. The design hour traffic volumes in PCUs / hour are given in the following table :*

From	*N*			*E*			*S*			*W*		
To	*E*	*S*	*W*	*S*	*W*	*N*	*W*	*N*	*E*	*N*	*E*	*S*
PCUs/hr	*40*	*800*	*70*	*60*	*500*	*50*	*60*	*660*	*60*	*70*	*680*	*60*

Calculate the optimum cycle time for a fixed time installation. Sketch the phasing diagram for each phase. Draw a diagram showing the timings for all three aspects for a complete cycle.

Make suitable assumptions for Amber and for the pedestrian interval.

Solution. The width roadway being 12 metres, the time taken for pedestrians to cross the street, is 10 seconds, with a speed of walk of 1.2 metres per second. Because of the large width of the streets, it is desirable to have a central pedestrian refuge of at least 1 m width. The time that will be needed by a pedestrian to reach the pedestrian refuge from the kerb will then be :

$$\frac{12-1}{2} \times \frac{1}{1.2} = 4.6, \text{ say 5 seconds.}$$

This will be the pedestrian clearance interval during which no signal is displayed to the pedestrians and those who have just left the kerb or the central refuge before the termination of the pedestrian green signal can reach safely the central refuge of the kerb as the case may be. The American practice (Ref. 2) is to display the flashing "DONT WALK" indication for this clearance interval. The pedestrian clearance interval is followed by Amber of the next vehicular phase and by the Red signal in the pedestrian phase.

A total length of 13 seconds will be provided for the pedestrian phase as follows :

Pedestrian green time = 8 seconds

Pedestrian clearance interval = 5 seconds

The width of the approach road from each direction is

$$\frac{12-1}{2} = 5.5 \text{ m}$$

and this will be used to calculate the saturation flow from the formula.

$$s = 525\,W \qquad \text{from (Eq. 15.5)}$$

Since the site is "average" and is level with parking prohibited, no corrections are needed for the saturation flow obtained from the above formula.

The effect of left-turning traffic will be accounted for if it constitutes more than 10 percent of the traffic by counting each left turner as equivalent to 1.25 straight ahead vehicles. Since no exclusive right turning lanes are provided, the effect of right turning traffic will be accounted for by counting each right turner as equivalent to 1.75 straight ahead vehicle.

The following tabulations indicate the sequence of calculations :

From	*N*			*E*			*S*			*W*		
To	*E*	*S*	*W*	*S*	*W*	*N*	*W*	*N*	*E*	*N*	*E*	*S*
Given flow PUCs/hr	40	800	80	80	500	52	60	660	60	70	680	60
Correction for left turners	–	–	–	+ 20	–	–	–	–	–	–	–	–
Correction for right turners	–	–	+ 60	–	–	+ 39	–	–	+ 45	–	–	+ 45
Total	40	800	140	100	500	91	60	660	105	70	680	105
q		980			691			825			855	
$s = 525\ W$		2900			2900			2900			2900	
$y = \frac{q}{s}$		0.34			0.24			0.28			0.29	
y(max)												
N–S		0.34										
E–W											0.29	

Provide Amber time (a) of 2 seconds

All red period after each vehicular phase = 4 seconds (given). The Intergreen period (I) will, therefore, consist of :

Amber = 2 × 2 = 4 seconds

All Red = 4 seconds

8 seconds

Pedestrian phase = 13 seconds

Amber following pedestrian phase = 2 seconds

Starting delay per vehicular phase (l) = 2 seconds

Lost time $L = (13 + 2) + \Sigma(I - a) + \Sigma l$

$= 15 + (8 - 2) + (8 - 2) + 2 + 2 = 31$ seconds

Optimum cycle time,

$$C_0 = \frac{1.5\,L + 5}{1 - Y} \qquad \text{from (Eq. 15.1)}$$

$$= \frac{1.5\,(31) + 5}{1 - 0.34 - 0.29}$$

$$= \frac{52}{0.37} = 140 \text{ seconds}$$

This should be restricted to 120 seconds maximum.

Effective green time per cycle available to the vehicular phases

$= C_0 - L$

$= 120 - 31 = 89$ seconds.

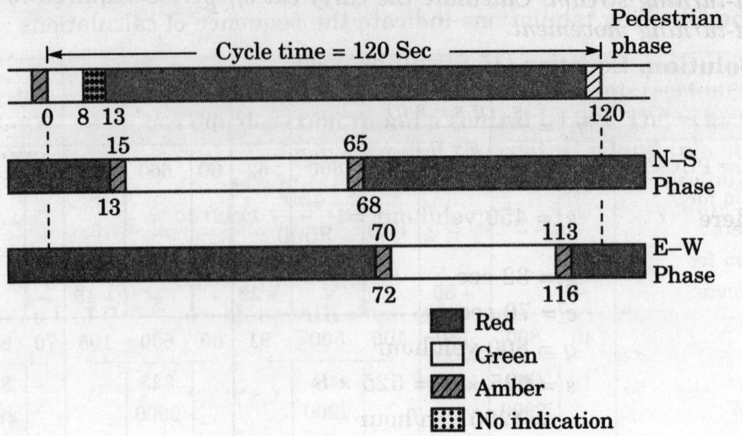

Fig. 15.12. Timing Diagram.

This will be apportioned between *N–S* and *E–W* phases as follows :

$$g_{NS} = \frac{y_{NS}}{Y}(C_0 - L)$$

$$= \frac{0.34}{0.63} \times 89 = 49 \text{ seconds}$$

$$g_{EW} = \frac{y_{EW}}{Y}(C_0 - L)$$

$$= \frac{0.29}{0.63} \times 89 = 40 \text{ seconds}$$

The timing diagram is given in Fig. 15.12.
The phasing diagram is given in Fig. 15.13.

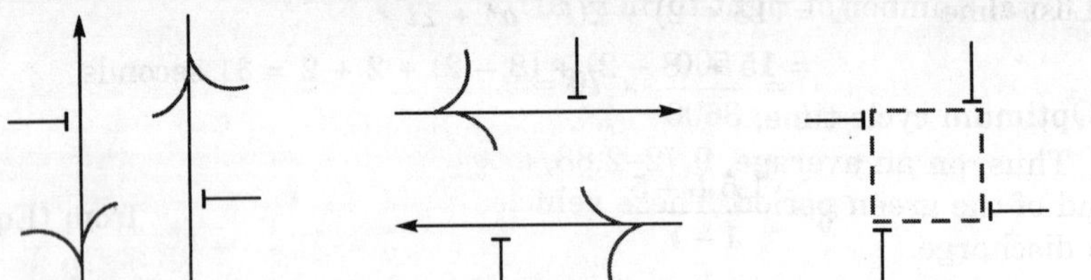

Fig. 15.13. Phasing diagram.

Problem 15.3. *An intersection controlled by traffic signals is formed by two roads running North to South and East to West. There is a heavy turning movement of 500 vehicles per hour from South to East. The flow from North to South is 800 vehicles per hour through an approach width at stop line of 8 m. The cycle time is 70 seconds and the effective green time for the flow from the North is 32 seconds. The effective right-turning saturation flow is 450 vehicles per hour for an opposing flow of 800 vehicles per hour, related to a minimum head-way of 2½ seconds for* *the right-turning stream. Calculate the early-cut-off period required to allow for the right-turning movement.*

Solution. Equation (15.8) will be used.

$$n_r = \frac{s_r\,(g\,s - c\,q)}{s - q}$$

Here $s_r = 450 \text{ veh/hour} = \dfrac{450}{3600} \text{ veh/sec}$

$g = 32 \text{ sec}$

$c = 70 \text{ sec}$

$q = 800 \text{ veh/hour}$

$s = 525 \times w = 525 \times 8$

$= 4200 \text{ veh/hour}$

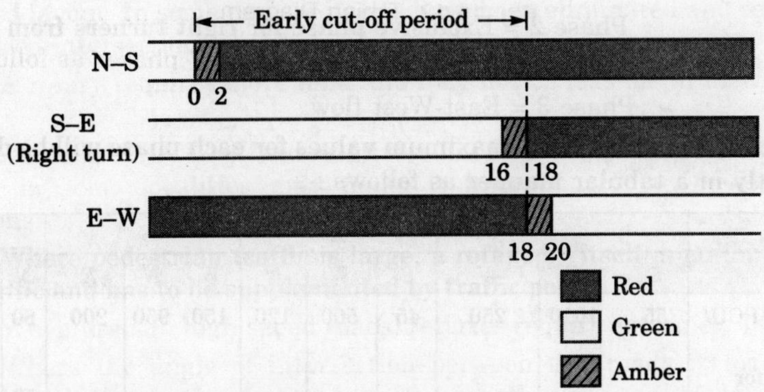

Fig. 15.14. Timing diagram.

$$n_r = \frac{450\,(32 \times 4200 - 70 \times 800)}{3600\,(4200 - 800)} = 2.88 \text{ veh/cycle.}$$

Total number of right-turners arriving per cycle

$$= \frac{500}{3600} \times 70 = 9.72$$

Thus, on an average, 9.72–2.88, *i.e.*, 6.84 or 7 vehicles will be left over at the end of the green period. These vehicles will take 7 × 2½ sec = 17.5 sec or 18 sec to discharge.

Thus the early cut-off period should be 18 sec. The timing diagram is indicated in Fig. 15.14

Problem 15.4. *A dual carriageway runs North and South and is intersected by a single carriageway running East and West. The layout of the junction has been designed such that there are two straight through lanes of 3.50 m width and an exclusive right-turn lane in each of the Northern and Southern approaches. The central reservation is reduced to 1.2 m at the junction to accommodate the exclusive rightturning lane. The radius of the right-turning stream is 15 m. The width of the East-West carriageway is 12 m. A fixed cycle three-phase signal is to be installed with an exclusive phase for right-turners from North and South. Assume lost time due to "starting up" and "tailing off" to be 2 seconds. Suitable values for the* *intergreen and amber times may be assumed. The traffic flow in PCUs/hour are given in the following table :*

From	*N*			*E*			*S*			*W*		
To	*E*	*S*	*W*	*S*	*W*	*N*	*W*	*N*	*E*	*N*	*E*	*S*
Flow PUCs/Hour	*55*	*1010*	*250*	*45*	*500*	*120*	*150*	*950*	*200*	*60*	*500*	*110*

Calculate the optimum cycle time and the actual green times to be set on the controller for each phase. Draw the phasing diagram.

Solution. **The following assumptions would be made :**

Intergreen period (*I*) = 4 seconds

Amber period (*a*) = 2 seconds

Let the three phase be designated as follows :

Phase 1 = North-South straight-ahead flow

Phase 2 = Exclusive phase for right turners from North and South

Phase 3 = East-West flow.

The calculation of the *y* maximum values for each phase will be done more conveniently in a tabular manner as follows :

From	*N*			*E*			*S*			*W*		
To	*E*	*S*	*W*	*S*	*W*	*N*	*W*	*N*	*E*	*N*	*E*	*S*
Given flow PCU/hour	**55**	**1010**	**250**	**45**	**500**	**120**	**150**	**950**	**200**	**60**	**500**	**110**
Correction for left turners (+ 25%)		**–**						**+ 38**				
Correction for right turners (+ 75%)						**+ 90**					**+ 83**	
Total flow Phase 1		**1065**						**1138**				
Saturation flow Phase 1		**525 × 2 × 3.5 = 3675**						**3675**				
y_1		$\frac{1065}{3675} = 0.29$						$\frac{1138}{3675} = 0.31$				
Saturation flow Phase 2			$\frac{1800}{1 + \frac{1.52}{15}} = 1620$						**= 1620**			
y_2			$\frac{250}{1620} = 0.15$						$\frac{200}{1620} = 0.12$			

From	N			E			S			W		
To	E	S	W	S	W	N	W	N	E	N	E	S
Total flow Phase 3					755						753	
Saturation flow Phase 3					$525 \times \frac{12}{2}$ = 3150						3150	
y_2					$\frac{755}{3150} = 0.24$						$\frac{753}{3150} = 0.24$	

It may be noted that the correction for right turners is not applied to flow from *N* and *S* since an exclusive phase is provided for them.

Maximum values of y : $y_1 = 0.31$

$y_2 = 0.15$

$y_3 = 0.24$

$Y = 0.31 + 0.15 + 0.24 = 0.70$

Lost time, $L = \Sigma(I - a) + \Sigma l$

$= 3 \times (4 - 2) + 3 \times 2 = 12$ seconds

Optimum cycle time, $C_0 = \frac{1.5\,L + 5}{1 + X} = \frac{1.5 \times 12 + 5}{1 - 0.7} = \frac{23}{0.3} = 77$ seconds

Effective green time = 77 – L = 77 – 12 = 65 seconds.

This will be apportioned between the phases as follows :

$$g_1 = \frac{y_1}{Y} \times 65 = \frac{0.31}{0.70} \times 65 = 29 \text{ seconds}$$

$$g_2 = \frac{y_2}{Y} \times 65 = \frac{0.15}{0.70} \times 65 = 14 \text{ seconds}$$

$$g_3 = \frac{y_3}{Y} \times 65 = \frac{0.24}{0.70} \times 65 = 22 \text{ seconds}$$

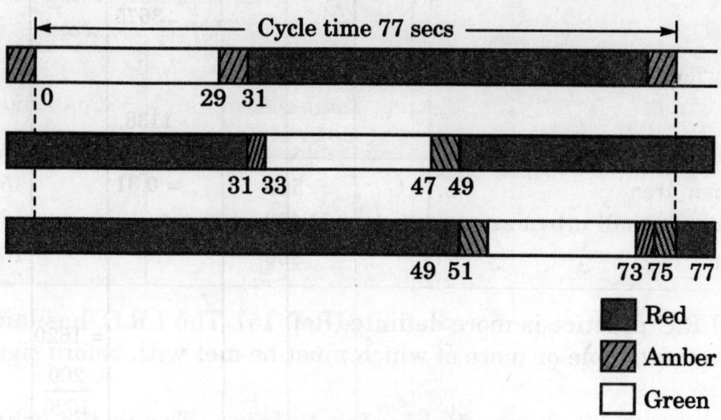

Fig. 15.15. Timing Diagram.

15.10. Warrants for Signals

In U.K., the installation of signals at a particular intersection is judged on the merits of the situation, there being no laid down warrants. Generally, the Department of the Environment have laid down certain criteria for deciding on signal installation (Ref. 7)

The minimum traffic flows for which signals are considered justified as per current U.K. practice are given below :

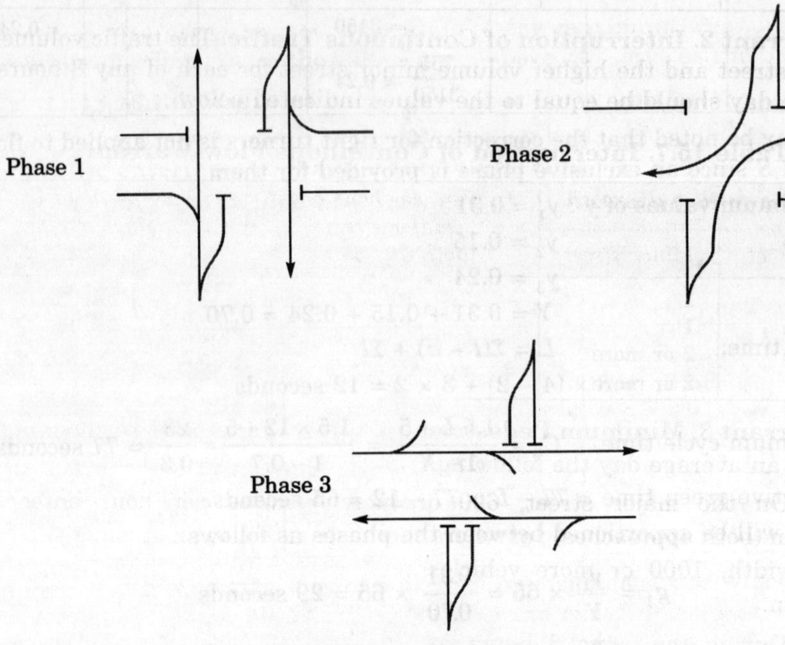

Fig. 15.16. Phasing timing.

Table 15.5. Minimum Traffic Flows for Signal Installation
(Ref. 7)

Type of Area	*Average hourly flow in V.P.H. to exceed*	
	Total entering intersection	*Contribution from side roads*
Large urban area	500	150
Suburban and small urban areas	400	125
Elsewhere	300	100

The I.R.C practice is more definite (Ref. 15). The I.R.C. has laid down the following warrants, one or more of which must be met with before signals can be installed :

Warrant 1. Minimum Vehicular Volume. The traffic volume on the major street and the higher volume minor street for each of any 8 hours of an average day should be equal to the values indicated below :

Table 15.6. Minimum Vehicular Volume Warrant

No. of lanes on each approach		*Veh. per hour on major street (both approaches)*	*Veh. per hour on minor street (one direction only)*
Major street	*Minor street*		
1	1	650	200
2 or more	1	800	200
2 or more	2 or more	800	250
1	2 or more	650	250

Warrant 2. Interruption of Continuous Traffic. The traffic volume on the major street and the higher volume minor street for each of any 8 hours of an average day should be equal to the values indicated below :

Table 15.7. Interruption of Continuous Flow Warrant

No. of lanes on each approach		*Veh. per hour on major street (both directions)*	*Veh. per hour on minor street (one direction only)*
Major street	*Minor street*		
1	1	1000	100
2 or more	1	1200	100
2 or more	2 or more	1200	150
1	2 or more	1000	150

Warrant 3, Minimum Pedestrians Volume. For each of any hour of any 8 hours of an average day the following traffic volume must exist :

(*i*) On the major street, 600 or more vehicles per hour enter the intersection (both approaches) ; or where there is a raised median island 1.2 m or more in width, 1000 or more vehicles per hour (both directions) enter the intersection.

(*ii*) During the same 8 hours as in (*i*) above, there are 150 or more pedestrians per hour on the highest volume cross-walk crossing the major street.

Warrant 4, Accident Experience. The accident experience warrant is satisfied when :

(*i*) Adequate trial of less restrictive remedies with satisfactory observance and enforcement have failed to reduce the accident frequency ;

(*ii*) Five or more reported accidents, of types susceptible of correction by traffic signal control have occurred within a period of 12 months, each accident involving personal injury or property damage to an apparent extent of ₹ 2000 or more.

(*iii*) The signal installation will not seriously disrupt traffic flow.

15.11. Co-ordinated Control of Signals

15.11.1. Need for co-ordinated control

Need for co-ordinated control of signals arises on a main traffic route when it is desirable to reduce delays and avoid main traffic from having to stop at every junction. When a signal indicates a stop aspect at a junction, a queue of vehicles is formed behind the stop line. When the signal changes to green, the vehicles start moving in a platoon. If this platoon is made to meet a green aspect at the next junction no delay is caused to the vehicles. This principle of linking adjacent signals so as to secure maximum benefits to the flow of traffic is called co-ordinated control of signals.

The co-ordination of signals is sought for with the following objectives in view :

(*i*) To pass the maximum amount of traffic without enforced halts.

(*ii*) To have minimum overall delay to traffic streams, both in the main and side roads.

(*iii*) To prevent the queue of vehicles at one intersection from extending and reaching the next intersection.

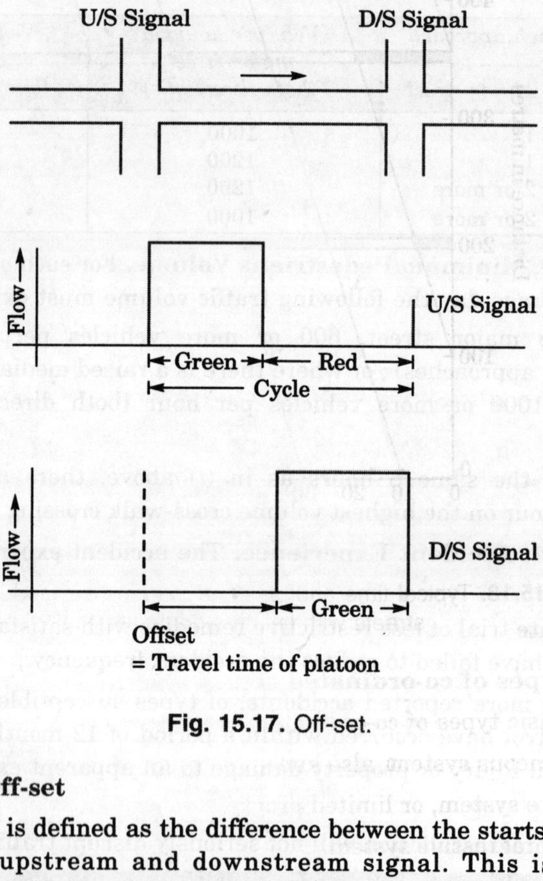

Fig. 15.17. Off-set.

15.11.2. Off-set

The *off-set* is defined as the difference between the starts of green time at the successive upstream and downstream signal. This is an important consideration in planning a system of co-ordination. If the start of green at the downstream signal is offset at a particular value such that the platoon, which starts at the upstream signal upon a green indication there, arrives at the downstream signal just in time for the green signal, the platoon has an unhindered movement (Fig. 15.17).

15.11.3. Time-and distance diagram

In planning a system of co-ordinated signal control, it is often expedient to indicate the system diagrammatically by what is known as a "time-and-distance" or "time-and-space" diagram. On this diagram, the time and signal settings are indicated along the horizontal axis to a suitable scale, whereas the distance travelled along the major route is plotted vertically to a suitable scale. A typical time and distance diagram for a one way street is shown in Fig. 15.18 and one for a two-way-street in Fig. 15.19.

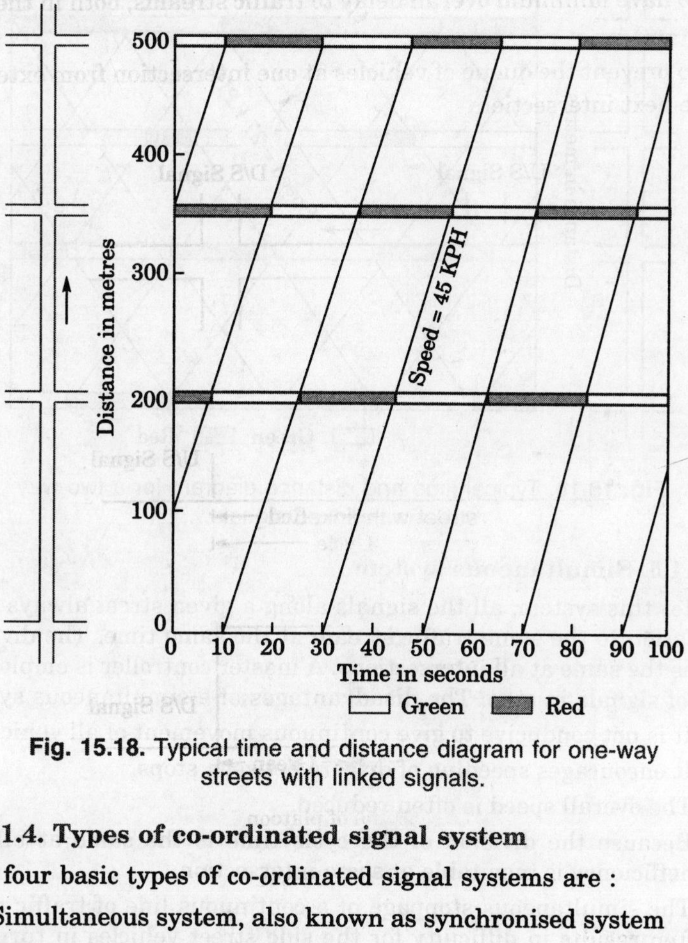

Fig. 15.18. Typical time and distance diagram for one-way streets with linked signals.

15.11.4. Types of co-ordinated signal system

The four basic types of co-ordinated signal systems are :

(*i*) Simultaneous system, also known as synchronised system

(*ii*) Alternate system, or limited progressive system

(*iii*) Simple progressive system.

(*iv*) Flexible progressive system.

These systems are described in detail below.

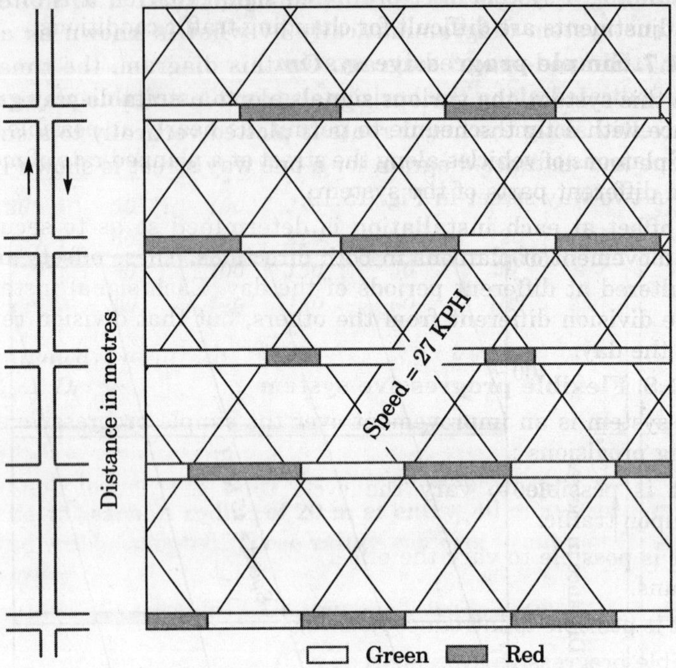

Fig. 15.19. Typical time and distance diagram for a two-way street with linked signals.

15.11.5. Simultaneous system

Under this system, all the signals along a given street always display the same indication to the same traffic stream at the same time. The division of the cycle time is the same at all intersections. A master controller is employed to keep the series of signals in step. The disadvantages of a simultaneous system are :

(*i*) It is not conducive to give continuous movement of all vehicles.

(*ii*) It encourages speeding of drivers between stops.

(*iii*) The overall speed is often reduced.

(*iv*) Because the division of the cycle time is the same at all the intersections, inefficiency is inevitable at some intersection.

(*v*) The simultaneous stoppage of a continuous line of traffic at all intersections often results in difficulty for the side street vehicles in turning into or crossing the main side street.

15.11.6. Alternate system (limited progressive system)

Under this system, consecutive signal installations along a given road show contrary indications at the same time. This permits the vehicles to travel one block in half the cycle time. This system operates efficiently where the blocks are of equal lengths. It also brings about a certain measure of speed control since speeding drivers are stopped at each signal. Some of the disadvantages of this system are :

(*i*) The green times for both the main and side streets have to be substantially equal, resulting inefficiency at most of the intersections.

(*ii*) In situations where the block lengths are unequal, the system is not well suited.

(*iii*) Adjustments are difficult for changing traffic conditions.

15.11.7. Simple progressive system

With this system, the various signals along a street display green aspects in accordance with a time schedule to permit, as nearly as possible, continuous operation of platoons of vehicles along the street at a planned rate of motion, which may vary in different parts of the system.

The offset at each installation is determined so as to secure the best continuous movement of platoons in both directions. These offsets are fixed and cannot be altered at different periods of the day. Each signal installation may have a cycle division different from the others, but that division remains fixed throughout the day.

15.11.8. Flexible progressive system

This system is an improvement over the simple progressive system with the following provisions :

(*i*) It is possible to vary the cycle time and division at each signal depending upon traffic.

(*ii*) It is possible to vary the offset, thus enabling two or more completely different plans.

(*iii*) It is possible to introduce flashing or shut down during off peak hours.

Flexible progressive systems require a master controller which keeps the local controllers at each intersection in step.

15.12. Signal Approach Dimensions

The approach dimensions have an influence on the design and performance of intersections. One method of improving the efficiency of signalised intersections is to modify the approach dimensions. Since the signals permit traffic movement from any approach for only a proportion of the time, it stands to reason that the approach roads in immediate vicinity of the intersection should have a wider roadway than the normal. Thus when planning an improvement in the geometric layout of the existing intersections or planning for new facilities, it helps to keep in view the improvement in flow conditions that can be achieved with wider approaches.

Webster and Newby (Ref. 8) have proposed the following simple rules for determining the approach dimensions :

(*i*) For a two phase cross-roads the approach widths should be proportional to the square roots of the flows. If q_1 and q_2 are the maximum flows on phase 1 and 2 respectively, the green times (g_1 and g_2), lengths widened (d_1 and d_2) and widths (w_1 and w_2) are related in the following way, vide Fig. 15.20.

$$\frac{w_1}{w_2} = \frac{g_1}{g_2} = \frac{d_1}{d_2} = \sqrt{\frac{q_1}{q_2}} \qquad ...(15.9)$$

If the approach width deduced from the above rule is less than that of the feeder road, it should be made equal to that of the feeder road and the green time made correspondingly less. The extra green time thus allocated to the other phase results in less widening being necessary on those approaches. The flow used should be the maximum flow on the two or more arms of the same phase.

(ii) With multi-phase intersections the above rule itself can be extended as follows :

$$w_1 : w_2 : w_3 \ldots w_n = \sqrt{q_1} : \sqrt{q_2} : \sqrt{q_3} : \ldots \sqrt{q_n} \quad \ldots(15.10)$$

$$= g_1 : g_2 : g_3 : \ldots g_n \quad \ldots(15.11)$$

$$= d_1 : d_2 : d_3 : \ldots d_n \quad \ldots(15.12)$$

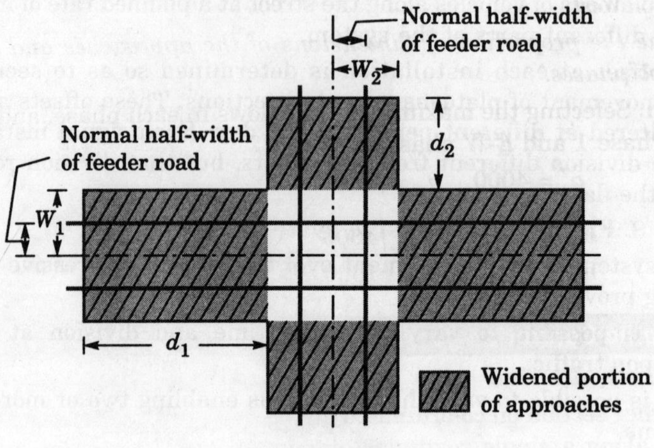

Fig. 15.20.

(iii) For T-junctions with 2 phase control, the ratios of widths, green times and lengths should be :

$$\frac{w_1}{w_2} = \sqrt{\frac{q_1}{2q_2}} \quad \ldots(15.13)$$

and

$$\frac{g_1}{g_2} = \frac{d_1}{d_2} = \sqrt{\frac{2q_1}{q_2}} \quad \ldots(15.14)$$

where the suffix 2 refers to the stem of the T-junction.

The examples given below illustrate the use of the above principles.

Problem 15.5. ***A T-junction has a design year flow of 300 vehicles per hour on the south arm and 2400 vehicles per hour on the east and west arms. What should be the ratio of approach widths, lengths and green times of the arms ?***

Solution. $\frac{w_{EW}}{w_S} = \sqrt{\frac{q_{EW}}{2q_S}} = \sqrt{\frac{2400}{600}} = 2$

$$\frac{g_{EW}}{g_S} = \frac{d_{EW}}{d_S}$$

$$= \sqrt{\frac{2q_{EW}}{q_S}} = \sqrt{\frac{4800}{300}} = 4$$

The east-west arm should be made twice as wide as the south arm. The green time apportioned to the east-west phase should be four times the green time apportioned to the south phase. The east-west approach should be widened over a length which is four times the length widened on the south arm.

Problem 15.6. *The following table gives the flows in the arms of an intersection where a two phase signal is to be designed.*

Arm	*Flow (vehicle / hour)*
North	*4000*
South	*3800*
East	*1000*
West	*900*

Determine the proportion of dimensions of the approaches and the green times for the two phases.

Solution. Selecting the maximum of the flows in each phase, and denoting *N-S* phase as Phase 1 and *E-W* phase as Phase 2.

$$q_1 = 4000, \quad q_2 = 1000$$

$$\frac{w_1}{w_2} = \frac{g_1}{g_2} = \sqrt{\frac{q_1}{q_2}} = \sqrt{\frac{4000}{1000}} = 2.$$

15.13. Area Traffic Control

15.13.1. Introduction

In an earlier section on co-ordinated signal systems, a description of simple linked systems along a single road was given. Area traffic control is a further extension of the same principles of coordination to include signals in a substantial area.

Area Traffic Control can be defined as a technique which provides for a centralised control of numerous signal installations distributed throughout an urban area, such that there is a planned coordination between signals at different junctions. The technique invariably employs digital computers for achieving the desired objective.

The earliest known scheme of Area Traffic Control was in Toronto begun in a modest scale in 1959. The success of the project prompted its application in many important cities of the world. In U.K. controlled experiments were initiated in certain areas of Glasgow and London to assess the amount of benefit that could be expected from different system of centralised control. By now, nearly 100 cities have some form of Area Traffic Control.

The subject has grown into a very interesting and promising field with extensive literature. In this section only a general introduction is sought to be given.

The objectives in an Area Control System are one or more of the following:

(*i*) Minimising journey time for vehicles

(*ii*) Minimising vehicular stops, resulting in less noise, less pollution and less consumption of fuel

(*iii*) Reducing accidents

(*iv*) Discouraging use of certain areas

(*v*) Minimizing person time.

Area Control System has proved to be a very efficient tool in tackling the serious problem of congestion at signalised intersection.

15.13.2. Traffic control methods

The following are the main types of methods in general use :

1. Fixed time plans based on historical data and calculated off line by a computerised optimizing technique. The information about vehicle movement is obtained manually or through detectors and fed to the computer, which then determines the signal settings, and transmits the settings to the signals. Examples of this type are the Combination Method and TRANSYT (Traffic Analysis Study Tool), which will be described briefly later.

2. Co-ordinated systems with local response at each signal. Example of this type are the FLEXIPROG (Flexible Progressive) and EQUISAT (Equally Saturated).

3. Fully responsive systems such as S.P.G., (Signal Plan Generation) and PLIDENT (Platoon Identification).

15.13.3. Combination method

This method has been developed by the Road Research Laboratory, UK and uses delay/offset relationship to obtain the relative timings or off-sets of traffic signals in a network (Ref. 9). It applies a rigorous optimization process to a reasonable model of the traffic and assumes that the cycle time, green times, flow and saturation flows are known and then chooses the offsets of the signals to minimise delay over the whole network. The technique can be applied on an area basis, subject to some restraints on the type of the network. It suffers from a serious restriction in assuming that the delay between two signals depends solely on the relative settings of the two signals, which is a good approximation only in heavily loaded conditions.

15.13.4. TRANSYT method (Traffic Analysis Study Tool)

The TRANSYT (Ref. 10), implemented by a digital computer programme, is a method of automatically determining fixed time signal plans to meet known network conditions. It is a widely used system, and as demonstrated by the Glasgow experiments, is an improvement over the Combination Method (Ref. 11). The traffic model makes allowance for flow intersection between successive sections of roads and represents average traffic behaviour more correctly than the Combination Method. It represents platoon dispersion effectively. The solution time is short and good convergency on the optimum signal settings is achieved by a "hill climbing" type of optimization procedure. The overall impedance to traffic is measured by a "performance index" that can be chosen with any desired balance between journey time and number of stops. The optimization procedure minimizes the performance index by altering the points within the signal cycle at which each stage starts. Thus, both signal offsets and green times are included in the optimization procedure.

15.13.5. FLEXIPROG (Flexible progressive)

This is a vehicle actuated progressive system and requires the use of detectors on the approach arms. With a continuous stream of traffic all over the detectors, the signals behave as a fixed time system. Under lighter traffic conditions, the signals can change after detecting a suitable gap in the traffic. Stages are missed if there is no demand for them. This system has not proved better than the combination method in the Glasgow experiments (Ref. 11).

15.13.6. EQUISAT (Equally Saturated)

Under this system, the cycle time and offset pattern are fixed. The allocation of green time is varied to equalise the degree of saturation of each stage.

Detectors on each approach arm and computer logic are used to measure or deduce both flow and saturation flow. This system too did not prove to be measurably better than the Combination method in the Glasgow experiments (Ref. 11).

15.13.7. PLIDENT (Platoon Identification)

This is a system developed by the Road Research Laboratory, U.K. (Ref. 12). The system identifies platoons of vehicles and operates signals to allow unimpeded passage to them on priority routes. The scheme has no fixed cycle or offsets, but adjusts the length of each stage to suit the various platoons. The effect of the scheme is to give the priority streets just the amount of green they require at times which avoid delay to the approaching platoons. This scheme, though successfully implemented in the Glasgow experiment, produced the largest average journey times there.

15.13.8. SPG (Signal Plan Generation)

This is a fully responsive system and generates cycle times, splits and offsets on line using measured data on traffic conditions. This type of system is installed in Madrid and Barcelona (Ref. 13).

15.13.9. SCOOT (Split Cycle Offset Options Technique)

The SCOOT is a new entrant into the field of traffic signal controlled network. It has been developed in the U.K. by the Transport and Road Research Laboratory and the British industry. A large number of detector loops are laid across the road which transmit the data on traffic to the central control system which adjusts the signal plans according to actual needs. Traffic delays can be substantially reduced (upto 12%) when compared to fixed time plans (Ref. 16, 17 and 18).

15.14. Delay at Signalised Intersections

15.14.1. An important consideration that has to be faced by an engineer in designing and operating a signalised installation is the delay to traffic. An understanding of the delay problem is important because the economic losses involved by vehicular delays at signals can be enormous.

15.14.2. Webster's (Ref. 5) investigations on the subject at the Road Research Laboratory, U.K. are based on queueing theory and computer simulation. The following formula has been presented by him for the average delay per vehicle on a particular intersection arm :

$$d = \frac{c\,(1-\lambda)^2}{2\,(1-\lambda x)} + \frac{x^2}{2\,q\,(1-x)} - 0.65\left(\frac{c}{q^2}\right)^{1/3} x^{(2\,+\,5\lambda)} \qquad \text{...(15.15)}$$

where, d = average delay per vehicle on the particular arm

λ = proportion of the cycle which is effectively green for the phase under consideration (*i.e.*, g/c)

x = the degree of saturation. This is ratio of the flow to the maximum possible flow under the given settings of the signals and equals q/λ_s

c = cycle time

g = effective green time

q = flow

s = saturation flow.

In the above formula, if c is in seconds, q should be in vehicles per second and delay will be obtained in seconds.

In order to simplify the calculations, ready made tables are available (Ref. 1, 5). A nomogram for the determination of delay based on the above equation has been presented by Dick (Ref. 14).

APPENDIX 15.1

Derivation of equation (15.8) for number of right turning vehicles which are able to discharge through gaps in the opposing flow.

Right turning vehicles can discharge through gaps in the opposing flow only during the period from the end of the saturated flow of the opposing stream to the beginning of the red period.

It is assumed that no vehicles remain in the opposing queue at the end of the green period.

Denoting the saturated green time for the opposing flow g_s, the effective red time by r and the opposing flow by q, the number of vehicles discharged during the saturated green time is : $(r + g_s)q$.

The time these vehicles take to discharge is :

$$g_s = \frac{(r + g_s)q}{s}$$

where s is the opposing saturation flow.

Thus $$g_s = \frac{rq}{s-q} = \frac{(c-g)q}{s-q} \qquad \text{...(15.16)}$$

where, c = cycle time = $r + g$

g = effective green time

If g_u is the unsaturated green time, then

$$g_u = g - g_s$$

Substituting from Eq. (15.16), $$g_u = \frac{gs - qc}{s-q}$$

The proportion of unsaturated green time, g_u / c is :

$$= \frac{gs - qc}{c\,(s-q)}$$

If s_r, is the effective right turning saturation flow through the opposing gaps, then the actual flow of right turners through gaps in the opposing stream, q_r, is given by :

$$q_r = s_r \left\{ \frac{gs - qc}{c\,(s-q)} \right\} \qquad \text{...(15.17)}$$

The number of right turning vehicles per cycle, n_r, is :

$$n_r = s_r \left\{ \frac{gs - qc}{c\,(s-q)} \right\} \qquad \text{...(15.18)}$$

I. MULTIPLE CHOICE QUESTIONS

1. The yellow junction box at the center of the intersection in the picture indicates no temporary stopping allowed
(*a*) right
(*b*) not on above
(*c*) wrong
(*d*) I dont know

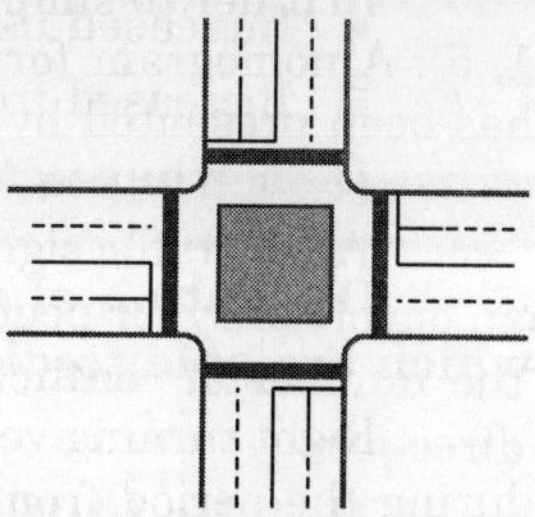

2. The hand signal of the traffic police in the picture is
(*a*) an auxiliary signal for stopping
(*b*) a signal for stopping
(*c*) a signal for turning
(*d*) a signal for pulling over

3. The hand signal of the traffic police in the picture is a signal for traffic approaching from behind to stop
(*a*) right
(*b*) wrong
(*c*) I don't know
(*d*) none of the above

Answers

1. (*a*) 2. (*b*) 3. (*a*)

II. SOLVED QUESTIONS AND ANSWERS

Q. 1. What do you mean by traffic signals? Explain its advantages.

Ans. Traffic signals control vehicle and pedestrian traffic by assigning priorities to various traffic movements to influence traffic flow. Properly designed, located and maintained traffic signals have one or more of these advantages:

- Provide for orderly movement of traffic;
- Increase traffic-handling capacity of an intersection;
- Reduce frequency and severity of certain types of crashes, especially right-angle collisions;
- Provide for continuous movement of traffic at a definite speed along a given route;
- Interrupt heavy traffic at intervals to permit other vehicles or pedestrians to cross.

Q. 2. What are the disadvantages of traffic signals?

Ans. Traffic signals are not a solution for all traffic problems at intersections, and unwarranted signals can adversely affect the safety and efficiency of traffic by causing one or more of the following:

- Excessive delay;
- Increased traffic congestion, air pollution and gasoline consumption;

- Disobedience of signals;
- Increased use of less-adequate streets to avoid traffic signals;
- Increased frequency of crashes, especially rear-end collisions.

Q. 3. What is the purpose of a traffic signal?

Ans. Traffic signals are designed to ensure an orderly flow of traffic, provide an opportunity for pedestrians or vehicles to cross an intersection and help reduce the number of conflicts between vehicles entering intersections from different directions.

Q. 4. Do traffic signals reduce intersection crashes?

Ans. Certain types of crashes can be reduced in number or severity by installing a signal, while other types of crashes will increase. Where signals are used unnecessarily, the most common result is an increase in total crashes, especially rear-end collisions.

Q. 5. How do traffic signals work?

Ans. Fixed-time signals follow a predetermined sequence of signal operation, always providing the same amount of time to each traffic movement, whether traffic is present or not. Actuated signals change the lights according to the amount of traffic in each direction. They use various types of sensors to detect vehicles, and adjust the length of the green time to allow as many vehicles as possible through the intersection before responding to the presence of vehicles on another approach.

Q. 6. When should signals be considered?

Ans. Traffic signals should be considered when they will alleviate more problems than they create. A warranted signal properly operated may provide for more orderly movement of traffic, and reduce the occurrence of certain types of collisions. However, traffic signals function by stopping traffic, and whenever a vehicle is stopped on the road, the potential for a crash is created. As a result, unwarranted signals can result in increased crashes, delays and congestion.

Q. 7. What happens once the signal is recommended?

Ans. Design plans are developed and reviewed. Environmental and right-of-way documents are prepared and, if necessary, public meetings are held. Once the design plans are completed, the signal installation project is publicly bid and awarded to an electrical contractor. The entire process may take two years or more.

A flashing yellow light means you should drive with caution when approaching and moving through the intersection.

Q. 8. Explain blank traffic lights.

Ans. During an electrical power loss, traffic lights at intersections will not work. Yield the right-of-way to vehicles in the intersection and to vehicles entering the intersection from your right. Go cautiously and use the intersection the same way you would use an intersection with all-way stop signs.

Q. 9. What are traffic beacons?

Ans. A traffic beacon is a single flashing light hung over an intersection or placed over signs or on obstacles in the road.

Q. 10. What is a flashing red beacon.

Ans. A flashing red beacon above an intersection or stop sign means you must come to a complete stop. Move through the intersection only when it is safe to do so.

Q. 11. Define flashing yellow beacon.

Ans. A flashing yellow beacon above an intersection, above a warning sign or on an obstruction in the road, warns you to drive with caution.

Q. 12. What are pedestrian signals?

Ans. Pedestrian signals help pedestrians cross at intersections with traffic lights. The signal for pedestrians to walk is a white walking symbol. A flashing or steady orange hand symbol means pedestrians must not begin to cross.

Q. 13. What is meant by pedestrian walk signal?

Ans. A pedestrian facing a walk signal may cross the road in the direction of the signal. While crossing, pedestrians have the right-of-way over all vehicles.

III. REVIEW QUESTIONS

1. What are the advantages and disadvantages of traffic signals?
2. Discuss signal indications.
3. Illustration of the signals.
4. What are fixed time signals and vehicle actuated signals?

REFERENCES

1. F.V. Webster and B.M. Cobbe, Traffic Signals. Road Research Laboratory (U.K.), Road Research Technical Paper No. 56, H.M.S.O. London, 1966.
2. Manual on Uniform Traffic Control Devices, Federal Highway Administration, U.S. Deptt. of Transportation, Washington, 1977.
3. Department of Environment, Pedestration Safety, H.M.S.O, London, 1973.
4. Specification for Road Traffic Signals, I.S. : 7537–1974, Indian Standards Institution, New Delhi, 1975.
5. F.W. Webster, Traffic Signal Settings Deptt. of Scientific and Industrial Research, Road Research Laboratory, Road Research Technical Paper No. 39, H.M.S.O., London, 1961.
6. A Method of Measuring Saturation Flow at Traffic Signals, Road Research Laboratory, Road Note No. 34, H.M.S.O., London, 1963.
7. Technical Memorandum, H1/73, Criteria for Traffic Light Signals at Junctions, Deptt. of Environment, U.K.
8. Webster F.V. and Newby R.F., Research into the Relative Merits of roundabouts and Traffic Signal Controlled Intersections, Proceedings of the Institution of Civil Engineers, London, 1964.
9. Hillier, A.J. The Area Traffic Control Experiments in Glasgow, Traffic Engineering and Control, 1965/66, 7 (8) and 7 (9), London.
10. D.L. Robertson, Transyt, Traffic Network Study Tool, RRL Report LR 253, Road Research Laboratory, Crowthorne, 1969.
11. Holroyd J. and Hillier J.A., Area Traffic Control in Glasgow, Traffic Engineering and Control, 11 (5), London, 1969.
12. J. Holroyd and J.A. Hillier : The Glasgow Experiment. Plident and after, RRL Report LR 384, Road Research Laboratory, Crowthorne, 1971.
13. OECD, Area Traffic Control Systems, Organization for Economic Co-operation and Development, Paris, 1972.

14. A.C. Dick, A Nomogram for solution of the RRL Traffic Signal Delay Formula, Traffic Engineering and Control, 1964, 6 (2), London.

15. Guidelines on design and installation of Road Traffic Signals, I.R.C : 93–1985, Indian Roads Congress, New Delhi, 1985.

16. Peck, C *et al*, Urban Traffic Control, Developing World Transport, Grosvenor Press International, London, 1991.

17. Robertson, D.I., and P.B. Hunt, A Method of Estimating the Benefit of Coordinating Signals by TRANSYT and SCOOT, Traffic Engineering and Control 23, 1982.

18. Hunt, P.B., Robertson, D.I., Bertherton, R.D. and R.I. Winton, SCOOT—A Traffic Responsive Method of Coordinating Signals, TRRL Report : LR 1014, Transport and Road Research Laboratory, Crowthorne, 1981.

16. Miscellaneous Traffic Control Aids and Street Furniture

16.1. Background

he term street furniture covers a wide variety of terms that are needed to support the main function of streets which is the movement of vehicles and pedestrians. The common types of street furniture found on the streets are :

Traffic Aids

1. Roadway delineators
2. Safety barriers
3. Barricades & channelisers
4. Railings
5. Traffic signs
6. Overhead traffic signs
7. Traffic signal posts
8. Traffic impact attenuators
9. Traffic aid posts
10. Hazard markers and Object markers
11. Speed breakers

Road Beautification

(*i*) Tree guards
(*ii*) Planters

Facilities

(*i*) Bus stands
(*ii*) Taxi stands
(*iii*) Roadside toilets
(*iv*) Roadside rest areas

Information Providers

Roadside Advertisements

The detailed coverage of each of these is beyond the scope of this book. However, this Chapter gives a brief description of some of the terms.

16.2. Principles of Street Furniture Design

The following are the general principles to be borne in mind while selecting and designing street furniture items (Ref. 1, 2) :

- Only those items which are absolutely necessary should be provided. Superflous items clutter the streetscape. Sometimes it is possible to combine light poles and signal posts.
- The initial cost and cost of maintaining the facilities must be given due consideration.
- The materials selected should be durable.

- The design should be vandal-proof.
- The designs should be aesthetically pleasing and match with the surroundings.

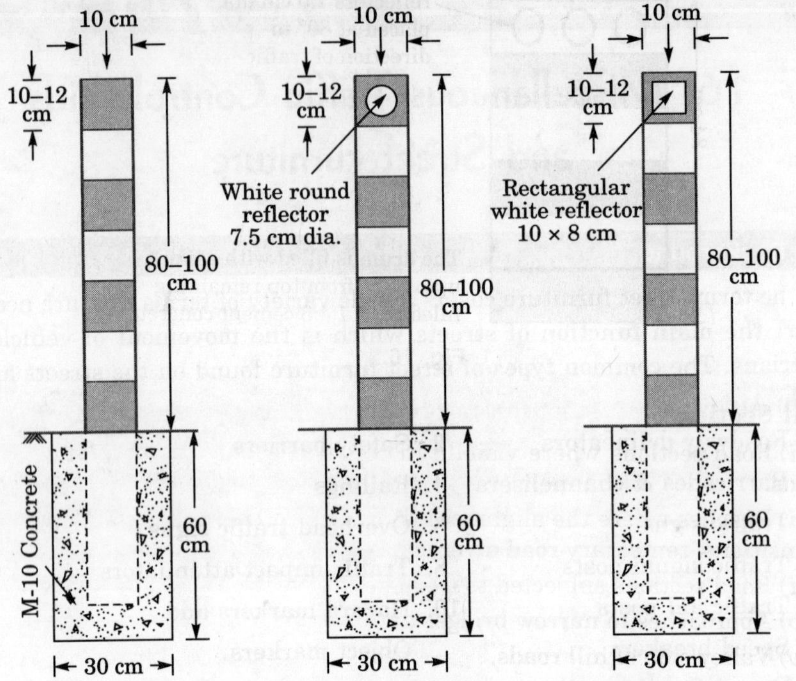

Fig. 16.1. Roadway Delineators.

16.3. Roadway Delineators

Roadway delineators are intended to provide visual aids to drivers concerning the alignment of the road, especially at night. They are effective in locations where the horizontal and vertical geometry change and during severe weather conditions such as rain, fog or snow. Generally, delineators are reflectionised for better visibility during night.

Roadway delineators are generally in the form of guide posts of metal, concrete, timber or cut stone, having a circular, rectangular or triangular cross-section.

The side facing the traffic should have a dimension not less than 10 cm. They should be 80–100 cm high. Typical designs are indicated in Fig. 16.1.

As an alternative, empty bitumen drums, suitably painted, may also be used, as in Fig. 16.2.

The use of roadway delineators is generally in rural highway sections under the following situation :

Curved Sections

(*i*) Horizontal curves of radius 1000 m or less.

(*ii*) Vertical curves with inadequate visibility.

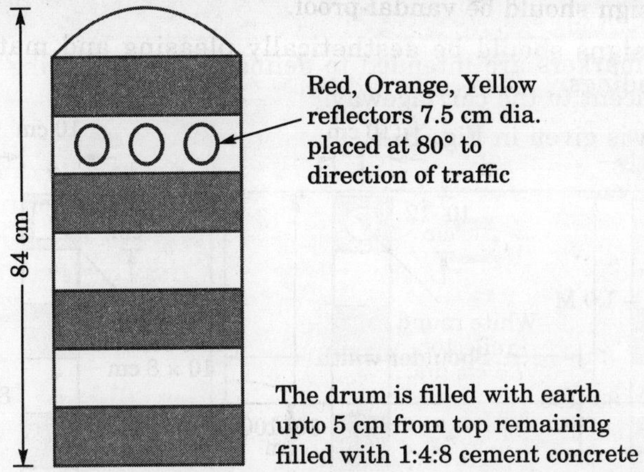

Fig. 16.2

Straight Sections

(*i*) Road sections where visibility is often poor due to mist, fog or snowy conditions.

(*ii*) Reaches where the alignment is uncertain to the driver, *e.g.*, pavement with transitions, temporary road diversions etc.

(*iii*) Road sections subjected to frequent submergence and ponding.

(*iv*) Approaches to narrow bridges and culverts.

(*v*) Valley side of hill roads.

(*vi*) Road embankments exceeding 3 m in height.

(*vii*) Approaches to important intersections.

(*viii*) Special problem points such as causeways and tunnels.

The spacing of delineators should be 50—70 m on straight sections, the posts being in pairs, one on each side of the roadway. On horizontal curves, the spacing should be fixed in relation to the radius of curve, as in Table 16.1. The location of delineators is indicated in Fig. 16.3.

Table 16.1. Recommended Spacing of Roadway Indicators on Horizontal Curves (*Source* : *Ref.* 3)

Radius in metres	*Spacing on curves in metres*
30	6
50	8
100	12
200	20
300	25
400	30
500	35
600	38
700	42
800	45
900	48
1000	50

16.4. Hazard Markers

Hazard markers are intended to define obstructions like guardrails and abutments adjacent to the carriageway.

A design is given in Fig. 16.4.

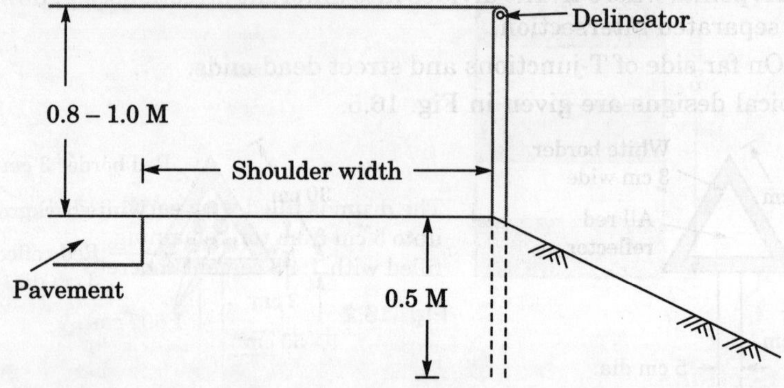

Fig. 16.3

Note : The stripes slope down at 45° towards the side on which traffic is to pass

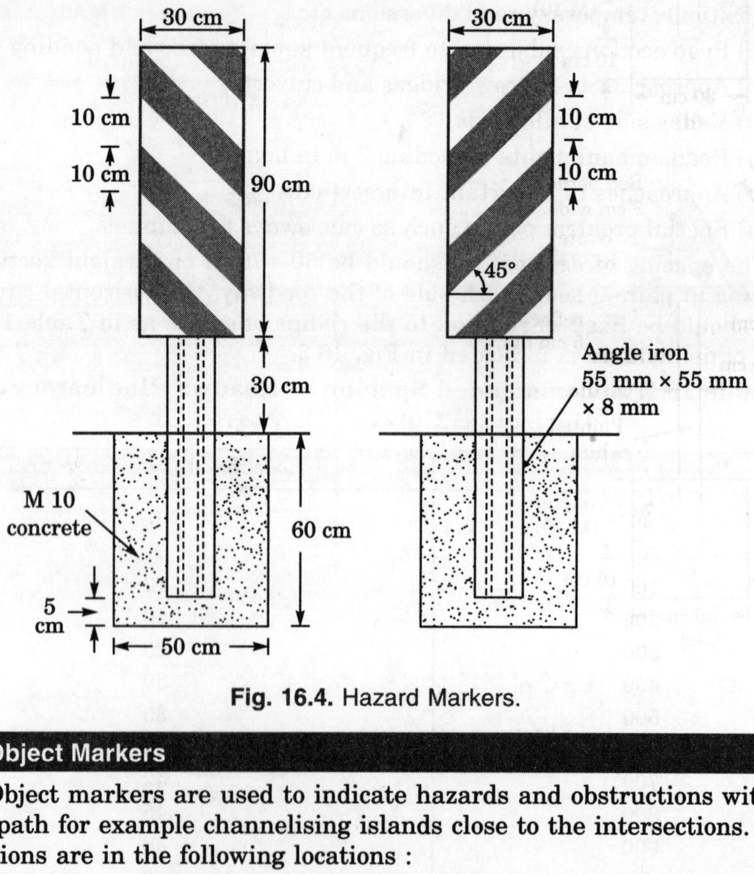

Fig. 16.4. Hazard Markers.

16.5. Object Markers

Object markers are used to indicate hazards and obstructions within the vehicle path for example channelising islands close to the intersections. Typical applications are in the following locations :

(*i*) Traffic islands at approaches to intersections.

(*ii*) Around periphery of rotary islands.

(*iii*) Median openings.

(*iv*) Facing approaching traffic at islands forming left infiltration lane.

(*v*) On medians/islands on far side of the intersections.

(*vi*) At points where traffic divides into different directions, *e.g.* down ramps of a grade separated intersection.

(*vii*) On far side of T-junctions and street dead-ends.

Typical designs are given in Fig. 16.5.

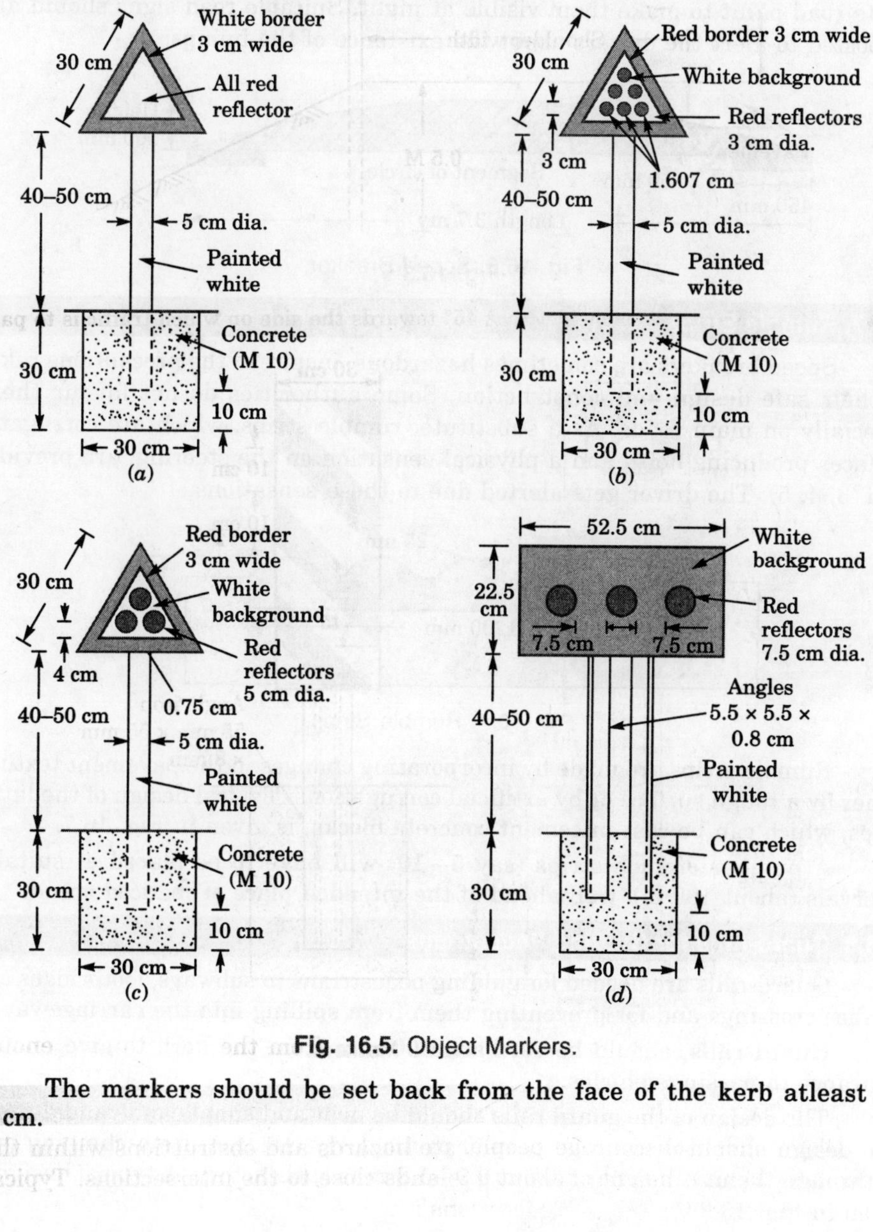

Fig. 16.5. Object Markers.

The markers should be set back from the face of the kerb atleast by 50 cm.

16.6. Speed Breakers

Speed breakers, also known as speed control humps, are commonly used on private access roads and near schools and universities in urban situations. They are to be disfavoured on through roads since they are dangerous to the vehicles. Their indiscriminate use is to be deprecated. In fact, in India, the Ministry of Surface Transport (Roads Wing) and the Indian Roads Congress have been opposing the provision of these obstacles on the road in an indiscriminate manner.

However, in some special situations as indicated above, they are used in the country. The design is shown in Fig. 16.6. The humps should be painted with white road paint to make them visible at night. Suitable road signs should also be posted to alert the drivers about the existence of the humps.

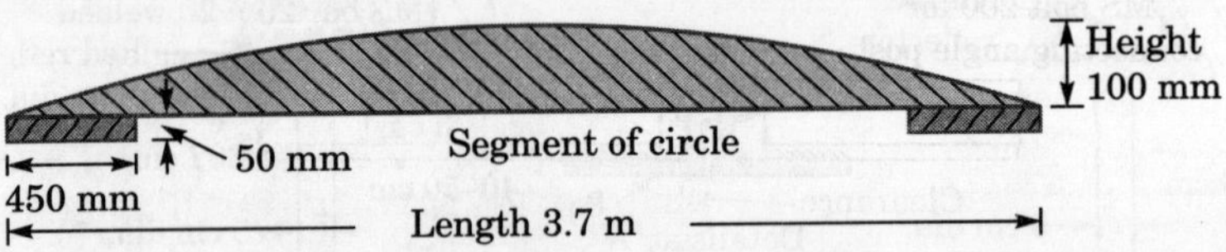

Fig. 16.6. Speed Breaker.

16.7. Rumble Strips

Speed breakers are sometimes hazardous, inspite of the precautions taken in their safe design and construction. Some authorities do not favour them, especially on main roads. As a substitute, rumble strips, which are corrugated surfaces producing noise and a physical sensation on the steering, are provided (Ref. 3, 4, 5). The driver gets alerted due to these sensations.

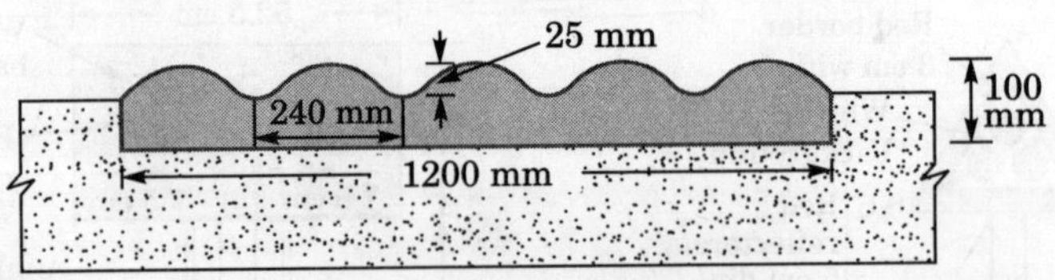

Fig. 16.7. Rumble Strips.

Rumble strips are made by incorporating changes in the pavement texture, either by a rough surface or by artificial corrugations. Typical design of the latter types, which can be cast in cement concrete blocks, is given in Fig. 16.7.

A number of such strips (say 5—10) will have to be placed at suitable intervals (about 100 m), well ahead of the intended place of reduced speed.

16.8. Guard-rails for Guiding Pedestrians

Guard-rails are needed for guiding pedestrians to subways, footbridges and surface crossings and for preventing them from spilling into the carriageway.

Guard rails, should be set back 500 mm from the kerb to give enough clearance to passing vehicles.

The design of the guard rails should be neat and simple and vandal-proof. The design should discourage people, particularly children, from climbing over or through them. A height of about 0.9—1.2 m is satisfactory. A typical design is given in Fig. 16.8.

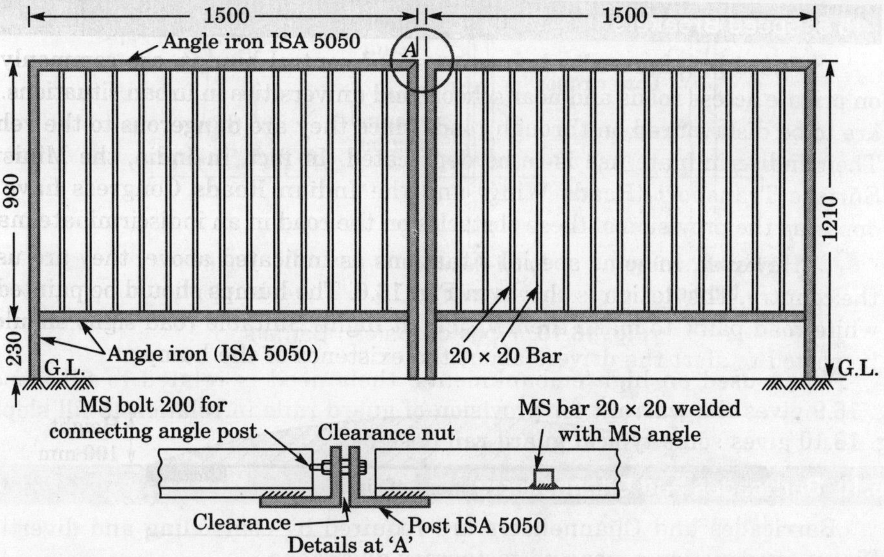

Fig. 16.8. Typical guard-rail.

16.9. Safety Barriers

Safety barriers (Ref. 4) for motor-traffic are intended to prevent vehicles accidentally leaving the highway. Typical locations are approaches to bridges and high embankments. They are designed to withstand the impact of a vehicle and deflect the same inwards.

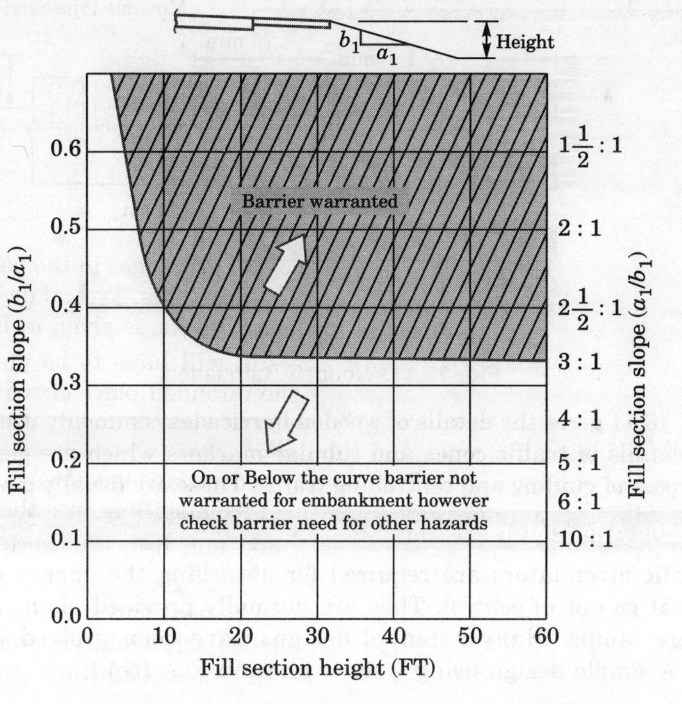

Fig. 16.9. Warrant for provision of Safety Barrier.

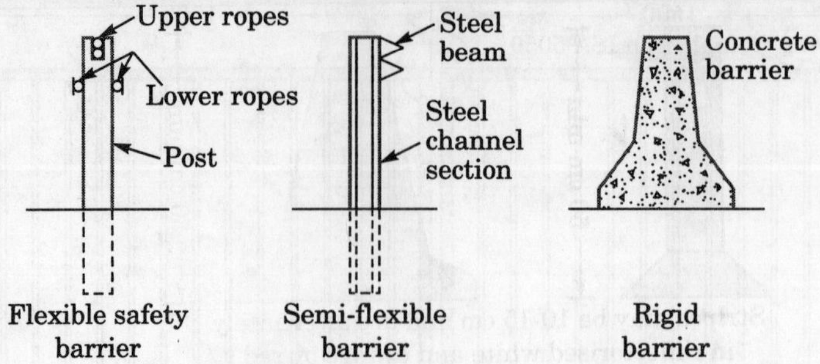

Fig. 16.10. Types of Safety Barriers.

When used on high embankments, their need is related to fill slopes. Fig. 16.9 gives the warrant for provision of guard-rails in relation to fill slopes. Fig. 16.10 gives some typical guard-rail designs.

16.10. Barricades and Channelisers

Barricades and Channelisers are required by controlling and diverting traffic on construction zones and temporary diversions.

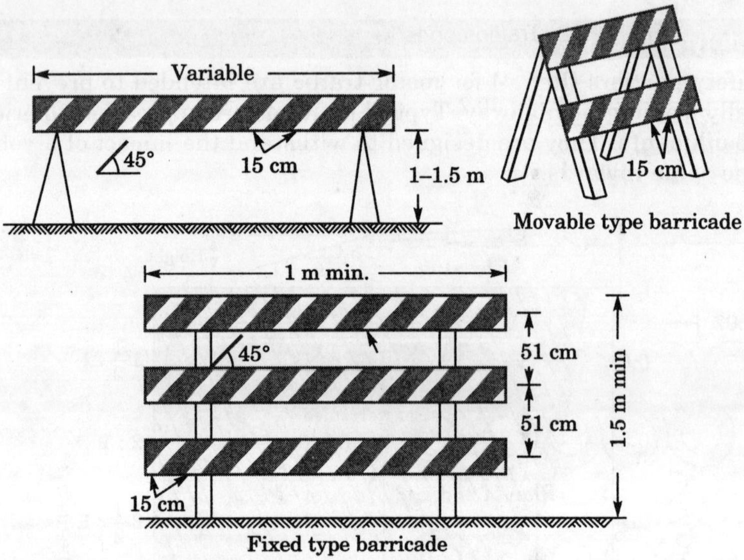

Fig. 16.11. Wooden Barricades.

Fig. 16.11 gives the details of wooden barricades commonly used. Fig. 16.12 gives the details of traffic cones and tubular markers which are generally used for the purpose of guiding and controlling traffic. These are usually made of plastic.

16.11. Traffic Attenuators

Traffic attenuators are required for absorbing the energy of impact of vehicles that go out of control. They are normally provided in the gore area of interchange ramps. Many patented designs have been evolved in advanced countries. A simple design using tyres is given in Fig. 16.13.

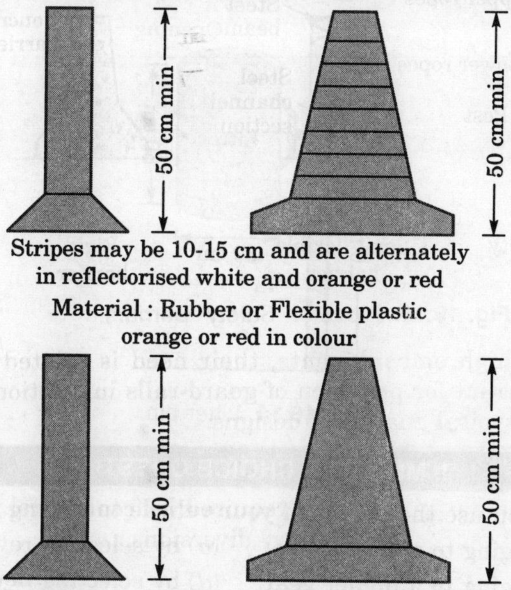

Fig. 16.12. Traffic cones and tubular markers.

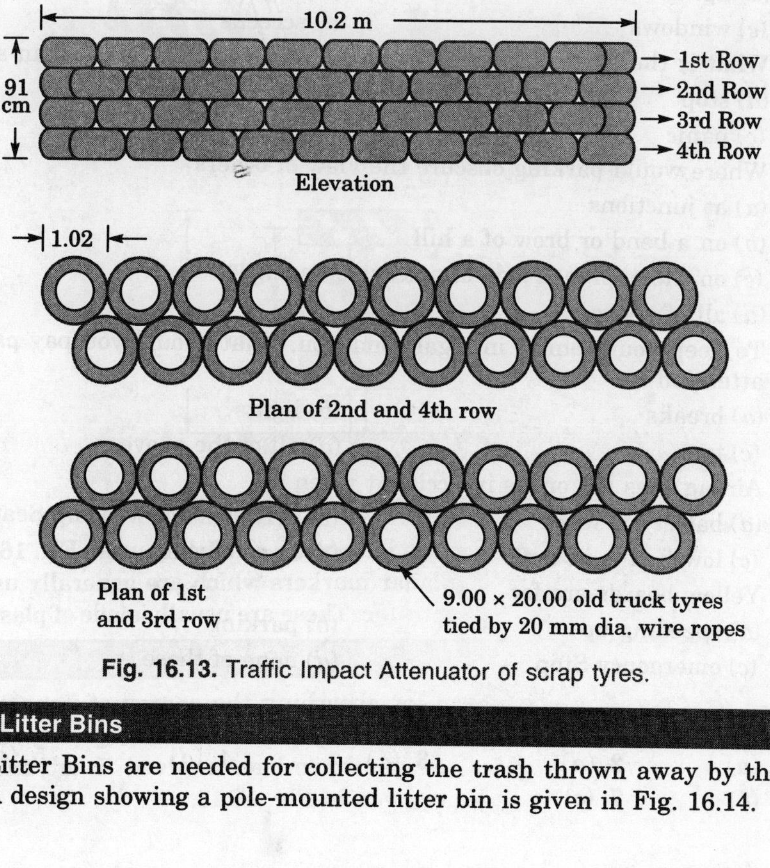

Fig. 16.13. Traffic Impact Attenuator of scrap tyres.

16.12. Litter Bins

Litter Bins are needed for collecting the trash thrown away by the road users. A design showing a pole-mounted litter bin is given in Fig. 16.14.

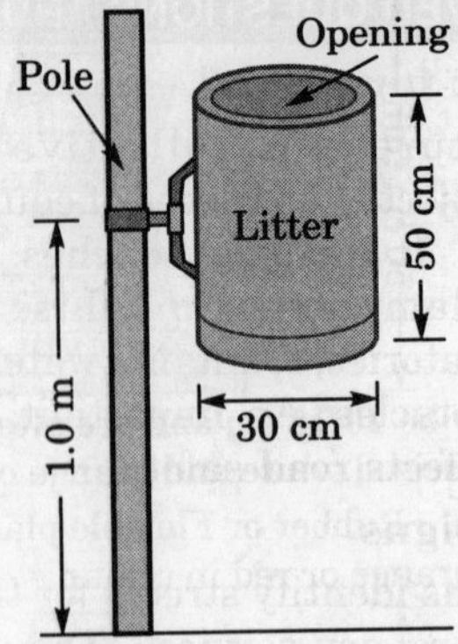

Fig. 16.14. Litter bin.

I. MULTIPLE CHOICE QUESTIONS

1. **How can you use the engine of your vehicle to control your speed?**
   (*a*) by changing to a lower gear **(*b*) by selecting reverse gear**
   (*c*) by changing to a higher gear **(*d*) by selecting neutral**
2. **What should always be kept clean on your vehicle**
   (*a*) lights (*b*) reflectors
   (*c*) windows (*d*) screens
3. **What is the first thing you should do after being involved in an accident**
   (*a*) stop (*b*) run
   (*c*) panic (*d*) all of the above
4. **Where would parking obscure the view of others**
   (*a*) at junctions
   (*b*) on a bend or brow of a hill
   (*c*) on the approach side of a zebra crossing
   (*d*) all of above
5. **To keep your vehicle in legal condition, what should you pay particular attention to**
   (*a*) breaks (*b*) lights
   (*c*) tyre (*d*) all of the above
6. **Airbag does not opens in accident when**
   (*a*) battery is weak (*b*) driver is not wearing Seat belt
   (*c*) low fuel is in tank (*d*) none of these
7. **Yellow boards are for**
   (*a*) speeding up (*b*) parking
   (*c*) emergency Stop (*d*) none of these

Answers

1. (*a*)	**2.** (*d*)	**3.** (*a*)	**4.** (*d*)	**5.** (*d*)
6. (*c*)	**7.** (*c*)			

II. SOLVED QUESTIONS AND ANSWERS

Q. 1. What is street furniture?

Ans. Street furniture is a collective term (used in the United Kingdom and Canada) for objects and pieces of equipment installed on streets and roads for various purposes. It includes benches, traffic barriers, bollards, post boxes, phone boxes, streetlamps, traffic lights, traffic signs, bus stops, tram stops, taxi stands, public lavatories, fountains watering troughs, memorials, public sculptures, and waste receptacles. An important consideration in the design of street furniture is how it affects road safety.

Q. 2. Street name signs.

Ans. Street name signs identify streets for the benefit of visitors, especially postal workers and the emergency services. They may also indicate the district in which a street lies.

Q. 3. A bench.

Ans. A bench is essentially a chair made for more than one person, usually found in central parts of settlements (such as plazas and parks). They are often provided by the local councils or contributors to serve as a place to rest and admire views. Armrests in between are sometimes provided to discourage lying down and/or unwanted closeness.

Q. 4. Bollards.

Ans. Bollards are posts, short poles, or pillars with the purpose of preventing the movement of vehicles onto sidewalks or grass. They can also protect buildings and other site features.

Q. 5. Litter bins.

Ans. Litter bins are also a form of street furniture. Litter bins are strategically placed to try and deter people from littering on the streets. Litter bins provide a place for people to dispose of their waste.

Q. 6. Speed bumps.

Ans. Speed bumps (or speed breakers) are the common name for a family of traffic calming devices that use vertical deflection to slow motor-vehicle traffic in order to improve safety conditions. Variations include the speed hump (or speed ramp), speed cushion, and speed table.

Q. 7. What are shoulder rumble strips?

Ans. There are four types of shoulder rumble strips. They differ primarily in how they are installed, their shape and size, and the amount of noise and vibration produced:

- Milled
- Rolled
- Formed
- Raised

Highway agencies often use all types of shoulder rumble strips, depending on the need and the material. The groove pattern can be installed intermittently or continuously. The groove pattern, depth, width, shape, and spacing may also change with the road agency.

Q. 8. What do you mean by rolled rumble strips?

Ans. Image of Rolled Rumble into hot asphalt pavements and shoulders when the constructed or reconstructed surface course is compacted. The strips are made by a roller with steel pipes welded to drums, which make the depressions as they pass over the hot pavement. The sound and vibration of rolled rumble strips is typically much less noticeable than milled rumble strips, but varies based on width, depth, and spacing.

Q. 9. Explain safety barrier.

Ans. A **safety barrier** is a component which prevents passage into a dangerous area. It is commonly used to mitigate risk in the Hazard-**Barrier**-Target model, as studied in **safety** science.

Q. 10. What are barricades?

Ans. Barricades are critical tools for diverting traffic away from active construction areas. Available in a variety of materials, colors and sizes, Plasticade barricades are designed with the safety of drivers and construction personnel foremost in mind. They not only increase visibility of construction areas for drivers, they also provide a guide for orderly and predictable movement of traffic, a safety benefit for drivers and workers alike.

III. REVIEW QUESTIONS

1. What is traffic aids and street furniture?
2. Discuss the principles of street furniture design.
3. Explain roadway delineators.
4. What do you mean by hazard markers?
5. Discuss about the speed breakers.
6. What are traffic attenuators?

REFERENCE

1. Development of Aesthetic Designs for Roadside Furniture, State of Art Report, Prepared for Ministry of Surface Transport by Dr. L.R. Kadiyali and Associates, New Delhi, 1996.
2. Street Furniture—A Design Council Catalogue, The Design Council of U.K., London, 1979.
3. Recommended Practice for Road Delineators, I.R.C : 79-1981, Indian Roads Congress, New Delhi, 1981.
4. Guidelines for Traffic Safaty Barriers, I.R.C : 119–2015, Indian Roads Congress, New Delhi, 2015.

SECTION 7
TRAFFIC REGULATIONS

17. Regulation of Traffic

17.1. Basic Principles of Regulation

17.1.1. Need for regulation of traffic

The motor vehicle is a machine in charge of a human being and this makes it necessary for the formulation of suitable regulations for safe operation of traffic and enforcement of these regulations. The regulations should be framed so as to achieve safe and efficient movement of traffic and pedestrians, without at the same time infringing unduly upon the individual rights of the road users. The regulations should evoke respect by the road users and not be disregarded. They should be flexible and should keep pace with the changing conditions and times. They should be reviewed and altered periodically in light of traffic conditions and accident data analysis. The traffic engineer should be conversant with the principles of regulation. The design of streets and facilities and the safe operation of traffic are vitally connected with the Traffic Regulations.

17.1.2. Scope of traffic regulations

Traffic regulations cover matters dealing with the control of vehicles, drivers and road users. The control of vehicles deals with the registration, weight, size, design, construction and maintenance. Driver regulations deal with the licencing and other aspects of operation of vehicles by drivers. Regulation of other road users deals with the rules regarding pedestrians, cyclists and motorcyclists. In India, traffic is mixed in character, and this brings in the need to regulate the movement of animal drawn vehicles, cycle rickshaws and hand carts.

17.1.3. Traffic laws

Most of the countries have uniform traffic laws operating in all parts of the cuntry so that the laws are understood consistently everywhere and are obeyed. In U.S.A., the Uniform Motor Vehicle Code (Ref. 1) serves as a pattern for the States to evolve their own system of traffic laws. Many States have laws in conformity with the Motor Vehicle Code, and some others have rules similar to those contained therein (Ref. 2). The Uniform Vehicle Code has the following five Acts :

Act I. Uniform Motor Vehicle Administration, Registration Certificate of Title and Anti-theft Act.

Act II. Uniform Motor Vehicle Operator's and Chaffeur's Licence Act.

Act III. Uniform Motor Vehicle Civil Liability Act.

Act IV. Uniform Motor Vehicle Safety Responsibility Act.

Act V. Uniform Act Regulating Traffic on Highways.

In U.K., the Road Traffic Act, 1972 (Ref. 3) provides the framework for rules regarding the traffic regulations. It is divided into the following parts :

Part I. Principal Road Safety Provisions.

Part II. Construction and Use of Vehicles and Equipment.

Part III. Licencing of Drivers of Vehicles.

Part IV. Licencing of Drivers of Heavy Goods Vehicles.

Part V. Driving Instruction.

Part VI. Third Party Liabilities.

Part VII. Miscellaneous and General.

In India, the Motor Vehicles Act. 1988 (Ref. 4), provides the basis for regulating vehicles, drivers, other road users and traffic. It contains the following Chapters :

Chapter I. Preliminary.

Chapter II. Licencing of Drivers of Motor Vehicles.

Chapter III. Licencing of Conductors of Stage Carriages.

Chapter IV. Registration of Motor Vehicles.

Chapter V. Control of Transport Vehicles.

Chapter VI. A. Special Provision Relating to State Transport Undertakings.

Chapter VII. Construction, Equipment and Maintenance of Motor Vehicles.

Chapter VIII. Control of Traffic.

Chapter IX. Motor Vehicles Temporarily Leaving or Visiting India.

Chapter X. Liability without Fault in Certain Cases.

Chapter XI. Insurance of Motor Vehicles against Third Party Risks.

The Act is applicable throughout India and provides for the State Governments to make rules for the purpose of carrying into effect the various provisions of the Act.

17.2. Regulation of Speed

17.2.1. Need for regulation of speed

High speeds are always associated with accidents, as discussed in Chapter 18 on Road Accidents. Regulation of the speed is one of the means of ensuring safer travel.

17.2.2. General principles governing application of speed limits

Speed limits selected should be 'realistic' so that they are not disregarded by the drivers. For this purpose, the limits should be appropriate to the conditions at site. Limits should be reviewed periodically based on accumulated experience and future needs.

The limits should be related to the volume and character of traffic, the presence of pedestrians, the general features of the road and safety built into it and the development that has taken place along the road.

17.2.3. Speed limits in urban areas

Speed limits in urban areas are governed by the type and volume of traffic. Traffic conditions in urban areas are such that the chances of accidents are increased. Pedestrians and cyclists use the urban roads and compete for space with vehicles. Mixed traffic conditions increase the chances of accident occurrence. Parking is frequent along the streets and this is another source of accidents. The need for speed limit under these conditions is well known and most of the countries in the world prescribe a speed limit on urban roads.

In U.K., the principal urban limit is 30 MPH (Ref. 5), though in some areas a limit of 40 MPH has been introduced (Ref. 5, 6). The definition of the urban area in U.K. is accepted as any area provided by street lighting.

In the U.S.A., the Uniform Vehicle Code (Ref. 1) prescribes an absolute speed limit of 30 MPH in any urban district. Many States observe speed limits of 25 to 30 MPH in residential districts and 20 to 25 MPH in business districts.

In India the traffic is mixed in character. The street pattern in many towns is the inheritance of the past and is characterised by narrow and winding alleys with buildings abutting the carriageway. Pedestrian traffic is heavy. The need for regulation of speed is paramount. The speed limits on urban traffic vary with the nature of congestion and the type of streets. Table 17.1, gives general guidelines (Ref. 8).

Table 17.1: Speed limits for different types of vehicles under mixed traffic condition (*Source : Ref. 8*)

Different categories of road and streets	*Speed limits in kilometers per hour*	
	Group I	*Group II*
	Light and Medium vehicle	*Heavy vehicles*
Major Roads of Arterial/Subarterial character mostly in open and thinly built-up areas	50	40
Roads with moderate traffic situated in semi-built-up areas.	40	30
Congested roads in built-up areas	30	20

17.2.4. Speed limits in rural areas

Or rural highways, the speed limits normally imposed are 80 K.P.H., 100 K.P.H. and 110 K.P.H. The Uniform Vehicle Code (Ref. 1) recommends a speed limit of 60 M.P.H. for locations other than urban areas during the day time. In U.K., certain rural sections of motorways used to have an upper speed limit of 70 M.P.H., till recently when the energy crisis forced the Government to reduce this. Single carriageway rural roads in U.K. operate on speed limits of 50 M.P.H. or 60 M.P.H.

In India, the speed limits generally followed are given in Table 17.2.

17.2.5. Problem of small villages along rural highways

Small villages located along rural highways have peculiar safety problems. Since people may live on either side of the road, frequent crossing of the road by pedestrians is unavoidable. Speeding vehicles can cause hazardous conditions for these pedestrians. In such situations, the speed limits will have to be set based on accidents records and the extent of local traffic.

17.2.6. Speed limits for different types of vehicles

Apart from prescribing speed limits for different classes of roads, limits for speed of different classes of vehicles are also important. In U.K., on all purpose roads, vehicles which carry goods are subjected to a speed limit of 40 M.P.H. and buses and coaches to 50 M.P.H. (Ref. 5). If trailers are towed, the speed limit is 10 M.P.H. lower. On motorways, as a rule, no speed limits for vehicle classes apply.

In India, the Eighth Schedule of the Motor Vehicles Act (Ref. 4) limits the speed of different types of vehicles to those given in Table 17.2.

Table 17.2. Speed limits generally followed in India for rural highways (*Source : Ref. 4*)

	Classes of Vehicles	*Maximum Speed K.P.H.*
1.	Vehicles without trailers	
	(*i*) Light motor vehicle or a motor cycle	No limit
	(*ii*) Medium or heavy passenger motor vehicle	60
	(*iii*) Medium or heavy goods motor vehicle	60
2.	Articulated heavy motor vehicle	50
3.	Vehicle drawing not more than one trailer	
	(*i*) Vehicle being a light motor vehicle, and trailer being two-wheeled having a laden weight not more than 800 kg	60
	(*ii*) Vehicle being a light motor vehicle, and trailer having more than two-wheels or a laden weight more than 800 kg	50
	(*iii*) Vehicle being a medium motor vehicle	50
	(*iv*) Vehicle being a heavy motor vehicle	40
	(*v*) Vehicle being a heavy motor vehicle used by the fire brigade	50
4.	Any case not covered by above	30

17.2.7. Speed limits at nights

Speed limits at nights are generally lower than during day time. This is obviously necessitated by greater risk hazards during night. In U.S.A., the Uniform Vehicle Code recommends a speed limit of 55 M.P.H. in rural locations during night time, as against 60 M.P.H. during day time (Ref. 1).

17.2.8. Criteria for application of speed limits of 50, 65, 80 and 100 K.P.H.

Before selecting any particular speed limit, the following factors should be considered (Ref. 5) :

(*i*) ***Speed of traffic.*** For selecting the speed limits, the 85th percentile speed of vehicles is normally selected (See Chapter 9 on Speed Studies). Speed survey would yield data to determine this.

(*ii*) ***Road conditions.*** The geometry of the road (its width and layout), gradient, sight distance, street lighting, superelevation, provision of proper signs and adequate safety measures for pedestrians are some of the road characteristics that should be studied before selecting the speed limit.

(*iii*) ***Environment of the road.*** The extent of roadside development, the frequency and nature of access points, the presence of schools, playing fields, industry, housing estates etc. are some of the environmental conditions governing the speed limit.

Table 17.3. Guidelines for selection of speed limit (*Source : Ref. 5*)

Appropriate Speed (M.P.H.) (1)	*Speed of traffic (before limit is applied or altered)* (2)	*Environment of road* (3)	*Traffic composition* (4)	*Character of road* (5)	*Casualty rate* (6)
30	85th percentile speed of cars not greater than 37 M.P.H.	(*i*) Built-up area on both sides (*ii*) Direct access from development on sides (*iii*) Presence of schools, recreational grounds and playing fields	(*i*) High proportion of 2-wheeled vehicles (*ii*) Large number of pedestrians	(*i*) Frequent junctions with visibility inadequate for speeds greater than 30 M.P.H. (*ii*) Pedestrian crossings (*iii*) Development close to footways	Above 5 per million veh. miles (average of last years for fatal, serious and slight injuroes).
40	85th percentile speed of cars not greater than 47 M.P.H.	(*i*) Built-up area as above (*ii*) Partially built-up, transition between built-up and rural (*iii*) Little or no frontage development	(*i*) As above, but with more severe parking restrictions (*ii*) 2-wheeled vehicles and pedestrians more numerous	(*i*) Important radial or ring roads with well designed footpaths on both sides (*ii*) Footways provided, roads of adequate capacity and well designed. (*iii*) Roads such as urban motor ways where geometric design is based on 40 M.P.H. speed.	(*i*) Above 3 per million veh. miles for all types of injury. (*ii*) Above 1 per million veh. miles for fatal injury. (*iii*) Above 1 per million veh. miles for pedestrians and drivers two wheeled vehicles.

(1)	(2)	(3)	(4)	(5)	(6)
50	85th percentile speed of cars not less than 40 M.P.H. and not greater than 57 M.P.H.	(*i*) Rural roads less developed than 40 (*ii*) but having above average casualty rates. Roads between sections having speed restrictions of 30 or 40 M.P.H. (*ii*) Urban or rural roads with little or no frontage development. (*iii*) Urban or rural roads with no frontage development		(*i*) Application to single carriage-way roads of two or three traffic lanes, not usually carrying less than 4000 v.p.d. (*ii*) Dual or single carriageway roads such as by-passes. (*iii*) New roads such as urban motorways where geometric design is based on speed of 50 M.P.H.	Fatal and serious casulty rate per million vehicle miles normally exceeding 1.
70	85th percentile speed of cars usually about or above 58 M.P.H.	Little or no frontage development		Dual carriageway or divided roads	As above for 50 M.P.H. limits.

(*iv*) ***Traffic.*** The nature and volume of traffic dictate the selection of speed limits.

(*v*) ***Accident rates.*** Past accident rates given positive indication of whether there is any necessity of posting any speed limits.

The broad guidelines in Table 17.3 indicate the condition appropriate to the selection of speed limits (Ref. 5).

17.2.9. Enforcement methods and instruments for detection of speed violators

For success of speed regulation measures, strict enforcement of law is needed. Motorists who violate the speed limit must be apprehended, given official summons or arrested, and penalties must be imposed. They should be tried before courts which are usually constituted to deal with traffic cases exclusively.

One of the common methods of detecting speed violation is to follow the violator in an automobile or motor cycle. These vehicles are made conspicuous by marking them with a distinctive colour. The speedometers in the checking vehicles should be accurate and for this purpose should be calibrated frequently.

Enoscopes and radar speedmeters, described in detail under Chapter 3 have become very popular in recent times with the police authorities for detecting speed violation. For accurate and unbiased estimate of speed, they should be located so as not to be conspicuous to the drivers.

17.2.10. Speed zoning

17.2.10.1. Speed zoning is the designation of various sections of highways for different values of safe speed limits. The limits, are selected after careful consideration of various factors that are conducive for the safety of traffic, and are posted by signs. Some of the techniques and studies used to determine the proper speed limits for speed zoning are (Ref. 9) :

(*i*) Complete engineering investigations.

(*ii*) Study of accident frequency.

(*iii*) Number of business establishments on street.

(*iv*) Volume study

(*v*) Speed study.

(*vi*) Observation by police department

(*vii*) Width of street

(*viii*) Pedestrian traffic volume.

17.2.10.2. Speed zoning at horizontal curves

The safe speed at which a vehicle can negotiate a curve is dependent upon the radius of the curve, the value of the superelevation and the permissible value of the coefficient of friction. The following formula relates these factors :

$$V^2 = 127\,R(e + f) \qquad ...(17.1)$$

where, V = Speed in kilometres per hour

R = Radius in metres

e = Value of superelevation, expressed as a fraction

f = Coefficient of friction.

The values of f recommended by AASTHO (Ref. 10) are given in Table 17.4.

Table 17.4. Coefficient of lateral friction recommended by AASTHO
(Adapted from Ref. 10)

Design speed K.P.H.	50	65	80	100	120	130
Maximum lateral friction	0.16	0.15	0.14	0.13	0.12	0.11

The Indian Road Congress recommends a constant value of 0.15 (Ref. 11).

It is also possible to arrive at the speed on curve which drivers consider as a reasonable maximum. If speed data of vehicles are collected at the curve, the 85th percentile speed could be used as the safe speed for posting on the curve.

17.2.10.3. Safe speeds at intersections

The safe speeds at intersections are governed by factors such as the sight distance available and the type of traffic sign that is provided (STOP or GIVE WAY sign). The most common methods available for determining the safe speeds prescribed by some of the U.S. bodies are (Ref. 12, 13) :

(*i*) National Safety Council method.

(*ii*) American Automobile Association method.

(*iii*) American Association of State Highway Officials method.

(*i*) ***National Safety Council Method.*** The following figure represents the basis of the method :

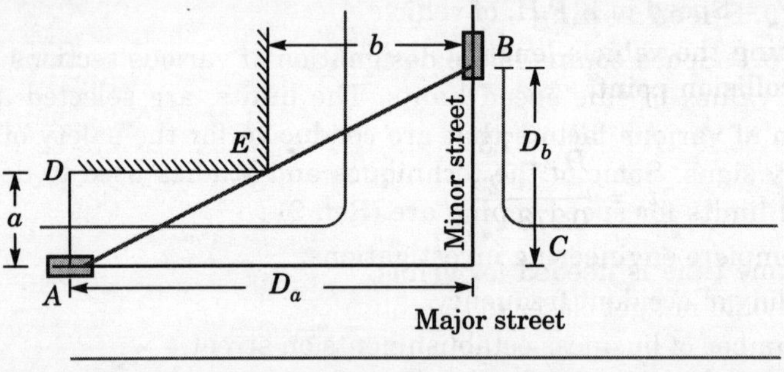

Fig. 17.1. Basis of National Safety Council Method of Speed Zoning.

A and *B* are two vehicles, *A* travelling on the major or preferential street and *B* travelling on the side street. The view at the intersection is restricted due to the presence of an obstruction. Point *C* represents the potential collision point and *CBA* represents the visibility triangle.

Since triangles *CBA* and *DAE* are similar,

$$\frac{D_b}{D_a} = \frac{a}{D_a - b} \quad \text{or} \quad D_b = \frac{a\,D_a}{D_a - b} \qquad ...(17.2)$$

D_a is taken as equal to the safe stopping distance of vehicle *A* travelling at a speed *V* K.P.H. plus a clearance of 5 m

The braking distance is given by :

$$d = \frac{v^2}{2gf} = \frac{V^2}{254 f} \qquad ...(17.3)$$

where, d = braking distance in metres

v = speed in metres per second

g = acceleration due to gravity, m/sec²
f = coefficient of friction
V = speed in K.P.H.

If f is assumed as 0.53,

$$d = \frac{V^2}{254 \times 0.53} = 0.0074\, V^2 \qquad ...(17.4)$$

The safe stopping distance S is the braking distance plus the distance travelled in the perception reaction time, t seconds.

$$S = 0.0074\, V^2 + \frac{V}{3.6}\, t$$

$$S = 0.0074\, V^2 + 0.278\, Vt \qquad ...(17.5)$$

The values of t for various values of V are given in Table 17.4. Table 17.4

Table 17.4. Perception-reaction time for different speeds

V (K.P.H.)	0	20	40	60	80	100
t (seconds)	0.76	0.80	0.95	1.15	1.55	2.0

$$D_a = \text{Safe stopping distance} + 5 \text{ m}$$
$$= 0.0074\, V_a^{\,2} + 0.278\, V_a t + 5 \qquad ...(17.6)$$

where V_a = Speed in K.P.H. of vehicle A.

Assuming the vehicle length to be 7.3 m, the time required for vehicle A to clear the collision point,

$$t_a = \frac{D_a + \frac{7.3}{2}}{0.278\, V_a} \qquad ...(17.7)$$

The same time is needed for vehicle B to travel a distance of D_b–5, since this vehicle should not get closer to the collision point than the clearance distance of 5 m.

Equating both,

$$\frac{D_a + 3.65}{0.278\, V_a} = \frac{D_b - 5}{0.278\, V_b}$$

or

$$V_b = \frac{(D_b - 5)\, V_a}{D_a + 3.65} \qquad ...(17.8)$$

The above method is applicable to streets intersecting at any angle, provided the measurements a and b are taken parallel to the paths of vehicles A and B respectively.

(*ii*) . The same principles as above are applied here, with the change that f is taken 0.50, the clearance distance is taken as 4.5 m and the perception reaction time uniformly taken 1.0 second. This gives :

$$D_a = 0.0074\, V_a^{\,2} + 0.278\, V_a + 4.5 \qquad ...(17.9)$$

Assuming that V_a is the legal speed limit, the speed of vehicle B can be calculated on the premise that the distance from B to the collision point should be equal to the safe stopping distance plus a clearance of 4.5 m.

$$D_b = \frac{a\,D_a}{D_a - b}$$

$$= 0.0074\,V_b^2 + 0.278\,V_b + 4.5 \quad \text{...(17.10)}$$

The above quadratic equation can be solved for V_b.

(*iii*) ***AASHO method*** **(Ref. 13) : In third method the coefficient of friction is taken as 0.4 and the perception reaction time is taken as 2.0 seconds. This gives :**

$$D_a = 0.0099\,V_a^2 + 0.556\,V_a \quad \text{...(17.11)}$$

$$D_b = \frac{a\,D_a}{D_a - b}$$

$$= 0.0099\,V_b^2 + 0.556\,V_b \quad \text{...(17.12)}$$

V_b can be calculated from this quadratic equation.

The use of the above methods can be illustrated by the following example.

Problem 17.1. *Assume V_a = 60 K.P.H., a = 15 m, b = 20 m. Calculate the speed limit on minor street.*

Solution. By NSC Method,

$$t = 1.15 \text{ sec. from Table 17.4}$$

$$D_a = 0.0074\,V_a^2 + 0.278\,V_a t + 5 \text{ (Eq. 17.6)}$$

$$= 0.0074\,(60)^2 + 0.278\,(60)\,(1.15) + 5$$

$$= 26.6 + 19.2 + 5 = 50.8 \text{ m}$$

$$D_b = \frac{a\,D_a}{D_b - b} \quad \text{from (Eq. 17.2)}$$

$$= \frac{15 \times 50.8}{50.8 - 20} = \frac{15 \times 50.8}{30.8} = 24.8 \text{ m}$$

$$V_b = \frac{(D_b - 5)\,V_a}{D_a + 3.65} = \frac{(24.8 - 5)\,60}{50.8 + 3.65}$$

$$= \frac{19.8 \times 60}{54.45} = 21.8 \text{ K.P.H.}$$

By AAA method $D_a = 0.0074\,V_a^2 + 0.278\,V_a + 4.5$

$$= 0.0074\,(60)^2 + 0.278\,(60) + 4.5$$

$$= 26.6 + 16.7 + 4.5 = 47.8 \text{ m}$$

$$D_b = \frac{15 \times 47.8}{47.8 - 20} = \frac{15 \times 47.8}{24.8} = 25.8 \text{ m}$$

$$25.8 = 0.0074\,V_a^2 + 0.278\,V_b + 4.5$$

$$\therefore \quad 0.0074\,V_b^2 + 0.278\,V_b - 21.3 = 0$$

$$\therefore \quad V_b = \frac{-0.278 + \sqrt{(0.278)^2 + 4 \times 21.3 \times 0.0074}}{2 \times 0.0074}$$

$$= \frac{-0.278 + \sqrt{0.077 + 0.63}}{0.0148}$$

$$= \frac{-0.278 + 0.84}{0.0148} = 38 \text{ K.P.H.}$$

By AASHTO Method.

$$D_a = 0.0099\,V_a^{\,2} + 0.556\,V_a$$
$$= 0.0099\,(60)^2 + 0.556 \times 60 = 35.6 + 33.2$$
$$= 68.8 \text{ m}$$
$$D_b = \frac{a\,D_a}{D_a - b} = \frac{15 \times 68.8}{68.8 - 20} = \frac{15 \times 68.8}{48.8} = 21.1 \text{ m}$$
$$\therefore \quad D_b = 0.0099\,V_b^{\,2} + 0.556\,V_b$$
$$\therefore \quad 0.0099\,V_b^{\,2} + 0.556\,V_b - 21.1 = 0$$
$$\therefore \quad V_b = \frac{-\,0.556 + \sqrt{(0.556)^2 + 4 \times 21.1 \times 0.0099}}{2 \times 0.0099}$$
$$= \frac{-\,0.556 + \sqrt{0.309 + 0.835}}{0.0198} = \frac{-\,0.556 + 1.07}{0.0198} = \frac{0.514}{0.0198}$$
$$= 26 \text{ K.P.H.}$$

17.2.11. Speed laws

Speed laws vary from country to country and even in a country they vary from State to State. In U.S.A., for example, the following categories of law exist (Ref. 14) :

(*i*) Those which require that the driver should drive at a speed which is safe and reasonable according to the existing conditions of the road, and the potential hazards that may be present on the road. The rule under these laws is called the *Basic Speed Rule*.

(*ii*) Those which prescribe that the driver should drive at a safe and reasonable speed, but in addition provide that any vehicle speed in excess of a designated limit is deemed to be a violation of the basic speed rule. If any question is raised, the burden of proof falls upon the driver to prove that he was not travelling at an unsafe speed. The designated limit is called the *prima facie* speed limit.

(*iii*) Those which prescribe an absolute limit above which it is always illegal to drive and below which it is always legal to drive irrespective of the conditions.

(*iv*) Those which prescribe special speed limits in given zones, indicated by signs.

17.2.12. General effect of speed limits

Though a large number of drivers disregard the speed limits, yet it cannot be denied that speed limits have an overall salutory effect (Ref. 15). They reduce the accident rate and bring down the number of those who drive at higher speeds. Special beneficiaries are the motor cyclists and pedal cyclists (Ref. 6). Speed limit is, therefore, a powerful tool in traffic regulation.

17.3. Regulation of Vehicles

17.3.1. The regulation of vehicles broadly covers the aspects indicated below :

(i) Vehicle registration

(*ii*) Construction and equipment of vehicles.

(*iii*) Size, weight and loads of vehicles

(*iv*) Lighting of vehicles.

(*v*) Inspection of vehicles.

(*vi*) Control of transport vehicles.

(*vii*) Insurance.

17.3.2. Vehicle Registration

Vehicle registration is a basic requirement and the data accumulated provides guidance to administrators, planners, traffic police, economists and traffic engineers on total number of vehicles in use, their different types and use, taxation, name of the owner and licence plate number. In India, the vehicles registered in different States bear different licence plate numbers, and the identity of the State can be established by the alphabets preceding the number (Ref. 4).

The practice in India in regard to the colour of the letters and the background plate is as follow :

(*i*) In case of transport vehicle.	In black on a white background.
(*ii*) In case of temporary registrations.	In red on a yellow background.
(*iii*) In case of registration marks allotted to dealers.	In white on a red background.
(*iv*) In case of private vehicles.	In black on a white background.
(*v*) In case of taxis.	In black on a yellow background.

In U.K. a system is prevalent by which after the licence number a letter (A, B, etc.) is inserted which signifies the year of registration. Thus it is possible to find out the age of the vehicle when reading the licence plate number.

17.3.3. Construction and equipment of vehicles

It is essential that the motor vehicles be constructed and equipped in such a manner as to promote safe and efficient traffic. For this purpose, the regulations usually cover the following aspects :

(*i*) the width, height, length and overhang of vehicles and trailers ;

(*ii*) the diameter, width and condition of tyres of vehicles and trailers ;

(*iii*) the maximum unladen and laden weight of vehicles and trailers, and the maximum axle weight ;

(*iv*) seating arrangements in public service vehicles and the protection of passengers against weather ;

(*v*) brakes and steering gear (*e.g.*, in India it is necessary to have right hand steering control unless the vehicle is equipped with a mechanical or electrical signalling device) ;

(*vi*) the use of safety glass ;

(*vii*) signalling appliances, lamps and reflectors ;

(*viii*) speed governors ;

(*ix*) the emission of smoke, visible vapour, sparks, ashes, grit, oil or pollutants ;

(*x*) noise caused by the vehicles ;

(*xi*) the use of trailers with motor vehicles ;

(*xii*) prohibiting the use of horns at specified locations ;

(*xiii*) the periodical testing and inspection of the vehicles.

In India, the State Governments are empowered under the Motor Vehicles Act (Ref. 4), to make rules for the above purpose.

17.3.4. Control of Transport Vehicles

In India, under the Motor Vehicles Act (Ref. 4), the State Governments have been empowered to frame rules to govern the use of commercial transport vehicles. These rules are intended to grant permit for the plying of goods vehicles on specified routes and regions, having due regard to :

(*i*) the advantages offered by the development of road transport ;

(*ii*) the desirability of co-ordinating road and rail transport ;

(*iii*) the desirability of preventing the deterioration of the road system ; and

(*iv*) the desirability of preventing uneconomic competition among motor vehicles.

Similarly, the State Governments can grant permits for vehicles carrying passengers. The permits issued under this rule indicate the route or area of operation, the manner in which the vehicle is to be used (Stage carriage, contract carriage, private or public carrier etc.), the carrying capacity, schedule for trips etc. A system of National Permits for goods vehicles is prevalent in India. Till recently, the number of such permits used to be limited, but it has now been decided to do away with such restrictions.

17.3.5. Insurance

Most of the countries have laws requiring the vehicles to be insured against third party risks. The insurance is intended to cover liability in respect of death or injury to persons or damage to any property of a third party arising out of the use of the vehicle. In India, the Motor Vehicles Act contains suitable provisions in this respect.

17.4. Regulations Concerning the Driver

17.4.1. Regulations concerning the driver cover the following aspects :

(*i*) Licensing of the driver (*ii*) Requirements of physical fitness

(*iii*) Age of drivers

(*iv*) Disqualification and endorsement of licences.

17.4.2. Licencing of the Driver

Driving of a motor vehicle without the driver having a valid driving licence is an offence. The driving licence legally confers upon the driver the privilege to drive the vehicle. The licence is usually granted after the driver passes a test of competence. The Third Schedule of the *Motor Vehicle Act*, in India prescribes the criteria that should be satisfied by determining the competence of the driver in a test. Among other things, the driver will have to demonstrate his ability in driving the vehicle and should be conversant with the general traffic rules and regulations.

17.4.3. Physical Fitness

In India, the following diseases and disabilities are deemed to absolutely disqualify a person from obtaining a licence :

(*i*) Epilepsy (*ii*) Lunacy

(*iii*) Heart disease likely to produce sudden attacks of giddiness or fainting

(*iv*) Inadequate perception

(*v*) Deafness preventing the hearing of ordinary sound signals

(*vi*) Inability readily to distinguish red and green colours

(*vii*) Night blindness.

Persons suffering from Leprosy are ineligible to drive a Public Service Vehicle.

17.4.4. Age of Drivers

In India, the minimum age to drive a motor vehicle is 18 years and the minimum age to drive a transport vehicle is 20 years (Ref. 4)

In U.K. the minimum age is as under (Ref. 3) :

Class of Vehicle	*Age*
(*i*) Motor cycle or invalid carriage	16
(*ii*) Motor car	17
(*iii*) Trailer used primarily for work on land in connection with agriculture	17
(*iv*) Heavy locomotive, light locomotive, motor tractor or heavy motor car	21

17.4.5. Disqualification and endorsement of Licence

Habitual drinking and dangerous driving render a driver to be disqualified from holding a driving licence. In India, the Regional Transport Authority and the courts have the power to disqualify. Similarly, such authorities have powers to endorse upon the driving licence and particulars of the disqualification.

17.4.6. Offences and Penalties

The drivers are expected not to drive at excessive speed and to avoid careless and dangerous driving. Failure to observe such safety precautions is deemed as an offence. Driving under the influence of liquor or drugs and driving when mentally or physically unfit are also punishable offences. The Motor Vehicles Act of India lays down different punishment for different offences.

17.5. Regulations Concerning Traffic

17.5.1. In India, the traffic is mixed in character and consists of a motley of motorised vehicles (*e.g.*, cars, buses trucks, tractors and trailers, scooters, motor-cycles, scooter-rickshaws and vans), animal drawn vehicles (bullock carts, horse-drawn vehicles and camelcarts), cycles, cycle-rickshaws and hand-carts. In urban streets, mixed traffic presents hazards and affects the capacity of the streets adversely. Any city or town will have to, therefore, develop a system of rules to regulate such traffic and enforce them strictly. Some of the regulations that are possible are outlined here.

17.5.2. Cycles

The following rules promote safety of cycle traffic (Ref. 20) :

(*i*) Cycling while under the influence of alcohol or drugs shall be considered an offence.

(*ii*) Riding of more than one person on a cycle shall be considered an offence.

(*iii*) Wherever separate cycle tracks are provided, the cyclists shall ride on such tracks compulsorily.

(*iv*) Not more than two cycles shall remain abreast except at separate cycle tracks.

(*v*) Cyclists shall not use foot-ways.

(*vi*) Cyclists shall not be allowed to be towed by any other vehicle.

(*vii*) Where separate cycle tracks are provided, the cyclists shall keep to the extreme left of the carriageway.

(*viii*) All cycles shall be provided with good brakes, night lamps, a red reflector at the rear and bells or horns.

(*ix*) Cycles shall be parked only at designated places.

17.5.3. Motor Cycles and Scooters

The following are some of the special provisions for regulating the safe movement of motor cycles and scooters :

(*i*) Motor cycle and scooter riders shall wear compulsorily crash helmets.

(*ii*) All motor cycles and scooters shall be provided with a rear view mirror.

(*iii*) All motor cycles and scooters shall have a screen provided on the rear wheel to prevent loose garments getting entangled in the spokes.

(*iv*) Not more than one rider and a pillion rider shall be carried on a motor cycle or scooter.

(*v*) A rider with a learner's licence shall not carry a pillion rider.

(*vi*) A footrest shall be compulsorily provided for the pillion rider.

(*vii*) The load projections on the sides shall be restricted to 0.5 m from the longitudinal middle of the motor cycle, on the front to about 0.6 m from the front wheel and on the rear to about 1.0 m from the rear wheel.

17.5.4. Rules for Pedestrian Traffic

The following are some of the simple safety rules which the pedestrians should observe (Ref. 20) :

(*i*) Pedestrians shall obey the traffic signals and any violations shall be considered an offence. For example, crossing a street while the signal indicates red shall be an offence.

(*ii*) Pedestrians shall use the footways wherever they are provided, the street being used only for crossing.

(*iii*) Pedestrians shall cross the streets only at cross-walks wherever they are provided.

(*iv*) Where footways are not provided separately, pedestrians shall walk on the right hand side of the road facing traffic.

17.5.5. Rules for animal vehicles

The following are some of the regulations that are possible for animal drawn vehicles :

(*i*) The animal drawn traffic shall be prohibited from carrying long pipes, rods, bamboos etc. more than 5 metres in length.

(*ii*) The height, length and width of loads shall be restricted to safe limits, depending upon the size of the carriage.

(*iii*) The animal drawn traffic shall invariably keep to the extreme left of the carriageway having mixed traffic.

17.5.6. Rules for animals on streets

The following rules are some of those possible for regulating the animals on streets :

(*i*) Driving of animals such as camels and elephants on the busy streets shall be prohibited.

(*ii*) Cattle in droves shall not be driven on busy streets at peak periods of traffic.

(*iii*) Cattle in droves shall be accompanied by sufficient number of attendants to completely keep them under control.

(*iv*) A horse rider shall keep to the extreme left of the carriageway as far as possible.

(*v*) A horse rider shall not use the cycle tracks and footways.

17.6. General Rules Concerning Traffic

17.6.1. In addition to the special regulations concerning the vehicles, drivers, speed and parking, there are some fundamental traffic laws which every road user must be conversant with. These laws are of a basic nature and are intended to be applied on a uniform basis throughout the country.

17.6.2. "Keep to the left" rule

India is one of the countries which still adhere to the "Keep to the left" rule. U.K. is another. Countries such as U.S.A. and Canada follow the "Keep to the right" rule.

Though no inherent advantage can be claimed to either of the above rules, the majority of the countries in the world today follow the "Keep to the right" rule. Some of those following the "Keep to the left" rule are, therefore seriously considering changing over to the "Keep to the right" rule. Sweden is one of those that have switched on the "Keep to the right" rule recently. The argument in favour of adopting a uniform pattern throughout the world is that drivers travelling in different countries will have little difficulty if the rules of the road were the same every where, and the motor vehicle industry need not manufacture two models—one for internal consumption and the other for export. It is true that changing over from one system to the other is extremely costly, because of the need to modify the costly highway structures, signs, signal installations and road markings, and the changes that will be necessitated in the vehicle manufacturing industry. It is equally true, however, that if a change is deemed to be necessary, the earlier it is effected the better. This is especially true of developing countries, which are just at the threshold of the widespread use of the motor vehicle.

According to the "Keep to the left" rule, the drivers must keep to the left and stay on the left side except when overtaking another vehicle or making a right turn. Keeping to the left is not however, required in one-way streets or multi-lane highways.

17.6.3. Overtaking rules

Normal rules in countries with the left side rule prescribe overtaking on the right. When there are multiple lanes, this rule is not usually enforced. Where overtaking is prohibited by signs, the above rules do not apply.

17.6.4. Turning rules

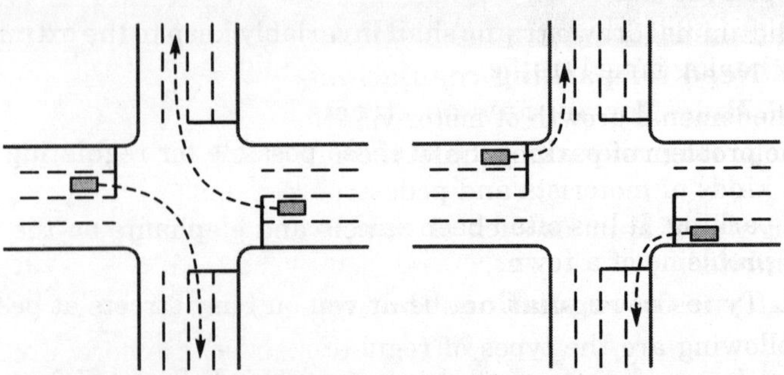

Fig. 17.2. Correct left and right turning manoeuvres.

In countries with the left side rule, left turning manoeuvres must be made from the lane nearest to the left hand side curb of the street and into the left lane of the street concerned nearest to the left side curb. Similarly, a right turning manoeuvre must be made from the lane nearest to the centre line of the street and into the left lane nearest to the centre line of the street concerned. The correct turning manoeuvres are indicated in Fig. 17.2.

17.6.5. Priority rules at intersection

The usual rule at an intersection is that the vehicle which arrives at an intersection first has the right of way. When two vehicles arrive simultaneously at an intersection, priority is given to the vehicle on the right. At intersections where the STOP sign is installed on any arm, all vehicles approaching the intersection on that arm must first stop and only then proceed.

17.6.6. Hand signals

In India, the following rules exist for hand signalling :

(*i*) ***Right turn.*** For turning to the right or for driving to the right in order to overtake another vehicle, a driver shall extend his right arm in a horizontal position outside of and to the right of his vehicle with the palm of his hand turned to the front.

(*ii*) ***Left turn.*** A driver shall extend his right arm and rotate it in an anti-clockwsie direction.

(*iii*) ***Slowing dawn.*** A driver shall extend his right arm with the palm downward and to the right of the vehicle and shall, move the arm so extended up and down several times.

(*iv*) ***Stopping.*** A driver shall raise his right forearm vertically outside of and to the right of the vehicle, palm to the front.

(*v*) ***Signal for allowing to be overtaken.*** The driver shall extend his right arm and hand horizontally outside of and to the right of the vehicle and shall swing the arm backwards and forwards in a semi-circular motion.

17.6.7. Rules intended to promote pedestrian movement on footways

The following are some of the rules that are enforced to promote pedestrian movement on footways :

(*i*) Prohibiting hawkers and pedlars from occupying footway space.

(*ii*) Preventing abutting shopkeepers from erecting temporary structures, projections, platforms on the footways.

17.7. Parking Regulations

17.7.1. Need for parking regulations

The phenomenal growth of motor vehicles in towns and cities has brought to the fore the problem of parking. Safety, comfort and convenience of road users, including all kinds of motorists and pedestrians, demand a proper and effective regulation of parking. It has often been said that proper parking policy is the key to the traffic problems of a town.

17.7.2. Types of regulations that are normally imposed.

The following are the types of regulations that are normally imposed on parking :

(*a*) **Space controls**

(*i*) Restrictions on loading and unloading of commercial vehicles.

(*ii*) Exclusive parking spaces for taxis.

(*iii*) Reservation of curb space for bus stops.

(*iv*) Restriction on parking near intersections.

(*v*) Limitations on curb parking and designation of parking places along the curb.

(*b*) **Time Controls**

(*i*) Peak hour parking prohibitions.

(*ii*) Parking control by fees.

17.7.3. Restrictions on loading and unloading of commercial vehicles.

Commercial goods transport vehicles consume considerable space and hinder the free flow of traffic. The following rules can be enforced to regulate loading and unloading :

(1) Making it necessary for the vehicles halting for the purpose of loading and unloading to draw up parallel to and alongside the curb and not athwart it or athwart a cross street.

(2) Restricting loading and unloading

(*i*) Within an intersection area and 75 metres on either side ;

(*ii*) On a pedestrian crossing or within 6 metres on either side of it ;

(*iii*) Near corners, bends, safety zones, and also entrances and driveways of properties ;

(*iv*) On side walks and cycle tracks, and

(*v*) Within 10 metres of a bus stop.

17.7.4. Exclusive parking spaces for taxis

Taxis serve a great need to city traffic and deserve to be patronised in the matter of parking spaces by the curb side. Exclusively reserved taxi parking spaces are common in most of the progressive cities.

17.7.5. Reservation of curb space for bus stops

Public transport is really the answer to the growing traffic dilemma in the cities. Designation of bus stops along the curb and prohibiting any other vehicle to use this space is a step in the direction of fostering of public transport.

17.7.6. Restrictions on parking near intersections

Safety requirements demand the prohibition of any kind of parking within an intersection area and 75 metres on either side of it. Similarly, parking should be prohibited on a pedestrian crossing or within 6 metres on either side of it. Corners, bends, safety zones, entrances and driveways of properties, side walks and cycle tracks are other locations where parking needs to be prohibited.

17.7.7. Limitations on curb parking—designation of parking places along the curb

Control on the curb parking is the answer to the problems posed by congested streets. If judiciously applied, it can increase the capacity and speed of traffic to a large extent, apart from bringing benefits due to reduced accident hazards. For this purpose, parking places are designated at desired locations alone, and no parking is permitted elsewhere. The measure is successful only if the streets are properly signed and the carriageway markings are adequate.

17.7.8. Peak hour parking prohibitions

If parking is found to be undesirable only during the peak hours, prohibition only during the peak hours can be resorted to. Proper indication of the prohibited hours on the road signs is necessary.

17.7.9. Parking control by fees

Regulation of parking by charging fees has the desirable effect of curbing long term parking while encouraging short term parking. The parking turnover is increased in the process. One method of collecting the charge is by installing parking meters.

17.8. Enforcement of Regulations

17.8.1. Importance of enforcement and punishment

The ultimate benefit of regulatory measures, to the community depends upon the extent to which these laws are enforced and observed. The enforcement effort put in by the police authorities and the administration determines in a large measure the degree of observance of the rules. The nature and type of penalties for nonobservance also has an effect on the degree of observance of the rules. Enforcement involves considerable deployment of police and administrative personnel and, by its very nature, is costly. If the resources are limited, it may be necessary to evaluate the cost of enforcement a developing country like India, where a good proportion of the road users are not educated, enforcement measures should be given a high priority (Ref. 18).

17.8.2. Goals and objectives

The major goals and objectives of traffic enforcement are :

(*i*) To achieve safe and efficient movement of traffic and prevent accidents :

(*ii*) To prevent violations of traffic laws.

(*iii*) To take presuasive action to prevent road users from violating traffic laws.

(*iv*) To punish those who violate traffic laws.

Enforcement measures can be preventive or punitive. Education programmes and traffic engineering techniques are preventive whereas punishment for offences is punitive.

17.8.3. Machinery for enforcement

The machinery for enforcement usually consists of police, courts or traffic authorities. In some countries, the police or traffic authorities deal with minor offences and impose penalties directly. The courts are usually called upon to deal with more serious traffic offences.

17.8.4. The police

The police forces have a vital role to play in traffic law enforce-ment. In many major cities, the traffic police is a specialist wing of the police forces in view of the complicated and peculiar nature of duties involved. In U.K., a system of traffic wardens is in operation to assist the traffic police in enforcing the traffic laws and in dealing with minor police functions pertaining to control and regulation of moving and parking vehicles.

The duties of the traffic police in enforcement are as below :

(*a*) Preventive duties.

(*i*) Control and direct the traffic flow so as to ensure safe movement ;

(*ii*) Supply information to the road users unfamiliar with the locality about the rules and regulations in force and their desired destinations.

(*iii*) Investigate accidents and build up data on them.

(*iv*) Intensive patrol in selected areas and accident prone locations to prevent recurrence of accidents.

(*v*) Initiate road safety propaganda and education campaigns.

(*b*) Persuasive duties.

(*i*) Issuing written warnings for traffic offences ;

(*ii*) Issuing verbal warnings for minor offences.

(*c*) Punitive duties.

(*i*) Issuing summons to the offender whenever a traffic law violation takes place.

(*ii*) Arresting a road user whenever the police have reasons to believe that the driver is endangering the lives of road users and property. Under the Motor Vehicles Act in India, the police in uniform have powers to arrest without warrant whenever specific offences are committed.

(*iii*) Imposing on-the-spot fines in minor offences to prevent the offenders being dragged unnecessarily to the courts.

(*iv*) Assisting the courts in dealing with the offenders brought before them.

(*v*) Seizing and detaining vehicles used unlawfully and without permits.

(*vi*) Impounding false documents found in the charge of drivers.

For the efficient discharge of the above duties, the police are provided with patrol vehicles, speed measuring devices and alcohol test devices.

17.8.5. The courts

The courts have the final say in the matter of punishing the offenders. They interpret the law, assess the guilt and impose penalties. In India, the Motor Vehicles Act prescribes the penalties for different traffic offences. For proper discharge of their duties, the Magistrates must have adequate knowledge of traffic laws and, in addition, should have specialized training to handle such offences.

17.8.6. In this section the existing Indian traffic laws will be examined and their shortcomings identified.

The advent and the increasing popularity of the motor vehicles in the thirties prompted the introduction of the Motor Vehicles Act in 1939. The Act, revised recently in 1988 is still the fountainhead of all traffic regulations in the country.

Though the Act was framed with the best of intentions and has been amended from time to time to suit new requirements, the experience of the road user and the enforcement authorities in its working has been that the law is deficient in many respects. A brief discussion of the deficiencies will now be made.

The Act was basically framed to codify the laws relating to Motor Vehicles. It is thus limited to only one of the constituents of traffic, namely, the motor vehicles, and is silent on the other important categories of road users, namely, the pedestrian, animal drawn transport and cyclists. The rules and regulations concerning these categories of traffic are left to be framed by the local authorities. This gives rise to fragmented pieces of regulatory measures, often halfhearted and more often contradictory, locally in force in different parts of the country. There is a need for a unified traffic law covering all categories of road traffic and applicable throughout the country. Even as regards certain aspects of motor vehicles, the practice varies from one part of the country to another. For example, the registered laden weight permitted on a truck of the same make and model is different in different States. The provision regarding periodic inspection of vehicles varies from one part of the country to another. The wide variation in laws in different parts of the country is not conducive to safety, since most of the road users do not have a clear conception of their obligations. In short, the Motor Vehicles Act, has not been conceived keeping in view the various safety aspects. The stress is on regulation of motor vehicles and is certainly not on prevention of accidents. There is a need to examine all the traffic laws framed in the country and to arrive at basic uniform traffic laws applicable throughout the country, and to supplement the same by additional provisions to control traffic in municipal areas. This has been highlighted in the recommendations of an Expert Study Group which went into the whole question of traffic safety in India recently (Ref. 19).

The Indian Motor Vehicles Act, does not cover the safety aspect of roads by the Highway authorities. The result is often that the roads are dug up indiscriminately and adequate warning light and signs are not installed. Diversion of traffic on substandard roads causes serious accidents. The law does not make it mandatory for the highway authority to provide minimum safety precautions.

A Highway Safety Code has now been brought out by the Indian Roads Congress (Ref. 20). The Code covers all types of road users including pedestrians, cyclists, motor vehicle drivers and passengers of public transport vehicles.

I. MULTIPLE CHOICE QUESTIONS

1. Traffic regulations may refer to:
   (*a*) motor vehicle traffic regulations (rules of the road)
   (*b*) highway code in the United Kingdom
   (*c*) traffic code in the United States
   (*d*) vienna convention on road traffic for international standard
2. Before a vehicle enters a curve of a mountain road, the driver .... if there is no vehicle coming in the opposite direction
   (*a*) should reduce speed, honk and drive on the left side
   (*b*) should drive along the outer side of the curve
   (*c*) may briefly borrow the opposite lane
   (*d*) none of the above
3. When passing an unmanned level crossing, the driver should ....
   (*a*) speed up and pass (*b*) reduce speed and pass
   (*c*) maintain the speed and pass (*d*) stop, look and pass
4. Unless stopping due to an abnormal situation on the expressway, the driver should select .... for stopping
   (*a*) the ramp (*b*) the acceleration lane
   (*c*) the deceleration lane (*d*) the service area
5. When a vehicle running on an expressway needs to stop when encountering an unexpected emergency, the driver may directly stop the vehicle in the carriageway
   (*a*) wrong (*b*) right (*c*) I don't know (*d*) none of them

Answers

1. (*a*) **2.** (*a*) **3.** (*d*) **4.** (*d*) **5.** (*a*)

II. SOLVED QUESTIONS AND ANSWERS

Q. 1. What are the rules for pedestrians traffic?

Ans. At all crossings. When using any type of crossing you should always check that the traffic has stopped before you start to cross or push a pram onto a crossing always cross between the studs or over the zebra markings. Do not cross at the side of the crossing or on the zig-zag lines, as it can be dangerous.

Q. 2. Zebra crossings.

Ans. Give traffic plenty of time to see you and to stop before you start to cross. Vehicles will need more time when the road is slippery. Wait until traffic has stopped from both directions or the road is clear before crossing. Remember that traffic does not have to stop until someone has moved onto the crossing. Keep looking both ways, and listening, in case a driver or rider has not seen you and attempts to overtake a vehicle that has stopped.

Q. 3. What do you mean by pelican crossings?

Ans. Pelican crossings are signal-controlled crossings operated by pedestrians. Push the control button to activate the traffic signals. When the red figure shows, do not cross. When a steady green figure shows, check the traffic has stopped then cross with care. When the green figure begins to flash you should not start to cross. If you have already started you should have time to finish crossing safely.

Q. 4. Explain puffin crossings.

Ans. Puffin crossings differ from pelican crossings as the red and green figures are above the control box on your side of the road and there is no flashing green figure phase. Press the button and wait for the green figure to show. When the road is congested, traffic on your side of the road may be forced to stop even though their lights are green. Traffic may still be moving on the other side of the road, so press the button and wait for the signal to cross.

Q. 5. What is meant by toucan crossings?

Ans. Toucan crossings are light-controlled crossings which allow cyclists and pedestrians to share crossing space and cross at the same time. They are push-button operated. Pedestrians and cyclists will see the green signal together. Cyclists are permitted to ride across.

Q. 6. Explain overtaking or passing.

Ans. Overtaking or passing is the act of one vehicle going past another slower moving vehicle, travelling in the same direction, on a road. The lane used for overtaking another vehicle is almost always a passing lane further from the road shoulder which is to the left in places that drive on the right and to the right in places that drive on the left.

Q. 7. What are the rules for animals vehicles?

Ans. Hand signals are given by cyclists and some motorists to indicate their intemations to other traffic. Under the terms of the Vienna Convention on Traffic, bicycles are considered to be vehicles and cyclists are considered to be drivers. The traffic codes of most countries reflect this.

In some countries (such as in the Czech Republic, Canada, and the United States), hand signals are designated not only for cyclists, but for every vehicle that does not have signal lights or has damaged signal lights. For example, drivers of older cars and mopeds may be required to make hand signals.The animal drawn traffic shall be prohibited from carrying long pipes, rods, bamboos etc more than 5 meters in length.

III. REVIEW QUESTIONS

1. What is the basic principles of regulation?
2. Discuss the speed regulation of vehicles?
3. Describe the general rules of concerning traffic.

REFERENCES

1. Uniform Vehicle Code, National Committee on Uniform Traffic Laws and Ordinances, U.S.A., 1968.
2. Traffic Engineering and Control in U.S.A., Organization for European Economic Cooperation, Paris, 1955.
3. Road Traffic Act, 1972, H.M.S.O., London.
4. The Motor Vehicles Act, 1988.
5. How fast ? Ministry of Transport, H.M.S.O., London, 1968.
6. Road Research Laboratory, Research on Road Traffic, H.M.S.O. London, 1965.
7. L.J. Pignataro, Traffic Engineering, Prentice-Hall, Inc., Englewood Cliffs, New Jersy, 1973.
8. Guidelines on Regulation and Control of Mixed Traffic in India I.R.C : 70–1977, Indian Roads Congress, New Delhi, 1977.
9. W.S. Smith and C.S. Le Craw, Traffic Speed Enforcement Policies, the ENO Foundation, Saugatuck, Connecticut, 1948.
10. A Policy on Geometric Design of Rural Highways. American Association of State Highway Officials, Washington, 1965.
11. Guidelines for Design of Horizontal Curves for Highways and Design Tables, I.R.C : 38–1988, Indian Road Congress, New Delhi, 1988.
12. H.K. Evans (Editor), Traffic Engineering Handbook, Institute of Traffic Engineers, New Haven, Connecticut, 1950.
13. AASHO, A Policy on Intersections at Grade, AASHO, Washington.
14. Traffic Engineering and Control in the U.S.A., Project No. 149, Organization for European Economic Cooperation, Paris, 1955.
15. Road Research Laboratory, Research on Road Safety, H.M.S.O, 1963.
16. Pocket Book of Transport in India. Transport Research Division, Ministry of Shipping and Transport, New Delhi, 1972.
17. Research on traffic law enforcement, Organization for Economic Cooperation and Development, Paris, 1974.
18. S. Dayashanker, Traffic Enforcement, National Seminar on Traffic Enforcement and Environment in Metropolitan Cities, Vol. I, Madras, 1972.
19. Report of the Study Group on Road Safety, Ministry of Shipping and Transport, New Delhi, 1972.
20. Highway Safety Code, Indian Roads Congress, New Delhi, 1996.

SECTION 8
TRAFFIC SAFETY

18. Road Accidents–Causes and Prevention

18.1. Road Accidents and the Traffic Engineer

The spectacular increase in the number of motor vehicles on the road has created a major social problem–the loss of lives through road accidents. The appalling human misery and the serious economic loss caused by road accidents demand the attention of the society and call for the solution of the problem. A multi-disciplinary approach is needed in understanding the problem and providing solutions. The traffic engineer is concerned because many features of the highway affect the safety of the vehicle and the other road users. The analysis of the accident statistics provides clues to the many factors that lead to the accidents and to improvements that may be desired. Based on the statistics the traffic engineer must devise ways to reduce the accidents through better planning, design, construction, maintenance and traffic operation. The traffic engineer is also concerned about the regulation and management of traffic to ensure safer travel. Accident data supply valuable information to control, regulate and manage the traffic more efficiently. The cost of traffic accidents helps the traffic engineer in evaluating an improvement scheme aimed at reducing the accidents.

18.2. Accident Situation in India

18.2.1. It is estimated that in the U.S.A, about 44,000 persons are killed and 3 million injured every year. This figure is really alarming. The accident situation is more serious in India because of the rapid growth of motor vehicles in the past few years and the inadequacy of many of our roads and street to cope up with this traffic. The mixed traffic conditions prevailing on the roads in India make the matters much worse.

18.2.2. The trend in the accidents and fatalities in India for the period 1960–97 is seen from Table 18.1 (Ref. 1).

Table 18.1. Road Accidents in India, 1970-2001 (Data Relates to Calendar Year)

(Source : Ref. 49, 63)

(In thousands)

Year	*Road Accidents*	*Persons Killed ('000)*	*Persons Injured ('000)*
1	2	3	4
1970	114.1	14.5	70.1
1971	120.2	15.0	70.7
1972	122.3	16.1	76.4
1973	121.6	17.6	79.3
1974	114.3	17.3	76.7
1975	116.8	16.9	77.0
1976	124.7	17.8	82.5
1977	135.4	20.1	95.6
1978	146.3	21.8	99.5
1979	144.4	22.6	102.9
1980	153.2	24.6	109.1
1981	161.2	28.4	114.0
1982	166.2	30.7	126.0
1983	177.0	32.8	134.1
1984	195.0	35.1	156.2
1985	207.0	39.2	163.4
1986	215.5	40.0	176.4
1987	234.0	44.4	189.0
1988	246.7	46.6	214.8
1989	270.0	50.7	229.7
1990	282.6	54.1	244.1
1991	293.4	56.4	255.0
1992	260.3	57.2	267.2
1993	280.1	60.7	287.8
1994	320.4	64.0	311.5
1995	348.9	70.6	323.2
1996	371.2	74.6	369.5
1997	373.7	77.0	378.4
1998	385.0	79.9	390.7
1999	386.4	82.0	375.0
2000	391.4	78.9	399.3
2001	405.6	80.9	405.2
2002	407.5	84.7	408.7
2003	406.7	86.0	435.1
2004	429.9	92.6	464.5
2005(P)	439.2	94.9	465.2

(P) : Provisional.

18.2.3. Type of road users killed and injured in accidents is given in Table 18.2.

Table 18.2. Type of Road Users Killed in Road Accidents in Delhi (1994) and USA (1995) and Thailand (1987)

Type of road users	*Percent of Persons killed*		
	Delhi	*USA*	*Thailand*
Pedestrians	42	13	47
Cyclists	14	2	6
Occupants of motorised two wheelers	27	5	36
Occupants of motorised four wheelers	12	79	12
Others	5	1	—

18.2.4. In 1968, the Government of India set up a study group to go into the accident situation in India and to recommend suitable preventive measures. The Report of the Study Group (Ref. 1) is a comprehensive document and touches almost all aspects of road safety in India. Another Committee on Road Safety submitted its report to the Government of India in 1983 (Ref. 52). The Prakash Narain Report for the Planning Commission (2001) has made several recommendations, pertaining to institutional set-up, trauma care etc. (Ref. 59). A more recent study of the National Highways Authority of India has made recommendations on legislation, financing and institutional reforms (Ref. 60).

18.3. International Comparison of Road Accidents

International comparison of road accident data is of great value but to be meaningful it should be based on agreed definitions of accidents, fatalities and injuries. Another difficulty in interpreting the results is the wide variation that exists in different countries as regards economic conditions, education and literacy, climate, types of vehicles, traffic conditions, legal measures in force, population density, degree of urbanisation and value of vehicle kilometers to which the statistics are usually related. Caution is, therefore, needed in drawing conclusions from such comparisons.

Most countries maintain accident statistics in some form or other. Organisations such as the World Health Organisation and Economic Commission for Europe publish statistics of road accidents. The Road Research Laboratory (U.K.) has assembled accident figures from a number of countries (Ref. 2).

Smeed and Jeffcoate (Ref. 3) have presented the accident figures from 68 countries (India being one of them), for the period 1960–67 and have fitted a relation between the number of motor vehicles (N) per population (P) and the number of deaths (D) per vehicle population (N). The relation is of the form :

$$\frac{D}{N} = 0.0003 \left(\frac{N}{P}\right)^{-2/3} \qquad ...(18.1)$$

and was earlier proposed by the Road Research Laboratory (Ref. 2).

Though this formula has its limitations, it provides a tolerably good procedure for estimating the number of road fatalities in a given country where the population and the number of vehicles are known. It will be seen that as the value of N/P increases the value of D/N decreases. This implies that as the vehicle ownership increases, the number of deaths per licensed vehicle decreases.

A further modification of the above relationship has been recently suggested by Jacobs and Hutchinson (Ref. 4), for developing countries, and is as follows :

$$\frac{D}{N} = 0.00077 \left(\frac{N}{P}\right)^{-2/5} \qquad ...(18.2)$$

This relationship also shows the tendency for fatalities per motor vehicle to decrease with increasing vehicle ownership.

Table 18.3 gives the fatality rates per 10,000 vehicles in some selected countries (Ref. 59).

Table 18.3. Fatality Rates from Road Accidents in Selected Countries (1992)

S. No.	*Country*	*Fatality Rate (Death per 10,000 vehicles)*	*S. No.*	*Country*	*Fatality Rate (Death per 10,000 vehicles)*
1.	Nigeria	160.0	2.	Tanzania	150.8
3.	Ghana	111.8	4.	Zambia	84.8
5.	Liberia	53.7	6.	Mauritius	36.2
7.	Ecuador	20.7	8.	China	26.2
9.	India	25.2	10.	Russian Federation	14.7
11.	Indonesia	12.6			
12.	Iran	6.1	13.	Malaysia	5.6
14.	Portugal	4.4	15.	Brazil	4.1
16.	Austria	2.9	17.	USA	2.1
18.	Japan	1.7	19.	Norway	1.2

The word average is about 10, with about 500,000 deaths every year and 500 million vehicles in use.

The inferences that can be drawn from Table 18.3 above are :

(*i*) The fatality rates are higher for less developed countries than for developed countries.

(*ii*) The reason for the above are that the vehicle ownership rates are very high in developed countries, and equation 18.2 shows that as the vehicle ownership rate increases, the fatality rate comes down.

(*iii*) The developed countries invest large amount of money on safety.

In India, the fatality rate had come down from 25.2 in 1994 to 12.8 in 2003 because of the increase in vehicle ownership. But the number of people killed in road accidents now is 140,000 per year and this is a matter of serious concern.

Another way of expressing fatality rates is in terms of deaths per 100,000 population. Such a comparison for the year 2003 (Source : various sources from the internet) is given below :

Table 18.4. Fatality Rates from Selected Countries (2003) in Terms of Deaths per 100,000 Population

S. No.	*Country*	*Fatality Rate (Deaths per 100,000 population)*	*S. No.*	*Country*	*Fatality Rate (Deaths per 100,000 population)*
1.	U.K.	5.8	2.	Norway	6.0
3.	Japan	7.0	4.	India	7.9
5.	Germany	8.0	6.	Canada	8.6
7.	France	9.5	8.	New Zealand	11.5
9.	Portugal	12.9	10.	USA	14.8
11.	China	18.4	12.	Russian Federation	24.8
13.	Malaysia	26.8			

Because of the large population in India, the fatality rate on the basis of population is comparable to some of the developed countries.

18.4. Collection of Accident Data

18.4.1. The usefulness of an accurate and comprehensive system of collection and recording accident data cannot be overemphasized. Such data serve to identify the basic causes of accidents and to suggest means for overcoming the deficiencies that lead to such accidents. The data, have a number of uses (Ref. 5), enumerated below :

(*a*) **Engineering uses.** 1. In determining the adequacy, size, shape and legibility of traffic signs.

2. In determining the justification for traffic control devices, such as traffic signals.

3. In determining and planning pedestrian safety features such as underpasses, overbridges, pedestrian barriers, refuge islands, pedestrian signals and street lighting.

4. In determining speed zoning and speed control.

5. In planning traffic regulation measures such as prohibition of on-street parking, one-way streets and prohibited turnings.

6. In designing safe and efficient street lighting.

7. In designing or redesigning intersection.

8. In designing and providing channelizing islands, central verges and refuge islands.

9. In planning safety measures for traffic during construction.

10. In identifying specific deficiencies in the maintenance procedures and specifications.

11. In improving the horizontal and vertical alignment.

12. In providing adequate sight distances.

13. In determining suitable width for pavement, shoulders and bridges.

14. In understanding the deficiency of the pavement surfaces and devising ways and means of improving them.

15. In determining any deficiencies in superelevation and camber and rectifying them.

16. In planning safety measures and guard rails.

(*b*) **Enforcement uses.** 1. In planning deployment of personnel for various duties such as manual control and street patrolling depending upon accident frequency at different locations and during different hours of the day.

2. In directing enforcement effort.

3. In controlling pedestrian behaviour.

4. In the safe and efficient operation of traffic control devices.

5. In enforcing curb parking regulations.

6. In planning and enforcing vehicle inspection measures.

7. In planning and enforcing cycle inspection measures.

8. In aiding prosecution of offenders in courts.

(*c*) **Administrative and policy issues.** 1. In initiating and administering Traffic Safety programmes.

2. In evaluating the success of Traffic Safety Programmes.

3. In determining the accident costs.

4. In identifying the need to amend the legislative measures in force and to take appropriate action to amend them.

(*d*) **Educational uses.** 1. In planning and organizing school safety educational Programmes.

2. In planning and organizing driver safety educational programmes.

(*e*) **Uses for the Motor Vehicle Administrator.** 1. In reviewing the procedures for licensing drivers.

2. In reviewing the procedures for registration and licensing of vehicles.

3. In reviewing the vehicle inspections requirement.

18.4.2. Requirements of accident records

(*i*) If accident records are to be of use, they should be accurate and comprehensive, instead of being vague and misleading.

(*ii*) Accidents must be reported on a standard form so that uniform procedure is followed.

(*iii*) The terms describing the accidents must be accurately defined.

(*iv*) For facility of being analysed by a computer, the data should be coded properly.

18.4.3. Standard definitions

The definition of some of the terms used in Accident Reporting Forms *A*-1 and 4 recommended by the I.R.C (Ref. 6) are given below :

Road accident. An accident (collision, overturning or slipping) which occurred or originated on a road open to public traffic resulting in either injury or loss of life, or damage to property, in which at least one moving vehicle was involved.

Person killed. Any person who was killed outright on the spot in the accident or whose death could be directly traced to the injury(ies) received in the accident.

Fatal accident. An accident in which one or more persons were killed.

Grievously injured person. A person who has received grievous injuries (in accident) such as fractures, concussions, internal lesions, crushing, severe cuts and lacerations, severe general shock requiring medical treatment and any other serious lesions requiring detention in hospital. Also those hurt grievously as defined in Indian Penal Code in accidents, as reproduced below :

(*i*) Emasculation

(*ii*) Permanent privation of the sight of either eye.

(*iii*) Permanent privation of hearing of either ear.

(*iv*) Privation of any member or joint.

(*v*) Destruction or permanent impairing of the powers of any member or joint.

(*vi*) Permanent disfigurement of the head or face.

(*vii*) Fracture or dislocation of a bone or tooth.

(*viii*) Any hurt which endangers life or which causes the sufferer to be, during the space of twenty days, in severe bodily pain, or unable to follow his ordinary pursuits.

Grievous injury accidents. Accidents in which person (s) were grievously injured. (These do not include fatal accidents for providing classification under item 2 of the Form).

Slightly injured persons. Persons who have sustained only minor injuries or bruises or sprains.

Minor injury accidents. Accidents in which persons received only minor injuries or bruises or sprains.

Non-injury accidents. Accidents in which no one was killed or injured.

Pedestrian. Any person other than a driver or passenger. Persons in or operating pedestrian conveyance such as perambulator, invalid chair without engine, push cart, etc. or pulling a cycle are pedestrians. Persons attending to a vehicle (*e.g.* for change of tyre, repairing engine, etc.) moving on roller skates etc., are also pedestrians.

Driver. Any person who drives a vehicle or rides a pack or saddle animal.

Passenger. Any person, other than a driver, who is in or on a vehicle.

Motor vehicle. Mechanically propelled road vehicle including tractors and cycles fitted with auxiliary engines.

Cycles. Two-wheeled or three-wheeled road vehicle fitted with pedals and using human energy as its sole means of propulsion.

Motor cycles. A motor operated vehicle with two wheels more than 450 mm in diameter without pedal operations.

Motor car. A free-wheeled self-propelled vehicle designed for the transportation of persons but limited in seating capacity to not more than 7 passengers.

Motor car (Taxi). A motor car constructed, adapted, or used to carry not more than six passengers, excluding the driver for hire or reward.

Auto-rickshaw. A motor operated vehicle with three wheels less than 450 mm in diameter without pedal operation.

Scooter. A motor operated vehicle with two wheels less than 450 mm in diameter without pedal operation.

Articulated vehicle. A unit made up of a road motor vehicle and a semi-trailer.

Nature of collision. Positions and directions of vehicles in each case are as indicated in Fig. 18.1.

(*i*) Head on collision. (*ii*) Rear end collision.
(*iii*) Brush/side swipe. (*iv*) Right angled collision.
(*v*) Right turn collision.

Types of insurance :

Comprehensive. Covers all risks.

Third party. This category includes all types, other than comprehensive insurance.

18.4.4. Standard accidents reporting forms

In India, the I.R.C. have prescribed (Ref. 6) two forms for accident reporting as follows :

Form A—1. Form for collecting the details of an accident.

Form 4. Summary of road accidents in a State during the year.

These forms are reproduced in Appendices 18.1 and 18.2.

18.4.5. Collision diagrams and condition diagram

A collision diagram is a schematic representation of all the accidents occuring at a particular location. Such a diagram greatly facilitates the understanding of the limitation of the site that often cause accidents. The diagram contains a large scale map of the location and the paths of the vehicles and the position of the pedestrians involved in each of the accidents. The map itself need

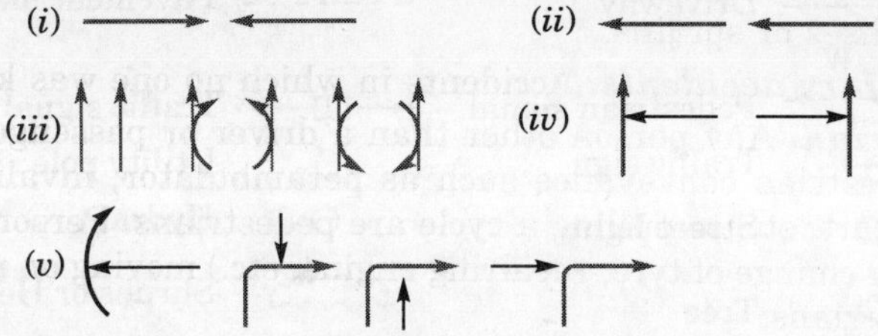

Fig. 18.1

not be necessarily to scale and the vehicle paths indicated therein are not the actual paths so as to avoid overlap confusion. A uniform system of symbols would be necessary to identify the full details and nature of the accidents. The symbols that are normally employed are indicated in Fig. 18.2.

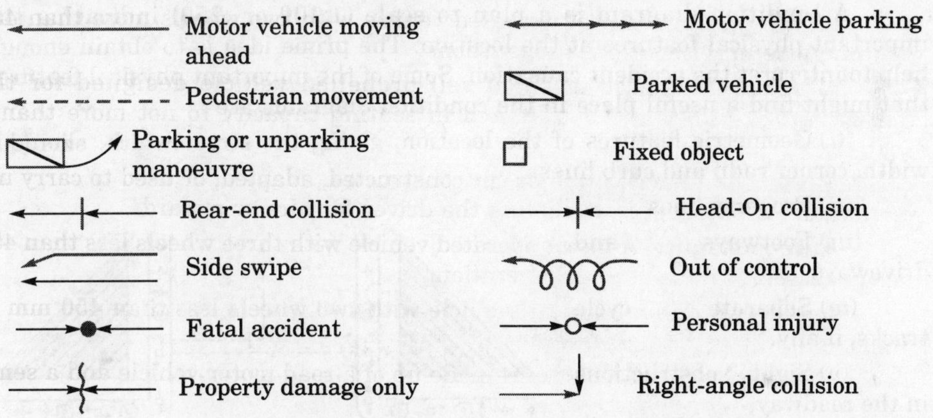

Fig. 18.2. Symbols employed in collision diagrams.

It may be seen that the pedestrian path is indicated by dotted lines whereas the path of the vehicle is indicated by solid lines, and the direction of travel is indicated by arrows. The date and time (to the nearest hour) can be conveniently indicated on one of the arrows. Any extraordinary weather or road-way conditions or details of the persons involved in the accident can also be noted such as 'fog', 'ice', 'drunken driver', etc. 'Before' and 'after' collision diagrams will be useful in assessing the efficacy of the measures undertaken to improve the location. A diagram of traffic flow at the junction can help provide valuable

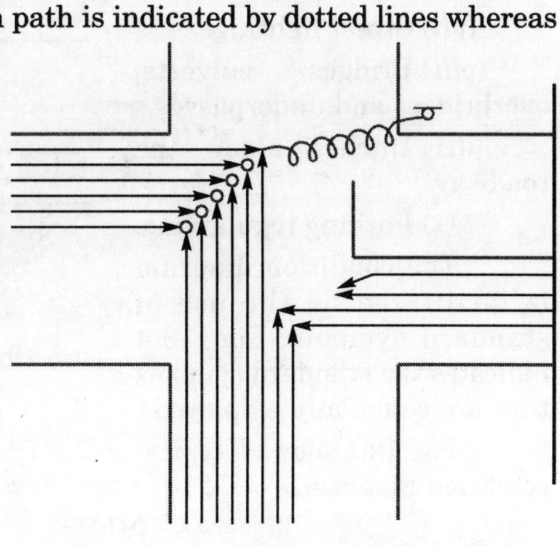

Fig. 18.3. Typical collision diagram at a junction.

clues to the accident causation.

A typical collision diagram is given in Fig. 18.3.

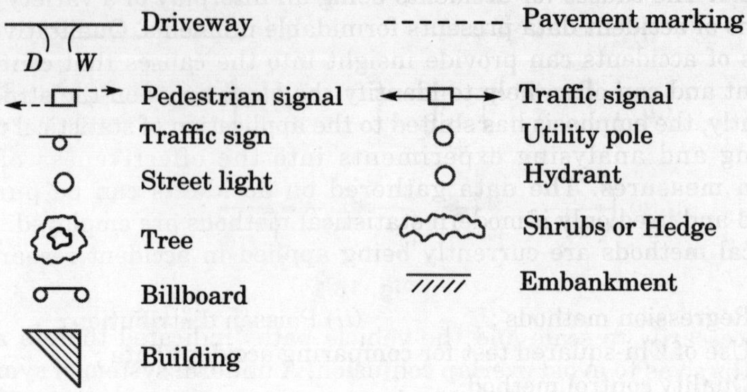

Fig. 18.4. Symbols for condition diagrams.

A condition diagram is a plan to scale (1:100 or :250) indicating the important physical features at the location. The prime idea is to obtain enough help to interpret the accident causation. Some of the important physical features that might find a useful place in the condition diagrams are :

(*i*) Geometric features of the location, giving the street width, shoulder width, corner radii and curb lines.

(*ii*) Property lines.

(*iii*) Footways and driveways.

(*iv*) Separate cycle tracks, if any.

(*v*) Sight obstructions in the roadway.

(*vi*) Traffic signs, signals and markings.

(*vii*) Street lighting.

(*viii*) Bridges, culverts, overbridges and underpasses.

(*ix*) Ditches along the roadway.

(*x*) Parking regulations.

The condition diagram is facilitated by the use of standard symbols. Fig. 18.4 indicates the standard symbols that are commonly employed.

Fig. 18.5 shows a typical condition diagram.

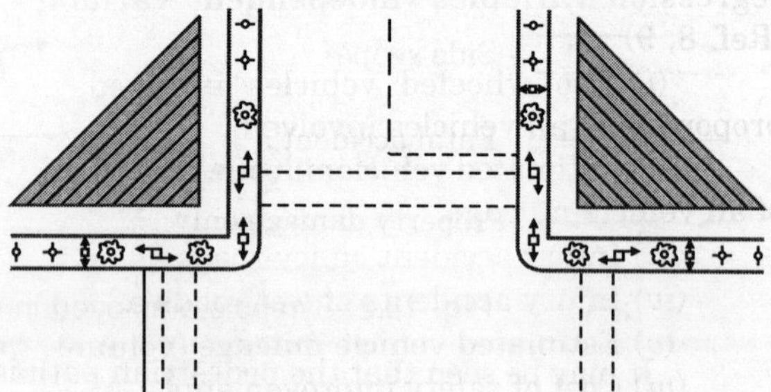

Fig. 18.5. Typical condition diagram.

18.5. Statistical Methods for Analysis of Accident Data

18.5.1. The causes for accidents being an interplay of a variety of factors, the analysis of accident data presents formidable problems. Qualitative methods of analysis of accidents can provide insight into the causes that contributed to the accident and can often help to identify the *black spots* on the street system. More recently, the emphasis has shifted to the application of statistical techniques in planning and analysing experiments into the effectiveness of accident prevention measures. The data gathered on accidents can be purposefully interpreted and used only if modern statistical methods are employed. A number of statistical methods are currently being applied in accident research. These include :

(*i*) Regression methods ; (*ii*) Poisson distribution ;
(*iii*) Use of Chi-squared test for comparing accident data ;
(*iv*) Quality control method.

18.5.2. Regression methods

The basic principle behind these methods is that the expected number of accidents, on a certain road system during a given time period, is dependent in a linear way on factors which are supposed to be of significance for the determination of accident frequency (Ref. 7). The number of accidents occurring on a certain day is itself assumed to be normally distributed, with a mean value being a linear function of regression variables, and a variance being constant and same for all days of a certain time period studied. The numbers of accidents on different days are assumed to be stochastically independent. Some of the regression variables (independent variables) that could be considered are (Ref. 8, 9) :

(*i*) Two wheeled vehicles involved in personal injury accidents as a proportion of all vehicles involved.

(*ii*) Estimated vehiclemileage travelled in hours of darkness as a proportion of all vehicle-mileage.

(*iii*) Injury accident on icy roads as a proportion of all injury accidents.
(*iv*) Injury accidents of wet roads as a proportion of all injury accidents.
(*v*) Estimated vehicle mileage (volume of traffic).
(*vi*) Cost of safety improvements.
(*vii*) Number of pedestrians.
(*viii*) Ratio of sidewalk to length.
(*ix*) Pavement width.
(*x*) Number of junctions per km length of road.
(*xi*) Curvature of the road, in degrees per km.

The regression model is of the following form :

$$Z = \alpha_0 + \alpha_1 x_1 + \alpha_2 x_2 + \alpha_3 x_3 + \alpha_4 x_4 + \ldots\ldots \alpha_n x_k \quad \ldots(18.3)$$

where. Z = accident rate

$\alpha_1, \alpha_2 \ldots \alpha_n$ = regression coefficients

$x_1, x_2 \ldots x_n$ = independent variables

α_0 = regression constant.

The regression equation is easy to determine by means of a computer programme by the method of least squares.

18.5.3. Poisson distribution in accident analysis

One of the commonly used statistical models for accident analysis is the Poisson distribution, which is very familiar to students of Statistics. In Chapter 8, a detailed treatment of this distribution is given. Poisson distribution is appropriate for accident study because accidents are governed by the laws of chance and the occurrence of accidents is a rare event in time or in distance or amongst drivers.

The mathematical formula for the Poisson distribution is :

$$P(r) = \frac{e^{-m}\, m^r}{r\,!} \qquad \text{...(18.4)}$$

where, $P(r)$ = Probability of occurrence of r events
m = average rate of occurrence of events
e = base of Naperian logarithms.

Applying the above formula to determine the probability of a driver causing an accident, let :

N be the number of drivers

M be the kilometres driven by each driver

p be the probability of having an accident per kilometre travelled

m be the average rate of occurrence of accidents in a length of travel of M kilometres ($m = p\,M$)

Then the probability of a driver having r accidents,

$$P(r) = \frac{e^{-m}\; m^r}{r\,!}$$

Also, the number of driver having r accidents

$$= N\,\frac{e^{-m}\; m^r}{r\,!} \qquad \text{...(18.5)}$$

In the above formula, the value of p is very small and the value of M is very large.

The above formula can also be used to determine the probability of accidents occurring in a given time span. To illustrate, let m be the average number of accidents occurring in a year at a given location.

Then $P(r)$ = probability of occurrence of r accidents during a year

$$= \frac{e^{-m}\; m^r}{r\,!}$$

The following examples illustrate the use of Poisson distribution in predicting accident occurrence.

Problem 18.1. *It is observed that on an average a vehicle driver drives 5000 km during the course of a year. The probability of having an accident is 100 per 200 million vehicle-kiometres. What is the probability of a driver having at least two accidents during his driving career extending to 25 years ?*

Solution. No. of accidents

= 100 per 200,000,000 vehicle kilometres

$\therefore$ p = 0.0000005 accidents per vehicle kilometre

$M = 5000 \times 25 = 125{,}000$ km

$m = pM = 0.0625$

$$\therefore \quad P(r) = \frac{e^{-0.0625}\; 0.0625^r}{r\,!}$$

$$P(r = \text{at least } 2) = 1 - P(r = 0) - P(r = 1)$$

$$= 1 - \frac{e^{-0.0625} \times 0.0625^0}{0!} - \frac{e^{-0.0625} \times 0.0625^1}{1!}$$

$$= 1 - e^{-0.0625} - e^{-0.0625} \times 0.0625$$

$$= 1 - e^{-0.0625} \times (1 - 0.0625)$$

$$= 1 - e^{-0.0625} \times 0.9375 = 0.1287.$$

Problem 18.2. *The accident records for three consecutive years at an uncontrolled junction indicate the following number of accidents :*

Year	*No. of accidents*
1972	*3*
1973	*6*
1974	*9*

Calculate the probability of 4 accidents occurring per year at the site.

Solution. Here, $m = \frac{3+6+9}{3} = 6$

$$\therefore \quad P(r) = \frac{e^{-6}\, 6^4}{r!},$$

$$\therefore \quad P(4 \text{ accidents}) = \frac{e^{-6}\, 6^4}{4!} = 0.1339.$$

Problem 18.3. *It has been found that on an average 1 in 100 drivers in a bus company are involved in an accident every year. If there are 500 drivers in the company, what is the probability that there are exactly 4 drivers who are involved in an accident during a year ?*

Solution. $p = \frac{1}{100} = 0.01$

$$n = 500$$

$$m = np = 500 \times 0.01 = 5$$

$$P(r) = \frac{m^r\, e^{-4}}{4!},$$

$$P(4) = \frac{5^4 \cdot e^{-4}}{4!} = \frac{625}{24} \times \frac{1}{54.60} = 0.48$$

18.5.4. Use of Chi-squared distribution for comparing accident frequencies

One of the situations a traffic engineer has to assess frequently is whether the safety measures adopted at a particular location or stretch of road have been really effective in reducing the number of accidents. *Before and after* data on accidents can be evaluated on statistical principles, and one of the handy tools in this direction is the Chi-squared test (Ref. 10, 11). Let b be the number of accidents before the improvements at a particular location and a the number after the improvements. Assuming that the improvements have no effect and the accident number is expected to increase due to the changes in traffic and weather then let $b\,C$ be the number of accidents expected if no improvements had been carried out, the factor C being called the *control ratio*. Then the value of the Chi-squared is :

$$\chi^2 = \frac{(a - bC)^2}{(a + b)\,C} \qquad \text{...(18.6)}$$

The null hypothesis H_0 stipulates that there is no real change due to the improvements.

Assuming a 5% level of significance, from Table 8.15, we find the value of χ^2 to be 3.841 with one degree of freedom. If $\chi^2 > 3.841$, we observe that the null hypothesis is unlikely to be true and that there is a real change. On the other hand, if $\chi^2 < 3.841$, we conclude that the null hypothesis is true and that there is no real change due to the improvements.

The following examples illustrate the above procedure.

Problem 18.4. *In an ordinary square junction of two roads there were 20 accidents in a year. After provision of traffic signals, the number of accidents dropped down to 8 per year. In the sector of the city where this junction is situated, the general trend observed was that number of accidents increased at a rate of 10 per cent during the period covered by the above two observations. Test whether the improvement in junction design has a significant effect at 5% significance level.*

Solution. Here $a = 8; \quad b = 20$

$$C = \frac{110}{100} = 1.1$$

$$\chi^2 = \frac{(a - bC)^2}{(a + b)\,C} = \frac{(8 - 22)^2}{28 \times 1.1}$$

$$= \frac{14 \times 14}{28 \times 1.1} = 6.36$$

From Table 8.15, χ^2 for 5% significance level and one degree of freedom = 3.841.

As χ^2, (observed) > 3.841, the data supplied provides strong evidence that the improvement in junction design was effective, and that the reduction in the number of accident was not merely due to chance alone.

Problem 18.5. *The accident data pertaining to a metropolitan city for the year 1965 and 1970 are given below :*

Year	*1965*	*1970*
Accidents	*300*	*400*
Vehicle-kilometre of travel	*250 million*	*300 million*

Test whether there is any significant increase in the accident rates in two years.

Solution. Here, the traffic has increased from 250 to 300 million vehicle kilometres, and thus the control rate C is

$$= \frac{300}{250} = 1.2$$

$$a = 400, \quad b = 300$$

Substituting in the formula :

$$\chi^2 = \frac{(a - bC)^2}{(a + b)\,C} = \frac{(400 - 360)^2}{700 \times 1.2} = \frac{1600}{840} = 1.90$$

For a 5% significance level, with 1 degree of freedom, χ^2, from Table 8.15.
$= 3.841$

χ^2 (observed) < 3.841

Conclude that the difference in the accident rates in the years might easily have arisen due to chance.

18.5.5. Use of Quality Control Approach

Statistical approach to Quality Control problem has been widely recognised as an effective tool, finding applications in diverse fields. An interesting use of such an approach is in the accident analysis field (Ref. 12). The method is basically built upon the Poisson distribution model for the occurrence of accidents. From past records on estimate of m, the expected average number of accidents, is first made. Thereafter confidence limits are determined for the number of accidents for any desired level, say 99 per cent. If the observed number of accidents falls beyond these limits, it can be deduced that there is a significant change in the accident causation and that the observed value is not merely due to chance.

18.6. Road and its Effect on Accidents

18.6.1. Accidents are rarely caused because of one single factor. Usually the interaction of a diverse set of factors causes the accidents, though one factor can be more responsible than the rest and can be easily identified. A car driver knocking down a school child–a very common occurrence in traffic accident statistics–can be considered as an example to illustrate. The car driver might have been speeding and driving recklessely. He might be even under the influence of liquor. The geometries of the road on which he was driving might be deficient. The street surface might have become slippery due to the use of very soft aggregates in the surfacing course, aggravated by the thick film of water on the road caused by a sharp shower. The driver on seeing the child suddenly crossing the street might have instantaneously applied the brakes. Mechanical failure of the brakes and the slippery condition of the surface might have rendered all his efforts of no avail, resulting ultimately in the unfortunate accident. The sequence of events leading to the accident in this case is rather involved and interlinked and no single factor can be isolated and identified as the sole cause of the accident. Complex as the problem is yet it cannot be denied that a study of the various causes of accidents is very vital for traffic safety. Such a study forms an important aspect of the work of a Traffic Engineer.

Broadly, the factors that cause accidents can be considered under the following headings :

(*i*) The road

(*ii*) The vehilce

(*iii*) The driver

(*iv*) The road user other than the motorist

(*v*) Environmental factors.

In this section detailed consideration of the road and its effect is given.

18.6.2. The characteristics of a road has a great influence on the causation of accidents. Improvement to the conditions of the road can bring about a measurable change in the accident. The Highway Engineer and the Traffic Engineer have an important duty in ensuring that the roads are designed to proper standards, built to the correct specifications and maintained adequately. It is being increasingly realised that safety can be built into the highways.

18.6.3. The safe speed at which a vehicle can negotiate a curve depends on the radius of curvature of the curve. Sharp curves permit low speeds and large radii curves cater to high design speeds. Geometric standards prescribed in different countries lay down the minimum and ruling radii for different speeds. The engineer should endeavour to provide as large a radius as possible, taking recourse to minimum and ruling radii only when constrained by site conditions.

18.6.4. Vertical alignment

The vertical alignment should be designed with the provision of suitable shock-proof vertical curves. The vertical curves should be made to blend homogeneously with the horizontal curves. It is a good practice to design the vertical curves to be fully contained within the horizontal curve.

18.6.5. Sight distance

Sight distance can be for (*i*) Safe stopping and (*ii*) Overtaking. All roads should be designed for the full length for safe stopping sight distance. As large a percentage of the length as is consistent with economy should be designed for overtaking sight distance. Wherever overtaking is prohibited, suitable signs should be erected.

18.6.6. Superelevation

For a vehicle to negotiate any curve at the desired speed, the road should have adequate superelevation. Deficiency in superelevation can cause serious accidents.

18.6.7. Carriageway width

The minimum carriageway width for a two-way road should be 7 m to cater for 2 lanes of traffic. Unfortunately, due to limited financial resources, most of the roads in the country are single lane. This is a major cause of accidents because of the inherent danger on such roads during overtaking and crossing. An intermediate width of 5.5 m is being advocated in this country for roads where the traffic is too heavy for a single lane carriageway (Ref. 13), but obviously this solution has its own limitation.

18.6.8. Width and conditions of shoulders

Whereas shoulders are needed for parking stopped vehicles, in India they are used for overtaking and crossing vehicles. This requirement demands that the shoulders be of adequate width and should be maintained in good condition, if accidents are to be avoided. They should have adequate supporting power even in the rainy season.

18.6.9. Deficiency in road signs and road markings

Properly designed and maintained road signs inform the driver of need for caution and can avert many an accident. Similarly road markings such as centre line marking, stop line marking and pedestrian crossing marking can prevent accidents.

18.6.10. Junction design

Any deficiency in the design of junctions can cause accidents. Elements of junction design such as turning radius, width at entry and exit and channelizing islands should receive proper attention. Improved *visibility* and staggered cross roads can reduce the accidents. Channelization reduces the number of collision points and hence promotes safety.

18.6.11. Pavement surface characteristics

Pavement surface characteristics determine to a large extent the safety of a vehicle when negotiating a curve or when suddenly applying brakes. The surface should have anti-skid properties even when the pavement is wet. Because of the extreme importance of this subject, a more detailed discussion is given later.

18.6.12. Formation delineators and guard rails

When driving along curves or on high embankments, the driver needs to be guided about the direction of the road. Formation delineators with reflectors aid in this requirement. Guard rails and safety barriers prevent vehicles from going off the roadway in the event of loss of control and are very useful in specific situations (See Chapter 16).

18.6.13. Narrow bridges and culverts

Narrow bridges and culverts can cause very serious accidents, especially when the roadway is wide. Widening of such narrow structures should be undertaken if accidents are frequently caused at the location. Posting proper warning signs can also save ugly accidents.

18.6.14. Median width

A wide median prevents glare from opposite stream of traffic and promotes safety. Strips of 15 m width seem to be desirable for safe operation of opposing streams of traffic (Ref. 14), but it is often difficult to obtain such a wide median and one has to do with 5 m or so In extreme situations, a width as low as 1.2 m is also resorted to.

18.6.15. Street Lighting

The importance of adequate street lighting for preventing road accidents needs no special emphasis. A British study reports that the average reduction in night accidents observed after the provision of improved lighting is about 30 per cent (Ref. 2). Besides reduction in the number of accidents in the night, good street lighting also reduces the proportion of the more severe type of accidents.

18.6.16. Results of research in India on accident rate and road features

A study in India (Ref. 50) has revealed the following relationships.

Mumbai–Poona Road

$$AR\ (3) = -\ 0.1526 + 0.0216\ RF + 0.0031\ CV + 0.4793\ J$$

Some selected roads in India

$$AR\ (1) = 0.2171 + 0.002884\ CV + 0.4126\ J - 0.3447\ W + 0.001274\ Q_{ADT}$$

In the above equations

AR (3) = Personal injury accident rate per million vehicle km per year

AR (1) = accident rate in number per km per year

AF = Rise and fall in m/km

CV = Curvature in deg/km

W = payment width in m

J = Number of junctions per km

Q_{ADT} = Total volume in terms of vehicles (all types) per day.

18.7. The Vehicle

18.7.1. It is being increasingly realised–thanks to consumer protection pioneers like *Ralph Nader*–that vehicles play a dominant role in road safety. One estimate puts that mechanical deficiencies cause about 5 per cent of all traffic accidents (Ref 3). The subject of vehicle safety is controversial : while on the one hand, the manufaturers maintain that the present day cars are safe provided they are run and maintained with care, the consumers argue that vehicles are the biggest culprits that cause serious accidents. The truth perhaps is that no vehicle is perfectly safe, and the researcher is constantly on the hunt for improving the quality of motor vehicles. Various aspects of vehicle design and maintenance are involved in making (*i*) accidents fewer in number, (*ii*) accidents less serious to other road user (*iii*) accidents less severe to vehicle occupants.

18.7.2. Braking system

Brakes are needed to bring the vehicles to a stop whenever a driver comprehends immediate danger. A good braking system should achieve this objective without resulting in the driver's loss of directional control of the vehicle. The interaction between the tyre and the road surface governs largely the ultimate braking performance of vehicles, although factors such as the braking system itself are significant. A detailed treatment of the subject of the interaction between the tyre and the road surface is given under the Section on Skidding. Elements of design of the braking system include the load distribution among the different wheels and the brake distribution between the front and rear wheels. One of the prime causes for accidents during braking is the wheels getting locked and the driver losing control of the vehicle consequently. Research is being undertaken to device ways of applying the brakes severely but at the same time to prevent the vehicles from being locked. This will be one of the major advances in designing vehicles for safety, and it is hoped that anti-skid brakes will soon be a feature of all future vehicles.

18.7.3. Vehicle lighting system

The lighting system of the vehicle consists of headlights, (driving main beam and the dipper beam), side lamps, parking lights, rear lights, direction indicators and stop lamps. An efficient and reliable system of lighting the vehicle is desirable for averting accidents.

The headlight should perform two functions : to provide a main beam for enabling the driver to see the road ahead for sufficiently long distance and to provide a dipper beam which must avoid glare to the opposing traffic, while at the same time providing sufficient visibility to the driver. Glare is a serious problem when two vehicles approach each other and can blind the driver's vision. It should be avoided by properly designing the dipper beam. Polarised headlights offer interesting break-through in this regard.

Rear lamps give indication to the driver following a vehicle about the presence of a vehicle in front of him. Their efficient design and functioning can avert front-to rear collisions.

Direction indicators give adequate notice of the intention of the driver to turn or to stop. The usual type of direction indicators are the flashing type. The position, type and colour of the indicators should be such as to be visible to the following drivers easily. The direction indicators are usually amber in colour, and preferably they should be mounted at the side of the vehicle approximately at the level of the driver's eye (Ref. 2). Semaphore arms may not be fully suitable to large vehicles, and in such cases an extra amber light may be installed at the rear (Ref. 15).

18.7.4. Vehicle body–its features

In the design of motor vehicles for safe driving, an aspect which has been receiving increasing attention is the features and dimensions of the vehicle body. Extensive work has been done in regard by the Harvard School of Public Health (Ref. 16).

Factors that need to be considered are :

(*i*) the shape and dimensions of the driver's seat

(*ii*) arrangement of dials on the dash board

(*iii*) positioning of controls in relation to the driver's seat

(*iv*) visibility of the driver from the seat

(*v*) noise levels in the vehicle

(*vi*) concentration of carbon monoxide inside the vehicle.

The seat of the driver should be comfortable and adjusted to give him a clear view all around and to secure easy access to all the controls. The controls themselves must be easily operated and the dials must be properly visible. The noise levels and carbon monoxide levels in the commercial vehicles can cause discomfort to drivers and should be within safe limits. From this point of view, vehicles with rear engines are safer than those with front engines.

The driver normally has a clear vision of the road to his front and sides, except for that portion obstructed by the bonnet and windscreen pillars. Yet these obstructions can be the cause of a lack of proper vision when a pedestrian crosses the road or a cyclist travels in front. The lesser these obstructions are, the better the design. Commercial vehicles suffer from the lack of visibility to the rear, with the result a number of accidents can be caused in reversing. External mirrors fitted to the sides eliminate this drawback partially, but for effectiveness should be of proper size and should be located at the most advantageous position. Windscreen wipers should be provided on all vehicles and should be maintained in good condition.

For safety of road users as well as the occupants, there should be as few protruding sharp points as possible both externally as well as internally. The knobs should preferably be recessed. The doors should have secure locks so that the occupants are not thrown out in accidents.

Another controversial issue is the horsepower of the vehicle. While one group holds the view that high horse-powered vehicles cause more accidents, the other group contends that basically high horse-power vehicles are safer ones and what causes more accidents with them is their misuse. Excessive speeding is possible with high powered vehicles and this probably contributes to more accidents. In such cases, speed governors are a great necessity.

18.7.5. Tyres

Tyres are an important aspect of vehicle design from the point of view of safety. The performance of tyres relative to puncture, blowouts, vulnerability to damage by sharp objects, braking and road surface characteristics are areas where considerable research has taken place. Skidding is a phenomenon which is governed by the interaction of the tyre, brakes, road surface, speed and the wetness of the road surface. Because of its extreme importance in traffic accident prevention, this aspect is treated in greater detail elsewhere.

18.7.6. Vehicle inspection and maintenance

Frequent inspection of vehicles and testing them is an important requirement. Such periodic tests identify the mal-functioning of the important parts and facilitate easy maintenance. This assumes added significance in India where old vehicles are used to their maximum and their life is made to stretch to the last. Most important items that need periodic inspection arc the brakes, lighting system and the steering mechanism. Additional features that could form part of routine inspection could be the tyres, suspension system and the windscreen. Legislative enforcement of an effective inspection system will yield good results in safety. Some countries have made it compulsory for testing the vehicle periodically, U.K. being one them (Ref. 17). Some States in the U.S.A. have also provision for compulsory periodic testing. The Motor Vehicles Act in India contains provision for the States to enact rules to require periodical testing and inspection of vehicles (Ref. 18).

18.8. The Driver

18.8.1. The driver is the key factor in most of the accidents. He is the human element in charge of the machine. He drives it, steers it, accelerates it, decelerates it, brakes it, and stops it. As such a study of driver behaviour offers a very deep insight into the causes of accidents and provides a valuable guidance for avoidance of accidents. His behaviour influences and controls the vehicle while his behaviour itself is controlled by his personality, training and attitudes. Alcohol and drugs affect his driving capability profoundly and are of special significance for traffic safety. Age and sex of drivers have also been found to be important factors in safety research.

18.8.2. The task of a driver in the modern busy street is really taxing. During the course of his normal driving he has to overtake many vehicles and be overtaken by some, perform numerous turning manoeuvres, comprehend dangerous situations as they present themselves at intersections, deal with a number of pedestrians who cross the streets at all sorts of odd places and must make many important decisions in a fraction of a second. It is this risk that a driver is exposed to that results in traffic accidents. It has been estimated by someone that a driver makes two billion spilt second decisions for every single decision that ultimately leads to fatality (Ref. 19). Driving today is no more a form of relaxing recreation, but is in fact the generator of an appreciable cardiovascular stress in a driver–this is shown by a study of 600 healthy drivers in U.K. (Ref. 20). Where exactly does hope lie for road safety in the sphere of activity controlled by the driver ? The present Section deals with some of these aspects likely to be of importance to the Traffic Engineer.

18.8.3. Driver judgment, skill and emotional make-up

When a driver perceives a danger, he immediately reacts to it. The time he takes to comprehend the danger depends upon his individual mental make-up and is called Perception time. Once the information is fed to him, some time must elapse before he can react to the stimulus. This time is called his Reaction time. The Perception time and the Reaction time vary from driver to driver. Some authorities also call the Perception and Reaction time as the PIEV time (Perception, Intellection, Emotion and Volition time–see Chapter 2). AASHO Policy (Ref. 21) assumes a perception time of 1.5 seconds and total perception and brake reaction time of 2.5 seconds for designing the sight distance. The reaction time for different events is also different.

After having assessed the situation mentally, the driver is now ready to decide on the course of action. His individual sense of 'judgment' is built upon his experience of past driving, driver education and his estimate of his own capacity to perform. Different drivers take different degrees of risks, and age and sex of the driver has an important effect in this regard.

By skill is meant the driver's capacity to control and manipulate his vehicle under situations he is called upon to deal with. His skill is partly inborn, but some skill is acquired out of training and experience.

An interesting approach is to study the relation between the physiological or psycho-physical make-up of a driver and his accident proneness. The Galvanic Skin Response (GSR) is one of the measurements suggested to determine the involuntary change in the electrical resistance of the skin (Ref. 54). It is noticed that the sweat glands in the skin, which are motivated by the autonomic nervous system increase their secretions during tension and excitement. The electrical resistance of the skin drops consequently.

18.8.4. Age of drivers

It is generally observed that young drivers cause more accidents than would be caused by mature drivers with considerable driving experience (Ref. 2, 22). Similarly, old drivers also seem to cause more accidents than middle-aged drivers. The minimum accidents are caused by the middle aged group, say 40–50 years. While inexperience is the cause of higher accident rate amongst the younger age group, physical illness leads to higher accidents amongst the older age group.

18.8.5. Sex of the drivers

It is generally seen that males have more accidents and convictions than females, but when miles driven are taken into account this difference essentially disappears (Ref. 23).

18.8.6. Marital status

Generally speaking, married persons (both male and female) have been observed to be better drivers (Ref. 23).

18.8.7. Training of drivers

It can be appreciated that systematic training of drivers should have a profound effect on their driving behaviour and performance. Professional drivers of long-distance haulage trucks and public transport buses, after receiving class-room training and driving instructions, are known to have inculcated good driving habits. Rewards from good and clean record of driving and deterrent penalties for negligent driving are some of the other methods adopted by organisations who have to deal with a large number of professional drivers. Suitable legislation for ensuring that driving instructions are given only by fully qualified personnel should be welcomed. Special courses for those drivers who already hold a driving licence, but have been summonned for infringing traffic regulations should be planned. The driving tests should be conducted with a view to prevent inept people from acquiring a licence and to ensure that those who secure a licence are reasonably competent to handle their vehicles and have learnt the basic principles of driving and traffic rules.

18.8.8. Alcohol and drugs and the driver

A subject of research on road safety that has received a large measure of attention and rightly so is the effect of alcohol and drugs on the driving capability. The judgment and skill of a driver can be exercised only so long as his sensory faculties are functioning unimpaired. Alcohol and drugs are known to be positively dangerous in this respect.

Study of accident figures during different parts of the day in U.K. (Ref. 20) has revealed that, apart from the evening rush hour, the period from 10 P.M. till midnight is very bad. The traffic on the roads at this hour is usually on the decrease, and yet why does the accident figures rise ? There cannot be any other explanation but alcohol. A similar conclusion is inevitable when one examines the traffic accident pattern by the days of the week in U.K. A study in that country revealed (Ref. 24) that Saturday is the worst day. The spurt in the rate on the week-end is probably due to the drinks people consume on that day.

The alcohol consumed by a person enters into the blood stream soon after it is consumed. The actual amount of alcohol in the blood depends on a variety of factors such as the amount consumed, the weight of the person, the kind of alcohol, the time interval between the drink and the test and whether the alcohol was consumed when the stomach was empty or when it was full. The amount of alcohol in the blood can be measured by simple clinical tests by taking a sample of blood. A quicker, and equally reliable, test is to measure the proportion of alcohol in the breath. The "Breathalyser" is one such apparatus used in U.K. The blood-alcohol content at which the driver may be considered to be unfit to drive a vehicle is probably above 0.5 parts per 1000 by weight, which corresponds roughly to 50 mg/100 ml.

The effect of various amounts of alcohol on the driver behaviour has been studied in a study at Grand Rapids, Michigan U.S.A. (Ref. 25). Table 18.3 gives a summary of the findings.

Table 18.3. Blood Alcohol Concentration and Accident Risk

Blood alcohol Concentration	*Accident risk*
0–20 mg/100 ml	In some subjects there is deterioration in special skills resembling driving.
20–40 mg/100 ml	Progressive deterioration in special skills is noticed.
40–50 mg/100 ml	Significant increase in accident involvement is noticed as the blood alcohol level approaches and passes 50 mg/100 ml.
50–80 mg/100 ml	Overall accident risk increases rapidly. At a concentration of 80 mg/100 ml, alcohol emerges as the dominant factor in accidents.
Greater than 80 mg per 100 ml	The accident probability of drivers at a blood alcohol concentration of 100 mg/100 ml is 6–7 times that of drivers with less than 10 mg/100 ml.

Many authorities around the world have laid down their own standards for the maximum blood alcohol content. In U.S.A. many states consider it a legal offence for a person to drive a motor vehicle if the blood alcohol content is 100 mg/100 ml or more. In Norway and Sweden the limit is 50 mg/100 ml. In U.K. the Road Safety Act, 1988 makes it a legal offence for a person to drive a motor vehicle if the blood alcohol content is 80 mg/100 ml (Ref. 17). In India, the Motor vehicle Act (Sec. 185) (Ref. 18) provides for suitable punishment for persons driving under the influence of alcohol and drugs to such an extent as to be incapable of exercise proper control over the vehicle. But no limit has been prescribed for the alcohol content. The Traffic Engineering Committee of the I.R.C. (Ref. 25) have recommended to the Government to amend the Motor Vehicles Act in such a way that persons having a blood alcohol content of more than 0.08 per cent should be made liable to fine of Rupee one thousand or imprisonment for a period of three months. The punishment provided in the M.V. Act now is six months imprisonment or a fine upto two thousand rupees or both.

Drugs which have a profound influence on the central nervous system cause changes in the behaviour of drivers to an appreciable extent. Drugs which fall into this category are hypnotics, tranquilising agents, psychotomimetic drugs, stimulating agents and similar others. More research is needed to establish the actual role of such drugs in road accidents, although it is believed that because of the very nature of their action, these drugs can be dangerous.

18.8.9. Fatigue

When a driver is tired, due to long hours of driving, he suffers usually from a lack of concentration, so vital for road safety. His experiences when he is tired may be linked with illusions or hallucinations, and the results can be disastrous. For instance, a tired driver may get the feeling of seeing a person on the road when there is none, and in order to avoid him may sverve away from the road and get involved in a serious accident. Long and monotonously straight reaches of a road have been known to induce drowsiness and lack of vigilance. On the other hand, alignment with a good blend of straights and curvature tends to keep the driver alert. Loss of sleep causes drowsiness and lapses in attention. Limiting the continuous period of driving by regulations is one of the methods to prevent accidents in case of professional drivers.

18.8.10. Use of crash helmets

Motorcyclists and pillion riders who are involved in an accident are prone to head injuries. Head injuries often cause fatality. Protection of the head against injuries is one of the ways of reducing fatalities. This can effectively be done by wearing a crash helmet, which absorbs the energy in a crash and reduces the impact on the head to a safe level which can be withstood without any detriment to the safety. It has been indicated that wearing a protective helmet reduces by 30 to 40 per cent the risk of injury to that part of the head covered by the helmet (Ref. 2). The helmet is usually made of fibre glass and is well padded inside to effectively absorb the energy of crash. Wearing of helmets is now compulsory in the U.K. where a British Standard is also available for the helmets (Ref. 27). In India, I.R.C. have already made a suggestion to the Government to make wearing of helmets compulsory for two-wheeler drivers and pillion riders. The Study Group of Road Safety appointed by the Government have also suggested that the use of crash helmets should be encouraged (Ref. 1) Sec. 129 of the M.V. Act makes it compulsory to wear helmets (Ref. 18).

18.8.11. Use of safety belts and air bags

One of the common causes of serious accidents is that drivers, when travelling at a high speed and being suddenly forced to apply the brakes, are thrown forward and hit the windscreen or the steering wheel. Another common occurrence is that the car door opens out in an accident and the driver is thrown out of a vehicle causing him serious injuries. In both the above types of occurrences if the driver is firmly held to his seat, the chance of serious injuries can be minimised to a great extent. This is the principle behind the safety belt which is fastened to the seat and worn around his body to restrain him from movement. Safety belts are known to have saved many lives and are becoming increasingly popular in many countries. For successful functioning, such belts should be strong enough to bear the forces that will act on them. They should also be convenient and comfortable to wear. Automatically **inflatable air bags** reduce the driver fatalities considerably. The Indian vehicles do not provide this safety feature, but in countries abroad, many vehicles have this. Combination of air-bags and seat belts gives additional safety to drivers.

18.9. Skidding

18.9.1. Road accidents are very commonly caused by pavement slipperiness which results in skidding of the vehicle. Accident statistics show that in U.K. 27 per cent of the accidents involving vehicles was attributable to skidding (Ref. 2). Considerable research effort has been expended on the subject of pavement slipperiness and skidding in countries abroad, especially in U.S.A., U.K., and some countries in the continent. In India, not much attention seems to have been paid so far to this subject, though skidding can occur on our roads in the rainy season (Ref. 29).

18.9.2. The phenomenon of skidding

When a driver moving at speed applies the vehicle brakes suddenly, the prime factor that governs the stability of the vehicle is the friction that develops at the tyre-pavement interface. If the friction that can be mobilised is greater than the decelerating force, the vehicle comes to a safe halt within a certain distance. If, however, the pavement is wet and slippery and the tyres bald, the friction that is developed is much smaller than the decelerating force, and the vehicle cannot be controlled by the driver. The wheels of the vehicle get locked due to sudden braking and skidding results if the frictional resistance offered is too low.

In the U.S.A., the skid resistance is denoted by the Skid Number (SN) which is 100 times the friction factor (Ref. 30). In U.K., however, the measure of skid resistance is the coefficient of friction itself. If μ is the coefficient of friction developed at the time of skidding, an 'α' is the acceleration or deceleration of the vehicle, the following simple relation can be proved :

$$\mu = \frac{\alpha}{g} \qquad ...(18.7)$$

where g is the acceleration due to gravity.

18.9.3. Factors determining skid resistance

The following factors determine the skid resistance developed at the pavement interface :

1. Pavement. Different types of pavements have different skid resistance properties. For instance, surfaces like open textured permix carpet have a better skid resistance property than mastic asphalt. The texture of the surface is provided by the (*i*) macro-texture which determines the large scale roughness provided by the size and the shape of the coarse aggregates and (*ii*) the micro texture which refers to the fine scaled roughness caused by the asperities on the individual aggregates. The macro-texture helps in providing a good drainage of the surface and also in deforming the rubber of the tyres, which brings into play the hysteresis component of the friction. The micro-texture aids in puncturing the thin film of water on the surface and in mobilising the adhesion component of friction.

Aggregates which polish under traffic into a smooth shape are not desirable when resistance to skidding is being sought. For this purpose, limestones are regarded with suspicion. The Ministry of Transport (U.K.) (Ref. 31) specify the following minimum Polishing Stone Values (PSV) when measured by the Portable Tester (Ref. 32) :

Description of site	*Recommended minimum value of PSV*
1. Difficult sites	62
2. Average	59

2. Effect of water on the surface. The greatest danger of skidding is where the pavement is wet and the frictional resistance drops down consequently. The action of the film of water can be understood from Fig. 18.6.

The tyre surface contact strip is divided into three zones. Zone 1 represents the zone of bulk displacement of water. Zone 2 represents the zone where the water film is much thinner and where the water carries a portion of the load of the tyre. In Zone 3, the water has been squeezed out and the tyre is in contact with the surface. As the speed of the vehicle increases, not enough time is available for the water to be squeezed out and the boundaries of zones move backwards. When the speed is very high a stage is reached when Zones 2 and 3 completely vanish and only Zone 1 is present. This stage represents the condition when a full film is present between the tyre and the road surface and the phenomenon is called "aquaplaning" or "hydroplaning".

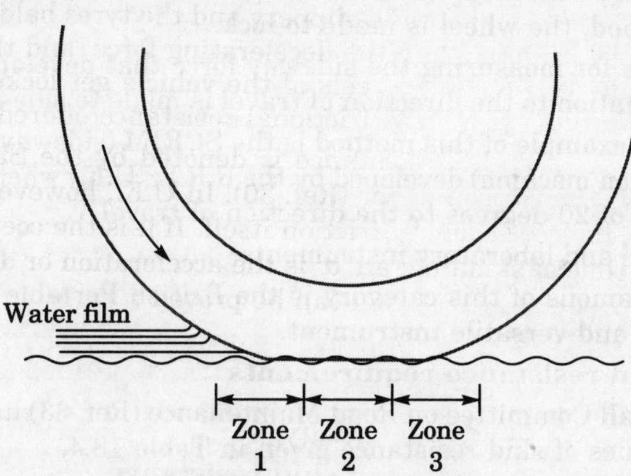

Fig. 18.6. Tyre-road surface-water film interaction.

The remedy is to drain off the water from the road surface as quickly as possible. This is achieved by providing adequate camber on the road and a rough surface texture which has numerous channels for the quick exit of water.

3. Speed. As speed increases, the skid resistance drops down markedly. Fig. 18.7 indicates the general trend.

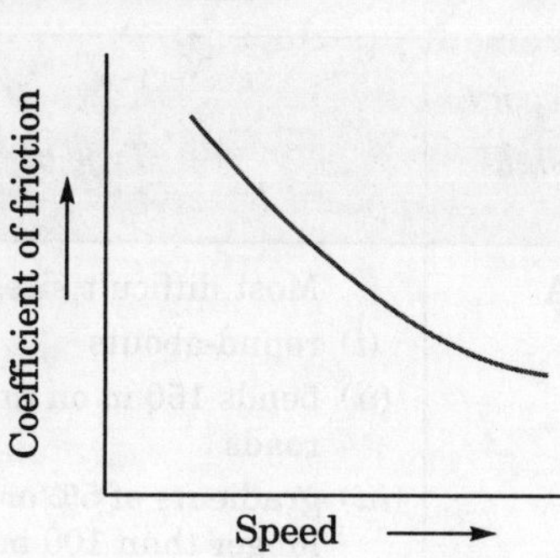

Fig. 18.7. General trend indicating drop in skid resistance as speed increases.

4. Tyre characteristics. The friction which develops between the tyre and the road surface is governed to a great extent by the tyre characteristics. There are two components of this friction :

(*i*) adhesion component, and (*ii*) hysteresis component.

The former is caused by the shear forces induced at the tyre pavement interface, and is important on dry roads. The hysteresis component is developed by the energy losses caused by the deformation of the rubber when it comes into contact with the irregularities of the surface. Hysteresis becomes important on wet roads and is attributable to macro-texture. The composition of rubber is important in determining the extent of mobilization of friction.

The tread depth and pattern of tyres influence the skid resistance to a great extent. Drainage of surface water is facilitated by the grooves in the tyre.

18.9.4. Measurement of skid resistance

Skid resistance is measured by a number of methods, which can be grouped as follows :

(*i*) Methods employing the stopping of test vehicles by braking and measuring the stopping distance or the deceleration caused.

Though simple, these methods involve considerable risk if braking is done at higher speeds.

(*ii*) Methods employing braking of trailers towed by vehicles.

In the N.A.S.A. method, the trailer wheel is made to rotate under slip. In the A.S.T.M. method, the wheel is made to lock.

(*iii*) Methods for measuring the sideway force that develops when a wheel placed at an inclination to the direction of travel is made to side-slip.

The classic example of this method is the SCRIM (sideway force coefficient routine investigation machine) developed by the R.R.L., U.K., where the fifth wheel is set at an angle of 20 degrees to the direction of travel.

(*iv*) Portable and laboratory instruments.

The most famous of this category is the British Portable Tester (Ref. 32) which is a handy and versatile instrument.

18.9.5. Skid resistance requirements

The Marshall Committee on Road Maintenance (Ref. 33) has recommended the minimum values of skid resistance given in Table 18.4.

Table 18.4. Minimum Values of Skid Resistance

(*Source* : Ref. 33)

Category of sight	*Type of Site*	*Skidding Resistance*		
		Test K.P.H.	*Speed (MHP)*	*Sideway force Coefficient*
A	Most difficult sites, *e.g.*, (*i*) round-abouts (*ii*) bends 150 m on unrestricted roads (*iii*) gradients of 5% or steeper or longer than 100 m (*iv*) approaches to traffic signals on unrestricted roads	50	(30)	0.55
B	Average sites, *e.g.*, (*i*) motorways and other high speed roads (*i.e.* speeds in excess of 95 km/hr or 60 MPH)	50 80	(30) (50)	0.50 0.45

	(*ii*) Trunk and principal roads and other roads with more than 2,000 vehicles per day in urban areas	50	(30)	0.50
C	Other sites, *e.g.*, Straight roads with easy gradients and curves, without junctions and free from any feature, such as mixed traffic, especially liable to create conditions of emergency.	50	(30)	0.40

18.9.6. Road construction practices for obtaining skid resistant surfaces

The following are some of the methods practised to obtain satisfactory skid-resistant surfaces :

(*i*) Stones, which polish under traffic and tend to become smooth (*e.g.*, limestones) are avoided in the surfacings.

(*ii*) In cement concrete pavements, the fine aggregates provide adequate skid resistance and must be selected with care.

(*iii*) Excess of bituminous binder, known as fatting, should be avoided.

(*iv*) Smooth and rounded aggregates are inferior to broken and angular aggregates.

(*v*) In cement concrete mixes, the water cement ratio is very important for satisfactory performance of the surface.

(*vi*) In cement concrete mixes, the surface finish should be selected to give a good texture depth.

(*vii*) Artificially, calcined bauxite aggregates hold much promise in anti-skid specifications (Ref. 34).

(*viii*) Methods employed for restoring the skid resistance characteristics of cement concrete surfaces are : acid etching, roughening the surface, grooving and bonding additional layer. (Ref. 64) The disadvantage with grooving is the noise created by traffic on grooved roads.

18.10. Speed in Relation to Safety

18.10.1. It is generally known that the frequency and severity of accidents increase as the speed increases. The following factors are responsible for this :

(*i*) The distance needed to bring a vehicle to a stop increases as the speed increases.

(*ii*) The minimum safe separation distance between vehicles increases with speed, and as many drivers do not maintain these distances, the risk of accident increases at greater speeds.

(*iii*) With higher speeds, the distance travelled during the driver's reaction time increases.

(*iv*) The skid resistance of wet roads decreases as the speed increases.

(*v*) When vehicles are approaching at fast speed, the judgment of the pedestrian can be less reliable.

(*vi*) Risk of tyre bursts is more at higher speeds.

(*vii*) Risk of mechanical failure due to fatigue and overheating is greater at higher speeds.

18.10.2. Whenever speed limits have been enforced, a sure reduction in the accident rate results. For instance, the imposition of the 50 MPH limits on certain week ends in England produced a 15% reduction is all injury accidents. (Ref. 35) After the imposition of a 40 MPH speed limit in some sections of roads in the London area, an apparent reduction of 20 percent in all injury accidents and of 30 percent in fatal and serious accidents has been observed. (Ref. 2)

Considerations in prescribing and enforcing speed limits are discussed in Chapter 17.

18.11. Weather and its Effect on Accidents

Elements of weather having an affect on accidents are (*i*) rain, (*ii*) snow or ice and (*iii*) fog.

Rain water causes the pavement to be slippery and can contribute to skidding accidents. Analysis of accident statistics for various months in U.K. (Ref. 2) suggests that whenever the rainfall was heavier, the accident rate was correspondingly higher.

Under conditions of snow and ice, accidents tend to be more frequent (Ref. 2). Slipperiness is the usual cause. Studded tyres were thought once to be the answer to such driving conditions, but they cause considerable damage to the road surface and tend to wear them.

Fog is a dangerous condition to drive with because visibility can get reduced to just a few metres under thick fog. But the usual observation is that the traffic volume itself drops down under adverse fog conditions and people tend to drive carefully under such situations. (Ref. 14).

18.12. Pedestrian Safety

18.12.1. Seriousness of the pedestrian safety problem

A large number of road accidents involve pedestrians. In the U.S.A., two-thirds of road deaths are of car occupants, and less than one fifth are pedestrians. (Ref. 36) In Europe one-third are pedestrians and the remaining one-third are riders of two wheelers. In Delhi, the pedestrians constitute 46 per cent of the fatalities from road accidents. (Ref. 55) These figures indicate the seriousness of the pedestrian safety problem.

18.12.2. Trends in pedestrian accident pattern

18.12.2.1. Age

Age has a significant effect on the accident pattern involving pedestrians. Most of the research findings have concluded that children and the very old are more likely to be in greater risk than the other age groups. This is easy to understand : the active adults are able to cross the road safely, whereas the very young and the very old are not. Children are frequently involved in crossing the road on their way to or from the school. They also get involved in accidents when playing in the streets or nearby. In U.K. it is estimated (Ref. 37) that nearly half of all the fatal and serious pedestrian casualties are children under 13 years, although this age group represents less than 25% of the population. The old people get involved in accidents due to impaired hearing and vision and decreased perception and reaction.

18.12.2.2. Sex

The influence of the sex of a person on the accident has been investigated by many. The general conclusion is that women are more careful pedestrians than men. According to Moore and Older, (Ref. 38) the risk of accident to pedestrians per 10,000 inhabitants is 5.1 for men and only 3.6 for women, in U.K. This is probably because men take greater risk in crossing roads at places other than those intended for pedestrians. Women, on the other hand, use the intended crossing places more frequently. Women have also been known to wait for more faviourable gaps in vehicle stream when crossing than men.

18.12.2.3. Social conditions

Social status influences pedestrian behaviour. For instance, it has been noticed that the underprivileged and low income groups figure prominently in pedestrian casualties. (Ref. 28). This is probably because of lack of adequate training and education. General environmental and living condition also have an influence on the accident rate. (Ref. 2).

18.12.2.4. Driving experience

If a person is known to have driving experience, his behaviour is expected to be more careful as a pedestrian. This is easily explained by the fact that a person with driving experience understands fully the limitations of the vehicle in charge of a human being.

18.12.2.5. Familiarity with the locality

Not much evidence is available whether the accident risk is influenced by the pedestrian's familiarity with the locality. Knowledge of the location of the crossing points and rules about restricted turning movements and one-way streets should lead to lesser accident rate among local pedestrians than with outsiders. More research is needed to understand this aspect.

18.12.2.6. The drunken pedestrian

The effect of alcohol and drugs on a driver has already been dealt with separately. The pedestrian is no different under the influence of liquor and drugs. Many of the pedestrian accidents can be traced to the action of alcohol and drugs.

18.12.3. The road and its influence on the pedestrian accidents

18.12.3.1. Road characteristics that can influence pedestrian accidents include geometric features of the road, design of intersections, traffic control devices, street lighting, colour of the road surface and footway design. It is at times difficult to ascribe the cause of an accident to any of the above single factors, since an interplay of driver behaviour, vehicle features, road characteristics and pedestrian behaviour brings about the accident.

18.12.3.2. Geometric features of the road

A divided carriageway with central verge is less accident prone than a two-way carriageway having the same number of lanes. This is because the pedestrian, when choosing a safe gap to cross, has only to reckon with unidirectional traffic in the former case. The greater the width of the road to be crossed, the greater is the risk of pedestrian accidents. (Ref. 40) One way streets are safer than two-way streets for pedestrians.

18.12.3.3. Design of intersections

Many features of the intersection, whether signalised or otherwise, have an effect on pedestrian safety. Properly designed channelising islands, which also serve as pedestrian refuges, offer an advantage to the pedestrians crossing a road at the intersection. The location of the stop line and the pedestrian cross-walk at the intersection should be determined with a view to ensure safe pedestrian crossing.

18.12.3.4. Pedestrian crossing design

Many types of pedestrian crossings have been tried with varying degrees of success. They fall under the following categories :

(1) Uncontrolled crossings (2) Controlled crossings

A typical example of the uncontrolled crossing is the "Zebra" crossing introduced in 1951 in U.K. This consists of a sufficient width of cross-walk painted with black and white longitudinal strips so as to be conspicuous to the drivers as well as the pedestrians. As soon as a pedestrain steps on to the striped cross-walk he has the right of way and all vehicles are required to stop legally. The design of these crossings is standardised in U.K. Apart from the striped zebra makings in black and white, the crossing consist of a zig-zag area (about 19 metres on both sides) to increase the driver's visibility of the crossing, black-and-white beacon posts, yellow flashing globe and studs marking the limits of crossing. A rough sketch of the U.K. Zebra crossing is given in Fig. 18.8.

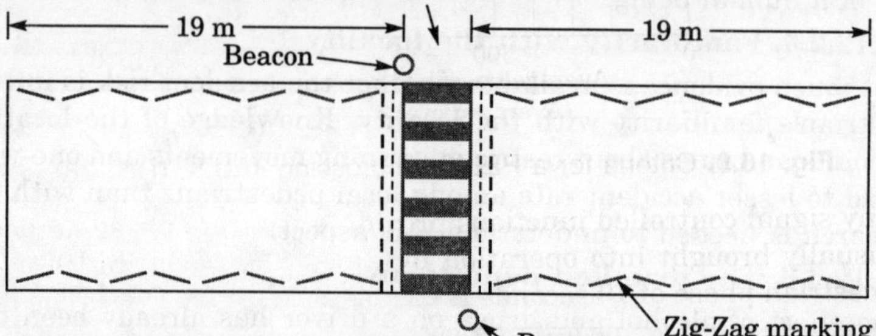

Fig. 18.8. Zebra crossing in U.K.

The experience in that country with such crossing has been very satisfactory. If the risk of crossing a road anywhere is arbitrarily taken as 1, the relative risk of accidents on a zebra crossing is 0.42. (Ref. 24) Uncontrolled crossings are normally provided where the vehicular traffic is not heavy. If the vehicular traffic is heavy, such crossings are likely to delay the vehicles to a great extent. In such cases, signal controlled crossings should be thought of.

The simplest form of controlled crossings is one controlled by police or traffic wardens. The assistance rendered by such guides in allowing the school children to cross at safe places has increased the safety to a large extent.

Signal controlled crossings are the safest of all. If the risk of crossing a road at any place is arbitrarily taken as 1, the risk of crossing at a signal controlled crossing is as 0.17 (Ref. 39). One of the most popular types of pedestrian controlled crossing in U.K. is the PELICAN crossings. These are operated by push button control, by which operation the waiting pedestrian demands a signal phase to cross. The criteria for the provision of PELICAN crossing U.K. in terms of pedestrian and vehicular volumes are given in Fig. 18.11.

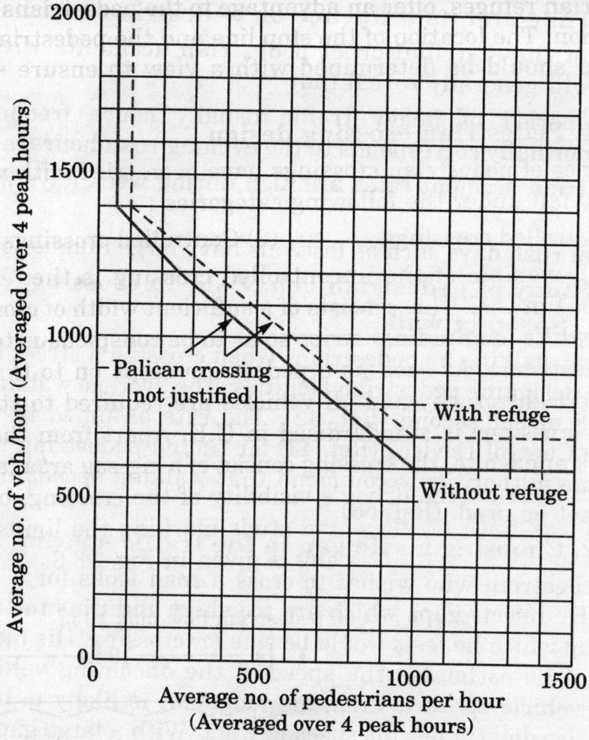

Fig. 18.9. Criteria for a PELICAN crossing in U.K. (Ref. 38)

Many signal controlled junctions provide a separate phase for pedestrians to cross, usually brought into operation by a push button installed on a signal post. A pedestrian phase of 10 seconds is commonly used (Ref. 27). The intergreen period can also be used for pedestrian crossings.

18.12.4. Road lighting and vehicle lighting

Both road lighting and vehicle lighting play an important role in ensuring that the driver observes clearly the pedestrian in the darkness. Improving the road lighting standard at pedestrian crossing should receive proper attention. The position of the lantern should be decided so that the illumination provided reveals the pedestrians crossing the street. More detailed treatment of the subject is given under Chapter 19 on Street Lighting.

If the pedestrian himself wears light coloured dresses or reflecting materials, the perceptibility to the driver is vastly improved.

18.12.5. Pedestrian footways

An adequately designed pedestrian footway goes a long way in segregating the vehicular and pedestrian traffic and affords a sense of safety and protection to the pedestrians. For being effective, the footways should be raised above the road level and must have a surface which will attract the pedestrians to use it. Encroachments by hawkers and vendors must be prohibited. The minimum width of a footway should be 1.5 m (Ref. 41). The capacity of footways is given in Table 11.39.

18.12.6. Time and its influence on pedestrian accidents

A study of the distribution of pedestrian accidents with reference to different times will generally reveal that :

(*i*) Dark period of heavy traffic usually cause frequent pedestrian accidents. This normally corresponds to the evening rush hour traffic in winter.

(*ii*) Pedestrian accident rates are high during week days when the traffic flow is heavy.

(*iii*) Special rush days such as festivals have large number of accidents.

18.12.7. Study of pedestrian behaviour in crossing

18.12.7.1. Speed of walk

The rate of walking by pedestrians when crossing a street is an important consideration in designing pedestrian facilities. The average walking speeds range from 0.75 to 1.8 m per second (Ref. 42). The rate assumed in the Manual on Uniform Traffic Control Devices (Ref. 43) for timing pedestrian signals is 1.2 m per second. Some authorities recommend that walking speeds below 0.9 m per second should not be used. (Ref. 58).

18.12.7.2. Choosing a safe gap in the traffic stream

Every pedestrian who wishes to cross a road looks for a safe gap in the traffic stream. He rejects gaps which are too short and thus too risky and waits for a suitable gap which he feels would be safe for crossing. His judgment is likely to be justified if he estimates the speed of the oncoming vehicle accurately. Usually, if the vehicle speed is high, his judgment is likely to be falsified. The gap he accepts also determines his speed of walk. With a large gap, say 7 seconds, the pedestrian can cross a 6 m carriageway safely (Ref. 2). If the gap is less than 4 seconds, he needs to hurry up. For a 2 seconds gap, he has to virtually run to avert being run over.

18.12.8. Pedestrian subways

Pedestrian subways are provided where large connection of pedestrian flow occurs. The subways should be attractive, well lit and maintained. A gentle gradient of say 1 in 10 or flatter, should be provided for the ramps. The width of the subways should be sufficient to meet the peak pedestrian flows.

18.12.9. Measures to increase pedestrian safety

The earlier discussion on the causes of pedestrian accidents has already provided some of the answers to the pedestrian safety question. These and other safety measures are summarised below :

1. Segregation of pedestrians and vehicles by the provision of raised footways decreases the pedestrian accidents.

2. Provision of railing barriers on busy roads to channelise the pedestrian flow in safe direction and to prevent pedestrians jay walking has yielded good results in many Indian cities. (Ref. 44).

3. Special attention should be paid to mark the crossings. Black and white stripes are known to be particularly satisfactory. Proper signing will also help. Near schools, warning signs of children would be needed.

4. Provision of dual carriageways, wherever feasible, should be thought of.

5. Pedestrian refuges should be provided at junctions.

6. Where pedestrian traffic is very heavy at intersections, separate pedestrian phases should be provided at signals.

7. Pedestrian actuated signals of the PELICAN type have proved successful.

8. Zebra crossings can bring down the accident rate and are recommended for locations where the vehicular traffic is not heavy.

9. Overbridges and subways for pedestrians should be thought of where the pedestrian traffic is concentrated and heavy, especially near busy intersections and generators of pedestrian traffic such as railways stations and bus stands.

10. Complete protection to the pedestrian can be afforded by converting some of the busy streets in the shopping centres to Pedestrian Precincts. Encouraging results by this measure have been experienced in many cities in the world. (Ref. 45)

11. If one way streets are being planned as part of traffic management, an incidental advantage is the reduction in the accident rate.

12. Adequate street lighting and vehicle lighting measures ensure pedestrian safety. Special attention needs to be paid for lighting pedestrian cross-walks and junctions. Increased visibility of the pedestrians themselves can prevent many accidents.

13. Children near schools need special patrol or guides for crossing the streets.

14. Establishment of play-grounds for children to play results in fewer accidents in the streets.

15. A legislative measure to force pedestrians to walk facing the on-coming traffic can prevent many an accident.

16. Road safety education, especially for children, propaganda and enforcement of traffic rules and regulations deserve careful consideration.

18.13. Cyclists

Cycles are a major mode of traffic in Indian towns and cities and even in the rural areas. Their number will continue to grow with rapid urbanisation and rural prosperity. In the developed world too, cycles have suddenly become popular in recent times, mostly because of renewed interest in the environmental movement. Cycle consumes no energy and occupies very much less space than an automobile. It makes no noise and emits no pollutants.

An unhappy feature of the cycle is that a large number of accidents are associated with it. In Delhi, 14 per cent of the road fatalities are to cyclists (1994 (Ref. 59). This is primarily because the vehicle itself offers no protection to the rider in the event of a collision with a motor vehicle.

The types of accidents usually are collision and falls. The latter, though more common, are less serious. The usual causes for cycle accidents are :

- Making improper turns.
- Disregarding traffic signals, signs and markings.
- Double-riding.
- Improper maintenance of roads (pot-holes etc.)

The ideal solution for preventing cycle accidents is to provide separate cycle tracks. The design criteria and guidelines for the provision of separate cycle tracks are given elsewhere.

It is desirable that cycle traffic be governed by rules intended to promote safety. Some of the common rules adopted by many city authorities are :

1. Not more than two cycles shall remain abreast except at separate cycle tracks.

2. Cyclists should not use foot-ways.

3. Cyclists should compulsorily use separate cycle tracks wherever they are provided.

4. Cyclists should not allow themselves to be towed by any other vehicle.

5. Double-riding shall not be allowed.

6. Where no separate cycle tracks are provided, the cyclists shall keep to the extreme left of the carriageway.

7. All cycles should be provided with good brakes, night lamps and red reflector at the rear.

18.14. Motorcycle and Scooter Riders

The number of motorcycle and scooters on the streets of Indian towns and cities is increasing enormously in the past years. This trend is likely to be repeated in many developing countries, where few can afford the luxury of a car, with the steep increase in the price of petrol. The rider of a motor cycle or a scooter is very much vulnerable to accidents because he is totally unprotected.

A study in Delhi has estimated that the chances of fatality is fourteen times as great for motor cycle and scooter riders as for car drivers. (Ref. 28). In 1994, 27 per cent of the road fatalities were in respect of two-wheeler riders.

Another cause for concern is that the two wheeler accident victims generally succumb to head injuries. This points to the usefulness of crash helmets for these riders.

Other regulatory measures aimed at safety of motor-cycle and scooter riders are :

(*i*) Compulsory provision of rear view mirrors.

(*ii*) Compulsory provision for a screen to cover the rear wheel so that loose garments (especially dhoties and sarees worn by Indians) do not get entangled in the spokes of the wheels.

(*iii*) Restricting the number of riders to the driver and one pillion rider.

(*iv*) Prohibiting a rider with a learner's licence to carry a pillion rider.

(*v*) Compulsory provision of a proper foot-rest for the pillion rider.

(*vi*) Restricting the speed of vehicles to a lower value when the rider is not wearing a crash helmet.

18.15. Parking and its Influence on Accidents

Parking of vehicles is one of the major contributors to accidents. Accidents can occur in the following ways :

1. Parking and unparking manoeuvres can cause accidents to pedestrians, cyclists or to other motorists.

2. Opening of car doors on parking can suddenly cause a cyclist or a motorist to be involved in an accident.

3. Pedestrians can appear from between parked vehicles or in front of a parked vehicle, unnoticed by a speeding vehicle.

4. Parking generally reduces the street space and increases the congestion, thereby indirectly causing accidents.

There is no doubt that a suitable policy of regulating on-street parking will have a salutary effect on the accident situation. Regulation of parking is considered in detail in Chapter 17.

18.16. Traffic Management Measures and Their Influence in Accident Prevention

Some of the traffic management measures such as one-way streets and prohibited turns can minimise the accident causation considerably. A detailed account of these measures is reserved for Chapter 20.

One way streets help to reduce accidents in the following ways :

1. The intersection conflict points are reduced-those between vehicles and other vehicles and those between vehicles and pedestrians.
2. Head-on and side-swipe collisions are eliminated.
3. Head light glare is eliminated.
4. Parking manoeuvres tend to be less dangerous and obstructive when all vehicles are facing in the same direction.

When left-turns are prohibited, a safer crossing for the pedestrians results.

Right-turns involve the greatest degree of hazard and conflict to traffic, apart from causing reduction in junction capacity. Prohibition of right-turns can, therefore, result in a reduction in the hazard.

18.17. Legislation, Enforcement, Education and Propaganda

18.17.1. It has been said that the three 'E's of road safety programme are : Engineering, Education and Enforcement.

Legislation and enforcement have brought in significant reductions in the accident rate.

18.17.2. Legislative measures that are possible

A variety of legislative measures are possible and different countries have adopted different measures. Some of them are listed below :

1. To stipulate age limits for drivers.
2. To introduce penalties of fine, imprisonment, disqualification or endorsements on licences for careless driving.
3. To enable police to check the drivers for their drunkenness and to impose suitable penalties.
4. To prescribe maximum hours of work for drivers of commercial vehicles and buses to prevent them from fatigue.
5. To prescribe uniform road signs throughout the country and provide for penalties for the non-observance of the same.
6. To lay down rules for pedestrians when crossing streets and to impose penalties for their non-observance.
7. To prescribe rules for cyclists.
8. To prescribe rules for motor cycle and scooter riders.
9. To prescribe rules for the maximum size and weight (axle loads) of vehicles.
10. To prescribe minimum standards for the design of vehicles.
11. To prescribe minimum standards for the inspection and maintenance of vehicles.

12. To prescribe rules for the regulation of mixed traffic.
13. To lay down rules for parking of vehicles.
14. To control the loading and unloading of goods by the curbside.
15. To prescribe speed limits and to provide for punishments for their infringements.
16. To make third party insurance compulsory.

18.17.3. Enforcement

Legislation by itself cannot achieve its objective unless it is enforced rigidly. The laws are flouted by many motorists and other road users, and unless cases of such offences are detected and the offenders brought to book, matters cannot be improved. The enforcement is in the hands of the Police Department in co-operation with the magistrates or special traffic courts. The police force should be adequately strengthened to deal with detection. In U.K., Traffic Wardens supplement the work of the police. The work of the police is rendered easy in certain cases of detection of offences, such as exceeding the speed limit, by mechanical aids. The radar speed measuring instrument can record the speed instantaneously from an inconspicuous position by the roadside. Breath analysers can detect the drunken driver. Police patrols in vehicles equipped with wireless are of great help.

18.17.4. Education

Road safety education occupies an important place in the prevention of accidents. School children who are particularly liable to accidents, can be easily imparted the necessary training in the school about the rules of the road and related safety aspects. Since the attitudes and personality of the children are in their formative stage, a lasting impression can be left in their minds about the precautions. The society can reap the benefits when the children grow up and can expect them to be law-abiding citizens in observing the traffic laws.

Training of professional drivers of commercial vehicles and buses and personnel of the armed forces can be of great value. Refresher courses for drivers can serve to educate them on the needs of traffic safety and how to use their motor vehicles in a safe and efficient manner.

In the U.S.A., high-school driver education is an integral part of the curriculum for students approaching the legal driving age. Under this programme, the potential drivers at a young age are imparted necessary instructions to learn fundamental driving skills and to achieve desirable traffic safety pattern.

18.17.5. Propaganda

Propaganda and exhortation are powerful tools in a road safety campaign. The Slough Experiment in U.K has proved the value of these. (Ref. 47) Propaganda and exhortation can be achieved in many possible ways. Some of those with proven value are :

1. Poster campaigns
2. Campaigns in local press
3. Short films exhibited on the television
4. Broadcasts in the radio
5. Use of imaginative and catching slogans.

18.18. Cost of Road Accidents

In any exercise involving the economic appraisal of highway, traffic or transportation project, it becomes necessary to assess the cost of road accidents. A detailed consideration of this subject is given in Chapter 41.

18.19. Road Safety Audit

One of the measures employed for preventing or reducing road accidents is to subject an existing road or a future road project to a formal examination, known as Road Safety Audit (RSA). The objective of RSA is to identify any aspects of a highway design that give rise to road safety concerns, and to suggest amendments in the design that would enhance safety. In U.K., (Ref. 61), RSA is required to be carried out at the following stages :

(*i*) on completion of preliminary Design
(*ii*) on completion of detailed Design
(*iii*) on completion of construction, *i.e.*, pre-opening.

The Auditors are independent and qualified experts, not connected with the project, and address the problem areas using the experience gained from accident reduction, schemes, accident investigation and research. RSA is also carried out on existing roads, when it becomes an accident reduction exercise, it is then also referred as Road Safety Audit Review (RSAR).

In India, recently some RSA studies have been completed on National Highways. They contain recommendations to improve road signages, road marking, highway lighting, delineators, intersection designs, and to eliminate roadside hazards like trees, hoardings etc. (Ref. 62).

18.20. Traffic Calming

Traffic calming is a set of engineering measures to reduce speeds and volumes of motor vehicles in local areas, the thereby increase safety.

Some of the measures include :

- Speed breakers or humps (see section 16.6)
- Rumble strips (see section 16.7)
- Crash barriers (see section 16.11)
- Road narrowing or chicanes (see section 20.3.7)
- Provision of separate cycle tracks (see section 11.15)

APPENDIX 18.1

Road Accident Form A-l (*Source* : Ref. 6)

1. Identification Particulars

(*i*) State (*ii*) District/Taluk
(*iii*) City/Town/Village (*iv*) Police zone/Station
(*v*) Accident number allotted by
Police Station
FIR Number
MVA accident number

2. Location

(*i*) Name of road (*ii*) Classification of road
1. National Highway 2. State Highway
3. Other Road

(*iii*) Location of area

1. Urban 2. Rural

(*iv*) If intersection, specify the names of roads

(*v*) Details of location (the details under this should be such as to give precise location)

(*vi*) Type of area

1. Near school or college 2. Near or inside a village
3. Near a factory/industrial area
4. Near a religious place
5. Near a recreation place/cinema
6. In bazar 7. Near office complex
8. Near hospital 9. Residential area
10. Open area 11. Near bus stop
12. Near petrol pump 13. At pedestrian crossing
14. Affected by encroachments

(*vii*) Narrow bridge or culverts

1. Yes 2. No

3. Date, Day and Time

(*i*) Date Month Year

(*ii*) Day of the week

(*iii*) Time A.M. P.M.

(*iv*) Holiday 1. Yes 2. No

4. Light Conditions

1. Daylight 2. Twilight
3. Dark hours with good street light
4. Dark hours with poor street light
5. Dark hours with no street light

5. Weather Conditions

1. Fine 2. Mist/Fog 3. Cloudy
4. Light rain 5. Hail/sleet 6. Snow
7. Strong wind 8. Dust storm 9. Very hot
10. Very cold
11. Other extraordinary weather conditions (specify).

6. Classification of Accident

1. Fatal 2. Grievous injury
3. Minor injury 4. Non-injury.

7. Type of Vehicles with Registration Numbers and Objects Involved

7.1. Vehicles involved

	Type of vehicle	*Registration number*	*Year or manufacture*
Vehicle 1	...	...	...
Vehicle 2	...	...	...
Vehicle 3	...	...	...

7.2. Pedestrian, Animal and other objects involved (specify type)

S. No.

1 ....

2 ....

3 ....

8. Nature of the Accident

8.1. Type

1. Overturning	2. Head on collision
3. Rear end collision	4. Collision brush/side swipe
5. Right angled collision	6. Skidding
7. Right turn collision	8. Others (describe)

8.2. Hit and Run 1. Yes 2. No

9. Details of drivers of vehicles involved

9.1. Name, Sex, Age, Education and Address of the Driver(s)

	Name of Driver	*Sex M/F*	*Age*	*Highest class passed*	*Address*
Vehicle 1	–	–	–	–	–
Vehicle 2	–	–	–	–	–
Vehicle 3	–	–	–	–	–

9.2. Person driving the vehicle

	Veh. 1	Veh. 2	Veh. 3
1. Owner of private vehicle			
2. Owner of public/commercial vehicle			
3. Paid driver			
4. Other			

9.3. Type of licence

	Veh. 1	Veh. 2	Veh. 3
1. Regular licence			
2. Learner's licence			
3. Without appropriate licence			

9.4. Licence Number and Other Details

	Veh. 1	Veh. 2	Veh. 3
1. Licence number			
2. Date of issue			
3. Date of expiry			
4. Place of issue			

9.5. Type of Manoeuvre

	Veh. 1	Veh. 2	Veh. 3
1. Diverging			
2. Merging			
3. Crossing			
4. Stationary			
5. Temporarily held up			
6. Parked			
7. Stopping			
8. Starting from near side			
9. Starting from off side			
10. Turning right			
11. Turning left			
12. Making U-turn			
13. Going ahead, overtaking			
14. Going ahead, not overtaking			
15. Using private entrance			
16. Reversing			

9.6. Responsibility of Driver

	Veh. 1	Veh. 2	Veh. 3
1. Consumption of alcohol or drugs			
2. Exceeded lawful speed			
3. Did not give right of way to vehicle			
4. Did not give right of way to pedestrian			
5. Followed too closely			
6. Overtaking on hill			
7. Overtaking on curve			
8. Cut in sharply after overtaking			
9. Other improper overtaking			
10. On wrong side of road			
11. Failed to give signal			
12. Gave improper signal			
13. Improper turn			
14. Disregarded Police Officer			
15. Disregarded traffic light signal			
16. Disregarded stop sign			
17. Improper starting from parked position			
18. Wrong/improper parking			
19. Asleep or fatigued or sick			
20. Inattentive or attention diverted at the moment			
21. Improper use of head-lights causing glare			
22. Other improper actions (describe)			

10. Particulars of Vehicle Involved

10.1. Particulars of vehicles

	Veh. 1	Veh. 2	Veh. 3
(*a*) Load :			
(*i*) Overloaded/Overcrowded			
(*ii*) Load protruding			
(*b*) Left hand drive			
(*c*) Vehicular defect :			
(*i*) Defective brakes			
(*ii*) Defective steering			
(*iii*) Punctured or burst tyres			
(*iv*) bald tyre			
(*v*) Other serious mechanical defect (describe)			

10.2. Certificate of fitness in the case of commercial vehicles

	Veh. 1	Veh. 2	Veh. 3
1. In force			
2. Not in force (expired)			

10.3. Particulars of Insurance

	Veh. 1	Veh. 2	Veh. 3
(*a*) Name of insurance company			
(*b*) Number of insurance			
(*c*) Type of insurance :			
1. Comprehensive			
2. Third party 3. Not insured			

11. Details of Pedestrian or Person other than the Driver involved in Accident

11.1. Age of Person/Pedestrian

S. No.	*Male/Female*	*Age*
1	...	...
2	...	...
3	...	...
4		

11.2. Particulars of Person/Pedestrian

1. Intoxicated or drugged
2. School child
3. Pedestrian deaf, blind or otherwise infirm
4. Crossing at intersection with signal
5. Crossing at intersection-against signal

6. Crossing at intersection-diagonally
7. Crossing at pedestrian crossing
8. Crossing not at intersection/pedestrian crossing
9. Coming from behind parked vehicle
10. Walking on road–with traffic, side-walk not available
11. Walking on road–with traffic, side-walk available
12. Walking on road–against traffic, side-walk not available
13. Walking on road–against traffic, side-walk available
14. Pushing or working on vehicle
15. Others working on road
16. Playing on road
17. Hanging on the vehicle
18. Sleeping or squatting on the road
19. Dismounting or mounting moving vehicle
20. Dismounting or mounting stationary vehicle
21. Not on road (explain)
22. Other actions (specify)

12. Details of Cyclists involved

1. Double riding
2. Overloading
3. Not keeping to the left
4. Without light at night
5. Going on carriageway-cycle track available
6. Cycling in the lane of fast moving vehicle
7. Cutting in the flow of traffic or zigzag moving
8. Turning right carelessly/without giving signal
9. Towing himself with other vehicle
10. Not observing traffic rules
11. Confused by traffic
12. Rider inexperienced
13. Loss of control
14. Skidding on wet/slippery road
15. Others (specify)

13. Type of Persons and Animals Killed or Injured

13.1. Type and Number of Persons Killed or Injured

Type	*No. Killed*		*No. Injured*			
			Grievously		*Slightly*	
	M.	*F.*	*M.*	*F.*	*M.*	*F.*
1. Pedestrian						
2. Bicycles						
(*i*) Drivers						
(*ii*) Passengers						

(Contd...)

3. Motor cycles
   (*i*) Drivers
   (*ii*) Passengers
4. Scooters
   (*i*) Drivers
   (*ii*) Passengers
5. Mopeds
   (*i*) Drivers
   (*ii*) Passengers
6. Autorickshaws
   (*i*) Drivers
   (*ii*) Passengers
7. Cars, taxis, vans and other light and medium motor vehicles
   (*i*) Drivers
   (*ii*) Passengers
8. Trucks
   (*i*) Drivers
   (*ii*) Passengers
9. Buses
   (*i*) Drivers
   (*ii*) Passengers
10. Other motor vehicles
   (*i*) Drivers
   (*ii*) Passengers
11. Animal driven vehicles
   (*i*) Drivers
   (*ii*) Passengers
12. Cycle rickshaws
   (*i*) Drivers
   (*ii*) Passengers
13. Hand carts and rickshaws
   (*i*) Pullers
   (*ii*) Passengers
14. Other persons

13.2. Number of animals killed/injured

Killed ......

Injured ......

14. Type of Damage to Vehicles and Property

Vehicle/Property	*Type of damage (specify)*
1. Vehicle 1	
2. Vehicle 2	
3. Vehicle 3	
4. Other property	

15. Road Condition

(*i*) *Horizontal features of the road*

1. Straight road 2. Slight curve 3. Sharp curve

(*ii*) *Vertical features of the road*

1. Flat road 2. Gentle incline 3. Steep incline
4. Hump 5. Dip

15.2. Type of surface

1. Surfaced (black topped/concrete)
2. Metalled 3. Kutcha

15.3. Condition of surface

1. Dry 2. Wet

15.4. Nature of surface

1. Good surface 2. Loose surface
3. Rutted/and or potholed 4. Road under repair/construction
5. Corrugated or wavy road 6. Slippery surface
7. Snowy 8. Muddy
9. Oily 10. Speed breaker
11. Others (specify)

16. Road Features

(*i*) Carriageway

1. Single lane 2. Two lanes
3. Three lanes or more without central divider (median)
4. Four lanes or more with central divider

(*ii*) Cycle track provided 1. Yes 2. No
(*iii*) Foot path provided 1. Yes 2. No
(*iu*) Pucca shoulder provided 1. Yes 2. No

17. Intersection Type and Control

17.1. Type of junction (Refer sec. 21 for figures)

1. T-junction 2. Y-junction
3. Four arm junction 4. Staggered junction
5. Junction with more than 4 arms

6. Roundabout junction
7. Manned rail crossing
8. Unmanned rail crossing

17.2. Type of traffic control

1. Traffic light signal
2. Police controlled
3. Stop sign
4. Flashing signal/blinker
5. Uncontrolled

18. Traffic Regulations

1. One-way street
2. Entry of heavy vehicles prohibited
3. Entry of slow moving vehicles prohibited
4. Speed restrictions
5. Parking prohibited
6. Any other (specify)

19. Names of Persons (with User's Classification such as Pedestrian Driver, etc.)

Name	User classification	Occupation	Monthly income (₹)
	Killed		
	Grievously injured		
	Slightly injured		

20. Cause of Accident

1. Fault of driver of motor vehicle
2. Fault of cyclist
3. Fault of driver of other vehicle
4. Fault of pedestrian
5. Fault of passenger
6. Defect in mechanical condition of motor vehicle
7. Poor light condition (including street light)
8. Defect in road condition
9. Result of weather conditions
10. Stray animal
11. Other causes (specify)

21. Diagrammatic sketch of accident site and brief description of the accident

22. Result of investigations and Court Proceedings

APPENDIX 18.2

(*Source* : Ref. 7)

Road Accident Form-4

Statement containing particulars of road accidents in the State of.......for the year ending December.........

1. Total number of accidents classified according to month of the year

Month	*Accidents*				
	Fatal (F)	*Grievously injured (GI)*	*Minor injured (MI)*	*Non-injury (NI)*	*Total*
1. January					
2. February					
3. March					
4. April					
5. May					
6. June					
7. July					
8. August					
9. September					
10. October					
11. November					
12. December					
Total					

17.2. Accidents classified according to time, area and type

Time	*Number of accidents*									
	URBAN					*RURAL*				
	F	*GI*	*MI*	*NI*	*Total*	*F*	*GI*	*MI*	*NI*	*Total*
0600–0700										
0700–0800										
0800–0900										
0900–1000										
1000–1100										
1100–1200										
1200–1300										
1300–1400										

Time	Number of accidents									
	URBAN					RURAL				
	F	GI	MI	NI	Total	F	GI	MI	NI	Total
1400–1500										
1500–1600										
1600–1700										
1700–1800										
1800–1900										
2000–2100										
2100–2200										
2200–2300										
2300–2400										
2400–0100										
0100–0200										
0200–0300										
0300–0400										
0400–0500										
0500–0600										

17.3. Accidents Classified According to Classification of Road

Classification of Road	Number of accidents				
	Fatal	Grievous injury	Minor injury	Non-injury	Total
1. National Highway					
2. State Highway					
3. Other Road					

4. Accidents Classified According to Location

Location	Number of accidents				
	Fatal	Grievous injury	Minor injury	Non-injury	Total
1. Near school or college					
2. Near or inside a village					
3. Near a factory/ industrial area					
4. Near religious place					
5. Near a recreation place/cinema					

6. In bazar					
7. Near office complex					
8. Near hospital					
9. Residential area					
10. Open area					
11. Near bus stop					
12. Near petrol pump					
13. At pedestrian crossing					
14. Affected by encroachments					
15. Narrow bridge or culvert					

5. Accidents Classified According to Weather conditions

Weather condition	*Number of accidents*				
	Fatal	*Grievous injury*	*Minor injury*	*Non-injury*	*Total*
1. Fine					
2. Mist/Fog					
3. Cloudy					
4. Light rain					
5. Heavy rain					
6. Hail/sleet					
7. Snow					
8. Strong wind					
9. Dust storm					
10. Very hot					
11. Very cold					
12. Other extra ordinary weather conditions					

6. Accidents Classified According to Type of Vehicles and Objects involved

Type of Vehicle	*Number of accidents*				
	Fatal	*Grievous injury*	*Minor injury*	*Non-injury*	*Total*
(A) Motor Vehicles :					
1. Motor cycle/scooter					
2. Moped					
3. Autorickshaw					
4. Motor car					
5. Jeep					
6. Taxi					
7. Bus					
8. Truck					
9. Tempo					

10. Articulated vehicle					
11. Tractor					
12. Other motor vehicles					
(B) Other Vehicles and Objects :					
1. Cycle					
2. Cycle rickshaw					
3. Hand drawn vehicle					
4. Animal drawn vehicle					
5. Pedestrian					
6. Animal					
7. Tree					
8. Level crossing					
9. Other fixed objects					

7. Accidents Classified According to Age of Vehicle

Age of vehicle (in year)	*Number of accidents*				
	Fatal	*Grievous injury*	*Minor injury*	*Non-injury*	*Total*
Less than 1 year					
1–2					
2–4					
4–6					
6–8					
8–10					
10 and above					

8. Accidents Classified According to Nature of Accident

Nature of accidents	*Number of accidents*				
	Fatal	*Grievous injury*	*Minor injury*	*Non-injury*	*Total*
1. Overturning					
2. Head on collision					
3. Rear end collision					
4. Collision brush/ side swipe					
5. Right angled collision					
6. Skidding					
7. Right turn collision					
8. Hit and run					
9. Others					

9. Accidents Classified According to Cause

Cause of accident	*Number of accidents*				
	Fatal	*Grievous injury*	*Minor injury*	*Non-injury*	*Total*
1. Fault of driver of motor vehicle					
2. Fault of cyclist					
3. Fault of driver of other vehicle					
4. Fault of pedestrian					
5. Fault of passenger					
6. Defect in mechanical condition of motor vehicle					
7. Poor light condition					
8. Defect in road condition					
9. Result of weather conditions					
10. Other causes					
11. Cause not known					

10. Accidents Classified According to Details of Drivers of Vehicles involved

	Number of accidents				
	Fatal	*Grievous injury*	*Minor injury*	*Non-injury*	*Total*
(A) Sex					
Male					
Female					
(B) Age					
1. Under 18 years.					
2. 18–21 yrs.					
3. 21–30 yrs.					
4. 31–40 yrs.					
5. 41–50 yrs.					
6. 51–60 yrs.					
7. 61–70 yrs.					
8. 71 and above					
(C) Educational Qualification of Drivers Involved					
Upto Standard 8					
Standard 8–10					
Standard 10 and above					
(D) Persons Driving the Vehicle					

1. Owner of private vehicle					
2. Owner of public/ commercial vehicle					
3. Paid driver					
4. Other					
(E) Type of Licence					
1. Regular licence					
2. Learner's licence					
3. Without appropriate licence					

11. Accidents Classified According to Type of Manoeuvre

Type of manoeuvre	*Number of accidents*				
	Fatal	*Grievous injury*	*Minor injury*	*Non-injury*	*Total*
1. Diverging					
2. Merging					
3. Crossing					
4. Stationary					
5. Temporarily held up					
6. Parked					
7. Stopping					
8. Starting from near side					
9. Starting from off side					
10. Turning right					
11. Turning left					
12. Making U-turn					
13. Going ahead overtaking					
14. Going ahead, not overtaking					
15. Using private entrance					
16. Reversing					

12. Accidents Classified According to Responsibility of Driver

Fault of driver	*Number of accidents*				
	Fatal	*Grievous injury*	*Minor injury*	*Non-injury*	*Total*
1. Consumption of alcohol or drug					
2. Exceeded lawful speed					
3. Did not give right of way to vehicle					
4. Did not give right of way to pedestrian					
5. Followed too closely					
6. Passed on hill					
7. Passed on curve					
8. Cut in sharply after passing					

Fault of driver	*Number of accidents*				
	Fatal	*Grievous injury*	*Minor injury*	*Non-injury*	*Total*
9. Other improper passing					
10. On wrong side of the road					
11. Failed to give signal					
12. Gave improper signal					
13. Improper turn					
14. Disregarded Police Officer					
15. Disregarded traffic light signal					
16. Disregarded stop sign					
17. Improper starting from parked position					
18. Wrong or improper parking location					
19. Asleep or fatigued or sick					
20. Inattentive or attention diverted at the moment					
21. Improper use of head-lights					
22. Other improper actions					

13. Accidents Classified According to Particulars of Vehicle involved

Particulars	*Number of accidents*				
	Fatal	*Grievous injury*	*Minor injury*	*Non-injury*	*Total*
(A) Load					
(*i*) Overloaded/ overcrowded					
(*ii*) Load protruding					
(B) Left Hand Drive					
(C) Vehicular Defect					
(*i*) Defective brakes					
(*ii*) Defective streering					
(*iii*) Punctured or burst tyres					
(*iv*) Bald tyre					
(*v*) Other serious mech-anical defect					
(D) Certificate of Fitnees in the case of Commercial Vehicles					
(*i*) In force					
(*ii*) Not in force (expired)					

14. Accidents Classified According to Age of Persons other than Drivers involved in Accidents

Age	Number of accidents				
	Fatal	*Grievous injury*	*Minor injury*	*Non-injury*	*Total*
Less than 5 years					
5–10 years					
11–15 years					
16–20 years					
21–25 years					
26–40 years					

15. Accidents Classified According to type of Persons and Animals Killed or Injured

	Number of persons									
	Male					Female				
	F	*GI*	*MI*	*NI*	*Total*	*F*	*GI*	*MI*	*NI*	*Total*
	(*1*)	(*2*)	(*3*)	(*4*)	(*5*)	(*6*)	(*7*)	(*8*)	(*9*)	(*10*)
(A) Type of Person killed or Injured										
1. Pedestrian										
2. Bicycles										
(*i*) Drivers										
(*ii*) Passengers										
3. Motor cycles										
(*i*) Drivers										
(*ii*) Passengers										
4. Scooters										
(*i*) Drivers										
(*ii*) Passengers										
5. Mopeds										
(*i*) Drivers										
(*ii*) Passengers										
6. Autorickshaws										
(*i*) Drivers										
(*ii*) Passengers										
7. Cars, taxis, vans and other light and medium motor vehicles										
(*i*) Drivers										
(*ii*) Passengers										

	Number of persons									
	Male					*Female*				
	F	*GI*	*MI*	*NI*	*Total*	*F*	*GI*	*MI*	*NI*	*Total*
	(*1*)	(*2*)	(*3*)	(*4*)	(*5*)	(*6*)	(*7*)	(*8*)	(*9*)	(*10*)
8. Trucks										
(*i*) Drivers										
(*ii*) Passengers										
9. Buses										
(*i*) Drivers										
(*ii*) Passengers										
10. Other motor vehicles										
(*i*) Drivers										
(*ii*) Passengers										
11. Animal drawn vehicles										
(*i*) Drivers										
(*ii*) Passengers										
12. Cycle rickshaws										
(*i*) Drivers										
(*ii*) Passengers										
13. Hand carts and rickshaws										
(*i*) Pullers										
(*ii*) Passengers										
14. Other persons										
(B) Animals	*Killed*					*Injured*		*Total*		
Number of animals										

16. Accidents Classified According to Road Condition

Road condition	*Number of accidents*				
	Fatal	*Grievous injury*	*Minor injury*	*Non-injury*	*Total*
(A) Type of Surface					
1. Surfaced					
2. Metalled					
3. Kutcha					
(B) Condition of Surface					
1. Dry					
2. Wet					

Road condition	*Number of accidents*				
	Fatal	*Grievous injury*	*Minor injury*	*Non-injury*	*Total*
(C) **Nature of Surface**					
1. Good surface					
2. Loose surface					
3. Rutted and/or pot hold					
4. Road under repair construction					
5. Corrugated or wavy road					
6. Slippery surface					
7. Snowy					
8. Muddy					
9. Oily					
10. Speed breaker					
11. Others					
(D) **Horizontal Features**					
1. Straight road					
2. Slight curve					
3. Sharp curve					
(E) **Verticle Features**					
1. Flat road					
2. Gentle incline					
3. Steep incline					
4. Hump					
5. Dip					

17. Accidents Classified According to Road Features

Carriageway width	*Number of accidents*				
	Fatal	*Grievous injury*	*Minor injury*	*Non-injury*	*Total*
1. Single lane					
2. Two lanes					
3. Three lanes or more without central divider (median)					
4. Four lanes or more with central divider					

18. Accidents classified According to Junction Type and Traffic Control

Carriageway width	*Number of accidents*				
	Fatal	*Grievous injury*	*Minor injury*	*Non-injury*	*Total*
(A) Type of Junction					
1. T-junction					
2. Y-junction					
3. Four arm junction					
4. Staggered junction					
5. Junction with more than four arms					
6. Rounda bout junction					
7. Manned rail crossing					
8. Unmanned rail crossing					
(B) Type of Traffic Control					
1. Traffic light signal					
2. Police controlled					
3. Stop sign					
4. Flashing signal/ blinker					
5. Uncontrolled					

19. Accidents Classified Accroding to Type and Number of Vehicles and Persons Involved

Type of accident	*Number of accidents*	*Number of vehicles involved*	*Number of persons involved*
Fatal			
Grievous			
Minor injury			
Non-injury			
Total			

I. MULTIPLE CHOICE QUESTIONS

1. In children ages 10-14, the leading cause of hospitalization is:
   (*a*) mental disorders (*b*) respiratory illnesses
   (*c*) injury (*d*) car accidents
2. You have been asked to write a handout that offers safety guidelines for staff supervising young children on the playground. Which of these statements includes information that you should include?
   (*a*) the average temperature for each month
   (*b*) developmental physical milestones for each grade level

(*c*) suggestions for several outdoor games
(*d*) a brief checklist for inspecting playground equipment

3. Your local parent group has asked you to present on accident prevention. Which of these topics is an important one for parents of elementary school children?
(*a*) common flu symptoms and safeguards
(*b*) safety precautions for pedestrians
(*c*) food allergies
(*d*) common allergy medications and risks associated with them

4. Which of these statements is true?
(*a*) children are at greatest risk when they are in a school swimming pool
(*b*) children are at greatest risk when they are playing a contact sport
(*c*) children are at greater risk travelling to and from school than at any other time in their school day
(*d*) children are at greatest risk at night

5. Your principal has invited you to offer training on the prevention of asthma-related emergencies in your school. Which of these symptoms is related to asthma?
(*a*) lips or fingernails are grey or blue
(*b*) coughs constantly
(*c*) trouble talking
(*d*) all of the above

6. Which of these notions is a myth about suicide?
(*a*) using alcohol can increase one's risk of suicidal behaviour
(*b*) suicidal behaviour is contagious
(*c*) teens may write about suicidal behaviour in papers
(*d*) teens who are suicidal usually do not tell others about their feelings

7. Which of these could be considered risky behaviour?
(*a*) choking game (*b*) huffing
(*c*) chubby bunny (*d*) all of the above

8. Which of the following statements is true?
(*a*) the choking game is played primarily by children with cognitive disabilities.
(*b*) a suicide attempters' lunch bunch support group is a helpful way for teachers to support students.
(*c*) asthma is one of the most common chronic childhood diseases
(*d*) children are safer outside of school than inside school

9. The school health Index is:
(*a*) an excellent resource but may be too expensive for some schools
(*b*) an annual report from the federal government that describes hospitalization rates
(*c*) a report on illnesses and accidents published by each state department of education
(*d*) a free assessment and planning tool

10. Which of the following statements is not true?

(*a*) the school health index is an assessment and planning tool, so it does not incorporate training.

(*b*) the school health index has eight modules for assessment and planning.

(*c*) the school health index includes training materials.

(*d*) the school health index includes information on the health of staff as well as children.

Answers

1. (*a*)	**2.** (*d*)	**3.** (*b*)	**4.** (*c*)	**5.** (*d*)
6. (*d*)	**7.** (*d*)	**8.** (*c*)	**9.** (*d*)	**10.** (*a*)

II. SOLVED QUESTIONS AND ANSWERS

Q. 1. What are the causes of road accidents?

Ans. Causes of road accidents

- Excessive speeding
- Driver's fatigue and reckless driving
- Equipment failure
- Poor roadway design and roadway maintenance
- Night driving
- Improper turning manner
- Driving on a wrong way
- Improper use of electronic gadgets while driving
- Disobedience to traffic instructions
- Impatient and tailgating
- Potholes on the highway
- Drinking before or while driving

Q. 2. Explain the preventions against the road accidents.

Ans. (*i*) Maintain regular check on general vehicle devices such as steering, lights, windshield wipers, horns, warning signals, side or rearview mirrors, restraint systems, tyre.

(*ii*) Avoid tailgating especially during heavy traffic.

(*iii*) Practice precaution when carrying passengers.

(*iv*) Avoid drinking of alcohol before and while driving.

(*v*) Always obey traffic regulation and road signs such as traffic lights etc.

(*vi*) Avoid over speeding

(*vii*) Avoid using cell phones and other technological gadgets while driving.

(*viii*) Avoid loading beyond the vehicle designed load limit.

(*ix*) Follow the established emergency precautions and constant awareness campaign against automobile accident by the safety authorities such as National Emergency Management Agency (NEMA), Federal Road Safety Corps (FRSC).

(*x*) Avoid slowing down suddenly to check your car or look at an unusual situation happening on the road or nearby.

(*xi*) Attend regular mandatory driver education and training courses, annual driver examinations to upgrade your driving skills.

(*xii*) Always obey the advocated speed limits and preventive techniques instituted by the Federal Road Safety Corps (FRSC), NEMA against car accident

Q. 3. How different factors of roads contribute in accidents?

Ans.

- *Drivers:* Over-speeding, rash driving, violation of rules, failure to understand signs, fatigue, alcohol.
- *Pedestrian:* Carelessness, illiteracy, crossing at wrong places moving on carriageway, Jaywalkers.
- *Passengers:* Projecting their body outside vehicle, by talking to drivers, alighting and boarding vehicle from wrong side travelling on footboards, catching a running bus etc.
- *Vehicles:* Failure of brakes or steering, tyre burst, insufficient headlights, overloading, projecting loads.
- *Road Conditions:* Potholes, damaged road, eroded road merging of rural roads with highways, diversions, illegal speed breakers.
- *Weather conditions:* Fog, snow, heavy rainfall, wind storms, hail storms.

Q. 4. What are preventive measures for accidents?

Ans. (*a*) Education and awareness about road safety

(*b*) Strict enforcement of law

(*c*) Engineering

(*d*) Vehicle design (*e*) Road infrastructure

Q. 5. Discuss the direct consequences of accidents:

Ans. Fatality (death)

- Injury
- Property damage

Q. 6. What are the causes of collisions?

Ans. The majority of road crashes are caused by human error. Research has shown that driver error accounts for over 80% of all fatal and injury crashes on Irish roads.

III. REVIEW QUESTIONS

1. What is skidding?
2. Discuss the speed in relation of safety.
3. Briefly explain the weather and its effect on accidents.
4. What is pedestrian safety?

REFERENCES

1. Report of the Study Group on Road Safety, Ministry of Transport, New Delhi, 1972.
2. Research on Road Safety, Road Research Laboratory, H.M.S.O., London, 1963.
3. Smeed, R.J. and G.O. Jeffcoate, The effect of changes in motorisation in various countries on the number of road facilities, Symposium on the use of statistical methods in the analysis of road accidents, Organization for Economic Co-operation and Development, Paris, 1970.
4. Jacobs, G.D. and P. Hutchinson, A study of accident rates in developing countries, TRRL Report, LR 546, Transport and Road Research Laboratory, Crowthorne, 1973.
5. Uses of Traffic Accident Records, ENO Foundation for Highway Traffic Control, Saugatuck, Connecticut, 1947.
6. Road Accident Forms A–1 and 4, The I.R.C., New Delhi, 1982.

7. J. Gustavsson, On the use of Regression Models in the study of road accidents, Symposium on the use of statistical methods in the analysis of road accidents, Organization for Economic Co-operation and Development, Paris, 1970.
8. G. Morlat and A. Ropars, Multiple Regression Applied to Growth of road accidents, *Ibid*.
9. Y. Kogane, An empirical study of motor vehicle accidents on National Routes in Japan, *Ibid*.
10. F. Garwood and R.F. Newby, Use of Chi-Squared Distribution for comparing accident frequencies, *Ibid*.
11. Road Research Laboratory, (U.K.), Research on Road Traffic, H.M.S.O., London, 1965.
12. H.K. Sinha, Methods in Accident Studies, Indian Highways, Vol. 3, No. 7, New Delhi, 1965.
13. Kadiyali, L.R., A study of the problems of single-lane pavements in India and their improvements, Paper No. 296, Journal of the I.R.C. XXX–3, New Delhi, 1973.
14. A.D. Little, Inc., The State of the Art of Traffic Safety, Paraeger publishers, New York, 1970.
15. Austin, M., Accident Black Spot, Penguin Books, Harmondsworth, 1966.
16. McFarland and others, Human body size and capabilities in the design and operation of vehicular equipment, Boston, 1953 (Harvard School of Public Health).
17. Ministry of Transport (U.K.), Road Traffic Act, 1988, H.M.S.O. London.
18. The Motor Vehicles Act, 1988 New Delhi, 1988.
19. Robert F. Baker, The Highway Risk Problem, Wiley–Interscience, 1971, New York.
20. Road Accidents, 1965, H.M.S.O., London, 1967.
21. A Policy on Geometric Design of Rural Highways, American Association of State Highway Officials, Washington, 1965.
22. Ernst, On the Effects of driver's age and experience on traffic, accidents, Symposium on the use of statistical methods in the analysis of road accidents, O.E.C.D, Paris, 1970.
23. Forbes, T.W., Human Factors in Highway Traffic Research, Wiley-Interscience, New York, 1972.
24. Cohen, J. and B. Preston, Causes and prevention of road accidents, Faber and Faber, London, 1968.
25. Goldberg, L, and J.D.J. Havard, Research on the effects of alcohol and drugs on driver behaviour, O.E.C.D., Paris.
26. Traffic Engineering Committee of the, I.R.C., Meeting held on the 5th and 6th Sept, 1975, New Delhi.
27. British Standards Institution, B.S. 2001 : 1956, Protective helmets for motor cyclists, London, 1956.
28. Safety of motorcycle and scooter riders, Editorial, Transport Communications Monthly Review, No, 254, New Delhi, 1973.
29. Pavement Slipperiness and Skid Resistance : State of Art., I.R.C. Highway Research Board, Special Report 2, New Delhi, 1976.
30. National Co-operative Highway Reseach Programme, Synthesis of Highway Practice, No. 14, "Skid Resistance", Highway Research Board, Washington, 1972.
31. Ministry of Transport, Specification for Road and Bridge Works, H.M.S.O., London, 1969.
32. British Standards Institution, B.S. 812–1967, London.
33. Report of the Committee on Highway Maintenance, (Marshall Report), H.M.S.O., London, 1970.
34. James, J.C, Calcined bauxite and other artificial polish resistant madstones, R.R.L. Report LR 84, Crowthorne, 1967.

35. How Fast ?, Ministry of Transport (U.K.), H.M.S.O. London, 1968.
36. Bull, J.P., International Comparison of Road Accident Statistics, Symposium on the use of Statistical methods in the analysis of road accidents, OECD, Paris, 1970.
37. Department of the Environment, Pedestrian Safety, H.M.S.O., London, 1973.
38. Moore, R.L., and S.J. Older, Pedestrians and motor vehicles are compatible in today's world, Traffic Engineering, 35 (12), 1965.
39. Biehl B.M., S.J. Older and D.J. Griep, Pedestrian Safety, OECD, Paris, 1970.
40. Mackie, A.M. and S.J. Older, Study of Pedestrian risk in crossing busy roads in London inner suburbs, Traffic Engineering and Control, 1965 7(6).
41. Guidelines for Pedestrian facilities, Indian Roads Congress, IRC : 103-1988, New Delhi, 1988.
42. AASTHO, A Policy on design of Urban Highways and Arterial Streets, Washington, 1973.
43. Manual on Uniform Traffic Control Devices, Federal Highway Administration, U.S. Dept. of Transportation, Washington, 1971.
44. Khanna T.S., Railing barriers on busy road intersections, Journal of the I.R.C., Vol. XXIII–1, New Delhi, 1958–59.
45. Streets for people, O.C.E.D., Paris.
46. Aaron, J.E., and M.K. Strasser, Driver and Traffic Safety Education, McMillan Company, New York, 1966.
47. Ministry of Transport and Civil Aviation, the Slough Experiment 1955–57, H.M.S.O., London, 1957.
48. Dawson, R.F.F., Current costs of road accidents in Great Britain, RRL Report LR 396, Crowthome, 1971.
49. Motor Transport Statistics, various issues, Transport Research Division, Ministry of Shipping and Transport, Govt, of India, New Delhi.
50. Road User Cost Study, Final Report, Central Road Research Institute, New Delhi, 1982.
51. Motor Transport Statistics, 1981–82 and 1982–83, Transport Research Division, Ministry of Transport, New Delhi, 1985.
52. Road Safety–A Committee Report, Ministry of Shipping and Transport New Delhi, 1983.
53. World Road Statistics, International Road Federation, Washington, 1984.
54. Babkov, V.F., Road Conditions and Traffic Safety, MIR Publishers, Moscow, 1975.
55. Sarna, A.C., *et al,* Road accidents on rural highways in India International Seminar on Road Safety, Indian Road Congress New Delhi, 1986.
56. Pocket Book on Transport in India, 1989, Transport Research Division, Ministry of Surface Transport, New Delhi, 1989.
57. Special Issue on Motor Vehicle Accident Statistics in India, 1986, Transport Research Division Ministry of Surface Transport, New Delhi, 1986.
58. Transportation, Planning Handbook, The Institute of Transportation Engineers, Prentice Hall, Englewood Clifts, 1992.
59. Report of the working group on Road Accidents, Chairman : Prakash Narain, Planning Commission, New Delhi, 2001.
60. Definition of Road Safety Policy and Development of Action Plans, Report of Study for Ministry of Road Transport and Highways and National Highways Authority of India, Span Consultants, New Delhi, 2003.
61. Road Safety Audit, H.D. 19/03, Design Manual for Roads and Bridges, U.K. (available on the internet).
62. Sarin, S.M., *et al,* Need for Road Safety Audit on Indian Roads, Vol. 60–3, Journal of the Indian Roads Congress, New Delhi, 1999.
63. MORT&H website morthnic.in/motorstat/mt3.pdf.
64. Handbook on Concrete Roads, Cement Manufacturers' Association, New Delhi, 2000.

SECTION 9

STREET LIGHTING

19. Street Lighting

19.1. Need for Street Lighting

Street lighting is primarily intended to enable the road users (motorists, cyclists and pedestrians) to see accurately and easily the carriageway and the immediate surroundings in darkness. A large proportion of the road accidents are caused in the night and one of the chief reasons is the unsatisfactory lighting. Driving in the night is facilitated by the headlights, but it is always not possible to depend on headlights. For instance, in city streets if all vehicles were to drive with the full headlights on, the glare problem would become acute and severe accidents could be caused. In city streets, therefore, there is a need to illuminate the carriageway and its immediate surroundings so that the use of headlights can be avoided or minimised. On the other hand, it will be rather too costly to install lighting in rural sections of highways and the driver has to use the headlights for night driving.

Improved visibility at night by means of artificial lighting lessens the strain on driving and ensures comfort. The drivers feel more confident in their manoeuvres and avoid jerky and dangerous driving.

Since the speed of traffic generally improves with better lighting, traffic flow conditions also improve. Favourable headway and lateral placements are brought about and the ability of the road to handle traffic is vastly improved.

Indirect benefits include reduction in crime, promotion of the aesthetic appearance of the surroundings and extension of business hours after sunset.

Street lighting has, therefore, become a very important part of the work of a traffic engineer, in association with the illumination engineer.

19.2. Definition of Common Terms

Some of the common terms met with in street lighting are defined here, as commonly understood (Ref. 1, 2).

1. ***Luminous flux (F).*** This is the radiant power given by light source.

2. ***Lumen (lm).*** This is the unit of luminous flux. It is the flux emitted in a solid angle of one steradian by a source having a uniform point source of one candela.

3. ***Steradian.*** A steradian is the unit measure of a solid angle. It is equal to the solid angle subtended at the centre of a sphere by unit area of its surface. The whole space surrounding a point subtends a solid angle of 4π.

4. ***Luminous intensity (I).*** Luminous flux emitted per unit solid angle, dF/dO.

5. ***Candela.*** The unit of luminous *intensity*. The candela is the basic practical photometric unit, and is defined in terms of the luminance of a black body at a temperature of freezing of platinum.

6. ***Candle power.*** Luminous intensity expressed in candelas.

7. ***Illumination, E.*** Luminous flux incident per unit area upon a point on a surface $E = dF/dA$.

The unit of illumination in the metric system is lux or lumen per square metre.

8. ***Luminance, B.*** The luminous intensity emitted or reflected per unit area, of a surface or its equivalent, orthogonally projected in the direction of observation.

$$B = dI/dA.$$

This unit of measurement is the candela per square metre.

9. ***Luminosity.*** Luminosity is the brightness sensation experienced by an observer when viewing a bright light or object. It is not a measurable quantity.

10. ***Lighting Installation.*** The entire equipment provided for lighting a street section, and consists of lanterns, means of support and the electrical and other auxiliaries.

11. ***Lighting system.*** An array of lanterns having a characteristic light distribution, sited in a manner concordant with this distribution. Lighting systems are commonly designated by the name of the characteristic light distribution *e.g.*, cut-off, semi-cut-off, etc.

12. ***Lamp.*** Bulb or light source.

13. ***Lantern (American term : Luminaire).*** A house for one or more lamps, comprising a body and any refractor, reflector, diffuser or enclosure associated with the lamp (*s*).

14. ***Outreach.*** The distance measured horizontally between the centre of a lantern mounted on a bracket and the centre of the column or wall face (Fig. 19.1).

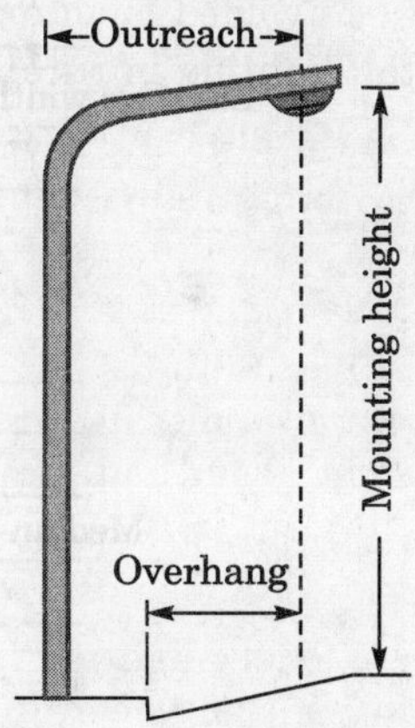

Fig. 19.1. Installation terms.

15. ***Overhang.*** The distance measured horizontally between the centre of a lantern mounted on a bracket and the adjacent edge of a carriageway (Fig. 19.1).

16. ***Mounting height.*** The vertical distance between the centre of the lantern and the carriageway (Fig. 19.1).

17. ***Spacing.*** The distance measured parallel to the centre line of the carriageway, between successive lanterns in an installation (Fig. 19.2), the successive lanterns being on the same or opposite side of the carriageway.

18. ***Span.*** That part of the highway lying between successive lanterns in an installation.

19. ***Arrangement.*** The pattern according to which lanterns are located on plan. The following arrangements are possible :

Staggered. An arrangement in which lanterns are placed alternately on either side of the carriageway (Fig. 19.2).

Opposite. An arrangement in which lanterns are placed on either side of the carriageway opposite to one another (Fig. 19.2).

Central. An arrangement in which lanterns are placed in an axial line close to the centre of the carriageway (Fig. 19.2).

Single-sided. An arrangement in which lanterns are placed on one side only of the carriageway (Fig. 19.2).

20. ***Diffuse reflection.*** Reflection such that light incident upon a surface is reflected from every part of the surface in many directions.

21. ***Preferential reflection*** **(Ref. 3).** Reflection of light characteristic of unpolished or partly polished surfaces, in which the reflected light is not distributed according to the laws of optical reflection, but is diffused, so that the luminance of the surface exhibits one or more maxima.

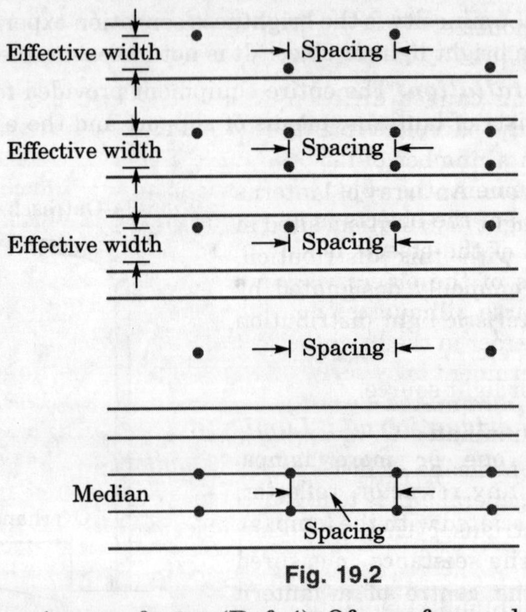

Fig. 19.2

22. ***Luminance factor*** **(Ref. 4).** Of a surface under specified conditions of illumination and observation : the ratio of the luminance of the surface to the luminance of a perfect white diffuser under same conditions.

19.3. Laws of Illumination

The light from a source received on a plane normal to the incident light will be inversely proportional to the square of the distance (d) of the plane from the source (Fig. 19.3).

If the plane is inclined at an angle θ to the normal, the amount of light received is proportional to $\cos \theta$ (Fig. 19.4).

These two can be combined together and expressed as below :

$$E = \frac{I \cos \theta}{d^2}$$

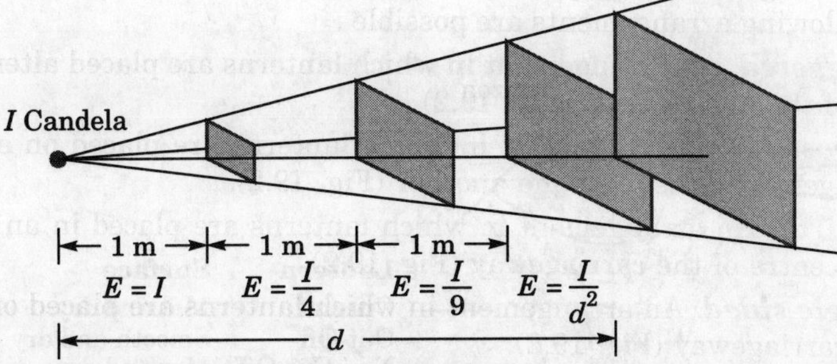

Fig. 19.3

19.4. Discernment by Artificial Lighting

Discernment by artificial lighting can be possible in three ways :

(*a*) by silhouette

(*b*) by reverse silhouette

(*c*) by surface detail.

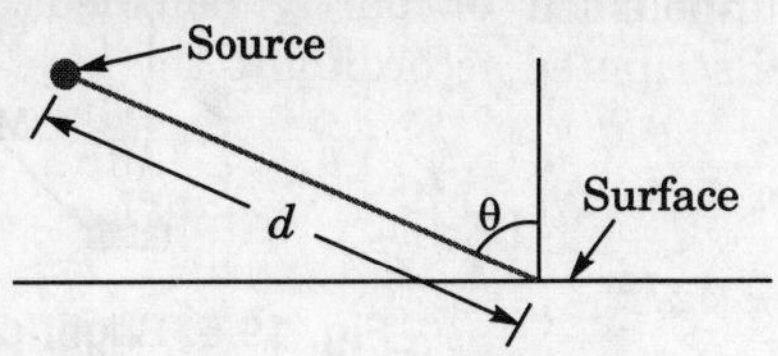

Fig. 19.4

Artificial lighting causes different degrees of brightness in the pavement and the objects, depending upon a number of factors associated with the pavement, the objects and the lantern. If the conditions are such that the brightness of the objects is less than that of the background (*i.e.,* pavement) discernment of the objects is said to be by silhouette. On the other hand, if the brightness of the objects is more than that of the background, discernment is by reverse silhouette. Both these means of discernment are dependent upon the contrast in the brightness of the objects and the background. The third means of discernment is by surface detail wherein a portion of the object facing the observer is illuminated by a high order of illumination, the other portions being not so illuminated.

The most common means of discernment adopted in road lighting is by silhouette, where in the general aim is to produce a bright surface on the pavement against which the pedestrians, cyclists and other objects which are comparatively not so bright can be easily seen.

19.5. Appearance of Lighted Pavement

A single lantern produces a bright patch on the pavement. The typical shape of the patch is roughly a Tee when viewed by the road user, vide Fig. 19.5. The head of the Tee stretches across the road and the tail extends towards the observer. The diagram also show how the shape of the patch changes with the condition of the surface and the type of the lantern. The head of the *T* is a zone of diffuse reflection whereas the tail is a zone of preferential reflection. With a highly polished surface approximating a mirror, the surface becomes a "specular or mirror reflector". Under this condition, the single lantern appears at a single point on the road surface which is the mirror image of the source. A smooth and wet surface has a long and highly luminous tail extending very near to the observer but with a less pronounced head. As the surface becomes rough and dry, the head is more pronounced and there is hardly any tail. The proportions of the bright area are also dependent upon the distance of the lantern from the observer,

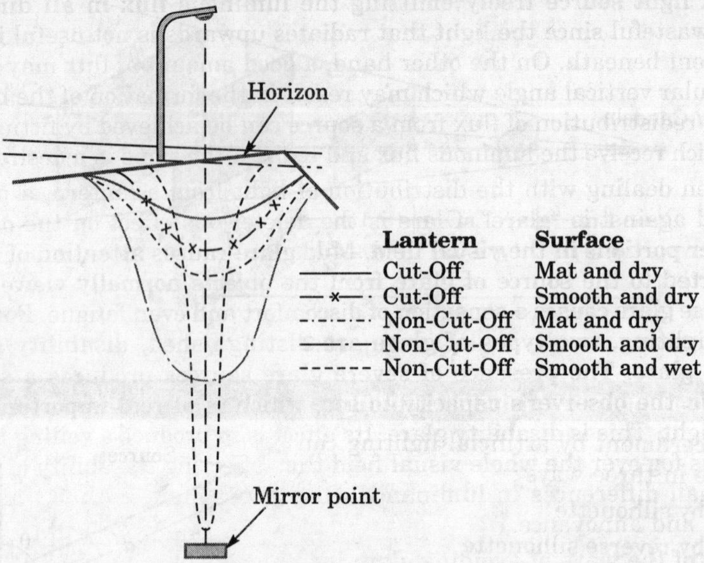

Fig. 19.5. Bright patches formed by a single lantern under varying conditions. (Source : Ref. 2)

the height of the light source, the distribution of the light from the lantern and the height of the observer. With a complete lighting installation composed of individual laterns, the bright patches formed by the individual lanterns build up the lighted pavement. The key to a successful lighting installation is to provide lanterns having such a distribution of light, and at such a height, that uniform bright areas are formed on the pavement when normally viewed by the motorist. Some of these factors will now be considered in detail.

19.6. Type of Surface

It has been briefly indicated earlier that the road surface controls the shape of the bright patch. This is because of the varying reflection properties of different surfaces. Smooth surfaces and those that have been polished due to wear are conducive to preferential reflection that produces a bright tail in the patch. Rough surfaces with projections help to diffuse the light and to obstruct preferential reflection. Examples of such type are freshly laid surface dressings of crushed stone. Wet surfaces tend to make the patches narrow-almost to "streaks" - leaving dark areas between them, in which objects are difficult to be seen. Conditions can be improved if the surface texture is sufficiently coarse. The colour of the road surface affects the strength of the diffuse reflection but not that of preferential reflection. Rough surfaces, which are desirable from skid resistance considerations, are not so desirable from lighting considerations. Highly matt surfaces, which are also dark in colour, therefore, present problems in lighting. Light coloured surfaces (*e.g.*, concrete) offer good reflection characteristics.

19.7. Distribution of Light from the Lantern—The Glare Problem

The pattern of distribution of light from the lantern has a profound effect upon the pattern of illumination of the pavement. A desired pattern of illumination of the pavement can be obtained by regulating the lighting distribution in the lantern. A light source freely emitting the luminous flux in all directions is obviously wasteful since the light that radiates upwards is not useful in lighting the pavement beneath. On the other hand, a good amount of flux may be needed at a particular vertical angle which may result in the formation of the best bright patch. The redistribution of flux from a source can be achieved by fitting suitable devices which receive the luminous flux and redirect the same in a desired pattern.

When dealing with the distribution of light from a lantern, a problem to be guarded against is "glare". Glare is the deleterious effect on the observer of the brighter portions in the visual field. Mild glare causes attention of the driver to be diverted to the source of glare from the objects normally viewed by him. More intense glare causes a sensation of discomfort and even fatigue. For purposes of street lighting, two types of glare are distinguished, disability glare and discomfort glare. The presence of severe glare sources produces a significant reduction in the observer's capacity to see, which is of real importance in safe travel at night. This is disability glare. Its effect is to produce a veiling luminance or luminous fog over the whole visual field thus affecting the ability of the eye to discern small differences in luminance. Discomfort glare is that which causes distraction and annoyance.

One of the ways of avoiding glare is to increase the height of the lantern mountings. Another is to locate the lanterns in such a way that they are removed from the driver's normal line of sight. It is generally seen that light intensities emitted at angles less than about 70° from the downward vertical cannot reach the driver's eyes (because of the shield provided by the roof of the vehicle) and so cannot cause glare (Fig. 19.6).

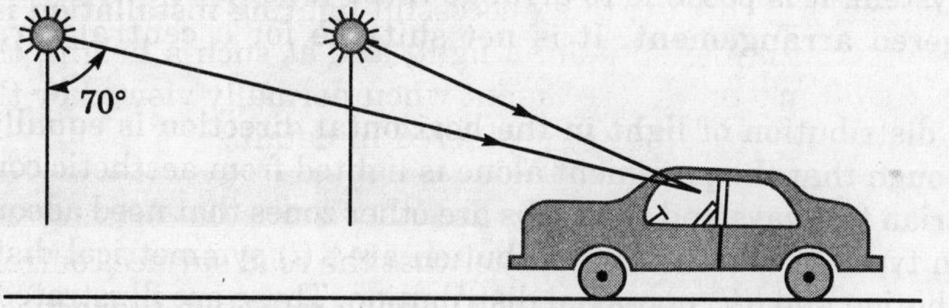

Fig. 19.6. Light rays emitted at angles less than about 70° from the downward vertical cannot cause glare.

All rays above this angle cause glare and it is desirable to limit the intensities above this angle.

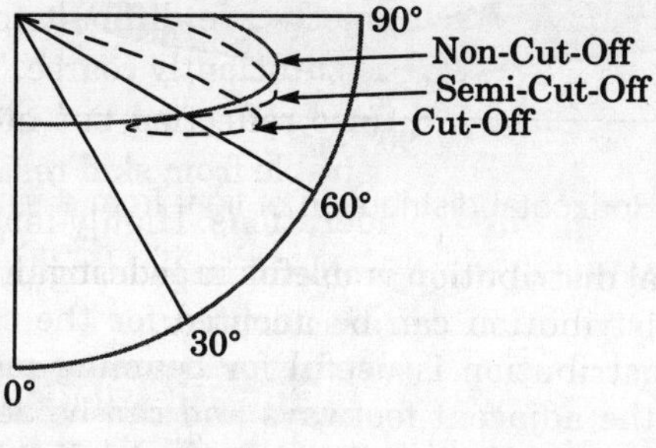

Fig. 19.7. Types of vertical light distribution.

The vertical distribution of light in street lighting systems follows one of the three types—cut-off, non-cut-off medium angle beam and non-cut-off high angle beam (also known as the semi-cut-off system). The principle behind the differentiation can be understood by a reference to Fig. 19.7.

In the cut-off system, the light is reduced as completely and as rapidly as possible above a 70° angle with the downward vertical, the main beam being usually at this angle. Hardly any light is emitted above 80°. This system eliminates glare. In the non-cut-off high angle beam system, the peak intensity is at about 80°, and there is considerable output near the horizontal plane. With this distribution, a certain amount of glare is inevitable, and modern lighting practice does not favour the adoption of this system. The non-cut-off medium angle beam system (semi-cut-off) is in between the above two types, with a peak intensity directed at an angle of 75°

The advantages of a cut-off system are absence of glare and a superior performance with matt surfaces. With this system, the tail of the bright patch is almost absent and the reliance is heavily on the head. The spacing has to be closer to achieve the desired brightness over the road, thus resulting in extra cost. However, the complete absence of glare makes it the superior type where cost is not much the deciding criterion. Because of its reliance on the head of a Tee, it is more suitable when the mounting is central. This solution is warranted when there are large trees on the edges of the pavement preventing location of the edge mountings.

The non-cut-off medium angle beam system, on the other hand, results in a certain amount of glare. It performs better with smoother surfaces, and results in longer lantern spacing. Since the tail of the bright patch is sufficiently long with this system, it is possible to arrange these tails to lie alongside one another by a staggered arrangement. It is not suitable for a central arrangement of lanterns.

The distribution of light in the horizontal direction is equally important. It is not enough that the pavement alone is lighted from aesthetic considerations. The pedestrian footways and the curbs are other zones that need adequate lighting. Three main types of horizontal distribution are : (*i*) symmetrical distribution, (*ii*) axial distribution and (*iii*) non-axial distribution. These are illustrated in Fig. 19.8.

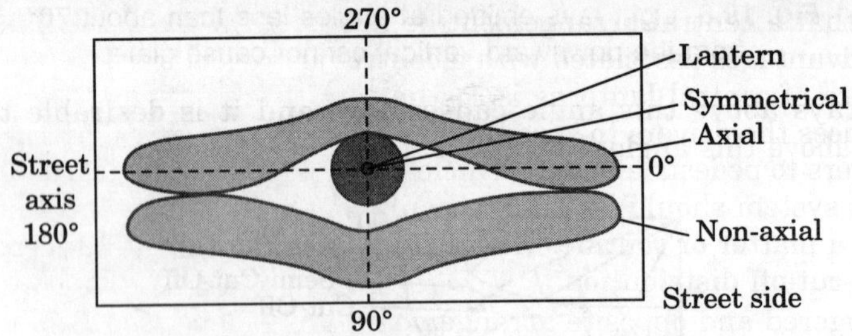

Fig. 19.8. Horizontal distribution of light from a single lantern.

The symmetrical distribution is useful in pedestrian precincts and the ends of cut-de-sac. Axial distribution can be adopted for the central arrangement of lanterns. Non-axial distribution is useful for beaming most of the light on the carriageway than on the adjacent footways and can be adopted for the lanterns mounted near the road curbs. The axial and non-axial distribution throw two main beams, one up and one down the road.

19.8. Mounting Height

A brief mention has already been made earlier that a high mounting reduces glare. In addition, a high mounting makes the luminance more uniform than if the mounting is low. A low mounting increases the illumination of objects on the road and thus hampers discernment by silhouette. On the other hand, the higher the mounting the costlier the installation becomes. The general practice is to have mountings of 7.5–12 m for traffic routes and 3–8 m for other streets. As more powerful lamps are used, the mounting height generally needs to be increased to avoid glare.

19.9. Spacing

Spacing of lanterns controls the minimum illumination on the pavement. The spacing is determined by the shape, and in particular the length, of the bright patch and the extent to which it is desired that individual patches should overlap. The length of the tail depends upon whether a cut-off system is used or a semi-cut-off. With the former, the tail is likely to be shorter than the latter. The spacing has also to be satisfactory from the point of view of pedestrian requirements. In general, the spacing should not exceed 55 m, and should preferably be 35-45 m on important traffic routes. As a rough guide, the spacing should be between 3 and 5 times the mounting height.

19.10. Lantern Arrangements

The different lantern arrangements have been defined earlier. These are :

(*i*) Single-sided. (*ii*) Staggered.
(*iii*) Central. (*iv*) Opposite.
(*v*) Combination of (*iii*) and (*ii*) or (*iv*).

Single-sided arrangement, through simple to install, is not desirable since it leaves one side of the road dark. It can be used only for narrow roads.

Central mounting has the advantage that more light reaches the carriageway and footways than when the lanterns are side mounted. The bright effect is a uniformly lighted roadway centre, with the luminance falling off gradually to the sides. Central mounting becomes unavoidable if there are large trees or buildings near the curb lines. If cut-off distribution is used, it is almost necessary that a central arrangement be adopted. On the other hand, there are many disadvantages associated with a central arrangement. The installation and maintenance of central lanterns is costlier and more difficult. The uniformly lit centre induces the drivers to speed. The comparatively poor lighting on the sides poses dangers to pedestrians and vehicles emerging from the sides. On the whole, the central system should be reserved only for special situations and should not be used as a matter or routine. Whenever it is used, it should incorporate a cut-off or semi-cut-off distribution.

Staggered and opposite arrangements are widely used. The latter results in a uniform luminance over the full pavement width but at greater cost.

With a side-mounting, sometimes the lanterns are located a little away from the curb-line. The distance of the lantern from the curb-line is known as the overhang (see definitions in para 19.2). The purpose of the overhang is to produce a luminance pattern somewhat near to the central arrangement, with a uniformly bright centre and a somewhat darker sides. The overhang is related to the mounting height. A maximum overhang of 0.25 of the mounting height is permitted by the British Code. (Ref. 1).

The out-reach of the lantern mounting determines its exact location with reference to the curb line (See definitions in para 19.2). The set-back from the curb line has an effect on safety. If the lamp posts are located too close to the curb line, vehicles which leave the roadway are liable to crash against the posts. The number of colli-sions generally decreases with the increase in the set-back distance. A distance of 0.9—1.5 is generally considered safe. In situations where there is a speed limit of 50 k.p.h in force, and it is not possible to adhere to these minimum values, a distance of 0.5 m may be considered. If there is a pedestrian footway close to the carriageway, lamp post should be sited at the back of the footway.

19.11. Types of Lamps

The principle energy sources employed for street lighting are gas and electricity. Gas used to be a popular source some years ago, but the versatility and convenience of electricity has made it the widely used energy source of the present day.

There is a large variety of electric lamps available for street lighting purpose. Considerations such as initial cost, life., colour rendering, wattage, brightness, efficiency etc. govern the choice. The following are the main types of lamps :

(*i*) Tungsten filament.
(*ii*) Tubular fluorescent.
(*iii*) Sodium vapour discharge.
(*iv*) High pressure mercury discharge.

Tungsten filament (bulb type) is the most familiar and the simplest light source, commonly used for domestic lighting. Its use for street lighting is limited to minor residential streets and pedestrian walk-ways. The lighting is caused by the passage of electric current through a tungsten wire inside a gas filled bulb. It has a very low light-producing efficiency, with about 8.14 lumens per watt. It also has a short-life, say about 1000 hours. It has got good colour rendering properties and gives almost a white coloured light. Its initial cost is low and hence it becomes the natural choice where initial cost has to be kept low.

Tubular fluorescent lamp consists of a long narrow tube with its interior coated with a fluorescent powder. With passage of electric current an electric discharge takes place at low pressures within the tube causing the excitation of the fluorescent powder. An excellent white light is emitted as a result. The wattage of the tube is low, say upto 120-watts. The lighting efficiency is high, with about 50–75 lumens per watt. The life of the tube is also high, about 6000 hours. It emits a while light which has also good colour rendering properties. It is, therefore, used popularly for shopping streets where the colour rendering and appearance are important requirements.

Sodium vapour discharge lamps operate with an electrical discharge in a vapour of sodium. There are two variations possible : low pressure discharge and high pressure discharge.

The low pressure discharge lamp gives a characteristic monochromatic yellow light, in which no colours can be distinguished. Its use in shopping areas is, therefore, undesirable. Its advantage is the absence of dazzle. The high pressure discharge lamp gives a white colour, and can illuminate all colours. The wattage for a low pressure lamp is about 120–160, the efficiency being about 120 lumens per watt. The life is about 6000 hours. For a high pressure lamp, the wattage is about 400, the efficiency being about 100 lumens per watt. Its life is about 4000 hours. High efficiency, longer life plus pleasing light make sodium vapour lamps ideal for street lighting. These lamps are now very popular in India.

Mercury vapour discharge lamps operate by an electric discharge between two electrodes causing the mercury to evaporate. The discharge is of a brilliant greenish white. The earlier designs are designated MA (mercury arc). MB type (mercury bulb) has a better colour than the MA type. With the incorporation of a fluorescent powder coated on the interior of the bulb, MBF (mercury bulb fluorescent) lamps have excellent colour rendering properties. The life of the mercury lamps is about 7,500 hours, the wattage is about 400 and the efficiency is about 23-60 lumens per watt. The latest development is the MBI lamps where Indiom is used. It is the most efficient, but also very expensive.

The characteristics of some of the common types of lamps are summarised in Table 19.1. Broad indications are also given therein about their suitability.

19.12. Quantity of Illumination Needed on Different Streets

The levels of illumination needed on different streets, based on IES-ASA Practice (America), are given in Table 19.2 (Ref. 5, 6).

Table 19.1. Lamp Characteristic, their Properties and Uses

Lamp type	*Tungsten filament*	*Fluore-scent tubular*	*Low Pressure sodium*	*High pressure sodium*	*Mercury bulb (MB)*	*Mercury bulb Fluorescent (MBF)*
Wattage	1000	120	60	400	400	400
Life (hours)	1000	6000	6000	4000	7500	7500
Design lumens	18,000	7300	20,500	38,500	38,500	22,200
Colour of light	white	white	yellow	white	blue-white	white
Colour rendering	good	good	bad	good	bad	excellent
Suitability rating of lighting	0 = un-suitable	1 = fair	2 = good	3 = excellent		
Rural expressway	0	0	3	3	1	3
Urban expressway	0	0	2	3	1	3
Complex junctions	0	0	3	3	1	3
Urban traffic routes	0	0	3	3	0	3
Rural traffic routes	0	0	3	3	1	3
Residential roads	0	3	0	3	0	3
Shopping streets	1	2	0	3	0	3
Squares etc.	1	2	0	3	0	3
Aesthetic qualities	2	3	0	3	0	3

Table 19.2
Recommended Average Illumination (lumens/sq. m.)
(Adapted from Ref. 5 and 6)

Roadways other than expressways

Roadways classification	*Area classification*		
	Urban downtown	*Intermediate*	*Outlying urban and rural*
Major	22	13	10
Collector	13	10	6
Local or minor	10	6	2*

*Residential

Expressways and freeways

	Expressways	*Freeways*
Continuous urban	15	6
Continuous rural	11	6
Interchange urban	22	6
Interchange rural	15	6

The I.S.I. recommendations are as given in Table 19.3. (Ref. 9)

Table 19.3. Classification of Lighting Installation and Levels of Illumination

Classification of lighting installation	*Type of Road*	*Average level of illumination on road*	*Ratio Minimum / Average illumination*	*Transverse variation of illumination*	*Type of Luminaire*	
					Preferred	*Permitted*
(1)	(2)	(3)(lux)	(4)	(5)	(6)	(7)
Group A1	Important traffic routes carrying fast traffic	30	0.4	33	cut-off	semi-cut-off
Group A2	Other main roads carrying mixed traffic, like main city streets, arterial roads through-ways, etc.	15	0.4	33	cut-off	semi-cut-off
Group B1	Secondary roads with considerable traffic like princi-local traffic routes, shopping streets etc.	8	0.3	20	cut-off or semi-cut-off	non-cut-off
Group B2	Secondary routes with light traffic	4	0.3	20	cut-off or semi-cut-off	non-cut-off

The average illumination in lumens per sq. m. is found from : Average illumination of street in lumens per sq. m.

$$= \frac{\text{(Lamp lumens)} \times \text{(Coefficient of utilization)} \times \text{(Lantern maintenance factor)}}{\text{(Spacing between lanterns in m)} \times \text{(Width of roadway in m)}}$$

19.13. Lamp Installation, at "T" Junctions and Cross-roads

A high standard of lighting is needed at junctions and many accidents can be prevented with adequate illumination.

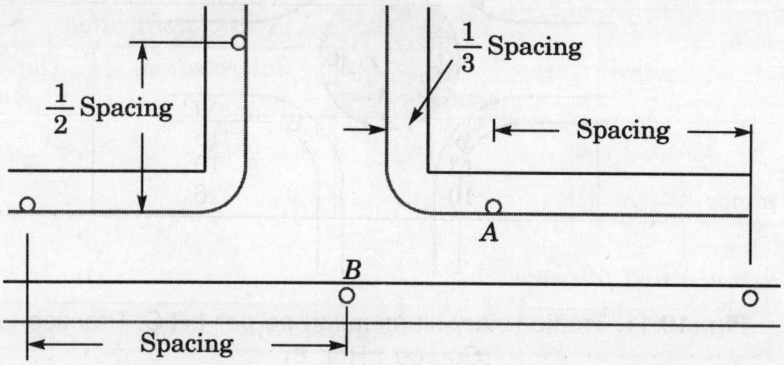

Fig. 19.9. Typical lamp installation : *T* junction.

In the case of *T* junctions, it is customary (Ref. 7) to locate a lantern (*B*) in the major road, just beyond the entrance from the side road, Fig. 19.9, to indicate to the driver on the side road that he is approaching a *T* junction. Lantern *A* serves to show the break in the curb-line to drivers on the main road and approaching from the left.

In the case of cross-roads, an arrangement shown in Fig. 19.10 may be adopted. The basic idea is to locate one lantern a little beyond the intersection in each road, so as to show up the curb line as dark shadow to all users of the junction.

19.14. Illumination of Traffic Rotaries

Traffic rotaries require special treatment. The driver requires to be forewarned about the presence of the traffic hazard, and once he enters the rotary he should be able to see his path in it and other vehicles.

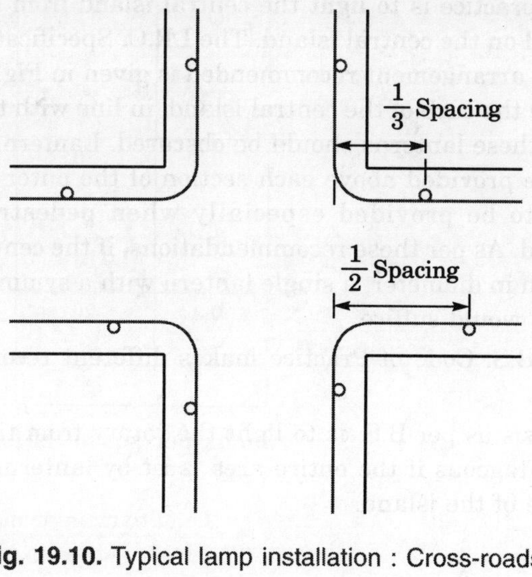

Fig. 19.10. Typical lamp installation : Cross-roads.

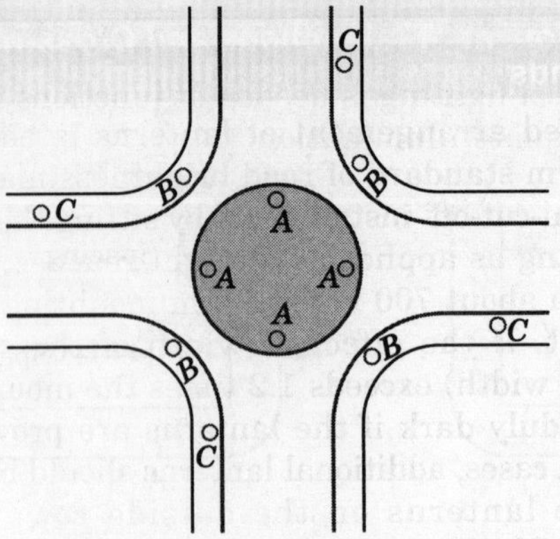

Fig. 19.11. Traffic rotary illumination as per I.R.C. Practice
(*Source* : Ref. 8)

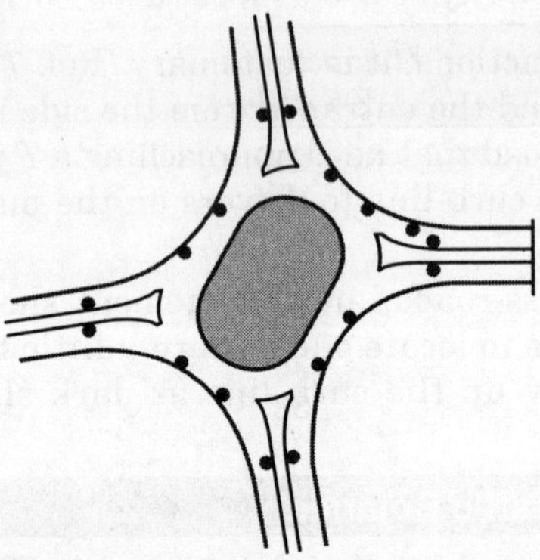

Fig. 19.12. Traffic rotary illumination as per current British practice
(*Source* : Ref. 7)

A common practice is to light the central island from lanterns supported by columns erected on the central island. The I.R.C. Specification (Ref. 8) follows this principle. The arrangement recommended is given in Fig. 19.11. Lanterns *A* are provided above the curb of the central island, in line with the approach traffic lane. The back of these lanterns should be obscured. Lanterns *B*, one or more in number, should be provided above each section of the outer curb of the rotary. Lanterns *C* are to be provided especially when pedestrians cross at the channelising island. As per these recommendations, if the central island is small, *viz.*, less than 20 m in diameter, a single lantern with a symmetrical distribution centrally mounted would suffice.

The latest B.S. Code of Practice makes different recommendations, vide Fig. 19.12 (Ref. 7).

The emphasis as per B.S. is to light the rotary from the outer perimeter. It would be advantageous if the entire area is lit by lanterns located on a high mast at the centre of the island.

19.15. Lighting at Bends

When a staggered arrangement of lanterns is employed in the straight length of a road, uniform standard of road brightness may be achieved on bends in both 'cut-off' and 'semi-cut-off' installations by sitting the lanterns on the outside of the bends. The spacing as applicable straight reaches, may be adopted where the radius is more than about 700 m for a 9 m mounting height or 1000 m for a 12 m mounting height. If the effective width of the road (see Fig. 19.2 for explanation of effective width) exceeds 1.2 times the mounting height, the inside of the bend may be unduly dark if the lanterns are provided only on the outer side of the bend, in such cases, additional lanterns should be provided on the inside of the bend, opposite lanterns on the outside row. Table 19.4 gives the recommended spacing of lanterns for smaller radii curves.

Table 19.4. Recommended Spacing of Lanterns on the Outside of Bends
(in metres)

Radius of curve (m)	*Mounting height 7.5 or 9 m*	*Mounting height 10.5 or 12 m*
150	21	26
180	24	29
210	26	30
240	28	33
270	29	35
300	30	36
360	33	40
430	36	43
500	38	45
550	41	49
600	43	52
670	–	55
730	–	56

19.16. Lighting Dual Carriageways

Generally each carriageway is treated separately except that wherever possible the columns on the central reserve are fitted with double arms to reduce the number of columns on the central reserve. If the central reserve is very wide, say wider than 3 m, single arm columns may be considered for each carriageway.

19.17. Lighting of Roads Carrying Only Local Traffic

For roads carrying only local traffic, lighting is essentially intended for the pedestrians and the motorist is expected to use dipped/dimmed head lights. The mounting height can be of the order of 5 to 6 m. The spacing can be a maximum of 36 m for lanterns with good optical control. Otherwise, the spacing can be reduced to 24–30 m.

19.18. Lighting Bridges

Bridges present special problems because it may not be possible to have standard mountings of 9 m or so. The lanterns may have to be mounted on parapets, the height of the lantern rarely exceeding 3.5 m above the deck. This necessitates the luminance to be restricted if glare is to be avoided. 'Cut-off' distribution is generally the answer.

19.19. Tunnel Lighting

The problem of tunnel lighting has to be approached with a view to provide a brightness pattern which produces automatic adaptation of a driver's eye as he approaches, passes through and leaves the tunnel. He should be able to drive safely in visual comfort either by night or by day. The brightness scheme by day will be different from that provided by night because in day-light the need is for the driver to adjust from the higher external brightness to a much lower level in the tunnel, while at night the brightness outside will be less than that inside. This calls for an installation pattern capable of being switched on to give two quite separate lighting schemes. In the day-time scheme, the lighting at the entrances should be such as to bring the driver's adaptation level down as he approaches the tunnel. Inside the tunnel itself, the level of illumination should be comfortable enough. During night-time, a uniform illumination is generally provided from portal to portal, the level being a little higher than that of normal main road lighting.

The light fittings should be relatively easy to open quickly for maintenance and must have lamps that do not need to be replaced at quick intervals. For cleaning tunnel surface, water jetting is commonly resorted to, and in order to withstand this treatment the fitting should be robust.

Exact spacing is difficult to give. However, as a general guide, the spacing of fittings may be 1.75–2 m at the entrances where a high standard of brightness is demanded. This spacing may be increased to 3–4 m inside the tunnel.

19.20. High Mast Lighting

High Mast Lighting is advantageous in locations such as Toll Plazas, Grade Separators, car parks and shopping malls. It eliminates the need for having too many lighting poles. The height of the lighting system varies from 15 to 45 m. A typical sketch is given in Fig. 19.13. (Ref. 10).

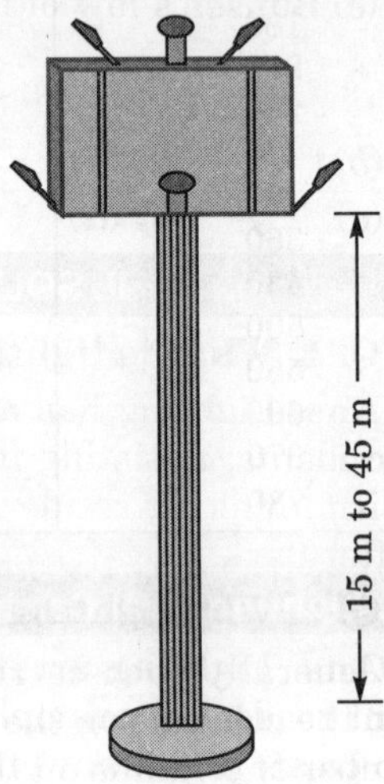

Fig. 19.13. Continuously tappered high mast pole.

19.21. Maintenance of Lighting Instations

For ensuring safety, a very high standard of maintenance of lighting installations is called for. Cleaning of lamps and replacement of faulty lamps should be carried out at regular intervals.

I. MULTIPLE CHOICE QUESTIONS

1. Radiant efficiency of the luminous source depends on
   (*a*) shape of the source (*b*) temperature of the source
   (*c*) wavelength of light rays (*d*) all of the above.

2. Light waves travel with a velocity of
(*a*) 3×10^{10} cm/s (*b*) 3×10^{12} cm/s
(*c*) 3×10^{15} cm/s (*d*) 3×10^{18} cm/s.
3. Carbon arc lamps are commonly used in
(*a*) domestic lighting (*b*) street lighting
(*c*) cinema projectors (*d*) photography.
4. The unit of solid angle is
(*a*) solid angle (*b*) radian (*c*) steradian (*d*) candela.
5. Candela is the unit of
(*a*) luminous flux (*b*) luminous intensity
(*c*) wavelength (*d*) none of the above.
6. The unit of luminous flux is
(*a*) steradian (*b*) candela (*c*) lumen (*d*) lux.
7. The illumination is directly proportional to the cosine of the angle made by the normal to the illuminated surface with the direction of the incident flux. Above statement is associated with
(*a*) Planck's law (*b*) Macbeth's law of illumination
(*c*) Bunsen's law of illumination (*d*) Lambert's cosine law.

Answers

1. (*b*) **2.** (*a*) **3.** (*c*) **4.** (*c*) **5.** (*b*)
6. (*c*) **7.** (*d*)

II. SOLVED QUESTIONS AND ANSWERS

Q. 1. What is lighting? Discuss its importance.

Ans. Lighting has an important role to play in: Reducing risks of night time accidents; Assisting in the protection of property; Discouraging crime and vandalism; Making residents feel secure; and Enhancing the appearance of the area after dark.

Q. 2. Why lighting arrangements are made?

Ans. Lighting arrangements may be used to identify the functions of different roads. For example, two lamp columns at the entrance to a road can be used to create a gateway effect. Adequate lighting should be provided in all parts of the layout to enhance safety and security for drivers, pedestrians and cyclists.

Q. 3. Why lighting is necessary?

Ans. Lighting is necessary to illuminate bends and traffic calming features, to enable road users to see potential obstacles and each other after dark and to reduce the fear of crime. 11.2.5 Sizes of trees and shrubs when mature and their location in the layout must be considered in relation to street lighting.

Q. 4. What is the importance of colour of lighting?

Ans. The colour of lighting is an additional consideration. Where pedestrians are concerned, white lighting is preferable, rather than orange. White lighting allows better discernment of street features, including the facial expressions of other street users, which can be an important factor in allaying personal security concerns.

Q. 5. What is illuminance?

Ans. The density of luminous flux incident on a surface; it is the quotient of the luminous flux by the area of the surface when the latter is uniformly illuminated, expressed in lumens per square meter.

Q. 6. What is a lamp?

Ans. It is a source of light. The device within a luminaire which converts the electrical energy to light.

Q. 7. Discuss light-loss factor.

Ans. A depreciation factor which is applied to the calculated initial average footcandle (lux) to determine the value of depreciated average illumination at a predetermined time in the operating cycle, usually just prior to relamping, and which reflects the decrease in effective light output of a lamp and luminaire during its life.

Q. 8. What is lumen?

Ans. The unit of luminous flux (time rate of flow of light). A lumen is defined as the luminous flux emitted by a point source having a uniform luminous intensity of one candela.

Q. 9. What is luminaire?

Ans. A complete lighting fixture consisting of a lamp or lamps together with the ballast, reflector, refractor, photocell when required, and the housing.

Q. 10. What is interchange lighting?

Ans. Interchange lighting places lights in the merging traffic and gore areas in the same locations as partial interchange lighting. In addition, it places lights along the ramps, on the through roadway through the interchange, and on the crossroad between the ramp terminals.

Q. 11. What is bridge lighting?

Ans. The roadway on a bridge is normally treated the same as other parts of the roadway. If there is no lighting on the adjacent roadway, there is normally no need for lighting on the bridge. An exception is a very long bridge, which may be lit even though the roadway is not lit at other locations. Where lights are to be installed on a bridge, the desirable locations for the lighting units are at abutments and at pier locations, or at distance from an abutment or pier not to exceed 25 percent of the length of the span. This placement of the lighting units reduces the effects of vibration. The light poles should utilize davit type mast arms so that there are no joints to be weakened by vibration.

Q. 12. Explain pole placement.

Ans. Pole placement is an engineering decision which should be based upon geometry, character of the roadway, physical features, environment, available maintenance, economics, aesthetics, and overall lighting objectives. Physical roadside conditions may require adjustment of the spacing determined from the base levels of illumination, indicated in the **AASHTO** Guide.

Q. 13. What are luminaires?

Ans. A luminaire consists of a lamp, reflector, refractor, ballast if required, photocell if required, and housing. Several factors have influenced the choice of the type of luminaire that the state currently uses. The efficiency of a lamp in converting electrical energy to light, the ability of the lamp to maintain its light output over the course of the lamp life, the length of the lamp life, the colour of the light, and the distribution of the light are all factors which affect the cost and effectiveness of installing, operating, and maintaining the lights, and, hence, affect the choice of light source. The luminaires should be a standard type that is maintainable by and approved by **OTST's** Electrical Services Section.

Q. 14. What is signal system?

Ans. A signal system is defined as any cabinet containing traffic signal, traffic management, or traffic recording equipment. The signal system refers to any cables that lead into or out of the signal cabinet. Power shall not be obtained from inside the signal cabinet. Power can be obtained from the service equipment/ service cabinet or from the unmetered lighting conductors in the signal bases.

III. REVIEW QUESTIONS

1. What is the need for street lighting?
2. Discuss the discernment by artificial lighting.
3. What are the types of lamps?
4. What is the quantity of illumination needed on different streets?

REFERENCES

1. British Standard Code of Practice : Street Lighting. C.P. 1004, Parts 1 and 2 : 1963, British Standard Institution, London, 1963.
2. Waldram, J.M., International recommendations for public thoroughfares, Traffic Engineering and Control, 7. No. 12 London, 1966.
3. Waldram, J.M., Street Lighting, E. Arnold and Co., London, 1952.
4. Research on Road Safety, Road Research Laboratory, (U.K.), Her Majesty's Stationery Office, London, 1963.
5. Design and application of roadway lighting, Street and Highway Safety Lighting Bureau, New York.
6. American Standard Practice for Roadway Lighting, Illumination Engineering Society, 1964.
7. British Standard Code of Practice, Street, Lighting, C.P. 1004, Parts 4, 6 and 8 : 1967, British Standards Institution, London, 1967.
8. Recommended Practice for Traffic Rotaries, I.R.C : 65–1976. Indian Roads Congress, New Delhi, 1976.
9. Code of Practice for lighting of public thoroughfares, I.S. : 1944 (Part I and I)-1970, Indian Standards Institution, New Delhi, 1970.
10. Development of Aesthetic Designs for Roadside Furniture, Research Scheme R-68, Guidelines, Planning Norms and Typical Design, Ministry of Surface Transport, prepared by L.R. Kadiyali & Associates, New Delhi, 2000.

SECTION 10
TRAFFIC MANAGEMENT

20. Transportation System Management

20.1. Introduction

As the traffic on the existing road system in cities grows, congestion becomes a serious problem. Medium and long term solution like widening roads, providing elevated fly-overs and constructing bypasses and urban expressways are costly. Simple and inexpensive solutions can tide over the crisis for some time. Transportation System Management (TSM) is a package of short term measures to make the most productive and cost-effective use of existing transportation facilities, services and modes (Ref. 10). TSM also embraces Travel Demand Management (TDM).

20.2. Travel Demand Management

TDM techniques are aimed at reducing the traffic flows, especially during the peak hour. Some of the techniques commonly adopted are :

1. Car pooling and other ride-sharing programmes
2. Peripheral parking schemes (described in 12.7)
3. Chartered buses (Institutional buses) to serve areas of trip origins to common work place
4. Staggering of office hours and flexible time of work
5. Internal shuttle service in the CBD
6. Parking restraint (described in 42.2.3)
7. Road pricing (described in 42.3)
8. Entry fee (described in 42.2.4)
9. Priority for buses in traffic.
10. Restrictions on entry of trucks during day-time.

20.3. Traffic Management

20.3.1. Scope of Traffic Management Measures

Many of the urban streets carry traffic volumes for which they were simply not designed. The inevitable result is delay, congestion and accidents. The resultant ills can be get over to some extent by controlling the traffic, imposing regulatory measures and enforcing management techniques, so as to make the most economic use of the streets. Traffic control measures include traffic signals and these have been already considered. Regulatory measures include restrictions on speed, parking, size of vehicles and so on, and these also have been discussed separately. The third of the set of measures available to the traffic engineer are collectively known as traffic management measures. These measures also form part of TSM.

The fundamental approach in traffic management measures is to retain as much as possible existing pattern of streets but to alter the pattern of traffic movement on these, so that the most efficient use is made of the system. In doing so, minor alternations to traffic lanes, islands, curbs etc. are inevitable, and are part of the management measures. The general aim is to reorient the traffic pattern on the existing streets so that the conflict between vehicles and pedestrians is reduced.

Some of the well-known traffic management measures are :

(*i*) Restrictions on turning movements
(*ii*) One-way streets
(*iii*) Tidal-flow operations
(*iv*) Exclusive bus-lanes
(*v*) Closing side-streets.

20.3.2. Restrictions of turning movements

20.3.2.1. The problem posed by turning traffic

At a junction, the turning traffic includes left-turners and right-turners. Left-turning traffic does not usually obstruct traffic flows through the junctions, but right-turning traffic can cause serious loss of capacity. At times, right-turning traffic can lock the flow and bring the entire flow to a halt. One way of dealing with heavy right-turning traffic is to incorporate a separate right-turning phase in the signal scheme, or to introduce an early cut-off or late start arrangement. These schemes have their limitations and result in a long signal cycle. Another solution is to ban the turning movement altogether.

20.3.2.2. Prohibited right-turning movement

Prohibition of right-turning movement can be established only if the existing street system is capable of accommodating an alternative routing. Depending upon the existing layout of the street system, three methods are available :

(*i*) Diversion of the right-turning traffic to an alternative intersection further along the road where there is more capacity for dealing with a right-turn. This scheme is shown in Fig. 20.1 (*a*), and is known as a *T* turn. It is often useful for dealing with a difficult right turn from a minor road into a major road. The right turn gets shifted to a minor-minor junction.

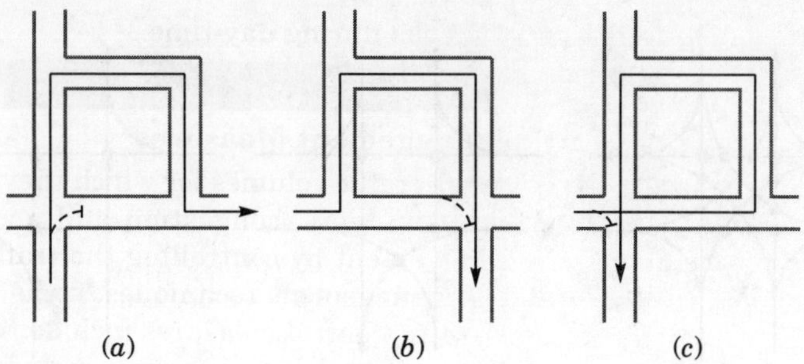

Fig. 20.1. Prohibited right turning schemes.

(*ii*) Diversion of the right-turning traffic to the left before the junction. This scheme is shown in Fig. 20.1 (*b*) and is known as a *G* turn. It is useful for a right-turn from a major road, since it is converted to a left turn from the major road and a straight-over movement at the original junction.

(*iii*) Diversion of the right-turning traffic beyond the junction. This scheme is shown in Fig. 20.1 (c), and is known as a *Q* turn. This entails three left turns and requires the driver to travel twice through the original junction, thus increasing the total volume of traffic handled by the junction. Since only left turns are involved it is considered the least obstructive.

Prohibition of right-turning movement is known to increase the saturation flow and the capacity of the junction. However, it is also desirable to remember that the prohibition of a right turn at a junction will not altogether make problems disappear. In fact it may compound the problems at others.

20.3.2.3. Prohibited left-turning movements

Left-turning movement is not obstructive to traffic and it is rare they are prohibited. However, such prohibition may be needed to provide a safe crossing for pedestrians, especially when the pedestrian traffic across the minor road is heavy.

20.3.3. One-way Streets

20.3.3.1. Purpose

As the name itself imples, one-way streets are those where traffic movement is permitted in only one direction. As a traffic management measure intended to improve traffic flow, increase the capacity and reduce the delays, one-way streets are known to yield beneficial results. They afford the most immediate and the least expensive method of alleviating the traffic conditions in a busy area. In combination with other methods such as banned turning movements, installation of signals and restrictions on loading and waiting, the one-way street system is able to achieve great improvement in traffic conditions of congested areas.

20.3.3.2. Advantages of one-way streets

(*i*) ***A reduction in the points of conflict.*** Traffic movements at junctions involve a number of points of conflict. These generate delay, congestion and accident hazards. Any scheme where the points of conflict are reduced in number is thus conducive to better safety and less delay. Fig. 20.2 gives the point of conflict a vehicles at adjunction of two 2-lane 2-way streets (*a*), and how the same can be reduced by making one street one-way (*b*) and both the streets one-way (*c*).

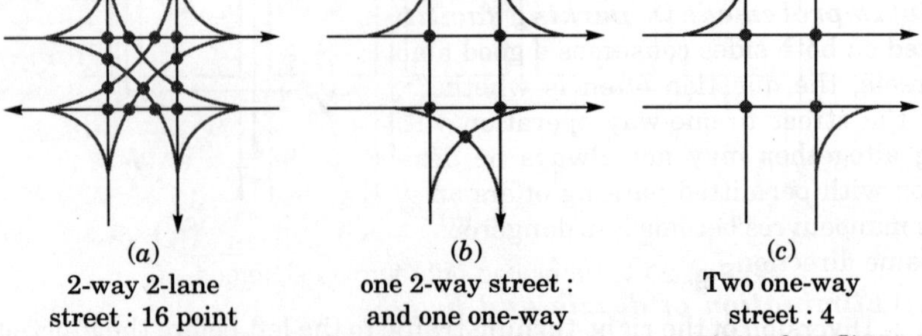

(*a*) 2-way 2-lane street : 16 point p.f of conflict

(*b*) one 2-way street : and one one-way street : 7 points

(*c*) Two one-way street : 4 points of conflict

Fig. 20.2. Vehicular points of conflict with 2-way streets and one-way streets.

The points of conflict of pedestrians and vehicles are similarly reduced vide Fig. 20.3.

(*ii*) ***Increased capacity.*** The removal of opposing traffic and the reduction of intersection points of conflict results in a marked increase in the capacity of a one-way street. However, this will not always be the case and broad generalisations on the relative efficiency of an individual approach operating one-way as compared to two-way, without consideration of the entire system can be misleading.

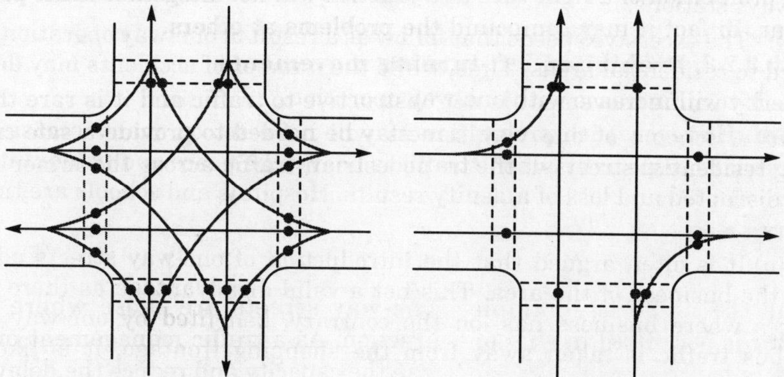

(*a*) Two 2-way streets : 24 points of conflicts

(*b*) Two one-way streets : 12 points of conflicts

Fig. 20.3. Points of conflict of pedestrians and vehicles.

Another consideration that leads to increased capacity is the utilization of the odd-traffic lane-under one-way street operation.

(*iii*) ***Increased speed.*** Since the opposing traffic is eliminated, drivers can operate at higher speeds. This is further facilitated by the more efficient operation of the traffic signal system that is possible under one-way street operation. With increased speeds, delays and journey times get reduced. An overall reduction of 20 percent in journey time has been reported in some London streets which were changed to one-way system.

(*iv*) ***Facilitating the operation of a progressive signal system.*** The use of one-way street operation offers advantages in designing a system of signals for an entire area. Progressive system design is easy with a pair of one-way streets. With progressive system, the flow becomes smoother and the safety is increased.

(*v*) ***Improvement in parking facilities.*** A two-way street with parking permitted on both sides consumes a good amount of street width. In narrow two-way streets, the question often is whether to eliminate parking altogether or convert the street to one-way operation with one-side parking. Elimination of parking altogether may not always be desirable. In such situations, one-way operation with permitted parking offers an alternative. As an added advantage, parking manoeuvres become less dangerous and obstructive when all vehicles face in the same direction.

(*vi*) ***Elimination of dazzle and head-on collision.*** The elimination of dazle in one-way streets is conducive to greater safety.
Similarly, the elimination of head-on collision results in reduction in accident severity.

20.3.3.3. Disadvantages of one-way street working

A number of disadvantages are associated with one-way street working. Some of these are :

(*i*) Although the journey times and delays are reduced, the actual distances to be covered by drivers increase.

(*ii*) Where buses operate on the streets, the stops will have to be relocated and in many instances the passengers will have to walk extra distances. It may happen that the stops will be away from important generators of bus traffic, such as railways stations.

(*iii*) The excessive speeds that follow as a result of one-way operation may be a hazard to residential areas. Thus, while the number of accidents may decrease, the severity will increase with one-way operation,

(*iv*) A scheme of one way streets will have to, on many occasions, use existing residential streets. If the traffic is heavy, the peace and tranquility of the area is disrupted and loss of amenity results. Hospitals and schools are the worst sufferers.

(*v*) It is often argued that the introduction of one-way streets adversely affects the business of the area. This not a valid disadvantage as there may be instances where business has, on the contrary, benefited by one-way streets. When bus traffic is taken away from the shopping frontage, it is likely that business drops.

(*vi*) In the initial stages of its introduction, confusion is likely to be created amongst motorists and pedestrians. With the passage of some time, familiarity sets in.

(*vii*) One of the important drawbacks of a one-way street operation is the prerequisite need for the availability of a street system that can be easily modified to suit that new scheme. Not all the existing patterns are amenable to a change-over. A gridiron pattern lends itself easily to the change. Irregular street patterns are difficult to adapt. The difficulties become large at the ends of the oneway roads. The ideal situation would be one where the two complementary roads converge to form a *Y*-junction, but it is rare to expect this in practice. Most often, the two roads will have to end in a cross-street, which will then have to handle a great amount of traffic. The two junctions may pose problems in design because of heavy turning movements.

(*viii*) Emergency vehicles (fire services, ambulances etc.) may be blocked by cars in all lanes at intersections waiting for signals to change.

20.3.3.4. Need for proper signing

Whenever a system of oneway streets is introduced, it is imperative that proper signs should be put up to foster safe and efficient traffic. 'No entry' signs are needed at all terminal points of the one-way streets. At the entrances and exits of all intersections within the scheme, 'one-way' and/or 'two-way' traffic signs should be displayed. It may be necessary to put up 'No left turn' and 'No right turn' signs at some junctions.

20.3.4. Tidal flow operation

20.3.4.1. Description

One of the familiar characteristics of traffic flow on any street leading to the city centre is the imbalance in directional distribution of traffic during the peak hours. For instance, the morning peak results in a heavy preponderance of flow towards the city centre, whereas the evening peak brings in heavier flow away from the city centre. In either case, the street space provided for the opposing traffic will be found to be in excess. This phenomenon is commonly termed as "tidal flow". One method of dealing with this problem is to allot more than half the lanes for one direction during the peak hours. This system is known as "tidal flow operation", or "reverse flow operation".

20.3.4.2. Methods

The principle of tidal flow operation can be translated into practice in two ways :

(*i*) The first is to apportion a greater number of lanes in a multilane street to the in-bound traffic during the morning peak and similarly a great number of lanes to the out-bound traffic during the evening peak.

(*ii*) The second requires the existence of two separate streets parallel to each other and close to each other, so that the wider of the two can be set apart for the heavier traffic both during the morning peak and the evening peak. In this case, the two streets will operate as one-way streets.

20.3.4.3. Favourable conditions for operation

For undivided streets, tidal flow operation is best done with the number of traffic lanes allocated to the two directions roughly corresponding with the directional split in traffic flow. Generally, tidal flow operation is justified where 65 per cent or more of traffic moves in one direction during peak periods. It is also necessary that the remaining lanes for the lighter flow are adequate for that traffic. With a three-lane street, two lanes can be reserved for heavier flow and one lane for the smaller flow. With a four-lane street, a 3 lane to 1 lane distribution may not correspond with the directional split in traffic flow. In such cases, it may be more prudent to allocate the entire street to one direction and designate a parallel street nearly to be opposing traffic. With 3 lane and 4 lane streets, there is a disadvantage that the lesser flow has to move only in one lane and this may not be convenient. With a five lane road, a 3 lane to 2 lane distribution becomes very convenient. On a 6 lane street, 4 lane to 2 lane distribution is convenient. During off-peak hours a 5 lane facility can be worked with 2 lane for each direction, with no traffic allowed in the central lane. A six lane facility can be worked with 3 lanes for each direction during off-peak period.

20.3.4.4. Special measures needed

With a tidal flow operation, special signing and additional control devices are needed. The assignment of traffic to proper lanes can be achieved by placing overhead signs. Better results are possible,, with signals to regulate the flow in each lane. If these signals are properly designed, an additional advantage can be gained by the progressive timing of these.

It is desirable to restrict parking on the side of the major flow during the peak periods. Direct rightturns by traffic in the major direction should be forbidden at important intersections because of the restrictions they place on the free moment of traffic.

Stanchions or traffic cones are desirable to separate the opposing traffic. 'No right turn' and 'Keep Left' signs on pedestals and stanchions are also necessary.

20.3.5. Closing Side-streets

20.3.5.1. Method

A main street may have a number of sidestreets where the traffic may be very light. In such situations, it may be possible to close some of these sidestreets without affecting adversely the traffic, and yet reap a number of benefits.

20.3.5.2. Advantages

The following are some of the advantages of closing sidestreets :

(*i*) Since interference from the traffic from side streets is eliminated, the speed increases and journey time reduces.

(*ii*) For the same reason as above, the accidents get reduced.

(*iii*) If the side streets are too many and at close intervals, it is difficult to formulate a scheme for the progressive system of signals. A spacing of 275 m between signalised intersections is desirable, and this necessitates the closure of all intermediate side streets.

(*iv*) The side-streets which are closed can be utilised for parking of vehicles, if there is an acute shortage of parking space in the area.

(*v*) The side-streets, closed for traffic from the main streets, can be easily converted to a pedestrian precinct, thereby enhancing the safety, comfort and convenience of pedestrians.

20.3.5.3. Disadvantages

The disadvantages of closing the side-streets can be :

(*i*) Closure of a number of cross-streets may increase the flow to and from the remaining cross-roads. This may necessitate signal control and other measures at these junctions.

(*ii*) When a number of side-streets are closed, the immediate effect is an increase in the parking of vehicles on the main street itself. Those vehicles which used to park on the side streets are now deprived of their parking space. This may create congested conditions on the main road and lead to delays and lower speeds.

20.3.6. Exclusive Bus Lanes

A recent innovation in traffic management practice in some of the cities is to reserve a lane of the carriageway exclusively for bus traffic. This is possibly only in situations where the carriageway is of adequate width and a lane can be easily spared for the buses. This implies that there should be atleast 3 lanes in each direction. For reasons of convenience of alighting and embarking passengers at the curb, the exclusive lane has to be adjacent to the curb.

Exclusive bus lanes running against heavy one-way flow (Contra-flow lanes) are also very common. (Ref. 1) One experience suggests that such an arrangement nearly halves the journey time. (Ref. 2) A good measure of enforcement is needed if serious accidents have to be avoided in this system.

Bus priority measures are a cheap and easy way to provide some aid to bus services. The journey time can be considerably reduced (Ref. 3, 4) and bus journey can be made more attractive. Regularity of the buses can be improved. Viewed in the present context of traffic dilemma caused by the increase in car ownership and the need to patronise public transport, this is a welcome strategy.

To be successful, the bus-lanes should be created for a good length of the road instead of in small bits. An effective enforcement policy is a prime requisite for safety. Satisfactory signing and marking of the lanes (Ref. 5) and adequate publicity are also needed. Recessed bus bays must be provided wherever reserved bus lanes are to be introduced.

In some of the advanced countries, as a further measure of patronage of the buses, traffic signals have been installed with special bus detectors at street intersections. Thereby it is possible to give preferential treatment for buses at traffic signals, minimizing delays (Ref. 6, 7).

The capacity of the exclusive bus lanes may be computed by assigning a PCU of 1.6 to each bus. (Ref. 8) For example, if the road-way lane has a capacity of 1500 PCUs per hour, the exclusive bus lane can be deemed to have a capacity of 1940 buses per hour. The width of the bus lane should be a minimum of 2.8 m and if possible 3.5 m. (Ref. 9).

'Bus only' streets can also be established, prohibiting all traffic except buses. This is usually enforced in shopping areas (Ref. 5) to facilitate pedestrians reach the buses within short walking distances.

A *Bus Rapid Transit* (BRT) corridor has been constructed in New Delhi, wherein two lanes (one for each direction) have been carved in the centre of a multi-lane road. Though it has increased congestion in the lanes meant for cars, buses are able to run at a fast speed. BRT corridors are also being planned in cities like Vishakhapatnam, Pune, Ahmedabad and Bangaluru.

20.3.7. It is some times necessary to reduce the carriageway width of a road to earmark some space for parking, such as in busy shopping centers. This can be achieved by chicanes (Fig. 20.4). Suitable signing that the carriageway is narrowed is necessary.

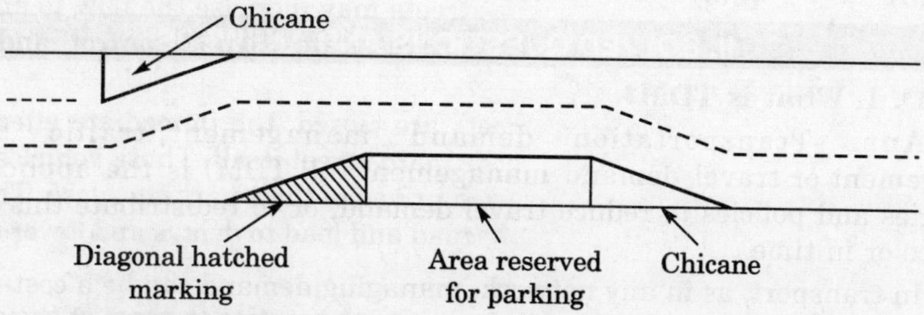

Fig. 20.4. Traffic calming by chicane (reduced lane width).

I. MULTIPLE CHOICE QUESTIONS

1. Who is responsible for the construction and maintenance of the National Highways?
   (*a*) state government (*b*) central government
   (*c*) federal government (*d*) national government
2. Who is responsible for the State highways and major district roads?
   (*a*) state government (*b*) central government
   (*c*) federal government (*d*) national government
3. When was the Border Roads Development Organisation (BRO) set up?
   (*a*) April, 1951 (*b*) April, 1952
   (*c*) April, 1955 (*d*) May, 1960
4. Which among the following does not belong to the scheduled airlines in the public sector?
   (*a*) Air India (*b*) Alliance Air
   (*c*) Air India Charters Ltd. (Air India Express)
   (*d*) Aryan Cargo Express Private Limited
5. Which was the first Indian private airline to launch flights to China?
   (*a*) Jet Airways (*b*) Sahara Airlines
   (*c*) Kingfisher Airlines (*d*) IndiGO

6. When was Pawan Hans Ltd. introduced?
(*a*) May, 1960 (*b*) October, 1985
(*c*) October, 1986 (*d*) April, 1995

7. When was International Airport Authority of India (IAAI) set up?
(*a*) 1960 (*b*) 1972
(*c*) 1985 (*d*) 1986

8. When was the Airports Authority of India proposed?
(*a*) September 18, 1992 (*b*) April 1, 1995
(*c*) June 1, 2004 (*d*) March 31, 2005

9. When was Shipping Corporation of India Ltd. established?
(*a*) October 2, 1961 (*b*) October 27, 1986
(*c*) September 18, 1992 (*d*) January 26, 1998

Answers

1. (*b*)	**2.** (*a*)	**3.** (*d*)	**4.** (*d*)	**5.** (*a*)
6. (*b*)	**7.** (*b*)	**8.** (*b*)	**9.** (*a*)	

II. SOLVED QUESTIONS AND ANSWERS

Q. 1. What is TDM?

Ans. Transportation demand management, traffic demand management or travel demand management (all TDM) is the application of strategies and policies to reduce travel demand, or to redistribute this demand in space or in time.

In transport, as in any network, managing demand can be a cost-effective alternative to increasing capacity. A demand management approach to transport also has the potential to deliver better environmental outcomes, improved public health, stronger communities, and more prosperous cities.

TDM techniques link with and support community movements for sustainable transport.

Q. 2. What is transportation management system? Explain.

Ans. A transportation management system (TMS) is a subset of supply chain management concerning transportation operations and may be part of an enterprise resource planning system. A **TMS** usually "sits" between an **ERP** or legacy order processing and warehouse/distribution module.

Transportation systems management strategies are low-cost but effective in nature, which include, but are not limited to:

- Intersection and signal improvements
- Freeway bottleneck removal programs
- Data collection to monitor system performance
- Special events management strategies

Q. 3. What do you mean by traffic signal and intersection improvements?

Ans. Traffic signal and intersection improvements include such elements as:

- signal timing optimization
- controller/cabinet and signal head upgrades

- vehicle detectors repair/replacement
- communication with a central system
- turning lanes
- grade separations
- pavement striping
- lane assignment changes
- signage and lighting

Q. 4. What is freeway and arterial bottleneck removal?

Ans. Freeway and arterial bottleneck removal consists of identifying congested locations and improving such elements as:

- insufficient acceleration/deceleration lanes and ramps
- weaving sections
- sharp horizontal/vertical curves
- narrow lanes and shoulders
- inadequate signage and pavement striping
- other geometric deficiencies

The identification and elimination of traffic bottlenecks can greatly improve travelling conditions and enhance system capacity, reliability, and safety, especially during peak periods. **TSM** projects can complement the major capacity improvements and infrastructure by providing improved traffic flow on arterials and local streets.

Q. 5. What is the purpose of through fare assessment program (TAP)/Traffic signal integration and monitoring program.

Ans. The purpose of this program is to improve traffic flow and enhance the capacity of existing arterial systems by implementing new signal timing and low-cost operational improvements along selected corridors, which include but are not limited to:

- lane assignment changes
- vehicle detector upgrades
- addition of pedestrian push buttons
- signal head upgrades
- controller and cabinet upgrades
- communication with central computer
- addition of GPS clocks
- re-striping

Q. 6. What is A one-way street?

Ans. A one-way street is a street either facilitating only one-way traffic, or designed to direct vehicles to move in **one** direction. One-way streets typically result in higher traffic flow as drivers may avoid encountering oncoming traffic or turns through oncoming traffic.

Pedestrian experience One-way streets influence the pedestrian's experience on two key things—the comfort of pedestrians as it relates to the direction of travel of adjacent vehicles and the predictability of vehicular movement at intersections. On a two-way street, pedestrians always have the choice of walking facing the oncoming traffic. On a one-way street, pedestrians moving toward the same direction as the traffic will always have adjacent traffic coming behind them regardless of which side of the street they choose to walk on.

III. REVIEW QUESTIONS

1. What do you mean by travel demand management? Explain.
2. Write a brief note on traffic management.

REFERENCES

1. Levinson, H.S. and D.B. Sanders. Reserved Bus Lanes On Urban Freeways : A Macromodel, Transportation Research Record No. 513, Transportation Research Board, Washington, 1974.
2. Bendixson, T., Results of Reading Bus Lane experiment, Traffic Engineering and Control, Vol. 12, No. 4, London, 1970.
3. Lane, R. and D.H. Hodgkinson, Evaluating a Bus Segregation Network in Central London, Proceedings of a Symposium held at TRRL, TRRL Report LR 570, Crowthrone, 1973.
4. Webster F.V. Priority to Buses as Part of Traffic Management. TRRL Report LR 448, Crowthorne, 1972.
5. Blows J.W., Proceedings of a Symposium held at TRRL, TRRL Report LR 570, Crowthorne, 1973.
6. Cecilia, F., How to Improve Conditions by Restraining Private Traffic Preference of Public Surface Transport, Proceedings of the Symposium on Techniques of Improving Urban Conditions by Restriant of Road Traffic, Organization for Economic Co-operation and Development, Paris, 1973.
7. Robertson, D.I and R.A. Vincent, Bus Priority in a Network of Fixed Time Signals, TRRL Report 666, Crowthorne, 1975.
8. Highway Capacity Manual, Special Report No. 87, Highway Research Board, Washington, 1965.
9. Coste, J.F., Bus Lane Experiments in Marseilles, Proceedings of a Symposium held at TRRL, TRRL Report LR 570, Crowthorne, 1973.
10. Transportation Planning Handbook, Institute of Transportation Engineers, Practice Hall, Englewood Cliffs, 1992.

SECTION 11

HIGHWAY CAPACITY

21. Highway Capacity

21.1. General

Capacity of a facility is defined as the maximum hourly rate at which persons or vehicles can reasonably be expected to traverse a point or uniform section of a lane or roadway during a given time period under prevailing roadway, traffic and control conditions (Ref. 1, 17).

The time period normally used for expressing capacity is the hour, and capacity is, therefore, expressed in so many vehicles or persons per hour. Sometimes it is considered more expedient to bring all the vehicles to a common type, usually the passenger car. The capacity is then expressed in passenger car units (PCUs) per hour.

The words "under prevailing roadway and traffic conditions" in the above definition are highly significant. These conditions affect the ability of a roadway to accommodate traffic, and hence it is essential to state these conditions whenever capacity is being expressed.

The earliest edition of the Highway Capacity Manual (1950) had defined three types of capacity (Ref. 2) :

- ***Basic capacity.*** The maximum number of passenger cars that can pass a point on a lane or a roadway during one hour under the most nearly ideal roadway and traffic conditions which can possibly be attained.
- ***Possible capacity.*** The maximum number of vehicles that can pass a given point on a lane or roadway during one hour, under prevailing roadway and traffic conditions.
- ***Practical capacity.*** The maximum number of vehicles that can pass a given point on a lane or roadway during one hour, without the traffic density being so great as to cause unreasonable delay, hazard or restriction to the driver's freedom to manoeuvre under prevailing roadway and traffic conditions. This term is frequently referred to as the design capacity.

From the above definitions it is seen that "capacity" in the 1985 Manual (Ref. 1) and in the 2000 Manual (Ref. 17) are the same as the "possible capacity" in the 1950 Manual. The "basic capacity" of the 1950 Manual has now been replaced with the phrase "capacity under ideal conditions". The earlier "practical capacity" has been replaced now by several "service volumes", related to a group of desirable operating conditions collectively termed as "level of service".

21.2. Importance of "Capacity" in Highway Transportation Studies

An understanding of the capacity of a roadway in highly important to a highway transportation engineer. The following are some of the applications :

(*i*) The design of a highway facility is possible only when capacity is related to the projected requirements of traffic. The design features governed by capacity are the highway type, number of lanes, width of lanes, intersections and weaving sections.

(*ii*) By comparing the present traffic volume with the capacity of existing highway networks, their adequacy or deficiency can be assessed. Transportation planning studies can be benefitted by this assessment.

(*iii*) Improvements and changes in the geometric features, junction features, traffic control devices and traffic management measures can be planned effectively if capacity studies are considered.

21.3. Early Capacity Studies and Theoretically Derived Formulae—Historical Perspective

Early attempts to determine the capacity were made on the basis of the following theoretical formula :

$$C = \frac{1000\,V}{S} \qquad \text{...(21.1)}$$

where, C = Capacity in vehicles per hour per lane

V = Speed, in K.P.H.

S = Average spacing in metres of moving vehicles.

The value of S, the headway distance, is known from actual observations or can be calculated from considerations of perception time braking distance and length of vehicle. The following formula is generally used for determining S :

$$S = L + \frac{tV\,.\,1000}{3600} + \left(\frac{V\,.\,1000}{3600}\right)^2 \cdot \frac{1}{2gf} = L + 0.278\,Vt + \frac{V^2}{254\,f} \qquad \text{...(21.2)}$$

where, S = Spacing in metres, L = Length of vehicle in metres

V = Speed in K.P.H.

t = Perception-reaction time in seconds.

f = Friction factor

g = Acceleration due to gravity, m/sec^2.

A number of early studies were concerned with measurement of S in an actual traffic stream and values of V. Pioneering in this areas was Greenshields (1933), who used a time lapse photographic technique (Ref. 3). The relation obtained by him connecting S and V was :

$$S = 21 + 1.1\,V$$

where S = Spacing in feet ; V = Speed in M.P.H.

The Road Research Laboratory, U.K., measured the headway between vehicles at various speeds and found the following relation to hold good in practice (Ref. 4) :

$$S = 17.5 + 0.8\,v + 0.004\,v^2$$

where, S = Headway in feet ; v = Speed in feet/sec.

The approach outlined above has not been found to be satisfactory for the following reasons :

(*i*) The above approach is based on the assumption that all vehicles travel at the same speed and are spaced at equal intervals. In actual practice this is rarely fulfilled.

(*ii*) The calculations made by different authorities for capacity at a given speed have been observed to vary widely, thus confusing rather than enlightening the engineers (Ref. 5).

Realising the inadequacy of the approach, the Highway Research Board Committee on Highway Capacity under the chairmanship of late O.K. Normann, continued the search for a rational and practical method of determination of highway capacity. This culminated in the publication of the Highway Capacity Manual in 1950, wherein the modern ideas of capacity were crystallized (Ref. 2). This manual provided a basis for measuring capacity in relation to speed and spacing, relying mainly on the empirical approach wherein relationships between two or more variables were developed from field observations. The 1965 edition was a further refinement in the concepts of capacity, embodying the empirical approach as well as the theoretical or experimental methods of mathematically expressing traffic flow (Ref. 16). The 1985 edition (Ref. 1) and the 2000 edition (Ref. 17) are further refinements.

21.4. Passenger Car Units (PCUs)

In British practice it is usual to express capacity in terms of "passenger car units". This system is also being followed in India. The basic consideration behind this practice is that different types of vehicles offer different degrees of interference to other traffic and it is necessary to bring all types to a common unit. The common unit adopted is the Passenger Car Unit (PCU).

If the addition of one vehicle per hour in a traffic stream reduces the average speed of the remaining vehicles by the same amount as the addition of say, x cars per hour, then one vehicle of that type is considered to be equivalent to x PCUs (Ref. 4).

The passenger car unit of a vehicle type has been found to depend upon the size and speed of the vehicle, type and kind of road environment (*e.g.*, rural road, urban street, roundabout, traffic signal). They are generally not dependent on the flow and road width.

The PCU equivalents are generally derived from observations, though rough approximations can also be derived by theoretical considerations (Ref. 6, 15).

The following table gives the British practice on PCUs.

Table 21.1. Passenger Car Unit Equivalent–British Practice

Class of vehicle	*Urban standards (Ref. 7)*	*Rural standards (Ref. 8)*	*Round-about (Ref. 7)*	*Traffic signals (Ref. 9)*
1. Private car, taxi, motor-cycle combinations, light goods vehicle (upto 30 cwt unladen).	1.00	1.00	1.00	1.00
2. Motor cycle (solo), motor scooter, moped.	0.75	1.00	0.75	0.33
3. Medium or heavy goods vehicle (over 50 cwt unladen), unladen), horse-drawn vehicle.	2.00	3.00	2.80	2.25
4. Bus, coach, trolley bus, tram.	3.00	3.00	2.80	2.25*
5. Pedal cycle.	0.33	0.50	0.50	0.20

*2.50 for tram.

In India, the following PCU equivalents have been recommended for rural condition (Ref. 10).

Table 21.2. PCU Equivalents as per Indian Practice

S. No.	*Vehicle Type*	*PCU Equivalency factor*
1.	Motor cycle or scooter	0.50
2.	Passenger car, pick-up van or auto-rickshaw	1.00
4.	Truck or bus	3.00
5.	Truck-Trailer, agricultural Tractor-trailer	4.50
6.	Cycle	0.50
7.	Cycle rickshaw	2.00
8.	Hand cart	3.00
9.	Horse-drawn vehicle	4.00
10.	Bullock cart	8.00*

*For smaller bullock carts, a value of 6 will be appropriate.

Based on recent research in India (Ref. 18) the following values of PCU equivalents are proposed in the draft revision (Ref. 19) of I.R.C Guidelines for capacity.

Table 21.3. Proposed PCU Equivalents in the Draft Revision of I.R.C Guidelines for Capacity

Vehicle Type	*PCU Single lane and two lane roads*	*Multilane roads*
Passenger car or pick up van	1.0	1.0
Bus or Truck	4.0	4.5
Multi-Axled Truck	5.0	6.0
Light Commercial Vehicle	2.5	2.8
Tractor	3.0	3.0
Tractor-Trailer	4.5	4.5
3-wheeler (Auto-rickshaw)	1.4	1.6
Motorized two wheeler	0.4	0.5
Pedal Cycle	0.6	0.6
Pedal rickshaw	2.0	2.5
Horse Cart	4.0	4.0
Bullock Cart	7.0	7.0

21.5. Capacity for Uninterrupted Flow Conditions in the HCM Manual (1985 and 2000)

The Highway Capacity Manual (1985) defines ideal conditions for uninterrupted flow as follows (Ref. 1, 17) :

(*i*) Uninterrupted flow, free from side interferences of vehicles and pedestrians.

(*ii*) Passenger cars only in the traffic stream.

(*iii*) Traffic lanes 3.6 m wide, with adequate shoulders and no lateral obstructions within 1.8 m of the edge of the pavement.

(*iv*) For rural highways, horizontal and vertical alignment satisfactory for average highway speeds of 100 K.P.H. for multi-lane highways.

The above ideal conditions are rarely met with in actual roads and the capacity figures need to be reduced. Guidance for reducing the capacity are given separately in Sec. 21.7.

The uninterrupted flow capacities under ideal conditions taken from the Highway Capacity Manual are given in Table 21.4.

Table 21.4. Uninterrupted Flow Capacities Under Ideal Conditions (Ref. 1, 17)

Highway Type	*Capacity (Passenger vehicles per hour)*
Freeways	2,400 per lane with free speeds of 120 km/hr. 2,300 per lane with free speeds of 110 km/hr.
Multi lane Highways	2,200 per lane width with freespeeds of 100 km/hr.
Two lane	Two-way 2800 total, both directions.

The above are the average maximum values that have been determined on the basis of extensive studies.

21.6. Level of Service Concept in the HCM Manual

When a road is carrying a traffic equal in volume to its capacity under ideal roadway and traffic conditions, the operating conditions become poor. Speed drops down and the delay and frequency of stops mount up. The service which a roadway offers to the road user can vary under different volumes of traffic. The Highway Capacity Manual (Ref. 1, 16) has introduced the concept of "Level of Service" to denote the level of facility one can derive from a road under different operating characteristics and traffic volumes. The concept of levels of service is defined as a qualitative measure describing the operational conditions within a traffic stream, and their perception by motorists and/or passengers. The following are the factors which might be considered in evaluating the level of service :

1. Speed and travel time, including the operating speed and overall travel time consumed in travelling over a section of roadway.

2. Traffic interruptions or restrictions, with due consideration to the number of stops per mile, delays involved and the speed changes necessary to maintain pace in the traffic stream.

3. Freedom to manoeuvre to maintain the desired operating speeds.

4. Driving comfort and convenience reflecting the roadway and traffic conditions.

5. Economy, with due consideration operating cost of the vehicle.

Even though it is desirable to consider all the above factors in identifying a particular level of service, it is difficult to incorporate all these in the absence of accurate data. The HCM, therefore, utilises (*i*) travel speed and (*ii*) the ratio of the service volume to capacity, depending upon the particular problem. The latter is often referred to as "*v/c* ratio" in the Manual. As regards the travel speed, the Manual recommends the use of operating speeds on those types of highway carrying generally uninterrupted flow, such as in rural areas. For urban locations, the Manual recommends the use of average overall travel speed.

The operating conditions for the six levels of service selected by the Manual are given below, Level *A* representing the highest and Level *F* representing the lowest :

Level of Service A. Free flow, with low volumes and high speeds. Traffic density is low, with speeds controlled by drivers' desired speed limits and physical roadway conditions. Little or no restriction in manoeuvrability due to presence of other vehicles, and drivers can maintain their desired speeds with little or no delay.

Level of Service B. Zone of stable flow, with operating speeds beginning to be restricted somewhat by traffic conditions. Drivers still have reasonable freedom to select their speed and lane of operation. Reductions in speed are not unreasonable. The lower limit (lowest speed, highest volume) of this level of service has been associated with service volumes used in the design of rural highways.

Level of Service C. Still in the zone of stable flow, but speeds and manoeuvrability are more closely controlled by higher volumes. Most of the drivers restricted in the freedom to select their own speed, lane changing or overtaking manoeuvres. A relatively satisfactory operating speed is still obtained with service volumes perhaps suitable for urban design practice.

Level of Service D. Approaches unstable flow, with tolerable operating speeds being maintained though considerably affected by changes in operating conditions. Fluctuations in volume and temporary restrictions to flow may cause substantial drops in operation speeds. Drivers have little freedom to manoeuvre : comfort and convenience are low, but conditions can be tolerated for short periods of time.

Level of Service E. Can not be described by speed alone, but represents operations at even lower operating speeds than in level *D*, with volumes at or near the capacity of the highway. At capacity, speeds are typically but not always in the neighbourhood of 50 K.P.H. Flow is unstable, and there may be stoppages of momentary duration.

Level of Service F. Forced flow operations at low speeds, where volumes are below capacity. Conditions result from queues of vehicles backing up from a obstruction downstream. The section under study will be serving as a storage area during parts or all of the peak hour. Speeds are reduced substantially and stoppages may occur for short or long periods of time, because of downstream congestion. In the extreme, both speed and volume can drop to zero.

Fig. 21.1 gives a typical curve showing the relationship generally found between operating speed and volume/capacity ratio. In this figure are depicted the zones where the different levels of service are generally found to occur. The diagram pertains to a multilane highway.

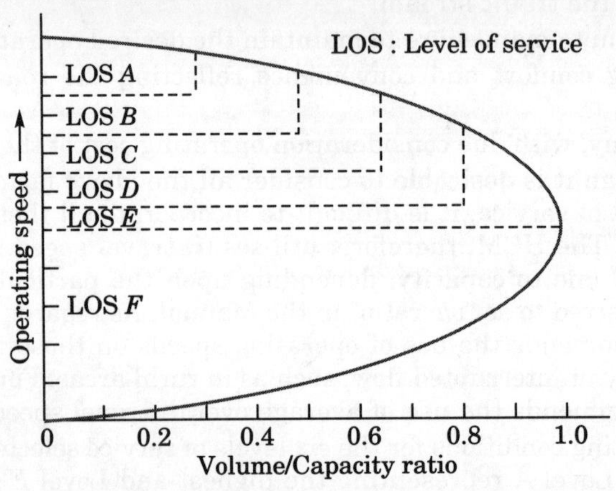

Fig. 21.1. Typical Speed Flow Curve for Multilane Highways.

The HCM terms the traffic volumes that can be served at each level of service as the *service volume*. After having selected a particular level of service for design purposes, the corresponding service volume then logically becomes the *design capacity,* which signifies that if the traffic volume using the road exceeds that value, the operating conditions will fall below the level of service selected for design.

The HCM contains no recommendations regarding the applicability of the different levels of service for the design of different types of highway facility. The choice is left to the designer, who will select an appropriate level of service based on considerations such as economy and need of a particular type of highway facility.

21.7. Factors Affecting Capacity and Level of Service

The factors which affect the capacity and level of service can be considered under the following two main categories :

(*i*) Roadway factors (*ii*) Traffic factors.

Roadway Factors

Roadway factors pertain to restrictive physical features of a road such as lane width, lateral clearance, shoulders, auxiliary lanes, surface conditions, alignment and grade.

Lane width. A lane width of 3.65 m is considered as the defined ideal, and the capacities of lanes narrower than this as given by the Highway Capacity Manual are given in Tables 21.5 and 21.6.

Lateral clearance. Lateral obstructions such as retaining walls, abutments, signposts, light poles and parked cars, located closer than 1.83 m from the edge of a traffic lane reduce the capacity. Tables 21.5 and 21.6 give the effect of lateral clearance on capacity.

Mountable curbs and vertical curbs 15 cm or less in height have insignificant influence in traffic operations.

Shoulders. Shoulders of adequate width are necessary for maintaining traffic continuously. The Highway Capacity Manual reckons that for lanes less than 3.65 m wide, paved shoulders of 1.2 m or more width increase the effective width of the adjacent traffic lanes by 0.3 m.

Table 21.5. Effect of Lane Width and Lateral Clearance on Capacity of Multi-lane Highways

(*Values given are multiplying factors*)

Distance from edge of travelled way to obstruction (m)	*Obstructions on one side* — *Lane width (m)*				*Obstructions on both side* — *Lane width*			
	3.65	*3.35*	*3.05*	*2.75*	*3.65*	*3.35*	*3.05*	*2.75*
	Four lane divided highway							
1.83	1.00	0.97	0.91	0.81	1.00	0.97	0.91	0.81
1.22	0.99	0.96	0.90	0.80	0.98	0.95	0.89	0.79
0.61	0.97	0.94	0.88	0.79	0.94	0.91	0.86	0.76
0	0.90	0.87	0.82	0.73	0.81	0.79	0.74	0.66
	Six lane divided highway							
1.83	1.00	0.96	0.89	0.78	1.00	0.96	0.89	0.78
1.22	0.99	0.95	0.88	0.77	0.98	0.94	0.87	0.77
0.61	0.97	0.93	0.87	0.76	0.96	0.92	0.85	0.75
0	0.94	0.91	0.85	0.74	0.91	0.87	0.81	0.70

Table 21.6. Effect of Lane Width and Lateral Clearance on Capacity of Two Lane Roads (*Values given are multiplying factors*)

Usable shoulder width (m)	*3.65 m lanes*		*3.35 m lanes*		*3.05 m lanes*		*2.75 m lanes*	
	LOS A-D	*LOS E*	*LOS A-D*	*LOS E*	*LOS A-D*	*LOS E*	*LOS A-D*	*LOS E*
1.83	1.00	1.00	0.93	0.94	0.84	0.87	0.70	0.76
1.22	0.92	0.97	0.85	0.92	0.77	0.85	0.65	0.74
0.61	0.81	0.93	0.75	0.88	0.68	0.81	0.57	0.70
0	0.70	0.88	0.65	0.82	0.58	0.75	0.49	0.66

Surface condition. A deteriorated and poorly maintained pavement adversely affects level of service but the Highway Capacity Manual states that adequate data are not available to develop suitable adjustment factors to reflect the effect of surface condition on capacity.

Alignment. The factors under alignment that affect capacity are the sharpness of curves and the percentage of road length with substandard overtaking sight distance. The Highway Capacity Manual contains tables and charts to account for the influence of these factors.

Grades. Grades adversely affect the speed of vehicles, especially trucks, and thus influence the capacity. The Highway Capacity Manual combines the effect of grades and the percentage of trucks in the traffic stream by giving suitable passenger car equivalent factors. The PCU of trucks and buses recommended for the analysis of multi lane highways are given in Table 21.7.

Table 21.7. Passenger Car Equivalents of Trucks and Buses on Multi lane Highways (*Ref.* 1)

Vehicle	*Terrain*		
	Plain	*Rolling*	*Mountainous*
Truck	1.7	4.0	8.0
Bus	1.5	3.0	5.0

Table 21.8 gives the average generalised passenger car equivalents of trucks and buses of 2-lane highways, over extended section lengths. It will be seen that the table takes into account the combined effect of the gradient and the presence of trucks and buses.

Table 21.8. Average Generalised Passenger Car Equivalent of Trucks and Buses on 2 lane Highways Over Extended Section Lengths (*Adapted from Ref.* 1)

Type of vehicle	*Level of service*	*Passenger car equivalents for*		
		Level terrain	*Rolling terrain*	*Mountainous terrain*
Trucks	*A*	3.0	4.0	7.0
	B and *C*	2.2	5.0	10.0
	D and *E*	2.0	5.0	12.0
Buses	*A*	1.8	3	5.7
	B and *C*	2.0	3.4	6.0
	D and *E*	1.6	2.9	6.5

The U.K. practice to account for the presence of different vehicles in the traffic stream is to assign suitable PCU factors. Similar practice is being followed in India.

21.8. Capacity of Freeways and Expressways in the Rural Areas

The distinguishing feature of freeways and expressway is the control over access provided in their design and operation. Through traffic is favoured exclusively and local traffic and service to the abutting properties are absent. These facilities are designed for a high design speed (usually 120 K.P.H.) and are meant to cater for a high volume of traffic. They have ideal in-built geometric features.

The levels of service and the related operating characteristics for freeways are given below :

Level of Service A. Free-flow conditions prevail in which a vehicle is not affected by other vehicles and driver is free to select the speed based on individual choice and design features. Operating speeds at or greater than 100 K.P.H. Service volume at this level is 700 passenger cars per hour per lane one direction.

Level of Service B. Higher speed range at stable flow. Operating speeds at or greater than 95 K.P.H. Service volume in one direction is around 1100 passenger cars per hour per lane.

Level of Service C. At this level operation is still stable. Operating speed of atleast 90 K.P.H. are needed. Service flow rate in one direction not exceeding 75 percent of the capacity rate *(i.e.,* 1550 passenger cars per hour per lane).

Level of Service D. This level represents the lower speed range of stable flow. Traffic operation approaches instability and becomes very susceptible to change in operating conditions. Operating speeds are generally in the range of 75 K.P.H. Service flow rates are about 0.90 of capacity rates *(i.e.,* 1850 passenger cars per hour per lane).

Level of Service E. This level represents unstable flows. Overall operating speeds of 50–60 K.P.H. Volumes approaching and at capacity of about 2000 passenger cars per lane per hour. This level is found in many freeways under peak period conditions, but design should never be attempted at this level.

Level of Service F. Forced flow conditions with operating speeds ranging from 50 K.P.H. to stop and go operation. Expressway acts as a storage for vehicles backing up from a downstream bottleneck.

Rural freeways and expressways are normally designed for level of service *B* giving a design service volume of 1100 passenger cars per hour per lane in one direction. In urban situations a level of service *C* may be more appropriate, giving a design service volume of 1500 passenger cars per hour per lane in one direction. (Ref. 11).

In the design of capacity of freeways, other critical elements that need evaluation are the ramp terminals and weaving sections. The Highway Capacity Manual gives detailed procedure for determining the level of service under such conditions.

The U.K. Rural Motorways Standards for carriageway capacities sections of such highways where the average journey length is more than 40 km are as given in Table 21.9. (Ref. 6).

Table 21.9. Design Capacity of Rural Motorways
(U.K. Practice—Ref, 8)

Type of carriageway	*Daily design capacity PCUs/day*	*Peak hourly design Capacity per lane PCUs/hour*
Dual 2-lane	33,000	1000
Dual 3-lane	50,000	1000
Dual 4-lane	66,000	1000

The design capacity values recommended by AASHTO on multilane highways in rural and suburban areas are given in Table 21.10. (Ref. 12).

Table 21.10. Design Capacity of Multi-lane Highways in Rural Areas as per AASHTO Practice

Type of highway	*Design capacity PCUs*
1. Freeways, (Controlled Access)	1,120 per lane
2. Multi-lane Highways without control of access	1,200 per lane

21.9. Capacity of Two lane Rural Highways, Without Access Control

Two lane rural highways without access control are a common part of the highway system of any country. Since the distribution of traffic by direction has practically no effect on operating conditions, the capacity and service volumes of a two lane highway are expressed in total vehicles per hour, in both directions. The service volumes are governed to a large extent by the availability of adequate overtaking sight distance.

The levels of service and the related operating conditions are given below :

Level of Service	*Operating Characteristics*
A	Represents free flow conditions. Operating speeds must be higher than 100 K.P.H. If overtaking sight distances are always available, volumes may reach 420 passenger cars per hour, both directions. About 75 percent of the desired overtaking manoeuvres can be made with little or no delay.
B	This level of service marks the beginning of stable flow, with operating speeds 90 K.P.H. or above. Service volumes may reach 750 passenger cars per hour, both directions.
C	The flow is still stable, but operating speeds are 80 K.P.H. or above. Service volume may reach 1200 passenger cars per hour, both directions.
D	Approaching unstable flow, with operating speeds falling down to 60 K.P.H. Service volume may reach 1800 passenger cars per hour, both directions.
E	This represents the capacity level with volumes reaching 2800 passenger cars per hour, both directions. Operating speeds usually in the neighbourhood of 50 K.P.H.
F	Forced flow conditions, operating speeds less than 50 K.P.H. and volumes are under 2000 passenger cars per hour, both directions.

For design of rural sections of two-lane highways, it is appropriate to adopt a level of service *B*. This gives a design service volume of 750 passenger cars per hour, both directions. AASTHO practice is given below in Table 21.11. (Ref. 12)

Table 21.11. Possible Design Capacities of 2-Lane Rural Highway as per AASHTO Practice *(Source : Ref. 12)*

Possibility capacity	*Design capacity for average running speeds of*		
	60–65 K.P.H.	65–75 K.P.H.	75–84 K.P.H.
2000 veh. per hour	1,500	1,150	900

In India, the I.R.C recommendations for design service volume are given in Tables 21.12, 21.14 and 21.23. (Ref. 10 and 19). A major research project to prepare an Indian Highway capacity manual is under progress in India, being implemented by the Central Road Research Institute.

Table 21.12. Suggested Design Service Volume for Indian Conditions for Two Lane Roads

S. No.	*Terrain*	*Curvature (Degrees per kilometre)*	*Suggested design service Volume in PCU/day*	
			Ref. 10	*Ref. 19*
1.	Plain	Low (0–50)	15,000	21,400
		High (above 51)	12,500	
2.	Rolling	Low (0–100)	11,000	15,700
		High (above 101)	10,000	
3.	Hilly	Low (0–200)	7,000	10,000
		High (above 201)	5,000	

Table 21.13. Suggested Design Service Volumes for Indian Conditions for Intermediate Lane (5.5) Roads (Ref. 10)

S. No.	*Terrain*	*Curvature (Degrees per kilometre)*	*Suggested design service volume in PCU/day*
1.	Plain	Low (0–50)	6,000
		High (above 51)	5,800
2.	Rolling	Low (0–100)	5,700
		High (above 101)	5,600
3.	Hilly	Low (0–200)	5,200
		High (above 201)	4,500

Table 21.14. Suggested Design Service Volumes for Indian Conditions for Single Lane Roads

S. No.	*Terrain*	*Curvature (Degrees per kilometre)*	*Suggested design service volume in PCU/day*	
			Ref. (10)	*Ref. (19)*
1.	Plain	Low (0–50)	2000	5000
		High (above 51)	2000	
2.	Rolling	Low (0–100)	1800	5000
		High (above 101)	1800	
3.	Hilly	Low (0–200)	1800	4000
		High (above 201)	1400	

The U.K. Standards provide for a design capacity on 9000 PCUs per day for a two-lane road (7.30 m width), which corresponds to a level of service *B*, the peak hour volume being 10 percent of the daily volume. (Ref. 8) The combined effect of non-standard lane width and lateral obstructions is taken into account by reducing the design capacity as per values given in Table 21.15.

Table 21.15. Capacity of Roads Restricted in Width and/or Clearance as Percentage of Standard Capacities
(U.K. Practice : Ref. 8)

Type of road	Carriage way width in metres	Obstruction on one side of road — Distance from carriageway in metres: 0	0.5	1.0	1.5	2.0 or more	Obstruction on both sides of road — Distance from carriageway in metres: 0	0.5	1.0	1.5	2.0 or more
Dual 3 lane road	11	94	97	98	99	100	90	95	97	99	100
	10	91	93	94	95	96	87	91	93	95	96
	9	85	87	88	88	89	81	84	86	88	89
	8.5	77	78	79	80	81	72	76	78	80	80
Dual 2 lane road	7.3	90	96	98	99	100	81	92	97	99	100
	6.75	87	93	95	97	97	79	89	94	96	97
	6.0	82	87	89	90	91	74	84	88	90	91
	5.5	73	78	79	89	81	66	75	78	80	81
2 lane single carriage way	7.3	85	90	95	98	100	70	79	88	96	100
	6.75	73	77	81	85	86	60	68	76	83	87
	6.0	66	69	72	76	77	54	61	68	74	78
	5.5	60	68	66	69	70	49	55	62	68	71

The U.K. Standards also indicate the reduction in capacity due to non-availability of overtaking sight distance, vide Table 21.16. (Ref. 8).

Table 21.16. Effect on Design Capacity and Speed where Overtaking Sight Distances are not Provided

Percentage of road length with sub-standard overtaking sight distance	0	20	40	60	80	100
Percentage of standard design capacity.	100	90	80	65	50	30
Estimated reduction in average speed km/hour of two-lane road carrying 900 PCUs per hr.	0	6	12	18	24	30

21.10. Capacity of Multi-lane Rural Highways Without Access Control

Multi-lane rural highways without access control are those which are undivided or which lack significant control of access features, or both. They are generally designed for a speed of 100 K.P.H.

The levels of service and the related operating conditions for such type of facilities are given below :

Level of Service	*Operating Characteristics*
A	Operating speed of 100 K.P.H. or more. Service volume is about 33 percent of capacity, *i.e.,* 650 passenger cars per hour per lane.
B	Beginning of stable flow area, operating speed about 90 K.P.H. Service volume will not exceed 50 percent of capacity, *i.e.,* 1000 passenger cars per hour per lane.
C	Stable flow, with service volume not exceeding 65 percent of capacity, *i.e.,* 1300 passenger cars per hour per lane. Operating speed at least 75 K.P.H.
D	Approaching unstable flow, with service volume upto 80 percent of capacity, *i.e.,* 1600 passenger cars per hour per lane. Operating speed about 60 K.P.H.
E	Flow at capacity, *i.e.,* 2000 passengers cars per hour per lane. Operating speed about 50 K.P.H.
F	Forced flow, congested conditions, with operating speed less than 50 K.P.H.

Level of service *B* is usually adopted for determining the design service volumes. The AASHTO practice for design capacity of multi-lane highways with interference from cross traffic and roadsides may be seen in Table 21.9 given earlier.

In U.K. practice, the capacity of dual carriageway roads other than motorways in the rural area is determined on the basis of a design service volume of 1000 PCUs/hour, per lane in the direction of heavier traffic (Ref. 8). The assumption made is that one carriageway carries 60 percent of the total flow in both directions. The design capacities recommended are given in Table 21.16.

Table 21.16. A Design Capacity of Dual Carriageways other than Motorways in U.K. (*Ref. 8*)

Carriageway width	*Design capacity (PCUs/day)*
Dual 14.6 m	66,000
Dual 11.00 m	50,000
Dual 7.30 m	33,000

Table 21.15 given earlier indicates the reduction in the above design capacities due to restricted widths and are lateral clearances.

In India, I.R.C recommends a design service volume of 40,000 PCUs per day for fourlane divided carriageway under mixed traffic conditions (Ref. 10).

The proposed revision of I.R.C Guidelines (Ref. 19) gives the following values of capacity.

Table 21.17. Recommended Design Service Volumes for Four-lane divided roads (Ref. 19)

Terrain	*Shoulder Type*	*Suggested design service volume in PCUs/day at*	
		LOSB	*LOSC*
Plain	Paved	75,000	105,000
	Earthen	67,000	95,000
Rolling	Paved	64,000	90,000
	Earthen	57,000	80,000
Hilly	Paved	53,000	75,000
	Earthen	47,000	66,000

21.11. Capacity of Urban Streets

Urban streets can be basically of 2 types :

(*i*) Arterial streets (*ii*) Downtown streets.

Urban streets are characterised by intersections at frequent intervals (signalised or uncontrolled), speed regulations, extensive roadside development, mixed traffic conditions and a high degree of congestion. Arterial streets have longer spacing of intersection and lesser roadside development than downtown streets. Broadly the intersections control the capacity of the street system as a whole and it, therefore, becomes critical to evaluate the capacity of intersections.

The levels of service for urban and suburban arterials can be related to the flow conditions, average overall travel speed, load-factor at intersections, peak-hour factor, and service volume capacity ratio. The peak hour factor is defined as the ratio of the volume occurring during the peak hour to the maximum rate flow during a given time period within the peak hour (say 5 minutes, 10 minutes or 15 minutes) (Ref. 1). The load factor is defined as the ratio of the total number of green signal intervals that are fully utilised by traffic during the peak hour to the total number of green intervals for that approach during the same period. The characteristics of different levels of service for urban and suburban arterials are summarised below :

Level of Service	*Operating Characteristics*
A	The average overall travel speed is 80 K.P.H. or more. Relatively free flowing, with service volume capacity ratio of 0.60 or less. Load factor at intersections about 0.0. Peak hour factor at 0.70 or less.
B	Stable flow. Average overall travel speed drops down to 40 K.P.H. or more. Slight delay is common. Service volume/ capacity ratio is 0.70 or less. Load factor about 0.1 or less, arid peak hour factor 0.80 or less.
C	Stable flow, with acceptable delays. Average overall travel speed drops down to 30 K.P.H. or more. Service volume/ capacity ratio is 0.80 or less. Load factor about 0.3 or less, and peak hour factor is 0.85 or less.
D	Approaching unstable flow, with tolerable delay. Average overall travel speed drops down to 25 K.P.H. Service volume/capacity ratio about 0.90 or less. Load factor 0.7 or less and peak hour factor 0.90 or less.
E	Unstable flow with congestion and intolerable delay. Average overall travel speed about 25 K.P.H. Service volumes are at capacity, or near

abouts. Load factor at intersection 1.0 or less, and peak hour factor 0.95 or less.

F Forced flow, with jammed conditions. Average overall travel speed below 15 K.P.H. Demand volume/capacity ratio may well exceed 1.0. Intersections overloaded.

For design purposes, a level of service of *C* is perhaps suitable for urban streets (Ref. 1, 11).

In the case of downtown streets the primary function is to afford service to local business, residences and similar abutting properties. The emphasis is more on circulatory flow rather than on through movements. Pedestrian volumes are rather heavy. Bus operations present impediments to the traffic flow. Many of the traffic operations in the downtown streets might appropriately fall under level of service *F* if measured on the same rating scale for higher type urban facilities. Considering the complex nature of traffic conditions existing and the absence of reliable and accurate data, the HCM recommends the following levels of service for such streets :

Level of Service	*Traffic Conditions*
A	Free flow relatively, with some stops likely to occur. Average overall speed 40 K.P.H. or more.
B	Stable flow, with delay, not unreasonable. Average overall speed 30 K.P.H. or more.
C	Stable flow, with significant delays which may be acceptable. Average overall speed 25 K.P.H.
D	Approaching unstable flow, with tolerable delays. Average overall speed 15 K.P.H. or more.
E	Unstable flow, with average overall speed below 15 K.P.H. but moving.
F	Forced flow, jammed conditions. Stop and go movement.

It may be noted that the level *E* cannot be considered as the capacity for downtown urban streets as in the case of other types of facilities ; the capacity is primarily governed by that of controlling intersections or other interruptions.

The AASHTO Policy on Urban streets recommends the design capacities given in Table 21.18 for arterial streets and urban highways. (Ref. 11).

Table 21.18. Design Capacities for Arterial Streets and Urban Highways

Type of Highway	*Design capacity (Passenger cars per hour per 3.65 m lane)*
1. Major suburban highway with moderate interference from cross traffic and roadsides, level of service *C*.	700-900
2. Major suburban highway, considerable interference from cross traffic and roadsides, level of service *C*.	500-700
3. Arterial street, traffic signals average 1.6 K.M. or more apart, parking prohibited and refuge provided for stalled vehicles, level of service *C*.	400-600
4. Arterial street, traffic signals average less than 1.6 K.M. apart, parking prohibited.	As governed by capacity of critical intersections

The practical capacities recommended by the U.K. Standards for Urban Roads are given in Table 21.19. (Ref. 7).

Table 21.19. Practical Capacities of Two-way Urban Roads

(Adapted from Ref. 7)

Effective width of carriageway (m)	*2-lane*			*3-lane*		*4-lane*			*6-lane*
	6.00	*6.75*	*7.30*	*9.00*	*10.00*	*12.0*	*13.5*	*14.6*	*22.0*
Description	*Capacity in PCUs per hour (Both directions of flow)*					*Capacity in PCUs per hour (One direction of flow)*			
1. Urban motorways with grade separation and no frontage access							3000		4500
2. All purpose roads with no frontage access, no standing vehicles permitted and negligible cross traffic	1200	1350	1500	2000	2200	2000	2400	2200	3600
3. All purpose street with high capacity junctions and 'no waiting' restrictions	800	1000	1200	1600	1800	1200	1500	1350	2500
4. All purpose street and capacity restricted by waiting vehicles with junctions	300 to 500	450 to 600	600 to 750	900 to 1100	1100 to 1300	800 to 900	1000 to 1200	900 to 1000	1600 to 2200

21.12. Capacity of Rotary Intersection U.K. Practice

Rotary Intersections (Round-abouts) are common in U.K. In India many intersections in urban locations are of this type. The general considerations in their design and layout are discussed in Chapter 11. Here the U.K. Practice for determining their capacities will be discussed.

Based on a number of experiments, the maximum capacity of the weaving sections in a rotary has been computed. The practical capacity is taken to be 80 percent of the maximum capacity and can be calculated from the following empirical formula, which is well known as the Wardrop formula (Ref. 7, 8, 13) :

$$Q_p = \frac{280\, w \left(1 + \dfrac{e}{w}\right)\left(1 + \dfrac{p}{3}\right)}{1 + \dfrac{w}{l}} \qquad \text{...(21.5)}$$

Further discussions on this formula are contained in Chapter 11.

21.13. Capacity of Weaving Sections-Highway Capacity Manual Approach

21.13.1. Weaving is defined as the crossing of streams of traffic moving in the same direction accomplished by merging and diverging. Weaving, occurs in a number of situations like interchanges, intersection freeway sections and ordinary highways. A knowledge of the capacity of the weaving section is, therefore, of considerable importance. Based on extensive research findings in the U.S.A., the Highway Capacity Manual outlines an approach to determine the capacity. (Ref, 1).

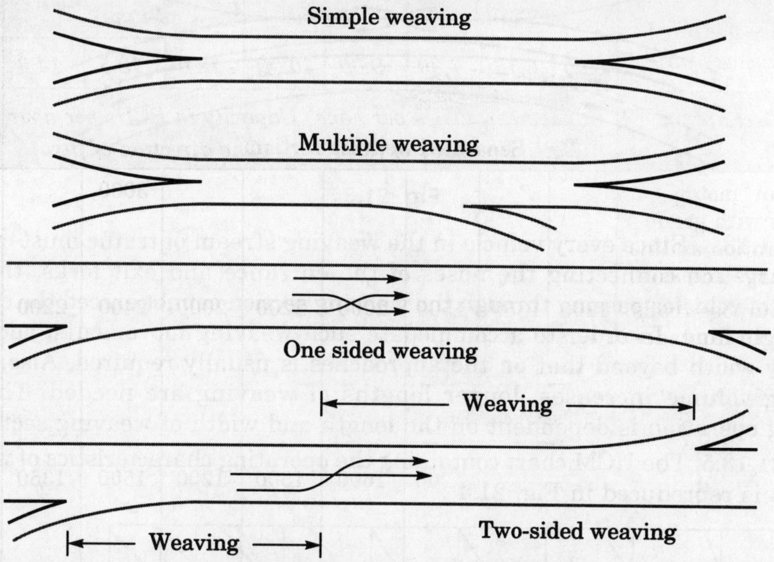

Fig. 21.2. Forms of weaving.

21.13.2. Some of the forms of weaving are illustrated in Fig. 21.2. Fig. 21.3 gives some weaving arrangements.

21.3.3. Traffic through a weaving section consists of weaving traffic and non-weaving traffic. These two groups tend to separate themselves and this enables each class to be examined and analysed separately. Non-weaving traffic is treated in the usual way as a lane of a highway.

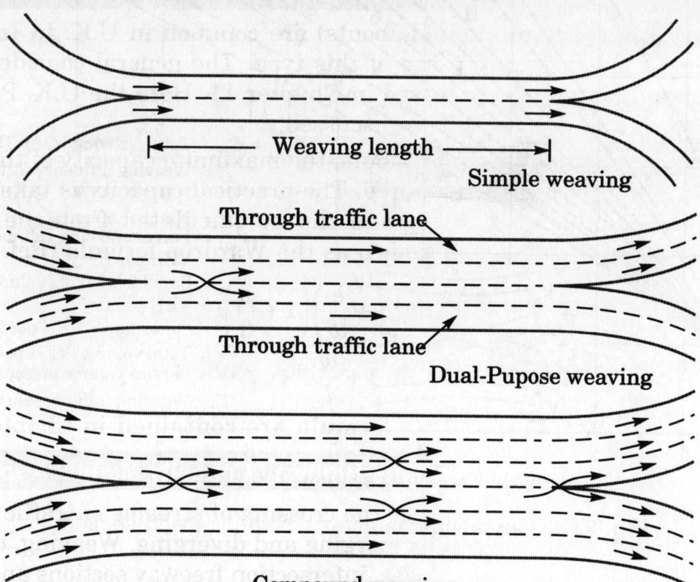

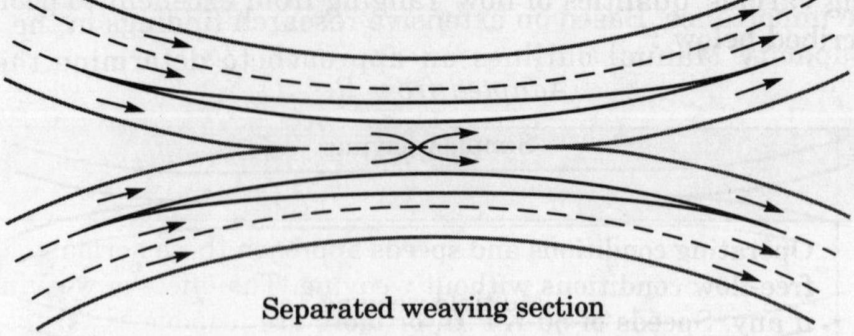

Fig. 21.3

21.13.4. Since every vehicle in the weaving stream of traffic must cross an imaginary line connecting the noses of the entrance and exit forks, the total number of vehicles passing through the weaving section cannot exceed the capacity of a single lane. In order to accommodate such weaving movements, additional roadway width beyond that on the approaches is usually required. Also, as the weaving volume increases, longer lengths of weaving are needed. Thus the weaving operation is dependent on the length and width of weaving section.

21.13.5. The HCM chart containing the operating characteristics of weaving sections is reproduced in Fig. 21.4 :

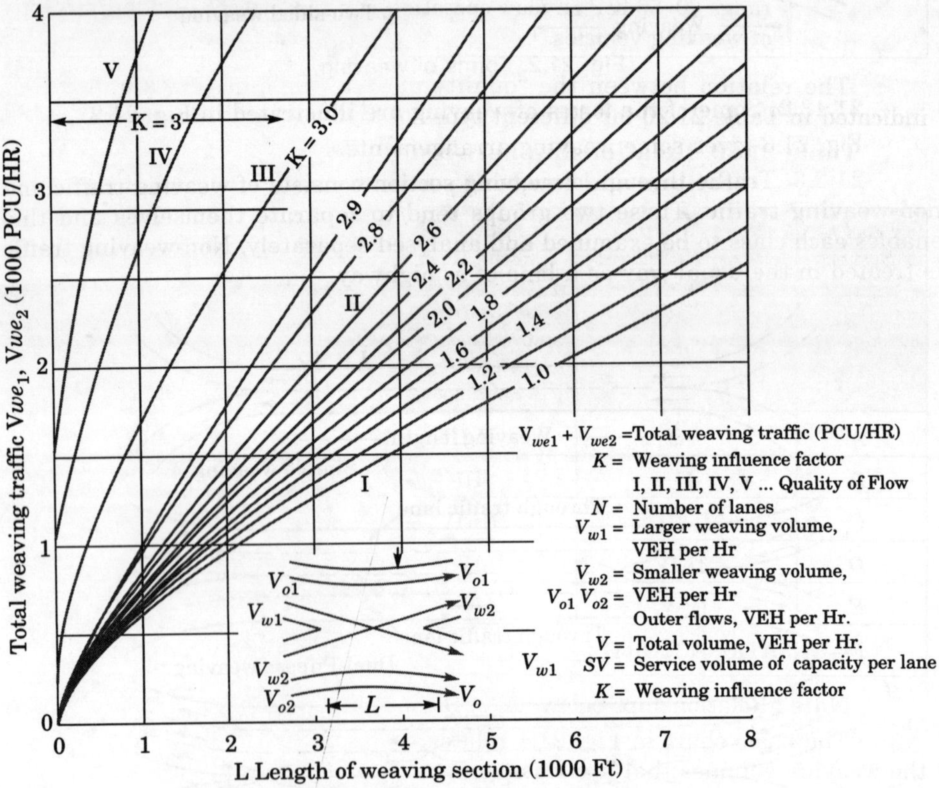

Fig. 21.4. Operating characteristics of weaving section.

The above chart contains a family of curves designated *I* through *V* representing various 'qualities of flow' ranging from excellent to poor. These are briefly described below :

(Adapted from Ref. 1)

Quality of flow	*Description of Operation*
I	Operating conditions and speeds approach those normally found under free-flow conditions without weaving. The effect of weaving is slight, if any. Speeds of 80 K.P.H. or more are feasible.
II	Operating conditions and speeds are only slightly more restricted than under free-flow conditions. The effect of weaving is slight to normal. Speeds of 75–80 K.P.H. can be achieved.
III	Drivers affected by other vehicles in the stream to a greater extent than normal under free-flow conditions. Speeds of 65–75 K.P.H. can be maintained by weaving vehicles. Suitable for approach roads with operating speed of 80 K.P.H.
IV	Weaving vehicles can maintain operating speeds of about 50-60 K.P.H. Occasional showdowns and restriction of manoeuvr ability. Suitable for operating speeds not exceeding 65 K.P.H.
V	This level represents the capacity of weaving section. Speeds in the range 30–50 K.P.H. Slow operation and turbulence, including stopping of weaving vehicles.

The relation between the "quality of flow" and the levels of service is indicated in Table 21.20 for different types of highways and streets :

Table 21.20. Relationship between quality of flow of weaving sections and levels of service (*Source : Ref. 1*)

Level of Service	*Freeways and Rural Highways Proper*	*Multi-lane Highways Connecting collector distributor roads and other interchange roadways*	*Two lane Rural Highway*	*Urban and Sub-urban Arterials*
A	I-II	II-III	II	III-IV
B	II	III	II-III	III-IV
C	II-III	III-IV	III	IV
D	III-IV	IV	IV	IV
E	IV-V	V	V	V
F—UNSATISFACTORY				

Note : Relationships below heavy illness not normally considered in design.

The basic chart in Fig. 21.4 relates the length of the weaving section to the weaving volumes that are feasible at different 'quality of flow' levels.

The first step in the solution of a weaving section is to determine the weaving length required from the known weaving volume and the desired quality of flow from the above chart. The determination of the width is made by calculating the widths separately for non-weaving flow and weaving flow. In the terminology adopted in Fig. 21.4, the number of lanes required to accommodate the nonweaving flow is $(V_{o1} + V_{o2})/SV$, where SV is the lane service volume. For the weaving flow, the number of lanes is calculated using the same value of lane service volume, but the flow is taken as $V_{w1} + k \cdot V_{w2}$, where V_{w1} is the larger weaving volume, V_{w2} is the smaller weaving volume and k is a factor (weaving influence factor) having values from 1.0 to 3.0. These values are indicated in the family of curves in Fig. 21.4. The maximum value (3.0) is applicable to the shorter weaving sections. The formula for the total number of lanes thus becomes :

$$N = \frac{V_{w1} + K \cdot V_{w2} + V_{o1} + V_{o2}}{SV} \quad ...(21.6)$$

If $V_{w1} + V_{w2} + V_{o1} + V_{02} = V$ (the total volume), then the above formula can also be written as :

$$N = \frac{V + (k-1)V_{w2}}{SV} \quad ...(21.7)$$

The Service Volume (SV) used in the above equations can be taken from the Table 21.21.

Table 21.21. Values of Service Volume (SV) for various quality of flow levels

Quality of flow	*Maximum lane SV value (Passenger cars per hr.)*
I	2,000
II	1,900
III	1,800
IV	1,700
V	1,600

As the length of the weaving section increases, the effect of weaving gets more and more dissipated and nullified. The *H.C.M* gives the guidelines given in Table 21.22 for determining the sections out of the realm of weaving.

Table 21.22. Volume-length combinations considered out of realm of weaving (*Adapted from Ref. 1*)

Volume Weaving (Passenger cars per hour)	*Minimum length of section (m)*
500	300
1000	750
1500	1200
2000	1800

The following example illustrates the use of the above approach for a simple weaving problem.

Problem 21.1. *AB and CD are two rural sections of a highway intersecting with a weaving section. It is desired to provide a level of service B for the traffic volumes indicated in the diagram below. The maximum service volume is 1000 passenger cars per hour per lane. Determine the minimum length and width of the weaving section.*

Solution. From Table 21.20, the level of serviced B corresponds to the quality of flow II.

Total weaving volume

$$= 700 + 400 = 1100 \text{ V.P.H.}$$

From Fig. 21.4, for a weaving volume of 1100 V.P.H. and quality of flow II, length of weaving section required is 450 m. Value of k for quality of flow II is 2.6.

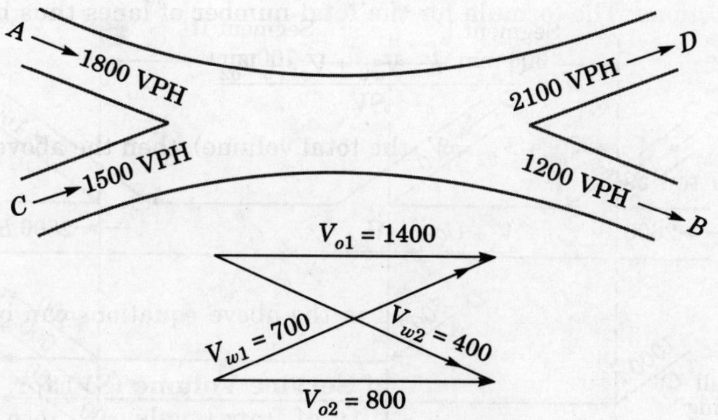

Fig. 21.5

$$\text{No. of lanes} = \frac{V + (k-1)V_{w2}}{SV} = \frac{3300 + (2.6 - 1)\,400}{1000}$$

$$= \frac{3300 + 640}{1000} = \frac{3940}{1000} = 3.94 \quad \text{Use 4 lanes.}$$

21.13.6. Multiple Weaving Sections

Multiple weaving sections are made up of overlapping simple weaving sections, each of which can be analysed separately. The relationships indicated in Fig. 21.4 equally apply to multiple sections as well. The following example illustrates the procedure to be adopted in analysing multiple weaving sections.

Problem 21.2. *Fig. 21.6 indicates the layout of a freeway having two entry ramps followed by an exit ramp. The weaving lengths of segment I and segment II are 300 m and 700 m respectively. It is required to determine the number of lanes required for the weaving sections in segment I and segment II on the basis of level of service C and service volume of 1200 vehicles for hour per lane.*

Solution. The volume of weaving that can take place in both the segments is apportioned between the segments on the basis of their weaving lengths.

The total weaving movement CB is 400 vehicles per hour and the total weaving movement AE is 700 vehicles per hour. These are apportioned between segments I and II as follows :

Segment I

$$\text{Weaving movement } CB = 400 \times \frac{300}{1000} = 120 \text{ V.P.H.}$$

$$\text{Weaving movement } AE = 700 \times \frac{300}{1000} = 210 \text{ V.P.H.}$$

Segment II

Weaving movement $CB = 400 - 120 = 280$ V.P.H.

Weaving movement $AE = 700 - 210 = 490$ V.P.H.

The above weaving movements are indicated in the incremental weaving diagram in Fig. 21.6.

It should be noted that in Segment II, the traffic which weaves with DB (500 V.P.H.) is 700 + 600 V.P.H., 700 from A and 600 from C.

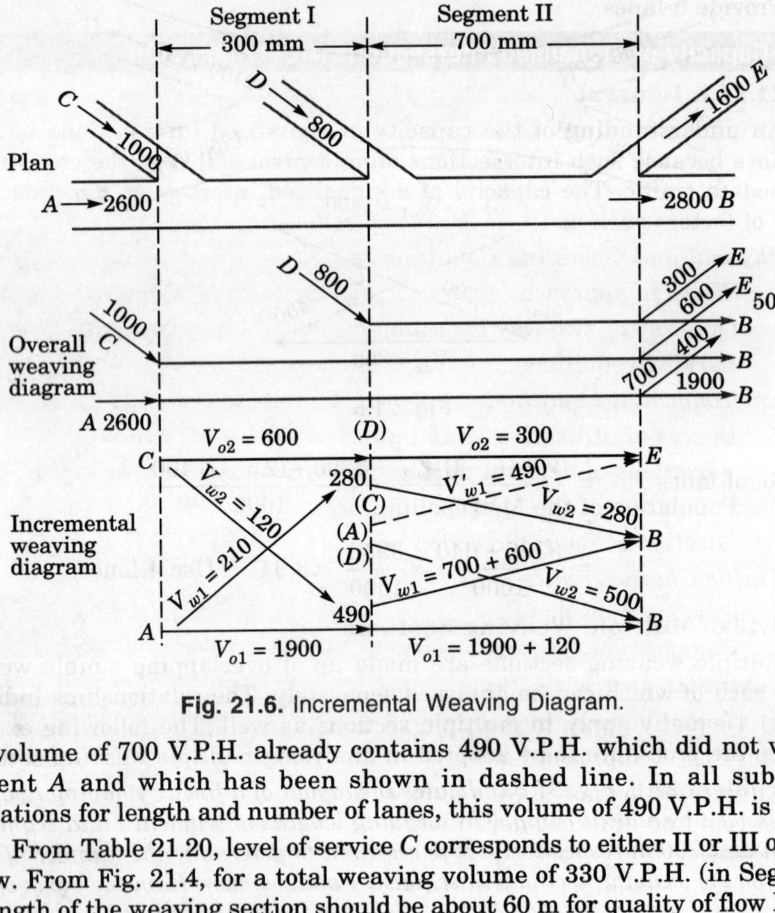

Fig. 21.6. Incremental Weaving Diagram.

This volume of 700 V.P.H. already contains 490 V.P.H. which did not weave in segment A and which has been shown in dashed line. In all subsequent calculations for length and number of lanes, this volume of 490 V.P.H. is omitted.

From Table 21.20, level of service C corresponds to either II or III of quality of flow. From Fig. 21.4, for a total weaving volume of 330 V.P.H. (in Segment I), the length of the weaving section should be about 60 m for quality of flow II, which is satisfied. For a weaving legnth of 200 m and a weaving volume of 330 V.P.H., from the same figure $k = 1.8$

$$\text{No. of lanes,} \quad N = \frac{V + \Sigma(k-1)\, V_{w2}}{SV} = \frac{3600 + 0.8 \times 120}{1200} = \frac{3696}{1200}$$

$$= 3.1 \text{ lanes.}$$

Use 4 lanes.

For Segment II, the total weaving volume is 700 + 600 + 500 + 280 = 2080 V.P.H. From Fig. 21.4, for quality of flow II, the length of weaving should be at least 900 m, and for quality of flow III the minimum length should be 400 m. The available length of 700 m will, therefore, give quality of flow intermediate between II and III and is satisfactory. From the same Fig. 21.4, for a length of weaving of 700 m and a weaving flow of 2080, $k = 2.9$. Similarly for a weaving flow of 490 + 280 = 770 V.P.H., and length of 700 m, $k = 1.8$.

∴ Total no. of lanes

$$= \frac{V + \Sigma(k-1)\, V_{w2}}{SV}$$

$$= \frac{4400 + 1.9 \times 500 + 0 \times 280}{1200} = \frac{5350}{1200} = 4.5$$

Provide 5 lanes.

21.14. Capacity of Signalised Intersections (HCM Approach)

21.14.1. General

An understanding of the capacity of signalised intersections is of vital importance because such intersections often control ability of the city streets to accommodate traffic. The capacity of a signalised intersection depends upon a number of factors such as :

Physical and Operating Conditions :

Width of approach.

One-way or two-way operation.

Parking conditions.

Environmental Conditions :

Degree of utilization of an individual approach

Variations of demand during the peak hour

Population of the Metropolitan area

Location of the intersection within the Metropolitan area

Traffic Characteristics :

Turning movements

Composition (Trucks, buses, cycles etc.)

Control Measures :

Traffic signals, including cycle length, green time to cycle time ratio

Marking of approach lanes.

An intersection may have a number of approaches, and the capacity of each approach could be different depending on the above factors. In actual practice, the term 'intersection capacity' is used to signify the individual approach capacity.

21.14.2. Normann method of determining signalised intersection capacity

Much of the work which formed the basis for the design of intersections contained in the 1965 edition of the Highway Capacity Manual was done under the direction of the late O.K. Normann and hence the procedure is named after him.

Two terms, *viz.,* 'load factor' and 'peak-hour factor' need to be defined before the method can be described. The term 'load factor' is measure of the degree of utilization of an approach roadway during one hour of peak flow. It is the ratio of the number of green phases that are loaded or fully utilised by traffic to the total number of green phases available for that approach during the same period. It can have a value ranging from 0.0 to 1.0, the former representing excellent or very good operating conditions and the latter representing very poor and jammed conditions. The term 'peak hour factor' is a measure of the variation in demand during the peak hour and is defined as the ratio between the number of vehicles counted during the peak hour and four times the number counted during the highest 15 consecutive minutes. The peak hour factor can range from 0.25 to 1.00, the former value representing the extreme peaking condition and the latter representing a uniform flow during the peak hour.

The method identifies six levels of service *A* to *F*, with capacity occurring at level *E*, as shown in Table 21.23.

Table 21.23. Levels of Service for Signalised Intersection Capacity
(Source : Ref. 1)

Level of Service	*Traffic Flow Description*	*Load factor*
A	Free flow, no approach phase fully utilized by traffic, approaches quite open, drivers find freedom of operation.	0.0
B	Stable flow, an occasional approach phase is fullly utilised, many drivers, begin to feel somewhat restricted. Suitable for rural design purposes.	$\lesssim$ 0.1
C	Stable operation continues, loading is still intermittent, occasional drivers may have to wait through more than one red signal indication, most drivers feel somewhat restricted. Suitable for urban design practice.	$\lesssim$ 0.3
D	Approaching unstable flow, delays to approaching vehicles may be substantial. Excessive back-ups do not occur.	$\lesssim$ 0.7
E	Represents capacity. Unstable flow condition prevaltions prevail. Full utilization of every cycle is seldom attained, long queues of vehicles likely and delays may be great.	$\lesssim$ 1.0
F	Represents forced flow. Jammed conditions prevail. Heavy back-ups.	Not applicable.

The method presents six fundamental capacity charts, enabling the determination of service volumes and capacity on a per hour of green basis, for different approach widths, load factors, peakhour factors, size of the metropolitan area and location within the metropolitan area. These are reproduced as Fig. 21.6 to 21.11.

Apart from the adjustments indicated in the above charts, additional adjustments are needed for the G/C ratio (ratio between the green time and the total cycle length), turning manoeuvres and presence of trucks and buses. The Manual presents charts and tables these adjustments, and a reference may be made to the Manual for these. A set of nomographs for ready solution of problems has been presented by Leisch (Ref. 14)

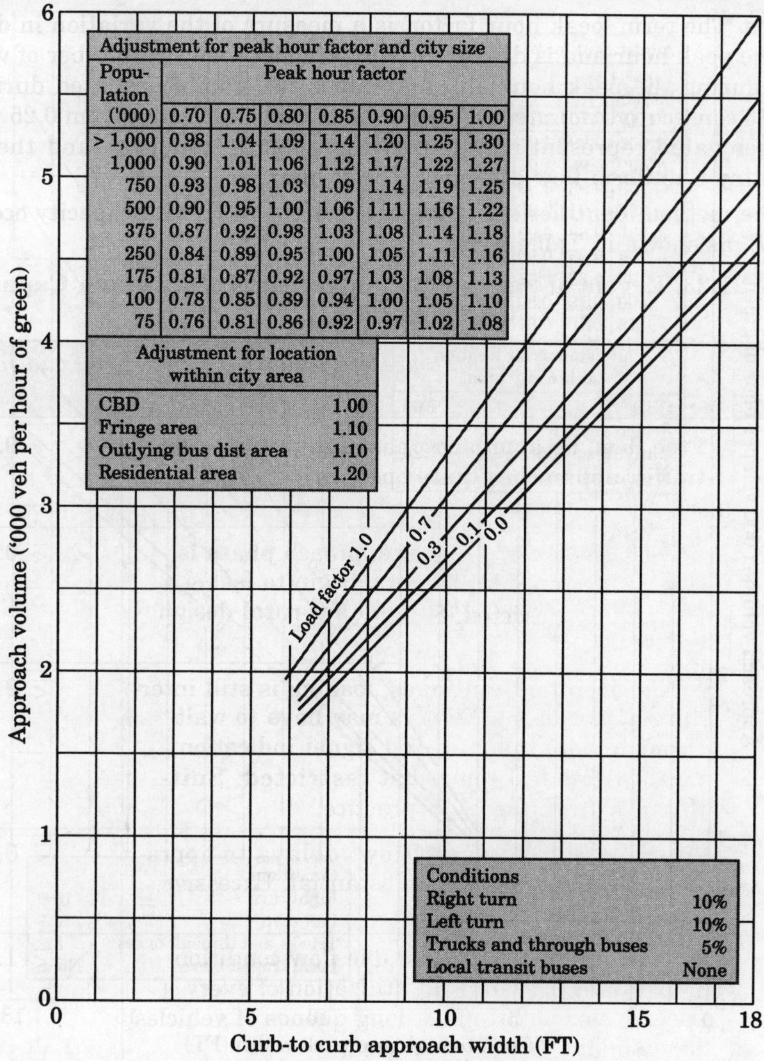

Adjustment for peak hour factor and city size							
Population ('000)	Peak hour factor						
	0.70	0.75	0.80	0.85	0.90	0.95	1.00
> 1,000	0.98	1.04	1.09	1.14	1.20	1.25	1.30
1,000	0.90	1.01	1.06	1.12	1.17	1.22	1.27
750	0.93	0.98	1.03	1.09	1.14	1.19	1.25
500	0.90	0.95	1.00	1.06	1.11	1.16	1.22
375	0.87	0.92	0.98	1.03	1.08	1.14	1.18
250	0.84	0.89	0.95	1.00	1.05	1.11	1.16
175	0.81	0.87	0.92	0.97	1.03	1.08	1.13
100	0.78	0.85	0.89	0.94	1.00	1.05	1.10
75	0.76	0.81	0.86	0.92	0.97	1.02	1.08

Adjustment for location within city area	
CBD	1.00
Fringe area	1.10
Outlying bus dist area	1.10
Residential area	1.20

Fig. 21.7. Urban intersection approach service volume, in vehicles per hour of green signal time, for one-way streets with no parking.

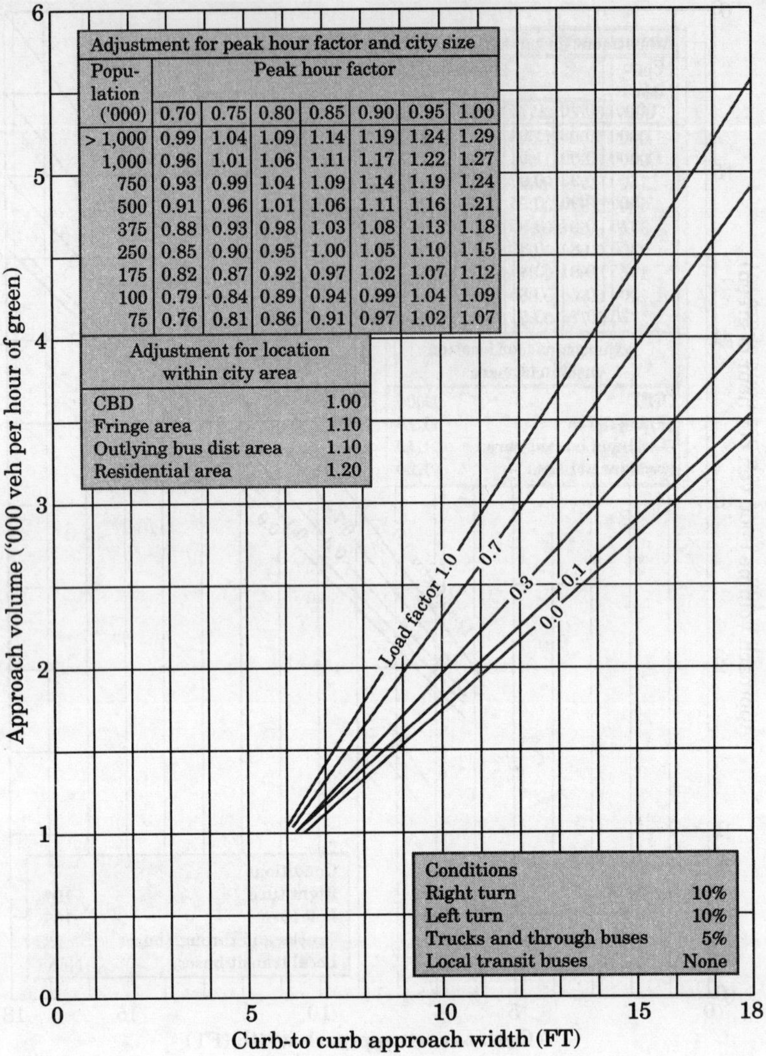

Fig. 21.8. Urban intersection approach service volume, in vehicles per hour of green signal time, for one-way streets with parking one side.

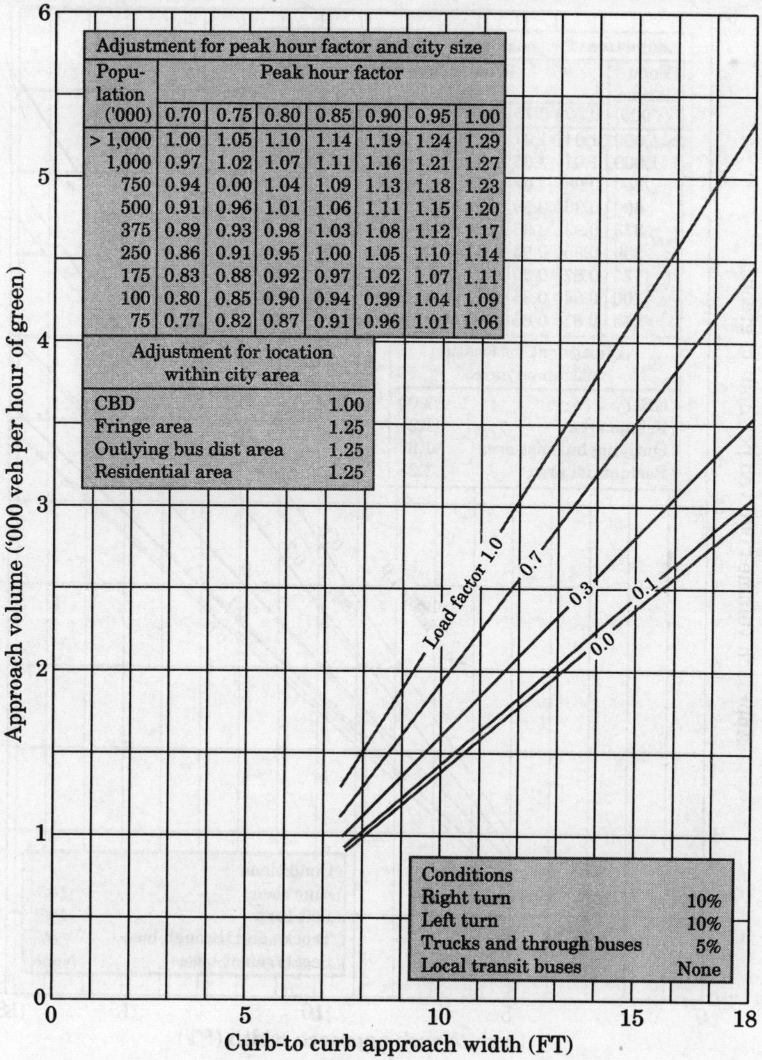

Fig. 21.9. Urban intersection approach service volume, in vehicles per hour for green signal time, for one-way streets with parking both sides.

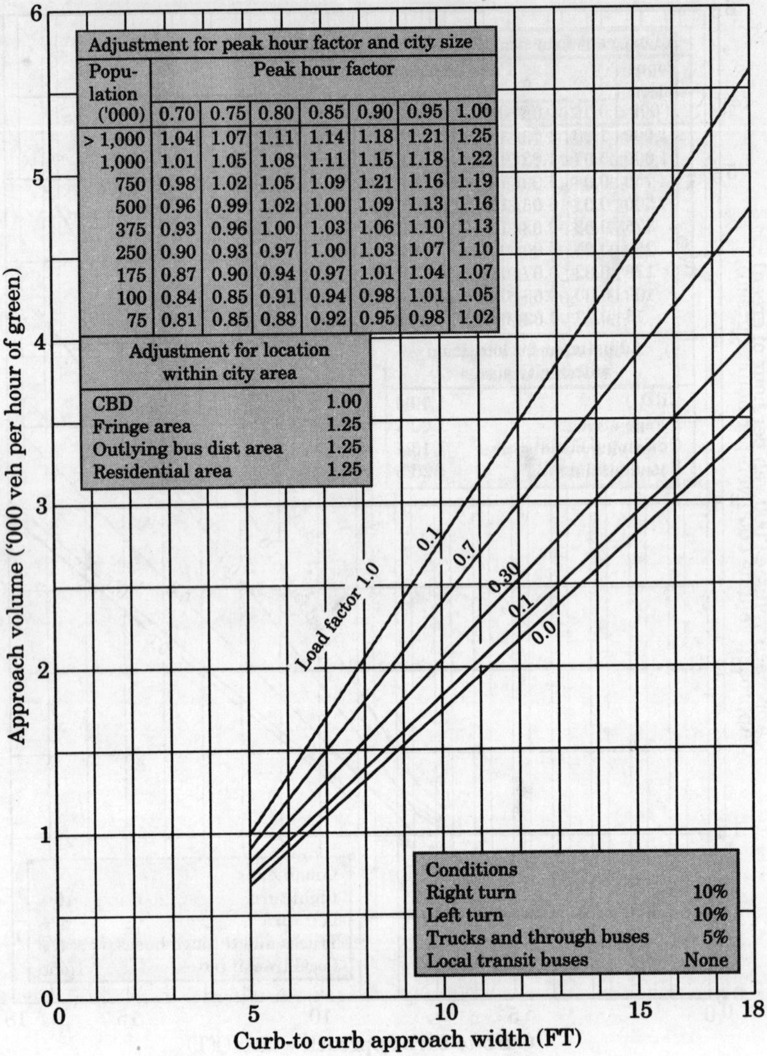

Fig. 21.10. Urban intersection approach service volume, in vehicles per hour of green signal time, for two-way streets with no parking.

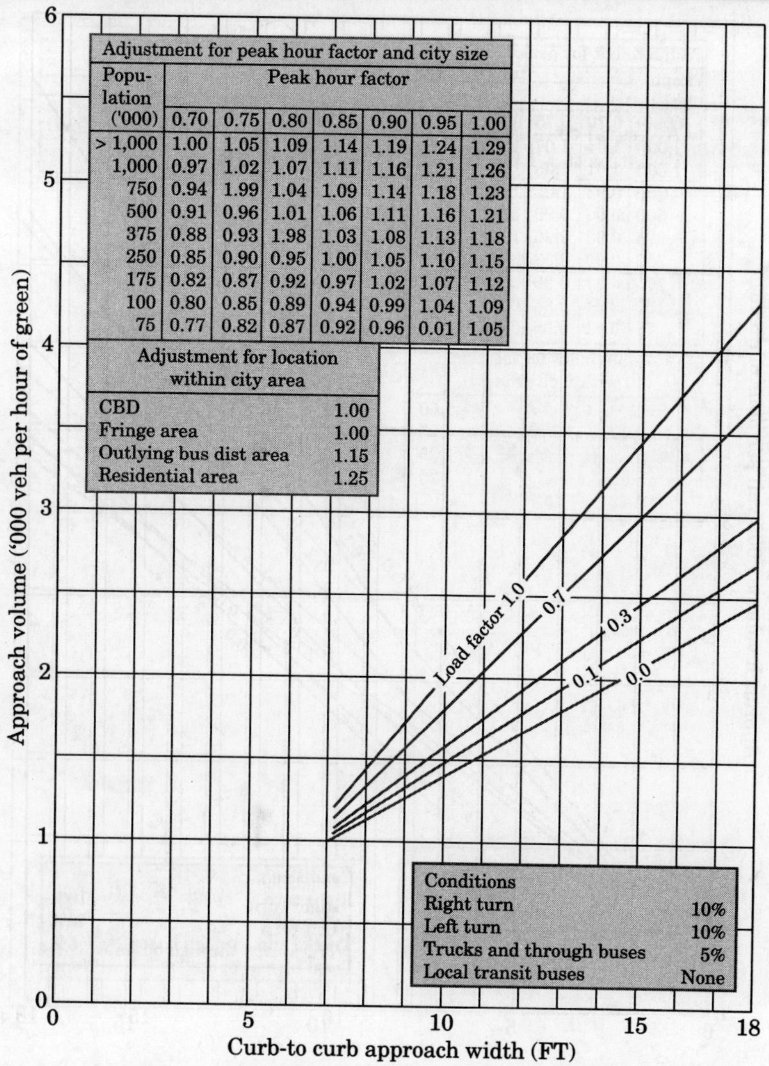

Fig. 21.11. Urban intersection approach service volume in vehicles per hour of green signal time, for two-way streets with parking.

A simple example will be considered here to illustrate the general basis of the procedure.

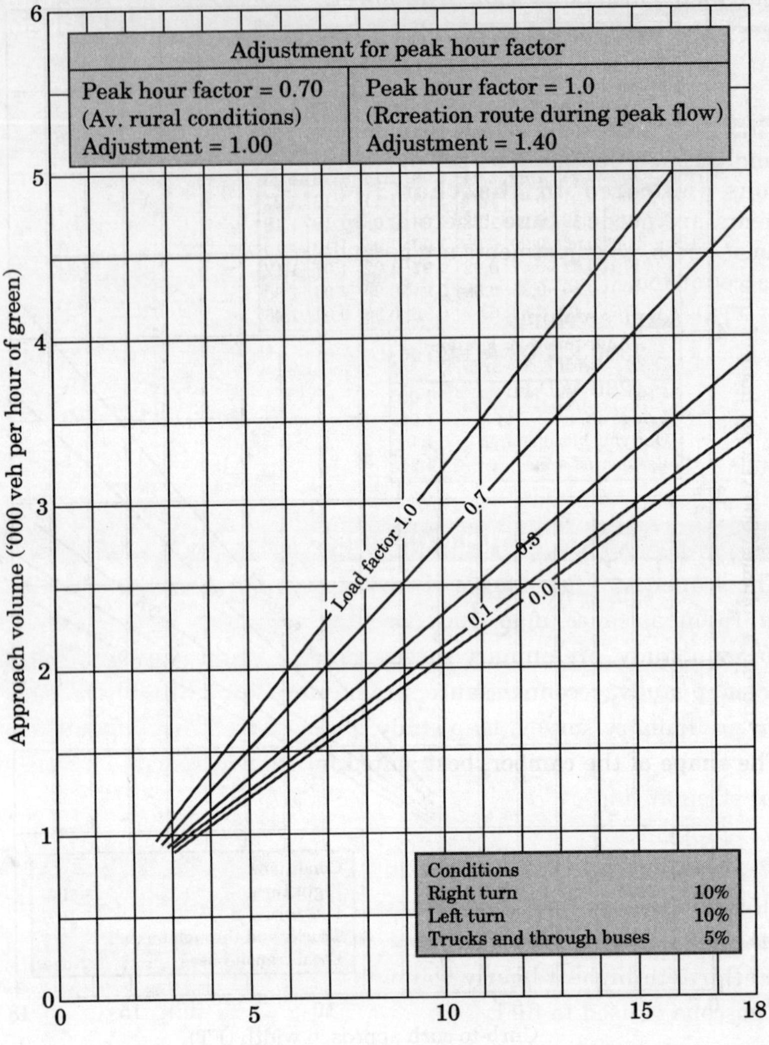

Fig. 21.12. Rural intersection approach service volume, in vehicles per hour of green signal time, for two-way highways with no parking on the travelled way.

Problem 21.3. *Fig. 21.13 represents a signalised intersection containing a 2-way street, with an approach width of 10 m, with no parking, and located in an outlying business district with population of 500,000. The load factor is 0.3 and the peak hour factor is 0.80. The signal has a cycle length of 60 seconds and the green time is 25 seconds. The percentage of left turns is 10%, the percentage of right turns is 10%, the percentage of trucks and through buses is 5% and there are no local transit buses. Determine the service volume of the intersection approach.*

Solution. For an urban 2-way street Fig. 21.10 is applicable. For a width approach of 10 m, and a load factor of 0.3, the volume per hour of green is 2300 vehicles, under average conditions. From the table in the upper portion of the chart, the adjustment factor needed for a city with a population of 500,000 and peak hour factor of 0.80 is 1.02. The lower table gives an adjustment factor of 1.25 for an outlying 25 business district. The G/C is $\frac{25}{60}$ = 0.42. Since the percentage of left turns and trucks and through buses are as per average conditions indicated in the chart, no adjustments are needed. Since there are no local transit buses, no adjustments are needed on this account too.

∴ The service volume

= 2300 × 1.02 × 1.25 × 0.42

= 1230 V.P.H.

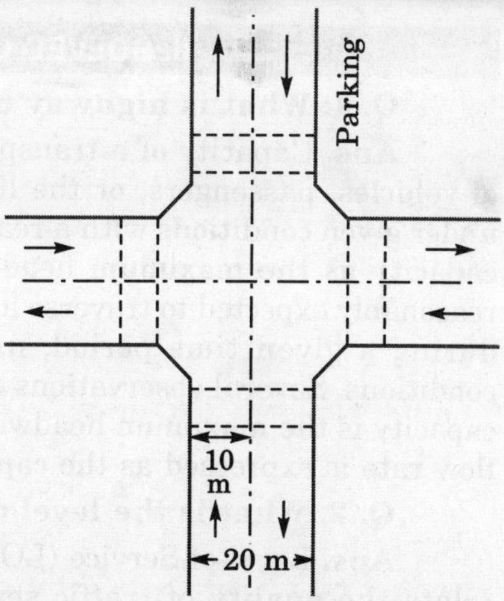

Fig. 21.13

I. MULTIPLE CHOICE QUESTIONS

1. The sequence of four stages of survey in a highway alignment is
   (*a*) reconnaissance, map study, preliminary survey and detailed survey
   (*b*) map study, preliminary survey, reconnaissance and detailed survey
   (*c*) map study, reconnaissance, preliminary survey and detailed survey
   (*d*) preliminary survey, map study, reconnaissance and detailed survey
2. The shape of the camber, best suited for cement concrete pavements
   (*a*) straight line (*b*) parabolic
   (*c*) elliptical
   (*d*) combination of the straight and parabolic line
3. Highway facilities are designed for
   (*a*) annual average hourly volume (*b*) annual average daily traffic
   (*c*) thirtieth highest hourly volume (*d*) peak hourly volume of the year.
4. Enoscope is used to find:
   (*a*) average speed (*b*) spot speed
   (*c*) space mean speed (*d*) time-mean speed
5. For highway geometric design purposes the speed used is
   (*a*) 15th percentile (*b*) 50th percentile
   (*c*) 85th percentile (*d*) 98th percentile
6. In highway construction, rolling starts from:
   (*a*) sides and proceed to center (*b*) center and proceed to sides
   (*c*) one side and proceed to other side
   (*d*) any of the above.

Answers

1. (*c*) 2. (*a*) 3. (*c*) 4. (*b*) 5. (*d*)
6. (*a*)

II. SOLVED QUESTIONS AND ANSWERS

Q. 1. What is highway capacity?

Ans. Capacity of a transport facility is defined as the maximum number of vehicles, passengers, or the like, per unit time which can be accommodated under given conditions with a reasonable expectation of occurrence. It defines the capacity as the maximum headway rate at which persons or vehicles can be reasonably expected to traverse a point or a uniform segment of a lane or roadway during a given time period, under prevailing roadway, traffic and control conditions. Several observations can be made from the above definition. Although capacity is the maximum headway rate, in many situations the break 15 minute flow rate is expressed as the capacity.

Q. 2. What is the level of service for the traffic facility?

Ans. Level-of-Service (LOS) of a traffic facility is a concept introduced to relate the quality of traffic service to a given flow rate. Level-of-service is introduced to denote the level of quality one can derive from a local under different operation characteristics and traffic volume.

Q. 3. What is interrupted facility?

Ans. Interrupted facilities include un-signalized intersection, signalized intersection, and arterials or corridors. They have respectively control delay, total delay and average travel speed as the measure of effectiveness. Other facilities may include pedestrian pathways, bicycle tracks, bus-transit system, rail-transit system and air-transportation system. Each of them have facility specific measure of effectiveness.

III. REVIEW QUESTIONS

1. What is meant by the capacity of two lane rural?
2. Explain the capacity of multilane rural highways.
3. Discuss the capacity of urban streets.
4. Discuss the capacity of signalised intersections.

REFERENCES

1. Highway Capacity Manual, Transportation Research Board Special Report 209, Washington, 1985.
2. Highway Capacity Manual (1950), U.S. Govt. Printing Office, 1950.
3. Greenshields B.D. The Photographic method of Studying Traffic behaviour, Proceedings of the Highway Research Board Vol. 13, 1932.
4. Research on Road Traffic, Road Research Laboratory, U.K. Traffic H.M.SO. London, 1965.
5. H.K. Evans (Editor), Traffic Engineering Handbook, Institute of Traffic Engineers, New Haven, Connecticut, 1950.
6. L.R. Kadiyali, A Study of the Problems of Singlelane Pavements in India and their Improvements, Paper No. 296, Journal of the I.R.C., Vol. XXXV 3, New Delhi, 1973.
7. Roads in Urban Areas, Ministry of Transport, H.M.S.O., London, 1966.
8. Layout of Roads in Rural Areas, Ministry of Transport, H.M.S.O., London, 1968.
9. F.V. Webster and B.M. Cobbe, Traffic Signals, Road Research Technical Paper No. 56, H.M.S.O., London, 1966.
10. Guidelines on Capacity of Roads in Rural Areas, I.R.C. : 64-1990, I.R.C., New Delhi, 1990.
11. A Policy on Design of Urban Highways and Arterial Streets, AASHTO Washington, 1973.

12. A policy on Geometric Design of Highways and Streets, AASHTO, Washington, 1994.
13. F.V. Webster and J.C. Wardrop, Capacity of Urban Intersections, Sixth International Study Week of Traffic Engineering, Salzbury, 1962.
14. J.E. Leisch, Capacity Analysis for Design of Signalised Intersections, Public Roads, August 1967 and October 1967, Vol. 34, Nos. 9 and 10, Washington.
15. Kadiyali, L.R. PCU factors under Indian conditions, some tentative findings, RUCS Technical Paper No. 37, Eleventh Quarterly Report, Central Road Research Institute, New Delhi 1980.
16. Highway Capacity Manual, Highway Research Board Special Report 87, Washington, 1965.
17. Highway Capacity Manual, HCM-2000, Transportation Research Board, Washington DC, 2000.
18. Chandra, S, Passenger Car Units for Indian Roads, Journal of the Indian Roads Congress, Vol. 69-2, New Delhi, 2008.
19. Guidelines for Capacity of Roads in Rural Areas, I.R.C : 64-2009, Proposed Draft Second Revision, Indian Roads Congress, New Delhi, 2008.

SECTION 12

THEORY OF TRAFFIC FLOW

22. Scope, Definitions and Basic Diagram of Traffic Flow

22.1. Scope

The theory of traffic flow can be defined as a mathematical study of the movement of vehicles over road network. The subject is a mathematical approach to define, characterise and describe different aspects of vehicular traffic.

The subject has virtually grown from the measurements in the field of the various characteristics of the traffic and the urge to describe these observed characteristics in precise mathematical language, with a view to understand traffic behaviour better. The study is of great importance to a traffic engineer as it provides him with a comprehensive knowledge of vehicular traffic, leading to improved techniques for the control, regulation and management of traffic.

The development of the subject has taken inspiration from various branches of knowledge. Statistics, applied mathematics, psychology and operations research are some of those. The subject is greatly being researched upon and newer and newer theories are emerging.

The first subject that will be covered will be the relationship between speed, flow and concentration. A number of approaches are available for arriving at this relationship. Of these, the carfollowing theory and the hydrodynamic analogy are important and will be discussed in detail.

Road traffic inevitably causes congestion. Queueing and delay occur in all congested situations. Delay causes economic loss since time means money in the modern age. A study of queueing and delay is, therefore of great relevance to a traffic engineer.

The distribution of cars linearly along a road and the study of headways and gaps between vehicles will aid the traffic engineer in designing intersections and operating the traffic facilities. The headways and gaps are governed by the arrival pattern of vehicles and the flow.

The subject derives heavily from Statistics. A knowledge of the theory of probability and statistics is essential for understanding the treatment.

22.2. Some Definitions

Traffic flow theory is mainly concerned with three measurable characteristics of road traffic, *viz.,* speed, flow and concentration.

Speed has already been defined in Chapter 3. The definitions of various terms associated with speed are :

Speed (v) is the rate of movement of traffic or of specified components of traffic and is expressed in metric units in kilometres per hour (K.P.H.).

Time mean speed (v_t) is the average of the speed measurements at one point in space over a period of time.

Space-mean speed (v_s) is the average of the speed measurements at an instant of time over a space.

Flow, also known as volume, (Q), is the number of vehicles passing a specified point during a stated period of time. It is usual to express it in vehicles per hour.

Concentration, also known as density, (K), is the number of vehicles present in a stated length of road at an instant. It is usually expressed in vehicles per kilometre length of road per lane.

Space Headway, (s), is the distance between the fronts of successive vehicles. It is measured in metres.

Time Headway, or simply headway, (h), is the time interval between the passage of the fronts of successive vehicles at a specified point. It is measured in seconds.

22.3. Relationships between the Variables

In Chapter 9, it was shown (Eq. 9.20) that the space mean speed, flow and density are related by the following equation :

$$\bar{v}_s = \frac{Q}{K} \qquad \text{...(22.1)}$$

Since concentration (K) is the reciprocal of the space headway $\left(\frac{s}{1000} \text{ when expressed in kilometres}\right)$,

$$\bar{v}_s = \frac{Q \times s}{1000} \qquad \text{...(22.2)}$$

s, the space headway, is related to h, the time headway as follows :

$$s = \frac{h}{3600} \times \bar{v}_s \times 1000 \qquad \text{...(22.3)}$$

22.4. Fundamental Diagram of Traffic Flow

In Eqn. 22.1 above, the concentration K has a theoretical maximum value when vehicles are packed from end-to-end. The theoretical maximum density is called the jamming concentration, K_j, which is the reciprocal of the length of a car. If the average length of cars in a stream is 5 m, the jamming concentration,

$$K_j = \frac{1000}{5} = 200 \text{ veh/km}$$

When the vehicles are packed from end-to-end, the flow Q is obviously zero. So also, when the concentration is zero, there are no cars, hence the flow is zero. As the concentration slowly increases from zero, the flow also increases and a point is reached when the flow is maximum. The maximum flow that is capable of being accommodated in a road is the capacity of the road. The general form of the Flow-Concentration curve, usually known as the Fundamental Diagram of Traffic Flow, is indicated in Fig. 22.1.

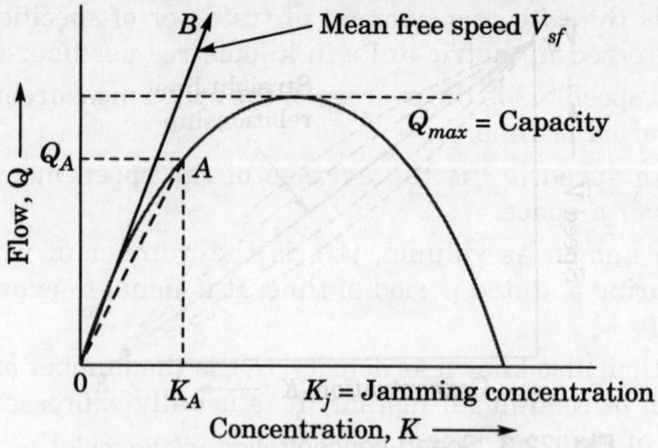

Fig. 22.1. Fundamental diagram of road traffic.

The exact shape of the flow-concentration curve will depend upon a number of factors. In fact, the curve is a characteristic of a particular road section at a particular time with a particular population of drivers (Ref. 1). Since all these can change and assume innumerable combinations, a large number of curves are possible.

Since $\quad \bar{v}_s = \dfrac{Q_A}{K_A}$ (Eq. 22.1),

the slope of the line joining the origin O to a point A on the curve gives the space mean speed associated with a concentration K_A and a volume Q_A.

The slope of the line OB tangential to the curve at the origin represents the mean free speed $\bar{v}_{sf}$, which is the speed the drivers would adopt if there were no interference from other vehicles.

For simplicity, it is often assumed that the shape of the curve is a parabola. The assumption is true for a straight line relationship between speed and concentration. More about this later.

The curve expressing the relationship between speed and flow will be of the form shown in Fig. 22.2. The general shape is the basis of describing the levels of service when dealing with the capacity of a road and has been utilised in Fig. 22.1. The curve should obviously start at the origin zero, since the speed is zero when the flow is zero. The curve should also pass through v_{sf} when again the flow is zero and speed is the maximum corresponding to free flow conditions. For intermediate values of speed, the volume varies and has a maximum value, Q max corresponding to the capacity of the road.

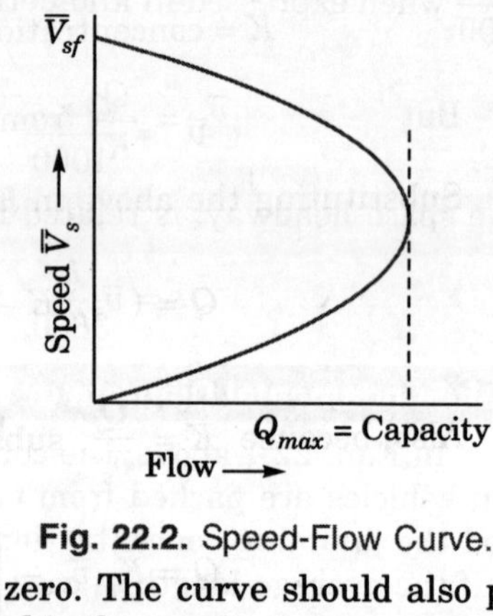

Fig. 22.2. Speed-Flow Curve.

The relationship between speed and concentration can similarly be deduced. When the speed is zero, the concentration is maximum and this corresponds to K_j, the jamming concentration. Also, when the concentration is zero, speed is maximum and is the free speed, v_{sf}. The exact shape of the curve joining these two points, depends on a number of factors. A straight-line relationship yields the parabolic shape of the curve relating flow and concentration. Typical speed-concentration curves are given in Fig. 22.3.

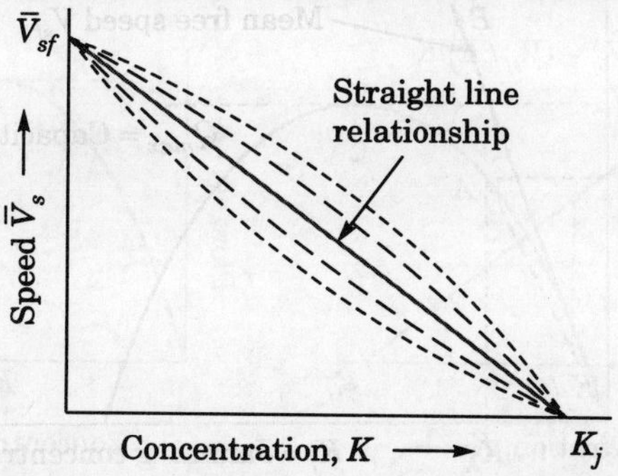

Fig. 22.3. Speed-concentration relationship.

22.5. Linear Relationship between Speed and Concentration

It has been indicated above that a linear relationship between speed and concentration is one of the possibilities. As early as in 1934, Greenshields (Ref. 2) analysed the data pertaining to rural roads in Ohio and found a linear relationship between speed and concentration of the following form :

$$\bar{v}_s = \bar{v}_{sf} - \left(\frac{\bar{v}_{sf}}{K_j}\right) K \qquad \text{...(22.4)}$$

where, $\bar{v}_s$ **= space mean speed**

$\bar{v}_{sf}$ **= space mean speed for free flow conditions**

K_j **= jamming concentration**

K **= concentration**

But $\bar{v}_s = \dfrac{Q}{K}$ **from Equation 22.1.**

Substituting the above in Equation 22.4,

$$Q = (\bar{v}_{sf}) K - \left(\frac{\bar{v}_{sf}}{K_j}\right) K^2 \qquad \text{...(22.5)}$$

Also, because $K = \dfrac{Q}{\bar{v}_s}$**, substituting in Equation 22.4**

$$Q = K_j\ \bar{v}_s - \left(\frac{K_j}{\bar{v}_{sf}}\right) \bar{v}_s^2 \qquad \text{...(22.6)}$$

Differentiating Equation 22.5 with respect to concentration the concentration when flow is a maximum is obtained. Thus

$$\frac{dQ}{dK} = \bar{v}_{sf} - 2 \times \bar{v}_{sf} \times \frac{K}{K_j} = 0$$

$$K = K\ \text{max.} = \frac{K_j}{2} \qquad \text{...(22.7)}$$

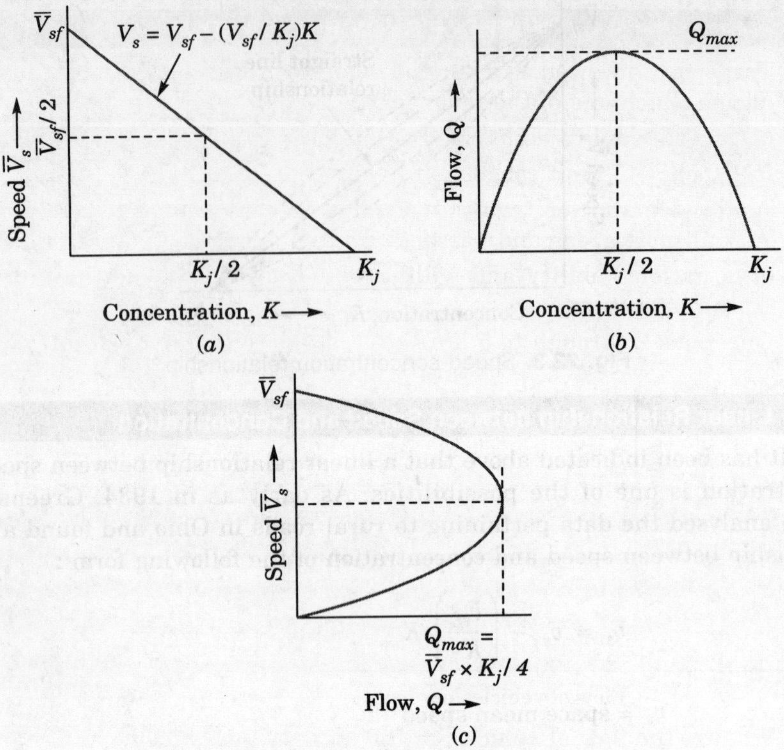

Fig. 22.4. Speed-flow-concentration curves when the speed-concentration relationship in linear.

To obtain the sped when the flow is maximum, Equation 22.6 is differentiated with respect to u_s. Thus :

$$\frac{dQ}{d\overline{u}_s} = K_j - \frac{2 \times K_j \times \overline{v}_s}{\overline{v}_{sf}} = 0$$

$$\therefore \qquad \overline{v}_s = \overline{v}_s \text{ max.} = \frac{\overline{v}_{sf}}{2} \qquad \text{...(22.8)}$$

Substituting the maximum values of K and v, in Equation 22.1, Q maximum is obtained.

$$Q \text{ maximum} = \overline{v}_s \text{ maximum} \times K \text{ maximum}$$

$$= \frac{\overline{v}_{sf}}{2} \times \frac{K_j}{2} = \frac{\overline{v}_{sf} \times K_j}{4} \qquad \text{...(22.9)}$$

These relations are shown in the curves in Fig. 22.4.

It may be mentioned that the assumption of a linear relationship between speed and concentration is only approximately true.

I. MULTIPLE CHOICE QUESTIONS

1. **Transport is viewed as a derived demand. Derived demand in this context means which one of the following?**

(*a*) an individual's ability and willingness to buy transport services

(*b*) a policy to limit congestion in city centres

(*c*) the demand for transport services is determined by the demand to transport goods and passengers from one place to another

(*d*) a nation's ability and willingness to provide an adequate transport infrastructure

2. **Questions 2 through to 5 are based on the following diagram.**

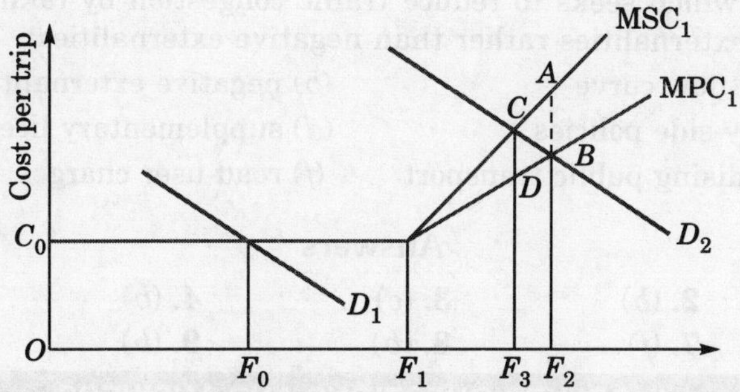

Flow of vehicles per hour along a particular route

What is the flow of vehicles in the off peak period?

(*a*) F_1 (*b*) F_0 (*c*) F_3 (*d*) F_2

3. **In the peak period the external cost of motoring is represented by the distance:**

(*a*) CD (*b*) DF_3 (*c*) AB (*d*) BF_2

4. **In order to achieve an efficient allocation of resources a road user charge is required of:**

(*a*) AB (*b*) CD (*c*) BF_2 (*d*) DF_3

5. **New road building will affect the diagram in which two of the following ways?**

1. Shift demand D_2 to the left
2. Reduce the gap between the MPC and MSC curves
3. Result in the demand curve D_2 becoming more elastic
4. Move the demand curve D_1 to the left
5. Shift the MPC and MSC curves to the right

(*a*) 1 and 3 (*b*) 2 and 3 (*c*) 2 and 4 (*d*) 3 and 5

(*e*) 1 and 5

6. **For the following five questions, match each description with its correct term.**

Describes a situation where the road user imposes costs on others (*e.g.*, slower journey times) for which he/she is not charged.

(*a*) negative externality (*b*) road-user charges

(*c*) supply-side policies (*d*) speed-flow curve

(*e*) supplementary licenses (*f*) subsidising public transport

7. An attempt to solve congestion problems by building more roads.
   (*a*) road-user charges (*b*) negative externality
   (*c*) supplementary licenses (*d*) subsidising public transport
   (*e*) speed-flow curve (*f*) supply-side policies

9. Shows the movement of vehicles along a particular road and indicates how motorists interact and impose delays and costs on each other.
   Road-user charges
   (*a*) negative externality (*b*) speed-flow curve
   (*c*) subsidising public transport (*d*) supply-side policies
   (*e*) supplementary licenses

10. A policy which seeks to reduce traffic congestion by taking into account positive externalities rather than negative externalities.
   (*a*) speed-flow curve (*b*) negative externality
   (*c*) supply-side policies (*d*) supplementary licenses
   (*e*) subsidising public transport (*f*) road-user charges

Answers

1. (*c*)	**2.** (*b*)	**3.** (*c*)	**4.** (*b*)	**5.** (*d*)
6. (*a*)	**7.** (*f*)	**8.** (*b*)	**9.** (*b*)	**10.** (*e*)

II. SOLVED QUESTIONS AND ANSWERS

Q. 1. What is traffic flow?

Ans. Traffic flow is the study of the movement of individual drivers and vehicles between two points and the interactions they make with one another. Unfortunately, studying traffic flow is difficult because driver behaviour is something that cannot be predicted with one-hundred percent certainty. Fortunately, drivers tend to behave within a reasonably consistent range and, thus, traffic streams tend to have some reasonable consistency and can be roughly represented mathematically.

Q. 2. What are the characteristics to represent traffic flow?

Ans. There are three main characteristics: (1) flow, (2) density, and (3) velocity. These relationships help in planning, design, and operations of roadway facilities.

Q. 3. Illustrate time-space diagram. Explain how does it show trajectories of vehicles over time and space?

Ans. Time-space diagram showing trajectories of vehicles over time and space. Traffic engineers represent the location of a specific vehicle at a certain time with a time-space diagram. This two-dimensional diagram shows the trajectory of a vehicle through time as it moves from a specific origin to a specific destination. Multiple vehicles can be represented on a diagram and, thus, certain characteristics, such as flow at a certain site for a certain time, can be determined.

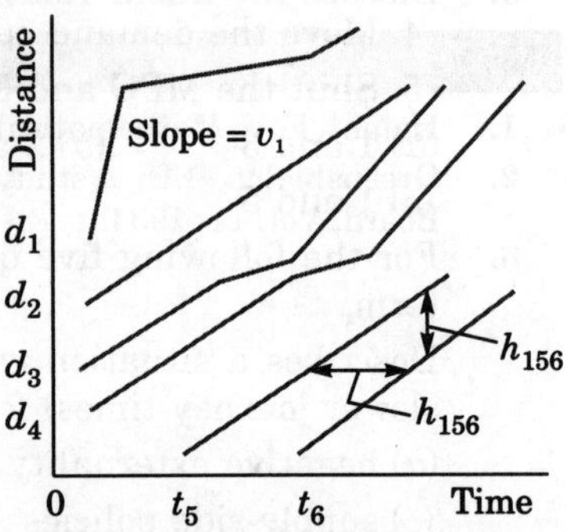

Q. 4. What is speed?

Ans. Measuring speed of traffic is not as obvious as it may seem; we can average the measurement of the speeds of individual vehicles over time or over space, and each produces slightly different results.

Q. 5. What do you mean by time mean speed?

Ans. Time mean speed is the arithmetic mean of speeds of vehicles passing a point.

Q. 6. What is space mean speed?

Ans. Space mean speed is defined as the harmonic mean of speeds passing a point during a period of time. It also equals the average speeds over a length of roadway.

Q. 7. Explain relating time and space mean speed.

Ans. Note that the time mean speed is average speed past a point as distinct from space mean speed which is average speed along a length. The two speeds are related as a rule of thumb time mean speed is about 2% more than space mean speed *i.e.,*

Q. 8. What is time headway?

Ans. Time headway is the difference between the time when the front of a vehicle arrives at a point on the highway and the time the front of the next vehicle arrives at the same point (in seconds).

Q. 9. What is space headway?

Ans. Space headway is difference in position between the front of a vehicle and the front of the next vehicle (in meters).

Q. 10. The spot speeds (expressed in km/hr) observed at a road sections are 66, 62, 45, 79, 32, 51, 56, 60, 53 and 49. The median speed (expressed in km/hr) is ......

Ans. Median speed is the speed at the middle value in series of spot speeds that are arranged in ascending order. 50% of speed values will be greater than the median 50% will be less than the median. Ascending order of spot speed studies are 32, 39, 45, 51, 53, 56, 60, 62, 66, 79.

III. REVIEW QUESTIONS

1. What is the relationship between variables?
2. Represent the fundamental diagram of traffic flow.
3. Discuss the linear relationship between speed and concentration.

REFERENCES

1. Haight, F.A., Mathematical Theories of Traffic Flow, Academic Press, London, 1963.
2. Greenshields, B.D., A study of traffic capacity, Proceedings of the Highway Research Board, Vol. 14, 1934.

23. Lighthill and Whitham's Theory

23.1. General

One of the attempts to understand traffic flow has been to draw an analogy between the flow of fluids and the flow of traffic. Lighthill and Whitham (Ref. 1) have contributed to this topic by their theory which is based on kinematic waves. The limitation of the theory is that it is based on the 'continuous flow' approach in fluid dynamics and it thus represents the limiting behaviour of a stochastic process for a large 'population', which in this case is the total number of vehicles. Therefore, it is applicable to large-scale problems only, and principally to the distribution of traffic on long, crowded roads.

23.2. Assumptions

The assumptions in the derivation of the equations of the theory are :

(*i*) The equation of continuity, *i.e.*, law of the conservation of vehicles, holds good. Thus :

Inflow = outflow ± storage

(*ii*) At any point on the road, the flow Q is a function of the concentration K.

23.3. Theory

Consider a length of road AB with a stationary observer O at the mid-point, Fig. 23.1.

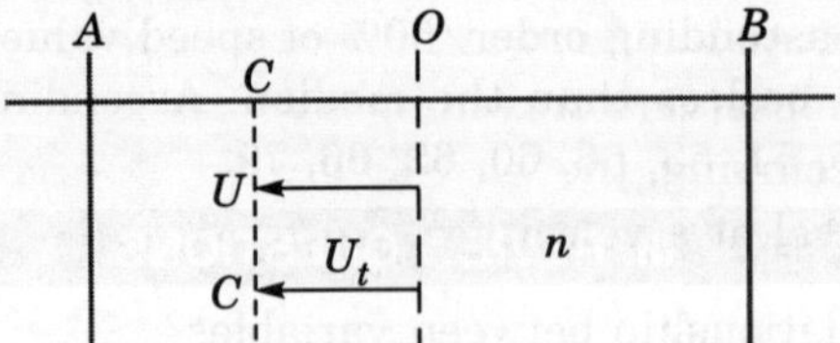

Fig. 23.1

There are n number of vehicles in the length OB when the observer starts counting the number passing him. The count stops at the expiry of time t when all the n vehicles have passed the observer.

At the same instant when the stationary observer starts counting, a moving observer starts at O and travels with the stream at speed U, which is less than $\overline{u}_s$, the space-mean speed of the stream, and counts the vehicles passing him. The count stops at the expiry of time t when all the n vehicles have passed O. The moving observer has travelled a distance OC in time t given by $OC = U.\ t$. His count will be less than the count of the stationary observer by the number of vehicles in the length OC.

Number counted by the moving observer

= Number in length l–Number in length OC

$$= Q.t - \frac{U.t \cdot n}{l}; \text{ but } \frac{n}{l} = K$$

$$= Q.\, t - K\, .\, U\, .\, t$$

Flow relative to the moving observer

$$= Q - K\, .\, U.$$

Suppose two moving observers are travelling at speed U at time t apart.

Suppose the flow and concentration change with time, the changes being relatively small ; but the observers have been told to adjust their speeds U so that the vehicles which pass them minus the vehicles which they pass is, on the average, the same for each. That is :

$Q - K.U$ is the same for each.

For observers 1 and 2,

$$Q_1 - K_1\, U = Q_2 - K_2\, U$$

$$\therefore \qquad U = \frac{Q_1 - Q_2}{K_1 - K_2} \qquad ...(23.1)$$

In the above equation, Q_1 and Q_2 are the flows and K_1 and K_2 the concentration observed by the observers at time t apart.

Calling the changes in flow and concentration as ΔQ and ΔK respectively,

$$U = \frac{\Delta Q}{\Delta K} \qquad ...(23.2)$$

But the number of vehicles between the observers must remain the same. And the number of vehicles passing any point between the times at which the observers pass it is Qt. As t is fixed, it follows that the flows remain unchanged along the path of the observers travelling with speed U. In other words, when changes in flow are occurring, the waves which carry such changes through the stream of vehicles travel at a velocity given by the equation :

$$U = \frac{\Delta Q}{\Delta K}$$

This velocity, relative to the road, may be positive or negative, but it never exceeds $\overline{v}_s$, the space-mean speed.

If Q and K *are* functionally related, then small changes in flow are propogated at the speed :

$$\phi = \frac{dQ}{dK} \qquad ...(23.3)$$

On the fundamental diagram of traffic flow, this implies that $\frac{dQ}{dK}$ is the slope of the tangent to the curve, Fig. 23.2.

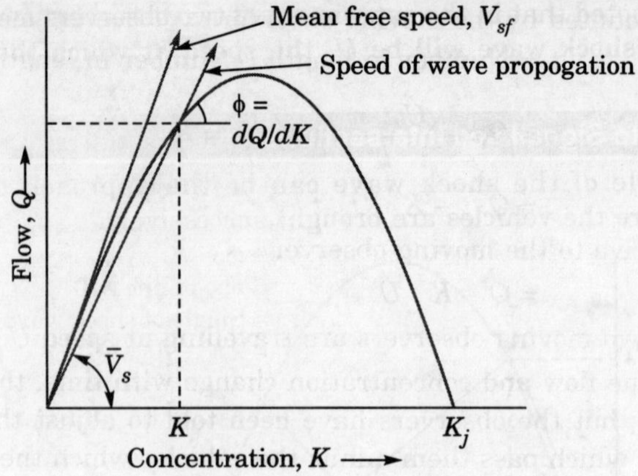

Fig. 23.2. Flow-concentration curve.

But $\quad Q = K \cdot v_s$

$$\therefore \quad \frac{dQ}{dK} = K \cdot \frac{d\bar{v}_s}{dK} + \bar{v}_s$$

$$\therefore \quad \phi = \bar{v}_s + K \cdot \frac{d\bar{v}_s}{dK} \qquad \text{...(23.4)}$$

It may be noted that the speed of the wave at the crest is zero.

Consider a stretch of road where the traffic is dense in front (A) and less dense behind (B), whose fundamental diagram of traffic flow is given in Fig. 23.3. The slope of the curve is steeper at B than at A. The speed of wave propagation of Q_B at B is greater than the speed of propagation of Q_A at A. A shock wave will now be generated if the flow changes discontinuously from A to B. The speed of the shock wave

$$= \frac{\Delta Q}{\Delta K} = \frac{Q_A - Q_B}{K_A - K_B}$$

$$= \text{slope of the chord } AB \qquad \text{...(23.5)}$$

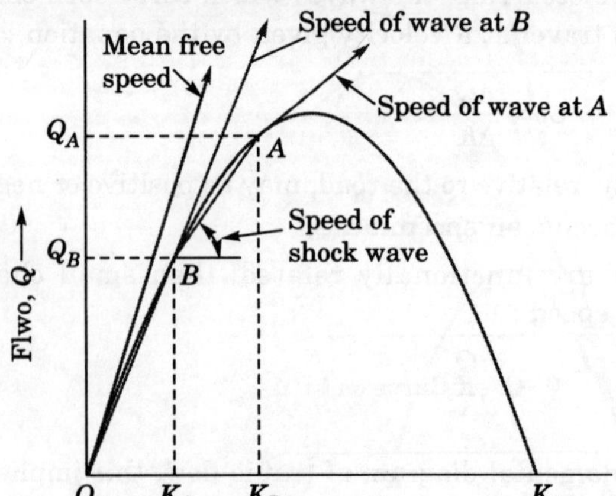

Fig. 23.3

It may be noted that in the experiment of two observers mentioned earlier, the speed of the shock wave will be U, the speed at which the two observers travelled.

23.4. Approach to Signallised Intersections

An example of the shock wave can be the approach to a signalised intersection, where the vehicles are brought successively.

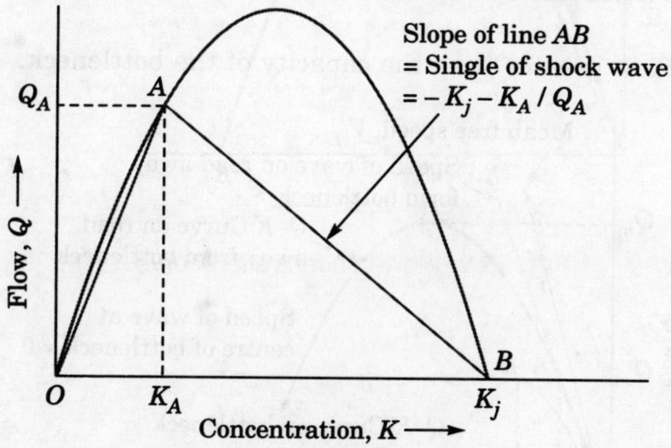

Fig. 23.4

to a halt, from a flow Q_A at a concentration K_A. In the region of the halted vehicles, the flow is zero and the concentration is the highest, almost reaching the jamming value. This region of halted flow and high concentration is propagated backward by means of a shockwave, whose speed is given by the slope of the line AB in Fig. 23.4.

Another example of the application of the above theory is in dealing with bottlenecks, say when one lane of a, four lane divided carriageway is closed for repairs. Three possibilities can be considered :

Case 1. Flow less than capacity of the bottleneck.

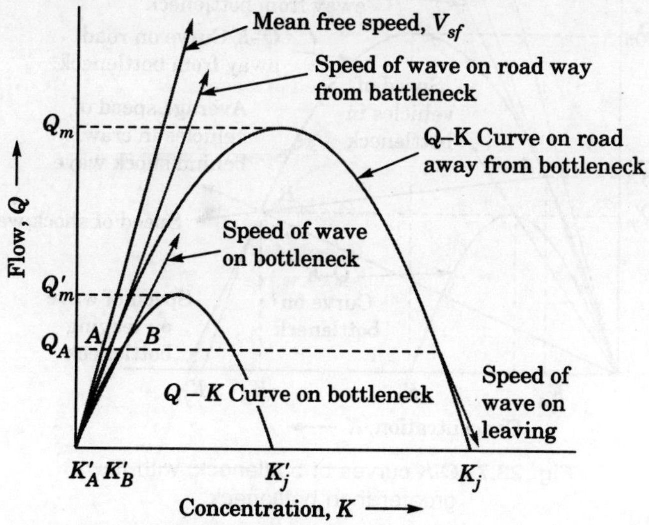

Fig. 23.5. *Q-K* curve of bottleneck, with flow less than bottleneck capacity.

Let Q_A represent the flow, which is less than Q'_m, the capacity of the bottleneck. On entering the bottleneck, the concentration increases from K_A to K_B. The mean speed of vehicles in the road, Q_A/K_A drops down to Q_A/K_{AB}. No shock wave is formed.

Case 2. **Flow equal to the capacity of the bottleneck.**

This is similar to case 1, except that the flow is equal to the maximum capacity of the bottleneck. The speed of the wave at the centre of the bottleneck is zero.

Case 3. **Flow greater than the capacity of the bottleneck.**

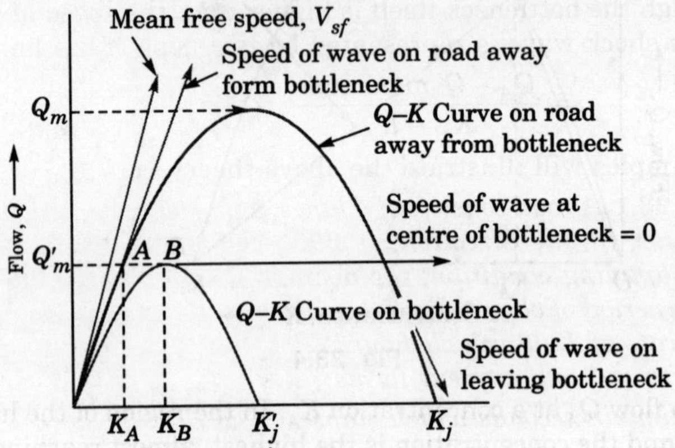

Fig. 23.6. *Q-K* curves of bottleneck with flow equal to bottleneck capacity.

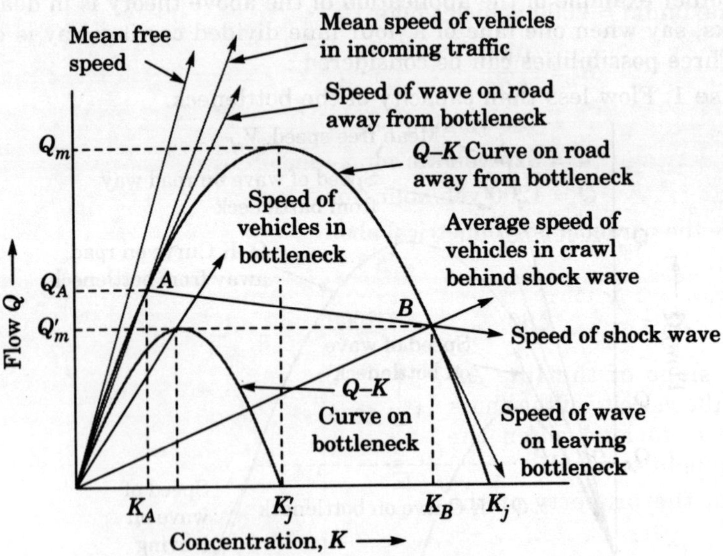

Fig. 23.7. *Q-K* curves of bottleneck, with flow greater than bottleneck.

The point A represents a traffic condition with flow Q_A greater than the bottleneck capacity Q maximum. The speed of the vehicles through the bottleneck drops from $\frac{Q_A}{K_A}$ to $\frac{Q' \text{max.}}{K_j'}$. The points B represents the flow condition on the second half of the $Q - K$ curve of the road away from the bottleneck, with a concentration equal to the bottleneck capacity Q' maximum. The crawl speed of traffic behind the shock wave is represented by $\frac{Q' \text{max.}}{K_B}$ which is very much lower than the speed of vehicles through the bottleneck itself. This shows that the speed of vehicles through the bottleneck itself is higher than the speed of crawl behind. The speed of the shock wave is represented by the slope of the line AB

$$= \frac{Q_A - Q' \text{ maximum}}{K_B - K_A}$$

A few examples will illustrate the above theory.

Problem 23.1. *A road consists of 4 lanes, 2 in each direction. The maximum capacity of 2 lanes in one direction, is 2000 vehicle/hour. When vehicles are stationary in a jamming condition, the average length occupied by a vehicle is. 6.25 m. During a period of observation, the actual volume of traffic in one direction is steady at the rate of 1200 vehicle/hour. This flow is brought to a halt when a traffic signal turns red and a queue forms.*

Find the time in seconds which elapses from the moment the signal turns red until the stationary queue reaches another intersection 75 m from the signal. Assume a linear relationship between speed and concentration.

Solution. Since the relationship between speed and concentration is linear, the Q-K curve will be parabolic. Fig. 23.8 indicates the Q-K diagram.

Q maximum = 2000 vehicles/hour (for 2 lanes)

$$K_j = \frac{1000}{6.25}$$

$= 160$ veh/km per lane

$= 320$ veh/km for 2-lanes.

$Q = 1200$ veh/hour (for 2 lanes)

Since the parabola is symmetrical about the optimum value of concentration K_0,

$$K_0 = \frac{K_j}{2} = 160.$$

The slope of the line AB represents the velocity of the shock wave that is formed when the traffic is brought to a halt.

From the property of the parabola,

$$\frac{x^2}{(160)^2} = \frac{y}{2000}$$

$$= \frac{800}{2000}$$

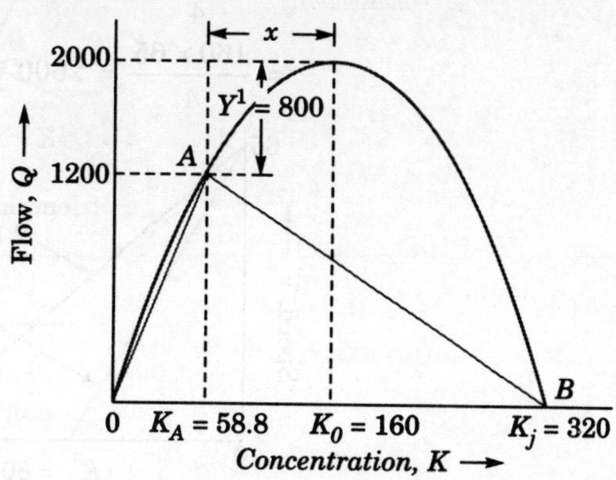

Fig. 23.8. Q-K curve.

$\therefore \quad x = \sqrt{\dfrac{800 \times 160 \times 160}{2000}} = 101.2$

$\therefore \quad K_A = 160 - 101.2 = 58.8$

$\therefore$ Slope of the line AB

$$= \frac{120}{320 - 58.8} = \frac{1200}{261.2} = 4.6.$$

$\therefore$ Speed of the shock wave = 4.6 K.P.H.

$\therefore$ Time in seconds to reach a distance of 75 m

$$= \frac{75 \times 3600}{4.6 \times 1000} = 58.7 \text{ sec.}$$

Problem 23.2. *One lane of a 2 lane one-way carriageway is closed for repairs. The maximum mean free speed under conditions of low flow in the 2 lane portion is 65 K.P.H. Under conditions of low flow, observations show that the maximum mean free speed in the bottleneck is also 65 K.P.H. The average space headway when vehicles are stationary is 12.5 m. The volume of traffic on the 2 lane road is 1800 veh/hour. Assuming that the speed-concentration relation is linear, find :*

(i) the mean speed of traffic through the bottleneck caused by the closure of one lane ;

(ii) the mean speed of traffic in the congested conditions immediately on the approach to the bottleneck ;

(iii) the mean speed of traffic on the carriageway clear of the influence of the bottleneck ;

(iv) the rate at which the queue of the congested traffic entering the bottleneck grows.

Solution. The speed-concentration curves and the flow-concentration curves for the 2 lane road and the bottleneck are shown in Fig. 23.9.

$$K_j = \frac{1000}{12.5} \times 2 = 160 \text{ veh/km for 2 lanes}$$

$$K_j = \frac{1000}{12.5} = 80 \text{ veh/km for single lane}$$

$$Q_{(maximum)} = \frac{K_j \times \bar{u}_{sf}}{4} \quad \text{(Eq. 23.1)}$$

$$= \frac{160 \times 65}{4} = 2600 \text{ veh/hour}$$

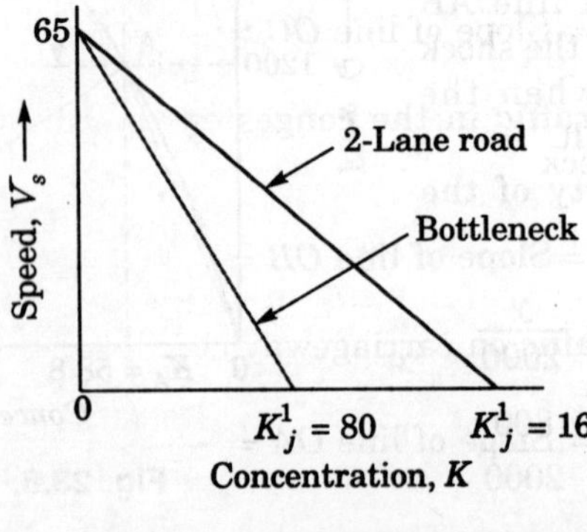

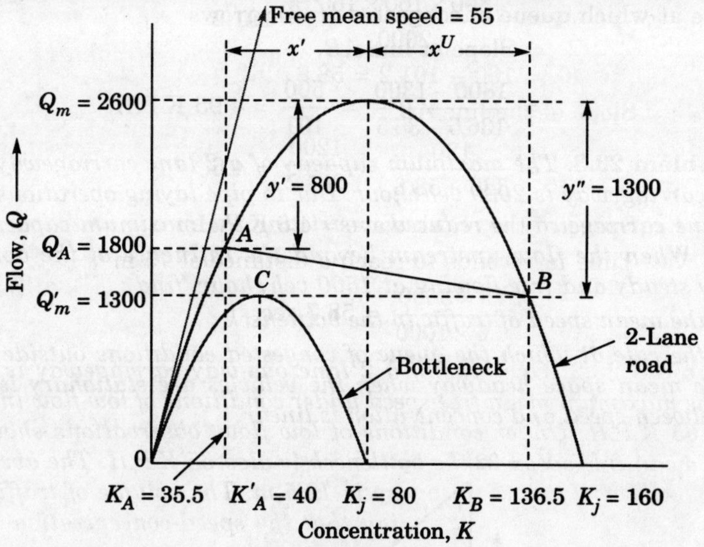

Fig. 23.9

$$Q_{(maximum)} = \frac{K_j \times \bar{v}_{sf}}{4} = \frac{80 \times 65}{4} = 1300 \text{ veh/hour.}$$

The observed volume is 1800 veh/hour and is indicated by point A on the Q-K curve. The point B corresponds to the flow corresponding to the bottleneck capacity.

$$\frac{(x'')^2}{(80)^2} = \frac{1300}{2600}$$

$$\therefore \quad x'' = 80\sqrt{0.5} = 56.5$$

$$\therefore \quad K_B = 80 + 56.5 = 136.5$$

$$\left(\frac{x'}{80}\right)^2 = \frac{800}{2600} = 0.308$$

$$\therefore \quad x' = 80 \times \sqrt{0.308} = 44.5$$

$$\therefore \quad K_A = 80 - 44.5 = 35.5.$$

Mean speed of traffic through bottleneck

$$= \text{Slope of line } OC = \frac{1300}{40} = 32.5 \text{ K.P.H.}$$

Mean speed of traffic in the congested conditions immediately on the approach to the bottleneck

$$= \text{Slope of line } OB = \frac{1300}{13.6} = 9.5 \text{ K.P.H.}$$

Mean speed of traffic on carriageway clear of the influence of bottleneck

$$= \text{Slope of line } OA = \frac{1800}{35.5} = 50.7 \text{ K.P.H.}$$

Rate at which queue of congested traffic grows

$$= \text{Slope of line } AB$$

$$= \frac{1800 - 1300}{136.5 - 35.5} = \frac{500}{101} = 4.95 \text{ K.P.H.}$$

Problem 23.3. *The maximum capacity of a 2 lane carriageway of a four-lane dual carriageway is 2000 veh/hour. Due to pipe-laying operations the width of the 2 lane carriageway is reduced, restricting the maximum capacity to 1100 veh/hour. When the flow upstream beyond the influence of the bottleneck is reasonably steady and free-flowing at 1500 veh/hour, find :*

(i) the mean speed of traffic in the bottleneck ;

(ii) the rate at which the queue of congested conditions outside bottleneck grows. The mean space headway when the vehicles are stationary is 8 m. The relation between speed and concentration is linear.

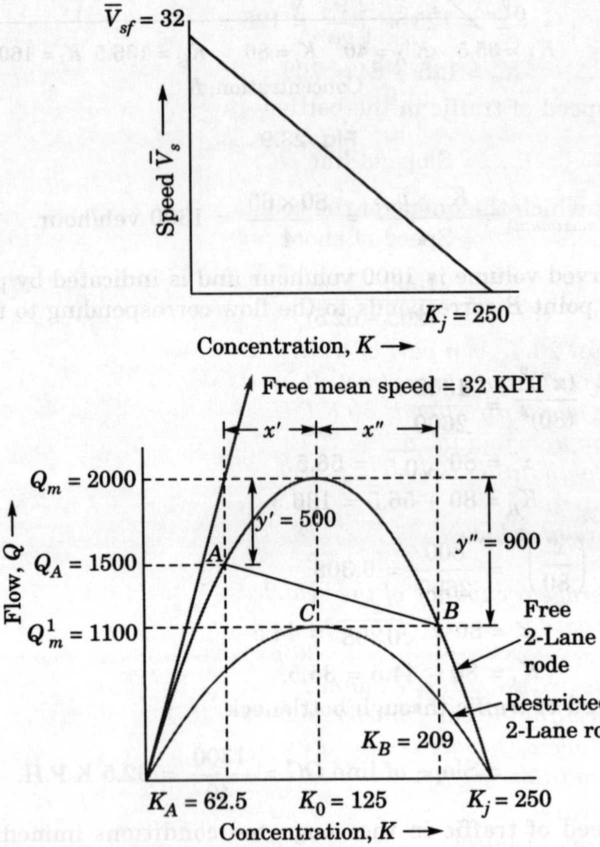

Fig. 23.10. Q-K and $\bar{v}_s$– K curve for the conditions in Problem 23.3.

Solution. Since the speed-concentration relationship is linear, the flow-concentration curve will be parabolic. These are indicated in Fig. 23.10. It may be noted that the pipe-laying operations have just reduced the width of the 2 lane carriageway and it is not indicated that one entire lane is closed. So it will be assumed that the bottleneck will be of 2 lanes, but with reduced width.

$$K = \frac{1000}{8} \times 2 = 250 \text{ veh/km.}$$

$$Q_{(max)} = \frac{K_j \times \bar{v}_{sf}}{4}$$

$$\therefore \quad 2000 = \frac{250 \times \bar{v}_{sf}}{4}$$

$\therefore$ $\bar{v}_{sf}$ the free mean speed

$$= \frac{2000 \times 4}{250} = 32 \text{ H.P.H.}, \left(\frac{x'}{125}\right)^2 = \frac{500}{1000}$$

$$\therefore \quad x' = 125 \times \sqrt{\frac{1}{4}} = 62.5$$

$$\left(\frac{x''}{125}\right)^2 = \frac{900}{2000}$$

$$\therefore \quad x'' = 125 \times \sqrt{\frac{9}{20}} = 125 \times \frac{3}{4.47} = 84$$

$$\therefore \quad K_B = 125 + 84 = 209.$$

Mean speed of traffic in the bottleneck

$$= \text{Slope of line } OC = \frac{1100}{125} = 8.8 \text{ K.P.H.}$$

Rate at which the queue of the congested conditions grows

= Speed of shock wave = Slope of line AB

$$= \frac{(1500 - 1100)}{(209 - 62.5)} = \frac{400}{146.5} = 2.73 \text{ K.P.H.}$$

Problem 23.4. *As a part of traffic management measure, it is proposed to ban the private car from a city street. It has been observed that the maximum free speed of traffic in one direction is 50 K.P.H., and this speed is unaffected for values of concentration upto and including 20 veh/km. The present flow is 1200 veh / hour per lane, the composition being (in percentage terms) :*

Private cars	:	*80*
Commercial vehicles	:	*18*
Buses	:	*2*

The average occupancy of the vehicles is :

Private cars	:	*1.5 persons per car*
Buses	:	*30 persons per bus.*

After the private cars are banned, it is expected that one-half of the persons previously travelling by car transfer to buses, the average occupancy remaining the same. The average space occupied by each type of vehicle when stationary in the jammed condition is :

Private cars	:	*7 metres*
Commercial vehicle	:	*12 metres*
Buses	:	*10 metres*

Assuming the relation between speed and concentration to be linear find :

(i) the maximum capacity of the street before the imposition of the ban on private cars ;

(ii) the maximum capacity of the street after the imposition of the ban on private cars ;

(iii) the mean speed of traffic in km/hour before the imposition of the ban on private cars ;

(iv) the mean speed of traffic in km/hour after the imposition of the ban on the private cars.

Solution. ***Before the imposition of the ban on private cars :***

The weighted spacing of vehicles

$$= 7 \times 0.8 + 12 \times 0.18 + 10 \times 0.02$$
$$= 5.6 + 2.16 + 0.2 = 7.96 \text{ m}$$

Jamming concentration before,

$$K_j = \frac{1000}{7.96} = 126 \text{ veh/hr.}$$

The v_s–K curve and Q-K curve before the imposition of the ban are given in Fig. 23.11.

$$Q_{(max)} = \frac{K_j \times \bar{v}_{sf}}{4} \quad \text{(Eq. 22.9)}$$

$$= \frac{126 \times 50}{4} = 1575 \text{ veh/hour/lane.}$$

For a flow corresponding to 1200 veh/hour, it is required to find the mean speed. This is given by the slope of the line *OA*.

$$\therefore \quad \left(\frac{x}{63}\right)^2 = \frac{375}{1575}$$

$$x = 63 \times \sqrt{\frac{375}{1575}} = \frac{63}{\sqrt{4.2}} = \frac{63}{2.05} = 30.8$$

$$\therefore \quad K_A = 63 - 30.8 = 32.2$$

$\therefore$ **Mean speed of traffic**

$$= \text{Slope of line } OA = \frac{1200}{32.2} = 37.3 \text{ K.P.H.}$$

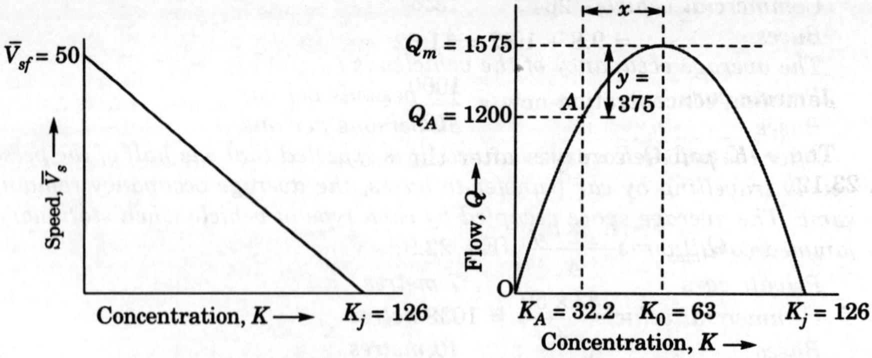

Fig. 23.11. v_s-K and Q-K curves for problem 23.4 before the imposition of ban on private cars.

After the imposition of the ban on private cars :

The number of persons travelling by car before the ban

$$= 1200 \times 0.80 \times 1.5 = 1440 \text{ persons/hour.}$$

The number of persons now switching over to buses

$$= 50\% \text{ of } 1440 = 720 \text{ persons/hour}$$

$\therefore$ No. of extra buses needed $= \dfrac{720}{30} = 24/\text{hour}$.

No. of buses originally present $= 1200 \times \dfrac{2}{100} = 24/\text{hour}$

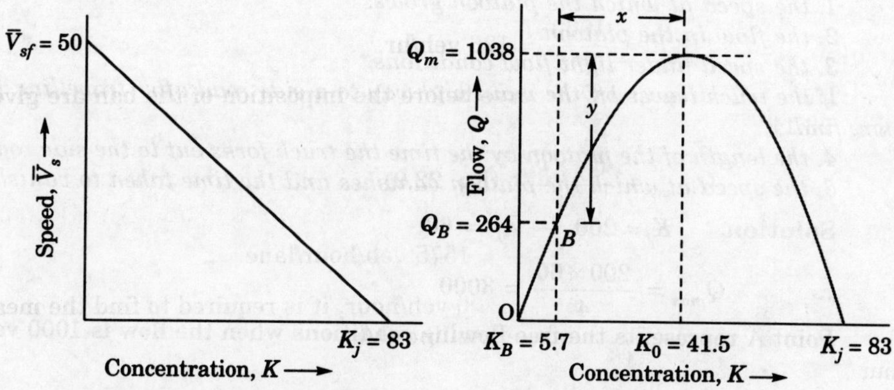

Fig. 23.12. v_s-K and Q-K curves for problem 23.4 after the imposition of ban on private cars.

Total No. of buses after the ban = 48/hour

No. of commercial vehicles

$$= 1200 \times 0.18 = 216/\text{hour}$$

Total flow $= 48 + 216 = 264$ vehicles/hour.

Weighted spacing of vehicles now :

$$= \frac{12 \times 216}{264} + \frac{10 \times 48}{264}$$

$$= 9.8 + 1.82 = 11.62, \text{ say } 12 \text{ m.}$$

Jamming concentration now $= \dfrac{1000}{12} = 83$ veh/km/lane.

The v_s-K and Q-K curves after the imposition of the ban are given in Fig. 23.12.

$$Q_{(max)} = \frac{K_j \times u_{sf}}{4} \quad \text{(Eq. 22.9)}$$

$$= \frac{83 \times 50}{4} = 1038 \text{ veh/hour/lane}$$

$$\left(\frac{x}{41.5}\right)^2 = \frac{1038 - 264}{1038} = \frac{774}{1038}$$

$$\therefore \quad x = \frac{41.5}{\sqrt{1.34}} = \frac{41.5}{1.16} = 35.8$$

$$\therefore \quad K_B = 41.5 - 35.8 = 5.7$$

∴ Mean speed of traffic after imposition of ban on private cars

$$= \text{Slope of line } OB = \frac{264}{5.7} = 46.5 \text{ K.P.H.}$$

Problems 23.5. *On a lane of a highway, the free speed of traffic is observed to be 60 K.M.P.H. and the jamming concentration is observed to be 200/km. Under light flow conditions, it is observed that there are 1000 vehicles per hour. A large truck travelling at 12 K.M.P.H. enters the stream, forcing the vehicles behind to queue up and move in a platoon. Find :*

1. the speed at which the platoon grows.
2. the flow in the platoon.
3. the speed under light flow conditions.

If the truck leaves off the main highway to a side road after travelling for 2 km, find ;

4. the length of the platoon by the time the truck forks out to the side road.
5. the speed at which the platoon vanishes and the time taken to vanish.

Solution. $K_j = 200, \quad \overline{u}_{sf} = 60$

$$\therefore \quad Q_{max} = \frac{200 \times 60}{4} = 3000$$

Point A represents the free-flowing conditions when the flow is 1000 veh/hour

$$\left(\frac{x'}{100}\right)^2 = \frac{2000}{3000} = 0.67$$

$$\therefore \quad x' = 100\sqrt{0.67} = 81.9$$

$$\therefore \quad \text{Speed under light flow conditions} = \frac{1000}{100 - 81.9}$$

Let B represents the congested flow conditions when the speed is 12 K.M.P.H. Let y represent the flow at B.

$$\frac{3000 - y}{3000} = \left(\frac{x''}{100}\right)^2 \quad \text{Also} \quad \frac{y}{100 + x''} = 12$$

$$\therefore \quad y = 1200 + 12\,x''$$

Substituting,

$$\frac{3000 - 1200 - 12x''}{3000} = \left(\frac{x''}{100}\right)^2$$

$$\therefore \quad 1800 - 12\,x'' = 0.3\,x''^2$$

Solving the quadratic equation, we have

$$x'' = \frac{-12 \pm \sqrt{144 + 4 \times 1800 \times 0.3}}{2 \times 0.3} = \frac{-12 \pm 48}{0.6} = \frac{36}{0.6} = 60$$

$$\therefore \quad y = 1200 + 12 \times 60 = 1920.$$

∴ Flow in the platoon = 1920 veh/hour.

Speed at which the rear of the platoon moves forward

= speed of shock wave = slope of AB

$$= \frac{1920 - 1000}{160 - 18.1} = \frac{720}{141.9} = 5.07 \text{ K.M.P.H.}$$

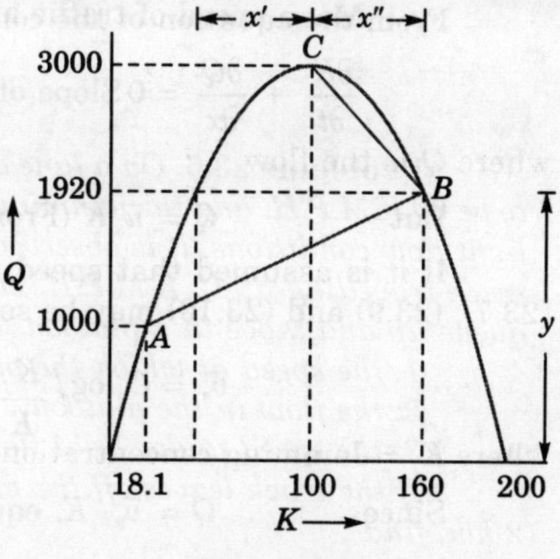

Fig. 23.13

Thus the front of the platoon moves forward at a speed of 12 K.M.P.H. whereas the rear moves forward at a speed of 5.07 K.M.P.H.

The platoon grows in size at a speed of :

12 – 5.07 = 6.93 K.M.P.H.

In the 2 km trip, the platoon grows to a length of :

= 2.12 × 6.93 km = 1.16 km.

When the truck forks out to the side road, the vehicles will travel at the maximum capacity, which is 3000 veh/hr. This point is represented by *C*. The front of the platoon then moves at a speed given by the slope of the line *CB*.

$$\text{Slope of line } CB = \frac{3000 - 1920}{160 - 100} = \frac{1080}{60} = 18 \text{ K.M.P.H.}$$

With the front of the platoon moving to the rear at this speed and the rear of the platoon moving forward at a speed of 5.07 K.M.P.H., the platoon will vanish at a speed of

– 18 + 5.07 = – 12.93 K.M.P.H.

Since the length of the platoon is 1.16 km, this is achieved in

(1.16/12.93) = 0.0897 hr = 5.382 minutes.

23.5. Greenberg's Extension of the Law of Continuity

23.5.1. The law of continuity which forms the basic assumption in Lighthill and Whitham's Theory has also been applied by others. Among them, the work of Greenberg deserves mention (Ref. 2, 3). Greenberg's approach yields very useful equations describing the relations between speed, flow and concentration.

23.5.2. The equation of motion for a one-dimensional continuous fluid is :

$$\frac{d\bar{v}_s}{dt} = -\frac{C^2}{K} \times \frac{\partial K}{\partial x} \qquad \text{...(23.6)}$$

where, $\bar{v}_s$ = Space mean speed
K = Concentration
x = Distance
t = Time to travel a distance x
C = Constant.

Since speed is a function of time (t) and distance (x),

$$\frac{d\bar{v}_s}{dt} = \frac{\partial \bar{v}_s}{\partial x} + \frac{dx}{dt} + \frac{\partial \bar{v}_s}{\partial t} \times \frac{dt}{dt}$$

$$= \frac{\partial \bar{v}_s}{\partial x} \times \bar{v}_s + \frac{\partial \bar{v}_s}{\partial t} \qquad \text{...(23.7)}$$

Substituting Eq. (23.6) in Eq. (23.7)

$$\frac{\partial \bar{v}_s}{dx} \times \bar{v}_s + \frac{\partial \bar{v}_s}{\partial t} + \frac{C^2}{K} \times \frac{\partial K}{\partial x} = 0 \qquad \text{...(23.8)}$$

From the equation of the continuity of flow,

$$\frac{\partial K}{\partial t} + \frac{\partial Q}{\partial x} = 0 \quad \text{...(23.9)}$$

where Q is the flow.

But $Q = \overline{u}_s K$ (From Eq. 22.1) ...(23.10)

If it is assumed that speed is only a function of concentration, the Eqs. (23.7), (23.9) and (23.10) may be solved to yield :

$$\overline{v}_s = C \log_e \frac{K_j}{K} \quad \text{...(23.11)}$$

where K_j = Jamming concentration.

Since $Q = \overline{v}_s, K$, equation 23.11 gives,

$$Q = CK \log_e \frac{K_j}{K} \quad \text{...(23.12)}$$

23.5.3. The above equation can also be obtained by the boundary condition approach. $Q = K \,.\, \overline{v}_s$

Differentiating with respect to K.

$$\frac{dQ}{dK} = K \times \frac{d\overline{v}_s}{dK} + \overline{v}_s \times \frac{dK}{dK}$$

$$\therefore \quad \frac{dQ}{dK} = K \times \frac{d\overline{v}_s}{dK} + \overline{v}_s \quad \text{...(23.13)}$$

Let $\quad K \,.\, \frac{dv}{dK} = C$ ...(23.14)

Then from Eq. (23.13). $\frac{dQ}{dK} - \overline{v}_s = C$

This expresses the hypothesis that the difference between the speed of propagation of the wave [Lighthill and Whitham's (Ref. 1) Kinematic wave] and the space-mean speed is a constant. With this assumption, Equation (23.14) can be re-written as :

$$\frac{dv}{dK} = \frac{C}{K} \quad \text{...(23.15)}$$

Integrating both sides with respect to K,

$$\frac{dv}{dK}\, dK = \int_{K=K}^{K=K_j} \frac{C}{K}\, dK$$

$$\therefore \quad [\overline{v}_s\ K = K_j - \overline{v}_s\ K = K] = C\,[\log_e K_j - \log_c K] \quad \text{...(23.16)}$$

But when $\quad K = K_j, \quad \overline{v}_s = 0$

$$\therefore \quad -\overline{v}_s = C \log_e \frac{K_j}{K} \quad \text{...(23.17)}$$

It is possible to rewrite Eq. (23.17) as below if the sign of the constant C is properly taken care of :

$$\overline{v}_s = \frac{K_j}{K} \quad \text{...(23.18)}$$

Multiplying by K yields :

$$Q = CK \cdot \log_e \frac{K_j}{K} \quad \text{...(23.19)}$$

This is the equation of the fundamental diagram of traffic flow. It can be shown that the value of C is equal to the stream velocity for maximum flow.

23.5.4. Plotting $\frac{Q}{CK_j}$ against $\frac{K}{K_j}$, the curve in Fig. 23.14 is obtained, which has a maximum value of $= \frac{Q}{CK_j} = \frac{1}{e}$ at a value of $\frac{K}{K_j} = \frac{1}{e}$.

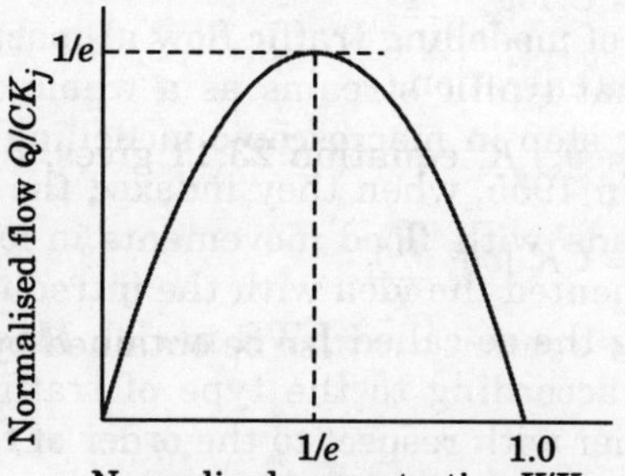

Fig. 23.14. Curve of $\frac{Q}{CK_j}$ against $\frac{K}{K_j}$.

I. SOLVED QUESTIONS AND ANSWERS

Q. 1. What is lighthill and Whiteham's theory?

Ans. The LWR model, named after the authors in LW55 and Ric56, is a macroscopic one-dimensional traffic model. The conservation law for traffic in one dimension is given by

$$\frac{\partial}{\partial t}\rho(t, x) + \frac{\partial}{\partial x} f(t, x) = 0$$

In this equation ρ is the traffic density (vehicles or pedestrians) and f is the flux which is the product of traffic density and the traffic speed v, *i.e.*, $f = \rho v$. There are many models researchers have proposed for how the flux should be dependent on traffic conditions. This relationship is given by the fundamental diagram.

Q. 2. Explain Greenshield's model.

Ans. Greenshield's model uses a linear relationship between traffic density and traffic speed by

$$v(\rho) = v_f \left(1 - \frac{\rho}{\rho_m}\right)$$

where v_f is the free flow speed and ρ_m is the maximum density. Free flow speed is the speed of traffic when the density is zero. This is the maximum speed. The maximum density is the density at which there is a traffic jam and the speed is equal to zero. The flux function is concave as can be confirmed by noting the negative sign of the second derivative of flow with respect to density, *i.e.*, $\delta_2 + \delta\rho_2$

< 0. The fundamental diagram refers to the relationship that the traffic density ρ, traffic speed v and traffic flow f have with each other.

Q. 3. What is a macroscopic traffic flow model?

Ans. A Macroscopic traffic flow model is a mathematical traffic model that formulates the relationships among traffic flow characteristics like density, flow, mean speed of a traffic stream, etc. Such models are conventionally arrived at by integrating macroscopic traffic flow models and converting the single-entity level characteristics to comparable system level characteristics.

Q. 4. When was the method of modelling traffic flow at macroscopic level originated?

Ans. The method of modelling traffic flow at macroscopic level originated under an assumption that traffic streams as a whole are comparable to fluid streams. The first major step in macroscopic modelling of traffic was taken by Lighthill and Whitham in 1955, when they indexed the comparability of 'traffic flow on long crowded roads' with 'flood movements in long rivers'. A year later, Richards (1956) complemented the idea with the introduction of 'shock-waves on the highway', completing the so-called LWR model. Macroscopic modelling may be primarily classified according to the type of traffic as homogeneous and heterogeneous, and further with respect to the order of the mathematical model.

Q. 5. What is diffusion model?

Ans. Diffusion model diffusion model is an extension of the Greenshield's model where the traffic speed depends not only on the traffic density but also on the density gradient. This models the driver behaviour where changes in traffic density in the x-direction affect the traffic speed. The model is given by

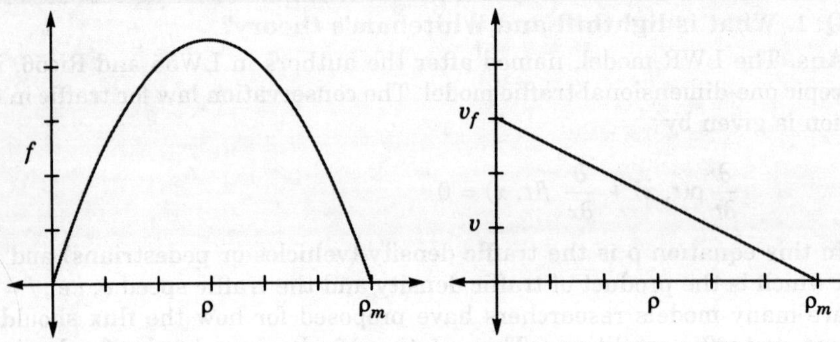

Fig. Fundamental diagram using Greenshield model

Q. 6. What is payne-Whitham model?

Ans. PW model or the Payne-Whitham model was proposed in the 1970s independently in Pay71 and Whi74. It uses two PDEs to represent the traffic dynamics. In its most general form, the model takes the following form

$$\rho_t + (\rho v)_x = 0$$

$$v_t + v\, v_x = \frac{V(\rho) - v}{\tau} - \frac{(A(\rho))_x}{\rho} + \mu \frac{v_{xx}}{\rho}$$

II. REVIEW QUESTIONS

1. **Discuss signalized intersections.**
2. **What is meant by bottleneck? Explain.**
3. **Discuss Greenberg's extension of the law of continuity.**

REFERENCES

1. **Lighthill, M.J. and G.B. Witham, on Kinematic Waves—II, A Theory of Traffic Flow on Long Crowded Roads, Proceedings of the Royal Society, Vol. 229, No. 1178, London 1955.**
2. **Greenberg, H., A Mathematical Analysis of Traffic Flow, Tunnel Traffic Capacity Study, the Port of New York, 1958.**
3. **Greenberg, H., An Analysis of Traffic Flow, Operations Research, Vol. 7, No. 1, 1959.**

24. Car-following Theory

24.1. Principles

In the car-following theory approach the vehicle driver system is visualised as one, where the driver of a following vehicle responds to the stimuli caused by the motion of the vehicle in front. The driver adjusts his speed, acceleration and spacing between the vehicle in front and his own to suit the motion of the vehicle in front. In this process, a certain time lag between receiving the stimulus and reacting to it is inevitable and has to be considered. This interesting theory has been built and enriched by the study of many researchers in the last 30 years (Ref. 1, 2, 3, 4). It may be noted that the theory applies to single lane dense traffic with no overtakings.

24.2. Derivation of Equations

In psychological terms, the car-following theory may be expressed in the form of the equation :

$$\text{Response} = \text{Sensitivity} \times \text{Stimulus} \qquad ...(24.1)$$

Response is taken as the acceleration of the following vehicle, since this is the quantity directly under the control of that driver. The stimulus could be taken as the relative speed between the following vehicle and that ahead. The stimulus can also depend upon the spacing between the two vehicles.

Mathematically, the stimulus-response equation can be put in the following form :

$$\ddot{x}_{(n+1)}(t+T) = \lambda\ [(\dot{x}_n t) - \dot{x}_{(n+1)}(t)] \qquad ...(24.2)$$

where $\ddot{x}$ = acceleration

$\dot{x}$ = speed

x = position on the road

t = time

T = time lag of response to stimulus (usually 1 to 2.2 seconds) (Ref. 4)

λ = a proportionality coefficient.

In the above equation, the subscript (n) refers to the nth vehicle, $(n + 1)$ to the $(n + 1)$th vehicle, the subscript (t) refers to the position at time t and the subscript $(t + T)$ refers to the position at time $t + T$.

The value of the sensitivity, λ, in Eq. (24.2) is intended to take care of the driver's awareness of the stimuli from the vehicle ahead and the intensity of his reaction to it. It is unlikely that λ is a constant and, in fact, it may be given a number of values such as :

(*i*) $\lambda = c$, a constant

(ii) $\lambda = a$, a constant where $s \leq s_c$, s being the spacing, defined as

$s = x_n - x_{n+1}$, and s_c is the critical spacing

and $\lambda = b$, a constant, where $s \leq s_c$

(iii) $\lambda = c/s$, c being a constant

(iv) $\lambda = \dfrac{\dot{x}_{(n+1)}}{s^2}$

(v) $\lambda = c/s^2$, c being a constant.

The values (iii) to (v) above can be included in a general expression of the form :

$$\lambda = \frac{c\dot{x}_{(m+1)}(t+T)}{\left[x_n(t) - x_{(n+1)}(t)\right]^l}$$

where, l and m are constants and take different values to obtain the three cases.

Two characteristics of traffic flow are important and need consideration : stability and the steady state of flow. Stability is important because of its application to accident studies, but unfortunately these stability criteria are difficult to establish except for the easiest case of $\lambda = C$. For this simple case, the car-following law may be expressed as follows :

$$\ddot{x}_{(n+1)}(t+T) = C\left[\dot{x}_{(n)}(t) - \dot{x}_{(n+1)}(t)\right] \qquad ...(24.3)$$

In the steady state, let each car be moving with a speed of a, so that we may write $x = u$, and since we are in a steady state condition we may drop T, the timelag of response. This implies that the stability conditions which involve T have been satisfied.

Eq. (24.3) can be then written as :

$$\frac{du(n+1)}{dt} = C\left[u(n) - u(n+1)\right] \qquad ...(24.4)$$

Now, since $x_{(n)} - x_{(n+1)} = S_{(n+1)}$

$$= \frac{1}{K_{(n+1)}} \qquad ...(24.5)$$

where K is the concentration.

$$u_{(n)} - u_{(n+1)} = \frac{ds_{(n+1)}}{dt}$$

$$= -\frac{1}{K^2_{(n+1)}}\frac{dK_{(n+1)}}{dt} \qquad ...(24.6)$$

Omitting subscripts, and combining Eqs. (24.4) and (24.5)

$$\frac{du}{dt} = \frac{C}{K^2}\frac{dK}{dt} \qquad ...(24.7)$$

Integrating Eq. (24.7) . $u = \dfrac{C}{K} + A$ ...(24.8)

However, if it is assumed that all changes of concentration are governed by this equation and that when the concentration is K_j (jamming concentration), the flow and, therefore, the velocity is zero, we can evaluate A by using the boundary conditions, $u = 0$ at $K = K_j$, giving $A = -C/K_j$

Thus, $$u = C\left[\frac{1}{K} - \frac{1}{K_j}\right] \quad ...(24.9)$$

Also, in the steady state, $Q = uK$

and hence, $$Q = C\left(1 - \frac{K}{K_j}\right) \quad ...(24.10)$$

Eq. (24.10) is the flow-concentration relationship.

Now consider the case of reciprocal spacing, when

$$\lambda = \frac{C}{x_{(n)} - x_{(n+1)}} \text{ or } \lambda = \frac{C}{s}$$

The car-following Eq. (24.2) now becomes :

$$\ddot{x}_{(n+1)}(t+T) = \frac{C\left[\dot{x}_{(n)}(t) - \dot{x}_{(n+1)}(t)\right]}{x_{(n)}(t) - x_{(n+1)}(t)} \quad ...(24.11)$$

In the steady state, Eq. (24.11) can be written as :

$$\frac{du\,(n+1)}{dt} = \frac{C\left[u_{(n)} - u_{(n+1)}\right]}{x_{(n)} - x_{(n+1)}} \quad ...(24.12)$$

From Eq. (24.5), we have $u_{(n)} - u_{(n+1)} = -\dfrac{1}{K^2}\dfrac{dK_{(n+1)}}{dt}$

and from Eq. (24.6), we have $x_{(n)} - x_{(n+1)} = \dfrac{1}{K_{(n+1)}}$

Substituting these in Eq. (24.12), we have

$$\frac{du}{dt} = -\frac{C}{K}\frac{dK}{dt} \quad ...(24.13)$$

Integrating, and using boundary conditions,

$u = 0$ at $K = K_j$, we get

$$u = C \log \frac{K_j}{K} \quad ...(24.14)$$

In the steady state, $Q = u\,K$, and hence

$$Q = CK \log_e \frac{K_j}{K} \quad ...(24.15)$$

Eq. (24.15) is the same as Eq. (23.19), which was derived from different conditions.

It may be noted that Eqs. (24.10) and (24.15) are not valid near $K = 0$ and hence they apply only to dense traffic.

24.3. A number of experiments on the 'follow-the leader' concept have been carried out in vehicular tunnels in New York City, in particular in the Holland

and Lincoln Tunnels (Ref. 3). These experiments have proved that the mathematical theory behind car-following approach gives a reasonably accurate description of some of the simple properties of traffic flow.

I. MULTIPLE CHOICE QUESTIONS

1. You should leave at least how many feet of space between your vehicle and the bicycle?

   (*a*) 5 feet (*b*) 2 feet
   (*c*) 3 feet (*d*) 4 feet

2. Which **THREE** pieces of information are found on a vehicle registration document?

   (*a*) Registered keeper (*b*) Make of the vehicle
   (*c*) Service history details (*d*) Date of the **MOT**
   (*e*) Type of insurance cover (*f*) Engine size

3. How can you use the engine of your vehicle to control your speed?

   (*a*) By changing to a lower gear (*b*) By selecting reverse gear
   (*c*) By changing to a higher gear (*d*) By selecting neutral

4. You are driving down a steep hill. Why could keeping the clutch down or rolling in neutral for too long be dangerous?

   (*a*) Fuel consumption will be higher
   (*b*) Your vehicle will pick up speed
   (*c*) It will damage the engine
   (*d*) It will wear tyres out more quickly

5. On a three-lane dual carriageway the right hand lane can be used for road

   (*a*) overtaking only, never turning right
   (*b*) overtaking or turning right
   (*c*) fast-moving traffic only
   (*d*) turning right only, never overtaking

6. What basic rule applies when you're using a motorway?

   (*a*) Use the lane that has the least traffic
   (*b*) Keep to the left-hand lane unless overtaking
   (*c*) Overtake on the side that's clearest
   (*d*) Try to keep above 50 mph to prevent congestion

7. In which of these situations should you avoid overtaking?

   (*a*) Just after a bend (*b*) In a one-way street
   (*c*) On a 30 mph road (*d*) Approaching a dip in the road

8. The **MAIN** benefit of having four-wheel drive is to improve:

   (*a*) road holding (*b*) fuel consumption
   (*c*) stopping distances (*d*) passenger comfort

Answers

1. (*a*) **2.** (*a*, *b*, *f*) **3.** (*a*) **4.** (*b*) **5.** (*b*)
6. (*b*) **7.** (*d*) **8.** (*a*)

II. SOLVED QUESTIONS AND ANSWERS

Q. 1. What is car following theory?

Ans. According to this theory, a driver accelerates (or decelerates) in response to the velocity changes of the vehicle immediately downstream. Each such acceleration of the driver occurs following a time lag. Much of the work related to this theory was concerned with how the values of the time lag affect stability; *i.e.,* the amplification of disturbances propagating through traffic.

Q. 2. What is Newell's car following model?

Ans. In traffic flow theory, Newell's car-following model is a method used to determine how vehicles follow one another on a roadway. The main idea of this model is that a vehicle will maintain a minimum space and time gap between it and the vehicle that precedes it.

Thus, under congested conditions, if the leading car changes its speed, the following vehicle will also change speed at a point in time-space along the traffic wave speed, $-w$.

Q. 3. What does car following theories describe?

Ans. Car following theories describe how one vehicle follows another vehicle in an uninterrupted flow. Various models were formulated to represent how a driver reacts to the changes in the relative positions of the vehicle ahead.

Q. 4. What is the basic assumption of pipe's model ?

Ans. The basic assumption of this model is a good rule for following another vehicle at a safe distance is to allow yourself at least the length of a car between your vehicle and the vehicle ahead for every ten miles per hour of speed at which you are travelling" According to Pipe's car-following model, the minimum safe distance headway increases linearly with speed. A disadvantage of this model is that at low speeds, the minimum headways proposed by the theory are considerably less than the corresponding field measurements.

Q. 5. What is Forbes' model?

Ans. In this model, the reaction time needed for the following vehicle to perceive the need to decelerate and apply the brakes is considered. That is, the time gap between the rear of the leader and the front of the follower should always be equal to or greater than the reaction time. Therefore, the minimum time headway is equal to the reaction time (minimum time gap) and the time required for the lead vehicle to traverse a distance equivalent to its length. A disadvantage of this model is that, similar to Pipe's model, there is a wide difference in the minimum distance headway at low and high speeds.

Q. 6. Explain the general motors' model.

Ans. The general motors' model is the most popular of the car-following theories because of the following reasons:

- Agreement with field data; the simulation models developed based on General motors' car following models shows good correlation to the field data.
- Mathematical relation to macroscopic model; Greenberg's logarithmic model for speed density relationship can be derived from General motors car following model.

Q. 7. What is the basic philosophy of car following theory?

Ans. The basic philosophy of car following model is from Newtonian mechanics, where the acceleration may be regarded as the response of a matter to the stimulus it receives in the form of the force it receives from the interaction with other particles in the system.

Q. 8. What is the concept of optimal velocity model?

Ans. The concept of this model is that each driver tries to achieve an optimal velocity based on the distance to the preceding vehicle and the speed difference between the vehicles. This was an alternative possibility explored recently in car-following models.

REFERENCES

1. Herman, R., The Mathematical Theory of Traffic flow, Institute of Trafllc Engineers, 1960.
2. Herman, R., E. Montroll, R, Potts and R. Rothery, Traffic dynamics : Analysis of stability of car following, Operation Research, Vol. 7, 1959.
3. Herman, R., and R.B. Potts, Single Lane Traffic theory and Experiments, Proceedings of the Symposium on Theory of Traffic Flow, Detroit, 1959 : Elsevier Publishing Company, 1961.
4. Montroll, E.W. and R. Potts, Car Following and Acceleration Noise, Highway Research Board Special Report 79, Washington, 1864.
5. Gazis, D.C., R. Herman and R. Potts, Car following Theory of Steady State Traffic Flow, Operations Research, Vol. 7, 1959.

25. The Queueing Theory and Its Application to Traffic Engineering Problems

25.1. Introduction

In many daily life situations, one is faced with the problem of providing services for randomly originating demands. Take the simple case of motorists arriving at a toll booth. If the arrival rate is heavy and the rate of servicing is not able to cope up with it, it is inevitable that a queue builds up. Delays occur as a consequence. A study of these aspects is attempted in this Chapter. The pioneer, investigator of the queueing theory was A.K. Erlang (1909). Since then the subject has been approached by many investigators. Queueing is one of the problems studied in the subject of Operations Research which has been developed in recent times. The literature available is very vast. For a beginner, an elementary book on Operations Research (Ref. 1) should serve the purpose of an introduction. Exhaustive presentations are also available (Ref. 2).

25.2. Applications to Traffic Engineering

Most traffic engineering problems concern themselves with the provision of adequate capacity for the average flow of vehicles in the system, so as to avoid congestion and delay under the average conditions. With the capacity designed only for the average conditions, it is inevitable that congestion occurs in most of the peaking situations. Examples of the occurrence of queues in traffic flows are: signalised intersections, car parks, toll booths, bottleneck situations and so on. Because of the easy applicability of the queueing theory to these problems, traffic engineers have been evincing great interest in these subject recently. A useful introduction would be Cleveland and Capelle's excellent summarization of the queueing theory approaches to traffic engineering problems (Ref. 3) and a T.R.B. publication (Ref. 4).

25.3. Queueing Approach

In the queueing approach, the following are some of the variety of questions for which the answers are sought :

1. What is the average (mean) and the distribution of :
   (*i*) The queue length ?
   (*ii*) The waiting time (time spent in the queue plus the service time) ?
   (*iii*) The time spent in the queue ?
   (*iv*) The number of vehicles in the system ?
2. What is the probability of :
   (*i*) There being n units in the system ?
   (*ii*) There being more than n units in the system ?
   (*iii*) The waiting time being greater than t ?

(*iv*) The waiting time being between t and $t + \delta t$?

(*v*) A customer having spent time d is the system ?

(*vi*) The service channel being idle ?

25.4. Solving Any Queueing Problems

In solving any queueing problem, a number of elements that go as inputs to the process need to be understood and precisely defined. They are the

(*i*) Arrival pattern

(*ii*) Service facility characteristics

(*iii*) Queue discipline characteristics

(*iv*) Order of service

(*v*) Service time characteristics

(*vi*) Training and familiarity and

(*vii*) Service discipline.

The arrival pattern can be :

(*i*) Regular—with a rectangular distribution.

(*ii*) Random—with a Poissonian, Erlangian or normal distribution.

In actual practice, it is more likely that a Poissonian arrival pattern is encountered, the number of customers being infinitely large.

To describe the arrival pattern fully, we need to know the average rate of arrival and the statistical distribution of the arrivals.

The characteristics of the service facility can cover the following :

(*i*) Single channel

(*ii*) Multiple channels, with one queue for each channel

(*iii*) Single queue feeding into multiple channels

(*iv*) Number of queues greater than one, but less than the number of channels

(*v*) Multiple channels in series

(*vi*) Combination to multiple service channels, some in series, some in parallel.

The queue discipline characteristics can be the following :

(*i*) Join a single queue and wait.

(*ii*) The queue is of a fixed capacity and further customers are turned away.

(*iii*) Not to join the queue at all (baulking).

(*iv*) Join the queue and leave, because not served before a certain time (reneging).

(*v*) In a system of multiple queues, to change from one queue to another (jockeying).

The order of the service can be any one of the following :

(*i*) First in first out.

(*ii*) Last in first out.

(*iii*) No regular system of priority (*e.g.* a telephone operator).

(*iv*) A system of priority.

The service time can have the following types :

(*i*) Rectangular distribution

(*ii*) Random, exponential, Poisson distribution.

The training and familiarity determine whether the service is affected by the number of customers in the queue and whether there is any difference between different operators.

The service discipline can mean serving only one customer at a time or serving many customers at a time.

It is thus apparent from the above that the problems that arise in queueing approach can be of varied types, each requiring a particular solution.

25.5. Assumptions Made in a Simple Queueing Approach as Applied to Traffic Flow

The queueing theory will be developed on the basis of the following assumptions as applicable to traffic flow :

(*i*) The system is in a steady state and has "settled down". This assumption is valid only when the arrival and the service patterns are sustained for indefinitely long periods of time, and not for "peaking situations" or "transient behaviour". This implies that the traffic intensity (ρ), which is defined as

$$\frac{\text{Average rate of arrival } (\lambda)}{\text{Average rate of service } (\mu)} \text{ is less than 1. Thus}$$

$$\rho = \frac{\lambda}{\mu} < 1 \qquad \text{...(25.1)}$$

(*ii*) The number of customers is "discrete".

(*iii*) The population of potential customers is infinite.

(*iv*) The arrivals are random in nature and Poissonian distribution applies, the mean rate of arrival being λ.

(*v*) There will be no simultaneous arrivals.

(*vi*) There will be a single service channel. Separate modifications of the approach are available for multiple channels.

(*vii*) The queue is single and is of infinite capacity.

(*viii*) The order of service is first in first out.

(*ix*) There is a single follow-on service discipline.

(*x*) The service times vary and follow an exponential distribution, with a mean rate μ.

25.6. Derivation of Formulae

Notation :

λ = Mean rate of arrival per unit time (veh/unit time)

μ = Mean rate of service (veh/unit time)

$$\rho = \text{traffic intensity} = \frac{\lambda}{\mu} < 1$$

$P_n(t)$ = Possibility that the queueing system has *n* items at time *t*.

From Poisson distribution,

$$\text{Prob. } (x \text{ arrivals in time } dt), = \frac{(\lambda dt)^x \, e^{-\lambda dt}}{x\,!}$$

$P(1)$ = Prob. (1 arrival in time *dt*)

$$= \lambda dt\, e^{-\lambda dt} = \lambda dt \left[1 - \lambda dt + \frac{(\lambda dt)^2}{2!} - \ldots\ldots\right]$$

$= \lambda dt$, neglecting terms of dt^2 and more than the second order, dt being small ...(25.2)

$P(0)$ = Prob. (0 arrivals in time dt)

$$\frac{\lambda dt^0\, e^{-\lambda dt}}{0!} = e^{-\lambda dt} = \left[1 - \lambda dt + \frac{(\lambda dt)^2}{2!} - \ldots\ldots\right]$$

$= (1 - \lambda dt)$ for the same reason as above ...(25.3)

In the stable state, (*i.e.* $\rho = \lambda/\mu < 1$) the state of the system is completely described by the number of units in the system.

Let $P_n(t)$ denote the probability that the queueing system has n items at time t.

Consider the situation at time $t + \delta t$, where δt is so short that only one vehicle can enter or leave the system during this time.

The ways in which a state $(n - 1)$ can be reached in time $t + \delta t$ are indicated in Table 25.1.

Table 25.1

	Situation at time t way		*Arrivals in δt*		*Completion of service in δt*	
	No. in system	*Prob.*	*No.*	*Prob.*	*No.*	*Prob.*
1.	$n - 1$	$P_{n-1}(t)$	0	$1 - \lambda\delta t$	0	$1 - \mu\delta t$
2.	$n - 1$	$P_{n-1}(t)$	1	$\lambda\delta t$	1	$\mu\delta t$
3.	n	$P_n(t)$	0	$1 - \lambda\delta t$	1	$\mu\delta t$
4.	$n - 2$	$P_{n-2}(t)$	1	$\lambda\delta t$	0	$1 - \mu\delta t$

All other possible ways, including 2 in Table 25.1 involve two or more events in the short time dt and their probabilities are of the order of dt^2 or more, and can be neglected.

$\therefore\ P_{(n-1)}(t + dt)$ = Sum of ways No. 1, 3 and 4 of reaching the state $(n - 1)$

$$= P_{n-1}(t)(1 - \lambda\delta t)(1 - \mu\delta t) + P_n(t)(1 - \lambda\delta t)(\mu\delta t) + P_{n-2}(t)(\lambda\delta t)(1 - \mu\delta t)$$

Neglecting the terms in δt^2,

$$P_{n-1}(t + \delta t) = P_{n-1}(t)(1 - \mu\delta t - \lambda\delta t) + P_n(t)(\mu\delta t) + P_{n-2}(t)(\lambda\delta t)$$

Rearranging and letting $\delta t \to 0$,

$$\frac{P_{n-1}(t + \delta t) - P_{n-1}(t)}{\delta t} = -(\lambda + \mu)P_{n-1}(t) + P_n(t)\mu + P_{n-2}(t)\lambda$$

$$\therefore \quad \frac{dP_{(n-1)}(t)}{dt} = -(\lambda + \mu)P_{n-1}(t) + P_n(t)\mu + P_{n-2}(t)\lambda$$

in which, $n = 1, 2, 3, \ldots\ldots$ *i.e.*, > 0 ...(25.4)

The probability of the system being in state 0 at time $(t + dt)$ can come about in two ways :

(1) There are no units at time (t) and none arrives in the interval δt

or (2) There is one unit at time t and one unit departs in the interval δ and none arrives.

Thus, $P_0(t + \delta t) = P_0(t)(1 - \lambda\delta t) + P_1(t)\,[(\mu\delta t)(1 - \lambda\delta t)]$

Neglecting second order infinitesimals,

$$\frac{P_0(t + \delta t) - P_0(t)}{\delta t} = \mu P_1(t) - \lambda P_0(t)$$

Letting $\delta t \to 0$.

If $n = 1$, the term $P_{n-2}(t)$ disappears.

$$\therefore \quad \frac{dP_0(t)}{dt} = \lambda P_0(t) + P_1(t)\,\mu \qquad \text{...(25.5)}$$

Equations (25.3) and (25.4) can be expressed as differential difference equations whereby the steady state solutions are obtained by setting :

$$\frac{d\,P_n(t)}{dt} = 0$$

From Eq. (25.4), $\mu P_n = (\lambda + \mu)\,P_{n-1} - \lambda P_{n-2}$ ...(25.6)

From Eq. (25.5), $P_1 = \dfrac{\lambda P_0}{\mu}$

$\therefore \quad = \rho P_0$ ...(25.7)

Dividing Eq. (25.6) throughout by μ,

$$P_n = \left(\frac{\lambda + \mu}{\mu}\right) P_{n-1} - \frac{\lambda}{\mu} P_{n-2} \qquad \text{...(25.8)}$$

$$= (1 + \rho)\,P_{n-1} - \rho P_{n-2} \text{ for } n > 1 \qquad \text{...(25.9)}$$

$$P_2 = (1 + \rho)\,P_1 - \rho P_0 \qquad \text{...(25.10)}$$

From Eq. (25.7), $P_1 = \rho P_0$

Substituting the above in Eq. (25.10),

$$P_2 = \rho(1 + \rho)\,P_0 - \rho P_0 = \rho^2 P_0 \qquad \text{...(25.11)}$$

The general solution is

$$P_n = \rho^n P_0 \text{ for } n > 0 \qquad \text{...(25.12)}$$

The sum of probabilities from $n = 0 \to \infty$ must be 1

$$\therefore \quad 1 = \sum_0^{\infty} P_n = P_0 \sum_0^{n} \rho^n$$

$$= P_0\{1 + \rho + \rho^2 + \rho^3 + \ldots\rho^n\} = P_0\,\frac{1}{1 - \rho}$$

$$\therefore \quad P_0 = 1 - \rho \qquad \text{...(25.13)}$$

Substituting Eq. (25.13) in Eq. (25.12)

$$P_n = \rho^n (1 - \rho) \qquad \text{...(25.14)}$$

Expected number of customers in the system

$$n = \sum_{n=0}^{n=\infty} nP_n = 0 + P_1 + 2P_2 + 3P_3 + ... + nP_n$$

$$= P_0 (\rho + 2\rho^2 + 3\rho^3 + ... + n\rho^n)$$

$$= (1 - \rho)\rho \frac{1}{(1-\rho)^2} = \frac{\rho}{1-\rho} \qquad ...(25.15)$$

Variance of the number in the system :

$$\text{Var. } (n) = \sum_{n=0}^{n=\infty} \left[n - E(n)^2\right] P_n,$$

where, $E = (n)$ = Expected value.

The above can be proved to be equal to :

$$\text{Var. } (n) = \frac{\rho}{(1-\rho)^2} \qquad ...(25.16)$$

Average time a unit spends in the system

$$d = \frac{1}{(\mu - \lambda)} = \frac{1}{\mu\left(1 - \frac{\lambda}{\mu}\right)}$$

$$= \frac{1}{\mu(1-\rho)} \qquad ...(25.17)$$

q = Average queue length = Expected number of customers in the system – Expected number in the service channel

$$= \frac{\rho}{1-\rho} - \rho = \frac{\rho^2}{1-\rho} \qquad ...(25.18)$$

$\overline{w}$ = Average waiting time spent in queue

$= \overline{d}$ – Time spent in service

$$= \overline{d} - \frac{1}{\mu} = \frac{1}{(\mu - \lambda)} - \frac{1}{\mu}$$

$$= \frac{1}{\mu(1-\rho)} - \frac{1}{\mu} = \frac{\rho}{\mu(1-\rho)} \qquad ...(25.19)$$

Prob. (number of units in system $> n$)

$= P(> n)$

$$= 1 - \sum_{0}^{n} P(r) = 1 - \sum_{0}^{n} \rho^r (1 - \rho)$$

$$= 1 - (1 - \rho) \sum_{0}^{n} \rho^r = 1 - (1 - \rho)(1 + \rho + \rho^2 ......)$$

$$= 1 - (1 + \rho + \rho^2 + \rho^3 ... + \rho^n - \rho - \rho^2 - \rho^3 ... - \rho^{n+1})$$

$$= \rho^{n+1} \qquad ...(25.20)$$

Prob. (waiting time is zero) = $P_{(0)} = 1 - \rho$ ...(25.21)

Prob. (of a queue existing)

$$= P(n > 1) = 1 - P_{(0)} - P_{(1)} = \rho^2 \quad \text{...(25.22)}$$

25.7. Problems

Some simple problems will now be solved to illustrate the above principles.

Problem 25.1. *A toll booth at the entrance to a bridge can handle 120 veh/hour, the time to process a vehicle being exponentially distributed. The flow is 90 veh/hour with a Poissonian arrival pattern. Determine :*

(i) the average number of vehicles in the system ;

(ii) the length of the queue ;

(iii) the average time spent by the vehicle in the system ;

(iv) the average time spent by the vehicle in the queue.

Solution. In this problem,

$$\lambda = \frac{90}{3600} = \frac{1}{40} \text{ veh/second,}$$

$$\mu = \frac{120}{3600} = \frac{1}{30} \text{ veh/second}$$

$$\rho = \frac{\lambda}{\mu} = \frac{30}{40} = \frac{3}{4}$$

Expected number of customers in the system

$$\bar{n} = \frac{\rho}{(1-\rho)} = 3$$

Average length of the queue, which is the number of vehicles in the queue

$$\bar{q} = \frac{\rho^2}{1-\rho} = 3 \times \frac{3}{4} = 2.25.$$

Average time spent by the vehicle in the system

$$\bar{d} = \frac{1}{\mu(1-\rho)} = \frac{1}{\frac{1}{30} \times \frac{1}{4}} = 120 \text{ seconds.}$$

Average time spent by the vehicle in the queue

$$= \bar{w} = \frac{\rho}{\mu(1-\rho)} = \frac{3}{4} \times 120 = 90 \text{ seconds.}$$

Problem 25.2. *The off-peak traffic flow arriving at random at toll booth facility is 90 V.P.H. and the peak flow is 180 V.P.H. The service rate, exponentially distributed, at the booth is 3.5 per minute. What is the average number of customers in the queue for each flow ?*

Solution. Off-peak flow : $\lambda = \frac{90}{60} = 1.5$ veh/minute

$$\mu = 3.5 \text{ veh/minute}$$

$$\rho = \frac{\lambda}{\mu} = \frac{1.5}{3.5} = \frac{3}{7}$$

Average number of vehicles in the queue

$$= \bar{q} = \frac{\rho^2}{1-\rho} = \frac{3}{7} \times \frac{3}{7} \times \frac{7}{4} = \frac{9}{28}$$

Peak flow: $\lambda = \frac{180}{60} = 3.0$ veh/minute

$$\mu = 3.5 \text{ veh/minute}, \rho = \frac{3}{3.5} = \frac{6}{7}$$

Average number of vehicles in the queue

$$= \bar{q} = \frac{\rho^2}{1-\rho} = \frac{6}{7} \times \frac{6}{7} \times \frac{7}{1} = \frac{36}{7} = 5 \text{ (approx).}$$

Problem 25.3. *There is a single toll booth in operation at a motorway where motorists are charged before being allowed to cross a bridge. The booth can handle 800 V.P.H. and the service times may be considered exponential. The peak flow is 720 V.P.H., the vehicle arrivals being random. Calculate :*

(i) the average number of vehicles in the system ;

(ii) the average time a vehicle is in the system ;

(iii) the average time a vehicle is in the queue ;

(iv) the chances of there being more than 4 vehicles in the system ;

(v) the percentage of time a toll booth operator is free ;

(vi) the probability that there is no vehicle in the system ;

(vii) the probability that the number of vehicles in the system is three.

Solution. $\lambda = \frac{720}{3600} = \frac{1}{5}$ veh/second.

$$\mu = \frac{800}{3600} = \frac{2}{9} \text{ veh/second.}$$

$$\rho = \frac{\lambda}{\mu} = \frac{1}{5} \times \frac{9}{2} = \frac{9}{10}$$

Average number of vehicles in the system

$$= \bar{n} = \frac{\rho}{(1-\rho)} = \frac{9}{10} \times \frac{10}{1} = 9$$

The average time a vehicle is in the system

$$\bar{d} = \frac{1}{\mu(1-\rho)} = \frac{1}{\frac{2}{9} \times \frac{1}{10}} = 45 \text{ seconds.}$$

The average time a vehicle is in the queue

$$\bar{w} = \frac{\rho}{\mu(1-\rho)} = 45 \times \frac{9}{10} = 40.5 \text{ seconds}$$

Probability of there being more than 4 vehicles in the system

$$= \rho^{n+1} = \left(\frac{9}{10}\right)^{4+1} = \left(\frac{9}{10}\right)^5 = 0.59$$

Percentage of time toll booth operator is free

$$= \frac{800 - 720}{800} = 10 \text{ per cent.}$$

Probability that there is no vehicle in the system

$$P_0 = (1 - \rho) = \frac{1}{10} = 0.1$$

Probability that there are three vehicles in the system

$$= P_{(3)} = (1 - \rho)\rho^3 = \left(\frac{9}{10}\right)^3 \frac{1}{10} = \frac{9^3}{10000}$$

$$= 0.0729.$$

I. MULTIPLE CHOICE QUESTIONS

1. Time study is used to
   (*a*) determine standard costs
   (*b*) determine the number of machines a person may run
   (*c*) provide a basis for setting piece price or incentive wages
   (*d*) all of the above
2. CPM has following time estimate
   (*a*) One time estimate (*b*) Two time estimate
   (*c*) Three time estimate (*d*) Four time estimate
3. Simplex method is the method used for
   (*a*) value analysis (*b*) network analysis
   (*c*) linear programming (*d*) queuing theory
4. Time study is
   (*a*) the appraisal, in terms of time, of the value of work involving human effort
   (*b*) machine setting time
   (*c*) time taken by workers to do a job
   (*d*) method of fixing time for workers
5. Service time in queuing theory is usually assumed to follow
   (*a*) normal distribution (*b*) poisson's distribution
   (*c*) erlang distribution (*d*) exponential law

Answers

1. (*d*) 2. (*a*) 3. (*c*) 4. (*a*) 5. (*d*)

II. SOLVED QUESTIONS AND ANSWERS

Q. 1. What is queueing theory? Explain.

Ans. Queueing theory is the mathematical study of waiting lines, or queues. In queueing theory, a model is constructed so that queue lengths and waiting time can be predicted. Queueing theory is generally considered a branch of operations research because the results are often used when making business decisions about the resources needed to provide a service.

Queueing theory has its origins in research by Agner Krarup Erlang when he created models to describe the Copenhagen telephone exchange. The ideas have since seen applications including telecommunication, traffic engineering, computing and the design of factories, shops, offices and hospitals as well as in project management.

Q. 2. What is mean arrival rate (λ)?

Ans. It is rate at which customers arrive at a service facility. It is expressed in flow (customers/hr or vehicles/hour in transportation scenario) or time headway (seconds/customer or seconds/vehicle in transportation scenario). If inter arrival time that is time headway (h) is known, the arrival rate can be found out from the equation:

$$\lambda = \frac{3600}{h}$$

Q. 3. What do you mean by FiFO?

Ans. First in first out (FIFO): If the customers are served in the order of their arrival, then this is known as the first-come, first-served (FCFS) service discipline. Prepaid taxi queue at airports where a taxi is engaged on a first-come, first-served basis is an example of this discipline.

Q. 4. What is meant by FiLO?

Ans. First in last out (FILO): Sometimes, the customers are serviced in the reverse order of their entry so that the ones who join the last are served first. For example, assume that letters to be typed, or order forms to be processed accumulate in a pile, each new addition being put on the top of them. The typist or the clerk might process these letters or orders by taking each new task from the top of the pile. Thus, a just arriving task would be the next to be serviced provided that no fresh task arrives before it is picked up. Similarly, the people who join an elevator first are the last ones to leave it.

Q. 5. What is SIRO?

Ans. Served in random order (SIRO): Under this rule customers are selected for service at random, irrespective of their arrivals in the service system. In this every customer in the queue is equally likely to be selected. The time of arrival of the customers is, therefore, of no relevance in such a case.

Q. 6. What is priority service?

Ans. Under this rule customers are grouped in priority classes on the basis of some attributes such as service time or urgency or according to some identifiable characteristic, and **FIFO** rule is used within each class to provide service. Treatment of **VIPs** in preference to other patients in a hospital is an example of priority service.

Q. 7. What is processor (or time) sharing?

Ans. The server is switched between all the queues for a predefined slice of time (quantum time) in a round-robin manner. Each queue head is served for that specific time. It doesn't matter if the service is complete for a customer or not. If not then it'll be served in it's next turn. This is used to avoid the server time killed by customer for the external activities (e.g., preparing for payment or filling half-filled form).

III. REVIEW QUESTIONS

1. What are applications to traffic engineering?
2. What is the queuing characteristics?
3. What are the characteristics of queuing?

REFERENCES

1. Saaty, T.L. Mathematical Methods in Operations Research, McGraw Hill Book Company, New York, 1959.
2. Saaty, T.L., Elements of Queueing Theory, McGraw Hill Book Company, New York, 1961.
3. Cleveland, D.E. and Capelle, D.G., Queueing Theory Approaches, Highway Research Board Special Report 79, Washington, 1964.
4. Traffic Flow Theory—A Monograph, Special Report 165, Transportation Research Board, Washington, 1975.

26. Vehicle Arrivals, Headways and Gaps

26.1. Probabilistic Aspects of Traffic Flow

Traffic along highways can be described by their probabilities or statistical aspects in addition to theoretical models such as the car following or hydrodynamic models which deal with deterministic characteristics of traffic flow. Many variables of traffic, such as arrival pattern, headways, gaps, speed, delay and volume can be conveniently dealt with by the ordinary statistical techniques. These techniques find valuable applications in the practical problems that confront a traffic engineer. Typical situations are : study of intersections, gap acceptance, timing signal installations, provision of adequate storage space for separate turning lanes, operation of toll booths where queue gets formed and many others. The present chapter deals with the application of statistical techniques to deal with the probabilistic nature of traffic. Queueing Theory, which is also an allied subject, is however, dealt with separately in an earlier Chapter.

26.2. Spacing and Headway Characteristics

An important variable that describes the traffic flow is the spacing and headway between individual vehicles in a traffic stream. The terms spacing and headway have been used freely to denote the position of vehicle in the traffic stream, though a subtle difference between the two terms is made out by the Highway Capacity Manual, which defines spacing as the *distance measure* and headway as the *time measure* from head to head of each of successive vehicles. The two terms are related to each other by the speed of the vehicles. Thus:

$$\text{Headway (sec)} = \frac{\text{Spacing (m)}}{\text{Speed (m/sec)}} \qquad \text{...(26.1)}$$

The concentration (sometimes called density) is the number of vehicles in a unit length of road (say a kilometre) and is obtained from :

Concentration (veh/km)

$$= \frac{1000\ \text{(m)}}{\text{Average Spacing (m)}} \qquad \text{...(26.2)}$$

The study of spacing and headway is very important as these variables almost determine the operating characteristics of a headway and its *level of service.* The level of service is governed to a large extent by the ease and facility with which individual drivers can drive, manoeuver and overtake on a highway with comfort and safety. The spacing and headway limit the drivers in the choice of their speeds and the decisions they make in weaving, merging, cutting, overtaking or just following behind a car. They enable the driver to judge the availability of suitable gaps in traffic streams when they can make right-turns or left-turns. The study of Highway Capacity has in fact been built up on a study of spacing and headway distributions.

26.3. Study of Vehicle Arrivals

The distributions of vehicles on a highways can be studied in space or in time. The first case refers to a certain length of road with vehicles moving on at different locations and we obtain the true picture of the traffic at any instant in time by an aerial photography covering the entire stretch. The spacing obtained then describes the distribution on vehicles in space. On the other hand, if a stationary observer positions himself at a fixed point and notes down the time of arrival of each vehicle that passes him, one can obtain the distribution of headway over a period of time. Though aerial photography has been used in many studies to determine the distribution of vehicles in space, yet it is the distribution of headways in time that is studied more commonly.

The simplest procedure for finding the distribution of headways in time is to note the time of arrival at a point by means of an accurate stop-watch. Two observers can be deployed, one to read out the time from the stop-watch which is kept continuously running and the other to record the readings.

Such a simple expedient may not be entirely satisfactory when the headways are too short. Moreover, the headways have to be found in the office by successive subtraction of timings of vehicles. A slight innovation consists of three stop-watches mounted on the same recording board but connected by a suitable lever mechanism. Initially all the three watches are set to zero and the lever is pressed when a vehicle crosses the point. This sets the first stop-watch on motion. When the next vehicle arrives, the lever is again pressed, and this results in the hand of the first watch stopping at that instant and in starting the second watch. The recorder thus directly reads the first stop-watch giving him the first headway. When the third vehicle arrives, the lever is again pressed, resulting in the hand of the first watch returning to zero, the hand of the second watch stopping at that instant, and the third watch starting. The reading on the second watch now gives the second headway directly. The procedure gives accurate measurements without straining either the recorder or the person who presses the lever when a vehicle passes the fixed point.

An ordinary tape recorder can be used for recording the arrival of vehicles. If the recorder is later on played, the accurate time of arrival can be obtained. It is necessary to ensure that the speed of rotation of the tape is constant and the same speed obtains when recording as when playing. A simple check is to record the stop-watch timings at every minute simultaneously with the recording of the arrival of vehicles.

A more sophisticated instrument is a multi-channel event recorder. A graduated strip of paper moves at a constant speed between two spools, and the pressing of any of the buttons causes a pen to record the event by a distinct mark on the moving paper strip. Any number of channels can be possible, and each channel can be allotted to a particular lane of traffic. For example, in a simple T-junction, the straight ahead vehicles on the road can be allotted two channels, the left turners and right turners from the main road to the side can be allotted two more channels and left turners and right turners from the side road can be allotted two more channels, thus making a total of six channels and covering all possible movements.

A simplified method is to count the number of vehicles in short intervals of time, say 10 seconds each. The stop-watch is kept running continuously and one observer counts the number of vehicles by means of hand tallies. A second observer indicates the timing at every 10 seconds and a third observer records the number of vehicles counted in each 10 second interval. This method is known as the counting distribution as against the methods earlier mentioned which yield the gap distribution. Gap distribution methods are more frequently resorted to in urban traffic situations since they give directly the data regarding gap availability which is needed for study of intersections.

26.4. Poisson Distribution of Vehicle Arrivals

The distribution of vehicles in space or in time may assume various mathematical forms. In general, it is likely that the spacing between vehicles will be distributed in a random and haphazard manner with gaps of various sizes occurring and no apparent regularity being discernible. The word random here implies the following conditions :

(*i*) An arrival of a vehicle at a given point is completely independent of the occurrence of any other event, *i.e.* the number of vehicles that arrived during any previous interval.

(*ii*) Equal intervals of time or space are equally likely to contain equal number of events or vehicles.

When the above conditions are fulfilled, one of the statistical models to describe the distribution is the Poisson distribution, which has already been described in Chapter 8.

It is well to remember that in addition to the above conditions, the Poisson distribution applies to a population having a discrete variable. This is obvious because the number of vehicles will have to be a whole number or zero and cannot be a continuous variable.

The road traffic can be considered as random when the flow is free and is not subjected to the influence of any signals or intersections nearby. This also implies that the flow of vehicles should be very light (such that the occurrence of an event is independent of any other event) and vehicles can overtake freely. There should be no bunching of vehicles (Ref. 1) and no impediments due to the geometry of the road. When restrictions do occur, more complicated mathematical models have been proposed.

The Poisson distribution has been indicated in Chapter 8. It is of the following form :

$$P(r) = \frac{(\lambda t)^r \, e^{-\lambda t}}{r!} \qquad \text{...(26.1)}$$

where, $P(r)$ = Probability of arrival of r vehicles in any interval of t sec.

λ = average rate of arrival (vehicle per unit time)

t = time interval

e = base of natural logarithms

The above expression can be directly derived as follows :

Let $P_1(t)$, $P_2(t)$ .... $P_i(t)$ denote the probability of 1, 2,..........i arrivals upto the time t.

Consider a small incremental period δt, such that the probability of more than one arrival in this time δt is negligible.

Let λ be the average rate of arrival per unit time.

$\therefore$ $\lambda \delta t$ = probability of one arrival in the incremental period δt. (This implies that the probability of one arrival in a short time is proportional to the length of interval, the proportionality constant being λ) ...(26.2)

And $1 - \lambda\delta t$ = probability that of no arrivals in the incremental periods δt ...(26.3)

If $P_i(t + \delta t)$ = probability that i arrivals have taken place upto the time $(t + \delta t)$.

Then this can occur in two ways :

(*i*) either i arrivals have taken place upto time t and no arrivals can take place in time δt and

(*ii*) $(i - 1)$ arrivals have taken place upto time t and one arrival takes place in time δt.

This gives :

$$P_i(t + \delta t) = P_i(t) \times P_0(\delta t) + P(i-1)(t) \times P_1(\delta t) \quad ...(26.4)$$

$$= P_i(t) \times (1 - \lambda\delta t) + P(i-1)(t) \times \lambda\delta t$$

$$= \lambda\delta t\ [P(i-1) - (t)\ P_i(t)] + P_i(t) \quad ...(26.5)$$

$$\therefore \quad \frac{P_i(t+\delta t) - P_i(t)}{\delta t} = \lambda[P_{(i-1)}(t) - P_i(t)]$$

or

$$\frac{\delta P_i(t)}{\delta t} = \lambda[P_{(i-1)}(t) - P_i(t)] \quad ...(26.6)$$

Now, $P_{-1}(t) = 0$ (this implies that the possibility of less than zero arrivals is zero) ...(26.7)

$P_0(0) = 1$ (this implies that there are no arrivals upto time $t = 0$, which is the initial condition) ...(26.8)

$P_i(0) = 0$ for $i \geq 1$ (This implies that the probability of i arrivals at time $t = 0$ is zero, which is also an initial condition) ...(26.9)

Setting $i = 0$ in Eq. (26.6) above

$$\frac{dP_0(t)}{dt} = \lambda[P_{-1}(t) - P_0(t)]$$

$$\frac{dP_0(t)}{P_0(t)} = -\lambda dt, \ \ln P_0(t) = -\lambda t + c$$

Taking anti-logs,

$$P_0(t) = e^{-\lambda t + c}$$

Since $P_0(0) = 1$ (from Eq. 26.7), $1 = e^{0 + c}$

$\therefore$ $c = 0$

$\therefore$ $P_0(t) = e^{-\lambda t}$ ...(26.10)

Setting $i = 1$ in Eq. (26.6) above

$$\therefore \quad \frac{dP_1(t)}{dt} = \lambda[P_0(t) - P_1(t)]$$

$$= \lambda[e^{-\lambda t} - P_1(t)], \text{ substituting from (26.10)}$$

$$\frac{dP_1(t)}{dt} + \lambda dP_1(t) = \lambda e^{-\lambda t}$$

Solving this as a differential equation, one gets :

$$P_1(t) = (\lambda t)\ e^{-\lambda t} + C_2 e^{-\lambda t}$$

Since $P_1(0) = 0$, $C_2 = 0$

$\therefore \quad P_1(t) = (\lambda t)\, e^{-\lambda t}$

Similarly $\quad P_2(t) = \dfrac{(\lambda t)^2 e^{-\lambda t}}{2!}$ and $P_3(t) = \dfrac{(\lambda t)^3\, e^{-\lambda t}}{3!}$

and $$P_r(t) = \frac{(\lambda t)^t\, e^{-\lambda t}}{r!} \quad \text{...(26.11)}$$

Putting the above in the standard form used earlier, one gets

$$P(r) = \frac{(\lambda t)^t\, e^{-\lambda t}}{r!} \quad \text{...(26.12)}$$

The validity of the above distribution can be checked by simple observations of the vehicular counts. The following example illustrates the technique.

Problem 26.1. *In order to check the validity of the Poissonian arrival in a lightly trafficked street free from the influence of intersections or signals nearby a count of vehicles passing an observer was taken. The number of vehicles arriving in intervals of 20 sec each was counted and recorded in table below :*

Number of vehicles in time interval of 20 seconds	0	1	2	3	4	5	6	7	8	9	10	11	12
Number of vehicles intervals with stated number of vehicles	–	4	13	23	26	24	14	20	6	2	3	3	1

Analyse the above data and fit a Poisson distribution function. Test the goodness-of fit by means of a χ^2 test and comment on the results.

Solution. Total number of vehicles counted during the study

$= 0 \times 1 + 4 \times 1 + 13 \times 2 + 23 \times 3 + 26 \times 4 + 24 \times 5 + 14 \times 6$
$+ 20 \times 7 + 6 \times 8 + 2 \times 9 + 3 \times 10 + 3 \times 11 + 1 \times 12$
$= 0 + 4 + 26 + 69 + 104 + 120 + 84 + 140 + 48 + 18 + 30 + 33 + 12 = 687$

Total intervals

$= 4 + 13 + 23 + 26 + 24 + 14 + 20 + 6 + 2 + 3 + 3 + 1 = 139$

$$m = \text{average arrivals per 20 second interval} = \frac{687}{139} = 4.96$$

$$P(r) = \frac{m^r e^{-m}}{r!},\ P(0) = e^{-4.96} = 00070$$

Theoretical number of intervals with 0 arrivals

$I_0 = 0.0070 \times 139 = 1$
$P(1) = P(0)\, m = 0.0347 \quad \therefore \quad I_1 = 5$
$P(2) = P(1)\, m/2 = 0.086 \quad \therefore \quad I_2 = 12$
$P(3) = P(2)\, m/3 = 0.142 \quad \therefore \quad I_3 = 20$
$P(4) = P(3)\, m/4 = 0.176 \quad \therefore \quad I_4 = 25$
$P(5) = P(4)\, m/5 = 0.175 \quad \therefore \quad I_5 = 25$
$P(6) = P(5)\, m/6 = 0.144 \quad \therefore \quad I_6 = 20$

$P(7) = P(6)\ m/7 = 0.102 \qquad \therefore \quad I_7 = 14$

$P(8) = P(7)\ m/8 = 0.063 \qquad \therefore \quad I_8 = 9$

$P(9) = P(8)\ m/9 = 0.035 \qquad \therefore \quad I_9 = 5$

$P(10) = P(9)\ m/10 = 0.0173 \qquad \therefore \quad I_{10} = 2$

$P(11) = P(10)\ m/11 = 0.0079 \qquad \therefore \quad I_{11} = 1$

$P(12) = P(11)\ m/12 = 0.0036 \qquad \therefore \quad I_{12} = \text{negligible, say } 0$

No. of vehicles	*0*	*1*	*2*	*3*	*4*	*5*	*6*	*7*	*8*	*9*	*10*	*11*	*12*
Observed values of 20 sec. intervals with stated number of vehicles (0)	–	4	13	23	26	24	14	20	6	2	3	3	1
Expected values of 20 sec. intervals with stated number of vehicles (*E*)	1	5	12	20	25	25	20	14	9	5	2	1	–

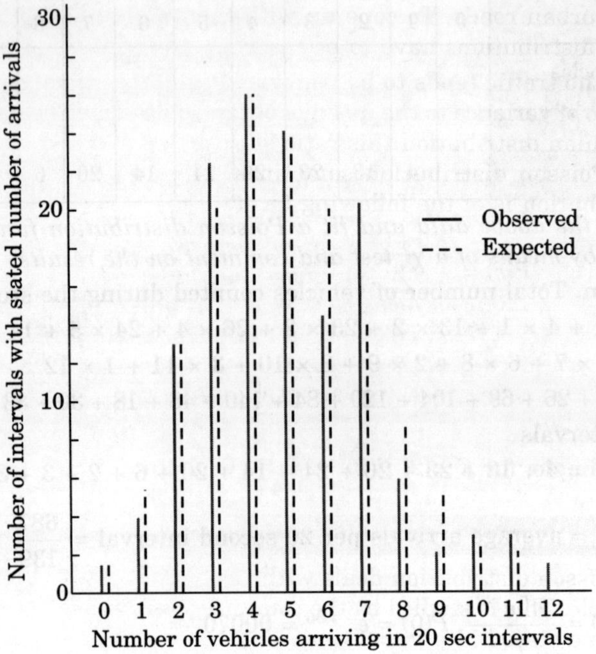

Fig. 26.1

$$\chi^2 = \Sigma \frac{(0-E)^2}{E} = \frac{(4-6)^2}{6} + \frac{(13-12)^2}{12} + \frac{(23-20)^2}{20} + \frac{(26-25)^2}{25} + \frac{(24-25)^2}{25} + \frac{(14-20)^2}{20} + \frac{(20-14)^2}{14} + \frac{(6-9)^2}{9} + \frac{(9-8)^2}{8}$$

$$= 0.67 + 0.08 + 0.45 + 0.04 + 0.04 + 1.8$$
$$+ 2.57 + 1.0 + 0.13 = 6.78$$

[**Note** : Readings less than 5 are clubbed]

$$c = \text{Number of classes} = 9$$

$\therefore$ Degrees of freedom = 9 – 2 = 7 (see Table 8.16).

For a significance level of 5% and 7 degrees of freedom, χ^2 from Table 8.15 is 14.07.

Since $\chi_{obs}{}^2 < \chi_c{}^2$ we conclude that the Poisson distribution is reasonably likely to represent the observed distribution.

The results are plotted in Fig. 26.1.

26.5. Limitations of the Poisson Distribution

The Poisson distribution holds good only when the conditions of traffic mentioned earlier are fulfilled. In particular, the flow should be light, and the behaviour of the vehicles should not be inhibited by other vehicles, or by the geometry of the road. There should be no intersections nearby and the flow should be free. No bunching of vehicles should occur.

It is rare that all the above conditions are fulfilled in traffic situations, especially on urban roads. To cope up with the complexities then arising, more sophisticated distributions have to be resorted to. These are dealt with later.

When the traffic tends to be heavy and the presence of one car influences other, the ratio of variance to the mean of observed data may be less than 1. Under a pure Poissonian distribution this ratio should be equal to 1. In such cases, the generalised Poisson distribution sometimes gives a good fit. The generalised Poisson distribution is of the following form :

$$P(r) = \sum_{j=Kr}^{K(r+1)-1} \frac{\lambda^j e^{-\lambda}}{j!} \qquad \text{...(26.12)}$$

$$= \sum_{i=1}^{K} \frac{e^{-\lambda} (\lambda)^{rK+i-1}}{(rK+i-1)!}$$

For example, for $K = 2$, $P(2) = \dfrac{\lambda^4 e^{-\lambda}}{4!} + \dfrac{\lambda^5 e^{-\lambda}}{5!}$

26.6. Gap and Headway Distribution

The Poisson distribution deals with the arrival rate of vehicles which is a discrete variable. This is studied by the counting distribution technique and noting the arrivals in a stated interval of time. The other technique indicated earlier in the Chapter is the gap distribution where in the gap in seconds (headway in seconds) between successive vehicles is recorded.

The gap or headway distribution function can be directly derived from Poisson distribution function mentioned earlier. Thus,

$$P(r) = \frac{(\lambda t)^r e^{-\lambda t}}{r!}$$

If in a particular interval of time t seconds, there is no arrival of a vehicle, then it can be concluded that the gap between the previous vehicle (which must have passed in the period prior to the beginning of the interval of time (t) and the next vehicle which can only arrive after the expiry of the interval of time (t) must be t seconds or more.

This implies that $P(0)$ is the probability of a gap equal to or greater than t seconds.

Expression the above relationship in the language of probability.

$$P(h \geq t) = P(0) = \frac{(\lambda t)^0 \, e^{-\lambda t}}{0\,!} = e^{-\lambda t}$$

[since $(\lambda t)^0$ and $0\,!$ are both equal to 1] ...(26.13)

In the above equations, h is the headway in seconds, which is also called the gap.

The above equation is called the negative exponential distribution though it is also called simply the 'exponential distribution'. Headways are thus governed by the exponential distribution curve, which unlike the counting distribution, is a continuous function. This is so because t can assume any value between 0 and infinity.

Extending Eq. (26.13) further, the probability of a headway being less than t, $P(h \leq t)$, is given by :

$$P(h \leq t) = 1 - e^{-\lambda t} \quad \text{...(26.14)}$$

In Eqs. (26.13) and (26.14) above, λ can be substituted by the mean time headway $\bar{t}$ as follows :

Mean time headway

$$\bar{t} = \frac{1}{\lambda} \quad \text{and} \quad \lambda = \frac{1}{\bar{t}}$$

$$\therefore \quad P(h \geq t) = e^{-(t/\bar{t})} \quad \text{...(26.15)}$$

$$\text{and} \quad P(h \leq t) = 1 - e^{-(t/\bar{t})} \quad \text{...(26.16)}$$

The mean time headway $\bar{t}$ is found by noting the number intervals (N) and the total period of these intervals (T).

$$\bar{t} = \frac{T}{N} \text{ seconds.}$$

A simple example will illustrate the use of these formulae.

Problem 26.2. *An observation on an Expressway yielded a count of 200 vehicles in a period of half an hour. Calculate the number of headways in this count :*

(i) greater than 4.5 seconds and (ii) less than 18 seconds.

Solution. The mean time headway

$$\bar{t} = \frac{T}{N} = \frac{1}{\lambda} = \frac{1800}{200} = 9 \text{ seconds}$$

$$P(h \geq i) = e^{-(t/\bar{t})}$$

$$P(h \geq 4.5) = e^{-(4.5/9)} = e^{-0.5} = 0.6065$$

Total number of headways counted = 200 – 1 = 199

Number of headways greater than 4.5 seconds

= 199 × 0.6065 = 120

$$P(h \le t) = 1 - e^{-(t/\bar{t})}$$

$$\therefore \quad P(h \le 18) = 1 - e^{-(18/9)}$$

$$= 1 - e^{-2} = 1 - 0.1353 = 0.8647$$

$\therefore$ Number of headways less than 18 seconds

$$= 199 \times 0.8647 = 172$$

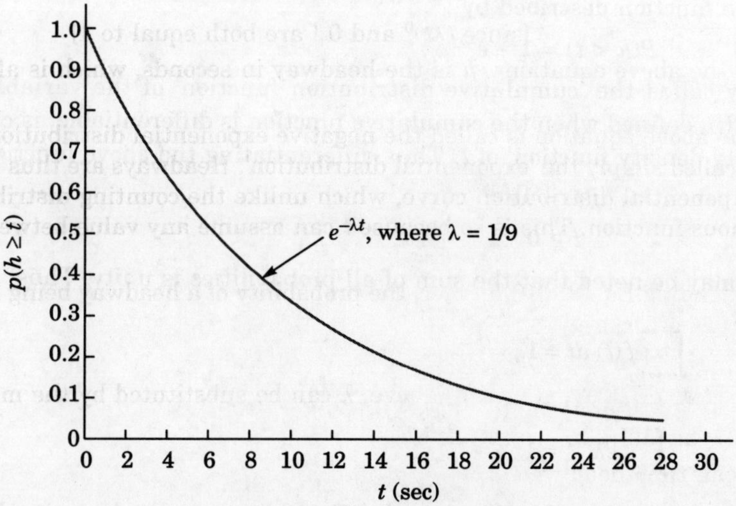

Fig. 26.2

The graphs of the two functions, $P(h \ge t) = e^{-\lambda t}$

And $P(h \le t) = 1 - e^{-\lambda t}$ for the values given in the above example are given in Figs. 26.2 and 26.3 respectively.

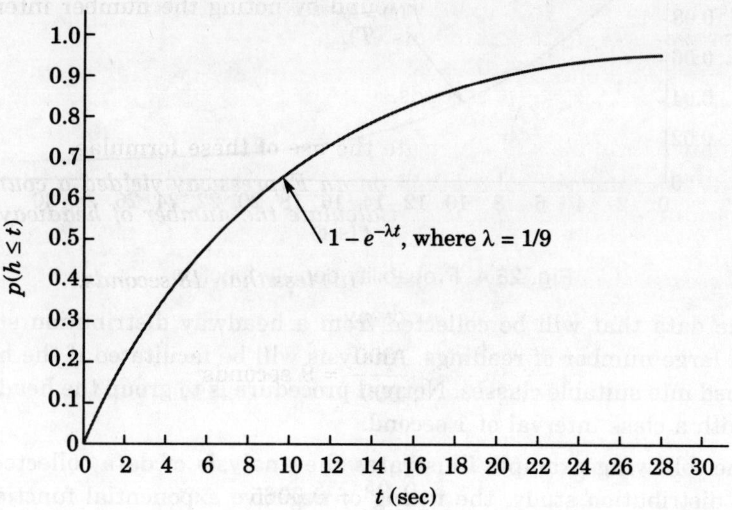

Fig. 26.3

It may be noted that because the functions are continuous, a smooth line has been drawn connecting different values of $P(h \geq t)$ and $P(h \leq t)$. In the case of the counting distribution, the probability values were merely indicated as ordinates, Fig. 26.1, without the values being connected by a smooth line. This was because of the discrete values for r, the number of arrivals or number of vehicles.

The function described by :

$$P(h < t) = 1 - e^{-\lambda t}$$

is usually called the 'cumulative distribution function' of the variable t. The function $f(t)$, defined when the cumulative function is differentiable, is called the 'probability density function' of t. Thus differentiating the above function,

$$f(t) = \lambda e^{-\lambda t}$$

for $$t \geq 0 \qquad \text{...(26.17)}$$

It may be noted that the sum of all probabilities is unity. Thus:

$$\int_{-\infty}^{+\infty} f(t)\, dt = 1,$$

$$\therefore \qquad \int_{-\infty}^{+\infty} e\lambda^{-\lambda t} = 1$$

The probability density function for the example above is plotted in Fig. 26.4.

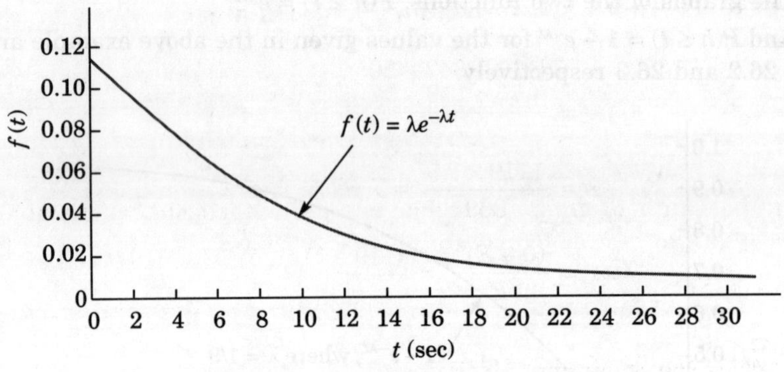

Fig. 26.4. Probability density function.

The data that will be collected from a headway distribution study will contain a large number of readings. Analysis will be facilitated if the headways are grouped into suitable classes. Normal procedure is to group the headway into classes with a class interval of 1 second.

The following example illustrates the analysis of data collected from a headway distribution study, the fitting of negative exponential function to the observed data and the testing of the goodness of fit by means of a χ^2 test.

Problem 26.3. *The distribution of headways by mean of a multi-watch is given in Table 26.1.*

Table 26.1

Headway class *t sec.*	*Observed frequency of* *headways* (f_i)
0—2.9, say 3	119
3—6	114
6—9	41
9—12	41
12—15	13
15—18	7
18—21	11
21—24	5
24—27	7
27—30	3
30—33	4
33—36	3
36—39	1

Test the goodness-of fit of a χ^2 test.

Solution. The calculations are conveniently carried out in a tabular form, vide Table 26.2.

λ in above Table is calculated thus :

$$\text{Mean headway} = \bar{t} = \frac{\begin{matrix}[119\times1+114\times3+41\times5+41\times7+13\\ \times9+7\times11+11\times13+5\times+7\times17+3\\ \times19+4\times21+3\times23+1\times21]\times1.5\end{matrix}}{\Sigma f_i}$$

$$= \frac{\begin{matrix}[119+342+205+287+17+143\\ +75+119+57+84+69+25]\times1.5\end{matrix}}{369}$$

$$= \frac{1719\times1.5}{369} = 7 \text{ sec}, \lambda = \frac{1}{\bar{t}} = \frac{1}{7}$$

$$\chi^2_{obs} = \frac{(0-E)^2}{E} = \frac{(119-115)^2}{115} + \frac{(114-103)^2}{103} + \frac{(41-51)^2}{51}$$

$$+ \frac{(41-34)^2}{34} + \frac{(13-22)^2}{22} + \frac{(7-14)^2}{14}$$

$$+ \frac{(11-11)^2}{11} + \frac{(5-8)^2}{8} + \frac{(18-12)^2}{12}$$

$$= 0.14 + 1.17 + 1.96 + 1.44 + 3.69 + 3.50 + 0 + 1.13 + 3.00 = 16.03$$

Note that the last five rows have been combined to give a value of $E > 5$

c = number of classes = 9

Degree of freedom = $c - 2$ (Refer. Table 8.16) = 7

From Table 8.15, χ_c^2 for 5 per cent significance level and 7 degrees of freedom is 14.07.

Table 26.2

Headway class (t sec) (1)	*Observed frequency of headways in the class (0)* (2)	*Observed frequency greater than lower class limit in Col. 1* (3)	*Column (3) expressed as percentage of total headways* (4)	*Theoretical percentage frequency of headway greater than lower class limit 100 × $e^{-\lambda t}$* (5)	*Theoretical frequency of headways greater than lower class limit Col.* (5) (6)	*Theoretical frequency in the class (E)* (7)
0—3	119	369	100	100	369	115
3—6	114	250	68	66	254	103
6—9	41	136	37	41	151	51
9—12	41	95	26	27	190	34
12—15	13	54	15	18	66	22
15—18	7	41	11	12	44	14
18—21	11	34	9	8	30	11
21—24	5	23	6	5	19	8
24—27	7 } 18	18	5	3	11	4 } 12
27—30	3	11	3	2	7	3
30—33	4	8	2	1.5	5	2
33—36	3	4	1	1	4	1
36—39	1	1	0	0.6	2	2

Since the $\chi^2_{observed}$ is higher than χ_c^2, there is a significant difference between the observed and theoretical headway distributions, showing a lack of good fit.

26.7. Shifted Exponential Distribution

The exponential distribution described above is a very simple function and can be used with great facility for fitting observed headways. Its one disadvantage, just as in the case of Poisson distribution, is that it is valid for light flows when the traffic is truly random and Poissonian. As the flow increases, some of the vehicles will be constrained in their movements and the flow can no longer be termed Poissonian. The other disadvantage is that the vehicles themselves have a definite length and are not merely points. This physical property limits the occurrence of small-sized headways, generally below 1 second. In order to overcome these drawbacks, a 'shifted exponential' curve is often fitted into the observed data.

The procedure lies in shifting the starting point of the curve from 0 to the right to a satisfactory minimum gap of at least 1 second, vide Fig. 26.5.

The shift of the starting point by τ results in disturbing all other points by this amount, and the equation of the curve then becomes (Ref. 2, 3, 4) :

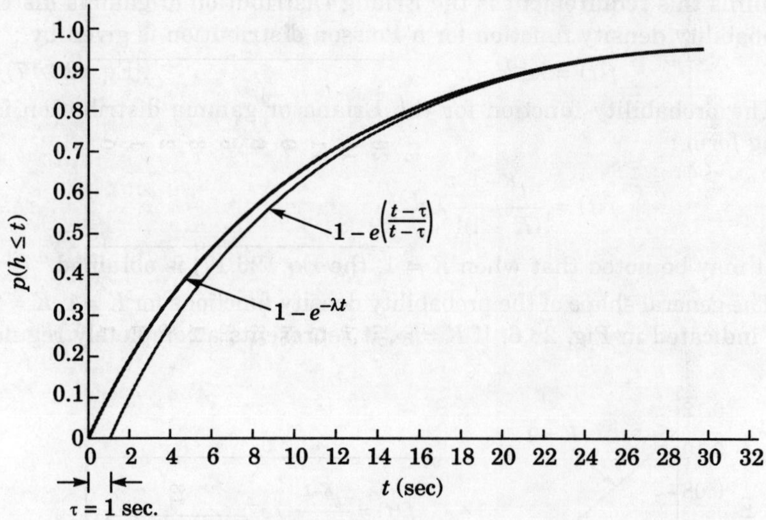

Fig. 26.5. Shifted exponential function.

$$P(h \leq t) = 1 - e^{\frac{-(t-\tau)}{(\bar{t}-\tau)}} \quad \text{...(26.18)}$$

Schuhl (Ref. 5) has proposed an ingenious way of avoiding the error in the curve by its bodily shifting. His proposition is that road traffic can be considered as forming two distinct sub-sets, one freemoving and the other retarded, each having distinct mean spacings between vehicles and each obeying some Poissontype distribution law. The resulting distribution is a composite Poisson type distribution of the following form :

$$P(h < t) = (1-r)\left[1 - \exp\left(\frac{-t}{\bar{t}_1}\right)\right] + r\left[1 - \exp\left(-\frac{t-\tau}{t_2-\tau}\right)\right] \quad \text{...(26.19)}$$

where r = proportion of vehicles in the retarded group.

μ = minimum headway of vehicles in the retarded group etc.

t_1 = average headway of free-following vehicles, sec.

$\bar{t}_2$ = average headway of vehicles in the retarded group, sec.

t = headways, sec.

e = base of natural logarithms.

The use of this equation requires the determination of parameters λ, $\bar{t}_1$ and $\bar{t}_2$. Some researchers [Kell (Ref. 6), Grecco and Sword (Ref. 7)] have studied how the parameters can be found. Kell assumes that there are separate values for free-flowing vehicles and retarded group vehicles.

26.8. Erlang Distribution

Random distribution has marked deficiencies as a model for vehicular arrangement (Ref. 1). Under this distribution, the probability of a gap between two fixed values increases as these values move towards zero. Moreover, small headways should be considered improbable rather than impossible. The curve which fulfils this requirement is the Erlang Distribution or gamma distribution. The probability density function for a Poisson distribution is given by :

$$f(t) = \lambda e^{-\lambda t} \quad \text{...[Eq. (26.17) earlier]}$$

The probability function for the Erlang or gamma distribution is of the following form :

$$f(t) = \frac{t^{K-1}}{(K-1)!} \lambda e^{-\lambda t} \quad \text{...(26.20)}$$

It may be noted that when $K = 1$, the Eq. (26.17) is obtained.

The general shape of the probability density functions for $K = 1$, $K = 2$ and $K = 3$ are indicated in Fig. 26.6. If $K = \infty$, it represents a completely regular flow.

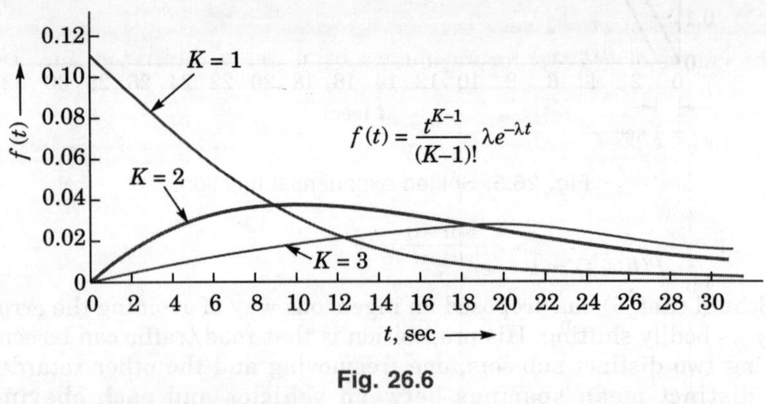

Fig. 26.6

The probability function $f(t)$ is equal to the probability of waiting time until the Kth arrival. Thus if K were to be equal to 2, then $f(t)$ would represent the probability of waiting time being t until two vehicles arrive. As a special cases, when $K = 1$, $f(t)$ would represent the probability of waiting time being t until the first vehicle arrives.

Corresponding to the generalised Poisson distribution mentioned earlier, the cumulative Erlang distribution is :

$$P(h \geq t) = \frac{\sum_{i=0}^{K-1} \left(\frac{Kt}{\bar{t}}\right)^i e^{-(K_t/\bar{t})}}{i\,!} \qquad \text{...(26.21)}$$

When, $K = 1$, the expression reduces to the exponential distribution :

$$P(h \geq t) = e^{-(t/\bar{t})} \qquad \text{(From Eq. 26.15)}$$

When, $K = 2$, the expression becomes :

$$P(h \geq t) = \left[1 + \left(\frac{Kt}{\bar{t}}\right)\right] e^{-(Kt/\bar{t})} \qquad \text{...(26.22)}$$

When, $K = 3$, the expression becomes :

$$P(h \geq t) = \left[1 + \left(\frac{Kt}{\bar{t}}\right) + \left(\frac{Kt}{\bar{t}}\right)^2 \times \frac{1}{2!}\right] e^{-(Kt/\bar{t})} \qquad \text{...(26.23)}$$

It may be noted that the value of K in the above can be approximately determined by :

$$K \approx \frac{\bar{t}^2}{s^2} \qquad \text{...(26.24)}$$

where, $\bar{t}$ = mean time headway from the observed data

and, s^2 = variance of the observed data.

Problem 26.4. *Fit a shifted exponential function to the data in Problem 26.3, assuming a minimum headway of 1 second.*

Solution. From Problem 26.3, $\bar{t} = 7$ sec.

$$P(h \leq t) = 1 - e^{-\frac{(t-\tau)}{(\bar{t}-\tau)}}$$

$$\tau = 1 \text{ sec.}$$

The values of $P(h \leq t)$ for various values of t are tabulated below :

t	$P(h \leq t)$
1	0
2	0.154
3	0.330
4	0.393
5	0.486
6	0.565
7	0.632
8	0.689
9	0.736
10	0.777
15	0.903
20	0.958

Problem 26.5. *The headways observed are as under :*
0.9, 2.1, 2.5, 3.8, 1.4, 1.1, 2.9, 1.8, 1.2, 1.0, 1.1, 2.0, 2.1, 2.5, 4.6, 1.1, 5.7, 2.1, 2.3, 3.5, 0.9, 5.6, 0.8, 1.7, 1.4.
Derive the Erlang distribution function. What is the probability that headways are greater than 4 seconds ?

Solution. The headway (t) and t^2 are tabulated.

Table

t	t^2
0.9	0.81
2.1	4.41
2.5	6.25
3.8	14.44
1.4	1.96
1.1	1.21
2.9	8.41
1.8	3.24
1.2	1.44
1.0	1.00
1.1	1.21
2.0	4.00
2.1	4.41
2.5	6.25
4.6	21.16
1.1	1.21
5.7	32.49
3.1	9.61
2.3	5.29
3.5	12.25
0.9	0.81
5.6	31.36
0.8	0.64
1.7	2.89
1.4	1.96
Total 57.1	191.71

$$\bar{t} = \frac{57.1}{25} = 2.204$$

Variance $$s^2 = \frac{1}{n-1}\left[\Sigma t^2 - n(\bar{t})^2\right]$$

$$= \frac{1}{24}\ [191.71 - 25 \times (2.204)^2] = 2.554$$

Thus, $$K = \frac{\bar{t}^2}{s^2} = \frac{(2.204)^2}{2.554} = 1.99, \text{ say } 2$$

$$P(h \geq t) = \left[1 + \frac{Kt}{\bar{t}}\right] e^{-(Kt/\bar{t})}$$

$$= \left[1 + \frac{2t}{\bar{t}}\right] e^{-(2t/\bar{t})} = \left[1 + \frac{2t}{2.204}\right] e^{-(2t/2.204)}$$

$$P(h \geq 4) = \left[1 + \frac{2 \times 4}{2.204}\right] e^{-(8/2.204)} = 4.63\ e^{3.63} = 0.123.$$

26.9. Pearson Type III Distribution

Another frequently used model for representing vehicle headways is the generalised Pearson Type III distribution. Karl Pearson has developed a number of distributions to cover a large variety of statistical data and of them Type III gives a good fit for observed headway data. The probability density function for this distribution is :

$$f(t) = \frac{\lambda}{T(k)} [\lambda(t - \alpha)]^{K-1} e^{-\lambda(t-\alpha)} \text{ for } t \geq \alpha \qquad ...(26.25)$$

where $T(K)$ is known as the gamma function and is given by

$$T(K) = \int_0^\infty x^{K-1} e^{-x} dx \qquad ...(26.26)$$

and λ, K and α are parameters of the function.

The Erlang distribution is, in fact, a simplified form of the Pearson Type III distribution, when the parameter K is an integer. In the generalised Pearson Type III distribution, K can have non-integer values also.

26.10. Gap Acceptance

26.10.1. Introduction

A common subject associated with the behaviour of driver and pedestrians in a traffic stream is their decision whether to accept or reject gaps. Gap acceptance or rejection is inherent in a variety of situations, some of which are :

(*i*) Ramp vehicles merging into the mainstream of a highway.

(*ii*) Merging, weaving and lane-changing on multilane highway facilities.

(*iii*) Pedestrians about to cross a road.

(*iv*) Unsignalised intersections, where the drivers from one leg of the intersection wait for a gap for entry into the stream in the other legs.

(*v*) Signalised intersections, where there is no separate phase for cutting across an on-coming stream of traffic.

(*vi*) Overtaking on two-lane, two-way roads.

In all the above situations, pedestrians or drivers will mentally analyse the traffic conditions they have to cope up with, and decide whether to accept or reject a particular gap offered.

The subject has been studied in great detail and a large amount of literature is available. Simulation techniques by the use of computers have facilitated the research.

26.10.2. Gap and Lag

Two terms are frequently met with in this subject, and it is necessary to know their definitions precisely. These terms are : Gap and Lag. These are defined below :

Gap is the time interval between the passing of the reference line by two successive vehicles in a lane of traffic.

Lag is that portion of the current gap remaining in a lane of mainstream of traffic when a vehicle intending to enter the mainstream arrives from any other stream. In other words, it is the elapsed time between the arrival of a vehicle intending to enter the mainstream and the arrival of the next vehicle in the mainstream.

The studies of gap acceptance in the past have covered both gaps and lags. Gaps and lags have their own individual characteristics.

26.10.3. Gap Acceptance Study Techniques

A multi-channel event recorder equipment is ideally suited to a study of gap acceptance. In this equipment a paper tape is made to run at a constant rate and the events (arrival or departure) on each traffic stream recorded on the tape by a pen recorder, when the relevant switch is pressed by the operator. Fig. 26.7 represents the

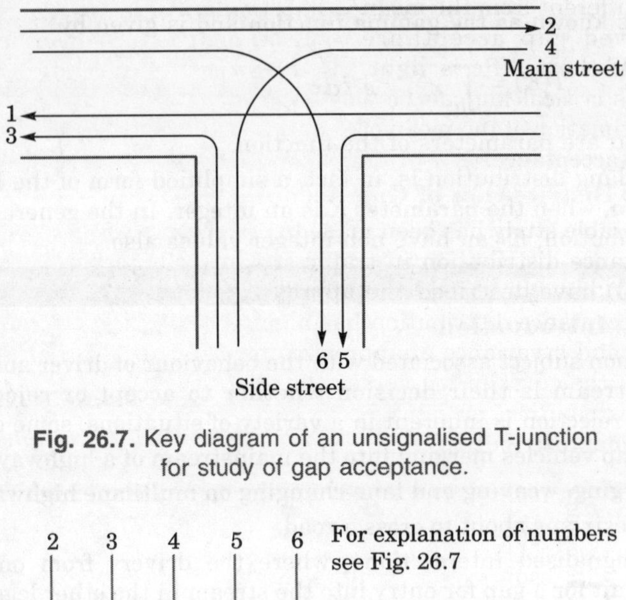

Fig. 26.7. Key diagram of an unsignalised T-junction for study of gap acceptance.

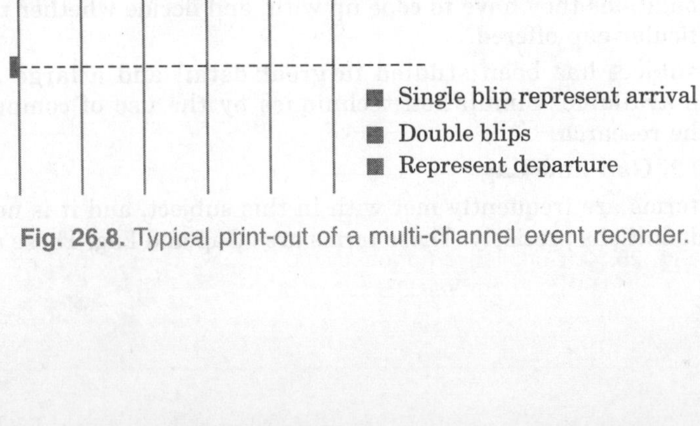

Fig. 26.8. Typical print-out of a multi-channel event recorder.

key diagram of an unsignalised T-junction and the flows studied by a 6-channel event recorder. The typical print-out of the tape is given in Fig. 26.8 ; one blip represents the arrival of a vehicle and two blips represents the departure of the vehicle.

26.10.4. Critical Lag

In a study of gap acceptance, one of the common method of analysis is by considering the 'critical lag'. This method is due to Raff (Ref. 9) who defines critical lag L as the lag which has the property that the number of accepted lags shorter than L is the same as the number of rejected lags longer than L. This is indicated diagramatically in Fig. 26.9.

It may be noted that Raff's critical lag is different from the mean of the observed gap acceptance distribution. If the traffic is light, however, Raff's critical lag can be a satisfactory estimation of the mean of the driver gap-acceptance.

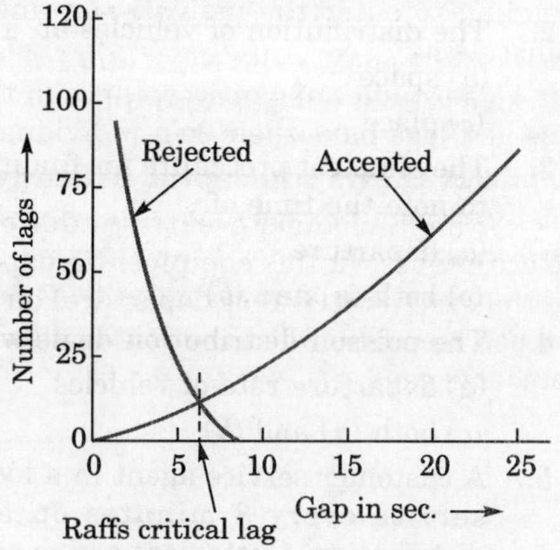

Fig. 26.9. Critical lag.

26.10.5. Distribution of Gap Acceptance

Considerable study has been made to understand the general character of the gap-acceptance distribution in traffic streams. Blunden et at (Ref. 11) and Miller (Ref. 10) have described the approach followed by various authors. In general, the acceptance distribution has a gamma or Erlang form, with a mean of 4.5 seconds and a variance of approximately 2 seconds (Ref. 12). As a rough approximation, the curve of lag versus the percentage acceptance can be taken as a normal distribution, as can be seen in Fig. 26.10. A better fit is obtained with a log-normal distribution.

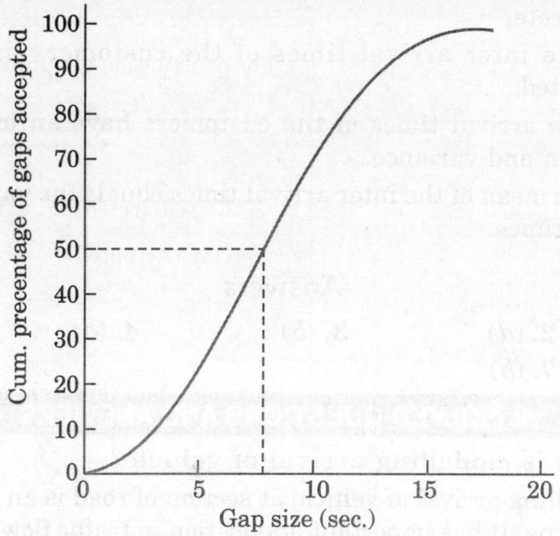

Fig. 26.10. Cumulative normal distribution of lag acceptance at an unsignalised intersection.

I. MULTIPLE CHOICE QUESTIONS

1. **The number of defects per 1,000 feet of extruded plastic pipe is best modeled with the ....**

(*a*) poisson distribution (*b*) pascal distribution

(*c*) binomial distribution (*d*) hypergeometric distribution

2. **The distribution of vehicles on a highway can be studied in**

(*a*) space (*b*) time

(*c*) place (*d*) both (*a*) and (*b*)

3. **The simplest procedure for finding the distribution of headways in time is to note the time of**

(*a*) departure (*b*) arrival

(*c*) both (*a*) and (*b*) (*d*) none of these

4. **The poisson distribution deals with the**

(*a*) departure rate of vehicles (*b*) arrival rate of vehicles

(*c*) both (*a*) and (*b*) (*d*) none of these

5. **A customer service agent in a local bank has a track record of completing service every 8 minutes on average, according to an exponential distribution. Customers arrive seeking service at the rate of six per hour, according to a Poisson distribution. What percentage of time is the customer service representative busy serving customers?**

(*a*) 60% (*b*) 75% (*c*) 85% (*d*) 80%

6. **A customer service agent in a local bank has a track record of completing service every 8 minutes on average, according to an exponential distribution. Customers arrive seeking service at the rate of six per hour, according to a Poisson distribution. On average, how many seats should be provided to assure adequate seating at the customer service area?**

(*a*) 4 (*b*) 5 (*c*) 6 (*d*) 3

7. **If customer arrivals are generated according to a Poisson distribution:**

(*a*) **then the mean of their arrival rate equals the standard deviation of the arrival rate.**

(*b*) **then the inter arrival times of the customers are exponentially distributed.**

(*c*) **then the arrival times of the customers have an identical standard deviation and variance.**

(*d*) **then the mean of the inter arrival times equals the variance of the inter arrival times.**

Answers

1. (*a*)	**2.** (*d*)	**3.** (*b*)	**4.** (*b*)	**5.** (*d*)
6. (*a*)	**7.** (*b*)			

II. SOLVED QUESTIONS AND ANSWERS

Q. 1. What is modelling arrival of vehicle?

Ans. Modelling arrival of vehicle at section of road is an important step in traffic flow modelling. It has important application in traffic flow simulation where vehicles are to be generated how vehicles arrive at a section. The vehicle arrival is obviously a random process. This is evident if one observe how vehicles are arriving at a cross section. Some times several vehicles come together, while at other times, they come sparsely. Hence, vehicle arrival at a section need to be characterized statistically.

Q. 2. What do you mean by vehicle arrivals?

Ans. Vehicle arrivals can be modelled in two inter-related ways; namely modelling what is the time interval between the successive arrival of vehicles or modelling how many vehicle arrive in a given interval of time. In the former approach, the random variables the time denoting interval between successive arrival of vehicle can be any positive real values and hence some suitable continuous distribution can be used to model the vehicle arrival.

In the later approach, the random variables represent the number of vehicles arrived in a given interval of time and hence takes some integer values.

Q. 3. What is headway modelling?

Ans. The inter-arrival time or the time headway is not constant due to the stochastic nature of vehicle arrival. A common way of modelling to treat the inter-arrival time or the time headway as a random variable and use some mathematical distributions to model them.

The behaviour of vehicle arrival is different at different flow condition. It may be possible that different distributions may work better at different flow conditions traffic state can be divided into three; namely low, medium and high flow conditions.

Q. 4. What is low volume flow?

Ans. Headway follow a random process as there is no interaction between the arrival of two vehicles. The arrival of one vehicle is independent of the arrival of other vehicle. The minimum headway is governed by the safety criteria. A negative exponential distribution can be used to model such flow.

Q. 5. What is high volume flow?

Ans. The flow is very high and is near to the capacity. There is very high interaction between the vehicle. This is characterized by *near* constant headway. The mean and variance of the headway is very low. A normal distribution can used to model such flow.

Q. 6. What is intermediate flow?

Ans. Some vehicle travel independently and some vehicle has interaction with other vehicles. More difficult to analyze, however, has more application in the field.

Q. 7. What is the Poisson distribution?

Ans. In probability theory and statistics, the **Poisson distribution** is a discrete probability distribution that expresses the probability of a given number of events occurring in a fixed interval of time and/or space if these events occur with a known average rate and independently of the time since the last event. The Poisson distribution can also be used for the number of events in other specified intervals such as distance, area or volume.

III. REVIEW QUESTIONS

1. What is Poisson distribution of vehicle arrivals?
2. What are the limitations of the Poisson distribution?
3. Discuss gap and headway distribution.

4. What is shifted exponential distribution?
5. Discuss the Erlang distribution.
6. What are Pearson type III distribution?
7. What is meant by gap acceptance? Explain.

REFERENCES

1. Ashton, W.D., The Theory of Road Traffic Flow, Nethuen & Co., London, 1966.
2. Gerlough, D.L., and F.C. Barnes, Poisson and Other Distributions in Traffic ENO Foundation for Transportation, Saugatuck 1971, Connecticut.
3. Gerlough, D.L., Traffic Input for Simulation on A Digital Computer, Proceedings of The Highway Research Board, Vol. 38, 1959.
4. Cleaveland, D.G. Capelle, Queueing Theory Approaches, "An Introduction to Traffic Flow Theory" H.R.B. Special Report 79, Washington 1954.
5. Schuhl Andre', Probability theory applied to vehicle distribution on two-lane highways, Poisson and other distributions in Traffic, ENO Foundation for Transportation, Saugatuck, 1971, Connecticut.
6. Kell, J.H., Analysing Vehicular delay at intersection, H.R.B. Bulletin 356, Highway Research Board, Washington, 1962.
7. Greeco, W.L. and E.C. Sword, Reduction of Parameters for Schuhl's Headway Distribution, Traffic Engineering, Vol. 38, 1968.
8. Haight, E.A. Whisler and W.W. Mesher, New Statistical Methods, for Describing Highway Distribution of Cars, Proceedings, Highway Research Board, Washington, 1961.
9. Raff, M.S., A Volume Warrant of Urban Stop Signs ENO, Foundation, Saugatuck, 1950.
10. Miller, A.J., Nine Estimators of Gap-acceptance Parameters, Proceedings of the Fifth International Symposium on the Theory of Traffic Flow and Transportation, American Elsevier Publishing Co., New York, 1971.
11. Blunden, W.R., C.M. Clissold and R.B. Fisher, Distribution of Acceptance Graps for Crossing and Turning Manoeuvres, Proceedings of the Australian Road Research Board, Vol. 1, Part I, 1962.
12. Blunden, W.R., The land-use transport system, Pergamon Press 1971.

27. Delay to Traffic at Uncontrolled Intersections

27.1. Introduction

The delay caused to traffic (including pedestrians) at intersections is a common problem in dealing with urban traffic. The problem is different in nature when signalised and unsignalised intersections are involved. The calculation of delay in signalised junctions has been briefly dealt with in Chapter 15. The present chapter deals with delays at unsignalised interactions, where priority is assigned to main-street traffic and side-street traffic is cautioned to slow or halt. The same situation more or less obtains when a driver on a ramp waits for a suitable gap to join the main stream of traffic.

27.2. Adam's Formula

A systematic study of the delay problem really begins with the classic paper of Adams (Ref. 1) dealing with the stochastic character of the traffic stream. This applies to arrival of pedestrians or a single vehicle on the side street. When traffic is light, the arrival pattern of vehicles can be assumed to be random (Poisson arrival), which gives rise to an exponential distribution of headways (Refer Chapter 26). It has been shown earlier that in such a case the probability of headway equal to or greater than t seconds is :

$$P(h \geq t) = e^{-\lambda t} \qquad \text{(From Eq. 26.13)}$$

where, e = base of natural logarithms

λ = average rate of arrival per unit time

= V/T veh/second.

t = time interval in seconds

h = headway in seconds (or gap)

V = Volume of vehicles arriving in time T seconds

T = Time interval in seconds.

Probability of delay occurring

$$= 1 - e^{\lambda t} \qquad \text{...(27.1)}$$

The proportion of time occupied by intervals greater than t seconds is given by :

$$P_t = e^{-\lambda t}(\lambda t + 1) \qquad \text{...(27.2)}$$

Thus, the proportion of time occupied by gaps less than t

$$1 - P_t = 1 - e^{\lambda t}(\lambda t + 1) \qquad \text{...(27.3)}$$

When a side-street vehicle arrives and finds in line, the probability that no main-street vehicle arrives during the next t seconds is $e^{-\lambda t}$. The waiting time then is zero.

The probability that a mainstreet vehicle arrives during the first t seconds, but none arrives during the t seconds following the arrival of the first mainstreet vehicle is $(1 - e^{-\lambda t})\, e^{-\lambda t}$. The waiting time for the sidestreet vehicle in this case is one interval. If the first two intervals in the main street are each less than t seconds, but the third is greater than t, the probability of this occurrence is $(1 - e^{-\lambda t})\, e^{-\lambda t}$, and the sidestreet vehicle will have to wait for two intervals, each less than t seconds. The same reasoning leads to the general expression for the probability of a sidestreet vehicle having to wait for n intervals, each less than t seconds, as :

$$(1 - e^{-\lambda t})^n\, e^{-\lambda t}$$

The expected number of intervals a sidestreet vehicle has to wait, (n) is given by :

$$\bar{n} = \sum_{n=0}^{\infty} n\,(1 - e^{-\lambda t})\, e^{-\lambda t}$$

$$= \frac{1 - e^{-\lambda t}}{e^{-\lambda t}} \qquad \text{...(27.4)}$$

The average length of intervals through which the sidestreet vehicles have to wait

$$= \frac{\text{Total time occupied by intervals less than } t}{\text{Number of intervals less than } t}$$

$$= \frac{T \times \text{Proportion of time occupied by gaps less than } t \text{ sec}}{V \times \text{Probability of an interval being less than } t \text{ sec}}$$

$$= \frac{T\,[1 - e^{-\lambda t}(\lambda t + 1)]}{V\,(1 - e^{-\lambda t})}, \text{ using Equation (27.3)}$$

$$= \frac{1}{\lambda} - \frac{te^{-\lambda t}}{1 - e^{-\lambda t}} \qquad \text{...(27.5)}$$

The expected mean time in service,

$\overline{ds}$ = the expected number of intervals a sidestreet vehicle has to wait × Average length of intervals through which side-street vehicle has to wait.

$$= \left[\frac{1 - e^{-\lambda t}}{e^{-\lambda t}}\right] \times \left[\frac{1}{\lambda} - \frac{te^{-\lambda t}}{1 - e^{-\lambda t}}\right]$$

$$= \frac{1 - e^{-\lambda t}}{\lambda e^{-\lambda t}} - t \qquad \text{...(27.6)}$$

$$= \frac{1}{\lambda}\left(e^{\lambda t} - 1\right) - t \qquad \text{...(27.7)}$$

In the above equation, t = critical lag.

Eq. (27.6) above includes vehicles which are not delayed at all. The mean time-in service of those vehicles which are delayed at least one interval

$$= \frac{\bar{d}}{1 - e^{-\lambda t}}$$

$$= \frac{1 - e^{-\lambda t}}{\left(\lambda e^{-\lambda t}\right)\left(1 - e^{-\lambda t}\right)} - \frac{t}{1 - e^{-\lambda t}} = \frac{1}{\lambda e^{-\lambda t}} - \frac{t}{1 - e^{-\lambda t}} \qquad \text{...(27.8)}$$

The following problem illustrates the above formulae :

Problem 27.1. *A two lane two way road in the E.W. direction is joined by a side-street from the South, the intersection being of the uncontrolled priority type, with all side-street vehicles being required to stop on arrival at the intersection. The peak-hour flow is as below :*

Direction	*Flow*
E to W	*900 veh/hour*
W to E	*600 veh/hour*
From S	*120 veh/hour*

The critical lag for left turning traffic from the side-street is 4 seconds, that for the right turning traffic from the side-street is 5 seconds. Find :

(i) the probability of delay to a vehicle arriving on the side street and intending to turn left ;

(ii) the probability of delay to a vehicle arriving on the side street and intending to turn right ;

(iii) the average delay to vehicles arriving on the side-street and intending to turn right ;

(iv) the average time-in-service of those left-turning vehicles which are delayed at least one interval.

Solution. *(i)* Probability of occurrence of delay to left turning vehicles :

$$= 1 - e^{-\lambda t} = 1 - e^{-\frac{900}{3600}\times 4} = 1 - e^{-1}$$

$$= 1 - \frac{1}{e} = 1 - 0.37 = 0.63.$$

(iii) Probability of occurrence or delay to right-turning vehicles :

$$= 1 - e^{-\lambda t}$$

$$= 1 - e^{-\frac{(600+900)\times 5}{3600}}$$

$$= 1 - e^{-2.08} = 1 - 0.12 = 0.88.$$

(iii) Average delay to left-turning vehicles :

$$= \frac{1}{\lambda}\left(e^{-\lambda t} - 1\right) - t = \frac{3600}{900}\left[e^{\frac{900}{3600}\times 4} - 1\right] - 4$$

$$= 4(e - 1) - 4 = 4(2.72 - 1) - 4$$

$$= 6.88 - 4 = 2.88 \text{ seconds.}$$

(iv) Average delay to right-vehicles $= \frac{1}{\lambda}\left(e^{-\lambda t} - 1\right) - t$

$$= \frac{3600}{900+600}\left[e^{\frac{900+600}{3600}\times 5} - 1\right] - 5$$

$$= \frac{3600}{1500}\left[e^{\frac{1500}{3600}\times 5} - 1\right] - 5$$

$$= \frac{12}{5}\left[e^{25/12} - 1\right] - 5 = 2.4\ (7.0) - 5 = 11.8 \text{ seconds.}$$

(v) Average time-in-service of those left-turning vehicles which are delayed at least one interval : $= \dfrac{1}{\lambda e^{-\lambda t}} - \dfrac{t}{1 - e^{-\lambda s}}$

$$= \frac{1}{\frac{900}{3600} \times e^{-\frac{900}{3600} \times 4}} - \frac{4}{1 - e^{-\frac{900}{3600} \times 4}}$$

$$= \frac{4}{e^{-1}} - \frac{4}{1 - e^{-1}} = 4 \times 2.72 - \frac{4}{1 - 0.368}$$

$$= 10.88 - \frac{4}{0.632} = 10.88 - 6.34 = 4.54 \text{ seconds.}$$

27.3. Tanner's Formulae

Tanner (Ref. 3) proposed a theoretical model to relate the various parameters connected with the delay problem in dealing with an intersection of a major road with a minor road. The notation used are :

q_1 = major road volume

q_2 = minor road volume

β_1 = minimum time headway between major road vehicles

β_2 = minimum time headway between vehicles emerging from the minor road

α = the average gap or lag in the major road traffic stream accepted by the minor road drivers when entering the major road traffic stream

w_2 = average delay to minor road vehicles.

The following formulae give the relationships between the parameters :

$$\overline{w_2} = \frac{\frac{1}{2} E(y)^2 / Y \, q_2 Y \exp.(-\beta_2 q_1)[\exp.(p_2 q_1) - \beta_2 q_1 - 1] q_1}{1 - q_2 Y[1 - \exp.(-\beta_2 q_1)]} \quad \text{...(27.9)}$$

$$E(y) = \frac{\exp.[q_1(\alpha - \beta_1)]}{q_1(1 - \beta_1)} - \frac{1}{q_1} \quad \text{...(27.10)}$$

$$E(y^2) = \frac{2 \exp.[q_1(\alpha - \beta_1)]}{q_1^2 (1 - \beta_1 q_1)^2} \{\exp.[q_1(\alpha - \beta_1) - \alpha q_1 (1 - \beta_1 q_1)]$$

$$- 1 + \beta_1 q_1 - \beta_1^2 q_1^2 - \frac{1}{2} \beta_1^2 q_1^2 / (1 - \beta_1 q_1)\} \quad \text{...(27.11)}$$

$$Y = E(y) + \frac{1}{q_1} \quad \text{...(27.12)}$$

$$q_2 \text{ (max.)} = \frac{q_1 (1 - \beta_1 q_1)}{\exp.[q_1(\alpha - \beta_1)][1 - \exp.(\beta_2 q_1)]} \quad \text{...(27.13)}$$

I. MULTIPLE CHOICE QUESTIONS

1. Near a pedestrian crossing, when the pedestrians are waiting to cross the road, you should:
   (*a*) sound horn and proceed. (*b*) Slow down, sound horn and pass.
   (*c*) Stop the vehicle and wait till the pedestrians cross the road and then proceed.
2. You are approaching a narrow bridge, another vehicle is about to enter the bridge from:
   (*a*) To cross the bridge as fast as possible.
   (*b*) Put on the head light and pass the bridge.
   (*c*) Wait till the other vehicle crosses the bridge and then proceed.
3. The pedestrians shall not cross the road at sharp bends or very near to a stopped vehicle. Why?
   (*a*) Inconvenience to other vehicles.
   (*b*) Inconvenience to other road users.
   (*c*) Drivers of other vehicles coming at a distance may not see persons crossing the road.
4. Maximum service volume of a road is defined as the total number of vehicles that
   (*a*) can pass a given point in a specified period of time
   (*b*) can be accommodated on a unit length of road
   (*c*) can pass a given point in unit time
   (*d*) None of these
5. For the movement of vehicles at an intersection of two roads without any interference, which type of grade separation is generally preferred?
   (*a*) delta (*b*) diamond (*c*) trumpet (*d*) cloverleaf

Answers

1. (*c*) **2.** (*c*) **3.** (*c*) **4.** (*c*) **5.** (*d*)

II. SOLVED QUESTIONS AND ANSWERS

Q. 1. What are uncontrolled intersections?

Ans. Uncontrolled intersections are the most common type of intersection usually occurs where the intersecting roads are relatively equal importance and found in areas where there is not much traffic. Uncontrolled intersections are the traffic junctions where there is no explicit traffic control measures are adopted.

The important aspects that will be covered in this chapter are: the concept of two-way stop controlled intersection, all-way stop controlled intersection, gap acceptance, critical gap, follow-up time, potential capacity, and delay determination. These concepts are primarily adopted from Highway Capacity Manual.

Q. 2. What is intersection?

Ans. An intersection is a road junction where two or more roads either meet or cross at grade. This intersection includes the areas needed for all modes of travel: pedestrian, bicycle, motor vehicle, and transit. Thus, the intersection includes not only the pavement area, but typically the adjacent sidewalks and pedestrian curb cut ramps.

Q. 3. Briefly explain about priority intersection.

Ans. Priority Intersection occur where one of the intersecting roads is given definite priority over the other. The minor road will usually be controlled by some form of sing marking, such as stop or yield sign; thus ensuring that priority vehicles travelling on the main street will incur virtually no delay.

Q. 4. What is space sharing intersection?

Ans. Space sharing intersection is intended to permit fully equally priority and to permit continuous movement for all intersecting vehicle flows; example would be rotaries and other weaving areas.

Q. 5. What is meant by time sharing intersection?

Ans. Time sharing intersection are those at which alternative flows are given the right of way at different point in time. This type of intersection is controlled by traffic signal or by police officer.

Q. 6. What is a two-way stop-controlled intersection?

Ans. It is one of the uncontrolled intersections with stop control on the minor street.

Q. 7. What are the characteristics of TWSC intersections?

Ans. At TWSC intersections, the stop-controlled approaches are referred to as the minor street approaches; the intersection approaches that are not controlled by stop signs are referred to as the major street approaches.

A three-leg intersection is considered to be a standard type of TWSC intersection if the single minor street approach is controlled by a stop sign. Three-leg intersections where two of the three approaches are controlled by stop signs are a special form of uncontrolled intersection control.

AWSC intersections require every vehicle to stop at the intersection before proceeding. Since each driver must stop, the judgment as to whether to proceed into the intersection is a function of traffic conditions on the other approaches. If no traffic is present on the other approaches, a driver can proceed immediately after the stop is made. If there is traffic on one or more of the other approaches, a driver proceeds only after determining that there are no vehicles currently in the intersection and that it is the driver's turn to proceed.

Q. 8. What are the characteristics of TWSC intersections?

Ans. Two geometric features are common to all intersections. The angle of intersection is formed by the intersecting streets' centerlines. Where the angle of intersection departs significantly (more than approximately 20°) from right angles, the intersection is referred to as a skewed intersection.

Q. 9. What are intersection legs?

Ans. Intersection legs are those segments of roadway connecting to the intersection. The leg used by traffic approaching the intersection is the approach leg, and that used by traffic leaving is the departure leg.

Q. 10. What is the pavement edge corner?

Ans. The pavement edge corner is the curve connecting the edges of pavement of the intersecting streets. In addition to the basic geometric design features, options may be added to improve service for various users.

Q. 11. Explain auxiliary lanes.

Ans. Auxiliary lanes are added at the intersection, usually to accommodate turning motor vehicles. They may also be used to add through lanes through an intersection. Channelizing and divisional islands may be added to an intersection to help delineate the area in which vehicles can operate, and to separate conflicting movements. Islands can also provide for pedestrian refuge.

Q. 12. What is a turning roadway?

Ans. A turning roadway is a short segment of roadway for a right turn, delineated by channelizing islands. Turning roadways are used where right-turn volumes are very high, or where skewed intersections would otherwise create a very large pavement area.

Q. 13. What are traffic control devices?

Ans. Traffic control devices assign right of way, to both motorized and non-motorized traffic and include traffic signals, pavement markings, STOP signs, **YIELD** signs, pedestrian signal heads and other devices (such as raised pavement markings, flashing beacons, and electronic blank-out signs).

Q. 14. How pedestrians are affected by traffic control?

Ans. Pedestrians are significantly affected by the type of traffic control installed at an intersection, the specific parameters of the control, and the resulting motor vehicle operations. At **STOP** controlled, **YIELD** controlled, and uncontrolled intersections, pedestrians' ability to cross the street and the delay experienced is influenced by the yielding behaviour of motor vehicles.

At signalized intersections, the length and frequency of time provided for pedestrian crossings, the clarity of information provided, conflicting turning movements, and motor vehicle yielding are key influences on pedestrians' ability to cross the street, and on delay.

Q. 15. What is road traffic control?

Ans. Road traffic control involves directing vehicular and pedestrian traffic around a construction zone, accident or other road disruption, thus ensuring the safety of emergency response teams, construction workers and the general public.

III. REVIEW QUESTIONS

1. What is Adam's formula?
2. Write a brief note on Tanner's formulae.

REFERENCES

1. Adams, W.F., Road Traffic considered as a random series, Journal of the Institution of Civil Engineers, 1936.
2. Tanner, J.C., A theoretical analysis of delays at an uncontrolled intersection, Biometrical, 49 (1962).

28. Simulation of Traffic

28.1. Introduction

28.1.1. Traffic flow phenomenon consists of a wide range of complex activities, embracing vehicle arrivals, speed of travel, lane discipline, overtaking and crossing logic, gap acceptance, acceleration and deceleration etc. The characteristics of traffic flow include speed, concentration and flow and these are governed by a variety of factors attributable to the road features, the vehicle performance characteristics and the road user behaviour. An understanding of the subject of traffic flow can be achieved by observing how the vehicles move in the stream of traffic, collect data and thereafter synthesise the flow characteristics through analytical or mathematical models. The speed density flow relationships, the hydrodynamic analogy model, car-following theory, queueing theory etc. are examples of the analytical approach. The science of traffic flow theory has been developed to a high degree with the help of this approach.

28.1.2. The analytical approach requires that field data be collected on a large scale. For example, if a proper understanding of the complex heterogeneous traffic on Indian roads is to be achieved, real life observations will have to be conducted on all types of roads met with in India (covering pavement widths, roughness, curvature and vertical profile) on a complete range of vehicle type and composition in the traffic and a full coverage of traffic volumes and speeds. Such a procedure is extremely tedious, time consuming and expensive. At the end of the effort the building up of suitable relationships will be often fraught with uncertainties, and it often happens that a reliable relationship does not emerge.

28.1.3. Traffic engineers are, therefore, taking recourse to a new technique known as Simulation Modelling which is finding application in a variety of situations other than traffic flow problems, Simulation is defined as creating a working analogy of real life problems into a computer-based model. It is a technique whereby some part of the real world is represented dynamically by building a computer model, running the model through time to solve problems having any given constraints and inputs.

28.2. Advantages of Simulation Techniques

In traffic engineering problems, Simulation techniques offer a number of advantages over conventional analytical methods.

1. Simulation is cheaper than many forms of field experimentation and analytical modelling, in terms of time, resources and cost.

2. Simulation is a powerful tool for comparing the consequences of a number of alternative strategies and improvement plans.

3. The task of developing the logic of various events and their occurrences involves the collection of pertinent data. The engineer gains an insight into the traffic characteristics and the way traffic interaction takes place during the

process of collection of data and modelling. This may ultimately lead to a better formation of an analytical model.

4. In real life situations, it is extremely difficult to obtain conditions in the field which are needed for building a better analytical formulation.

For example, in developing the speedflow relationships for rural highways, the extremely congested conditions are not easily found. A Simulation model can build up such situations on the computer and predict the performance of the system.

5. Simulation techniques can be employed to check uncertain analytical solutions.

6. Simulation techniques provide opportunity for controlled experimentation, altering one variable at a time or specific variables simultaneously, and the final effect can be observed. Traffic systems in particular, are highly complex systems, with many variables, interactions and sub-systems.

7. In the case of many analytical models certain assumptions have to be made to simplify the task. Often these assumptions are of a doubtful nature. Simulation models can overcome such deficiencies.

8. Simulation models are "transparent," in the sense that anyone who wishes to know how they work can see through the model.

28.3. Steps is Simulation

28.3.1. Certain sequential operations are involved in any problem wherein Simulation techniques are adopted. These are :

- Definition of the problem.
- Field studies to determine inputs needed for model formulation.
- Development of logic.
- Development of computer simulation programme.
- Calibration of model.
- Simulation runs covering actual conditions.
- Validation of the model.

28.3.2. Definition of the Problem

Simulation techniques can be used for solving a wide range of problems commonly met with in road traffic and transportation. Some of these are :

- Development of speed-flow relationships
- Prediction of fuel consumption of vehicles
- Gap acceptance problems in the design of intersections and interchange ramps
- Queueing problems
- Bus route scheduling
- Accident occurrence.

For a good model to be evolved, it is very necessary that the problem to be dealt with be accurately defined and understood by the modeller.

28.3.3. Field Studies Necessary to Provide Inputs to the Model

The model is a depiction of real life happenings. As such the formulation of a model involves a study of the real life behaviour and drawing of conclusions therefrom. These conclusions will thereafter be used as inputs to the various components of the model. To cite an example, the speeds which are adopted by drivers of cars on the road can cover a complete range, depending upon the characteristics of the vehicle, the road, the traffic stream and the driver himself. If the Simulation Model has to predict the speeds for a full range of possibilities the modeller has to study the actual speed distribution in real life under free-flow conditions as the starting point. Thereafter, he has also to study how speeds get affected by road characteristics, such as roughness, pavement width, curvature and sight distance. The results of field observations can thereafter be simplified into simple relationships which will be used in the Model.

28.3.4. Development of Logic

Certain activities in the model are interrelated through some logic which has to be established by the modeller. For example, given a *"T"* intersection problem, the vehicles arriving at the side road have to wait for suitable gaps in the traffic stream in the main road. Drivers accept gaps which are safe, though some drivers are extremely cautious and some others are prone to take risks. The gap acceptance phenomenon follows a certain pattern. The modeller has to establish the logic involved in such operations.

28.3.5. Development of the Computer Simulation Programme

The modeller is now in a position to write the Simulation Programme which will incorporate in inputs and logic he has determined. The actual programme language can be the choice of the modeller. The languages that are normally used are :

1. Fortran 2. Algol 3. Simula

The choice depends upon the computer system available and the ability of the program language to incorporate the requirements of the model as inexpensively as possible (in terms of computer time).

28.3.6. Calibration of the Model

Inspite of best efforts to incorporate all the needed inputs to the model, it is quite likely that certain essential features have been either missed or have been erroneously determined. The initial run of the model can detect such errors and give the modeller an opportunity to calibrate like model to eliminate any inaccuracies.

28.3.7. Simulation Runs

Once the model is calibrated it is ready for actual use. Before it can be universalised to cover the range of values of traffic, roadway, vehicle and driver characteristics, a number of Simulation runs are made covering the extremes of conditions.

28.3.8. Validation of the Model

The results of the Simulation runs are then compared with real life observations under similar conditions. If the model were to be reliable the results of the Simulation runs and predictions should converge with the actuals to a good degree of accuracy. If not, the modeller has to investigate the causes for divergence and recalibrate the model, till satisfactory convergence is achieved.

28.4. Scanning Techniques

Techniques adopted by the modeller to depict the sequence of events in the program can be of two types :

1. Event scanning 2. Time scanning.

Event scanning is the procedure by which the model assesses the conditions at the occurrence of every event. The model is allowed to run after the occurrence of one event till the occurrence of the next. Time scanning analyses the conditions at selected time intervals. Event scanning is economical in computer time but requires program sophistication. On the other hands, time scanning is simple to program, but may involve higher computer time and cost.

28.5. An Example of Simulation

28.5.1. In Traffic Engineering the determination speed-flow relationship is one of the commonly encountered problems. The analytical approach which was hitherto followed has inherent disadvantages, especially for Indian road and traffic conditions. Realising the potentialities of the Simulation Modelling approach in this field, a major Indo-Swedish modelling effort has been successfully concluded (Ref. 1). The Central Road Research Institute, the Indian Institute of Technology (Kanpur), the Ministry of Transport (Roads Wing) and the Swedish Road and Traffic Research Institute were involved in the study.

28.5.2. Problem Definition

The objectives, of the study were as under :

1. To develop a computer based Simulation Model for predicting the speeds or vehicles under :

(*a*) varying road widths commonly met with in India, with or without shoulders ;

(*b*) roads of different riding quality ;

(*c*) road sections having any desired vertical and horizontal geometry ;

(*d*) heterogeneous traffic conditions met with in India, covering a mixture of slow and fast moving traffic ; and

(*e*) free flow and congested conditions.

2. To develop a computer-based traffic Simulation Model with capabilities of predicting fuel consumption.

28.5.3. Stages Involved

The stages involved in the work were :

1. Problem definition
2. Assessing of the Swedish (VTI) Model as a useful take-off point
3. Field studies
4. Program development
5. Model calibration
6. Mode validation.

28.5.4. VTI Model

The Swedish (VTI) Model is microscopic in nature and operates on the basis of discrete events (Ref. 2). The behaviour of vehicles on roads is represented in a logical sequence which replicates the observed traffic flow conditions.

The model conceives the basic desired speed of vehicles as that speed which a driver would choose to adopt on a wide, straight, horizontal road with no speed restrictions. The following factors reduce the basic desired speed :

1. Narrow roads
2. Curves with radii less than 1000 m
3. Speed limits
4. Gradients.

The effect of each of the above factors is considered and the free speed is thereby obtained.

The effect of traffic is then considered. The overtaking phenomenon is modelled, encompassing the vehicle speeds, available sight distance, available gaps in the on-coming stream and probability of gap acceptance.

The model also predicts fuel consumption.

The model is valid for two lane roads and is written in the SIMULA language.

The output of the model consists of statistical distribution of speed, headways, journey time, number of overtakings, delay and fuel consumption.

Since the VTI Model had behind it nearly ten years of modelling effort it was decided that the Indian work would use the basic structure of the model and adapt it to Indian conditions. The adaptations that would be needed were identified as under :

1. The Indian vehicles have different horsepower, size and shapes.
2. The Indian roads have sub-standard widths (*e.g.*, single lane roads) poor riding quality and poor geometries.
3. The traffic on the Indian roads is heterogeneous with slow moving and fast moving vehicles occupying the same road space.

28.5.5. Field Studies

The following field studies were carried out to provide Indian inputs into the model :

1. Basic desired speed distribution of cars, trucks, buses and two wheelers (mean and standard deviation).
2. Effect of pavement width on vehicle speeds
3. Effect of shoulder type and condition on vehicle speeds
4. Effect of roughness on vehicle speeds
5. Effect of radius of curvature on vehicle speeds
6. Power weight ratio of Indian vehicles
7. Determination of the coefficient of air resistance and friction (Ref. 4).
8. Gap acceptance probabilities for various types of vehicle and roadway combinations.

For gap acceptance observations, the video equipment mounted on a travelling vehicle was used.

28.5.6. Model Programming and Calibration

The Indian Model has been developed in the SIMULA language and is structured on the lines similar to the VTI Model. After development of the model, it has been calibrated using typical Indian values of vehicle and traffic characteristics.

28.5.7. Model Validation

The model has been validated for selected road and traffic conditions. Validation involves the comparison of the model output with the observed values. The closer the agreement between the values the better is the model predictability.

The model has been validated for selected stretches of two lane and four lane roads in plain, rolling and hilly terrain. For single lane and intermediate lane roads, road sections in plain terrain were selected.

The values of predicted journey speeds, spot speeds, headways number of overtakings and fuel consumption were compared with observed values. Standard statistical tests were employed to check the agreement. It has been found that the model gives predicted values which are in close agreement with the observed values.

I. MULTIPLE CHOICE QUESTIONS

1. .... is considered to be a numerical computation technique used in conjunction with dynamic mathematical models.
   (*a*) analysis (*b*) system simulation
   (*c*) dynamic computation (*d*) none of the above.
2. Which system model applies computational procedures to solve equations:
   (*a*) dynamic model (*b*) static model
   (*c*) analytical Model (*d*) numerical model
3. Which system/model applies deductive reasoning of mathematical theory to solve a model:
   (*a*) dynamic model (*b*) static model
   (*c*) analytical model (*d*) numerical model.
4. Which model follows the changes over time that results from the system activities:
   (*a*) dynamic model (*b*) static model
   (*c*) analytic model (*d*) numerical model

Answers

1. (*b*) **2.** (*d*) **3.** (*c*) **4.** (*d*)

II. SOLVED QUESTIONS AND ANSWERS

Q. 1. What is traffic simulation?

Ans. Traffic simulation or the simulation of transportation systems is the mathematical modelling of transportation systems (*e.g.*, freeway junctions, arterial routes, round abouts, downtown grid systems, etc.) through the application of computer software to better help plan, design and operate transportation systems.

Q. 2. What are traffic models?

Ans. Simulation methods in transportation can employ a selection of theories, including probability and statistics, differential equations and numerical methods.

Q. 3. What is Monte Carlo method?

Ans. One of the earliest discrete event simulation models is the Monte Carlo simulation, where a series of random numbers are used to synthesise traffic conditions.

Q. 4. What is cellular automata model?

Ans. This was followed by the cellular automata model that generates randomness from deterministic rules.

Q. 5. Explain discrete event and continuous-time simulation.

Ans. Discrete event simulation models are both stochastic (with random components) and dynamic (time is a variable). Single server queues for instance can be modelled very well using discrete event simulation, as servers are usually at a single location and so are discrete (*e.g.* traffic lights). Continuous time simulation, on the other hand, can solve the shortcoming of discrete event simulation where the model is required to have input, state and output trajectories within a time interval. The method requires the use of differential equations, specifically numerical integration methods. These equations can range from simple methods, such as Euler's method, to higher order Taylor's series methods, such as Heun's method and Runge-Kutta.

Q. 6. What is car-following models?

Ans. A class of microscopic continuous-time models, known as car-following models, are also based on differential equations. Significant models include the Pipes, intelligent driver model and Gipps' model. They model the behaviour of each individual vehicle in order to see its implications on the whole traffic system.

Q. 7. What do you mean by systems planning?

Ans. The methods noted above are generally used to model the behaviour of an existing system, and are often focused around specific areas of interest under a range of conditions (such as a change in layout, lane closures, and different levels of traffic flow). Transport planning and forecasting can be used to develop a wider understanding of traffic demands over a broad geographic area, and predicting future traffic levels at different links (sections) in the network, incorporating different growth scenarios, with feedback loops to incorporate the effect of congestion on the distribution of trips.

III. REVIEW QUESTIONS

1. What are the advantages of simulation techniques?
2. What are the steps in simulation?
3. What is meant by scanning techniques?

REFERENCES

1. Traffic Simulation Modelling Study, Central Road Research Institute New Delhi, 1985.
2. Gynnerstedt, G. et al, A Model for the Monte Carlo Simulation of Traffic Flow along two-lane single carriageway rural roads, Meddelande 43, Swedish Traffic and Road Research Institute, Linkoeping, 1979.
3. Birtwistle, G.M. et al, Simula Begin Chart Well, Bratt Ltd., Bromley, 1982.
4. Kadiyali, L.R. et al, Rolling and air resistance factors for Indian Road conditions and vehicles, Highway Research Bulletin No. 19, Indian Roads Congress, New Delhi, 1982.
5. The Art of Simulation, by K.D. Techer, D. Van Nostrand Company, Princeton, 1963.
6. An Introduction to Traffic Flow Theory, H.R.B. special Report 79, 1986.
7. Workshop on Rural Traffic Simulation, Australian Road Research Board, 1983.

SECTION 13

TRANSPORT PLANNING

29. Transport Planning Process

29.1. Scope of the Subject

29.1.1. Transportation occupies a high place in modern life. Advancement in all spheres of life has been to a large extent influenced by transportation. Transport planning is a science that seeks to study the problems that arise in providing transportation facilities in an urban, regional or national setting and to prepare a systematic basis for planning such facilities. Since the developed countries where this science has evolved are mainly urban oriented the emphasis is more on urban transport planning. However, the principles of urban transport planning can be applied to regional or national transport planning as well with due changes wherever called for.

29.1.2. Human activities take place in adapted spaces linked by communications through channels. Town and country planning is a science that deals with the study of the urban or country "system" covering the interacting activities using adapted spaces linked by communications through channels. Transport planning is an important part of overall town and country planning, since it deals with the transport network which is an important channel of a communications. Any change in the transport system is reflected in number of impacts.

29.1.3. Though motor vehicles have revolutionised our life and brought comfort, pleasure and convenience, they have created problems of congestion, lack of safety and degeneration of the environment. The situation has already become unmanageable in many towns and cities. To understand the nature of these problems and formulate proposals for the safe and efficient movement of goods and people from one place to another is the subject of transport planning.

29.2. Interdependence of the Land Use and Traffic

29.2.1. In 1954, *Mitchell and Rapkin* (Ref. 1) made a statement that urban traffic was a function of land use. Though it appears now that there is nothing spectacular or startling in this statement, yet it paved the way for a new line of thinking in urban transportation and land-use planning. Till then, transportation planning was limited to the measurement of traffic using streets, identifying those sections where the present traffic had exceeded the capacity and undertaking improvement measures to relieve the congestion and bottlenecks in the smooth flow of traffic. This simplistic approach failed to deal with the crux of the transportation problem and only supplied the engineers with short term palliatives to deal with the transport malady. This approach has now been virtually abandoned.

29.2.2. Mitchell and Rapkin (Ref. 1) observed that various kinds of activities based on the land-called land use-"generated" different amounts and kinds of traffic. They concluded that though measures such as (*i*) regulation and control of traffic and (*ii*) provision and improvement of physical channels of movement where effective in dealing with urban traffic, the most basic level of action for a long-run solution of the traffic problems is the planning, guidance and control in the pattern of land use. More recently, Buchanan (Ref. 2) has also emphasised the interrelationship between traffic and buildings in a town. He states that in towns, traffic takes place because of buildings, and in fact all movements in a town have an origin and destination in a building. The pattern traced by traffic is thus closely related to the manner in which buildings are arranged. Commuter flows are closely dependent upon the location and size of the work-places and of the homeareas. School traffic is governed by the location of the schools and the home areas.

29.2.3. Just as transport is a function of land-use, the reciprocal statement that land-use is a function of transport is also true. As new systems of transport are built, the land-use pattern that follows has a close relation to the accessibility that has been made possible.

29.2.4. The above interdependence is the key-note of modern transport planning. The early Detroit Area Transportation Study demonstrated the empirical validity of the proposition that transport was a function of land use. The *Penn-Jersey Transportation Study* tested the reciprocal proposition that land use was a function of transportation. These concepts have by now been used in a number of important transportation studies in many of the principal towns and cities all over the world.

29.3. Systems Approach to Transport Planning

29.3.1. In the recent past, a new activity known as Operations Research has taken shape and is finding interesting applications in diverse fields. Transport Planning is one field where this approach has already been tried and found extremely useful. Operations Research is mainly concerned with optimising the performance of a "system". A "system" is defined as a complex whole, an organised whole, consisting of set of connected things or parts, whose components and interconnections are vital to the operation of the system.

29.3.2. The processes involved in the systems approach in transport planning can be represented by Fig. 29.1.

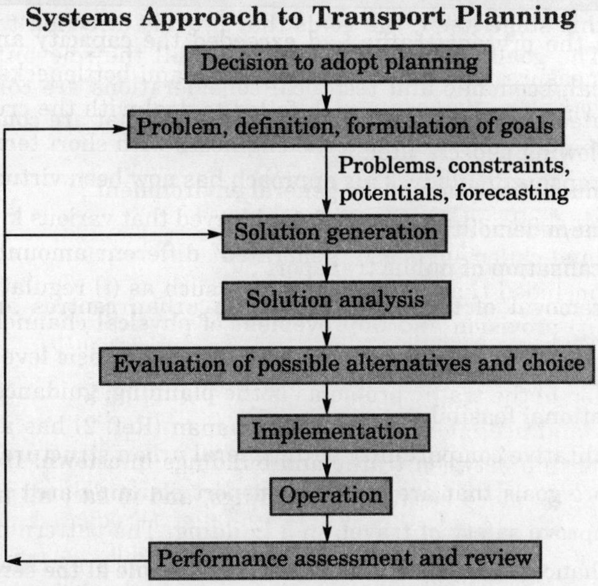

Fig. 29.1. Systems approach to Transport Planning.

29.3.3. The transport planning process starts with the decision to adopt planning as a tool for achieving certain desired goals and objectives. After the goals and objectives are defined, solutions are generated, taking due cognizance of problems, constraints, potentials and forecasting. These solutions are evaluated after thorough analysis. The best amongst them is chosen for implementation. After implementation, the system is studied in operation and its performance assessed. Based on this assessment it may be necessary to go back to certain stages of planning and repeat the sequence.

29.4. Stages in Transport Planning

The broad sequence of operations in the systems approach to transport planning has already been outlined. In this section, it proposed to identify in greater detail the various stages in the transport planning process.

Fig. 29.2, gives the flow diagram where the various stages have been indicated.

It will be seen that the transport planning process can be broken down to five important stages :

- Survey and analysis of existing conditions
- Forecast, analysis of future conditions and plan synthesis
- Evaluation
- Programme adoption and implementation
- Continuing study.

These stages are discussed in detail below.

29.5. Survey and Analysis of Existing Conditions

The survey and inventory starts with the definition of the survey area and division of the area into smaller units, called *zones*, to study the pattern of movement. At this stage itself, the goals that provide direction to the planning effort are set. The goals are so formulated that all the relevant social, legal, aesthetic political, economic and technical considerations are comprehensively weighed, evaluated and reconciled. Some of the goals that are commonly sought embrace the following aspects (Ref. 5) :

(*i*) Minimum disruption of the general environment ;

(*ii*) Minimum demolition of housing ;

(*iii*) Re-vitalisation of public transport ;

(*iv*) The removal of through traffic from urban centres and residential areas ;

(*v*) A high benefit/cost ratio ;

(*vi*) Operational feasibility ;

(*vii*) A qualitative compatibility with general urban structure.

Some more goals that are set for transport planning are :

- To improve safety of travel.
- To enhance the levels of mobility of people while at the same time reduce dependence on the use of personalized vehicles, conserve fuel and protect the environment.

An idea of the broad goals that are set out can be had from the following enunciation of goals by the U.S. Deptt. of Transportation (Ref. 4) :

"To provide that mix of transportation alternatives, including modal systems, related facilities and manpower, research and development, etc. which results in maximum benefits such as service, convenience, comfort, capacity and speed for a given cost."

"To increase the benefits derived from the preservation and enhancement of the environmental, aesthetic and social factors of transportation."

"To minimise the loss of human life and human suffering through injury from transportations-related accidents."

"To further all other objectives of the Federal Government whenever they affect transportation and the DOT can perform a particular task more effectively and efficiently."

The inventory of existing travel pattern includes :

(*i*) Collection of data on origins and destinations of journey by home-interview, road side interview, registration number plate survey, pre-paid post card survey, screenline and cordon surveys.

(*ii*) Collection of data on traffic volume on the various links in the existing network, including classification of vehicles, variation in traffic volumes during the days of the week and hours of the day.

(*iii*) Collection of data on movement of goods vehicles.

(*iv*) Collection of data on movement of public transport buses and coaches.

(*v*) Collection of data on movement by rail transit.

(*vi*) Parking characteristics (supply, usage, duration and method of charging).

Table 29.2. Stages in the Transport Planning Process

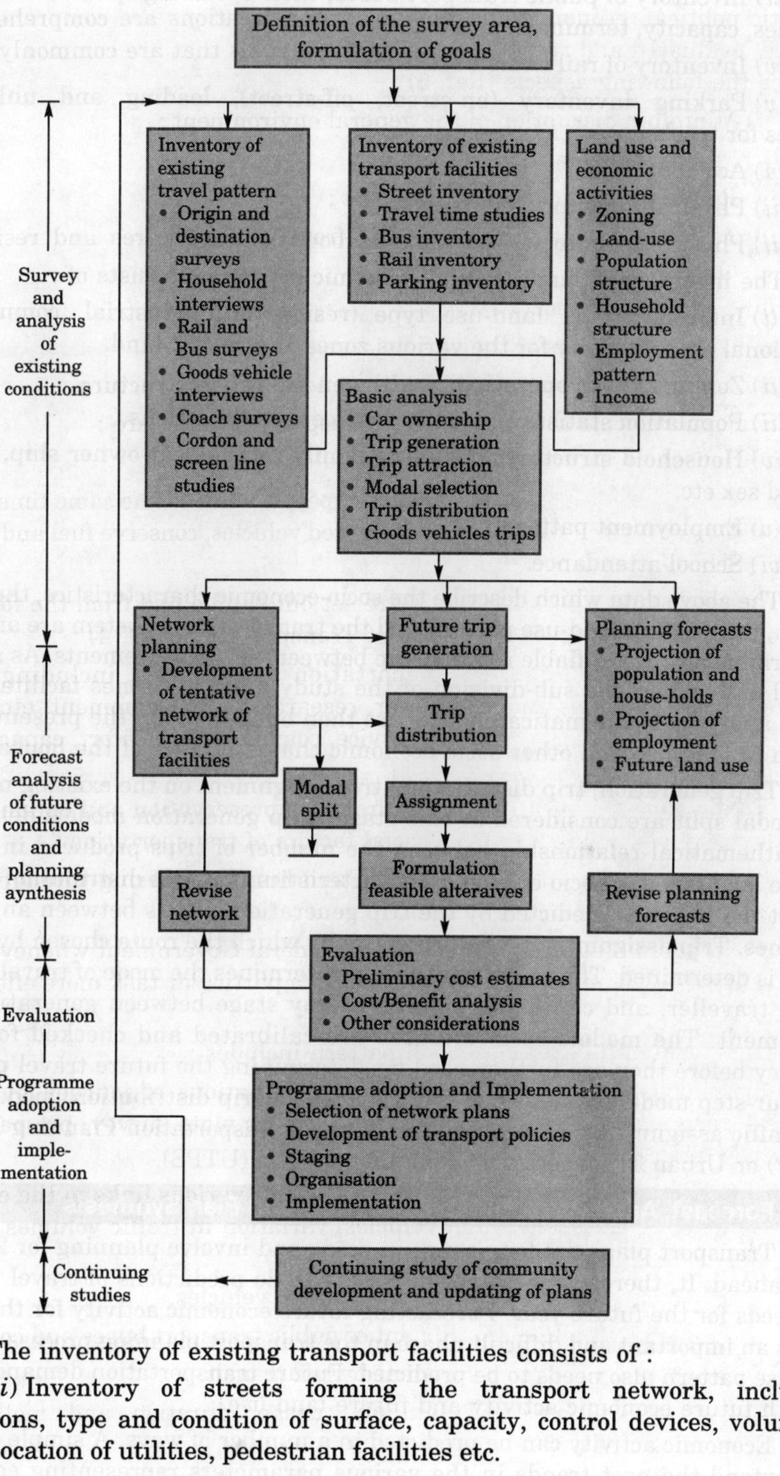

The inventory of existing transport facilities consists of :

(*i*) Inventory of streets forming the transport network, including dimensions, type and condition of surface, capacity, control devices, volume of traffic, location of utilities, pedestrian facilities etc.

(*ii*) Studies on travel time by different modes.

(*iii*) Inventory of public transport buses, their operating speeds, headways, schedules, capacity, terminals, passengers carried etc.

(*iv*) Inventory of rail transit facilities.

(*v*) Parking inventory (on-street, off-street), loading and unloading facilities for trucks.

(*vi*) Accident data.

(*vii*) Physical inventory of airports.

(*viii*) Physical inventory of port and harbour facilities.

The inventory of land-use and economic activities consists of :

(*i*) Information on land-use type (residential, industrial, commercial, recreational etc.) intensity for the various zones and vacant land.

(*ii*) Zoning laws in operation.

(*iii*) Population statistics, usually from census operations.

(*iv*) Household structure, including family income, car-owner ship, family size and sex etc.

(*v*) Employment pattern.

(*vi*) School attendance.

The above data which describe the socio-economic characteristics, the travel pattern, the existing land-use pattern and the transportation system are analysed to determine any quantifiable relationship between the measurements. As already pointed out earlier, the sub-division of the study area into zones facilitates the spatial analysis. Mathematical models are then built to relate the present travel pattern to land use and other socio-economic characteristics of the household.

Trip generation, trip distribution, trip assignment on the existing network and modal split are considered at this stage. Trip generation models determine the mathematical relationship between the number of trips produced in a zone and the land-use and socio-economic characteristics. The trip distribution models connect the tripends predicted by the trip generation models between any set of two zones. Trip assignment is the procedure by which the route chosen by a trip-maker is determined. The modal split model determines the mode of travel chosen by the traveller, and can be considered at any stage between generation and assignment. The models so formulated are calibrated and checked for their accuracy before they are further used for determining the future travel pattern. The four-step model consisting of trip generation, trip distribution, mode choice and traffic assignment is known as the Urban Transportation Planning Process (UTPP) or Urban Transportation Planning System (UTPS).

29.6. Forecast, Analysis of Future Conditions and Plan Synthesis

Transport plans are long-range in scope and involve planning for 20 to 25 years ahead. It, therefore, becomes necessary to do predictions of travel pattern and needs for the future year. Forecasting future economic activity for the study year is an important and difficult phase in the transport planning process. Future land-use pattern also needs to be predicted. Future transportation demand is tied up with future economic activity and future land-use.

Economic activity can be predicted in a number of ways. A simple method is to extend the past trends in the various parameters representing economic activity. More sophisticated methods such as the input-output model analysis used by the Chicago Area Transportation Study require detailed data, particularly on the interrelationships among the many independent sectors of economic activity.

Population forecast can be done by studies of the past trend. More detailed studies of the birth, death and migration data help in formulating population prediction models. Apart from an estimate of the total population, it is also necessary to know the age distribution, family size and number of house holds.

Estimate of future employment pattern follows from predictions of economic activity. Future level of car ownerships is an important parameter that influences future travel and mode choice.

Since the economy of a region and its transport are closely related, it is possible to establish econometric models relating the transport indicators and economic indicators of the following type (Ref. 6) : $\log_e T = A_0 + A_1 \log_e GNP$

where, T = Transport indicator, such as number of cars, for example

GNP = Gross National Product

A_0 = Regression constant

A_1 = Regression coefficient, also known as Elasticity coefficient.

Knowing the expected rate of growth in GNP, the expected rate of growth of T can be obtained by multiplying it by the Elasticity coefficient, A_1.

The predicted growth in land-use pattern such as residential, industrial, commercial recreational and open spaces can be determined by formulating suitable land-use allocation models. Considerations such as the policy of the administration towards urban renewal, zoning regulations, cost and availability of land etc. should be employed in building such models.

The future land-use activity arrangement and information concerning future income levels, family size, car ownership rates, employment, population and other economic activity factors are then fed as inputs to the previously formulated trip generation model. The output will be the future trip generation rates. The other stages such as trip distribution, assignment and modal split are gone through by using the future predicted parameters governing travel pattern and the respective models formulated for the base year. The data are fitted into a tentative network planned for the horizon year.

The output from the above stages yields the flow on each link of the network, and speed and level of service afforded by the planned facility.

29.7. Evaluation

In an urban system, a number of alternative transport plans are feasible for a given set of goals and policies. In order to select the best from these, it is necessary to evaluate each of the alternatives as to how it fulfils the desired objectives.

Cost/Benefit techniques are often used to evaluate the alternatives in economic terms.

At this stage, it may be necessary to revise the plans and go back to the initial stage of design to evolve further alternatives.

29.8. Programme Adoption and Implementation

The best alternative emerging from the evaluation study is selected for adoption and implementation. The stages in which the project is to be implemented are decided with the consideration for the financial resources. The necessary organisation for handling the project is built up and the work executed.

29.9. Continuing Study

Because transport planning is a dynamic and complex process, there cannot be any finality about the plan. The urban system and the people inhabiting it are not deterministic and are governed by random behaviour. Considerable uncertainty is always associated with such systems. Technology and preferences of people may change. Plans and policies which are relevant today may not remain so in the future context. This underlines the need for a continuous review and updating of the plan. The process is one of constant iteration and feedback. Periodic surveys should be carried out on the trends in travel patterns, journey times and other relevant factors and the plan readjusted, if need be.

29.10. Citizen Participation

Since the transport plans are essentially intended to serve the community and fulfil their aspirations, it is necessary to consult the affected people in the community and give them a voice in the formulation of decisions. The policies and goals may need to be redefined as a result of such consultations. It is desirable that a report with the community is established right from the start so that there will be a minimum of controversies and public criticism.

29.11. Difficulties in the Transport Planning Process

In view of the complexities involved, the process of transport planning is a difficult exercise. An interplay of numerous factors makes the process rather involved in computational procedures. Even though considerable advance has been made in model building to accurately synthesise the travel demand, the probabilistic nature of peoples' behaviour can make the plans go away. The dynamic behaviour of the urban system makes the whole process iterative with a continuous need for review and feed back. The conventional sequential four-stage UTPP models require large quantities of data which involve considerable time and resources to collect.

29.12. Relevance of Transport Planning Process to Less Developed Countries

Most of the research and modelling pertaining to the transport planning process has been carried out in the developed countries. These countries have a high car-ownership rate and a high per capita income. The degree of urbanisation in these countries is of a high order. The exact application of the processes and models developed for these conditions to the less developed countries is a matter of doubt.

Secondly, the conventional transport models were evolved in the U.S.A and Western Europe during the period 1950-1970. This was era of abundant availability of liquid fuel for propulsion of cars. The period was also one of economic stability. The energy crisis of the mid-seventies and the political instability in the oil-producing countries has altered the travel pattern even in the developed countries to a significant extent.

The advent of the computer, the internet and the mobile phone services has vastly altered the need to travel from one place to another. Hence, past trends in travel may not hold the key to predict future travel demand.

In the less developed countries, urbanisation is till now on a moderate scale; for example it is 28 percent in India, compared to 79 per cent in U.S.A. (2000-2001). Big cities in India have grown to mega-metropolises and it is doubtful whether such cities can absorb any more people. On the other hand, small and medium sized towns and cities have space and potential for absorbing the migration of the rural population. Under such circumstances, one has to be extremely careful to apply models developed under different conditions.

29.13. Computer Applications in Transport Planning

Many organisations have developed tbeir own "Soft ware" preprogrammed packages for dealing with standard problems met with Transport Planning.

The Bureau of Public Roads (U.S.A) and Federal Highway Administration (U.S.A) developed a number of packages such as :

(1) Gravity Model (Ref. 7) (2) Trip Generation (Ref. 8)
(3) Trip Distribution (Ref. 9) (4) Traffic Assignment (Ref. 10)
(5) Transportation Planning System (Ref. 11).

The efficient management of data on at a computer calls for meticulous care at every stage. The various stages involved in data management are :

(1) Data collection (2) Data verification
(3) Coding (4) Date entry
(5) Date validation (6) Report generation
(7) Creation of analysis files.

The advances in computing capabilities have simplified the effort involved. However, it needs hardly be emphasised that the data collection in specially designed forms is of prime importance.

Many micro-computer applications software for Urban Transportation Planning System (UTPS) are available such as MINUTP, TRANPLAN, MICROTRIPS, QRSII for Windows and TRIPS.

I. MULTIPLE CHOICE QUESTIONS

1. IT in supply chain transaction execution is concerned with:
   (*a*) collection, generation, and storage of vast data and tracking of the same through automated means.
   (*b*) enabling managers to process and evaluate SCM-related decisions using different optimization techniques.
   (*c*) measurement of the supply chain performance through data analysis tools.
   (*d*) cooperation and collaboration with suppliers, partners, etc. via the internet.
2. IT in supply chain decision support is concerned with:
   (*a*) enabling managers to process and evaluate SCM-related decisions using different optimization techniques.
   (*b*) collection, generation, and storage of vast data and tracking of the same through automated means.
   (*c*) measurement of the supply chain performance through data analysis tools.

(*d*) cooperation and collaboration with suppliers, partners, etc. via the internet.

3. IT in supply chain measurement and reporting is concerned with:
(*a*) measurement of the supply chain performance through data analysis tools.
(*b*) enabling managers to process and evaluate SCM-related decisions using different optimization techniques.
(*c*) cooperation and collaboration with suppliers, partners, etc. via the internet.
(*d*) collection, generation, and storage of vast data and tracking of the same through automated means.

4. Which of the following cannot be categorized under supply chain transaction execution?
(*a*) point of sale tracking system (*b*) manufacturing execution system
(*c*) warehouse management system
(*d*) customer relationship management
(*e*) inventory management system

5. An **ERP** system falls under:
(*a*) supply chain collaboration and coordination
(*b*) supply chain performance measurement and reporting
(*c*) supply chain transaction execution
(*d*) supply chain decision support

6. Which of the following does **CPFR** fall under?
(*a*) Supply chain transaction execution
(*b*) Supply chain collaboration and coordination
(*c*) Supply chain decision support
(*d*) Supply chain performance measurement and reporting
(*e*) both supply chain collaboration and coordination and supply chain decision support

7. Which of the following does not characterize **ERP II**?
(*a*) **ERP II** is concerned with optimization of supply chain through collaboration with trading partners.
(*b*) **ERP II** crosses all sectors and segments of business, including service, government and asset-based industries.
(*c*) In an **ERP II** system, the same information is available across the whole supply chain to the authorized participants.
(*d*) **ERP II** enables extended portal capabilities that help an organization involve its customers and suppliers to participate in the workflow process.
(*e*) **ERP II** systems are monolithic and closed.

8. **DSS** in a supply chain helps managers in taking decisions of:
(*a*) strategic level. (*b*) tactical level.
(*c*) operational level
(*d*) both strategic level and tactical level
(*e*) both strategic, tactical and operational level

Answers

1. (*a*)	**2.** (*a*)	**3.** (*a*)	**4.** (*d*)	**5.** (*c*)
6. (*b*)	**7.** (*e*)	**8.** (*e*)		

II. SOLVED QUESTIONS AND ANSWERS

Q. 1. What are the main stages of the transportation planning process? Explain.

Ans. In most of the countries transport planning is treated as a part of general economic planning and no special attention has been paid, but now not only developed countries but developing countries have also realised the need for separate planning for the transportation, not only for the existing system but for the future development also. The study of development and planning is basically a study of interaction between man, land and activity in the form of spatial organisation of economy.

After industrial revolution and rapid growth of urbanisation, development in the field of transport is enormous both in infrastructures, speed as well as in transport technology. Now a days every country of the world is having its own national transport system, not in isolation but as a part of international system of transportation. Transport now has, as ever, become an integral and essential part of the economy and requires a planned growth, which should be 'sustainable'.

In fact, transport planning is the process of regulating and controlling the provision of transport to facilitate the efficient operation of the economic, social and political life of the country at the lowest social cost. In practice, this means assuring adequate transport capacity and efficient operations to meet the needs generated by the nation's geographical array of activities.

Q. 2. What is aim of planning process?

Ans. The primary aim of transport planning is the identification and evaluation of the future transport needs. The four main stages of the transportation planning process are:

(*i*) Transportation survey, data collection and analysis;

(*ii*) Use of transportation model;

(*iii*) Future land use forecasts and alternative policy strategies; and

(*iv*) Policy evaluation.

Q. 3. What is the transportation model?

Ans. The second stage of the transportation planning process is to use the collected data to build up a transportation model. This model is the key to predicting future travel demands and network needs and is derived in four recognised stages, *i.e.*, trip generation, trip distribution, traffic assignment and model split.

Q. 4. Explain trip generation. What are the types of trip generation?

Ans. The first stage of model building process is that of trip generation. Trips are made for a variety of purposes and for various land uses. For convenience, trips are often split into two groups:

(i) Home-based trips: Such trips have one trip end at the home of the person making the trip, which may be either the origin or destination of the given trip.

(ii) Non-home-based trips: These have neither origin nor destination tripend at the home of the person making the trip. This initial part of the transport model expresses trip-making relationships in a mathematical form so that ultimately we can calculate the total number of trips-ends originating from the defined survey zones.

Q. 5. What is multiple regression?

Ans. Multiple regression technique are often used to calibrate a trip-generation model incorporating the above household variables. This model takes the following general form:

$$Y = a + b_1 x_1 + b_2 x_2 + \ldots + b_n x_n$$

where, Y = number of trips (by mode and purpose) generated in a given zone

a = constant term

$b_1 \ldots b_n$ = regression co-efficients relating to independent variables (*e.g.*, household income, car-ownership, household structure, etc.)

New estimates of the independent variables are made and inserted into the equation in order to estimate future levels of trips generation.

Multiple regression analysis, therefore, provides a suitable method for estimating future trip levels. Its main disadvantage, however, is that the original regression estimates have been established at a given point in time and are expected to remain constant over the period for which the forecast is required.

Consequently, a more recent approach to trip generation has been to use a technique known as 'category analysis'. The trip-generation stage of the planning process estimates the total number of trips originating in the survey area at one or more future dates.

Q. 6. What is trip distribution?

Ans. Trip involves on analysis of trips between zones. Lane states the function of this stage of the model:

It is the function of trip distribution to calculate the number of trips between one zone and another, given the previously determined numbers of trip ends in each zone together with further information on the transport facilities available between these zones.

For example, given that in zone I, g_i trip ends are generated and that in zone j, a_i trip ends are attracted, it is the purpose of the trip distribution model to determine the number of trips (t_{ij}) which would go from zone i to zone j. That is, the trip distribution model calculates the proportion of trip ends generated in zone i which would travel between i and j and so take up a certain proportion of the available attractions in zone j.

Q. 7. What is model split?

Ans. The term 'Model Split' is used by transport planners to describe the phase where the choice of travel mode is incorporated into the model. The positioning of this stage is neither fixed nor singularly definable since elements of model split are part of the other stages. Its position within the transportation model differs between studies. It is either used at the trip generation stage by stratifying the total trips or at the assignment stage of the model. The main purpose of the model split stage is to determine the trip shares of public, as against private, transport.

Q. 8. What is the transportation planning process?

Ans. Transportation planning is a cooperative, performance-driven process by which long and short-range transportation improvement priorities are determined. Metropolitan planning organizations (MPOs), States, and transit operators conduct transportation planning, with active involvement from the travelling public, the business community, community groups, environmental organizations, and freight operators.

Transportation planning typically follows the following steps:

- Engaging the public and stakeholders to establish shared goals and visions for the community.
- Monitoring existing conditions and comparing them against transportation performance goals.
- Forecasting future population and employment growth, including assessing projected land uses in the region and identifying major corridors of growth or redevelopment.
- Identifying current and projected transportation needs by developing performance measures and targets.
- Analyzing various transportation improvement strategies and their related trade-offs using detailed planning studies.
- Developing long-range plans and short-range programs of alternative capital improvement, management, and operational strategies for moving people and goods.
- Estimating how recommended improvements to the transportation system will impact achievement of performance goals, as well as impacts on the economy and environmental quality, including air quality.
- Developing a financial plan to secure sufficient revenues that cover the costs of implementing strategies and ensure ongoing maintenance and operation.

Q. 9. What is a regional transportation planning organization?

Ans. A regional transportation planning organization **(RTPO)** is a voluntary group of non-metropolitan area local officials and transportation system operators that States may assemble to assist in the Statewide and nonmetropolitan transportation planning process. **RTPOs** emphasize nonmetropolitan areas of the State. An **RTPO** may have additional representatives from the State, private businesses, transportation service providers, economic development practitioners, and the public.

Q. 10. What is a public transportation operator?

Ans. Public transportation operators are public agencies and governmentally chartered authorities that deliver transit services to the general public. As such, public transit operators cooperate with States and **MPOs** to carry out the Federally required transportation planning process in metropolitan areas. **MPOs** and States must include projects from public transit operators in **MTPs** and **TIPs** in order for those projects to receive Federal financial support.

Q. 11. How do agencies cooperate?

Ans. Transportation planning must be cooperative, because no single agency is responsible for the entire transportation system. Some roads that are part of the Interstate Highway System are subject to certain standards and are usually maintained by a State **DOT**. Other roads are county arterials or city streets that, along with bicycle and pedestrian facilities, are designed, operated, and maintained by counties or local municipalities.

Transit systems are often built, operated, and maintained by an entity or special regional authority that is not an agency of the State or local jurisdiction. There may be more than one public transit operator in a metropolitan area serving a network of separate but interdependent mobility needs.

Q. 12. What are the key products of the transportation planning process?

Ans. The key products are of the transportation planning process:

- Planning Work Programs, which include **UPWPs** prepared by **MPOs** and State Planning and Research Work Programs prepared by States.
- Transportation Plans, which include **MTPs** prepared by **MPOs** and **LRSTPs** prepared by states.
- Transportation Improvement programs, which include metropolitan **TIPs** prepared by **MPOs** and Statewide **TIPs** prepared by states.

Q. 13. What is the relationship between transportation and air quality?

Ans. The transportation system can be an influential factor affecting a region's air quality and estimated emissions of pollutants from motor vehicles are often a key consideration in transportation planning. Areas that have non attainment or maintenance air quality status must ensure that emissions from Federal transportation investments conform with levels set fourth in State air quality plans, and State **DOTs** and **MPOs** must understand air quality-related transportation planning requirements.

Non-attainment areas are geographic areas that do not meet the Federal air quality standards. Maintenance areas are that once violated but currently meet the Federal air quality standards. If no violations of air quality standards have been found, the area may be determined to be in compliance or attainment with a Federal air quality standard.

Q. 14. What are the major sources of air pollution?

Ans. The air quality of an area is affected by how pollutants interact with sunlight, topography, and weather patterns. Pollutants are emitted by motor vehicle operation and by a variety of other sources, including manufacturing, energy production, burning petroleum-based products, and even by small business activities, such as dry cleaning.

Stationary sources include relatively large, fixed facilities, such as power plants, chemical process industries, and petroleum refineries.

Q. 15. What is the congestion management process?

Ans. Congestion management is the term for applying strategies that improve transportation system performance and reliability by reducing the adverse impacts of congestion on the movement of people and goods. A congestion management process (CMP) is a systematic approach, defined by region, for managing congestion. The **CMP** provides accurate, up-to-date information on transportation system performance and assesses alternative strategies for congestion management that meet State and local needs. It works within the transportation planning process to move congestion management strategies into the funding and implementation stages.

Q. 16. What are the requirements for the CMP?

Ans. A **CMP** is required in **TMAs**—Transportation Management Areas. The **CMP** is intended to address congestion through a process that provides for effective transportation system management and operations, based on cooperatively developed travel demand reduction and operational management strategies. Even if a metropolitan area is not a **TMA**, the **CMP** represents good practice in monitoring, assessing, and resolving congestion issues. The **CMP** establishes a systematic method to identify and evaluate transportation improvement strategies, including operations and capital projects.

III. REVIEW QUESTIONS

1. What is meant by citizen participation?
2. Explain the difficulties in the transport planning process.
3. Discuss the relevance of transport planning process.
4. Write a brief note on the computer applications in transport planning.

REFERENCES

1. Mitchell R.B. and Rapkin C., Urban Traffic : A function of land use, Columbia University Press, New York, 1954.
2. Traffic in Towns, Buchanan Report, H.M.S.O., London, 1963.
3. A.G. Wilson, Bayliss D., Blackburn A.J. and B.J. Hutchinson, New Directions in Strategic Transportations Planning The Urban Transportation Planning Process, O.E.C.D., Paris, 1971.
4. U.S. Deptt. of Transportation, Goals for Urban Transportation, Washington, 1968.
5. The Urban Transportation Planning Process, O.E.C.D., Paris, 1971.
6. Guidelines for Traffic Prediction on Rural Highways, I.R.C: 108-1996, Indian Roads Congress, New Delhi, 1996.
7. U.S. Department of Commerce, Bureau of Public Roads, Calibrating and testing a gravity model for any size urban area, U.S., Govt. Printing Office, Washington, 1965.
8. U.S. Department of Transportation, Federal Highways, U.S. Government Printing Office, Washington, 1967.
9. U.S. Department of Transportation Federal Highway Administration. Trip distribution, and peripheral programmes, U.S. Government Printing Office, Washington, 1969.
10. U.S. Department of Transportation Federal Highway Administration. Assignment and peripheral programmes, U.S. Government Printing Office, Washington, 1969.
11. U.S. Department of Transportation, Urban Mass Transportation Administration, UMTA Transportation Planning System Reference Manual, U.S. Government Printing Office, Washington, 1972.

30. Transportation Survey

30.1. Introduction

The first stage in the formulation of a transportation plan is to collect data on all factors that are likely to influence travel pattern. The work involves a number of surveys so as to have an inventory of existing travel pattern, an inventory of existing transport facilities and an inventory of existing land use and economic activities. This stage of the planning process entails voluminous work and may take as much as two years for completion. The cost involved is also very high and this underlines the need for proper organization and careful planning. It is for this purpose that standardised procedures and guidelines have been established.

30.2. Definition of the Study Area

The study area for which transportation facilities are being planned is first of all defined. Transportation planning can be at the national level, the regional level or at the urban level. For planning at the urban level, the study area should embrace the whole connurbation containing the existing and potential continuously built-up areas of the city.

The imaginary line representing the boundary of the study area is termed as the 'external cordon'. The area inside the external cordon line determines the travel pattern to a large extent and as such is surveyed in great detail. The land-use pattern and the economic activities are studied intensively and detailed surveys (such as the home interview) are conducted in this area to determine the travel characteristics. On the other hand, the area outside the external cordon line is studied to a lesser degree of detail.

The selection of the external cordon line for an urban transportation study should be done carefully with due weightage to the following factors :

(*i*) The external cordon line should circumscribe all areas which are already built-up and those areas which are considered likely to be developed during the period of study.

(*ii*) The external cordon line should contain all areas of systematic daily life of the people oriented towards the city centre and should in effect be the 'commuter-shed'.

(*iii*) The external cordon line should be continuous and uniform in its course so that movements cross it only once. The line should intersect roads where it is safe and convenient to carry out traffic surveys.

(*iv*) The external cordon line should be compatible with previous studies of the area and of studies planned for the future.

30.3. Zoning

The defined study area is sub-divided into smaller areas called zones.

The purpose of such a sub-division is to facilitate the spatial quantification of land-use and economic factors which influence travel pattern. The data collected on individual household basis cannot be conveniently considered and analysed unless they are aggregated into small zones which reflect the average characteristics of the individual households. Sub-division into zones further helps in geographically associating the origins and destinations of travel. Zones within the study area are called internal zones and those outside the study area are called external zones. In large study projects, it is more convenient to divide the study area into sectors, which are sub-divided into smaller zones. Zones can themselves be sub-divided into sub-zones depending upon land-use.

A convenient system of coding of the zones will be useful for the study. One such system is to divide the study area into 9 sectors. The central sectors (CBD) is designated 0, and the remaining eight are designated 1 to 8 in a clockwise manner. The prefix 9 is reserved for the external zones. Each sector is sub-divided into 10 zones bearing numbers 0 to 9. Thus a system of three digits denotes a sub-zone. A sub-zone bearing the number 481 belongs to Sector 4 and to Zone 8 in that sector and is sub-zone 1 in that zone.

It would be helpful if the following points are kept in view when dividing the area into zones :

(*i*) Land-use is the most important factor in establishing zones for a transportation survey (Ref. 2). It is only when the origin and destination zones reflect properly the land use can traffic generated within the zones be predicted, measured and quantified accurately. The pre-dominant land use for establishing zones are : residential, industrial, commercial, recreational, open space, shopping, etc.

(*ii*) The zones should have a homogeneous land use so as to reflect accurately the associated trip-making behaviour.

(*iii*) Anticipated changes in land use should be considered when sub-dividing the study area into zones.

(*iv*) It would be advantageous if the sub-division follows closely that adopted by other bodies for data collection. This will facilitate correlation of data. Specially useful in this respect is the sub-division adopted for Census operations, since the census is a very useful source of much data needed for transportation planning.

(*v*) The zones should not be too large to cause considerable errors in data. At the same time, they should not be too small either, to cause difficulty in handling and analysing the data. As a general guide, a population of 1000–3000 may be optimum for a small area and a population of 5000–10000 may be optimum for a large area. In residential areas the zones should accommodate roughly 1000 households.

(*vi*) The zones should preferably have regular geometric form for easily determining the centroid which represents the origin and termination of travel.

(*vii*) Sectors should represent the catchment of trips generated on a primary route.

(*viii*) Zones must be compatible with screen lines and cordon lines.

(*ix*) Zone boundaries should preferably be water-sheds of trip-making.

(*x*) Natural or physical barriers such as canals, rivers etc. can form convenient zone boundaries.

In addition to the external cordon lines, there may be a number of internal cordon lines arranged as concentric rings to check the accuracy of survey data. Screen lines running through the study area are also established to check the accuracy of data collected from home interview surveys. Screen lines can be conveniently located along physical or natural barriers having few crossing points. Examples of such barriers are rivers, railway lines, canals etc.

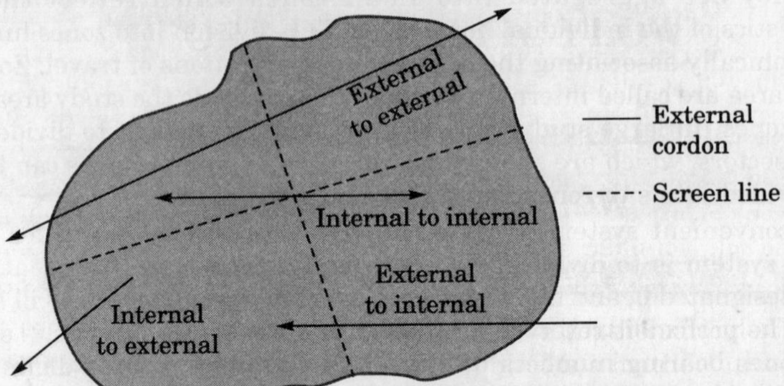

Fig. 30.1. Basic movements in a transportation survey.

30.4. Types of Surveys

30.4.1. The basic movements for which survey data are required are :

(*i*) Internal to Internal (*ii*) Internal to External

(*iii*) External to Internal (*iv*) External to External.

These are symbolically illustrated in Fig. 30.1.

30.4.2. For large urban areas, the internal to internal travel is heavy whereas for small areas having a small population (say less than 5000) the internal to internal travel is insignificant. The internal to internal travel is best studied by the home interview technique with checks by screen-line surveys. The internal-external, external-internal and external-external travels can be studied by cordon surveys. The internal to external travel can also be surveyed by home interview technique.

30.4.3. The surveys can collect the data

(*i*) at the home, (*ii*) during the trip and

(*iii*) at the destination end of the trip.

When collected at the home, the data can be wide-ranging and can cover all the trips made during a given period. The data collected during the trip is necessarily of limited scope since the procedure yields data only on the particular trip intercepted. At the destination ends, the direct interview types of surveys provide data on demand for parking facilities and on major traffic attractors such as factories, offices and commercial establishments.

30.4.4. The following are some of the surveys that are usually carried out :

(*i*) Home interview surveys (*ii*) Commercial vehicle surveys

(*iii*) Taxi surveys (*iv*) Road-side interview surveys

(*v*) Post card questionnaire surveys

(*vi*) Registration number surveys (*vii*) Tag surveys

(*viii*) Public transport surveys.

These are dealt in detail in subsequent sections.

30.5. Home-Interview Surveys

Home-interview survey is one of the most reliable type of surveys for collection of origin and destination data. The survey is essentially intended to yield data on the travel pattern of the residents of the household and the general characteristics of the household influencing trip-making. The information on travel pattern includes number of trips made, their origin and destination, purpose of trip, travel mode, time of departure from origin and time of arrival at destination and so on. The information on household characteristics includes type of dwelling unit, number of residents, age, sex, race, vehicle ownership, number of drivers, family income and so on. Based on these data it is possible to relate the amount of travel to household and zonal characteristics and develop equations for trip generation rates.

Because of the wide variety of data that can be collected by the home interview technique and the high cost involved, it is necessary to standardise the procedure for such surveys. One such well-known procedure is contained in the Bureau of Public Roads Manual (Ref. 3) which precisely describes how the interview is to be conducted and what information is to be collected. Excellent guidance is also available in the US Department of Transportation Publication (Ref. 10).

It is impractical and unnecessary to interview all the residents of the study area. Since travel patterns tend to be uniform in a particular zone, it is sufficient if a sampling procedure is employed. The size of the sample is usually determined on the basis of the population of the study area, and the standards given in Table 30.1 of the Bureau of Public Roads (B.P.R.) are often used (Ref. 3).

Table 30.1. B.P.R. Standards for Sampling Size for Home Interview Survey (*Source* : Ref. 3)

Population of study area	*Sample size*
Under 50,000	1 in 5 households
50,000–150,000	1 in 8 households
150,000–300,000	1 in 10 households
300,000–500,000	1 in 15 households
500,000–1,000,000	1 in 20 households
Over 1,000,000	1 in 25 households

The above standards are perhaps too costly to practice. In any case, the sample size should not be less than given in Table 30.2.

Table 30.2. Minimum Sampling Size for Home Interview Survey

Population of study area	*Minimum sample size*
Under 50,000	1 in 10 households
50,000 to 150,000	1 in 20 households
150,000 to 300,000	1 in 35 households
300,000 to 500,000	1 in 50 households
500,000 to 1,000,000	1 in 70 households
Over 1,000,000	1 in 100 households

In a recent survey in Mumbai (Ref. 9), 1.5 (*i.e.,* 1 in 66) per cent of the households in Greater Mumbai were covered.

The sample is selected in an unbiased way from the register of electors or valuation list.

A number of techniques are available for the home-interview survey. The full-interview technique involves interviewing as many members of the household as possible and directly recording all the information. In the home questionnaire technique, the interviewer collects only details of the household characteristics, leaving forms for household residents to complete in regard to travel information. The completed forms are collected by the interviewer after a day or two.

In both the above methods, it is necessary to send out a letter to the selected households prior to the proposed interview, explaining the nature, importance and objectives of the survey and eliciting their co-operation. In addition wide publicity is given to the survey in the local press, radio and television.

Usually the full-interview technique is more expensive, and it may be possible to collect the needed information only at the rate of eight interviews per eight hour day per interviewer (Ref. 4). The home-questionnaire technique is more speedy and it may be possible to cover about 20 households per day. While the full interview technique yields very accurate data, the same cannot be said of the home-questionnaire method.

Variations of the above techniques sometimes used in the past are the telephone interview and the postal survey. In the former the questionnaire is sent out by post before the survey data and the replies are elicited by telephone. This method can be successful only in areas of high telephone ownership. In the postal survey method, the questionnaire is mailed and the households are requested to send their replies by post in replypaid envelopes. It may often be found that response to the postal survey technique will be a low level inspite of wide prior publicity.

The information to be collected from the home-interview survey can broadly be classified under two groups : household information and journey data. The household information contains information such as address, size of household, age and sex structure of household, earning members, occupation, place of work, number of motor vehicles owned, household income, and so on. The journey data contains information on all journeys made during the previous 24 hour period, including the origin and destination of journeys, purpose of trip, mode of travel etc. The survey forms are generally standardised for this purpose and the questions are structured carefully to avoid ambiguity. The usual practice is to have the household information in the front of the form and the trip information on the back of the form. The form should designed so that the data can be easily coded. A typical design of the form is given in Tables 30.3 and 30.4 (Ref. 5).

30.6. Commercial Vehicle Surveys

Commercial vehicle surveys are conducted to obtain information on journeys made by all commercial vehicles based within the study area. The addresses of the vehicle operators are obtained and they are contacted. Forms are issued to drivers with a request that they record particulars of all the trips they would make. A typical questionnaire which elicits information on the origin and destination of each trip is given in Table 30.5.

30.7. Taxi Surveys

Large urban areas usually have a sizeable amount of travel by taxis. In such cases, a separate taxi survey is necessary. The survey consists of issuing questionnaires or log sheets to the taxi drivers and requesting them to complete the same.

30.8. Road Side Interview Survey

Road side interview survey is one of the methods of carrying out a screen-line or cordon survey. The road side interview survey can be done either by directly interviewing drivers of the vehicles at selected survey points or by issuing prepaid post cards containing the questionnaire to all or a sample of the drivers.

The survey points are selected along the junction of the cordon-line or screen-line with the roads. The cordons may be in the form of circular rings, radial lines of rectangular grids. For small towns, say with a population less than 5000, single circular cordon at the periphery of the town should suffice. The internal travel being light, the external cordon survey in that case will give the origin-destination data. In the case of medium sized cities, say with a population in the

Table 30.5. Typical Form for Commercial Vehicles Survey

COMMERCIAL VEHICLE SURVEY

Firm/Owner	Visit	Time	Date	Interviewer's Name	Sheet of
Address	1 2 3 4 5	Comments :			Ser. No. Zone No.
Person Interviewed					
Telephone No.					

Part I. Business Information

A. Nature of Business Male/female

B. Total Number of employees

C. Number of vehicles operating from address

Cars	Parked on premises	Parked elsewhere
Light Goods	Parked on premises	Parked elsewhere
Heavy Goods	Parked on premises	Parked elsewhere

D. Day of travel

E. Total number of trips made on day of travel

F. Type and number of vehicles interviewed

Vehicle No.

Vehicle Reg. No.

Part II. Vehicle Trip Information

G	*H*	*I*	*J*	*K*	*L*	
					Time of Trip	
Vehicle Type	Vehicle Number	Trip Number	Origin Address	Destination Address	Start AM PM	Finish AM PM
		Nature of business :		Nature of business:		
					AM PM	AM PM
		Nature of business :		Nature of business:		

range 5,000 to 75,000 two cordon lines are necessary, the external cordon at the edge of the urban development and the internal cordon at the limits of the central business district. Road side interviews at the intersection of roads with these two cordon lines should be able to fairly assess the patterns of travel in such cities. For large cities, the cordonlines and screen lines may be more complicated, and the home-interview technique cannot be dispensed with. Cordon line and screen line surveys by the road side interview technique serve to check the accuracy of the home interview survey data.

For dual carriageways or roads with very little traffic the traffic in both the directions is dealt with simultaneously. In other cases the traffic in two directions will be interviewed at different times. If the survey covers most of the

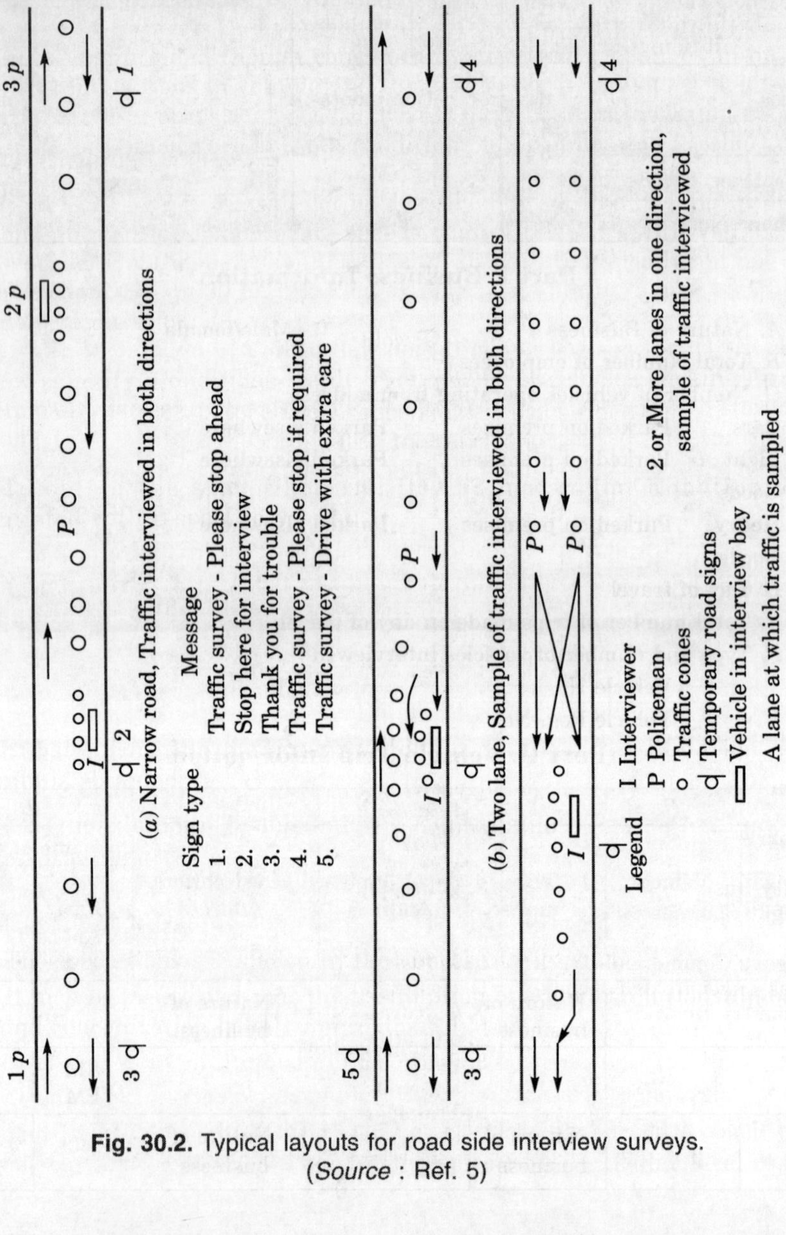

Fig. 30.2. Typical layouts for road side interview surveys.
(*Source* : Ref. 5)

day it may be sufficient to interview traffic in one direction only and to assume that the journeys in the opposite direction are the same as in the direction interviewed.

It is impractical to stop and interview all the vehicles. Sampling is, therefore, necessary. The number of samples depends on the number of interviewers and the traffic using the road. It may become necessary to vary the

Table 30.6. Sample Roadside Interview Form (*Source* : Ref. 6)

Day Date		Wet Dry	Separate Yes ... Count ? No ...	Recorded by		Sheet No.
Site				Private	Sample	Total
Direction				Bus		
Half-hour starting at				Goods		
Veh. Type	No. of occupants	Last stop	Next stop	Last land use	Next land use	Trip purpose

Vehicle Type	*Land use*	*Trip purpose*
0. Motor cycle	0. Residential	0. To work
1. Car	1. Hotel, guest house, Restaurant	1. Work to home
2. Taxi	2. School, College, University	2. Other to home
3. City Bus	3. Offices	3. Shopping
4. Light Commercial	4. Shops, markets	4. Business (own or employed)
5. Heavy Commercial (truck)	5. Industrial, docks, utilities	5. Personal affairs
6. Passenger Vehicle (truck)	6. Hospitals, doctors	6. School
7. Country Bus	7. Cultural recreational and other	7. Eat meal
	8. Police, military, fire stations	8. School/recreational
	9. Transport terminals, bus stops	9. Serve passenger

sampling rate at the traffic flow changes during the different parts of the day. Sampling methods should eliminate any element of bias. A convenient method is to sample one in a fixed number of vehicles, *viz.* every tenth, fifteenth or twentieth vehicle etc. Another simple method is to select the next vehicle as soon as each interview is completed.

Since interviews may last for several minutes, vehicles must be stopped in a interview bay so that traffic flow is not obstructed. Suitable advance warning signs should be erected. Some typical layouts for road side interview surveys are illustrated in Fig. 30.2.

The interviewers have no statutory powers to stop the vehicles and question the drivers. This makes it necessary to seek the help of the police to control and direct the traffic for being interviewed.

In addition to at least one interviewer at each survey station, an observer is posted to record the classified traffic flow concurrently with the period of interviewing.

The period and duration of the survey are important matters that need careful prior thought. A 24 hours count will not normally be needed, and the survey is often restricted to 16 hour (0600–2200 hours) or 12 hours (0700–1900 hours) in a day. For the remainder of the day, vehicular counts are, however, made. In order to eliminate bias due to unusual conditions on any particular day, it is the practice to obtain data for each week day (Monday-Friday).

For complete and reliable data to be obtained by the survey, it is necessary to frame and design the questions with care. The enumerators must also be given adequate training and instructions to avoid ambiguity in the answers and to ensure uniform pattern of data from different enumerators. Pre-printed forms are used to record the answers. A sample form used for the purpose is given in Table 30.6.

It will be seen from the sample form in Table 30.6 that coding of data will be an advantage.

The analysis of the data by computers will be easy. Since the interviews are done by sampling basis, expansion factors are needed to calculate the total number of trips. These expansion factors will be calculated separately for each class of vehicle and for different time periods (half-hour etc.). Thus if x_c represents the number of cars interviewed in a particular time period, and if X_c is the total number of cars counted during this time period, then the expansion factor is X_c / x_c.

Road side interview is an economic method of survey and yields accurate and reliable data. The disadvantage with the method is that vehicles are delayed when being interviewed.

30.9. Post-card Questionnaire

In this method, reply-paid questionnaires are handed over to each of the drivers or a sample of them at the survey points and requesting them to complete the information and return by post. The method avoids delay caused to the drivers by the direct roadside interview method but suffer from the disadvantage that the response may not be good. For this reason its use is not generally recommended for developing countries (Ref. 5). It is possible to get a good amount of information from this method. The method is simpler and cheaper than many others.

As regards the selection of survey points and questions asked, the method is similar to the direct roadside interview technique. A good amount of publicity is needed before the actual survey in order to get favourable response. It is reported that well planned and publicised post card questionnaire surveys have yielded returns of 50% or more. (Ref. 6).

30.10. Registration Number Plate Surveys

Registration number plate survey consists of noting the registration numbers of vehicles entering or leaving an area at survey points located on the cordon line. By matching the registration numbers of vehicles at the points of entry and exit from the area, one is enabled to identify two points on the paths of the vehicles. The method neither gives the origin or the destination of the trip nor yields any other useful information such as trip purpose. The advantage of the method is that the work does not interfere with the traffic in any way. However, a large number of observers are needed and the analysis of the results can be complicated.

First of all, the area to be surveyed is defined, and the roads intersecting the cordon line are identified. At each survey point, one or two observers are stationed to record the data in each direction of travel. If two observers are available at each point, one can call out the registration number of the vehicle and the other can record. Time should be recorded at regular intervals. If the actual times at entry and exit are noted, an estimate of the journey speed of the vehicles can also be had. The type of vehicles (car, commercial vehicle, bus etc.) as well as the full registration number (letters and figures) are noted. If some sampling is necessary because of heavy traffic, only numbers ending with pre-selected digits are recorded.

A sample form of the survey, using a 5 minute interval for recordings, is given in Table 30.7.

The analysis consists of tallying the numbers of vehicles at points of entry and exit. The procedure is cumbersome if done manually and hence the use of electronic computer is often sought. Manual method is organised by allotting data from each survey point to different persons who sit around a table. The registration numbers of vehicles are called in turn and the persons having the survey data at exit points match them with their entries. When a number is thus matched, the passage of that vehicle in between the two points now known is entered on a suitable form. If any number is not matched, the reason may be that the particular vehicle entered the survey area, but did not move out during the survey period,

Table 30.7. Sample Registration Number Plate Survey Form

	Sheet No.
Site : Okhla	Weather : Overcast
Observation point : Mathura Road	Observer : BNG
Movement Observed : Delhi-bound	Recorder : OPB
Sheet starting time : 8.00 A.M.	Date : 24-1-76

Sample :

Registration Numbers ending with : all even numbers, including zero. Vehicle class : All

Column Start Time			
8-00	8-05	8-10	8-15
DLQ 2660	DLX 742	JKN 528	UPS 584
DLM 2832	DLN 3856		PUN 4974
DLP 8486	HRP 9486		DLQ 948

or the number itself was recorded wrongly in the field. When the movements of all the sampled vehicles have been thus determined the true traffic movement is determined by applying the appropriate expansion factor. The expansion factor will be known by the method of sampling adopted. Thus if only numbers ending with an even number have been recorded, the expansion factor will be 2.

30.11. Tags on Vehicles

In this method, at each point where the roads cross the cordon line, vehicles are stopped and a tag is affixed, usually under a windscreen wiper. The tags for different survey stations have different shapes and/or colour to identify the survey station. The vehicles are stopped again at the exit points where the tags are removed. The times of entering and leaving the area may be marked on the tags in order to enable the journey times to be determined. If it is not possible to handle all the vehicles, sampling may be resorted to. Regular sampling (such as a every second, third, fifth ... vehicle) will make it easier to determine the expansion factor. Otherwise, an additional observer may be stationed at each survey point to take a classified count of the traffic. The survey can be conveniently divided into periods of say 15 or 30 minutes intervals. The analysis is simple and errors are not very large. The method can also be extended to cover vehicles that enter the area but are parked in the area during the rest of the study period. For this purpose, the parking areas are surveyed at the end of the study period and the tags on parked vehicles collected.

30.12. Public Transport Surveys

In order to assess the number of bus passengers passing through an external cordon, the survey can either be by direct interview with the passengers or by issuing post-card questionnaires. Direct interview is likely to result in large delays and requires a large number of interviews. In order to minimize the delays, the interviewer may enter the vehicle and carry out the interviews when the vehicle is in motion. Post-card questionnaires eliminate delays, but are likely to evoke poor response or contain an element of bias.

An external cordon rail survey can be carried out by interviewing the passengers on trains. Alternatively, pre-paid questionnaires may be distributed to persons residing at stations outside the survey area. These questionnaires may also be collected at the stations inside the survey area.

30.13. Inventory of Transport Facilities

The inventory of existing transport facilities should be undertaken to identify the deficiencies in the present system and the extent to which they need to be improved. The inventory consists of :

(*i*) Inventory of streets forming the transport network
(*ii*) Traffic volume, composition, peak and off-peak
(*iii*) Studies on travel time by different modes
(*iv*) Inventory of public transport buses
(*v*) Inventory of rail transport facilities
(*vi*) Parking inventory
(*vii*) Accident data.

30.13.1. Inventory of Streets

An understanding of the extent and quality of the road network is very important to formulate plans for the future. The inventory should cover details such as classification of the street system, length, cross-sectional dimensions, type and condition of the surface, capacity, intersections, control devices, structures, street furniture, etc.

30.13.2. Traffic Volume

Data pertaining to traffic volume and its composition will be needed to check on the survey data collected by the home-interview and cordon surveys. The variation of the traffic volume over different hours of the day, different days of the week and different months of the year is also needed. The techniques available for traffic volume counts have been discussed in detail under Chapter 4.

30.13.3. Travel Time Studies

An estimate of travel times between different zones by various modes is necessary for transport planning. Travel times are usually measured for the peak-hour conditions and non-peak hour conditions. The methods available for determining travel times are discussed in Chapter 3.

30.13.4. Inventory of Public Transport Buses

The inventory of public transport buses includes information on the total number of buses, their capacity, schedules, routes, operating speeds, terminals, number of passengers carried, economic picture of the public transport system and the fare structure.

30.13.5. Inventory of Rail Transport Facilities

The inventory of rail transport facilities should include the length, capacity, schedules, operating speeds, stations, number of passengers carried, economic picture of the rail transport undertaking and the fare structure.

30.13.6. Parking Inventory

The parking inventory should collect information on the existing on-street and off-street parking facilities, the parking demand and the utilization of existing facilities. Data on parking charges and the system used for charging should also be collected. Chapter 6 gives details of parking survey.

30.13.7. Accident Data

Accident data over the past years will help to understand the nature and extent of the hazards inherent in the present system and the need to improve the situation. The method of accident data collection is fully dealt with in Chapter 18.

30.14. Inventory of Land Use and Economic Activities

30.14.1. Inventory of Land Use

Since travel characteristics are closely related to the land-use pattern, it is of utmost importance that an accurate inventory of land-use be prepared. The zones are classified into land use activity such as residential, industrial, commercial, recreational, open space and so on. For this purpose, the Town and Country Planning authorities need to be consulted. If any master plan has been prepared for the survey area, the same should be given due consideration. Aerial photography has also been used as a source of land-use data (Ref. 7).

30.14.2. Inventory of Economic Activities

Data on economic activities should be collected to include the following :

1. Population of the survey area and the various zones
2. Age, sex and composition of the family
3. Employment statistics
4. Housing statistics
5. Income
6. Vehicle ownership.

Some of the data pertaining to economic activities will be already available from periodic census. A careful analysis of the census data will indicate the gaps that need to be filled in by home-interview surveys or some other means. The limitations of the census data should be clearly recognised before they are put to use. (Ref. 9).

The population data helps in the estimation of the future trip-making behaviour. Population maps indicating the density, school enrolment, institutional population and sociological factors will facilitate presentation of results and a better understanding of the travel pattern.

30.15. Expansion of Data from Samples

In order to derive the travel characteristics of the whole population from the data derived from sampling, certain expansion factors have to be used.

For the home-interview surveys, the expansion factor is calculated on a zonal basis as follows :

Expansion factor

$$= \frac{A - \dfrac{A}{B}\left(C + \dfrac{C}{B} \times D\right)}{B - C - D}$$

where, A = Total number of addresses in the original list.

B = Total number of addresses selected as original sample.

C = Number of sample addresses that are ineligible.

D = Number of sample addresses where no response is obtained.

Similar expansion factors can be used for the commercial vehicle surveys, but on an area-wise basis.

I. MULTIPLE CHOICE QUESTIONS

1. Traffic surveys aim to
   (*a*) capture data that accurately reflects the real-world traffic situation in the area
   (*b*) capture data of the real-world traffic situation in the area
   (*c*) achieve data of the real-world traffic situation in the area
   (*d*) none of these
2. A rubber tube that is placed across the lanes that uses pressure is
   (*a*) pneumatic road tube (*b*) piezo-electric sensor
   (*c*) inductive loop (*d*) none of these
3. A device that is placed in a groove cut into the roabed of the lane(s) being counted
   (*a*) piezo-electric sensor (*b*) pneumatic road tube
   (*c*) inductive loop (*d*) none of these
4. A wire embedded in the road in a square formation that creates a magnetic field that relays the information to a counting device at the side of the road.
   (*a*) inductive loop (*b*) piezo-electric sensor
   (*c*) pneumatic road tube (*d*) none of these
5. A sensor detecting the presence, speed and type of vehicles by measuring infra-red energy radiating from the detection area.
   (*a*) manual observation (*b*) microwave-Doppler/Radar
   (*c*) passive and active infra-red (*d*) none of these
6. A devices that sound waves or sound energy to detect vehicles. Those using ultrasound are placed overhead to record vehicle presence but can be affected by temperature and turbulence
   (*a*) manual observation (*b*) microwave-moppler/radder
   (*c*) ultrasonic and passive acoustic (*d*) none of these
7. The efficiency of the brakes of a vehicle depends upon
   (*a*) condition of road surface (*b*) condition of the tyres
   (*c*) presence of the show moisture (*d*) all the above
8. The type of transition curves generally provided on hill roads, is
   (*a*) circular (*b*) cubic parabola
   (*c*) lemniscate (*d*) spiral
9. Increase in traffic volume, due to increase in transport vehicles, is known as
   (*a*) development traffic (*b*) normal traffic growth
   (*c*) generated traffic growth (*d*) current traffic
   (*e*) none of these
10. A district road with a bituminous pavement has a horizontal curve of 1000 m for a design speed of 75 kmph. The superelevation is
    (*a*) 1 in 40 (*b*) 1 in 70 (*c*) 1 in 50 (*d*) 1 in 60
11. Traffic engineering only includes
    (*a*) design and application of control devices
    (*b*) traffic operation

(*c*) analysis of traffic characteristics
(*d*) all

12. First operation during the detailed survey of a hill road, is
(*a*) fixation of bench marks
(*b*) hydrological and soil surveys
(*c*) adjustment of alignment along with curves
(*d*) derivation of longitudinal and cross-sections

13. Width of vehicles affects the width of
(*a*) parking spaces (*b*) all
(*c*) shoulders (*d*) lanes

14. The usual width of parapet walls along Highways in hilly region, is
(*a*) 75 cm (*b*) 70 cm (*c*) 50 cm (*d*) 60 cm

15. Traffic surveys are carried out
(*a*) to determine the facilities to traffic regulations
(*b*) all
(*c*) to design proper drainage system
(*d*) to know the type of traffic

16. Road makers along roads from the edge of a curve should not be less than
(*a*) 40 cm (*b*) 50 cm (*c*) 55 cm (*d*) 60 cm

17. Design of horizontal and vertical alignments, superelevation, sight distance and grades, is worst affected by
(*a*) speed of the vehicle (*b*) length of the vehicle
(*c*) width of the vehicle (*d*) height of the vehicle

18. For the movement of vehicles at an intersection of two roads, without any interference, the type of grade separator generally preferred to, is
(*a*) clover leaf. (*b*) delta
(*c*) trumpet (*d*) diamond interchange

Answers

1. (*a*)	**2.** (*a*)	**3.** (*a*)	**4.** (*a*)	**5.** (*c*)
6. (*c*)	**7.** (*d*)	**8.** (*d*)	**9.** (*b*)	**10.** (*a*)
11. (*d*)	**12.** (*a*)	**13.** (*b*)	**14.** (*d*)	**15.** (*b*)
16. (*d*)	**17.** (*a*)	**18.** (*a*)		

II. SOLVED QUESTIONS AND ANSWERS

Q. 1. What is meant by zoning?

Ans. Zoning describes the control by authority which designates legal areas in a municipality to permit and prohibit land uses. Zoning may specify a variety of outright and conditional uses of land. It may also indicate the size and dimensions of land area as well as the form and scale of buildings. These guidelines are set in order to guide urban growth and development.

Q. 2. What is the role of surveys?

Ans. All persons involved in transport and land-use planning will at some stage be involved with data collection. Even if not directly concerned with the design and conduct of surveys, they will certainly wish to use data at some time, and at that stage they will realise what should have been done in the design and administration phases of the survey. Each individual's introduction to survey data may have widely differing emotional connotations.

Q. 3. What is the process of transport survey?

Ans. The conduct of a survey is not an informal procedure. Rather, it should follow a series of logical, interconnected steps which progress toward the final end product of the survey.

III. REVIEW QUESTIONS

1. Discuss the home interview surveys.
2. What is road side interview surveys?
3. What is post card questionnaire?
4. Discuss the registration number of plate surveys.
5. What are inventory of transport facilities?

REFERENCES

1. Srinivasan N.S., I.C. Agrawal, B.L. Suri and Anand Parkash, Intracity Origin–Destination Travel Survey of Delhi, Journal of the Indian Roads Congress, Vol. 36–2 New Delhi, 1975.
2. Kerr, Determination of O-D Zones by means of land-use data, H.R.B. Bulletin 347, Highway Research Board, Washington 1962.
3. Bureau of Public Roads, Procedure Manual : Conducting A Home Interview Origin-Destination Survey, Vol. 2 B. Washington, 1956.
4. Taylor, M.A., Travel Survey by Home Qestionnaire, Traffic Engineering and Control, London, July, 1965.
5. ECAFE, Manual on Traffic Surveys, United Nations, New York, 1971.
6. "Research on Road Traffic, Road Research Laboratory (U.K.). H.M.S.O., London, 1965.
7. Hadfield, S.M., An evaluation of land-use and dwelling unit data derived from aerial photography, H.R.B. Bulletin 357, Highway Research Board, Washington, 1962.
8. Fisher, R.J., and A.B. Sosslau, Census data source for urban transportation planning. Highway Research Record No. 141, Highway Research Board, Washington, 1966.
9. Planning road system for Bombay Metropolitan Region, Central Road Research Institute, New Delhi, 1983.
10. Urban Origin-Destination Studies, U.S. Department of Transportation, Federal Highway Administration, Washington D.C., 1973.

31. Trip Generation

31.1. Introduction and Definitions

The first phase of the transportation planning process deals with surveys, data collection and inventory. The next phase is the analysis of the data so collected and building models to describe the mathematical relationship that can be discerned in the trip-making behaviour. The analysis and model building phase starts with the step commonly known as Trip Generation.

Trip geneation is a general term used in the Transportation Planning Process to cover the field of calculating the number of trip ends in a given area. The objective of the trip generation stage is to understand the reasons behind the trip making behaviour and to produce mathematical relationships to synthesise the trip-making pattern on the basis of observed trips, land-use data and household characteristics.

Since a considerable confusion can occur in the meaning of the various terms used in trip generation, it is desirable to understand the extact meaning of the various terms. A *trip* is a one-way person movement by a mechanised mode of transport, having two *trip ends,* an *origin* (the start of the trip) and a *destination* (the end of the trip). Trips are usually divided into *home-based* and *non-home-based. Home-based trips* or those having one end of the trip (either origin or destination) at the home of the persons making the trip, while *non-home based trips* are those having neither end at the home of the person making the trip. The trip ends are classified into *generations* and *attractions. Ageneration* is the home end of any trip that has one end at the home (*i.e.,* of a home-based trip), and is the origin of a trip with neither end home based (*i.e.,* of a non-home-based trip). An *attraction* is the non-home end of a home-based tirip, and is the *destination* of a trip with neither end home-based (*i.e.,* of a non-home-based trip).

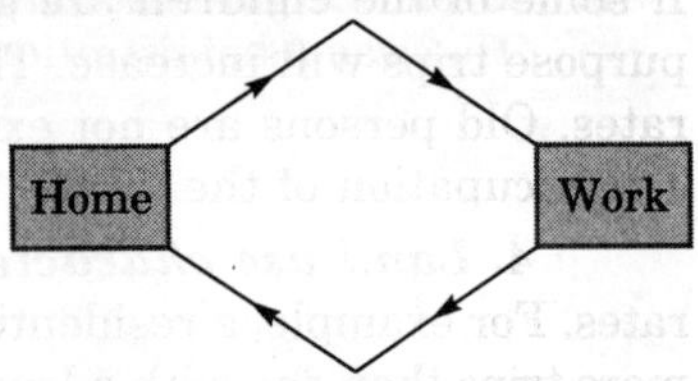

Fig. 31.1. Home-based trips.

The above definitions are further classified by the following examples. Consider a trip from home to work and the return trip from work to home, vide Fig. 31.1. Both these trips are home-based, because one end of the trips is the home. Both these trips are considered to have been generated at the home zone and attracted to the work zone. We thus have two work-purpose trip end generations in the home zone and two work-purpose attractions in the work zone. Consider another example of the trip from the place of work to shop and return to the place of work, as is usual during the lunch recess, vide Fig. 31.2. Both these trips are non-home-based, because neither end of the trip is the home of the person making the trip. Both these trips are considered to have been generated at the work zone and attracted to the shop zone. We thus have two shopping-purpose trip end generations in the work zone and two shopping-purpose attractions in the shopping zone.

By the above definitions, the total number of trip generations in any area should be equal to the total number of attractions.

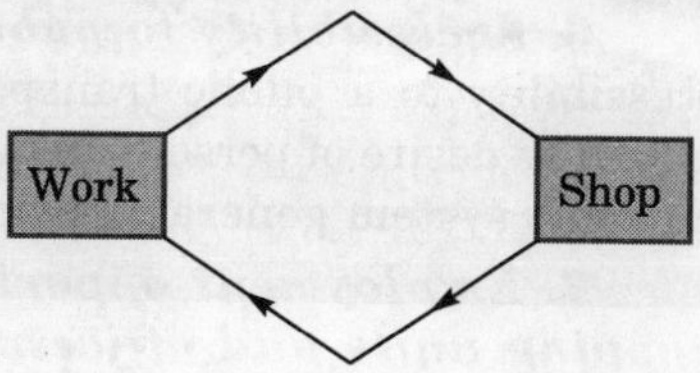

Fig. 31.2. Non-home-based trips.

31.2. Trip Purpose

Trips are made for different purposes and a classification of trips by purpose is necessary. The following are some of the important classes of trip purpose :

- Work, school, susiness
- Social or recreational, sports
- Others

The break-up of trips by purpose is normally done for the home-based trips which represent nearly 80–90% of the total trips.

31.3. Factors Governing Trip Generation and Attraction Rates

A number of factors govern the trip generation rates. These are discussed below :

1. ***Income.*** Obviously, family income which represents its ability to pay for a journey affects the number of trips generated by a household. A general trend is that the higher the income the higher is the trip generation rate.

2. ***Car ownership.*** A car represents easy mobility, and hence a car owning household will generate more trips than a non-car-owning household. By the same reasoning, the more cars there are in a household, the more the number of trips generated. Of course, number of cars owned is itself related to the income of the family, which has been listed earlier as a factor.

3. ***Family size and composition.*** The bigger the family, the more trips there are likely to be generated. Apart from the size, the composition of the family itself is important. For instance, if both the husband and wife are employed, the trips generated will be more than when only the husband is employed. If there are many school-going children, the number of schoolpurpose trips will be large. If some of the children are grown up and are employed, the number of work-purpose trips will increase. The age structure of the family also governs the trip rates. Old persons are not expected to generate as many trips as younger ones. The occupation of the family is also known to influence the travel pattern.

4. ***Land use characteristics.*** Different land uses produce different trip rates. For example, a residential area with a high density of dwellings can produce more trips than one with a low density of dwellings. On the other hand, low density areas may represent dwellings of the affluent society, which may produce a large number of private car trips. The rateable value of the dwelling and the type of the dwelling units affect the trip generation rates. The most important assumption made in transportation planning is that the amount of travel is dependent on land-use.

5. ***Distance of the zone from the town centre.*** The distance of the zone from the town centre is an important determinant of the amount of travel that people might like to make to the town centre. The farther the town centre, the less the number of trips are likely to be.

6. *Accessibility to public transport system and its efficiency.* The accessibility to a public transport system and its efficiency determine to some extent the desire of persons to make trips. An easily accessible and efficient public transport system generates more trips.

7. *Employment opportunities, floor space in the industrial and shopping units and offices, sales figures in shops etc.* The employment potentiality of an industrial or shopping unit or an office establishment directly governs the trip attraction rate (Ref. 1). Similarly, another factor to which the trip attraction rate can be related is the floor space in the premises of industries, shops and offices.

31.4. Multiple Linear Regression Analysis

31.4.1. Multiple linear regression analysis is a well-known statistical technique for fitting mathematical relationships between dependent and independent variables. This technique has been exploited fruitfully in a number of transportation planning studies carried out so far and has become a very powerful tool in the hands of the transportation planner. In the case of trip generation equations, the dependent variable is the number of trips and the independent variables are the various measurable factors that influence trip generation. These independent variables are the land-use and socioeconomic characteristics discussed earlier. The general form of the equation obtained is :

$$Y_p = a_1 X_1 + a_2 X_2 + a_3 X_3, \ldots, a_n X_n + U \qquad \ldots(31.1)$$

where, Y_p = number of trips for specified purpose p

$X_1, X_2, X_3, \ldots\ldots, X_n$ = independent variables relating to, for example, land-use socio-economic factors etc.

$a_1, a_2, a_3, \ldots, a_n$ = coefficients of the respective independent variable $X_1, X_2, X_3, \ldots\ldots X_n$, obtained by linear regression analysis

U = Disturbance term, which is a constant, and representing that portion of the value of Y_p not explained by the independent variables.

The equation of the above form is developed from the present day data pertaining to independent variables and the dependent variable, using the statistical technique of "least squares" fitting. The equation thus developed is used for determining the future values of trips, knowing the estimated future values of the independent variables.

As an example of the multiple linear regression analysis equation, the following equation developed from a study of Toronto (Ref. 2) is given :

Total trips (Generated at home on an average week day)

= (0.318)) × (population 5 years and older)

+ (0.458) × (number of households)

+ (0.890) × (number of cars owned)

Another example below is from a study of Modesto (Ref. 2) :

$$Y = 2.18 + 3.404\,A + 0.516\,H + 0.0119\,X_1 - 0.343\,X_2$$

where, Y = Average trips per occupied dwelling unit

A = Car ownership

H = Household size

X_1 = Social rank index

X_2 = Urbanization index.

31.4.2. Assumptions in Multiple Linear Regression Analysis and their validity in trip generation studies

The statistical theory of regression analysis is based on the following important assumptions :

(*i*) All the variables are independent of each other ;

(*ii*) All the variables are normally distributed ;

(*iii*) All the variables are continuous ;

(*iv*) A linear relationship exists between the dependent variable and the independent variable ;

(*v*) Influence of independent variable is additive, that is the inclusion of each variable in the equation contributes a distinct portion of the trip numbers.

It is difficult to ensure that the above basic assumptions are satisfied in most of the trip generation studies. Firstly, the so called independent variables in the regression equations are not truely independent of each other, and some sort of correlation normally exists among them. The variables such as car-ownership, family income, residential density etc. are all interrelated to a certain extent. Secondly, many of these variables are, strictly speaking, not normally distributed. Finally, some of them are not continuous variables, an example being the car ownership. The number of cars owned by a family can only be a discrete variable.

The above imperfections notwithstanding, the technique has gained popularity, one of the reasons for this being its easy adaptability to computer programming when handling voluminous data.

31.4.3. Aggregated and disaggregated analysis

Multiple linear regression analysis is of two types :

(*i*) Aggregated, or Zonal least-square regression, where each traffic zone is treated as one observation.

(*ii*) Disaggregated, or Household least-square regression, where each household is treated as an observation.

The aggregated analysis which is most widely used, is based on the assumption that contiguous households exhibit a certain amount of similarity in travel characteristics. This assumption allows the data in a zone to be grouped and the mean value of the independent variable used in further calculations.

Some of the disadvantages associated with the aggregated models are (Ref. 4, 5) :

1. The analysis masks the variations in the data.
2. The data is inefficiently utilised.
3. The zonal sample mean is not necessarily a reliable estimate of the population mean.
4. The model is dependent on the type of zoning system adopted.
5. The method is based on an important assumption that zones are to a large extent homogeneous with respect to travel and socio-economic characteristics. While care is taken normally to select the zone boundaries to fulfil the above assumption, the variations within the zone can sometimes be large.
6. In the process of making the zones as small as possible to make them truely homogeneous, the planner increases the complexity of analysis, notably trip distribution and assignment.

Disaggregated analysis, though not so widely used, treats each household as an observation. In this process, all the enormous amount of data is used more effectively resulting in a more meaningful description of the characteristics. As compared to aggregated analysis, disaggregated analysis produces better results (Ref. 4) and is considered more likely to be stable over time and to provide more reliable future estimates (Ref. 6).

31.4.4. Criteria for Evaluation of Regression Equations

The following criteria should generally be applied in evaluating and selecting a regression eqation (Ref. 7, 8) :

(*i*) The multiple correlation coefficient should have a value at least 0.75 or even higher. A value close to 1.0 shows a very good correlation.

(*ii*) The standard error of the estimate of the dependent variable should be sufficiently small.

(*iii*) The *F* test should be carried out to examine evidence of the degree of certainty that a meaningful relationship exists between the dependent and independent variables.

(*iv*) The equation should have accuracy, validity, simplicity, sharpness and constancy (Ref. 9).

31.4.5. Disadvantages of the Multiple-Linear Regression Analysis Technique

Some of the disadvantages associated with the assumptions made in the linear regression technique have already been discussed earlier. In addition, the following points deserve mention :

(*i*) The equation derived is purely empirical in nature and fails to establish a meaningful relationship between the dependent and independent variables.

(*ii*) The technique is based on the premise that the regression coefficients initially established will still remain unchanged in the future and can be used in the regression equation for predicting future travel. How far the prediction is valid in future is a moot question.

(*iii*) Difficulties arise in evaluating the effect of statistical problems relating to non-linearity of the response surface and high correlations amongst the explanatory variables.

31.5. Category Analysis

31.5.1. Category Analysis or cross-classification technique is a method developed by Wotton and Pick (Ref. 10) and has been used in some transportation studies in U.K. It is based on determining the average response or average value of the dependent variable for certain defined categories of the independent variables. A multidimensional matrix defines the categories, each dimension in the matrix representing one independent variable. The independent variables themselves are classified into a definite number of discrete class intervals.

31.5.2. Assumptions

The technique is based on the following assumptions :

(*i*) The household is the fundamental unit in the trip generation process, and most journeys begin or end in response to the requirements of the family.

(*ii*) The trips generated by the household depend upon the characteristics of that household and its location relative to its required facilities such as shops, school and work place.

(*iii*) Households with one set of characteristics generate different rates of trips from households with other set of characteristics.

(*iv*) Only three factors are of prime importance in affecting the amount of travel a household produces : car-ownership, income and household structure.

(*v*) Within each of the above three factors, a limited number of ranges can be established so as to describe the tripgenerating capacity of a household by a limited number of categories.

(*vi*) Trip generation rates are stable over a time so long as factors external to the household are the same as when the trips were first measured.

31.5.3. Categorisation of Households

As stated above, households are classified on the basis of three factors, *viz*., car ownerships, income and household structure. These are then classified into different ranges as indicated below :

(1) *Car ownership*–3 levels, (0 car, 1 car and more than 1 car)

(2) *Disposable Income*–6 classes :

(*i*) < £ 500 p.a. (*ii*) £ 500–1000 p.a.

(*iii*) £ 1000–1500 p.a. (*iv*) £ 1500–2000 p.a.

(*v*) £ 2000–2500 p.a. (*vi*) > £ 2500 p.a.

(3) *Household Structure–6 classes* :

(*i*) No employed residents and one non-employed adult.

(*ii*) No employed residents and 2 or more non-employed adults.

(*iii*) One employed resident and one or less non-employed adult.

(*iv*) One employed resident and 2 or more non-employed adults.

(*v*) Two or more employed residents and one or less non-employed adult.

(*vi*) Two or more employed residents and two or more non-employed adults.

The above system gives in all 3 × 6 × 6 = 108 categories.

In addition, it is also possible to consider 3 modes of travel *viz*., self-driven car, public transport, passenger in a car and 6 trip purposes (work, school, business, shopping, social-recreational and sport, others). Thus we have 18 mode purpose combinations.

31.5.4. Critical appraisal of the category analysis technique

The advantages that have been claimed (Ref. 11) for the technique are :

(*i*) The whole concept of household trip-making is simplified in this technique. The technique categorises the household according to certain socio-economic characteristics and this appears rational.

(*ii*) Unlike regression analysis technique, no mathematical relationship is derived between trip-making and household characteristics. This takes away many of the statistical drawbacks of the regression analysis.

(*iii*) Since data from the census can be used directly, it saves considerable effort, time and money spent on home-interview survey.

(*iv*) The computations are relatively simpler.

(*v*) Since disaggregate data are used, the technique simulates human behaviour more realistically than the zonal aggregation process normally employed in regression analysis.

The following are some of the disadvantages of the technique (Ref. 11) :

(*i*) It is difficult to test the statistical significance of the various explanatory variables.

(*ii*) The technique normally makes use of studies in the past made elsewhere, with broad corrections.

(*iii*) In the analysis it is assumed that income and car ownership increase in future. The categories of higher incomes and higher car ownership are, however, the ones which are the least represented in the base year. Moreover, they are the ones most likely to be used for future estimates of trip generations.

(*iv*) New variables cannot be introduced at a future date.

(*v*) Large samples are needed to assign trip rates to any one category.

I. MULTIPLE CHOICE QUESTIONS

1. Which one is not a public transport mode

(*a*) MRTS (*b*) BRTS (*c*) LRTS (*d*) CAR

2. ITS technology used in public transport

(*a*) Public transport travel information

(*b*) Electronic toll collection system

(*c*) Vehicle actuated signals

(*d*) Pelican signals

3. Which is not a urban road system

(*a*) Grid iron system

(*b*) Concentric and radial street system

(*c*) Polynomial street system

(*d*) Organic street system

4. An example of combination of rectangular with radial street system

(*a*) New Delhi (*b*) Bhubaneshwar

(*c*) Chandigarh (*d*) Gandhinagar

5. Estimate trip rate for a residential land use with 2744 thousands of square feet and 6574 person trips

(*a*) 2.4 (*b*) 10 (*c*) 11.2 (*d*) 7.6

6. If trip rates are 0.5 for H.H. size till 3 and 1.0 for H.H. size above Q. No. 4 estimate likely trip to

(*a*) 1100 (*b*) 1115 (*c*) 1000 (*d*) 2000

Answers

1. (*d*) **2.** (*a*) **3.** (*c*) **4.** (*a*) **5.** (*a*)

6. (*b*)

II. SOLVED QUESTIONS AND ANSWERS

Q. 1. What is trip generation?

Ans. Trip generation is the first stage of the classical first generation aggregate demand models. The trip generation aims at predicting the total number of trips generated and attracted to each zone of the study area. In other words this stage answers the questions to "how many trips" originate at each zone, from the data on household and socioeconomic attributes.

Q. 2. What are the factors that affect trip generation?

Ans. The main factors affecting personal trip production include income, vehicle ownership, house hold structure and family size. In addition factors like value of land, residential density and accessibility are also considered for modeling at zonal levels. The personal trip attraction, on the other hand, is influenced by factors such as roofed space available for industrial, commercial and other services.

At the zonal level zonal employment and accessibility are also used. In trip generation modelling in addition to personal trips, freight trips are also of interest. Although the latter comprises about 20 percent of trips, their contribution to the congestion is significant. Freight trips are influenced by number of employees, number of sales and area of commercial firms.

Q.3. Let the trip rate of a zone is explained by the household size done from the field survey. It was found that the household size are 1, 2, 3 and 4. The trip rates of the corresponding household is as shown in the table below.

Household size (x)

	1	2	3	4
Trips per day(y)	1 2 2	2 4 3	4 5 3	6 7 4
Σy	5	9	12	17

Solution. The linear equation will have the form, $y = bx + a$ where, y is the trip rate, and x is the household size, a and b are the coefficients. For a best fit, b is given by

$$b = \frac{n\Sigma xy - \Sigma x \Sigma y}{n\Sigma x^2 - (\Sigma x)^2}$$

$$a = \overline{y} - b\overline{x}$$

$$\Sigma x = 3 \times 1 + 3 \times 2 + 3 \times 3 + 3 \times 4 = 30$$

$$\Sigma x^2 = 3 \times (1^2) + 3 \times (2^2) + 3 \times (3^2) + 3 \times (4^2) = 90$$

Q. 4. What is input data?

Ans. A forecasting activity, such as one based on the concept of economic base analysis, provides aggregate measures of population and activity growth. Land use forecasting distributes forecast changes in activities in a disaggregate-spatial manner among zones. The next step in the transportation planning process addresses the question of the frequency of origins and destinations of trips in each zone: for short, trip generation.

Q. 5. What is poisson regression?

Ans. Poisson Regression Modelling Poisson regression is a non-linear modelling method that over comes some of the problems of OLS regression. It is particularly suited to count data. In the model, the number of events is modeled as a Poisson random variable with a probability of occurrence being.

Q. 6. What are the advantages of the Poisson regression model?

Ans. The Poisson model over comes some of the problems of the OLS model.

First, the Poisson model has a minimum value of 0. It will not predict negative values. This makes it idea l for a distribution in which the mean or the most typical value is close to 0. Second, the Poisson is a fundamentally skewed model; that is, it is non-linear with a long right.

Again, this model is appropriate for counts of rare events, such as crime incidents.

In other words, the variance is many times greater than the mean. Most real-world count data are similar to this; the variance will usually be a lot greater than the mean. What this means in practice is that the residual errors, the difference between the observed

Q. 7. What is Trip-rate trip rate analysis?

Ans. Trip-rate trip rate analysis refers various models those are based on the determination of the average trip-production or trip-attraction rates associated with the trip generators within the region.

III. REVIEW QUESTIONS

1. What are the factors governing trip generation and attraction rates?
2. Give multiple linear regression analysis.

REFERENCES

1. P.W. Shuldiner, Land use, activity and non-residential trip generation, Highway Research Record No. 141, Highway Research Board, Washington, 1966.
2. D.M. Hill and N. Dodd, Studies of trends of travel between 1954 and 1964 in a large metropolitan area, Highway Research Record No. 141, Highway Research Board, Washington, 1966.
3. P.W. Shuldiner, Trip generation and the home, H.R.B. Bulletin 347, Highway Research Board, Washington, 1962.
4. H. Kassoff and H.D. Deutschman, Trip Generation : A Critical Review, Highway Research Record No. 279, Highway Research Board, Washington, 1969.
5. G.M. McCarthy, Multiple-Regression Analysis of Household Trip Generation–A Critique, Highway Research Record No. 297, Highway Research Board, Washington, 1969.
6. A.A. Douglas and R.J. Lewis, Trip Generation Techniques, 3–House-hold least squares regression analysis, Traffic Engineering and Control, Vol. 12, No. 9, 1971.
7. G.M. Lamb, Introduction to Transportation Planning, 3–Trip Generatipn, Traffic Engineering and Control, Vol. 11, No. 11, London, March, 1970.
8. A.A. Douglas and R.J. Lewis, Trip Generation Techniques, 2–Zonal least squares regression analysis, Traffic Engineering and Control, Vol. 12, No. 8, 1970.
9. M.A. Taylor, Studies of Travel in Gloucester, Northampton and Reading, RRL Report L.R. 141, Road Research Laboratory, Crowthorne 1968.
10. Wootton, H.J. and G.W. Pick, A model for trips generated by households, Journal or Transport Economy and Policy, May, 1967.
11. A.A. Douglas and R.J. Lewis, Trip generation Techniques, 4 category Analysis, Traffic Engineering and Control, Vol. 12, No. 10, 1971.

32. Trip Distribution

32.1. What is Trip Distribution

After having obtained an estimate of the trips generated from and attracted to the various zones, it is necessary to determine the direction of travel. The number of trips generated in every zone of the area under study has to be apportioned to the various zones to which these trips are attracted. Thus, if g_i is the number of trips ends generated in zone i and a_j is the number of trip ends attracted to zone j, trip distribution stage determines the number of trips t_{i-j} which would originate from zone i and terminate in zone j.

The distribution of trips between zones can best be understood by a matrix, given in Fig. 32.1.

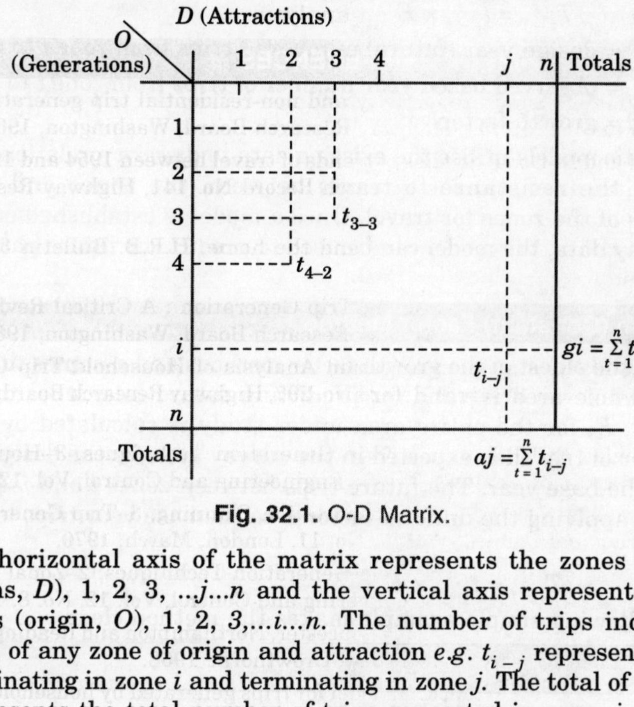

Fig. 32.1. O-D Matrix.

The horizontal axis of the matrix represents the zones of attractions (destinations, D), 1, 2, 3, ...j...n and the vertical axis represents the zones of generations (origin, O), 1, 2, 3,...i...n. The number of trips indicated at the intersection of any zone of origin and attraction *e.g.* t_{i-j} represents the number of trips originating in zone i and terminating in zone j. The total of any individual row, i, represents the total number of trips generated in zone, *i.e.* g_i. Similarly the total of any individual column j, represents the number of trips terminating in zone j, *i.e.*, a_j.

32.2. Method of Trips Distribution

There are two types of trip distribution methods, *viz*.,

(*i*) Growth factor methods.

(*ii*) Synthetic methods.

Growth factor methods have been used in earlier studies but have yielded place now to the more rational synthetic models. A description of the growth factor methods is, however, given here because they are still used in small studies and they serve to provide the historical background before the more recent methods are discussed.

The following are the important growth factor methods :

(*i*) Uniform factor method. (*ii*) Average factor method.

(*iii*) Fratar method. (*iv*) Furness method.

The synthetic methods that will be discussed here will be the following :

(*i*) Gravity model.

(*ii*) Tanner model.

(*iii*) Intervening opportunities model.

(*iv*) Competing opportunities model.

The growth factor methods are based on the assumption that the present travel patterns can be projected to the design year in the future by using certain expansion factors. This can be represented by the general formula :

$$T_{i-j} = t_{i-j} \times E \qquad \text{...(32.1)}$$

where, T_{i-j} = design year (future) number of trips from zone *i* to zone *j*.

T_{i-j} = observed based year number of trips from zone *i* to zone *j*.

E = growth factor.

Synthetic models utilise the existing data to discern a relationship between trip making, the resistance to travel between the zones and the relative attractiveness of the zones for travel. Once a model is established on the basis of the present day data, the model can be used to predict the future pattern-of travel between zones.

32.3. Uniform (Constant) Factor Method

This is the oldest of the growth factor methods and assumes that the growth rate for the whole area is valid for predicting future interzonal trips. A single growth factor, *E*, for the entire area under study is calculated by dividing the future number of trip ends expected in the survey area for the design year by the trip ends in the base year. The future trips between zones *i* and *j*, $T_{i-j,}$ are then claculated by applying the uniform factor *E* to the base year trips between zones *i* and *j*. Thus :

$$T_{i-j} = t_{i-j} \times E$$

The following simple example makes the method clear :

O \ D	1	2	3
1	60	100	200
2	100	20	300
3	200	300	20

The distribution of present trips among the zones 1, 2 and 3 are given in the O.D matrix above. The future trips generated in zones 1, 2 and 3 are expected to be 360, 1260 and 3120 respectively. It is required to distribute the future trips among the zones.

O \ D	1	2	3	t_i	T_i
1	60	100	200	360	360
2	100	20	300	420	1260
3	200	300	20	520	3120
				Total = 1300	4740

t_i in the above matrix represents the row totals, *i.e.* totals of the trips generated in the three zones at present. T_i represents the future trips generated in the three zones. The total trips generated at present is 1300 and the total trips generated in the design year is 4740, thus giving a value of :

$$E = \frac{4740}{1300} = 3.646$$

Multiplying the cells in the matrix by the uniform factor of $E = 3.646$, the following matrix results :

O \ D	1	2	3	T_i *Calculated*	T_i *(given)*
1	218	365	729	1312	360
2	365	73	1094	1532	1260
3	729	1094	73	1896	3120
				Total = 4740	4740

It will be seen that even though the trips have been distributed resulting in a total of 4740 trips, which is the same as the total of future trips as per data given, the totals the trips generated in each zone as now calcualted do not tally with the values given. This is because of the assumption of a uniform growth rate for all the zones. The method, therefore, suffers from certain disadvantages, which are enumerated below :

(*i*) The assumption of a uniform growth rate for the entire study is not correct, because each zone will have its own growth rate and the rate of growth of traffic movement between any two zones will be different.

(*ii*) The method underestimates movements where present day development is limited and overestimates movements where present day development is intensive.

(*iii*) If the present trip movement between any two zones is zero, the future trip movement also becomes zero as per this method. This may rarely be the case in reality.

32.4. Average Factor Method

In this method, a growth factor for each zone is calculated based on the average of the growth factors calculated for both ends of the trip. The factor thus represents the average growth associated both with the origin and the destination zones. The following mathematical relationship represents the principle employed :

$$T_{i-j} = t_{i-j}\left[\frac{E_j + E_i}{2}\right] \qquad ...(32.2)$$

where, T_{i-j} = future trips from zone i to zone j.

T_{i-j} = present trips from zone i to zone j

$E_i = \frac{P_i}{p_i}$ = Generated trips growth factor for zone i

$E_j = \frac{A_j}{a_j}$ attracted trips growth factor for zone j

P_i = future generated trips for zone i

p_i = present generated trips for zone i

A_i = future attracted trips for zone j

a_j = present attracted trips for zone j.

After the distribution has been completed on the above basis, it will be found that the sums of the trips from zone i will probably not agree with the projected trip ends in zone i and the sums of trips to zone j will not agree with the projected trip ends in zone j. This difficulty is overcome by an iterative process using new values for E_i and E_j calculated from :

$$E_i' = \frac{P_i}{p_i'}$$

and

$$E_j' = \frac{A_j}{a_j'}$$

where a_i' and a_j' are the total generations and attractions of zones i and j respectively obtained from the first stage of distribution.

The iteration is carried out till the growth factor approaches unity and the values balance within, say, plus or minus 1%.

The following simple example using data given earlier under para 32.3 illustrates the method :

The matrix is a symmetrical one, with the total origins and destinations at each zone being equal.

O \ D	1	2	3	p_i	P_i	$E_i = Pi/p_i$
1	60	100	200	360	360	1
2	100	20	300	420	1260	3
3	200	300	20	520	3120	6
a_j	360	420	520	1300	4740	
A_j	360	1260	3120			
$E_j = \frac{A_j}{a_j}$	1	3	6			

$$T_{1-1} = \frac{1+1}{2} \times 60 = 60$$

$$T_{1-2} = \frac{1+3}{2} \times 100 = 200$$

$$T_{1-3} = \frac{1+3}{2} \times 200 = 700$$

$$T_{2-1} = \frac{3+1}{2} \times 100 = 200$$

$$T_{2-2} = \frac{3+3}{2} \times 20 = 60$$

$$T_{2-3} = \frac{3+6}{2} \times 300 = 1350$$

$$T_{3-1} = \frac{6+1}{2} \times 200 = 700$$

$$T_{3-2} = \frac{6+3}{2} \times 300 = 1350$$

$$T_{3-3} = \frac{6+6}{2} \times 20 = 120$$

The matrix then becomes :

O \ D	1	2	3	p_i'	P_i'	$E_i' = \frac{P_i}{p_i}$
1	60	200	700	960	360	0.375
2	200	60	1350	1610	1260	0.783
3	700	1350	120	2170	3120	1.438
a_j'	960	1610	2170	4740		
j	360	1260	3120		4740	
$E_j' = \frac{A_j}{a_j'}$	0.375	0.783	1.438			

For the next interation, the values of interzonal movement are calculated with the new values E, shown as E in the last column of the above matrix.

The average factor method has the same disadvantages of the uniform factor method. The multiplying factor has no real significance and is only a convenient tool to balance the movements. There is no explanation of the movement between zones and the factors causing the movement. It has the additional disadvantage that a large number of iterations are required. As in the case of the uniform factor method, if t_{i-j} is zero, T_{i-j} also becomes zero. Because of these drawbacks the method is rarely used now-a-days except for updating existing stable and uniform data and for quick results.

32.5. Fratar Method

This method was introduced by T.J. Fratar (Ref. 1) and is based on predicting future interzonal movements by successive approximations.

According to this method, the total trips for each zone are distributed to the interzonal movements, as a first approximation, according to the relative attractiveness of each movement. Thus, the future trips estimated for any zone would be distributed to the movements involving that zone in proportion to the existing trips between it and each other zone and in proportion to the expected growth of each other zone. This may be expressed mathematically as follows :

$$T_{i-j} = t_{i-j} \times \frac{P_i}{p_i} \times \frac{A_j}{a_j} \times \frac{\sum^{k} t_{i-k}}{\sum^{k} \left[\frac{A_k}{a_k}\right] t_{i-k}} \quad ...(32.3)$$

where, T_{i-j} = Future trips from zone i to zone j

t_{i-j} = Present trips from zone i to zone j

P_i = Future trips produced at zone i

p_i = Present trips attracted to zone i

A_i = Future trips attracted to zone j

a_j = Present trips attracted to zone j

k = Total number of zones.

When the future traffic into and out of all zones is similarly distributed, each interzonal trip has been assigned two tentative values-one the result of the distribution for one of the zones involved and the other, the result of the distribution for the other zone involved. As a first approximation those pairs of tentative values are averaged. A new 'growth factor' for each zone is then calculated and the distribution process is repeated.

The following simple example from Fratar's original paper is given here to illustrate the actual procedure.

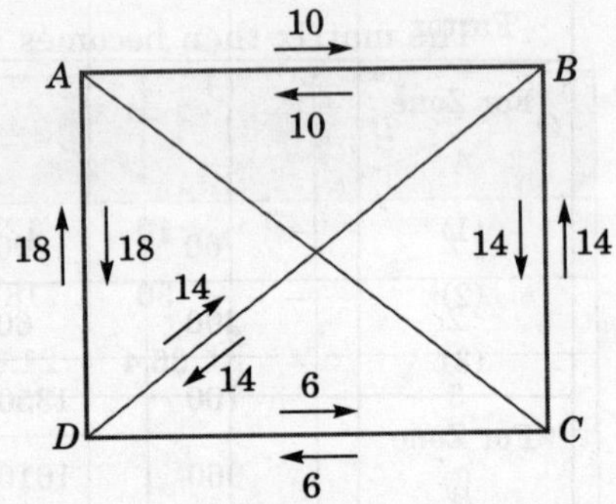

Fig. 32.2

O \ D	A	B	C	D
A	–	10	12	18
B	10	–	14	14
C	12	14	–	6
D	18	14	6	–
Present totals	40	38	32	38
estimated future totals	80	114	48	38
growth factor, E	2	3	1.5	1

$$T_{A-B} = t_{A-B} \times \frac{P_A}{p_A} \times \frac{A_B}{a_B} \times \frac{t_{AB} + t_{AC} + t_{AD}}{t_{AB}\, E_B + t_{AC}\, E_C + t_{AD}\, E_D}$$

$$= 10 \times \left[\frac{80}{40}\right] \times \left[\frac{114}{38}\right] \times \frac{40}{10 \times 3 + 12 \times 1.5 + 18 \times 1}$$

$$= 10 \times 2 \times 3 \times \frac{40}{10 \times 3 + 12 \times 1.5 + 18 \times 1} = 36.4$$

The computations are summarised below. Line 1 for each summary shows the existing trips for the indicated inter-zonal movement. Line 2 shows for each zone the interzonal trips multiplied respectively by the growth factor of the other zone involved. These products are summarised for each zone to provide a common denominator for the distribution of trips of that zone. This distribution is accomplished by dividing the total trips desired for the zone by the common denominator and multiplying the quotient by the products shown in the line 2. The new distribution shown in line 3, adds up to the total number of trips desired for the zone.

First Approximation

Zone	*A*	*B*	*C*	*D*	*Sum of products of trips and growth factors*	*Desired trips*	*Ratio of desired new total trips to sum of products*
Growth Factor	2	3	1.5	1			
For Zone *A*							
(1)	–	10	12	18			
(2)	–	30	18	18	66	80	1.21
(3)	–	36.4	21.8	21.8	80		
For Zone *B*							
(1)	10	–	14	14			
(2)	20	–	21	14	55	114	2.07
(3)	41.5	–	43.5	29	114		
For Zone *C*							
(1)	12	14	–	6			
(2)	24	42	–	6	72	48	0.667
(3)	16	28	–	4	48		
For Zone *D*							
(1)	18	14	6				
(2)	36	42	9	–	87	38	0.437
(3)	15.8	18.3	3.9	–	38		

The pairs of interzonal volumes obtained by these computations are averaged as shown below to obtain the first approximation for interzonal trips.

	A – B	A – C	A – D	B – C	B – D	C – D
	36.4	21.8	21.8	43.5	29.0	3.9
	41.5	16.0	15.8	28.0	18.3	4.0
Total	77.9	37.8	37.6	71.5	47.3	7.9
Average	39.9	18.9	18.8	35.7	23.6	4.0

The averages for the trips radiating from each zone are summarised below to determine the new growth factors :

	A	B	C	D
	39.0	39.0	18.9	18.8
	37.8	35.7	35.7	23.6
	37.6	23.6	4.0	4.0
New Totals	104.4	98.3	58.6	46.4
Desired Totals	80.0	114.0	48.0	38.0
New Growth Factors $= \dfrac{\text{Desired Totals}}{\text{New Totals}}$	1.04	1.16	0.82	0.82

The process is then repeated to obtain a second approximation on the same lines, but using new growth factors and new values of the inter zonal movement obtained from the first approximation.

The procedure is laborious except for simple problems, but can be conveniently tackled by a computer. It has the same drawbacks as other growth factor models and is unable to forecast trips for those areas which were predominantly under-developed during the base year. It does not take into account the effects of changes in accessibility for various portions of the study area.

32.6. Furness Method

The method devised by K.P. Furness (Ref. 2) is also iterative in nature. For this the estimates of future traffic originating and terminating at each zone are required, thus yielding origin growth factors and destination growth factors for each zone. The traffic movements are made to agree alternately with the future traffic originating in each zone and the estimated future terminating in each zone, until both these conditions are roughly satisfied.

The following example illustrates the method for a four-zone system :

O \ D	1	2	3	4	*Total present trips*	*Predicted future trips (Total)*	*Origin growth factor*
1	8	3	16	15	42	147	3.5
2	6	9	8	5	28	42	1.5
3	10	8	3	8	29	32	1.1
4	2	4	7	12	25	30	1.2
Total present trips	26	24	34	40	124		
Predicted future trips	39	24	68	120	251		
Destination growth factors	1.5	1.0	2.0	3.0			

The matrix gives the base-year distribution. From the values of the predicted future origins and destinations, the origin growth factors and destination growth factors are calculated for each zone. Each column of the matrix is then scaled by the appropriate destination growth factor, such that the column totals become equal to the predicted column totals. However, this results in the row totals not tallying with the predicted totals. A new set of origin growth factors is now computed for each zone by dividing the predicted trip total by the total obtained for the row by the column scaling mentioned earlier. The following position then results :

D / O	1	2	3	4	Total	Predicted trip total	New origin growth factor
1	12	3	32	45	92	147	$\frac{147}{92}$ = 1.60
2	9	9	16	15	49	42	0.86
3	15	8	6	24	53	32	0.60
4	3	4	14	36	57	30	0.5
Total	39	24	68	120	251		
Predicted total	39	24	68	120		251	

In the next step, each row is scaled by the new growth factor giving correct row totals. This, however, disturbs the column totals. As before, new destination growth factors are computed for each zone by dividing the predicted trip total by the total obtained by the row scaling. The following position then results :

D / O	1	2	3	4	Total	Present total trips
1	19.20	4.80	51.20	72.00	147.20	147
2	7.74	7.74	13.76	12.90	42.14	42
3	9.00	4.80	3.60	14.40	31.80	32
4	1.59	2.12	7.42	19.08	30.21	30
Total	37.53	19.46	75.98	118.38	251.35	
Predicted trip total	39	24	68	120		251
New destination growth factors		1.04	1.23	0.89	1.01	

The columns are now scaled again using the new destination growth factors and this iterative procedure is repeated until all growth factors are unity or sufficiently near to unity for the user to feel satisfied and confident that he has obtained the degree of accuracy he needs.

The Furness method gives results similar to Fratar, but requires less computations.

32.7. Criticism of Growth Factor Methods

The following are some of the disadvantages of the growth factor methods :

(*i*) Present trip distribution matrix has to be obtained first, for which large scale *O–D* studies with high sampling sizes are needed so as to estimate the smaller zone-to-zone movements accurately.

(*ii*) The error in original data collected on specific zone-to-zone movements gets magnified.

(*iii*) None of the methods provide a measure of the resistance to travel and all imply that resistance to travel will remain constant. They neglect the effect of changes in travel pattern by the construction of new facilities and new network.

Despite the above shortcomings, the growth factor methods are relatively simpler to use and understand. They can be used for studies of small areas and for updating stable and uniform data.

32.8. Gravity Model

32.8.1. In synthetic models of trip distribution, an attempt is made to discern the underlying causes of movement between places, and relationships are established between trips and measures of attraction, generation and travel resistance. Synthetic models have an important advantage that they can be used not only to predict future trip distributions but also to synthesise the base-year flows. The necessity of having to survey every individual cell in the trip matrix is thus obviated and the cost of data collection in reduced.

32.8.2. One of the well-known synthetic models is the Gravity Model. Based on Newton's concept of gravity, the model as proposed by Voorhees (Ref. 4) assumes that the interchange of trips between zones in an area is dependent upon the relative attraction between the zones and the spatial separation between them as measured by an appropriate function of distance. This function of spatial separation adjusts the relative attraction of each zone for the ability, desire or necessity of the trip maker to overcome the spatial separation. Whereas the trip interchange is directly proportional to the relative attraction between the zones, it is inversely proportional to the measure of spatial separation.

32.8.3. A simple equation representing the above relationship is of the following form :

$$T_{i-j} = \frac{KP_i\, A_j}{d_{ij}^n} \qquad \text{...(32.4)}$$

where, T_{i-j} = Trips between zones i and j

P_i = Trips produced in zone i

A_j = Trips attracted to zone j

d_{ij} = Distance between zone i and zone j, or the time or cost of travelling between them

K = A constant, usually independent of i and

n = An exponential constant, whose value is usually found to lie between 1 and 3 (Ref. 3).

k = Total number of zones.

The following formulation was also used in earlier studies dispensing with the proportionality constant :

$$T_{i-j} = P_i \times \frac{\dfrac{A_j}{(d_{i-j})^n}}{\dfrac{A_j}{(d_{i-j})^n} + \dots \dfrac{A_k}{(d_{i-k})^n}} \qquad ...(32.5)$$

where, T_{i-j}, P_i, A_j, d_{i-k} and n haye the same meaning as given earlier.

In order to simplify the computation requirements of the model, the following formulation has been frequently used (Ref. 5, 6, 7, 8) :

$$T_{(i-j)m} = \left[\frac{P_i\, A_{jm}\, F_{i-j}\, C_{i-j}}{\sum_{x=1}^{k} A_{km}\, F_{(i-k)} K_{(i-k)}} \right]_p \qquad ...(32.6)$$

where, T_{i-j} = Trips produced in zone i and attracted to zone j

P_i = Trips produced in zone i

A_j = Trips attracted to zone j

F_{i-j} = Empirically derived travel time factor which expresses the average area-wide effect of spatial separation on trip interchange between zones i and j

K_{i-j} = A specific zone-to-zone adjustment factor to allow for the incorporation of the effect on travel patterns of defined social or economic linkages not otherwise accounted for in the gravity model formulation

k = Total number of zones

m = Iteration number

p = Trip purpose.

The above relationship can be used for determining the trip interchange for each trip purpose and each mode of travel.

32.8.4. A few simple problems will now be solved to illustrate the above concepts.

Problem 32.2. *A self-contained town consists of four residential areas A, B, C and D and two industrial estates X and Y. Generation equations show that, for the design year in question, the trips from home to work generated by each residential area per 24 hour day are as follows :*

A	*1000*
B	*2250*
C	*1750*
D	*3200*

There are 3,700 jobs in industrial estate X and 4,500 in industrial estate Y. It is known that the attraction between zones is inversely proportional to the square of the journey times between zones. The journey times in minutes from home to work are :

Zones	*X*	*Y*
A	*15*	*20*
B	*15*	*10*
C	*10*	*10*
D	*15*	*20*

Caltulate and tabulate the inter zonal trips for journeys from home to work.

Solution.

$$T_{i-j} = \frac{P_i \dfrac{A_j}{(d_{i-j})^2}}{\sum \dfrac{A_i}{(d_{i-n})^2}}$$

$$T_{A-X} = \frac{1000 \times \dfrac{3700}{(15)^2}}{\dfrac{3700}{(15)^2} + \dfrac{4500}{(20)^2}} = \frac{1000 \times 16.5}{16.5 + 11.25} = 604$$

$$T_{A-Y} = \frac{1000 \times \dfrac{4500}{(20)^2}}{\dfrac{3700}{(15)^2} + \dfrac{4500}{(20)^2}} = \frac{1000 \times 11.25}{16.5 + 11.25} = 396$$

$$T_{B-X} = \frac{2250 \times \dfrac{3700}{(15)^2}}{\dfrac{3700}{(15)^2} + \dfrac{4500}{(10)^2}} = \frac{2250 \times 16.5}{16.5 + 45} = 604$$

$$T_{B-Y} = \frac{2250 \times \dfrac{4500}{(10)^2}}{\dfrac{3700}{(15)^2} + \dfrac{4500}{(10)^2}} = \frac{2250 \times 45}{16.5 + 45} = 1646$$

$$T_{C-X} = \frac{1750 \times \dfrac{3700}{(10)^2}}{\dfrac{3700}{(10)^2} + \dfrac{4500}{(10)^2}} = \frac{1750 \times 37}{37 + 45} = 790$$

$$T_{C-Y} = \frac{1750 \times \dfrac{4500}{(10)^2}}{\dfrac{3700}{(10)^2} + \dfrac{4500}{(10)^2}} = \frac{1750 \times 45}{37 + 45} = 960$$

$$T_{D-X} = \frac{3200 \times \dfrac{3700}{(15)^2}}{\dfrac{3700}{(15)^2} + \dfrac{4500}{(20)^2}} = \frac{3200 \times 16.5}{16.5 + 11.25} = 1980$$

$$T_{D-Y} = \frac{3200 \times \dfrac{4500}{(20)^2}}{\dfrac{3700}{(15)^2} + \dfrac{4500}{(20)^2}} = \frac{3200 \times 11.25}{16.5 + 11.25} = 1220$$

The results are tabulated in the matrix below :

	X	Y	T_{i-j} *for origin zones A, B, C, D. Total Productions*
A	604	396	1000
B	604	1646	2250
C	790	960	1750
D	1980	1220	3200
Total calculated attractions, C_j	3978	4222	8200
Total predicted attractions, A_j	3700	4500	8200

It will be seen that as a result of the distribution, the total attractions do not tally with the predicted attractions. An iterative procedure is necessary to balance out both the predicted attractions and the productions. For this purpose, the total attractions are first adjusted as follows :

$$A_{jk} = \frac{A_j}{C_{j(m-1)}} \times A_{j\,(m-1)} \qquad \text{...(32.7)}$$

For destination zones X and Y.

where, A_{jm} = Adjusted attraction factor, iteration m

A_j = Desired attraction

$A_{j(m-1)}$ = Attraction factor, iteration $m-1$

$C_{j(m-1)}$ = Actual attraction factor, iteration $m-1$.

For the second iteration,

$$m = 2$$

$$A_{j2} \text{ for zone } X = \frac{3700}{3978} \times 3700 = 3440$$

$$A_{j2} \text{ for zone } Y = \frac{4500}{4222} \times 4500 = 4800$$

Recalculating,

$$T_{A-X} = \frac{1000 \times \dfrac{3440}{(15)^2}}{\dfrac{3440}{(15)^2} + \dfrac{4800}{(20)^2}} = \frac{1000 \times 15.3}{15.3 + 12} = 560$$

$$T_{A-Y} = \frac{1000 \times \dfrac{4800}{(20)^2}}{\dfrac{3440}{(15)^2} + \dfrac{4800}{(20)^2}} = \frac{1000 \times 12}{15.3 + 12} = 440$$

$$T_{B-X} = \frac{2250 \times \dfrac{3440}{(15)^2}}{\dfrac{3440}{(15)^2} + \dfrac{4800}{(10)^2}} = \frac{2250 \times 15.3}{15.3 + 48} = 540$$

$$T_{B-Y} = \frac{2250 \times \frac{4800}{(10)^2}}{\frac{3440}{(15)^2} + \frac{4800}{(10)^2}} = \frac{2250 \times 48}{15.3 + 48} = 1710$$

$$T_{C-X} = \frac{1750 \times \frac{3440}{(10)^2}}{\frac{3440}{(10)^2} + \frac{4800}{(10)^2}} = \frac{1750 \times 34.4}{34.4 + 48} = 730$$

$$T_{C-Y} = \frac{1750 \times \frac{4800}{(10)^2}}{\frac{3440}{(10)^2} + \frac{4800}{(10)^2}} = \frac{1750 \times 48}{34.4 + 48} = 1020$$

$$T_{D-X} = \frac{3200 \times \frac{3440}{(15)^2}}{\frac{3440}{(15)^2} + \frac{4800}{(20)^2}} = \frac{3200 \times 15.3}{15.3 + 12} = 1790$$

$$T_{D-Y} = \frac{3200 \times \frac{4800}{(20)^2}}{\frac{3440}{(15)^2} + \frac{4800}{(20)^2}} = \frac{3200 \times 12}{15.3 + 12} = 1410$$

The results are tabulated in the matrix below :

O \ D	X	Y	T_{i-j} *for origin zones A, B, C, D (Total Production)*
A	560	440	1000
B	540	1710	2250
C	730	1020	1750
D	1790	1410	3200
T_{i-i} for destination zones X, Y (Total attraction)	3620	4580	8200
Total predicted attractions	3700	4500	8200

It will be seen that the results are now closer to the total predicted attractions. If more accuracy is needed, further iterations can be done.

Problem 32.3. *The total trips produced in and attracted to the three zones A, B and C of a survey area in the design year are tabulated as :*

Zone	*Trips produced*	*Trips attracted*
A	*2000*	*3000*
B	*3000*	*4000*
C	*4000*	*2000*

It is known that the trips between two zones are inversely proportional to the second power of the travel time between zones, which is uniformly 20 minutes. If the trip interchange between zones B and C is known to be 600, calculate the trip interchange between zones A and B, A and C, B and A, C and B.

Solution. Using the formula :

$$T_{i-j} = \frac{K.P_j\ A_j}{t^n}$$

$$T_{B-C} = \frac{K.P_B\ A_C}{t^n}$$

i.e.,
$$600 = \frac{K \times 3000 \times 2000}{20 \times 20}$$

$$K = \frac{600 \times 20 \times 20}{3000 \times 2000} = \frac{1}{25}$$

$$\therefore \quad T_{A-B} = \frac{1}{25} \times \frac{2000 \times 4000}{20 \times 20} = 800$$

$$T_{A-C} = \frac{1}{25} \times \frac{2000 \times 2000}{20 \times 20} = 800$$

$$T_{B-A} = \frac{1}{25} \times \frac{3000 \times 3000}{20 \times 20} = 900$$

$$T_{C-A} = \frac{1}{25} \times \frac{4000 \times 3000}{20 \times 20} = 900$$

$$T_{C-B} = \frac{1}{25} \times \frac{4000 \times 4000}{20 \times 20} = 1600.$$

Problem 32.4. *The details of four zones in a study are given below :*

Zone 1 : *Inter-zonal travel time of 3 minutes, 100 work trips attracted to it from all zones in the study area, produces 100 work trips and terminal time is 9 minutes.*

Zone 2 : *Inter-zonal travel time between zones 1 and 2 is 10 minutes, 70 work trips attracted to it from all zones in the study area and terminal time is 3 minutes.*

Zone 3 : *Inter-zonal travel time between zones 1 and 3 is 15 minutes, 60 work trips attracted to it from all zones in the study area and terminal time is 4 minutes.*

Zone 4 : *Inter-zonal travel time between zones 1 and 4 is 20 minutes, 500 work trips attracted to it from all zones in the study area and terminal time is 5 minutes.*

Distribute the 100 work trips to all other zones from zone 1. The travel time factor F_{ij} may be assumed to be related to the actual travel time as below :

Travel time (minutes)	*Travel time Factor F_{i-j}*
7	*100*
10	*85*
15	*65*
20	*50*

The socio-economic zone-to-zone adjustment factor may be taken as unity.

Solution. From the above data the following table can be easily constructed :

Zone	A_j	Driving time	Terminal time (Initial + Find)	Total travel time	F_{i-j}	A_jF_{i-j}	T_{i-j}
1	100	3	2 + 2 = 4	7	100	10,000	22
2	70	5	2 + 3 = 5	10	85	5950	13
3	60	9	2 + 4 = 6	15	65	3900	9
4	500	13	2 + 5 = 7	20	50	25,000	56
						$\sum_{j=1}^{4} A_jF_{i-j}$ = 44850	Σ = 100

From Eq. (32.6),

$$T_{i-j} = \frac{P_i\, A_i\, F_{i-j}\, K_{i-j}}{\sum_{x=1}^{n} A_x\, F_{(i-x)}\, K_{i-x}}$$

Since $K_{i-j} = 1$

the above formula can be rewritten as :

$$T_{i-j} = \frac{P_i\, A_j\, F_{i-j}}{\sum_{j=1}^{4} A_i\, F_{i-j}}$$

Thus

$$T_{1-2} = \frac{100 \times 70 \times 85}{44{,}850} = 13.3, \text{ say } 13$$

$$T_{1-4} = \frac{100 \times 500 \times 50}{44{,}850} = 55.7, \text{ say } 56$$

$$T_{1-1} = \frac{100 \times 100 \times 100}{44{,}850} = 22.3, \text{ say } 22$$

$$T_{1-3} = \frac{100 \times 60 \times 65}{44{,}850} = 8.7, \text{ say } 9.$$

These values have been shown in the last column of the above table.

Problem 32.5. *The number of trips produced in and attracted to the three zones 1, 2 and 3 are tabulated as :*

Zone	*1*	*2*	*3*	*Total*
Trips produced (P_i)	*14*	*33*	*28*	*75*
Trips attracted (A_j)	*33*	*28*	*14*	*75*

As a result of calibration, the friction factors to be associated with the impedance values between the various zones have been found to be as follows :

Impedance Units	*Friction Factor*
1	*82*
2	*52*
3	*50*
4	*41*
5	*39*
6	*26*
7	*20*
8	*13*

The impedance values between the various zones can be taken from the following matrix :

O \ *D*	*1*	*2*	*3*
1	8	1	4
2	3	6	5
3	2	7	4

Distribute the trips between the various zones.

Solution. $T_{(i-j)m} = \dfrac{P_i \, A_{jm} \, F_{i-j}}{\sum\limits_{j=1}^{n} (A_{jm} \, F_{i-j})}$ ...[Eq. 32.6]

First Iteration (m = 1)

$$T_{1-1} = \frac{P_1 \, A_1 \, F_{1-1}}{A_1 \, F_{1-1} + A_2 \, F_{1-2} + A_3 \, F_{1-3}}$$

$$= \frac{14 \times 33 \times 13}{33 \times 13 + 28 \times 82 + 14 \times 41} = \frac{6006}{3299} = 1.82$$

$$T_{1-2} = \frac{14 \times 28 \times 82}{3299} = 9.74$$

$$T_{1-3} = \frac{14 \times 14 \times 41}{3299} = 2.44$$

$$T_{2-1} = \frac{P_2 \, A_1 \, F_{2-1}}{A_1 \, F_{2-1} + A_2 \, F_{2-2} \times A_3 \, F_{2-3}}$$

$$= \frac{33 \times 33 \times 50}{33 \times 50 + 28 \times 26 + 14 \times 39} = \frac{54{,}450}{2924} = 18.62$$

$$T_{2-2} = \frac{33 \times 28 \times 26}{2924} = 8.22$$

$$T_{2-3} = \frac{33 \times 14 \times 39}{2924} = 6.16$$

$$T_{3-1} = \frac{P_3 \, A_1 \, F_{3-1}}{A_1 \, F_{3-1} + A_2 \, F_{3-2} + A_3 \, F_{3-3}}$$

$$= \frac{28 \times 33 \times 52}{33 \times 52 + 28 \times 20 + 14 \times 41} = \frac{48{,}048}{2850} = 16.86$$

$$T_{3-3} = \frac{28 \times 28 \times 20}{2850} = 5.50$$

$$T_{3-3} = \frac{28 \times 14 \times 41}{2850} = 5.64$$

The matrix then becomes :

D / O	1	2	3	Total P_i
1	1.82	9.74	2.44	14.00
2	18.62	8.22	6.16	33.00
3	16.86	5.50	5.64	28.00
Total $C_j(1)$	37.30	23.46	14.24	75.00

Second iteration

It will be seen that the total trip attractions do not equal the desired attractions. Further iterations are, therefore, necessary. The following formula can be used to adjust the attraction factors :

$$A_{jm} = \frac{A_j}{C_{j(m-1)}} \times A_{j(m-1)} \qquad \text{...(Eq. 32.7)}$$

where, A_{jm} = Adjusted attraction factor, iteration m

A_j = Desired attraction

$A_{j(m-1)}$ = Adjusted attraction factor, iteration $m - 1$

$C_{j(m-1)}$ = Actual attraction, iteration $m - 1$

In this case, for the second iteration, $m = 2$

	Zones		
	1	2	3
Desired attraction (total) A_j	33	28	14
Actual attraction $C_j(1)$	37.30	23.46	14.24
Adjusted attraction factor $A_j(2-1) = A_j(1)$	33	28	14
Adjusted attraction factor $A_{j(m)} = A_{j(2)}$	$\frac{33 \times 33}{37.30}$ = 20.19	$\frac{28 \times 28}{23.46}$ = 33.41	$\frac{14 \times 14}{14.24}$ = 13.76

Using the above values of $A_{j(2)}$.

$$T_{1-1} = \frac{14 \times 29.19 \times 13}{29.19 \times 13 + 33.41 \times 82 + 13.76 \times 41} = \frac{5312.58}{3683.25} = 1.44$$

$$T_{1-2} = \frac{14 \times 33.41 \times 82}{3683.25} = 10.41$$

$$T_{1-3} = \frac{14 \times 13.79 \times 41}{3683.25} = 2.15$$

The full matrix then becomes :

O \ D	1	2	3	Total
1	1.44	10.41	2.15	14.00
2	16.81	10.01	6.18	33.00
3	15.45	6.80	5.75	28.00
Actual attraction $C_{j(2)}$	33.70	27.22	14.08	75.00

Third iteration. It is seen that the attraction figures, though considerably closer to the desired values than the first iteration, need still further to be adjusted. The adjusted attraction figures, $A_{j(3)}$ are calculated as below :

	Zone 1	Zone 2	Zone 3
Desired attraction, A_j	23	28	14
Actual attraction, $C_{j(3-1)} = C_{(2)}$	33.70	27.22	14.08
Adjusted attraction factor $A_{j(3)}$	$\frac{33 \times 29.19}{33.70}$ = 28.58	$\frac{28 \times 33.41}{27.22}$ = 34.36	$\frac{14 \times 13.76}{14.08}$ = 13.63

The process is continued with new values of $A_{j(3)}$ till a satisfactory agreement is reached between the desired and actual attraction figures.

32.8.5. Calibration of the Gravity Model

Calibration is the process of determining the travel time factor F_{i-j} which expresses the area-wide effect of spatial separation on trip interchange between zones and the factor K_{i-j} which is the specific zone-to-zone adjustment factor to account for the social and economic factors influencing the travel pattern (Ref. 7, 8, 9, 10, 11).

The following are the phases involved in the calibration process :

Phase 1. In this phase, the basic data on the area's travel patterns and transportation facilities are processed to provide the three basic inputs to the gravity model formula, *i.e.*, zonal trip productions and attractions and the spatial separation between the zones as measured by travel times. The network is studied and driving time and terminal times are determined. The interzonal and intrazonal travel times are derived.

Phase 2. The basic survey trip data are analysed and a table of zone-to-zone movement for each trip purpose is built. The effect of trip length on trip making is studied by obtaining the trip length frequency distribution by one minute travel time increments, for each trip purpose being analysed.

Phase 3. This phase relates to the development of travel time factors. Initially, set of travel time factors are assumed, and this is facilitated by a pervious study of a city of comparable size. This trip interchanges are then calculated by using the gravity model formula on the basis of the data on zonal trip productions and attractions, minimum path travel times and assumed travel time factors. The initial estimate of the trip interchanges is then combined with the minimum time paths to obtain an estimated trip length frequency distribution for each trip purpose category. A comparison of the actual and estimated trip length frequency distributions is then made, and successive iterations are done to get close agreement between the two.

Phase 4. In this phase, zone-to-zone adjustment factors (K_{i-j}) are developed to account for social or economic linkages. Topographical barriers such as rivers etc. are now eliminated by adjustments to the minimum path travel times.

Phase 5. The calibrated gravity model is tested to make sure that it simulates accurately existing travel patterns.

Computer programmes are available to facilitate the above process, which is otherwise very complex.

32.9. Tanner's Model

Tanner (Ref. 4) has suggested that the inverse of nth power, $1/(d_{i-j})^n$ in the gravity model formula (Eq. 32.4) cannot give valid estimates at both very small and very large distances. In this place he proposes the function $e^{\lambda d}/d^n$, where λ and n are constants. The new formula suggested by him is of the form :

$$t_{1-2} = \frac{m\, P_1\, P_2\, e_{1-2}^{-\lambda d}}{d_{1-2}} \left[\frac{1}{C_1} + \frac{1}{C_2}\right] \qquad \text{...(32.8)}$$

where, t_{1-2} = number of journey per day between the two places 1 and 2

m = a constant

P_1 and P_2 = populations, or other measures of size of the two places

d_{1-2} = distance between places 1 and 2 or the time or cost of travelling between them

C_1 and C_2 = constants, one for each place, C_1 being defined by :

$$C_1 = \Sigma P_j e^{-di-j}$$

where the summation is over all places j.

32.10. Opportunity Model

32.10.1. Opportunity models are based on the statistical theory of probability as the theoretical foundation. The concept has been pioneered by Schneider and developed by subsequent studies (Ref. 12, 13). The two well-known models are :

(*i*) the intervening opportunities models ;

(*ii*) the competing opportunities model.

The opportunity models can be represented by the general formula :

$$T_{i-j} = O_i P(D_j) \qquad \text{...(32.9)}$$

where, T_{i-j} = predicted number of trips from zone i to j

O_i = total number of trips originating in zone i

$P(D_j)$ = calculated probability of a trip terminating in zone j.

D_j = total trip destinations attracted to zone j.

32.10.2. Intervening opportunities model

In the intervening opportunities model, it is assumed that the trip interchange between an origin and a destination zone is equal to the total trips emanating from the origin zone multiplied by the probability that each trip will find an acceptable terminal at the destination. It is further assumed that the probability that a destination will be acceptable is determined by two zonal characteristics : the size of the destination and the order in which it is encountered as trips proceed from the origin. The probability function in Eq. (32.9) above, $P(D_j)$, may then be expressed as the difference between the probability that the trip origins at i will find a suitable terminal in one of the destinations, ordered by closeness to i, up to and including j, and the probability that they will find a suitable terminal in the destinations up to but excluding j. The following equation represents mathematically this concept :

$$T_{i-j} = Q_i(e^{-LB} - e^{-LA}) \quad ...(32.10)$$

where, T_{i-j} = predicated number of trips from zone i to j

Q_i = total number of trips originating in zone i

L = probability density (probability per destination) of destination acceptability at the point of consideration

A = number of destinations between i and j (including j) when arranged in order of closeness

B = number of destinations between i and j (excluding j) when arranged in order of closeness.

It may be noted that :

$$A = B + D_j \quad ...(32.11)$$

The following problem illustrates the application of the above method.

Problem 32.6. *The number of trips produced in an attracted to the three zones 1, 2, and 3 are tabulated below :*

Zone	*1*	*2*	*3*	*Total*
Trips produced	*14*	*33*	*28*	*75*
Trips attracted	*33*	*28*	*14*	*75*

The order of closeness of the zones is including by the following matrix :

O \ *D*	*1*	*2*	*3*
1	1	2	3
1	2	1	3
3	2	3	4

The zonal L factors are given below :

Zone	*L Factors*
1	*0.04*
2	*0.02*
3	*0.04*

Distribute the trips between the zones.

Solution. Using Eq. (32.10)

$$T_{1-1} = 14[e^{-0.04} \times 0 \; _{-e}{}^{-0.04 \times 33}] = 10.26, \text{ say } 10$$

$$T_{1-2} = 14[e^{-0.04 \times 33} - e^{-0.04} \times (33 + 28)] = 2.52, \text{ say } 3$$

$$T_{1-3} = 14[e^{-0.04 \times (32 + 28)} - e^{-0.04 \times (33 + 28 + 14)}] = 0.52, \text{ say } 1$$

$$T_{3-1} = 33[e^{-0.02 \times 28} - e^{-0.04 \times (28 + 33)}] = 9.11, \text{ say } 9$$

$$T_{2-2} = 33[e^{-0.02 \times 0} - e^{-0.02 \times 28}] = 14.15, \text{ say } 14$$

$$T_{2-3} = 33[e^{-0.02 \times (33 + 28)} - e^{-0.02 \times (33 + 28\ 14)}] = 2.$$
$$T_{3-1} = 28[e^{-0.04 \times 14} - e^{-0.04 \times (14 + 33 + 28)}] = 12$$
$$T_{3-2} = 28[e^{-0.04 \times (14 + 33)} - e^{-0.04 \times (14 + 33 + 28)}] = 3$$
$$T_{3-3} = 28[e^{-0.04 \times 0} - e^{-0.04 \times 14}] = 12$$

The matrix can be constructed as below :

O \ D	1	2	3	Total
1	10	2	1	13
2	9	14	2	25
3	12	3	12	27
Destination total	31	19	15	65

It is thus seen that only 65 out of 75 trips have been distributed by this stage. Further iterations are needed. The destination total can be adjusted by the formula :

$$D_j(m) = \frac{D_j}{C_{j(m-1)}} \times D_{(m-1)} \qquad ...(32.12)$$

where, $D_j(m)$ = Adjusted destination total for iteration m

D_j = Desired destination total

$D_{j(m-1)}$ = Adjusted destination total for preceding iteration $(m - 1)$

$C_{j(m-1)}$ = Actual destination totals, iteration $(m - 1)$.

The iteration is carried on till a reasonable closeness is obtained between the total trips and the calculated trips.

32.10.3. Competing Opportunities Model

In the competing opportunities model, the adjusted probability of a trip ending in a zone is the product of two independent probabilities, viz., the probability of a trip being attracted to a zone and the probability of a trip finding a destination in that zone. A form of this model is given below (Ref. 14) :

$$T_{i-j} = \frac{P_j \dfrac{A_i}{\sum_j A_j}}{\sum \left[\dfrac{A_j}{\sum_j A_j} \right]} \qquad ...(32.13)$$

I. MULTIPLE CHOICE QUESTIONS

1. Gravity model is used in transportation planning process for
   (*a*) model split (*b*) trip distribution
   (*c*) trip generation (*d*) trip assignment
2. Traffic conflicts that may occur in a rotary intersection are
   (*a*) merging and diverging (*b*) crossing and merging
   (*c*) crossing and diverging (*d*) crossing, merging and diverging

3. For detecting whether a bitumen is cracked (over heated) or not the test/ performed is/are
   (*a*) spot test (*b*) float test
   (*c*) ductility test (*d*) all of these
4. If in a dorry abrasion test the loss in weight is 21 gms, then co-efficient of hardness is
   (*a*) 9.5 (*b*) 13
   (*c*) 17 (*d*) 21
5. If the number of lanes on the carriageway of a road is more than two, the total width of lane ways is equal to 3.0 m
   (*a*) + 0.60 m (*b*) + 0.80 x_n
   (*c*) + 0.90 m (*d*) + 0.70 m

Answers

1. (*e*) **2.** (*a*) **3.** (*a*) **4.** (*a*) **5.** (*d*)

II. SOLVED QUESTIONS AND ANSWERS

Q. 1. What is trip distribution?

Ans. Trip distribution is the second component (after trip generation, but before mode choice and route assignment) in the traditional four-step transportation forecasting model. This step matches trip makers' origins and destinations to develop a "trip table", a matrix that displays the number of trips going from each origin to each destination. Historically, this component has been the least developed component of the transportation planning model.

Q. 2. What are the types of trip distribution?

Ans. There are two trip distribution matrices that need to be distinguished. The first is the observed (or empirical) distribution. This is the actual number of trips that are observed travelling between each origin zone and each destination zone. In general, with crime data, such an empirical distribution would be obtained from an arrest record where the residence (or arrest) location of each offender is listed for each crime that the offender was charged with. In this case, the residence/arrest location would be considered the origin while the crime location would be considered the destination.

Q. 3. What is the gravity model?

Ans. A model that is usually used for trip distribution is that of the gravity function, an application of Newton's fundamental law of attraction. On journey to crime since there is a common theoretical basis. In the original Newtonian formulation, the attraction, F, between two bodies of respective masses M_1 and M_2, separated by a distance D, will be equal to.

III. REVIEW QUESTIONS

1. What are the methods of trip distribution?
2. What is furness method?
3. Explain gravity model.
4. Write a brief note on Tranner's model.

REFERENCES

1. T.J. Fratar, Vehicular Trip Distributions by Successive Approximations, Traffic Quarterly, Vol. VIII, No. 1. Saugatuck, 1954.
2. K.P. Furness, Time Function Iteration, Traffic Engineering and Control, 7, 1965.
3. A.M. Voorhees, A General Theory of Traffic Movement, Proceedings of the Institute of Traffic Engineers, 1955.
4. J.C. Tanner, Factors Affecting the Amount of Travel, Road Research Technical Paper No. 51, H.M.S.O., London, 1961.
5. W.G. Hansen, Evaluation of Gravity Model Trip Distribution Procedures, H.R.B. Bulletin No. 547, Highway Research Board, Washington, 1962.
6. R.R. Roberts, An Evaluation of the Terminal Times on Gravity Model Travel Time Factors, Highway Research Record No. 250, Highway Research Board, Washington, 1968.
7. R.J. Bouchare and C.L. Pyers, Use of Gravity Model for Describing Urban Travel, Highway Research Record No. 88, Highway Research Board, Washington, 1965.
8. B.L. Smith, Gravity Model Theory Applied to Small City Using a Small Sample of Origin-Destination Data, Highway Research Record No. 88, Highway Research Board, Washington, 1965.
9. K.E. Heanue, L.B. Hammer and R.M. Hall, Adequacy of Clustered Home interview Sampling for Calibrating a Gravity Model Distribution Formula, Highway Research Record No. 88, Highway Research Board, Washington, 1965.
10. Calibrating and Testing a Gravity Model for any Size Urban Area, U.S. Bureau of Roads, Office of Planning, 1963.
11. C. Ben, R.J. Bouchard and C.E. Sweet, Simplified Procedures for Determining Travel Patterns Evaluated for a Small Urban Area, Public Roads, Vol. 33, February, 1955.
12. A.R. Tomazinis, A New Method of Trip Distribution in an Urban Area, H.R.B. Bulletin 347, H.R.B. Washington, 1962.
13. R.W. Whitaker and K.E. West, the Intervening Opportunities Model, A Theoretical Consideration, H.R. Record No. 250, H.R.B. Washington, 1968.
14. W.R. Blunden, The Land-Use/Transport System, Pergamon Press, 1971.

33. Traffic Assignment

33.1. Purpose of Traffic Assignment

Traffic assignment is the stage in the transport planning process wherein the trip interchanges are allocated to different parts of the network forming the transportation system. In this stage (*i*) the route to be travelled is determined and (*ii*) the interazonal flows are assigned to the selected routes. The purpose of traffic assignment are (Ref. 1) :

(*i*) To determine the deficiencies in the existing transportation system by assigning the future trips to the existing system.

(*ii*) To evaluate the effects of limited improvements and additions to the existing transportation system by assigning estimated future trips to the improved network.

(*iii*) To develop construction priorities by assigning estimated future trips for intermediate years to the transportation system proposed for those years.

(*iv*) To test alternative transportation system proposals by systematic and readily repeatable procedures.

(*v*) To provide design hour traffic volumes on highway and turning movements at junctions.

Thus the assignment process is useful both to the transport planner and the highway designer; to the former, because of the need to evaluate how the proposed transport system will work, and to the latter, for geometric design of individual links and intersections.

The advent of the modern digital computers has facilitated the growth of assignment techniques, which involve computations too laborious for manual handling.

33.2. General Principles

All assignment techniques are based on route selection. The choice of the route is made on the basis of a number of criteria such as journey time, length, cost, comfort, convenience and safety. Journey time is often considered as the sole criterion since length and cost can be considered as functions of time in most cases. The route selection is made manually for small jobs but large jobs make use of an electronic computer for this purpose.

As a first step, the highway network is described by a system of links and nodes. A link is a section of a highway network between two intersections. A node is either the centroid of a zone or the intersection of two or more links. For computer analysis, the network description is coded, key punched and stored in the memory of the computer. The computer is then made to select the minimum path between the zones and assign predicted trips to these paths. Traffic volumes

are thus accumulated for each section of the network. As stated earlier, the minimum path may be that route of travel which has least accumulation of time, distance or other parameters. The sequence of nodes which defines the links comprising the minimum path between any two zone centroids is called the 'tree'. The tree is determined by the computer starting from the zone centroid and progressively selecting the shortest path to the terminal zone centroid. When traffic is accumulated for each link it may so happen that certain individual links get overloaded. In that event, certain adjustments will have to be made in accordance with travel-time flow relationships fed to the computer.

A procedure commonly employed in assignment studies is what is known as 'Moore's Algorithm'. Moore developed a method for dealing with telephone calls or the basis of shortest path, and this method has been exploited in many computer programmes designed to assign the traffic in a street network. A simple example below illustrates the method of building the minimum-path tree. The

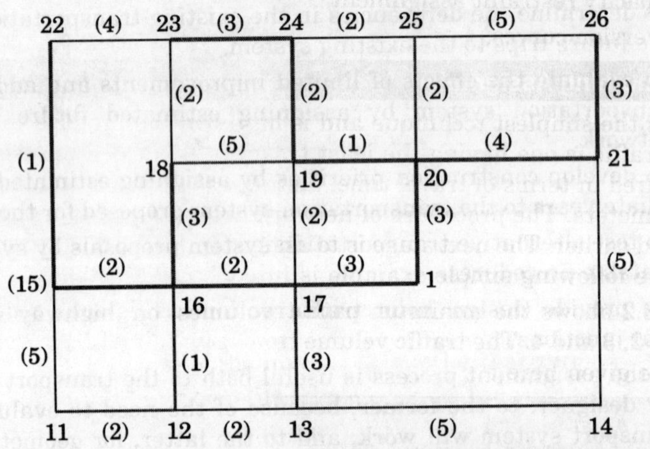

Minimum path tree is denoted by thick line

Fig. 33.1. Minimum path 'tree'

network shown in Fig. 33.1 consists of a zone centroid 1 and a number of links and nodes. The travel time on each link is indicated in figures in backets. It is required to build the minimum path tree from centroid 1.

Starting from Centroid 1, we go to each connecting node and note the time of travel to the node. Thus

$$T_{1-20} = 3$$
$$T_{1-17} = 3$$

The node close to Centroid 1 in time is considered next. In this example, the time taken to reach both the nodes being the same, the node having the lower number, *viz* Node 17 is taken up. From Node 17, the cumulative time to travel from Centroid connected directly to Node 17 is noted. Thus :

$$T_{1-17-19} = 5$$
$$T_{1-17-16} = 5$$
$$T_{1-17-13} = 6.$$

The next closest node to Centroid 1, *i.e.,* Node 20, is now considered and the cumulative time to travel from Centroid 1 to all nodes connected directly to it are noted. Thus :

$$T_{1-20-19} = 4$$
$$T_{1-20-25} = 6$$
$$T_{1-20-21} = 7.$$

It will be seen from the above that there are two possible routes to reach Node 19, *i.e.,* 1 – 17 – 19 and 1 – 20 – 19. The latter is shorter in time and is, therefore, chosen and the former is discarded. This process is repeated until all nodes have been covered by the shortest path. The minimum path 'tree' is indicated in Fig. 33.1.

33.3. Assignment Techniques

Though the broad principles of assignment have been discussed above, in practice different techniques of assignment are available. Some of these are :

(*i*) All-or-nothing assignment (Free assignment or Desire assignment)

(*ii*) Multiple route assignment

(*iii*) Capacity restraint assignment

(*iv*) Diversion curves.

33.4. All-or-nothing Assignment

This is the simplest technique and is based on the premise that the route followed by traffic is one having the least travel resistance. The resistance itself can be measured in terms of travel time, distance, cost or a suitable combination of these parameters. The procedure of network description and tree-building has been described earlier. The next stage is to assign the traffic flows to the minimum path tree. The following simple example is intended to illustrate the procedure.

Fig. 33.2 shows the minimum path tree connecting zone centroid 1 with zone centroid 2, 3 and 4. The traffic volume from zone centroid 1 to zone centroids 2, 3 and 4 are given here :

From Zone Centroid	*To Zone Centroid*	*Traffic Volume (vehicles/hour)*
1	2	2500
1	3	3000
1	4	4000

It is required to assign the flow from zone centroid 1 to zone centroids 2, 3 and 4.

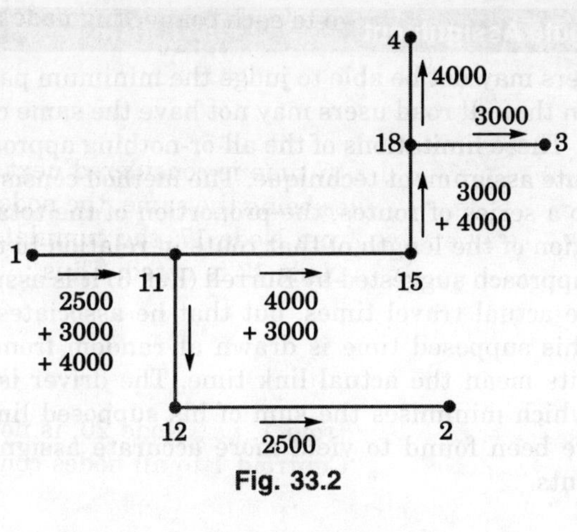

Fig. 33.2

It will be seen from the above figure, the traffic volumes assigned to the various links are :

Link	*Traffic flow (vehicles/hour)*
1—11	9500
11—12	2500
12—2	2500
11—15	7000
15—18	7000
18—3	3000
18—4	4000

Once the traffic volumes are assigned to the various links, a check is made to see that no link is loaded beyond its capacity. If overloading is found to exist, the journey times on this overloaded links are altered and the assignment is repeated.

Though simple, all-or-nothing assignment suffers from certain disadvantages. It is found to be the least desirable technique when compared with more accurate techniques (Ref. 1).

(*i*) If time alone is used as the governing factor for the choice of the minimum path, other equally important factors such as cost, reliability, convenience and safety will be neglected. But a driver may attach more value to these neglected factors, thus causing errors in the assigned flow. If cost is used as the sole governing factor, difficulties can arise because different persons value cost differently. It is also difficult to quantify all the costs involved in a journey, and some studies have restricted themselves only to quantifiable journey cost (Ref. 2).

(*ii*) Because of the very principle on which the technique is based, too many vehicles tend to be assigned to more attractive routes. This may cause increasing congestion on these routes and the technique takes no account of this factor. For the same reasons, all facilities in the network are not effectively utilised under this procedure.

(*iii*) If a superior facility is available, say, for example a motorway, people tend to prefer to use this facility for longer journeys. If the travel time or cost is the sole factor, the all-or-nothing assignment might not reflect this tendency.

(*iv*) Small differences in journey times by different routes between the same origin and destination can result in unrealistic journey paths.

33.5. Multiple Route Assignment

All road users may not be able to judge the minimum path for themselves. It may also happen that all road users may not have the same criteria for judging the shortest route. These limitations of the all-or-nothing approach are recognised in the multiple route assignment technique. The method consists of assigning the inter zonal flow to a series of routes, the proportion of the total flow assigned to each being a function of the length of that route in relation to the shortest route. In an interesting approach suggested by Burrell (Ref. 3) it is assumed that a driver does not know the actual travel times, but that he associates with each link a supposed time. This supposed time is drawn at random from a distribution of times, having as its mean the actual link time. The driver is then assumed to select the route which minimises the sum of his supposed link times. Multiple route models have been found to yield more accurate assignment than all-or-nothing assignments.

33.6. Capacity Restraint Assignment

33.6.1. Capacity restraint assignment is a process in which the travel resistance of a link is increased according to a relation between the practical capacity of the link and the volumes assigned to the link. This technique has been developed to overcome the inherent weakness of all-or-nothing assignment technique which takes no account of the capacity of the system between a pair of zones. The capacity restraint system, on the other hand, clearly restrains the number of vehicles that can use any particular corridor and, in fact, the whole system, if the assigned volumes are beyond the capacity of the network, and redistributes the traffic to realistic alternative paths.

33.6.2. Because of the iterative nature of the calculations involved, the capacity restraint technique is carried out entirely by an electronic computer. The procedure is similar to the all-or-nothing assignment as far as the initial data input are concerned. The additional data that is fed is the capacity of each link. The best paths are determined in the same way as in all-or-nothing technique by building the minimum path trees. Traffic is then assigned to the minimum paths, either fully or in stages, and as the assigned volume on each link approaches the capacity of the link, the new set of travel time on the link is calculated. This results in a new network with a different minimum path tree, differing significantly from the earlier minimum path tree. As a consequence, assigning the interzonal volumes to the new tree produces a new volume on each link. This iterative process is repeated until a satisfactory balance between volume and speed is achieved.

33.6.3. Some of the methods of capacity restraint are given below :

(1) Smock Method (Ref. 4). In this method, the all-or-nothing assignment is first worked out. In an iterative procedure, the link travel times are modified according to the function :

$$T_A = T_0 \, e^{\left(\frac{V}{c}-1\right)} \qquad \text{...(33.1)}$$

$$T_A \leq 5 \, T_0$$

where, T = original travel time or the travel time on a link when volume equals capacity.

T_A = adjusted travel time. e = exponential base.

V = assigned volume. c = computed link capacity.

In the second iteration, the adjusted travel times T_A are used to determine the minimum paths or trees. The resulting link volumes are averaged and these are again used to calculate the adjusted travel time for the next iteration.

(2) The Bureau of Public Roads (B.P.R.) Method. The formula used to update the link travel time is :

$$T_N = T_0 \left[1 + 0.15 \left(\frac{V}{c} \right)^4 \right] \qquad \text{...(33.2)}$$

where, T_N = link travel time at assigned volume

T_0 = base travel time at zero volume (equals 0.87 travel time at practical capacity).

V = volume, c = capacity.

In addition to the above two models, there are a few more which have been used in different studies (Ref. 5, 6, 7, 8). The Bureau of Public Road Assignment Manual is a useful reference guide (Ref. 9).

33.7. Diversion Curves

One of the frequently used assignment techniques is the diversion curves. These curves represent empirically derived relationships showing the proportion of traffic that is likely to be diverted on a new facility (bypass, new expressway, new arterial street etc.) once such a facility is constructed. The data collected from the pattern of road usage in the past serve to build up such curves. Diversion curves can be constructed using a variety of variables such as :

(*i*) Travel time saved. (*ii*) Distance saved.
(*iii*) Travel time ratio. (*iv*) Distance ratio.
(*v*) Travel time and distance saved. (*vi*) Distance and speed ratio.
(*vii*) Travel cost ratio.

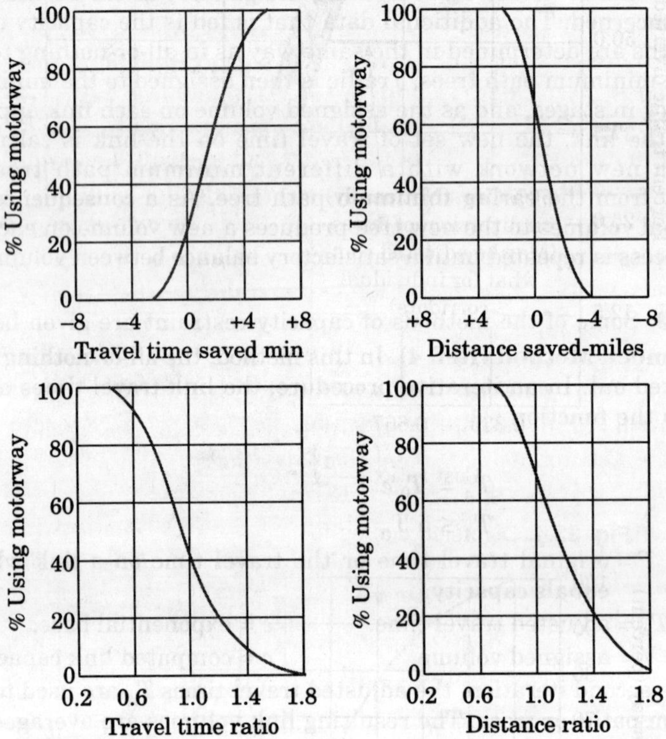

Fig. 33.3. Simple diversion curves using one variable.

Examples of diversion curves using single variable are given in Fig. 33.3.

The equations for diversion curves used in India for estimating the traffic that gets diverted to Expressways are given in Table 33.1 (Ref. 11).

Table 33.1. Diversion Curve Equations

Item	*Cost Ratio Interval*	*Equations*
Cars	< = 0.634	% Div. = 98.75 – (CR/0.634)*8.125
	0.634 < CR < 1.465	% Div. = 90.625 – ((CR – 0.634)/0.831)*84.375
	1.465 to 2.000	% Div. = 6.25 – ((CR – 1.465)/0.535)*5.25
Trucks & Buses	< = 0.750	% Div. = 100 – (CR)/0.75*5
	0.750 < CR < 1.250	% Div. = 95 – ((CR – 0.75)/0.5)*90
	1.250 to 2000	% Div. = (2 – CR)/0.75)*5

The curve is given in Fig. 33.4.

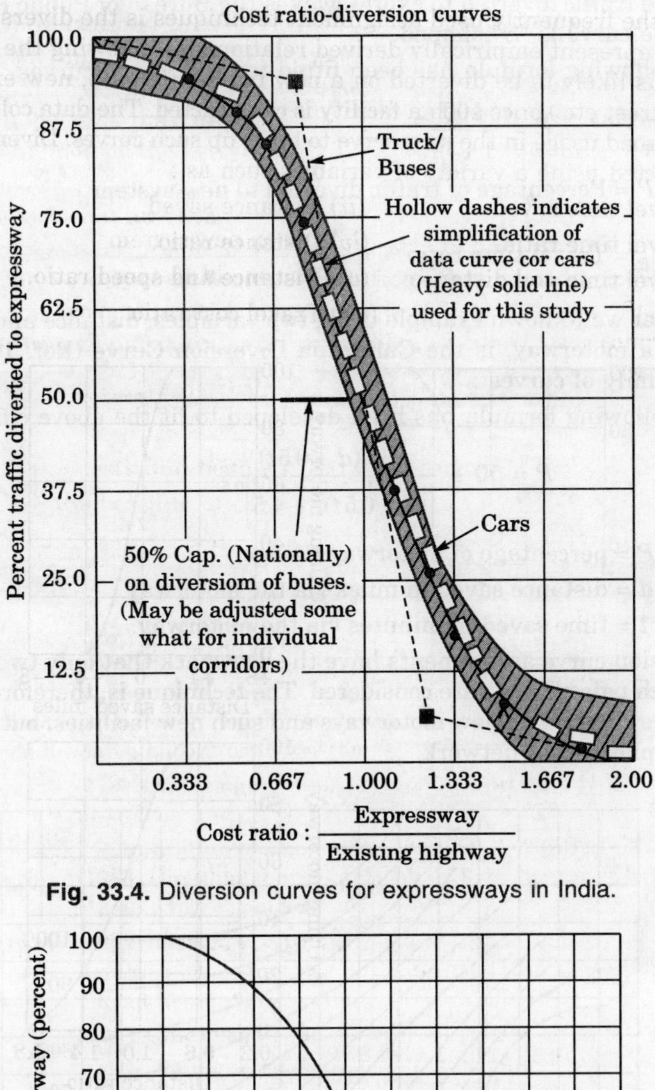

Fig. 33.4. Diversion curves for expressways in India.

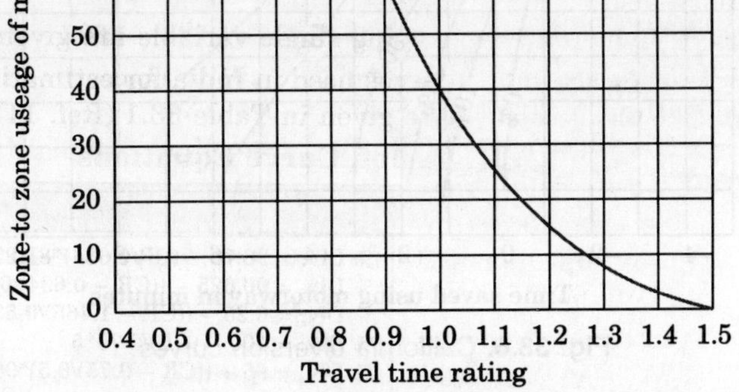

Fig. 33.5. Bureau of public roads diversion curve.

A well-known example of diversion curves using travel time ratio to determine the traffic diverted to expressway is the Bureau of Public Roads Curve, Fig. 33.5. The curve is *"S"* shaped.

The following formula has been fitted to this type of curves :

$$P = \frac{100}{1 + t_R^6} \qquad ...(33.3)$$

where, P = Percentage of traffic diverted to new system

$$t_R = \text{Travel time ratio \%} = \frac{\text{Time on new system}}{\text{Time on old system}}$$

Another well-known example using two variables, distance and travel time saved using a motorway, is the California Diversion Curve (Ref. 10). Fig. 33.6 shows the family of curves :

The following formula has been developed to fit the above curves.

$$P = 50 + \frac{50\,(d + 0.5t)}{\left[(d - 0.5t)^2 + 4.5\right]^{0.5}} \qquad ...(33.4)$$

where, P = percentage of motorway useage

d = distance saved in miles via the motorway

t = time saved in minutes via the motorway.

Diversion curve assignments have the drawback that only two alternative routes for each pair of zones are considered. The technique is, therefore, eminently suitable for new bypasses, new motorways and such new facilities, but is of limited use in a complex urban network.

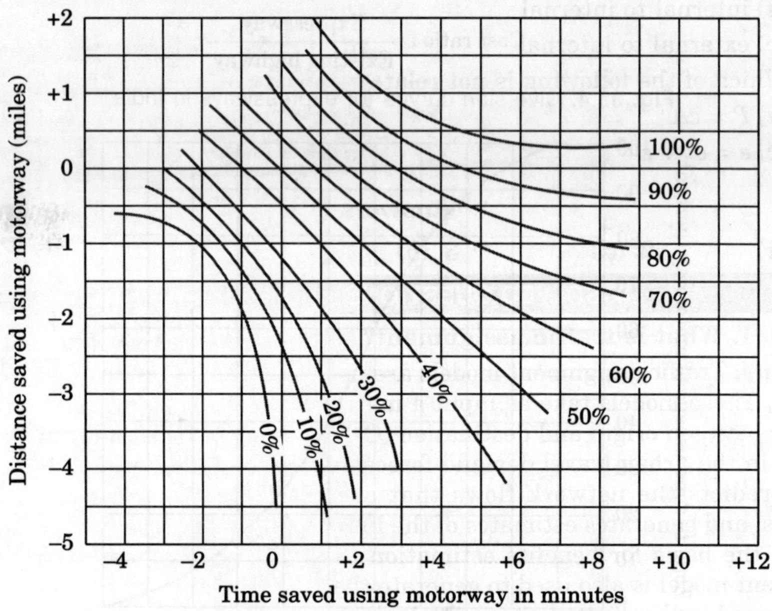

Fig. 33.6. California diversion curves.

Diversion curves reflect the travel resistance as measured by present day travel, but their use for future travel when the pattern can undergo radical changes is doubtful.

Problem 33.1. *In order to relieve congestion on an urban street network a motorway is proposed to be constructed. The travel time from one zone centroid to another via the proposed motorway is estimated to be 10 minutes whereas the time for the same travel via the existing streets is 18 minutes. The flow between the two zone centroids is 1000 vehicles per hour. Assign the flow between the new motorway and existing streets.*

Solution. Using Eq. (33.3).

$$P = \frac{100}{1+t_R^6}, \quad t_R = \frac{10}{18} = 0.56$$

$$P = \frac{100}{1+(0.56)^6} = 96.5\%.$$

Traffic diverted to the new motorway

$$= 1000 \times 96.5 = 965 \text{ veh/hour.}$$

and traffic using the existing streets = 35 veh/hour.

I. MULTIPLE CHOICE QUESTIONS

1. Utilities of two modes are 1.0 each. Estimate the probability of one of the modes.
   (*a*) 0.50 (*b*) 0.45
   (*c*) 0.52 (*d*) 0.60
2. Which movement is useful for planning a bye pass
   (*a*) internal to internal (*b*) internal to external
   (*c*) external to internal (*d*) external to external
3. Which of the following is not related to Lowry model
   (*a*) $P = eA$ (*b*) $e^s = p\,B$
   (*c*) $e = e^b + e^s$ (*d*) $e^p = e^b/e^s$

Answers

1. (*a*) 2. (*d*) 3. (*b*)

II. SOLVED QUESTIONS AND ANSWERS

Q. 1. What is traffic assignment?

Ans. Traffic assignment models are used to estimate the traffic flows on a network. These models take as input a matrix of flows that indicate the volume of traffic between origin and destination (O-D) pairs. Traffic assignment is a key element in the urban travel demand forecasting process. The traffic assignment model predicts the network flows that are associated with future planning scenarios, and generates estimates of the link travel times and related attributes that are the basis for benefits estimation and air quality impacts. The traffic assignment model is also used to generate the estimates of network performance that are used in the mode choice and trip distribution or destination choice stages of many models.

Q. 2. What is the aim of the traffic assignment?

Ans. The fundamental aim of the traffic assignment process is to reproduce on the transportation system, the pattern of vehicular movements which would be observed when the travel demand represented by the trip matrix, or matrices, to be assigned is satisfied. The major aims of traffic assignment procedures are:

1. To estimate the volume of traffic on the links of the network and obtain aggregate network measures.
2. To estimate inter zonal travel cost.
3. To analyze the travel pattern of each origin to destination (O-D) pair.
4. To identify congested links and to collect traffic data useful for the design of future junctions.

Q. 3. What is all-or-nothing assignment?

Ans. In this method the trips from any origin zone to any destination zone are loaded onto a single, minimum cost, path between them. This model is unrealistic as only one path between every O-D pair is utilized even if there is another path with the same or nearly same travel cost. Also, traffic on links is assigned without consideration of whether or not there is adequate capacity or heavy congestion; travel time is a fixed input and does not vary depending on the congestion on a link.

Q. 4. Explain incremental assignment.

Ans. Incremental assignment is a process in which fractions of traffic volumes are assigned in steps.In each step, a fixed proportion of total demand is assigned, based on all-or-nothing assignment. After each step, link travel times are recalculated based on link volumes. When there are many increments used, the flows may resemble an equilibrium assignment ; however, this method does not yield an equilibrium solution. Consequently, there will be inconsistencies between link volumes and travel times that can lead to errors in evaluation measures.

Also, incremental assignment is influenced by the order in which volumes for O-D pairs are assigned, raising the possibility of additional bias in results.

Q. 5. Wha is meant of capacity restraint assignment?

Ans. Capacity restraint assignment attempts to approximate an equilibrium solution by iterating between all or nothing traffic loadings and recalculating link travel times based on a congestion function that reflects link capacity. Unfortunately, this method does not converge and can flip-flop back and forth in loadings on some link.

Q. 6. Calculate the system travel and link flows by doing equilibrium assignment for the network in the given figure. Verify that the flows are at user equilibrium.

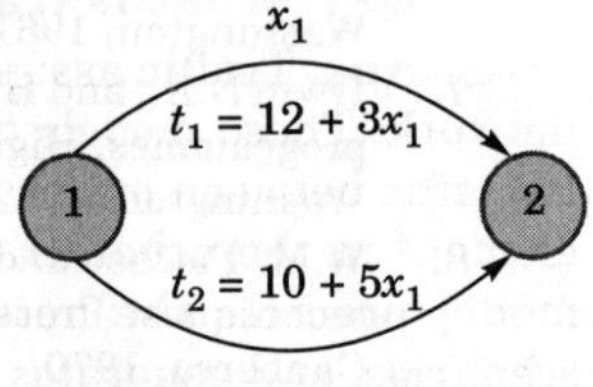

Fig. Two link problem with constant travel time.

Ans. Substituting the travel in the respective equation yield to

$$\min Z(x) = \int_0^{x_1} 12 + 3x_1 dx_1 + \int_0^{x_2} 10 + 5x_2 dx_2$$

$$= 12x_1 + \frac{3x_1^2}{2} + 10x_2 + \frac{5x_2^2}{2}$$

Subject to, $x_1 + x_2 = 12$

Substituting $x_2 = 12 - x_1$, in the above formulation will yield the unconstrained formulation as below:

$$\min Z(x) = 12x_1 + \frac{3x_1^2}{2} + 10(12 - x_1) + \frac{5(12 - x_1)^2}{2}$$

$$\min Z(x) = 4x_1^2 - 58x_1 + 480$$

Differentiate the above equation x_1 and equating to zero,

$$\frac{dz(x)}{dx} = 0 - 58 + 8x_1 = 0 \text{ or } x_1 = 7.25$$

Hence, $x_2 = 12 - 7, 25 = 4.75$

Travel times are,

$$t_1 = 12 + 3 \times 7.25 = 33.75 t_2 = 10 + 5 \times 4.75 = 33.75 \text{ i.e., } t_1 = t_2$$

It follows that the travel times are at user equilibrium.

III. REVIEW QUESTIONS

1. What is the purpose of traffic assignment?
2. Discuss the general principles of traffic assignment.
3. Explain the process of assignment technique.
4. Discuss diversion curves.

REFERENCES

1. M.J. Huber, H.B. Boutwell and U.K. Witheford, Comparative analysis of traffic assignment techniques with actual highway use N.C.H.R.P. Report 58, Highway Research Board, Washington, 1968.
2. T. Zakaria and J. Falcocchio, The traffic assignment process of the Delawars Valley Regional Planning Commission, Highway Research Record No. 250, Highway Research Board, Washington, 1968.
3. Burrell J., Multiple route assignment and its application to capacity restraint, 4th International Symposium on the Theory of Traffic Flow. Karlsruhe, 1968.
4. R. Smock, An iterative assignment approach to capacity restraint on arterial networks, Highway Research Board Bulletin 347, Highway Research Board, Washington, 1962.
5. M. Schneider, A direct approach to traffic assignment, Highway Research Record No. 6, Highway Research Board, Washington, 1963.
6. Irwin N.A. N. Dodd and Von Cube H.G,, Capacity restraint is assignment programmes, Highway Research Board Bulletin, Highway Research Board, Washington, 1961.
7. Irwin N.A., and H.G. Von Cube, Capacity restraint in multilateral mode assignment programmes, Highway Research Board Record No. 347, Highway Research Board, Washington, 1962.
8. W.M. Fischer and K.B. Davidson, Towards an improvement in traffic assignment techniques, Proceedings of the Australian Road Research Board, Vol. 5 Part 2, Canberra, 1970.
9. Bureau of Public Roads, Traffic Assignment Manual, U.S. Department of Commerce, 1964.
10. Moskowitz K., California method of assigning diverted traffic to proposed freeways, Highway Research Board Bulletin 130, Highway Research Board, Washington, 1956.
11. Development of Long Term Plan for Expressways in India, C.E.S. (India) Pvt. Ltd., Prepared for Ministry of Surface Transport, (Roads Wing), Govt, of India, 1991.

34. Modal Split

34.1. General Considerations

Modal split is the process of separating person-trips by the mode of travel. It is usually expressed as a fraction, ratio or percentage of the total number of trips. In general, modal split refers to the trips made by private car as opposed to public transport (road or rail).

An understanding of modal split is very important in transportation studies. Future transportation pattern can only be accurately forecast if the motivations that guide the traveller in his choice of the transportation modes can be analysed. Though the factors that govern the individual's choice of mode are complex, a study of the same is of great utility. Considerable attention is being given recent years to this aspect of transport planning.

34.2. Factors Affecting Modal Split

The factors that affect choice among alternative modes are heterogenous and numerous and it is convenient to discuss the factors under some broad categories as below :

1. Characteristics of the trip ;
2. Household characteristics ;
3. Zonal characteristics ;
4. Network characteristics.

These will now be examined in detail.

Characteristics the trip

(*i*) ***Trip purpose.*** The choice of mode is guided to a certain extent by the trip purpose. To give an example, home based school trips have a high rate of usage of public transport. On the other hand, home-based shopping journeys can have a higher rate of private car usage, for the simple reason that it is more convenient to shop when travelling in a personalised transport.

(*ii*) ***Trip length.*** The length can govern an individual's choice of a particular mode. A measure of the trip length is also possible by the travel time and the cost of travelling.

Household characteristics

(*i*) ***Income.*** The income of a person is a direct determinant of the expenses he is prepared to incur on a journey. Higher income groups are able to purchase and maintain private cars, and thus private car trips are more frequent as the income increases.

(*ii*) ***Car ownership.*** Car ownership is determined by the income and for this reason both income and car ownership are inter-related in their effect on modal choice. In general, families which own a car prefer private car trips, and in contrast families without car patronise public transport in the absence of any other alternative.

(*iii*) ***Family size and composition.*** The number of persons in the family, the number of school-going children, the number of wage earners, the number of unemployed, the age-sex structure of the family, and some other factors connected with the socio-economic status of the family profoundly influence the modal choice. Some of these factors are responsible for certain "captive" trips in public transport, such as those due to old age pensioners, school children, crippled and infirm persons and those who do not wish to drive. Normally, the principal family wage earner, usually the husband, takes priority in driving in the single car household, while the housewife and children must take to public transit, solicit a ride with friends or walk.

Zonal characteristics

(*i*) ***Residential density.*** The use of public transport increases as the residential density increases. This is because of the fact that areas with highest residential density are inhabited by persons with lower income, with lower levels of private car ownership. It is also found that higher density areas are served well by public transport system and such areas are oriented towards a better use of the public transport system.

(*ii*) Concentration of workers.

(*iii*) Distance from *CBD*.

Network characteristics

(*i*) ***Accessibility ratio.*** Accessibility ratio is a measure of the relative accessibility of that zone to all other zones by means of mass transit network and highway network.

(*ii*) ***Travel time ratio.*** The ratio of the travel time by public transport and travel time by private car gives a measure of the attractiveness or otherwise of public transport system. The travel time by public transport system is itself composed of :

(*a*) time spent walking to public transport vehicle at origin,

(*b*) time spent for waiting for public transport vehicle,

(*c*) time spent in public transport vehicle,

(*d*) time spent in transfer from one public transport vehicle to another, and

(*e*) time spent walking from public transport vehicle at destination.

Travel time by private car is composed of

(*a*) time spent driving the car,

(*b*) time spent in parking at destination, and

(*c*) time spent in walking from parked vehicle to destination.

It is generally found that as the travel time ratio increases, the usage of the public transport system falls down. Typical relations found (Ref. 5) are even in Fig. 34.1.

(*iii*) ***Travel cost ratio.*** The ratio of cost of travel by public transport and cost of travel by car is one of the most important factors influencing modal choice. In assessing the cost of travel by cars, the possibility of any arrangement of pooling cars for journey to work is also a consideration. The importance of travel cost is related to the economic status. People with high incomes are unmindful of cost and prefer more expensive modes (Ref. 4).

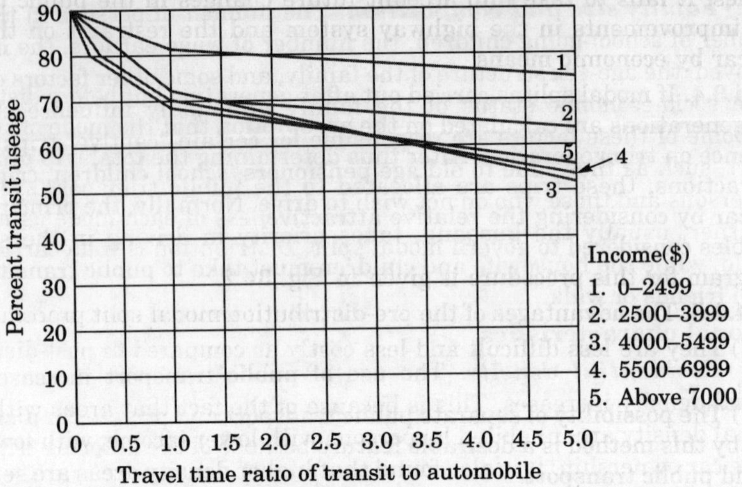

Fig. 34.1. Variation of modal split related to economic status and travel time ratios (Ref. 5).

Apart from the factors mentioned above, there are some others which are difficult of being quantified accurately. Comfort, convenience and prestige associated with various forms of travel fall under this category. Comfort is one of the main considerations in modal choice. Convenience which also includes seat probability in a public transport system and the presence of luggages in the hand, is another factor most frequently weighed by the travellers in deciding between private car or a public transport. The relative safety of travel is yet another factor which may be a consideration with some. The opportunity to read can be a factor favouring travel by rail. Waiting, transferring, overcrowding, standing and long walking are some considerations most unfavourable to public transport usage.

34.3. Modal Split in the Transport Planning Process

34.3.1. Modal split is considered in the transport planning process at a number of stages. Basically, two variations are possible :

(*i*) Pre-distribution modal split, in which the modal split is considered prior to trip distribution stages. This procedure is also known as trip end modal split procedure.

(*ii*) Post-distribution modal split, in which the modal split is considered after trip distribution stage. This procedure is also known as trip interchange modal split procedure.

34.3.2. In the pre-distribution modal split procedure, there are again two possibilities :

(*i*) At the trip generation stage itself ;

(*ii*) After trip generation, but before trip distribution.

34.3.3. If modal split is considered at the trip generation stage itself, it is necessary to derive separate multiple linear regression equation for each mode of transport, *viz.*, car, public transport, rail etc. The factors that are normally considered to influence modal choice in this type of procedure are the car ownership, residential density, distance of the zone of origin from the *CBD* and the relative accessibility of the zone of the origin to the transport facilities. Though the method reflects different trip generation patterns for different trip purposes and modes, it fails to take into account future changes in the public transport system, improvements in the highway system and the restraint on the use of private car by economic means.

34.3.4. If modal split is carried out after generation but before distribution, the trip generations are calculated on the assumption that the mode of travel has no influence on trip generation. After thus determining the total trip productions and attractions, these trips are allocated to the public transport system and private car by considering the relative attractiveness of each mode as measured by variables considered to govern modal split. Distribution is then carried out. A flow diagram for this procedure is given in Fig. 34.2.

34.3.5. The advantages of the pre-distribution modal split procedures are :

(*i*) They are less difficult and less costly as compared to post-distribution methods.

(*ii*) The possibility of separate public transport and private car distribution afforded by this method is a desirable feature because of the differing trip lengths by car and public transport.

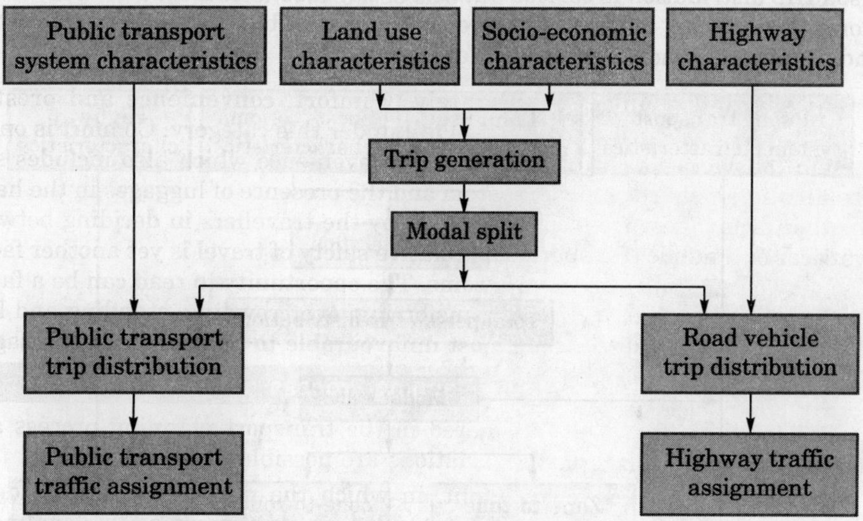

Fig. 34.2. Flow diagram for modal split carried out between trip generation and trip distribution.

(*iii*) This method reflects factors such as the income, car ownership, family structure, employment etc. (household and zonal characteristics) which are the characteristics affecting trip generation.

The disadvantages of pre-distribution methods are :

(*i*) Since these methods are strongly wedded to existing and historical levels of public transport service, they are inappropriate to studies involving planning of improvements to public transport systems where significantly different levels of service are contemplated.

(*ii*) In this method, the characteristics of the transportation system are fed on an average area-wide basis. As such, this procedure fails to reflect the particular zone-to-zone combination as precisely as tripinterchange models.

(*iii*) It does not consider the trip generation characteristics fully.

(*iv*) It is insensitive to future developments in inter-zonal travel.

34.3.6. In the post-distribution models, one possibility is to carry out modal split after distribution, but before assignment. In this procedure, the zone-to-zone home-based trips are known. Using this as input, the procedure determines the zone-to-zone public transport travellers on the basis of variables representing characteristics of the person making the journey, characteristics of the destination end of the journey and characteristics of the transport system—all measured on zone-to-zone basis. Relative travel times and relative travel costs are some of the usual parameters considered in the models. By subtracting the zone-to-zone public transport trips from the total zone-to-zone person-trips, the person trips mads by motor vehicle are derived. The assignment of these trips is carried out as the next stage.

A flow diagram for a model of the above type is given in Fig. 34.3.

34.3.7. The trip interchange model is more complex, especially if the number of zones is large. It also suffers from another drawback in that the total person trip distribution is carried out before any modal choice is considered. This ignores the differing lengths of journey by car and public transport. On the other hand, it has many advantages such as :

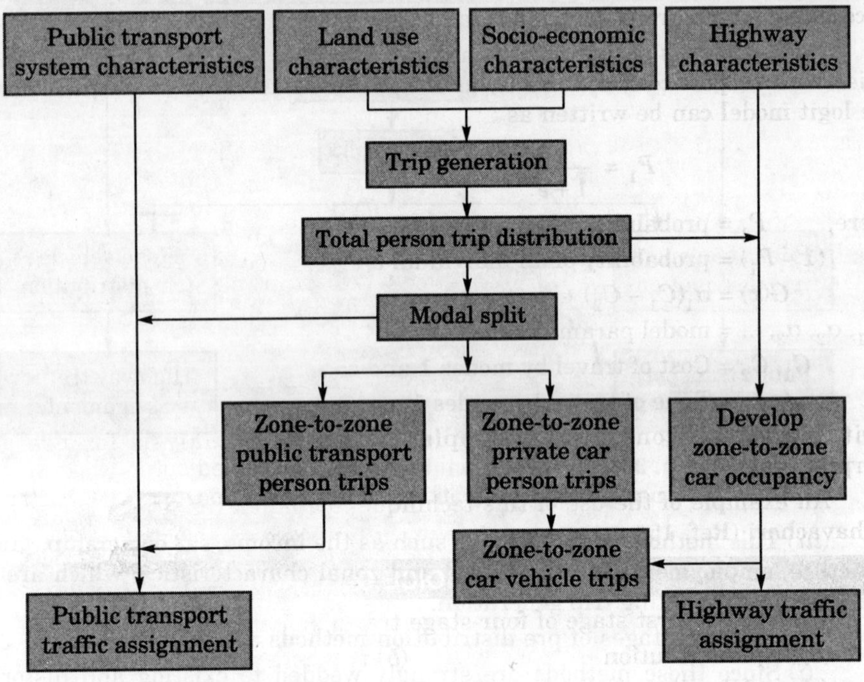

Fig. 34.3. Flow diagram for modal split carried out after trip distribution.

(*i*) It is useful in situations where serious consideration is given to public transport planning.

(*ii*) The method makes it possible to develop modal split relations based on a wide range of transport system variables influencing modal choice.

(*iii*) The method considers private car and public transport usage on a zone-to-zone basis instead of a zonal basis as in the pre-distribution methods.

34.4. Recent Developments in Modal Split Analysis

In vew of the complexities involved in determining modal split, a considerable amount of research has taken place in recent years to develop better techniques (Ref. 11, 12, 13, 14, 15).

The following are some of these :

(*i*) Probit analysis (*ii*) Logit analysis (*iii*) Discriminant analysis

Probit analysis is based on the principle that if members of population are subjected to a stimulus that can range over an infinite scale, the frequency of response to stimulus (assuming that the response is a 0–1 response) will be normally distributed.

In modal choice, the stimulus is assumed to be made up of the relative disutilities of travel on two modes and of the characteristics of the user. The probit equation can be written as :

$$Y = a_0 + a_1x_1 + a_2x_2 + \text{........}$$

where, Y = probit value for the probability of transit mode choice.

$x_1, x_2 ...$ = disutility variables.

The determination of the coefficients of the disutilities is done by calibration procedures, which are often lengthy and time-consuming.

Logit analysis assumes that the probability of the occurrence of an event varies with respect to function $F(x)$ as a sigmoid curve called the logistic curve. The logit model can be written as :

$$P_1 = \frac{1}{1 + e^{G(x)}}$$

where, P_1 = probability of an individual choosing mode 1.

$(1 - P_1)$ = probability of an individual choosing mode 2.

$G(x) = \alpha_1(C_1 - C_2) + \alpha_2(t_1 - t_2) + \alpha_3(\ ) +$

$\alpha_1, \alpha_2, \alpha_3, ...$ = model parameters

C_1, C_2 = Cost of travel by modes 1 and 2.

t_1, t_2 = Time of travel by modes 1 and 2

Logit analysis is considerably simpler than probit analysis for use and interpretation.

An example of the use of this technique for Indian conditions is given by Raghavachari (Ref. 15).

I. MULTIPLE CHOICE QUESTIONS

1. What is the first stage of four-stage travel demand modelling?

(*a*) trip distribution (*b*) trip generation

(*c*) modal split (*d*) traffic assignment

2. Model split aims to

(*a*) determine the number of trips on different modes.

(*b*) determine the speed of vehicles

(*c*) both (*a*) and (*b*)

(*d*) None of these

Answers

1. (*b*) 2. (*a*)

II. SOLVED QUESTIONS AND ANSWERS

Q. 1. What is meant by modal split?

Ans. This split was based on the predetermined relationships between public and personal transport being derived from the surveyed relationships and forecasts. The surveyed area (160 zones) was divided into three districts (the district between the administrative and master plan boundaries, the district between the master plan and central district boundaries, and the central district), as well as into eight corridors making up a geographic and functional entity.

In this way, 17 macro zones were obtained. Modal split parameters – public transport 40-80% and personal transport 20-60% – formed a 17 × 17 matrix

Q. 2. Write a brief note on modal split evaluation.

Ans. Modal split models used to simulate the selected mode are mostly based on official statistics (the number of inhabitants in the district, motorization degree, income, etc.) but may also include the data provided by transport studies (travel time or distance).

The relevance of these parameters in the analysis and modelling of the selected mode depends on input availability and quality. The data on relevant parameters are obtained from census records and some systemic local statistics which are periodically made, mostly not for transport studies. Most models rely on linear relationship between the variables on assumption that their effect is continuous, thus neglecting any stress, such as: the imposition of fuel consumption restrictions, an abrupt decline of living standards and similar.

Due to its simplified form, the models represent mutual influences only in part. The problem concerning aggregate four-step models involves the lack of a feedback between different steps. A change in transport conditions can modify not only the selected mode but also trip generation, as well as time and spatial split.

Q. 3. What models are used currently?

Ans. After surveying the modal split methods, it can be concluded that, in essence, three groups of models are in use:

- Methods by which the planners' aims are set and modal split is normatively made, assuming that the users will be influenced as desired, but without a clear recognition of the users' real behaviour.
- Analytical methods be which, on the basis of available data, a modal split model is developed, including relevant parameters which, during forecasting, are evaluated by engineering methods in real terms (time – travel time, distance, etc.)
- Methods which, to degree, rely on the users' real behaviour, including inertness (habits) with respect to a change of mode, inadequate information level, subjective evaluation and hostility towards a change, standing during the ride, etc.

III. REVIEW QUESTIONS

1. What are the general considerations in model split?
2. Describe the factors affecting modal spilt.
3. Explain the modal spilt in the transport planning process.
4. Discuss the recent developments in modal split analysis.

REFERENCES

1. Factors influencing modal trip assignment, NCHRP Report 57, Highway Research Board, Washington, 1968.
2. J.L. Schofer and A.M. Voorhees, Modal split model, Paper presented at O-D Committee meeting, 43rd Annual Metting, Highway Research Board, Washington, 1964.
3. L.E. Keefer, Characteristics of captive and choice transit trips in the Pittsburg Metropolitan Area, H.R.B. Bulletin 347, Highway Research Board Washington, 1962.
4. D.M. Hill and N. Dodd, Travel mode split in assignment programmes, H.R.B. Bulletin 347, Highway Research Board, Washington, 1962.
5. A.B. Sosslau. K.E. Heanue and A.J. Black, Evaluation of a new modal split procedure, Public Roads, Vol. 33, No. 1, Washington, 1964.
6. F.R. Wilson, The journey to work-Modal split, Maclaren 1967.
7. Quarmby, D.A., Choice of travel mode for journey to work, Journal for Transport Economics and Policy, Vol. 1. No. 3, September, 1967.
8. U.S. Dept. of Transportation, Bureau of Public Roads, Modal Split, Washington, 1966.
9. Bruton, M.J., Introduction to Transportation Planning, Hutchinson Technical Education, London, 1970.
10. Lane, A., Powell, T.J. and Smith P.R. Analytical Transport Planning, Duckworth, London, 1971.
11. McGillivrary, R.G., Demand and choice models for modal split, Journal of transport Economics and Policy, Vol. 4, No. 2, 1970.
12. Stopher. P.R., A probability model for travel mode choice for the work journey, Highway Research Record No. 283, Washington, 1969.
13. Warner, S.L., Stochastic choice of mode in urban travel : A study in binary choice, North Western University Press, Evanston Illinois, 1982.
14. Lave, C.A. A behavioural approach to modal split forecasting, Transport Research, Vol. 3, 1969.
15. Raghavachari, S., Modal split models forwork trip travel, Indian Highways, New Delhi, 1982.

35. Evaluation

35.1. Need for Evaluation

The transportation plans, which are prepared with specific goals, require to be evaluated to asses how far they meet the desired objectives. The plan which satisfies the requirements in the best possible manner is ultimately selected for implementation. Evaluation is, therefore, an important stage in the formulation of transportation plan.

35.2. Several Plans to be Formulated

Obviously, for achieving certain specific goals, there may be more than one solution. Some may be costly, and some may have other disadvantages and advantages. It is only by comparing the pros and cons of several alternatives that the best and optimum solution can be made to emerge. One of the alternatives is the "do nothing" situation, which serves as a convenient yardstick with which to measure up the remaining.

35.3. Testing

For comparing several alternatives, the initial step is to test the plan and assess how it would function in practice. The transport plan itself is based on a number of models which are designed to represent the various factors that determine the nature and amount of travel. This makes it incumbant on the part of the planner to test and assess how the plan is going to work. What are the actual flows in the network and what are the effects of these flow patterns on the environment, accident situation and comfort ? What is the impact of the plan on the parking requirements ? How do the intersections behave under the expected flows ? How is the land-luse development going to be influenced by the plan ? These are some of the questions that need to be answered by testing the several alternatives and by preparing a balance-sheet of advantages and disadvantages of each plan.

35.4. Considerations in Evaluation

Some of the considerations in evaluating alternatives are :

(*i*) System performance, including freedom of choice as to time and route of travel, comfort, convenience, safety and level of service provided in terms of reliability, travel time and delay.

(*ii*) Effect on the community environment including impact on the compatibility with adjacent land use, considering such factors as fumes, noise and disruption of neighbourhoods, effects on pattern of land development and aesthetics.

(*iii*) Economic factors, including annual capital costs, user operating costs (including accidents) and time costs.

(*iv*) Financial considerations as to the funds available to the community.

(*v*) Political acceptability considering the many governmental units within the typical metropolitan study area, many with differing and often opposing interests as to desirable transportation facilities.

35.5. Economic Evaluation

The qualitative answers obtained by testing the plans need to be translated into monetary terms so as to easily assess the comparative goods and evils of the plans. It is, therefore, very common to express the evaluation in economic terms. The idea is to use money as measure of the effectiveness of the several plans and to provide a common yardstick. In many situations it is possible to assign a monetary value to the impact of the plan, but there are many more where it is not possible, at the present stage of knowledge, to translate the impact in real money terms. Inspite of this drawback, it is convenient to evaluate the plans in economic terms and this practice has become common. A full discussion on economic evaluation is contained in Chapter 38.

I. MULTIPLE CHOICE QUESTIONS

1. Which one of the following equipments is useful in determining spot speed in traffic engineering?

(*a*) endoscope (*b*) periscope (*c*) radar (*d*) tachometer

2. Hot bitumen is sprayed over freshly constructed bituminous surface followed by spreading of 6.3 mm coarse aggregates and rolled. Which one of the following is indicated by this type of construction?

(*a*) surface dressing (*b*) gravel-bitumen mix

(*c*) liquid seal coat (*d*) seal coat

3. The walls which are necessary on the hill side of roadway where earth has to be retained from slipping is called

(*a*) retaining wall (*b*) breast wall

(*c*) parapet wall (*d*) none of these

4. When the path travelled along the road surface is more than the circumferential movement of the wheels due to rotation ,then it results in

(*a*) slipping (*b*) skidding

(*c*) turning (*d*) revolving

5. On a right angled road section with two way traffic the total number of conflicts point is

(*a*) 6 (*b*) 11 (*c*) 18 (*d*) 24

Answers

1. (*c*) **2.** (*a*) **3.** (*b*) **4.** (*b*) **5.** (*d*)

II. SOLVED QUESTIONS AND ANSWERS

Q. 1. What is Typical Economic Evaluation?

Ans. A typical economic evaluation involves the following general steps:

1. Describe each option, including a base-case and one or more alternatives.

2. Define the evaluation framework (described above), which identifies all impacts (costs and benefits) and objectives to be considered in the analysis. Classify impacts to avoid double-counting.

3. Quantify and monetize (measure in monetary value) impacts that are suitable for each option.

4. Calculate the total monetized benefits and costs for each year that is being considered (typically 10-20 years for a major investment project), and apply a discount value to future impacts. Sum the present value of benefits and costs to determine the Net Present Value.

5. Describe, and measure as much as possible, impacts that are unsuited for monetization (such as equity and effects on strategic community development objectives). Rates each alternative according to how much it supports or contradicts the objectives.

6. Conduct sensitivity analysis to determine how changes in key assumptions affect outcomes.

Q. 2. What are the main economic principles?

Ans. Efficient markets must reflect certain principles, including consumer choice, marginal cost pricing, economic neutrality, and competition). This is not to say that subsidies, under pricing and monopolies are always bad, they may be justified for various reasons (*e.g.*, to minimize transaction costs, for equity, to stimulate a certain type of economic development).

However, markets that reflect these principles tend to maximize productivity, consumer benefits and fairness. Transportation policies and programs that support market principles can provide additional economic benefits that are not usually reflected in conventional economic analysis. For example, funding road and parking facilities with user charges tends to be more efficient and fair than indirect funding, and variable fees that increase during congested periods are more efficient and equitable than flat fees.

Economic evaluation should indicate how a proposed policy or program impacts market principles, explicitly identify market distortions, and highlight opportunities to achieve transportation objectives in ways that reflect market principles and improve consumer choice.

Q. 3. Write a brief note on the impacts transportation on economic development.

Ans. Transportation made a great impacts on economic development. It is often intended to support economic development. This is particularly appropriate in undeveloped regions. If an area lacks basic access, paving a highway can significantly increase economic productivity and development. But once a basic road system exist, projects that provide marginal increases in traffic capacity or speed tend to provide small economic development benefits, and can even reduce local economic activity. Increased highway capacity tends to shift economic activity from one location to another (for example, favoring regional retail centers over neighborhood shops, and urban fringe housing over infill development), but often provides little increase in regional productivity or economic development. In most situations, the majority of economic benefits of transportation improvements can be measured in terms of direct user benefits, such as travel time and vehicle cost savings.

It is generally a mistake to include construction employment benefits (these are economic transfers, consisting of tax revenue collected throughout a jurisdiction devoted to a particular transportation project), property value increases, or macroeconomic benefits.

III. REVIEW QUESTIONS

1. What is the need for evaluation?
2. What consideration is taken in evaluation?

36. Land-use Transport Models

36.1. Introduction

In the previous Chapters, it has been shown how land-use determines transport demand and how this simple phenomenon can be made use of in building up the system of transport planning known commonly as the Urban Transport Planning Package (UTPP). The basic requirement in this technique is the land-use activities for the year being studied. A criticism of this technique has been that the travel demands estimated by it tend to be high in the principal travel corridors, with the result that highly costly rapid transit systems become necessary. Another weakness of the UTPP is that it requires a completely specified land-use allocation both at the production and at the attraction end of trips.

Attempts to overcome the above deficiencies have given rise to an entirely different transport planning technique, whereby land-use allocation and travel demands are determined simultaneously. The appeal of this technique is that land-use is not exogenously specified, but is on the contrary determined by the model itself, along with transport demand. The land-use transport models are sensitive to major development policy variables and thus give scope for manipulations in a variety of ways so as to select the best alternative.

An extensive review of land-use transport models is found in literature (Ref. 1, 2, 3, 4). In this Chapter, the principles of some of the popular models are discussed.

36.2. Selection of Land-use Transport Model

A variety of land-use transport models have been developed in the past 30 years. A few of them are "research models" having excellent capabilities for sensitive forecasting, while some are "operational models". The former group of models require extensive data collected through special surveys, whereas the latter need data which are collected routinely by planning departments. The Lowry derivative models fall into the latter group and are very popular.

While selecting a model, a number of considerations become important. These are :

1. Simplicity : The model should have a simple causal structure, which should be easy to comprehend. A simple model will generally consume less of time and resources.
2. Modest data requirements : Data requirements must be modest ; in fact, some of the good models make use of data routinely available with the planning department.

3. Adaptability : The model should be adaptable to any given location.

4. Comprehensiveness : The model should be comprehensive and should synthesise the relationship between activities, housing and transportation adequately well.

5. Operationality and rapidity : The model should be operational, capable of easy interpretation and should be able to test rapidly a wide range of policy options

6. Computer cost : The model should be operational at relatively cheap computer cost.

36.3. Lowry Derivative Models

36.3.1. The Lowry derivative models have many of the above attributes. They are simple to use, require modest data, are comprehensive and economical, have good response to change in input variables and have simple causal structure.

They have, therefore, been used extensively and successfully in a number of studies. Research carried out on the subject has yielded a variety of models, all based primarily on the original formulation.

36.3.2. Features of the Lowry Model

Lowry presented his model in 1964 for the spatial organisation of human activities in a metropolitan area (Ref. 5).

The fundamental structure of the model is illustrated in Fig. 36.1.

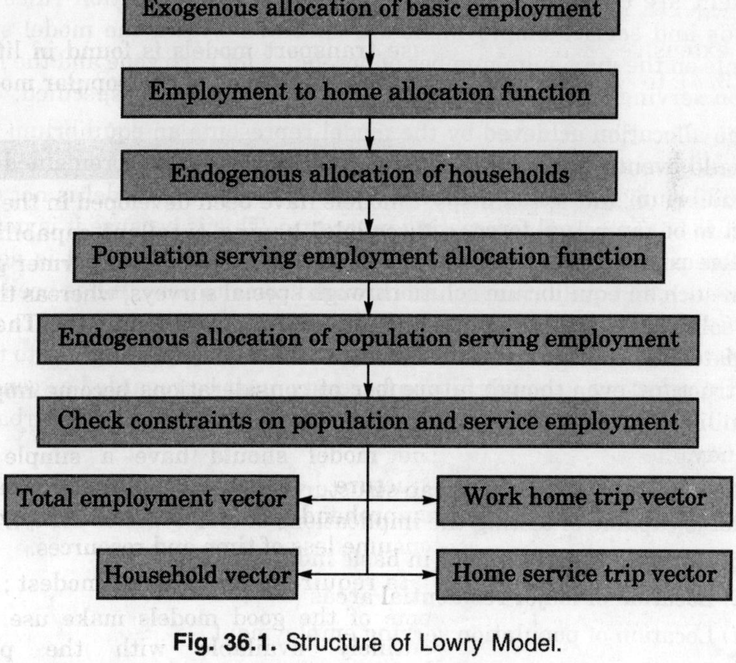

Fig. 36.1. Structure of Lowry Model.

The Lowry model relates the three principal components of the urban area, *viz* :

1. Population
2. Employment
3. Communication between population and employment.

Three broad sectors of activity are identified in the urban form :

1. Employment in basic industries
2. Employment in service (population serving) industries
3. Household or population sector.

The basic employment is defined as employment in those industries whose outputs and services are sold in markets external to the region under study. Following from this definition, the location of basic employment within a region is independent of the population and the service employment distribution of that region. Typical examples of basic industries are primary industries, manufacturing, national financial institutions, university etc.

Service employment is that category of employment which serves the population in a region. Thus, the location of service employment depends upon the population distribution of that region. Examples of service employment are the distribution and retail trade, utilities, personal services, elementary and high school employment etc.

The household sector consists of resident population.

In the model, the spatial distribution of basic employment is allocated exogenously. The spatial distribution of households and population serving employment are calculated by the model. The zonal allocation rules for both households and service employment are specified within the model structure. Constraints on the maximum number of hoseholds for each zone and the minimum population serving employment thresholds for zone are also specified.

The allocation achieved by the model represents an equilibrium situation which would eventually come to pass if all other factors had remained constant while equilibrium was being achieved. Consequently, the model is not (nor does it purport to be) an actual forecasting procedure. This is because it is not possible to associate points in real time with the model solutions since it is not clear if and when such an equilibrium solutions would come to pass. It has been observed that the solutions produced by the model bear a strong resemblance to reality or to anticipated realilty. As such, the model provides a useful insight into the urban spatial structure, even though future forecasts must be regarded as conditions of quasi-equilibrium, and not reflecting the development history of the urban system under study.

The use of the model for urban structure analysis is illustrated in Fig. 36.2. The model is capable of testing the implications of four types of policy variables :

(*i*) Location of employment in basic industries

(*ii*) Location of major residential areas

(*iii*) Location of population serving employment

(*iv*) Location of the principal transport system linking the population and the employment location.

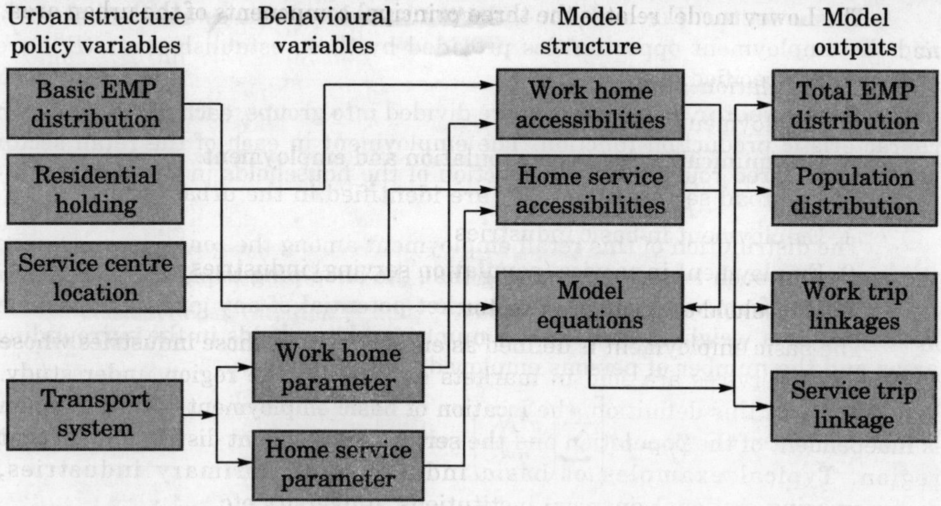

Fig. 36.2. Use of Lowry Land Use Transport Model for Urban Structure Analysis.

36.3.3. Model Equation System

The Lowry model can be represented in terms of nine simultaneous equations and three inequalities ;

The following notation will be used :

A = Area of land

E = Employment (number of persons)

N = Population (number of households)

T = Index of trip distribution

Z = Constraints

The following superscripts and subscripts are used in conjunction with the above notation :

U = Unusable land

B = Basic sector

R = Retail sector

H = Household sector

k = Class of establishments within the retail sector (groceries, clinics, primary schools, clothshops etc).

m = Number of classes of retail establishment ($k = 1........ m$)

n = Number of zones ($i = 1, 2, ...n$)

($j = 1, 2, ...n$)

Unspecified functions and coefficients are represented by lower- case letters, *a, b, c, d, e, f, g*.

Land Use. The total land available in each zone and the land therein not useable by any of the three activities are given. Thus :

$$A_j = A_j^U + A_j^B + A_j^R + A_j^H \qquad ...(36.1)$$

Basic sector. For each zone, the land used by basic establishments (A_j^B) and the employment opportunities provided by these establishments (E_j^B) are exogenously specified.

Retail sector. Retail sectors are divided into groups, each having its. own characteristic production function. The employment in each of the retail sector can be considered roughly to be a function of the households in the region.

$$E^k = a^k.N \qquad ...(36.2)$$

The distribution of this retail employment among the zones depends upon the market at each location. Assuming that the shop-ping trips originate either from home or from work places, the market potential of any given location can be defined as a weighted index of the number of households in the surrounding areas and the number of persons employed nearby. Thus :

$$E_j^k = b^k \left[\sum_{i=1}^{n} \left(\frac{c^k N_i}{T_{ij}^k} \right) + d^k E_j \right] \qquad ...(36.3)$$

The equation could be made more general. However, it has been assumed that none but short-range pedestrian trips originate from workplaces, so that only relevant origins are those in zone j. Those originating from home are often longer vehicular trips, but the likelihood of a shopping trip from i to j diminishes with intervening distance. The variable T_{ij}^k is a positive function of this distance, fitted from an analysis of home-based vehicular shopping trips. The coefficients c^k and d^k are measures of the relative importance of homes and work places as origins for a particular type of shopping. b^k is a scale factor which adjusts retail employment in each tract to the regional total determined in Equation 36.2.

$$E^k = \sum_{j=1}^{n} E_j^k \qquad ...(36.4)$$

Total Employment. The above procedure determines the amount of employment in any zone for each category of retail trade. The sum of these employment figures, plus the quantity of basic employment allocated to the zone is the total employment for the zone. Thus :

$$E_j = E_j^B \sum_{k=1}^{m} E_j^k \qquad ...(36.5)$$

Finally, with the aid of exogenously determined employment density coefficients (e^k) for each line of trade, we can determine the amount of land in each zone which will be occupied by retail establishments.

$$A_j^k = E_j^B \sum_{i=1}^{m} e^k E_j^k \qquad ...(36.6)$$

Household Sector. The region's population of households is a function of total employment. Thus :

$$N = f \sum_{i=1}^{n} E_j \qquad ...(36.7)$$

The population living in any zone j is also a function of the employment opportunities in the different zones and the measure of accessibility from zone j to each of these zones. Thus :

$$N_j = f \sum_{i=1}^{n} \frac{E_i}{T_{ij}} \qquad ...(36.8)$$

The coefficient g is a scale factor whose value is determined by the requirement that the sum of the zone populations must equal to the total population of the region as determined in Eqn. 36.7. Thus :

$$N = \sum_{i=1}^{n} N_j \qquad ...(36.9)$$

Constraints. In order to limit the dispersion of retail employment, a minimum size constraint (Z^k) is imposed, expressed in terms of employment. If the market potential of a particular location does not justify an establishment above this minimum size, the customers are sent elsewhere. Thus :

$$E_j^k \le Z^k, \text{ or else, } E_j^k = 0 \qquad ...(36.10)$$

In order to prevent the system from generating excessive population densities in locations with high accessibility indices, a maximum density constraint is imposed. The value of this constraint (number of households per unit area of residential space) may vary from zone to zone, as would be the case under zoning ordinances.

$$N_j \le Z_j^H \,.\, A_j^H \qquad ...(36.11)$$

And lastly, the amount of land set aside for retail establishments by Eqn. 36.6 must not exceed the amount available.

$$A_j^R < A_j - A_j^U - A_j^B \qquad ...(36.12)$$

Taken together with the accounting relationships expressed in Eqn. 36.1, this constraint also prevents the assignment of negative value to the residential land.

The solution of the system with the help of the above equations and inequalities is best done on a computer using an iterative approach.

36.4. Garin-Lowry Model

36.4.1. Garin (Ref. 6) has made very significant contribution to the development of the original Lowry model. Garin interpreted the economic base methodology explicitly, and expressed the Lowry algorithm in vector and matrix format. The vector and matrix expressions are described in section 36.5.

The Lowry model makes a distinction about population depen-dent on "basic" employment and that dependent on "service" employment. Thus if E is the total employment, B the basic employment and S the service employment,

$$E = B + S \qquad ...(36.13)$$

From the economic base theory, it is possible to relate the level of economic activity to the population level. Thus, a given number of total jobs E will support a certain number of people P. If we represent the number of people supported by one job as α, then

$$P = \alpha \,.\, E \qquad ...(36.14)$$

α is known as the population multiplier

$$\alpha = \frac{P}{E} \qquad ...(36.15)$$

Since $E = B + S, \quad P = \alpha\,(B + S)$

$$= \alpha\, B + \alpha\, S \qquad ...(36.16)$$

In the above equation, αB is the population depending upon basic employment and αS is the population depending upon service employment.

The second proposition of the economic base theory is that the levels of service employment are determined by the levels of population. To give a simple exmaple, the number of school teachers is dependent on the population. Thus,

$$S = \beta P \quad \text{...(36.17)}$$

Where β is termed the population serving ratio

Thus $$\beta = S/P \quad \text{...(36.18)}$$

It will now be possible to use the above concepts to forecast population and employment. Given the values of α and β and a forecast of the number of basic jobs, the population, dependent on the basic employment, $P(1)$, is found by multiplying the number of basic jobs by the population multiplier. Thus :

$$P(1) = \alpha B \quad \text{...(36.19)}$$

The above population is capable of generating a certain number of service employment. If we denote this by $D(1)$,

$$D(1) = \beta P(1) \quad \text{...(36.20)}$$

The service jobs will themselves have dependent population, $P(2)$, which can be found, thus :

$$P(2) = \alpha D(1) \quad \text{...(36.21)}$$

This additional population will again generate a demand for service employment, $D(2)$

$$D(2) = \beta P(2) \quad \text{...(36.22)}$$

The latest increment of service employment $D(2)$ will have an associated dependent population $P(3)$, which in turn will generate another increment of service employment. This incremental process will go on till each increment of service employment and service dependent population becomes smaller and smaller till they at last become insignificant. The sum of the increments of service employment $D(1) + D(2) + \ldots\ldots\ldots\ldots D(N)$ represents the total forecast of service employment and the sum of the increments of population is the total service dependent population. When the increments have become insignificant, the totals are said to have converged.

Garin has suggested adaptation to the population allocation procedure originally suggested by Lowry. The adaptation is based on the gravity model. The following expression is based on the gravity model :

$$T_{ij} = E_i A_i P_j^n d_{ij}^{-b} \quad \text{...(36.23)}$$

where, T = amount of activity allocated from zone i to zone j (*i.e.*, number of workers who work in zone i and live in zone j)

E_i = Total amount of activity to be allocated from zone i (*i.e.*, people employed in zone i)

P_i = a measure of attraction of zone j. This is generally the population of the zone.

d_{ij} = measure of spatial separation between the zone i and j.

$$A_i^n = \left(\sum_{j=1}^{n} P_j^{-b} di\, j \right)^{-1}$$

b = calibration constant

The total number of workers living in zone j is thus :

$$\sum_{j=1}^{n} T_{ij} \quad ...(36.24)$$

The total number of population living in any zone then becomes :

$$Pj = \alpha \sum_{j=1}^{n} T_{ij} \quad ...(36.25)$$

In the model, the location of the basic employment is an input. The first operation within the model is to allocate these basic employees to zones using the above gravity model formulation. Knowing E_i, the basic workers in zone i, the number of basic workers who work in zone i and live in zone j, T_{ij}, can be found by the gravity formulation. Multiplying T_{ij} by α, the population multiplier, one gets the population in j, p_j (1).

$$p_j(1) = \alpha \sum_{i} T_{ij} \quad ...(36.26)$$

The first increment of service employment is then :

$$D_j(1) = \beta p_j(1) \quad ...(36.27)$$

The service employment so generated is allocated to the various zones by the gravity formula :

$$S_{ji}(1) = D_j(1)\, B_j\, S_i\, d_{ji}^{-a} \quad ...(36.28)$$

where B_j = Service employment in zone i

a = calibration constant

d_{ij} = measure of accessibility

$$S_i(1) = \Sigma S_{ji} \quad ...(36.29)$$

$$S(1) = \sum_{i} \sum_{j} S_{ji} \quad ...(36.30)$$

where $S(1)$ is the first increment of total service employment in all zones.

The next operation is to find the increment of population dependent on $S_i(1)$ in Eqn. 36.23, which enables the location of service workers in zones of residences. Eqn. 36.27 then calculates the service dependent population by using the population multiplier. The whole sequence of operations is then repeated till the increments become insignificant and the totals converge.

36.4.2. Example of use of Garin-Lowry Model

The use of Garin-Lowry Model can be illustrated by a simple three zone example. Table 36.1 gives the employment and population in the zones and Table 36.2 gives the time/distance matrix. It will be assumed that a and b have a value of 2 from calibration.

Table 36.1. Employment and Population in Zones

Zone	*Basic employment*	*Service employment*	*Total employment*	*Total population*
1	2000	1000	3000	20000
2	5000	5000	10000	30000
3	10000	16000	26000	40000
Total	17000	22000	39000	90000

Table 36.2. Time/distance matrix

From \ To	1	2	3
1	2	7	5
2	7	3	4
3	5	4	3

$$\alpha = \frac{P}{E} = \frac{90{,}000}{39{,}000} = 2.308$$

$$\beta = \frac{S}{P} = \frac{22{,}000}{90{,}000} = 0.2444$$

$$T_{ij} = E_i\, A_i\, P_j\, d_{ij}^{-2}$$

Calculate, $A_i = \Sigma(P_j\, d_{ij}^{-2})$

Table 36.3

	$j = 1$	$j = 2$	$j = 3$	$\sum_{j=1}^{n} P_j d_{ij}^{-2}$
$i = 1$	$\frac{20000}{2^2} = 5000$	$\frac{30000}{7^2} = 612.24$	$\frac{40000}{5^2} = 1600$	7212.24
$i = 2$	$\frac{20000}{7^2} = 408.16$	$\frac{30000}{3^2} = 3333.33$	$\frac{40000}{4^2} = 2500$	6241.49
$i = 3$	$\frac{20000}{5^2} = 800$	$\frac{30000}{4^2} = 1875$	$\frac{40000}{3^2} = 4444.44$	7119.44

Table 36.4. Calculation of $(P_{ij}) = A_i^{\,n} P_j d_{ij}^{-2}$

	$j = 1$	$j = 2$	$j = 3$
$i = 1$	0.6932 $\left(= \frac{5000}{7212.24}\right)$	0.0848	0.2218
$i = 2$	0.0654	0.5340	0.4005
$i = 3$	0.1123	0.2633	0.6242

Table 36.5. Calculation of $T_{ij}(1) = E_j(P_{ij})$

	$j = 1$	$j = 2$	$j = 3$
$i = 1$	2000 × 0.6932 = 1386.4	2000 × 0.0848 = 169.6	2000 × 0.2218 = 443.6
$i = 2$	5000 × 0.0654 = 327.0	5000 × 0.5340 = 2670.0	500 × 0.4005 = 2002.5
$i = 3$	10000 × 0.1123 = 1123	10,000 × 0.2633 = 2633	1000 × 0.6242 = 6242
$\sum_{i=1}^{n} T_{ij}(1)$	2836.4	5472.6	8688.1
$P_j(1)$	$= \alpha\, \Sigma T_{ij}(1)$	$= 2.308 \times \Sigma T_{ij}(1)$	

Calculation of the basic popultion of each zone in turn

$$P_1(1) = 2.308 \times 2836.4 = 6546$$
$$P_2(1) = 2.308 \times 5472.6 = 12631$$
$$P_3(1) = 2.308 \times 8688.1 = 20052$$

Find the first increment of service employment demanded by the population resident in each zone

$$D_j(1) = \beta\, P_j(1) = 0.2444\, P_j(1)$$

For each zone in turn

$$D_1(1) = 0.2444 \times 6546 = 1600$$
$$D_2(1) = 0.2444 \times 12631 = 3087$$
$$D_3(1) = 0.2444 \times 20052 = 4901$$

Table 36.6. Calculation of $S_i d_{ij}^{-2}$

	j = 1	j = 2	j = 3
i = 1	$\frac{1000}{2^2} = 250$	$\frac{1000}{7^2} = 20.41$	$\frac{1000}{5^2} = 40$
i = 2	$\frac{5000}{7^2} = 102$	$\frac{5000}{3^2} = 555.56$	$\frac{5000}{4^2} = 312.5$
i = 3	$\frac{16000}{5^2} = 640$	$\frac{16000}{4^2} = 1000$	$\frac{16000}{3^2} = 1777.78$
$\sum_{i=1}^{n} S_i d_{ij}^{-2}$	992.0	1575.97	2130.28

Table 36.7. Calculation of $S_i d_{ij}^{-2} \left(\sum_{i=1}^{n} S_i d_{ij}^{-2}\right)^{-1}$

	j = 1	j = 2	j = 3
i = 1	250/992 = 0.252	20.41/1575.97 = 0.013	40/2130.28 = 0.019
i = 2	102/992 = 0.103	555.56/1575.97 = 0.353	312.5/2130.28 = 0.147
i = 3	640/992 = 0.645	1000/1575.97 = 0.635	1777.78/2130.38 = 0.835

Table 36.8. Allocation of service employment in each zone to zone of employment $S_{ji} = D_j(1)\, S_i d_{ij}^{-2} \left(\sum_{i=1}^{n} S_i d_{ij}^{-2}\right)^{-1}$

	j = 1	j = 2	j = 3	$\sum_{j=1}^{n} S_{ji}$
i = 1	1600 × 0.252 = 403	3087 × 0.013 = 40	4901 × 0.019 = 93	536
i = 2	1600 × 0.103 = 165	3087 × 0.353 = 1090	4901 × 0.147 = 720	1975
i = 3	1600 × 0.645 = 1032	3087 × 0.635 = 1960	4901 × 0.835 = 4092	7084

Table 36.9. Results of the first iteration

Zone	*Basic employment*	*Increment of population*	*Increment of service employment*
1	2000	6546	536
2	5000	12631	1975
3	10000	20052	7084

Table 36.10. Second Iteration $T_{ij}(2) = S_i(1) \,.\, (P_{ij})$

	$j = 1$	$j = 2$	$j = 3$
$i = 1$	536 × 0.6932 = 372	536 × 0.0848 = 45	536 × 0.2218 = 119
$i = 2$	1975 × 0.0654 = 129	1975 × 0.534 = 1055	1975 × 0.4005 = 791
$i = 3$	7084 × 0.1123 = 796	7084 × 0.2633 = 1865	7084 × 0.6242 = 4422
$\sum_{i=1}^{n} T_{ij}(2)$	1297	2965	5332

$$P_i(2) = \alpha\, \Sigma T_{ij}(2) \qquad = 2.308 \times \Sigma T_{ij}(2)$$
$$P_1(2) = 2.308 \times 1297 = 2993$$
$$P_2(2) = 2.308 \times 2965 = 6843$$
$$P_3(3) = 2.308 \times 5332 = 12306$$
$$D_j(2) = \beta \times P_j(2)$$
$$= 0.2444 \times P_j(2)$$
$$D_1(2) = 0.2444 \times 2093 = 731$$
$$D_2(2) = 0.2444 \times 6843 = 1672$$
$$D_3(2) = 0.2444 \times 11306 = 3008$$

Table 36.11. $S_{ji}(2)$

	$j = 1$	$j = 2$	$j = 3$	$\sum_{j=1}^{n} S_{ji}$
$i = 1$	731 × 0.252 = 184	1672 × 0.013 = 22	3008 × 0.019 = 57	263
$i = 2$	731 × 0.103 = 75	1672 × 0.353 = 590	3008 × 0.147 = 442	1107
$i = 3$	731 × 0.645 = 471	1672 × 0.635 = 1062	3008 × 0.835 = 2512	4045

Table 36.12. Results of the second iteration

Zone	*Increment of population*	*Increment of service employment*
1	2993	263
2	6843	1107
3	12306	4045

Table 36.13. Third Iteration $T_{ij}(3) = S_i(2)\ (P_{ji})$

	$j = 1$	$j = 2$	$j = 3$
$i = 1$	263 × 0.6932 = 182	263 × 0.0848 = 22	263 × 0.2218 = 58
$i = 2$	1107 × 0.0654 = 72	1107 × 0.5340 = 591	1107 × 0.4005 = 443
$i = 3$	4045 × 0.1123 = 454	4045 × 0.2633 = 4065	4045 × 0.6242 = 2525
$\sum_{i=1}^{n} T_{ij}(3)$	708	1678	3026

$$P_j(3) = \alpha \Sigma T_{ij}(3) = 2.308 \times \Sigma T_{ij}(3)$$
$$P_1(3) = 2.308 \times 708 = 1634$$
$$P_2(3) = 2.308 \times 1678 = 3873$$
$$P_3(3) = 2.308 \times 3026 = 6984$$
$$D_j = \beta \times P_j(3) = 0.2444 \times P_j(3)$$
$$D_1(3) = 0.2444 \times 1634 = 399$$
$$D_2(3) = 0.2444 \times 3873 = 947$$
$$D_3(3) = 0.2444 \times 6984 = 1707$$

Table 36.14. $S_{ji}(3)$

	$j = 1$	$j = 2$	$j = 3$	$\sum_j S_{ji}$
$i = 1$	399 × 0.252 = 101	947 × 0.013 = 12	1707 × 0.019 = 32	145
$i = 2$	399 × 0.103 = 41	947 × 0.353 = 334	1707 × 0147 = 251	626
$i = 3$	399 × 0.645 = 257	947 × 0.635 = 601	1707 × 0.835 = 1425	2283

Table 36.15. Results of the third iteration

Zone	*Increment of population*	*Increment of service employment*
1	1634	145
2	3873	626
3	6984	2283

Table 36.16. Fourth Iteration $T_{ij}(4) = S_i(3) \times (P_{ji})$

	$j = 1$	$j = 2$	$j = 3$
$i = 1$	145 × 0.6932 = 101	145 × 0.0848 = 12	145 × 0.2218 = 32
$i = 2$	626 × 0.0654 = 41	626 × 0.534 = 334	626 × 0.4005 = 251
$i = 3$	2283 × 0.1123 = 256	2283 × 0.2633 = 601	2283 × 0.6242 = 1425
$\sum_{i=1}^{n} T_{ij}(4)$	398	947	1708

$P_1(4) = 2.308 \times 398 = 919$
$P_2(4) = 2.308 \times 947 = 2186$
$P_3(4) = 2.308 \times 1708 = 3942$
$D_1(4) = 0.2444 \times 919 = 225$
$D_2(4) = 0.2444 \times 2186 = 534$
$D_3(4) = 0.2444 \times 3942 = 963$

Table 36.17. $S_{ji}(4)$

	$j = 1$	$j = 2$	$j = 3$	$\sum_{j=1}^{n} S_{ji}$
$i = 1$	225 × 0.252 = 57	534 × 0.013 = 7	963 × 0.019 = 18	82
$i = 2$	225 × 0.103 = 23	534 × 0.353 = 189	963 × 0.147 = 142	354
$i = 3$	225 × 0.645 = 145	534 × 0.635 = 339	963 × 0.835 = 804	1288

Table 36.18. Results of the fourth iteration

Zone	*Increment of population*	*Increment of service employment*
1	919	82
2	2186	354
3	3942	1288

Table 36.19. Fifth Iteration $T_{ij}(5) = S_i(4) \times (P_{ji})$

	$j = 1$	$j = 2$	$j = 3$
$i = 1$	82 × 0.6932 = 57	82 × 0.0848 = 7	82 × 0.2218 = 18
$i = 2$	354 × 0.0654 = 23	354 × 0.534 = 189	354 × 0.4005 = 142
$i = 3$	1288 × 0.1123 = 145	1288 × 0.2633 = 339	1288 × 0.6242 = 804
$\sum_{i=1}^{n} T_{ij}(5)$	225	535	964

$P_1(5) = 2.308 \times 225 = 519$
$P_2(5) = 2.308 \times 535 = 1235$
$P_3(5) = 2.308 \times 964 = 2225$
$D_1(5) = 0.2444 \times 519 = 127$
$D_2(5) = 0.2444 \times 1235 = 302$
$D_3(5) = 0.2444 \times 2225 = 544$

Table 36.20. $S_{ji}(5)$

	$j = 1$	$j = 2$	$j = 3$	$\sum_{i=1}^{n} S_{ji}$
$i = 1$	127 × 0.252 = 32	302 × 0.013 = 4	544 × 0.019 = 10	46
$i = 2$	127 × 0.103 = 13	302 × 0.353 = 107	544 × 0.147 = 80	200
$i = 3$	127 × 0.645 = 82	302 × 0.635 = 192	544 × 0.835 = 454	728

Table 36.21. Results of the fourth iteration

Zone	*Increment of population*	*Increment of service employment*
1	519	46
2	1235	200
3	2225	728

Table 36.22. Sixth Iteration $T_{ij}(6) = S_i(5) \times (P_{ji})$

	$j = 1$	$j = 2$	$j = 3$
$i = 1$	46 × 0.6932 = 32	46 × 0.0848 = 4	46 × 0.2218 = 10
$i = 2$	200 × 0.0654 = 13	200 × 0.534 = 107	200 × 0.4005 = 80
$i = 3$	728 × 0.1123 = 82	718 × 0.2633 = 192	728 × 6242 = 454
$\sum_{i=1}^{n} T_{ij}(6)$	= 127	= 303	= 544

$$P_1(6) = 2.308 \times 127 = 293$$
$$P_2(6) = 2.308 \times 303 = 699$$
$$P_3(6) = 2.308 \times 544 = 1256$$
$$D_1(6) = 0.2444 \times 293 = 72$$
$$D_2(6) = 0.2444 \times 699 = 171$$
$$D_3(6) = 0.2444 \times 1256 = 307$$

Table 36.23. $S_{ij}(6)$

	$j = 1$	$j = 2$	$j = 3$	$\sum_{j=1}^{n} S_{ji}$
$i = 1$	72 × 0.252 = 18	171 × 0.013 = 2	307 × 0.019 = 6	26
$i = 2$	72 × 0.103 = 7	171 × 0.353 = 60	307 × 0.147 = 45	112
$i = 3$	72 × 0.645 = 46	171 × 0.635 = 109	307 × 0.835 = 256	411

Table 36.24. Results of sixth iteration

Zone	*Increment of population*	*Increment of service employment*
1	293	26
2	699	112
3	1256	411

Table 36.25. Seventh Iteration $T_{ij}(7) = S_{ji}(6)\ (P_{ji})$

	$j = 1$	$j = 2$	$j = 3$
$i = 1$	26 × 0.6932 = 18	26 × 0.0848 = 2	26 × 0.2218 = 6
$i = 2$	112 × 0.0654 = 7	112 × 0.534 = 60	112 × 0.4005 = 45
$i = 3$	411 × 0.1123 = 46	411 × 0.2633 = 108	411 × 0.6242 = 257
$\sum_{i=1}^{n} T_{ij}(7)$	71	170	308

$$P_1(7) = 2.308 \times 71 = 164$$
$$P_2(7) = 2.308 \times 170 = 392$$
$$P_3(7) = 2.308 \times 308 = 711$$
$$D_1(7) = 0.2444 \times 164 = 40$$
$$D_2(7) = 0.2444 \times 392 = 96$$
$$D_3(7) = 0.2444 \times 711 = 174$$

Table 36.26. $S_i(7)$

	$j = 1$	$j = 2$	$j = 3$	$\sum_{j=1}^{n} S_i(7)$
$i = 1$	40 × 0.252 = 10	96 × 0.013 = 1	174 × 0.019 = 3	14
$i = 2$	40 × 0.103 = 4	96 × 0.353 = 34	174 × 0.147 = 26	64
$i = 3$	40 × 0.645 = 26	96 × 0.635 = 61	4 × 0.835 = 145	232

Table 36.27. Results of the seventh iteration

Zone	*Increment of population*	*Increment of service employment*
1	164	14
2	392	64
3	711	232

Table 36.28. Eighth Iteration $T_{ij}(8) = S_i(7)\ (P_{ij})$

	$j = 1$	$j = 2$	$j = 3$
$i = 1$	14 × 0.6932 = 10	14 × 0.0848 = 1	14 × 0.2218 = 3
$i = 2$	64 × 0.0654 = 4	64 × 0.534 = 34	64 × 0.4005 = 26
$i = 3$	232 × 0.1123 = 26	232 × 0.2633 = 61	232 × 0.6242 = 145
$\sum_{i=1}^{n} T_{ij}(8)$	40	96	174

$$P_1(8) = 2.308 \times 40 = 92$$
$$P_2(8) = 2.308 \times 96 = 221$$
$$P_3(8) = 2.308 \times 174 = 402$$
$$D_1(8) = 0.2444 \times 92 = 22$$
$$D_2(8) = 0.2444 \times 221 = 54$$
$$D_3(8) = 0.2444 \times 402 = 98$$

Table 36.29. $S_{ji}(8)$

	$j = 1$	$j = 2$	$j = 3$	$\sum_{i=1}^{n} S_i(8)$
$i = 1$	22 × 0.252 = 6	54 × 0.013 = 1	98 × 0.019 = 2	9
$i = 2$	22 × 0.103 = 2	54 × 0.353 = 19	98 × 0.147 = 14	35
$i = 3$	22 × 0.645 = 14	54 × 0.635 = 34	98 × 8.835 = 82	130

Table 36.30. Results of the Eighth iteration

Zone	*Increment of population*	*Increment of service employment*
1	92	9
2	221	35
3	402	130

Table 36.31. Ninth Iteration $T_{ij}(9) = S_i(8)\ (P_{ji})$

	$j = 1$	$j = 2$	$j = 3$
$i = 1$	9 × 0.6932 = 6	9 × 0.0848 = 1	9 × 0.2218 = 2
$i = 2$	35 × 0.0654 = 2	35 × 0.534 = 19	35 × 0.4005 = 14
$i = 3$	130 × 0.1123 = 15	130 × 0.2633 = 34	130 × 0.6242 = 81
$\sum_{i=1}^{n} T_{ij}(9)$	23	54	97

$$P_1(9) = 2.308 \times 23 = 53$$
$$P_2(9) = 2.308 \times 54 = 125$$
$$P_3(9) = 2.308 \times 97 = 224$$
$$D_1(9) = 0.2444 \times 53 = 13$$
$$D_2(9) = 0.2444 \times 125 = 31$$
$$D_3(9) = 0.2444 \times 224 = 55$$

Table 36.32. $S_{ji}(9)$

	$j = 1$	$j = 2$	$j = 3$	$\sum_{j=1}^{n} S_{ji}(9)$
$i = 1$	13 × 0.252 = 3	31 × 0.013 = 0	55 × 0.019 = 1	4
$i = 2$	13 × 0.103 = 2	31 × 0.353 = 19	55 × 0.147 = 8	20
$i = 3$	13 × 0.645 = 8	31 × 0.635 = 20	55 × 0.835 = 46	74

Table 36.33. Results of ninth iteration

Zone	*Increment of population*	*Increment of service employment*
1	53	4
2	125	20
3	224	74

Table 36.34. Tenth Iteration $T_{ij}(10) = S_i(9)\ (P_{ji})$

	$j = 1$	$j = 2$	$j = 3$
$i = 1$	4 × 0.6932 = 3	4 × 0.0848 = 0	4 × 0.2218 = 1
$i = 2$	20 × 0.0654 = 1	20 × 0.534 = 11	20 × 0.4005 = 8
$i = 3$	74 × 0.1123 = 8	74 × 0.2633 = 19	74 × 0.6242 = 46
$\sum_{i=1}^{n} T_{ij}(10)$	12	30	55

$P_1(10) = 2.308 \times 12 = 28$
$P_2(10) = 2.308 \times 30 = 69$
$P_3(10) = 2.308 \times 55 = 127$
$D_1(10) = 0.2444 \times 28 = 7$
$D_2(10) = 0.2444 \times 69 = 17$
$D_3(10) = 0.2444 \times 127 = 31$

Table 36.35. $S_{ji}(10)$

	$j = 1$	$j = 2$	$j = 3$	$\sum_{i=1}^{n} S_{ji}(10)$
$i = 1$	7 × 0.252 = 2	17 × 0.013 = 0	31 × 0.019 = 1	3
$i = 2$	7 × 0.103 = 1	17 × 0.353 = 6	31 × 0.147 = 5	12
$i = 3$	7 × 0.645 = 5	17 × 0.635 = 11	31 × 0.835 = 26	42

Table 36.36. Results of the 10th iteration

Zone	*Increment of population*	*Increment of service employment*
1	28	3
2	69	12
3	127	42

Table 36.37. Eleventh Iteration $T_{ij}(11) = S_i(10)\ (P_{rji})^p$

	$j = 1$	$j = 2$	$j = 3$
$i = 1$	3 × 0.6932 = 2	3 × 0.0848 = 0	3 × 0.2218 = 1
$i = 2$	12 × 0.0654 = 1	12 × 0.534 = 5	12 × 0.4005 = 5
$i = 3$	42 × 0.1123 = 5	42 × 0.2633 = 11	42 × 0.6242 = 26
$\sum_{i=1}^{n} T_{ij}(11)$	8	17	32

$P_1(11) = 2.308 \times 8 = 18$
$P_2(11) = 2.308 \times 17 = 39$
$P_3(11) = 2.308 \times 32 = 74$
$D_1(11) = 0.2444 \times 18 = 4$
$D_2(11) = 0.2444 \times 39 = 10$
$D_3(11) = 0.2444 \times 74 = 18$

Table 36.38. $S_{ji}(11)$

	$j = 1$	$j = 2$	$j = 3$	$\sum_{j=1}^{n} S_{ji}(11)$
$i = 1$	4 × 0.252 = 1	10 × 0.013 = 0	18 × 0.019 = 0	1
$i = 2$	4 × 0.103 = 0	10 × 0.353 = 4	18 × 0.147 = 3	7
$i = 3$	4 × 0.645 = 3	10 × 0.635 = 6	18 × 0.835 = 15	24

Table 36.39. Results of the 11th iteration

Zone	Increment of population	Increment of service employment
1	18	1
2	39	7
3	74	24

Table 36.40. Twelfth Iteration $T_{ij}(12) = S_i(11)\ (P_{ji})$

	$j = 1$	$j = 2$	$j = 3$
$i = 1$	1 × 0.6932 = 1	1 × 0.0848 = 0	1 × 0.2218 = 0
$i = 2$	7 × 0.0654 = 0	7 × 0.534 = 4	7 × 0.4005 = 3
$i = 3$	24 × 0.1123 = 3	24 × 0.2633 = 6	24 × 0.6242 = 15
$\sum_{i=1}^{n} T_{ij}(12)$	4	10	18

$P_1(12) = 2.308 \times 4 = 9$

$P_2(12) = 2.308 \times 10 = 23$

$P_3(12) = 2.308 \times 18 = 42$

$D_1(12) = 0.2444 \times 9 = 2$

$D_2(12) = 0.2444 \times 23 = 6$

$D_3(12) = 0.2444 \times 42 = 10$

Table 36.41. $S_{ji}(12)$

	$j = 1$	$j = 2$	$j = 3$	$\sum_{j=1}^{n} S_{ij}(12)$
$i = 1$	2 × 0.252 = 1	6 × 0.013 = 0	10 × 0.019 = 0	1
$i = 2$	2 × 0.103 = 0	6 × 0.353 = 2	13 × 0.147 = 1	3
$i = 3$	2 × 0.645 = 1	6 × 0.635 = 4	10 × 0.835 = 8	13

Table 36.42. Results of 12th iteration

Zone	Increment of population	Increment of service employment
1	9	1
2	23	3
3	42	13

Table 36.43. Thirteenth Iteration $T_{ij}(13) = S_i(12)\ (P_{ji})$

	$j = 1$	$j = 2$	$j = 3$
$i = 1$	1 × 0.6932 = 1	1 × 0.0848 = 0	1 × 0.2218 = 0
$i = 2$	3 × 0.0654 = 1	3 × 0.534 = 2	3 × 0.4005 = 1
$i = 3$	13 × 0.1123 = 1	13 × 0.2633 = 4	13 × 0.6242 = 9
$\sum_{i=1}^{n} T_{ij}(13)$	2	6	10

$P_1(13) = 2.308 \times 2 = 5$

$P_2(13) = 2.308 \times 6 = 14$

$P_3(13) = 2.308 \times 10 = 28$

$D_1(13) = 0.2444 \times 5 = 1$

$D_2(13) = 0.2444 \times 14 = 3$

$D_3(13) = 0.2444 \times 23 = 6$

Table 36.44. $S_{ji}(13)$

	$j = 1$	$j = 2$	$j = 3$	$\sum_{j=1}^{n} S_{ji}(13)$
$i = 1$	1 × 0.252 = 0	3 × 0.013 = 0	6 × 0.019 = 0	0
$i = 2$	1 × 0.103 = 0	3 × 0.353 = 1	6 × 0.147 = 1	2
$i = 3$	1 × 0.645 = 1	3 × 0.635 = 2	6 × 0.835 = 5	8

Table 36.45. Results of the 13th iteration

Zone	*Increment of population*	*Increment of service employment*
1	5	0
2	14	2
3	23	8

Table 36.46. Fourteenth Iteration $T_{ij}(14) = S_i(14)\ (P_{ji})$

	$j = 1$	$j = 2$	$j = 3$
$i = 1$	0	0	0
$i = 2$	2 × 0.0654 = 0	2 × 0.534 = 1	2 × 0.4005 = 1
$i = 3$	8 × 0.1123 = 1	8 × 0.2633 = 2	8 × 0.6242 = 3
$\sum_{i=1}^{n} T_{ij}(14)$	1	3	6

$P_1(14) = 2.308 \times 1 = 2$

$P_2(14) = 2.308 \times 3 = 7$

$P_3(14) = 2.308 \times 6 = 14$

$D_1(14) = 0.2444 \times 2 = 0$

$D_2(14) = 0.2444 \times 4 = 2$

$D_3(14) = 0.2444 \times 14 = 3$

Table 36.47. $S_{ji}(14)$

	$j = 1$	$j = 2$	$j = 3$	$\sum_{i=1}^{n} S_{ji}(14)$
$i = 1$	0	0	0	0
$i = 2$	0	1	0	1
$i = 3$	0	1	3	4

Table 36.48. Results of the 14th iteration

Zone	*Increment of population*	*Increment of service employment*
1	2	0
2	7	1
3	14	4

Table 36.49. Fifteenth Iteration $T_{ij}(15) = S_i(15)\ (P_{ji})$

	$j = 1$	$j = 2$	$j = 3$
$i = 1$	0	0	0
$i = 2$	0	1	0
$i = 3$	0	1	2
$\sum_{i=1}^{n} T_{ij}(15)$	0	2	2

$$P_1(15) = 2.308 \times 0 = 0$$
$$P_2(15) = 2.308 \times 2 = 15$$
$$P_3(15) = 2.308 \times 2 = 5$$
$$D_1(15) = 0.2444 \times 0 = 0$$
$$D_2(15) = 0.2444 \times 5 = 1$$
$$D_3(15) = 0.2444 \times 5 = 1$$

Table 36.50. $S_{ji}(15)$

	$j = 1$	$j = 2$	$j = 3$	$\sum_{j=1}^{n} S_{ij}(15)$
$i = 1$	0	0	0	0
$i = 2$	0	0	0	0
$i = 3$	0	1	1	2

Table 36.51. Results of the 15th iteration

Zone	*Increment of population*	*Increment of service employment*
1	0	0
2	5	0
3	5	2

Table 36.52. Results of the iterations of the lowry model

Iteration No.	*Service Employment* 1	2	3	Total	*Population* 1	2	3	Total
1	536	1975	7084	9595	6546	12631	20052	39229
2	263	1107	4045	5415	2993	6843	12306	22142
3	145	626	2283	3054	1634	3873	6984	12491
4	82	354	1288	1724	919	2186	3942	7047
5	46	200	728	974	519	1235	2225	3975
6	26	112	411	549	293	699	1256	2248
7	14	64	232	310	164	392	711	1267
8	9	35	130	174	92	221	402	715
9	4	20	74	98	53	125	224	402
10	3	12	42	57	28	69	127	224
11	1	7	24	32	18	39	74	131
12	1	3	13	17	9	23	42	74
13	0	2	8	10	5	14	23	42
14	0	1	4	5	2	7	14	23
15	0	0	2	2	0	5	5	10
Total	1130	4518	16368	22016	13275	28362	48387	90024

Table 36.53. The final output of the model

Zone	*Actual basic employment*	*Predicted service employment*	*Predicted total employment*	*Predicted total population*
1	2000	1130	3130	13,275
2	5000	4518	9518	28,362
3	10,000	16,368	26,386	48,387
Total	17,000	26,016	26,386	90,024

Table 36.54. Comparison of actual and predicted employment and population

Zone	*Service employment*		*Total employment*		*Total population*	
	Actual	*Predicted*	*Actual*	*Predicted*	*Actual*	*Predicted*
1	1000	1100	3000	3130	20,000	13,275
2	5000	4518	10,000	9518	30,000	28,362
3	16,000	16,386	26,000	26,386	40,000	48,387
Total	22,000	22,016	39,000	39,016	90,000	90,024

36.5. Matrix Operations for Simplifying Computations

The original Lowry algorithm can be expressed in vector and matrix format, thereby simplifying the computations and replacing the iterative process (Ref. 6, 7, 8).

The equation system is given below :

$$p = e^A \qquad \text{...(36.31)}$$

$$e^s = p^B \qquad \text{...(36.32)}$$

$$e = e^b + e^s \qquad \text{...(36.33)}$$

where p = a row vector of the population (or households) within each of the n zones

e = a row vector of the total employment in each zone

e^s = a row vector of the total population serving employment in each zone

e^b = a row vector of the basic employment in each zone

A = an $n \times n$ matrix of the workplace-to-household accessibilities

B = an $n \times n$ matrix of the household-to-service centre accessibilities

The A accessibility matrix may be expanded as follows :

$$A = (a_{ij}')\,(a_j) \qquad \text{...(35.34)}$$

where (a_{ij}') = an $n \times n$ square matrix of the probabilities of an employed working in i and living in j

a_j = an $n \times n$ diagonal matrix of the inverses of the labour participation rates, expressed either as population per employee, or households per employee.

The B accessibility matrix may be expanded as follows :

$$B = (b_{ij}')\,(b_j) \qquad \text{...(36.35)}$$

where (b_{ij}) = an $n \times n$ square matrix of the probabilities that the population in j will be serviced by the population serving employment in i.

(b_i) = an $n \times n$ diagonal matrix of the population serving employment-to-population ratios.

The a_{ij}' elements may be estimated empirically in the following way.

$$a_{ij}' = \frac{h_j\, f_{ij}^w}{\Sigma h_j\, f_{ij}^w} \qquad \text{...(36.36)}$$

where h_j = measure of attractivity of zone j for household location

f_{ji}^w = travel time factor between zone i and j which reflects the manner in which the spatial separation of zones influences the residential location choices of employees

It may be noted that the above equation is a gravity-type accessibility expression.

The b_{ij} elements may be estimated in a similar manner using the following expression :

$$b_{ij}^r = \frac{s_i^r\, f_{ij}^{sr}}{\sum_i s_i\, f_{ij}^{sr}} \qquad \text{...(36.37)}$$

where s_i^r = a measure of the attractivity of zone i for satisfying the service type r needs of the households

f_{ij}^{sr} = the travel time factor between zones i and j which reflects the manner in which the spatial separation of zones influences the type r service location choices of households.

The distribution of activities produced by the above set of equations should be such that the following constraint equations are satisfied :

$$p \leq p^c \quad ...(36.38)$$

$$e^{sr} \geq e^{sr\ min} \quad ...(36.39)$$

where p^c = a row vector of the population holding capacities of each of the the n zones

$e^{sr\ min}$ = a row vector of the population serving employment thresholds for the r service employment types considered to be viable for any zones.

If equations 36.38 and 36.39 are violated, then new accessibility matrices must be developed and the equation again solved using new matrices. This can be achieved by an approach similar to the attraction trip-end balancing procedure of the gravity model.

$$a_{ij}^{*} = \frac{h_i^{*} f_{ij}^{w}}{\sum_j h_i^{*} f_{ij}^{w}} \quad ...(36.40)$$

$$h_j^{*} = h_j \left(\frac{h_j}{p_j} \right) \quad ...(36.41)$$

$$h_{ji}^{*r} = \frac{s_i^{*r} f_{ij}^{sr}}{\sum s_i^{*r} i_{ij}^{sr}} \quad ...(36.42)$$

$$s_j^{*r} = s_i^r \text{ for zones in which } e_i^{sr} \geq e^{sr\ min} \quad ...(36.43)$$

$$= 0 \text{ for zones in which } e_j^{sr} \leq e^{sr\ min}$$

in which a_{ij}^{*} and b_{ij}^{*r} are new estimate of a_{ij} and b_{ij}^{r} to be used in the next iteration of the model.

The home-based work-trip matrix for any period (say the peak hour) may be derived from ;

$$T^w = WA' \quad ...(36.44)$$

where T^w = an $n \times n$ matrix of home-based work trips where the element t_{ij}^{w} represent the number of work trips between zones i and j

W = an $n \times n$ diagonal matrix of the work-trip generation rate per employee during the same period under consideration

A' = an $n \times n$ square matrix of work-to-home accessibilities, a'_{ij}.

The home-based service-trip matrices for any time period may be calculated in an analogous way.

$$T^{sr} = RB'^{r} \quad ...(36.45)$$

where T^{sr}= an $n \times n$ square matrix of the home-based service-trip interchanges i_{ij}^{sr} to type r population serving employment opportunities during the time period under consideration.

R = an $n \times n$ diagonal matrix of the type r service trip generation rate per household (or person) during the time period under consideration,

B'^r = an $n \times n$ square matrix of the home-to-type r service accessibilities $b_{ij}'^r$

The following simple problem illustrates the use of the equations 36.31 to 36.35.

Total employment vector e = [126, 177, 64, 216]

Basic employment vector e^b = [100, 150, 40, 200]

Journey to home function :

$$[a_{ij}'] = \begin{bmatrix} 0.35 & 0.30 & 0.20 & 0.15 \\ 0.25 & 0.35 & 0.20 & 0.20 \\ 0.15 & 0.10 & 0.35 & 0.40 \\ 0.10 & 0.25 & 0.20 & 0.45 \end{bmatrix}$$

Journey to shop function :

$$[b_{ij}'] = \begin{bmatrix} 0.50 & 0.25 & 0.10 & 0.15 \\ 0.30 & 0.45 & 0.15 & 0.10 \\ 0.15 & 0.20 & 0.40 & 0.25 \\ 0.20 & 0.25 & 0.35 & 0.20 \end{bmatrix}$$

Labour participation rate :
(households/employee)

$$[a_j] = \begin{bmatrix} 0.8 & 0 & 0 & 0 \\ 0 & 0.8 & 0 & 0 \\ 0 & 0 & 0.8 & 0 \\ 0 & 0 & 0 & 0.8 \end{bmatrix}$$

Service employment rato :
(service employment/household)

$$[b_i] = \begin{bmatrix} 0.2 & 0 & 0 & 0 \\ 0 & 0.2 & 0 & 0 \\ 0 & 0 & 0.2 & 0 \\ 0 & 0 & 0 & 0.2 \end{bmatrix}$$

Using equations 36.34 and 36.35,

$$A = \begin{bmatrix} 0.28 & 0.24 & 0.16 & 0.12 \\ 0.20 & 0.28 & 0.16 & 0.16 \\ 0.12 & 0.08 & 0.28 & 0.32 \\ 0.08 & 0.20 & 0.16 & 0.36 \end{bmatrix}$$

$$B = \begin{bmatrix} 0.10 & 0.05 & 0.02 & 0.03 \\ 0.06 & 0.09 & 0.03 & 0.02 \\ 0.03 & 0.04 & 0.08 & 0.05 \\ 0.04 & 0.05 & 0.07 & 0.04 \end{bmatrix}$$

The household vector may be calculated from Equation 36.31

$$[126, 177, 64, 216]\begin{bmatrix} 0.28 & 0.24 & 0.16 & 0.12 \\ 0.20 & 0.28 & 0.16 & 0.16 \\ 0.12 & 0.08 & 0.28 & 0.32 \\ 0.08 & 0.20 & 0.16 & 0.36 \end{bmatrix} = [95, 128, 101, 142]$$

The service employment vector may be calculated from Eqn. 36.32.

$$[95, 128, 101, 114]\begin{bmatrix} 0.10 & 0.05 & 0.02 & 0.03 \\ 0.06 & 0.09 & 0.03 & 0.02 \\ 0.03 & 0.04 & 0.08 & 0.05 \\ 0.04 & 0.05 & 0.07 & 0.04 \end{bmatrix} = [26, 27, 24, 16]$$

Equation 36.3 may be used to check if the household and employment allocations are internally consistent.

Original total employment vector e

$$= [126, 177, 64, 216]$$

$$e^b + e^s = [100, 150, 40, 200] + [26, 27, 24, 16]$$

$$= [126, 177, 64, 216]$$

The above result indicates that the co-distribution of employment and households is in an a state of equilibrium.

36.6. Application in India

Land-use transport modes have been applied on a very limited scale in India, through vast potentialities exist, Sarna's work (Ref. 8) on the application of the land-use transport model for Delhi is the first of its kind. The model disaggregates by socio-economic and spatial groups. The model has been applid to solve several horizon year planning problems in Delhi urban area. Alternative land-use transportation systems are generated by varying the locations of employment and population.

A limited study by Rao (Ref. 9) deals with a moderation of the Lowry land-use transport model for Indian conditions.

I. MULTIPLE CHOICE QUESTIONS

1. What happens to land values as you move away from the CBD?
   (*a*) They increase (*b*) They stay the same
   (*c*) They decrease
2. Which land use zone is being described below?
   "This area contains back to back terraced housing. The streets are laid out in a linear, grid iron pattern"
   (*a*) CBD (*b*) inner city
   (*c*) inner suburbs (*d*) outer suburb
3. Council estates are usually found in which land use zone?
   (*a*) CBD (*b*) inner city
   (*c*) inner suburbs (*d*) outer suburbs
4. Why is the concentric circle model (or Burgess model) useful?
   (*a*) It illustrates the layout of a city in a less economically developed country.
   (*b*) An examiner may ask a question about them.

(*c*) They are true to life and fit most cities.

(*d*) They provide a simplistic overview of the layout of an urban area

5. What problems are associated with the use of land use models?

(*a*) They only show the layout of a city in a less economically developed country and the location of shanty towns is incorrect.

(*b*) They only fit cities in the USA and France.

(*c*) They are too simplistic and rarely match a typical city.

6. Land is a convenient measure of space and land use provides a spatial framework

(*a*) for urban development and activities

(*b*) rural development.

(*c*) transport development.

7. Land use means

(*a*) spatial distribution or geographical pattern of the city,

(*b*) residential area, industry, commercial areas

(*c*) the space set for governmental, institution or recreational purposes

(*d*) all the above

Answers

1. (*a*)	**2.** (*a*)	**3.** (*c*)	**4.** (*a*)	**5.** (*a*)
6. (*a*)	**7.** (*d*)			

II. SOLVED QUESTIONS AND ANSWERS

Q. 1. How land use and transportation mutually interrelated?

Ans. Land use characteristics and transportation are mutually interrelated. The use of the term land use is based on the fact that through development, urban space put up a variety of human activities. Land is a convenient measure of space and land use provides a spatial framework for urban development and activities. The location of activities and their need for interaction creates the demand for transportation, while the provision of transport facilities influences the location itself. Land uses, by virtue of their occupancy, are supposed to generate interaction needs and these needs are directed to specific targets by specific transportation facilities.

Q. 2. What do you mean by land use transportation model?

Ans. Land use means spatial distribution or geographical pattern of the city, residential area, industry, commercial areas and the space set for governmental, institution or recreational purposes. Most human activities, economic, social or cultural involve a multitude of functions, such as production, consumption and distribution. These functions are occurring within an activity system where their locations and spatial accumulation. So, the behavioural patterns of individuals, institutions and firms will have an impression on the land use.

Q. 3. What is meant by land use system? What are its essential components?

Ans. The essential components of the land use system in terms of land use transport modelling are location and development. The urban land use is largely modelled by simulating the mechanisms that effect the spatial allocation of urban activities in the city. A number of other important economic concepts underpin land use transport models, serving as proxies for the complex interactions and motivations driving urban location. Among these are the ideas of bid rent, travel costs, inertia (stability of occupation of land), topography, climate, planning, and size.

Q. 4. What is transport system?

Ans. It is the traditional way of characterizing the transportation system in urban simulation models is a four stage process. The process begins with modelling travel demand and generating an estimate of the amount of trips expected in the urban system . Trip distribution allocates the trips generated in origin zones to destinations in the urban area. The third phase is modal split. Here trips are apportioned to various modes of transport. The four stage simulation processes concludes with trip assignment module that takes estimated trips that have been generated, distributed and sorted by mode and loads it on to various segments of the transport network.

Q. 5. What are the factors affecting transport land use relationship?

Ans. Factors affecting transport land use relationship

- Urban land development
- Dominance of private vehicle ownership
- Context of land use and transportation decision making
- Different time contexts for response.

Q. 6. Describe the classification of land uses.

Ans. The representation of this impression requires a typology of land use, which can be formal or functional as explained below: 135 CE-751, SLD, Class Notes, Fall 2006, IIT Mumbai Formal land use representations are concerned by qualitative attributes of space such as its form, pattern and geographical aspects and are descriptive in nature. Functional land use representations are concerned by the level of spatial accumulation of economic activities such as production, consumption, residence, and transport, and are mainly a socioeconomic description of space.

Q. 7. Write a brief note on land use and transportation?

Ans. The movement of people and goods in a city, referred as traffic flow, is the joint consequence of land activity and the capability of the transportation system to handle this traffic flow exactly like that of principle of demand and supply. There is a direct interaction between the type and intensity of land use and transportation facilities provided. Ensuring efficient balance between land use activity and transportation capability is primary concern of urban planning. Land use is one of the prime determinants of movement and activity i.e. trip generation which needs streets and transport systems for movement. This will lead to increased accessibility which further enhances value of land and land use.

Q. 8. What are the different land use models?

Ans. The purpose of land use transport models is to assess the policy impacts in terms of the implications of the future growth patterns on both land use and travel related issues. For this purpose, several researchers have developed various models with different theoretical backgrounds and data requirements. From the early developments of land use transport models to the latest state of art, can be broadly classified into three categories:

(*i*) Early models (*ii*) Intermediate era models

(*iii*) Modern era models.

Q. 9. What are the early land use of transport models?

Ans. There are several techniques which are representatives of earliest efforts in the development of urban development models and which continue to serve (either in original or modified form) a great number of transportation studies.These techniques are quite simple generally deal with aggregate relationships. These are developed primarily for location of residential activities. In addition many of these techniques can be applied without using computer or simple programs can be prepared for use on a computer. These simple techniques are considered most practical use in smaller urban areas because they require less time, cost and data.

Q. 10. What is Lowry land use model?

Ans. The Lowry model conceives of the major spatial features of an urban area in terms of three broad sectors of activity i.e., basic employment sector, the population serving employment and the household sector. The basic employment is employment whose products and services are utilized outside the study area. With Lowry model, spatial distribution of basic employment is allocated exogenously to the model while the other two activity sectors are calculated by the model by applying an iterative procedure, until the constraints, which are maximum no. of household for each zone and minimum population serving employment for any zone, are satisfied. The model views the spatial properties in terms of:

- Employment in basic industries
- Employment in population serving industries
- Household or population sector

Q. 11. Explain Lowry-Garin model.

Ans. Lowry-Garin Model Garin proposed a formulation of Lowry's model which prevents the need for the iterative solution to the equations described above. Garin has proposed a formulation of the Lowry model, which obviates the need for the iterative solution of to the equations. The following equations can be written:

$$
\begin{aligned}
P^b &= e^b\, A \\
e^{s(1)} &= P^b\, B = e^b\,(AB) \qquad \ldots(1) \\
p^{s(1)} &= e^{s(1)}\, A = e^b\,(AB)A \\
e^{s(2)} &= P^{s(1)}\, B = e^{s(1)}\,(AB) = e^b\,(AB)(AB) = e^b\,(AB)^2
\end{aligned}
$$

Successive iteration will yield:

$$
\begin{aligned}
e^{x(x)} &= e^b(AB)^x \\
p^{s(x)} &= e^b(AB)^x\, A
\end{aligned}
$$

Total employment and total population vectors are given by:

$$
\begin{aligned}
e &= e^b + e^{s(1)} + \ldots + e^{s(x)} + \ldots \\
&= e^b[I + AB + (AB)^2 + \ldots + (AB)^x + \ldots] \\
P &= P^b + P^{s(1)} + \ldots + P^{s(x)} + \ldots \\
&= e^b[I + AB + (AB)^2 + \ldots + (AB)^x + \ldots]A
\end{aligned}
$$

Q. 12. What is Sarna's land use model?

Ans. A critical problem in most Indian cities is the inadequacy of the transport infrastructure which is further aggravated by the increasing demands for intra city travel due to rapid growth in both population and employment. These demand based on travel forecasts led to recommendation for high capacity facility, requiring large capital as well as operating expenditure. The large investment required by the transport systems recommended for Indian metropolitan cities have simulated an approach to land use transport planning which attempts to minimize travel demands through the manipulation of land use. Dr Sarna's model considers Delhi as three districts as inner district, middle district, outer district and different socioeconomic groups according to their income.

Q. 13. What are the major functions of land use transport model?

Ans. The model used for this study is based on Lowry model according to which two major functions are given by

- It relates three elements of the urban/regional system, population, employment and transport and relate their interactions.
- It incorporates within its structure both allocation and forecasting procedures.
- It assumes an economic base mechanism where employment is divided into basic and non-basic (service) sectors.
- The basic employment sector includes those economic activities, the produce of which is utilized mostly outside the region e.g., manufacturing and other heavy government offices, the state head quarters, national financial institutions, university etc. All other are accounted as non-basic (sector population serving employment).
- The model assumes that the basic sector, both its location and magnitude is controlled exogenously.
- The model then determines the level and location of population and service (non basic) within the region.

Q. 14. What are the five stage of land use transport model?

Ans. The five-stage land use transport model has its decisions taken on instead of the conventional four-stage land use transport model, it is based on 167 CE-751, SLD, Class Notes, Fall 2006, IIT Mumbai:

- Destination
- Route
- Location
- Transport mode
- Mobility and

Q. 15. What are the types of surveys?

Ans. Surveys are of particular relevance to transport and land-use planning in several specific areas. Land-use surveys are an integral component of transport planning information requirements. This is due to the fact that travel is a so-called "derived demand." Thus, some surveys can be effective with only a large number of highly directed questions, while others need to explore a few topics in depth.

III. REVIEW QUESTIONS

1. Write a brief note on the selection of landuse transport models.
2. Discuss Lowry derivative models.
3. What is Garin-Lowry model?

REFERENCES

1. Ramana Murthy, P.V. and S. Raghavachari, A., review of various land use transport models, Indian Highways, New Delhi, May, 1981.

2. Urban Development Models, Highway Research Board Special Report 97, Washington, 1968.
3. Irwin, N.A., Review of existing land use forecasting techniques, Highway Research Record No. 88, Highway Research Board, Washington 1965.
4. Putman, S.H., Urban Landuse and transportation model-A State-of-Art Summary, Transport Research, Vol. 9, Pergamon Press, 1975.
5. Lowry, I.S., A model of metropolis, Technical Memorandum RM-4035-RC, The Rand Corporation, Santa Monica, California, 1964.
6. Garin, R.A., A matrix formulation of Lowry model for intrametropolitan activity allocation, Journal of American Institute of Planners 32, 1966.
7. Hutchinson, B.G., Principles of Urban Transport Systems Planning, Mc Graw-Hill Book Company, 1974.
8. Sarna, A.C., Application of land use transport model to Delhi, Journal of the Indian Roads Congress, Vol. 39-1, New Delhi, 1978.
9. Rao, R.L., Moderated Lowry land use transport model, Thesis, School of Planning and Architecture, New Delhi, 1978.

37. Transportation Planning for Medium and Small Size Cities

37.1. Introduction

The experience in India has been that metropolitan cities have been growing at an alarming rate and this unregulated growth is clearly harmful in many ways. To the traffic engineer and transportation planner, such growth poses serious problems. Mumbai, Kolkata, Chennai and Delhi are classic examples. It has been the thinking of the planners and administrators (Ref. 1, 2) that the growing urbanisation should be catered to by fostering the growth of medium and small sizes cities. The small and medium sized cities in India (1–10 lakh population) have grown in a haphazard manner over centuries. No thought has been given to the planning of the street network and transportation requirements of such towns. But in view of the likely shift in emphasis in favour of development of such cities, the transportation planner should have an idea of the needs of these cities, and how they can be met with.

37.2. Difficulties in Transport Planning for Small and Medium Cities

There are many difficulties in planning for the transportation needs of small and medium cities. Firstly, there is a total lack of experience in the country in devising suitable planning techniques. No comprehensive traffic studies have been made so far. Secondly, the data base for planning has not been built up. The trip-making behaviour, the relationship between population and employment, the speed of traffic of various types on the street etc. have not been investigated at all. Thirdly, the local authorities in charge of such cities do not possess the skill and expertise to plan effectively. They have been so far meeting with the situation on a purely tentative and *ad-hoc* manner and their decisions are unrelated to traffic requirements assessed on any scientific basis. Finally, the modern methods of transport planning are too sophisticated and costly to be used for medium and small sized cities. There is an urgent need to simplify the whole procedure, keeping the data requirements to the bearest minimum and making the solutions quick responsive.

37.3. Quick Response Techniques

Quick response transport planning techniques are those that depend upon the bearest minimum of data and a simplified approach to estimate travel demand (Ref. 3, 4, 5). For example, the travel demand is carried out at an aggregated level and a traffic zone is considered the basic unit for estimation. This can be achieved by a household survey which can yield the trip generation ratio. A ratio of 0.562 was found out for Gurgaon in the recent study (Ref. 5). If such studies are carried out in some selected Indian cities, a transferable and dependable value of per capita trip-generation rate can be established. This rate can be broken into home-based work trip, home-based non-work trip and non-home-based trips. Given the population for the horizon year, these trip generation rates will then yield the number of various types of trips from each zone. The trip-distribution can be achieved by gravity model, using simplified friction factors. The trips are then assigned to the network and the adequacy studied. The whole procedure can be handled manually without the use of computers. The appeal of this technique lies in its simplicity and relatively low cost. There is a vast scope for developing and standardising the technique for adoption in India.

I. MULTIPLE CHOICE QUESTIONS

1. Which of the following is not one of the features of urban community?
   (*a*) face to face relationships (*b*) complex life
   (*c*) materialistic (*d*) glamour in life
2. Which among the following is not a cause of growth of cities?
   (*a*) trade and commerce (*b*) industrialism
   (*c*) development of transport (*d*) hygienic outlook of people
3. The spatial feature of urbanisation in India has been
   (*a*) localised in nature (*b*) balanced
   (*c*) both (*a*) and (*b*) (*d*) none of the above

Answers

1. (*a*) **2.** (*d*) **3.** (*a*)

II. SOLVED QUESTIONS AND ANSWERS

Q. 1. What is the development status of electric bicycles?

Ans. Because of the small urban area and short trip distance, bicycle ownership used to be very large. But the rise of electric bicycles has sample survey, the electric bicycle (including moped) ownership was up to 88.1 per hundred households. In accordance with the city's population of 1.59 million households at the end of the year and 43.1% urbanization ratio of projection, the urban residents have approximately 603, 700 electric bicycles in all. The electric bicycle has brought great convenience to the residents, but also has a great influence on the composition of the urban transport characteristics of the trips.

Q. 2. Write a brief note on the development status of public transportation.

Ans. By the end of 2008, there had been 54 public cross-business lines, and the total length of line was 830.4 km, with an average length of 15.38 km. Public transportation network was mainly concentrated in the old urban area, the density of network was 1.22 km/km^2, the peripheral line network was much less.

It owned 763 various types of public transport vehicles, equivalent to 817 standard units; 300 meters radius of the bus stop coverage rate reached 47.55%, 500 meters radius of the bus stop coverage rate reached 76.06%, the annual passenger volume was 112 million people.

Q. 3. Why quick response technique was developed?

Ans. During the decades of the 1960s and 1970s, urban transportation planning techniques were developed to evaluate alternative transportation systems for entire regions. The same techniques were also applied by many agencies to non-regional issues such as site development, corridor analysis, and localized system changes. Large computers, considerable data for both model calibration and application and personnel with highly specialized expertise in Computers and modelling were required to utilize these planning techniques. Therefore, the procedures were costly, time-consuming, and not always responsive to the needs of decision makers. The increasing population in urban areas has resulted in increased travel demand thus leading to traffic congestion, environmental degradation and other infrastructure related problems.

When a transport system gets overloaded due to excess travel demand, its efficiency could be improved by providing additional capacity in the form of new modes of mass transportation or through demand management. This can be achieved by studying the travel behaviour with reference to choice of commuters in choosing the mode or in other words by 'Mode cope with demands. The problems and difficulties associated with moving about within the cities and towns Choice Modelling'. Keeping the above aspects points in mind, the principle objective of the study was to develop Quick Response Techniques (QRT) for travel demand estimation aimed at applying for small and medium and cities in India.

III. REVIEW QUESTIONS

1. What are the difficulties in transport planning for small and medium cities?
2. Discuss quick-response techniques.

REFERENCES

1. Report of Task Force for developing small and medium towns and cities, Vols. I and II, Ministry of Works and Housing, New Delhi, 1976.
2. National Transport Policy Committee, Planning Commission, New Delhi, 1980.
3. Quick-response technique and model parameters, Report 186, Transportation Research Board, Washington.
4. User's guide for quick response technique, Report 187, Transportation Reseach Board, Washington.
5. Technique for assessment of travel demand for small and medium cities (Case Study ; Gurgaon), School of Planning and Architecture, New Delhi, 1982.

SECTION 14

TRANSPORT ECONOMICS

38. Economic Evaluation of Transportation Plans

38.1. Need for Economic Evaluation

38.1.1. For a given set of goals and policies, it is possible to formulate a number of alternative transport plans. The cost of these plans may vary, and so also the benefits that are likely to accure from them. What then is the criterion for selecting a particular plan from among the alternatives ? Are there any accepted principles and procedures for assessing the real cost to a community of using resources to achieve a desired purpose ? The answers to these questions will be the subject of this Chapter.

38.1.2. Economic analysis has found a ready application in problems concerned with the evaluation of transport plans. An elementary treatment of the principles of engingeering economics is, therefore, not out of place in this book. In fact, most of the countries now follow some sort of economic project evaluation before any development plan is taken up for implementation. International lending institutions, such as the World Bank, attach the greatest importance to assuring themselves that a particular project for which a loan its required is not only economically and technically sound, but that its order of priority in relation to other possible projects has been carefully determined in the light of the overall development of the country concerned.

38.1.3. For any country, it is clear that the resources available are limited and are insufficient to meet all the demands. For a developing country, the need is greatest to husband the scarce resources and select the optimum plan from amongst those competing for the allocation of these scarce resources. Economic analysis is convenient procedure to select only those schemes that result in the greatest benefit from the resources available.

38.1.4. The following are some of the specific objectives in carrying out an economic evaluation :

(*i*) Whether the plan under consideration is worth investment at all.

(*ii*) To rank schemes competing for scarce resources in order of priority.

(*iii*) To compare mutually exclusive schemes and select the most economic.

(*iv*) To assist in phasing the programme over a time period depending upon the availability of resources.

38.2. Costs and Benefits of Transport Project

38.2.1. The basic principle behind any method of economic evaluation is to measure the costs of the project, determine the benefits that are likely to accrue and compare the two. Though the principle is simple, its actual application is not. The most difficult part of economic evaluation of a transport plan is the identification and quantification of the impact of the transport plan. The two components, the costs and benefits, are dealt with separately.

38.2.2. Cost. The costs can be considered broadly under the following categories :

(*i*) Capital cost of initial construction

(*ii*) Costs of delays to vehicles during the period of construction

(*iii*) Maintenance costs.

The capital cost of providing transport facility should be estimated accurately, and should include land costs and ancillary costs. Cost of traffic control and lighting installations and administration should also be included.

When a large programme of construction of a transport facility is undertaken, it is inevitable that a good deal of disturbance is caused to the operation of vehicles. Any significant delays caused to the vehicles in this respect should be added to the capital cost. If not significant they may be ignored (Ref. 1).

The maintenance costs are of a recurrent nature and represent the expenditure to keep the assets in a tolerably good condition in the future years. The impact of the new scheme on the existing transport network should be evaluated in determining the maintenance costs. For instance, the new proposals may siphon off a good amount of traffic from the existing facility, which then may need less expenditure on maintenance. It should also be ensured that future maintenance costs are discounted to the year in which the calculations are being made.

38.2.3. Benefits

38.2.3.1. Transportation plans are intended to bring benefits to the road user by providing cheaper, more efficient, quicker and safer travel. An evaluation of the benefits is, therefore, of prime concern in any economic analysis. Benefits usually represent the difference between the cost of operating on a new transport facility and the cost of operating on an existing facility. Thus, in order to determine the benefits, it is necessary to determine the cost of operation. Benefits can be grouped under the following heads :

(*i*) Benefits to the existing traffic by way of reduce operating costs, savings in travel time and reduction in accidents.

(*ii*) Benefits to the generated traffic.

(*iii*) Benefits to traffic diverted from other routes.

(*iv*) Benefits to the traffic operating on other roads (and railways) where reduction in traffic has been caused by the opening up of the new facility.

In addition, it is also necessary to consider the dis-benefits due to the scheme, such as environmental effects (pollution, fumes, noise, vibration, visual intrusions, congestion, loss of amenity, severance etc.). Other considerations are comfort, and convenience and some social benefits.

38.2.3.2. Vehicle operating costs

The subject of vehicle operating costs has been studied very extensively because these costs constitute a significant portion of the transport costs. Savings in operating costs can be brought about if the transport plans are prepared carefully. Savings in vehicle operating costs can be evaulated fairly accurately unlike some other kinds of benefits.

The subject is dealt with in detail in Chapter 39.

38.2.3.3. Travel time savings

Improved highways bring about significant reduction in travel time, thus benefitting passengers, vehicle crew, vehicle operators and consigners of goods. The subject is dealt with in detail in Chapter 40.

38.2.3.4. Accident costs

Bad roads cause high rate of accidents. Accidents involve costs to the economy. If accidents can be prevented, benefits to the economy are possible. This subject is dealt with in detail in Chapter 41.

38.2.3.5. Benefits to Generated Traffic

It is proper to consider the full benefits from highway improvements when dealing with traffic already using the highway. This follows from the reasoning that all those who travelled before the improvements were carried out must have placed a value on the trip at least equal to the cost of travel, and thus they benefit to the full extent when improvements are made. Generated traffic pertains to journeys which were not worthwhile before the improvements but are worthwhile after the improvements. Benefits to generated traffic are usually assessed at one-half the change in user costs (Ref. 10), vide Fig. 38.1.

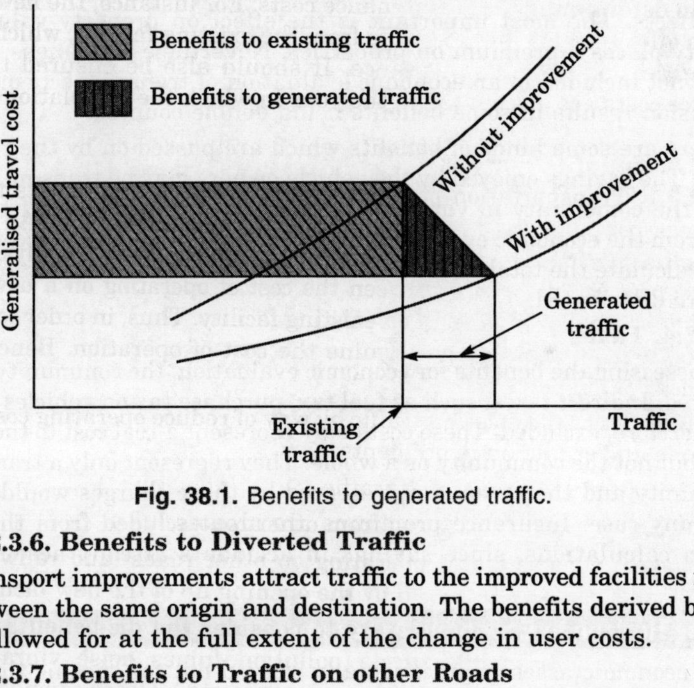

Fig. 38.1. Benefits to generated traffic.

38.2.3.6. Benefits to Diverted Traffic

Transport improvements attract traffic to the improved facilities from other routes between the same origin and destination. The benefits derived by diverted traffic is allowed for at the full extent of the change in user costs.

38.2.3.7. Benefits to Traffic on other Roads

Improvements to a road may cause reduction in traffic on other roads (and railways), thus resulting in lesser congestion. There may also be congestion on these roads as a result of the new scheme. These effects should be evaluated.

38.2.3.8. Environmental Effects

Traffic plans are likely to result in disbenefits caused by adverse effects on the environment. On the other hand, certain improvement schemes might be planned with the very objective of improving and preserving the environment. In all such cases, it is necessary to evaluate in monetary terms the impact of the scheme on the environment. The elements that need consideration are : noise, fumes, vibration, loss of amenity, severance, visual intrusion etc. Some of these are amenable to quantification in monetary terms, while some are not.

Noise is now considered a significant factor in environmental quality. The annoyance caused by noise is translated into noise costs by considering questions such as: "how much is an individual affected by noise prepared to pay to get relief from it" or "how much would the sufferer have to receive monetarily to reinstate his pre-noise level of satisfaction". The disbenefits can also be assessed by changes in rents and property values.

More research is needed before all the environmental effects can be considered in monetary terms.

38.2.3.9. Comfort and Convenience

Comfort and convenience represent the quality of service offered by a transport facility and are difficult to value. But yet they are important aspects since many road users are prepared to use a longer route just to derive in comfort and relaxation. The AASTHO guidelines (Ref. 9) recommended arbitrary values for the cost of discomfort and inconvenience.

38.2.3.10. Indirect Benefit

In addition to direct benefits to highway improvements there are many indirect effects. The most important is the effect on property values. A good accessibility places a premium on properties. Nevertheless, changes in property values are not included in an economic evaluation of transport schemes, because their inclusion results in some benefits being double counted.

There are some kinds of benefits which are passed on by the road user to the others. The savings enjoyed by the vehicle owners may be transmitted to other sectors of the community in various forms. Here also, the indirect benefits are excluded from the economic evaluation in order to avoid duplicate accounting. The idea is to calculate the total benefits to the community and not to assess how the benefits are distributed.

38.2.4. Taxes

In assessing the benefits for economic evaluation, the community as a whole is considered. Indirect taxes, such as fuel tax, purchase tax on vehicles and licence fees are therefore, excluded. These costs may represent a real cost to the individual road user but not the community as a whole. They represent only a transfer within the community and the revenue represented by these charges would have to be raised in any case. Insurance premiums are also excluded from the economic evaluation calculations, since savings in accidents already account for this element.

38.3. Time Horizon in Economic Assessment

An economic assessment is usually carried out for specific time horizon. For road schemes, the evaulation period commonly selected is 20 to 30 years. Even though it may be true that a scheme may have a life greater than this period, especially when properly maintained, it is difficult to keep the economic forecasts within realistic limits if longer time horizons are selected. For a railway line system, a higher time horizon is appropriate. A time period of 50 years was adopted for the London Victoria line study (Ref. 21).

At the end of the period selected for economic evaulation, the facility is likely to still remain an asset with some residual value. It is then appropriate to take the residual value into account duly discounting the same.

38.4. Basic Principles of Economic Evaluation

The fundamental principle on which all methods of economic evaluation are based is that money earns income over a period of time. For example, ₹ 100 today will be worth ₹ 672.75 at the end of 20 years if invested at 10 per cent compound rate of interest. Similarly, a sum of ₹ 672.75 which might become due to an individual after 20 years from today is worth only ₹ 100 at the present, assuming the same rate of interest. This leads to the necessity of devaluing the future benefits and costs of schemes to the present time in order to determine their present worth.

The following formulae are very useful in dealing with the problems in economic evaluation :

1. The amount S to which Re. 1 will increase in n years with a compound interest rate of r

$$S = (1 + r)^n \quad \text{...(38.4)}$$

2. The present value A of Re. 1 n years hence, when discounted at interest r

$$A = \frac{1}{(1+r)^n} = (1 + r)^n \quad \text{...(38.5)}$$

3. The present value A of an annuity of Re. 1 per year for n years when discounted at interest r.

$$A = \frac{1-(1+r)^n}{r} \quad \text{...(38.6)}$$

4. The amount P per annum for n years at interest rate r needed to accumulate Re. 1 at the end of n years. (Sinking fund factor).

$$P = \frac{r}{(1+r)^n - 1} \quad \text{...(38.7)}$$

5. The amount P per annum for n years at interest rate r needed for recovery of a capital of Re. 1 including interest. (Capital recovery factor).

$$P = \frac{r(1+r)^n}{(1+r)^n - 1} \quad \text{...(38.8)}$$

6. Present worth A for uniform payments of Re. 1 per year for n years at interest rate r (present worth factor).

$$A = \frac{r(1+r)^n - 1}{r(1+r)^n} \quad \text{...(38.9)}$$

Readymade tables giving the values of the above functions are available (Ref. 23, 24) and are handy to use.

38.5. Interest Rate

The rate of interest (or discount rate) is one of the most important elements in an economic analysis. The selection or rejection of a particular project is to a great extent dependent on the choice of an appropriate interest rate.

The choice of the interest rate is governed by a number of complex factors, and is dependent on the future availability of finance and the various opportunities for its use. Will the current generation prefer to consume the resources now or conserve it for future use by the current or future generation ? The answer to this question will give the "social time preference rate of interest", which will then determine the relative value that a government assigns to present consumption as against future consumption. Another approach is to find out the social yield that the resources employed by a marginal public project would have otherwise generated. This determines the "social opportunity cost rate of interest". In a truely competitive economy, the two rates of interest would be equal and investements and consumption would then be ideally allocated. Unfortunately, such an ideal situation is hard to find. The situation is especially so in developing countries where capital is very scarce.

The discount rate should be never less than the rate of borrowing or lending by the Government, or the market rate of interest. Since the interest rate represents opportunities forgone, the same is sometimes called the opportunity cost of capital.

A rate of 10–12% is often chosen when evaluating the transport sector schemes in developing countries.

38.6. Methods of Economic Evaluation

After having determined the costs and benefits of a scheme, a method has to be evolved for relating these two so as to arrive at an assessment of the soundness of a scheme in economic terms. A number of methods have been developed and the literature on them is voluminous. The following are the important methods :

A. Rate of return methods

1. Benefit cost (B/C) ratio method
2. First year rate of return method.

B. Discounting cash flow (DCF) methods

1. Net Present Value (NPV) method
2. Internal Rate of Return (IRR) method.

In the first group of methods, the benefits are expressed as the net benefits occurring in a single reference year and the costs are expressed as a net annual cost. In the second group of methods, the stream of costs and benefits over a period of time are compared on the basis of discounted cash flow analysis. The methods are described in detail in subsequent sections and a comparison of them is given in section 38.11.

38.7. Benefit-Cost Ratio Method

The benefit-cost ratio method is one of the widely used ones for evaluation of highway projects and is the basis of AASTHO Road-user analysis (Ref. 9). In this method, the ratio of the net annual benefits to the net annual costs is determined. The benefits are evaluated for a single reference year, which for convenience can be the first year of operation after construction or the median year of the analysis period. The costs are the equivalent annual charge representing equal amortization and interest payment (at a specified discount rate) spread over the economic life of the project. The benefit cost ratio for a particular project would be :

$$\frac{B}{C} = \frac{\text{Benefits in the reference year}}{\text{Annual costs}}$$

$$\text{Ratio} = \frac{\begin{array}{c}\text{(Road user cost for the existing road – Road user}\\\text{cost for the improved road) in the reference year}\end{array}}{\begin{array}{c}\text{Equivalent annual highway costs for the improved road}\\\text{– Equivalent annual highway costs for the existing road}\end{array}} \quad ...(38.10)$$

The numerator of the B/C ratio represents the benefits, which are really the reduction in user costs. The denominator represents the difference in annual highway costs between the new facility and the existing facility, including maintenance.

If a number of projects are being considered, the B/C ratio for each of these is determined in the same manner, comparing the conditions for the existing road with those for the improved road. A ratio greater than 1.0 indicates that the extra cost involved in the improvement is less than the benefits that are likely to accrue and the project is economically justified. If several projects are being considered, the ratios would enable them to be ranked in order of attractiveness, the one having the highest B/C ratio being the most attractive. When several major alternatives are under evaluation, a second analysis is then made, using the preferred alternative as the base to determine whether an added increment of investment might yield a proportionatley larger incease in road user savings when compared with another alternative.

The following problem illustrates the use of the benefit-cost ratio method.

Problem 38.1. *A single Lane road 50 km long is to be widened to two lanes at a cost of ₹ 8.0 lakhs per km, including all improvements. The cost of operation of vehicles on the single lane road is ₹ 1.20 per vehicle km, whereas it is ₹ 1.00 per vehicle km on the improved facility. The average traffic may be assumed 2500 vehicles per day over a design period of 20 years. The interest rate is 10 percent per annum. The cost of maintenance is ₹ 5,000 per km on the existing road and ₹ 10,000 per km on the improved road. Is the investment in the improvement scheme worthwhile ?*

Solution. Annual road user cost in the median year of analysis :

(*i*) *Improved road* ₹ 365 × 50 × 2500 × 1.00

= ₹ 455.00 lakhs

(*ii*) *Existing road*, ₹ 365 × 50 × 2500 × 1.20 = ₹ 545.00 lakhs.

The capital recovery factor for $n = 20$ and $r = 10$,

is :
$$A = \frac{r(1+r)^n}{r(1+r)^n - 1} = 0.117$$

Total cost of improvement = ₹ 50 × 8 lakhs = ₹ 400 lakhs

Present annual cost of improvement = ₹ 400 × 0.117 lakhs = ₹ 46.8 lakhs.

Difference in annual maintenance costs = ₹ 50 × 5000 = ₹ 2.5 lakhs.

$$\text{Benefit cost ratio} = \frac{₹\ (545 - 455)\ \text{lakhs}}{(46.8 + 2.5)\ \text{lakhs}} = 1.93$$

Since the ratio is greater than 1.0, the project is worthwhile.

38.8. First Year Rate of Return

The first year rate of return is the criterion followed by the Department of Environment, U.K. (Ref. 1). In this simple method, the benefits accuring in the first year of the scheme's operation alone are compared with the capital costs of construction. The result, expressing the benefits occurring in the first year as a percentage of the costs, in called the first year rate of return.

First Year Rate of Return

$$= \frac{\text{Benefits occurring in the first year}}{\text{Capital costs}} \quad ...(38.11)$$

The first year rate of return of a possible scheme gives an indication of its priority when compared with other schemes, and thus assists the selection of the most advantageous scheme. The priority for different schemes can also be decided on the basis of the first year rate of return.

The following example will make the principle clear.

Problem 38.2. *A new bypass is to be constructed at a busy town. The length of the bypass will be 5.2 km, and the length of the road through the town is 5.4 km. The cost of the project is likely to be ₹ 7,500,000. The speed of traffic through the town is 46.6 KPH. The predicted traffic after completing the bypass is 7,600 vehicles/day, out of which 50% will use the bypass. It is computed that if the bypass is not constructed the speed through the town will further drop to 44.1 KPH due to increased traffic, and the speed through the bypass is expected to be 77 KPH and that through the town will be 50.4 KPH. The travel costs at the three speeds are as below :*

Speed (K.P.H.)	*Travel costs per vehicle/KM (₹)*
44.1	*1.14*
50.4	*1.02*
77	*0.90*

It is expected that the construction of the bypass will bring down the accident rate from 1.75 per million vehicle km on the existing route to 0.60 per million vehicle km on the bypass. The cost of an accident can be taken as ₹ 15,000. The maintenance cost per km is ₹ 10,000. Calculate the First Year Rate of Return.

Solution.

(1)	*Existing Road* (2)	*Proposed Bypass* (3)	*Existing Road left open after construction of bypass* (4)
Travel Costs.			
(*i*) Travel cost per vehicle km (P) (₹)	1.14	0.90	1.02
(*ii*) Annual travel cost 365 × Vol × Length × P	365 × 7600 × 5.4 × 1.14 = ₹ 17,10,000	365 × 7600 × 5.2 × 0.90 × $\frac{50}{100}$ = ₹ 6,500,000	365 × 7600 × $\frac{50}{100}$ ×5.4 × 1.02 ₹ = 7,650,000

(1)	(2)	(3)	(4)
Accident Costs			
(*i*) Accident rate (J) per million vehicle km	1.75	0.60	1.75

(*ii*) Annual number of accidents (N)

$$= \frac{365 \times \text{Vol} \times \text{Length} \times J}{1{,}000{,}000} \quad 26.2 \quad 4.3 \quad 13.1$$

Extra Maintenance :

(*i*) Additional road length : 5.2 km

(*ii*) Additional maintenance cost

= ₹ 10,000 × 5.2 = ₹ 52,000

Net Benefits

(*i*) Saving in Travel costs :

₹ 6,500,000
+ ₹ 7,650,000
₹ 14,150,000

Savings ₹ 17,100,000
– ₹ 14,150,000
₹ 2,950,000

(*ii*) Savings in Accident costs : ₹ [26.2 – (4.3 + 13.1)] × 15,000

= ₹ 15,000 × 8.8

= ₹ 130,000

Net benefits ₹ 2,950,000
+ ₹ 130,000
₹ 3,080,000
– ₹ 52,000 (Addl. maintenance cost)
₹ 3,028,000

$$\text{First Year Rate of Return} = \frac{3{,}028{,}000 \times 100}{7{,}500{,}000} = 40.4\%$$

38.9. Net Present Value Method

Net present value (present worth) method is based on the discounted cash flow (DCF) technique. In this method, the stream of costs/benefits associated with the project over an extended period of time is calculated and is discounted at a selected discount rate to give the present value. Benefits are treated as positive and costs as negative and the net present value is found. Any project with a positive net present value is treated as acceptable. In comparing more than one project, a project with the highest net present value should be accepted.

The net present value is algebraically expressed as :

$$NPV_0 = (B_0 - C_0) + \frac{B_1 - C_1}{(1+i)} + \frac{B_2 - C_2}{(1+i)^2} \ldots + \frac{B_n - C_n}{(1+i)_n} \quad \ldots(38.12)$$

where, NPV_0 = net present value in the year 0.
B_t = value of the benefits which occur in the year t
C_t = costs which occur in the year t.
i = discount rate per annum.
n = the number of years for which the return is to be calculated.

The following problem illustrates the method.

Problem 38.3. *A section of road is at present having poor geometrics. The cost of upgrading this road is ₹ 100.00 lakhs. The road user costs with and without the improvements, the accident costs with and without the improvements and the extra maintenance costs with the improvements are tabulated below for a 20 year period after the execution of the improvement programme. Assuming a discount rate of 10% per annum, is the project economically justified ?*

Solution. Table 38.1 gives the calculations in a convenient form.

38.10. Internal Rate of Return Method

The internal rate of return is the discount rate which makes the discounted future benefits equal to the initial outlay. In other words, it is the discount rate which makes the stream of cash flows to zero. Eq. 38.12 can thus be modified as below, if B_0. is zero :

$$C_0 = \frac{B_1 - C_1}{(1+i)} + \frac{B_2 - C_2}{(1+i)^2} \ldots + \frac{B_n - C_n}{(1+i)^n} \qquad \ldots(38.13)$$

$$= \sum_{t=1}^{n} \frac{B_t - C_t}{(1+i)^t}$$

where, C_0 = Capital cost of the investment in the year zero
B_t = Value of the benefits which occur in the year t
C_t = Costs which occur in the year t
i = Internal rate of return
n = Number of year for which the analysis is done.

The solution to the above equation is rather tedious and is possible only by trial and error. With a computer programme the work is rendered very simple.

If the rate of return calculated from the above formula is greater than the rate of interest obtainable by investing the capital in the open market, the scheme is considered acceptable.

38.11. Comparison of the Various Methods of Economic Evaluation

The four methods of economic evaluation described earlier have their own advantages and disadvantages. Each one of them may be more appropriate than the other for a given sitution.

The benfit-cost ratio method, though widely used by highway engineers, suffers from the following drawbacks :

(*i*) It requires an assumption of the rate of interest, which should be somewhat related to the opportunity cost of capital. Unfortunately, the opportunity cost of capital is very often not known, or can be estimated only approximately (Ref. 10).

(*ii*) The significance of the B/C ratio is ambigous, (Ref. 22) and its relative value is difficult to understand and interpret. For example, if there are two proposals, one with a B/C ratio of 1.05 and the other with a ratio of 1.10, the difference is very difficult to appreciate.

Table 38.1. (All Amounts in ₹ Lakhs)

	Road user cost		Accident costs					
Year (t) (1)	With improvements (2)	Without improvements (3)	With improvements (4)	Without improvements (5)	Extra cost of maintenance (6)	Benefits = Col. (3) –Col (2)+ Col (5) –Col (4) = $B_t - C_t$ (7)	Benefit – cost (8)	Discounted value of Benefit–cost (9)
0	–	–	–	–	–	–	– 100.0	– 100.0
1	105.5	126.5	1.1	3.1	1.0	23.0	*Col.* (7)-Col. (6) 22.0	+ 20.0
2	110.3	132.2	1.1	3.1	1.0	24.1	22.0	+ 19.1
3	115.8	138.9	1.2	3.5	1.0	25.4	24.4	+ 18.4
4	121.6	145.8	1.2	3.7	1.0	26.7	25.7	+ 17.6
5	127.6	153.0	1.3	3.8	1.0	27.9	26.9	+ 16.7
6	134.0	161.0	1.3	4.0	1.0	29.7	28.7	+ 16.7
7	140.7	168.9	1.4	4.2	1.0	31.0	30.0	+ 15.4
8	147.8	177.0	1.5	4.4	1.0	32.1	31.1	+ 14.5
9	155.1	186.2	1.6	4.7	1.0	34.2	33.2	+ 14.1
10	162.9	195.2	1.6	4.9	1.5	35.6	34.6	+ 13.4
11	171.0	205.0	1.7	5.1	1.5	37.4	35.9	+ 12.6
12	179.6	214.6	1.8	5.4	1.5	38.6	37.1	+ 11.8
13	188.9	226.1	1.9	5.7	1.5	41.3	39.8	+ 11.5
14	198.0	237.5	2.0	5.9	1.5	43.4	41.9	+ 11.1
15	207.9	248.6	2.1	6.2	1.5	44.8	43.3	+ 10.4
16	218.3	262.1	2.2	6.6	1.5	48.2	46.7	+ 10.2
17	229.2	274.2	2.3	6.9	1.5	49.6	48.1	+ 9.5
18	246.7	288.5	2.4	7.2	1.5	52.6	51.1	+ 9.2
19	252.7	303.0	2.5	7.6	1.5	55.4	53.9	+ 8.8
20	256.3	318.0	2.7	8.0	1.5	58.0	56.5	+ 8.4
								Total = + 268.9 – 100.0

N.P.V. = + 268.9 – 100 = + 168.9

Since the N.P.V. is positive, the project is economically justified (*iii*) It is at times difficult to decide which items should be treated as costs and placed in the denominator and which as negative benefits and placed in the numerator.

The net present value method also suffers from the disadvantage that a discount rate has to be assumed initially. It is much simpler computationally when compared to the internal rate of return method.

The internal rate of return avoids the necessity for selecting a discount rate initially. The rate which is derived as a result of the computations can be easily compared with the market rate of interest, with which economists and financial experts are familiar. Thus, the method is considered to be more meaningful than the others. Its disadvantages are that the calculations are tedious and a solution can only be found by trial and error, and it may sometimes be misleading in comparing projects having different lives and different streams of benefits.

The first year rate of return method is quick to use, but has obvious shortcomings. More importantly, many projects may have attractive benefits initially but whose benefits taper off abruptly thereafter, and this factor will be lost sight of in the method.

I. SOLVED QUESTIONS AND ANSWERS

Q. 1. What is economic evaluation?

Ans. Economic evaluation (also called economic analysis or appraisal) refers to methods for determining the value of a policy, project or program. It is an important component of planning and management. Other important components include information collection and distribution, a decision-making process, public involvement, institution development, and negotiation between stakeholders. We frequently perform economic evaluation.

Virtually any consumer, business or public policy decision involves some sort of economic analysis. For minor decisions it may be sufficient to perform such analysis in your head or to use standard guidelines, but when making major decisions it is helpful to follow a more explicit and comprehensive process to insure that all significant impacts (costs, benefits and equity effects) are considered.

Economic evaluation can be performed well, or it can be performed badly. Transportation decisions are frequently criticized for inadequate and biased economic analysis. This paper describes techniques for performing comprehensive and objective evaluation, and identifies ways to avoid common problems.

Q. 2. Discuss the planning and evaluation techniques.

Ans. It is helpful to begin any planning process with a vision, goal or problem statement (these are essentially the same thing from different perspectives). Goals describe what you ultimately want to achieve, but are often difficult to measure. The next step is to establish specific, achievable objectives, with suitable evaluation criteria. For example, if the goal is to improve access, objectives may include a particular reduction in traffic congestion delay, an increase in transit trip speeds, or a reduction in barriers to non-motorized travel.

Evaluation criteria can include quantitative factors such as cost per unit of output (*e.g.,* dollars per additional passenger trip), and qualitative factors such as public acceptance. It is usually best to select several objectives that range from general (*e.g.,* improve travel choices for non-drivers) to specific (*e.g.,* a 50% reduction in potholes within two years).

Q. 3. How does measuring transportation affect evaluation process?

Ans. How transportation activities are measured can affect the evaluation process. There are three common ways to measure transport, representing different perspectives and assumptions (VTPI, 2006, "Measuring Transportation"):

- Traffic refers to vehicle movement. It is measured using vehicle-miles and vehicle-trips, roadway Level-of-Service (LOS), average traffic speeds, vehicle delay, and vehicle operating costs. It assumes that transport improvements consist of vehicle travel. It is the easiest way to measure transportation and is commonly used in transport planning.
- Mobility refers to the movement of people and goods. It is measured using person-miles, person-trips and ton-miles, average trip speeds, and cost per person or ton-mile. It recognizes automobile, transit, ridesharing, and to a lesser degree, non-motorized improvements to transportation problems. It is moderately difficult to measure.
- Accessibility refers to the ability to reach desired goods, services and activities. It is the ultimate goal of most transport (except the small portion of travel with no destination). It is measured using generalized cost (money, time and discomfort) per trip. Accessibility is relatively difficult to measure and so is not often used in transport planning.

Q. 4. What do you mean by economic analysis?

Ans. Economic analysis refers to the evaluation of economic impacts. These impacts are not limited to market resources (goods traded in a conventional market), they can be any scarce and valued resource, which may include money, land, time, health, safety, comfort, and environmental services such as clean air and water, as discussed below.

Economic analysis can be performed from various perspectives, such as an individual consumer, a particular group, a jurisdiction, a region, all of society (everybody), or current and future societies (which can be considered sustainability evaluation). For example, individual consumer generally evaluate a vehicle's safety from the perspective of occupants-risks to other road users are given less weight. Similarly, a particular group generally evaluates taxes and fees based on their cost burden, and may consider it desirable to shift costs to other groups.

Q. 5. What are the ways to evaluate equity analysis?

Ans. There are several ways to evaluate equity. Horizontal equity assumes that everybody should be treated equally. Vertical equity assumes that physically, economically or socially disadvantaged people should be favored compared with relatively advantaged people. Below are common equity objectives that can be used for evaluating policies and programs.

- Treats everybody equally. A policy does not arbitrarily favor one group over others.
- User-Pays Principle. Individuals bear the costs they impose unless a subsidy is specifically justified.
- Progressive with respect to income. Lower-income households benefit relative to higher-income households.

- Benefits transportation disadvantaged. Benefits people with disabilities, non-drivers, people who cannot afford a car, etc.
- Improves basic mobility. Helps satisfy basic mobility (travel that society considers valuable).
- Economic evaluation should indicate the distribution of costs and benefits (such as by income and need), and the degree to which options tend to achieve or contradict equity objectives.

Q. 6. What is cost effectiveness analysis?

Ans. Cost-Effectiveness is a relatively simple type of economic analysis. It measures the cost of achieving a specific objective, such as building a particular road or delivering a particular amount of freight. The quality of outputs (benefits) are held constant, so there is only one variable, the cost of inputs.

All relevant costs are converted into appropriate, comparable monetary units. For example, highway construction costs may be presented in costs per roadway-mile, cost per lane-mile, cost per additional vehicle-trip or passenger trip that uses the facility. An option is considered cost effective if its benefits outweight its costs. The lowest cost option for achieving an objective is considered most cost effective. For example, if there are two possible routes for a new highway, the cheapest option is most cost effective.

The costs of adding capacity to alleviate congestion should be assigned only to peak period travel (travel during congested periods) rather than all trips. Additional benefits from a project (such as increased safety or comfort) may be assigned to all users.

Q. 7. Write a brief note on benefit-cost and net benefit analysis

Ans. Benefit-cost analysis compares total incremental benefits with total incremental costs. Unlike cost-effectiveness analysis, benefit-cost analysis is not limited to a single objective or benefit. For example, potential highway routes may differ in construction costs and the quality of service (speed and safety) they provide. Benefit-cost analysis places a value on each incremental benefit and costs of each option. These are summed and compared. The results can be presented as a ratio, with benefits divided by costs (which is why it is often called "Benefit/Cost" or "B/C" analysis).

Q. 8. What is simple payback?

Ans. Simple Payback is similar to a Benefit-Cost Ratio. It indicates the time period required for benefits to repay an investment. However, these methods are affected by how benefits and costs are incorporated in the analysis. For example, some impacts can be considered either a reduction in benefits or an increase in costs, which is used will change the ratio.

Q. 9. What is meant by return on investment?

Ans. Return On Investment (ROI) indicates the annual rate of return an investment provides if all impacts are monetized. For example, a \$1 million investment that provides \$250,000 annual benefits has a four-year simple payback and an ROI of 25%. It is often used for business analysis.

Net benefits is defined as the sum of all benefits minus the sum of all costs. It provides an absolute measure of benefits (total dollars), rather than the relative measures.

Q. 10. What is optimization?

Ans. Optimization refers to solutions that provide the best balance between multiple, conflicting objectives. Transport planning is sometimes reductionist (evaluation that considers just one or two objectives), which can result in non-optimal solutions that may make society worse overall. For example, decision makers overwhelmed by the perceived complexity of considering multiple planning objectives sometimes ask planners to focus on just one or two problems. This can result in decisions that address certain problems (such as congestion or pollution) which exacerbate other problems (such as accidents and inadequate mobility for non-drivers), and tends to under value solutions that provide multiple benefits. More comprehensive optimization tends to be best for society overall. Fortunately, new computer-based analysis tools are available that allow more complex analysis.

II. REVIEW QUESTIONS

1. What is the need for economic evaluation?
2. Discuss the costs and benefits of transport project.
3. What are the benefits of diverted traffic?
4. What is the basic principles of economic evaluation?

REFERENCES

1. Department of the Environment (U.K.), The Economic Appraisal of Inter-urban Improvement Schemes, Technical Memorandum H/171.
2. Robinson R., H. Hide, J.W. Hodges. J. Rolt and S. Abaynayaka. A Road Transport Investment Models for Developing Countries TRRL Laboratory Report 674, Transport and Road Research Laboratory, Crowthorne, 1975.
3. Heggie, I.A., Transport Engineering Economics, McGraw Hill Book Company, London, 1972.
4. De Weille, J., Qualification of Road User Savings, World Bank Staff Occasional Papers No. 2, 1966 (Distributed by John Hopkins Press for the International Bank for Reconstruction and Development).
5. Hide H., S.W. Abaynayaka, I, Sayer and R.J. Wyatt, The Kenya Road Transport Cost Study : Research on Vehicle Operating Costs, TRRL Laboratory Report 672, Transport and Road Research Laboratory, Crowthrone, 1975.
6. Research on Road Traffic Road Research Laboratory, (U.K.) H.M.S.O., London, 1965.
7. Beesley, M.E., The Value of Tiome Spent in Travelling : Some New Evidence, Economica, 32, 1965.
8. Thomson, J.M. Modern Transport Economics, Penguin Books, Hamondsmorth, 1974.
9. AASTHO, Road User Analysis for highway improvements, Washington, 1960.
10. Adler, H.A., Sector and Project Planning in Transportation, World Bank Staff Occasional Papers No. 4. 1967, (Distributed by the John Hopkins Press for the International Bank for Reconstruction and Development).
11. Winfrey, R., Economic Analysis for Highways, International Textbook Company, Scranton, 1969.
12. NCHRP, Running Cost of Motor Vehicles as affected by Road Design and Traffic, Report No. 111, Highway Research Board, Washington, 1971.
13. Charlesworth, G. and J.L. Paisley. The Economic Assessment of return from Road Works, Proceedings, Institution of Civil Engineers, London, 1969.

14. Dawson, R.F.F., Vehicles Operating Costs in 1970, RRL Report LR 439, Transport and Road Research Laboratory, Crowthorne, 1972.
15. Reynolds, D.J., The Cost of Road Accidents, Journal of the Royal Statistical Society, Series A (General 119 (4), 1956.
16. Dawson, R.F.F., Cost of Road Accident in Great Britain, R.R.L. Report LR 79, Road Research Laboratory, Crowthrone, 1967.
17. Dawson, R.F.F. Current Costs of Road Accidents in Great Britain, R.R.L, Report LR 396, Road Research Laboratory, Crowthorne, 1971.
18. Srinivasan, N.S., D.V. Hingorani and B.M. Sharma, Economic Cost of Road Accidents, Journal of the IRC, Vol. 36–2, New Delhi, 1975.
19. Gwilliam, K.M. and P.J. Mackie, Economics and Transport Policy, George Allen and Unwin Ltd. London, 1975.
20. Dawson, R.F.F., The Economic Assessment of Road Improvement Schemes, Road Research Technical Paper No. 75. H.M.S.O., London, 1968.
21. Foster. R.D. and M.E. Beesley. Estimating the Social Benefit of Constructing an Underground Railway in London, Journal of the Royal Society, Vol. 126, London, 1973.
22. Wohl, M., and B.V. Martin, Evaluation of mutually exclusive design Projects-Special Report 92, Highway Research Board, Washington, 1967.
23. An Introduction to Engineering Economics, The Institution of Civil Engineers, London, 1969.
24. Manual of Economic Analysis for Highway Projects, Indian Roads Congress, New Delhi, 1995.

39. Vehicle Operating Costs

39.1. Introduction

The Road User Cost is composed of the following three components :

1. Vehicle operating cost (VOC)
2. Travel time cost
3. Accident cost.

The present Chapter deals with the vehicle operating cost (VOC). Chapter 40 deals with the travel time cost and Chapter 41 deals with the cost of road accidents.

39.2. Earlier Work Done Abroad

Considerable research has been done the world over on VOC. In fact, the interest in the same dates back to the thirties in the U.S.A. Ralph Moyer presented motor vehicle operating cost on untreated gravel and cement concrete road surfaces as early as 1939. (Ref. 1) Again, Moyer and Winfrey presented costs of operating vehicles on surfaced pavement and gravel and earthen roads in 1939 (Ref. 2). Winfrey did extensive work later on and he presented tables for motor vehicle runing costs for highway economic studies in 1963 (Ref. 3, 4). Claffey did extensive research on time and fuel consumption for highway user benefit study and presented the results in 1960. (Ref. 5, 6) The AASTHO Road User Benefit Analysis guide-book (Ref. 7) was published in 1960 and became a standard reference source for American conditions. Dawson's work reported in TRRL Report LR 439 (Ref. 8) gave the VOC in U.K. in 1970. Bonney and Stevens did considerable work on VOC on bituminous, gravel and earthen roads in East and Central Africa (Ref. 9). The World Banks' work (Ref. 10, 11) in this connection is worth mentioning. De Weille's tables for quantification of road user saviangs was a standard source of reference till early seventies. The work of the TRRL in Kenya on vehicle operating costs (Ref. 12, 13, 14) was the first scientific study on establishment of accurate relationships between the VOC components and factors likely to affect the same. The results of a larger study in Brazil (Ref. 15) are now available. The studies have enabled the development of a Highway Design Models, wherein the total transportation cost (cost of initial facility, cost of maintenance of the facility and the road user cost) can be determined for a number of alternative highway designs. The model can then select the most optimum combination of highway design and maintenance strategies.

39.3. Road User Cost Study in India

An ambitious research project, entitled Road User Cost Study, has been completed in India in 1982 (Ref. 16, 17, 18). The project, costing nearly ₹ 100 lakhs, was jointly sponsored by the Government of India and the World Bank and was implemented by the Central Road Research Institute. The project has yield valuable results on VOC, accident cost, travel time cost and speed-flow relationships under typical Indian road and traffic conditions. The work was updated in 1992 (Ref. 21). The work was further updated in 2001 (Ref. 22). The following paragraphs give some of the results of the study as far as VOC is concerned.

39.4. Components of VOC

The following are the components of the VOC :

Variable Cost

1. Fuel
   —Engine oil
   —Other oils
   —Grease
2. Lubricants
3. Tyre
4. Spare Parts
5. Maintenance Labour cost
6. Depreciation

Fixed Cost

1. Interest on capital
2. Insurance
3. Road tax
4. Other taxes
5. Registration fee
6. Garaging charges
7. Other fixed charges such as fines, tolls, octroi etc.
8. Permit charges
9. Loading and unloading charges (in respect of trucks only)
10. Commission on booking (in respect of trucks only)
11. Overhead Charges
    (*i*) Rent and water
    (*ii*) Clerical and managerial staff
    (*iii*) Audit fees
    (*iv*) Advertisement
    (*v*) Electricity
    (*vi*) Telephone and postage
    (*vii*) Stationery and printing
    (*viii*) Travelling expenses
    (*ix*) Freight and hire
    (*x*) Legal fees
    (*xi*) Bank charges
    (*xii*) Repairs and maintenance of buildings
    (*xiii*) Depreciation of fixed assets
    (*xiv*) Sundries
12. Crew costs

An idea of the relative proportion of the above components may be had from the following Table, which represents the cost of operation of a typical passenger bus in India (1990 prices)

Table 39.1. Cost of Operation of a Passenger Bus in India (1990)

S. No.	**A. Variable Costs** *Component*		*Cost in ₹ per km*
1.	Fuel		0.68
2.	Lubricants		
	(*i*) Engine oil	0.11	
	(*ii*) Other oil	0.01	
	(*iii*) Grease	0.01	
	Total 2	0.13	0.13
3.	Tyre		0.73
4.	Spare Parts		0.19
5.	Maintenance Labour cost		0.08
6.	Depreciation		0.19
	Total 'A'		2.00
	B. Fixed Costs		
7.	Crew Costs		0.79
8.	Other Fixed Costs		1.02
	Total 'B'		1.81
	Total 'A' and 'B'		3.81

39.5. Factors Affecting VOC

The following are the important factors affecting VOC :

A. Vehicle Factors

1. Age
2. Make
3. Horse power, engine capacity
4. Load carried
5. Condition of vehicle
6. Level of maintenance input
7. Type of fuel used (petrol/diesel/petrol-alcohol mixture)
8. Type of tyres used (rayon, nylon, radial, ply-rating, size etc.).

B. Roadway Factors

1. Roughness of surface
2. Type of surface
3. Horizontal curvature
4. Vertical profile
5. Pavement width
6. Type and condition of shoulder
7. Urban or rural location
8. Number of junctions per km.

C. Traffic Factors

1. Speed of travel
2. Traffic volume and composition.

D. Environmental Factors

1. Altitude
2. Rainfall
3. Temperature

39.6. Fuel Consumption Relationships

The data from vehicle Operators has yielded directly the relationships given in Table 39.2 (Ref. 18) :

In the above equations :

FC = Fuel econsumption in litres per 100 km
RF = Rise and fall in m/km
RG = Roughness in mm/km
V = Speed in km/hr
GVW = Gross vehicle weight in Tonnes
W = Pavement width in m
PW = Power/weight ratio in kw/tonne

Table 39.2. Fuel Consumptions Relationship from User Cost Survey
(*Source* : Ref. 18)

S. No.	*Vehicle*	*Equation*	R^2 *value*
1.	Bus	FC = Intercept + 0.001023 LK + 0.09177 RF $+ \dfrac{469.7}{V} + 0.006243\ V^2 + 0.5563\ GVW$ (The value of the intercept varies from operator to operator. For a typical operator, its value is –5.85)	0.709
2.	Truck	$FC = 7.170 - 0.9202\ PW + 0.1432\ RF - 0.3889\ W + \dfrac{567.1}{V} + 0.007868\ V^2$	0.526
3.	Car	$FC = 13.81 + 8.03669\ RF + 0.0003423\ RG + \dfrac{621.3}{V}\ 0.005005\ V^2$	0.177
4.	Jeep	$FC = -10.91 + 0.08823\ RF + 0.0005918\ RG + \dfrac{432.3}{V} + 0.005312\ V^2$	0.919

It is seen that fuel consumption generally :

(*i*) increases with load, and decreases with Power/Weight ratio.

(*ii*) increases with roughness.

(*iii*) increases with rise and fall.

(*iv*) decreases with increase in pavement width.

(*v*) follows the traditional U-shaped curve in relation to speed, exhibiting a low value at an optimum speed and higher values at speeds less than or more than the optimum.

In order to substantiate the above findings, the more reliable approach of controlled experiments was tried. The following equations emerged (Ref. 18).

Table 39.3. Fuel Consumption Equations from Controlled Experiments

S. No.	*Vehicle*	*Eqn.*	R^2
1.	Diesel Jeep	$FC = -3.083 + \dfrac{225.811^{**}}{V} + 0.00242\ V^{2**} + 0.00012\ RG^{**} + 0.12776\ RF^{**} - 0.05647\ FL^{**}$	0.73
2.	Ambassador Car	$FC = 1.031 + \dfrac{167.552^{**}}{V} + 0.00133\ V^{2**} + 0.00006\ RG + 0.13879\ RS^{**} - 0.10322\ FL^{**}$	0.880
3.	Premier Padmini Car	$FC = 4.984 + \dfrac{31.907^{**}}{V} + 0.00035 V^{2**} + 0.00019\ RG^{**} + 0.09416\ RS^{**} - 0.0677\ FL^{**}$	0.969
4.	Tata Truck	$FC = 8.507 + \dfrac{390.46^{**}}{V} + 0.00207\ V^{2**} + 0.00012\ RG^{**} + 0.3328\ RS^{**} - 0.1777\ FL^{**} - 0.6240\ PW^{**}$	0.932

[**Note**–* Denotes significant at 5 percent level]
**Denotes significant at 1 percent level]

In the above equations, the symbols have the same meaning as for symbols in Table 39.2. In addition:

RS = Rise in m/km FL = Fall in m/km

A comparison of the fuel consumption values predicted from Eqns. in Table 39.3 with actual consumption on *UCS* vehicles was made. It was observed that the actual fuel consumption was related to the experimental values as below :

Jeeps : Actual value = 1.53 × Experimental value

Cars : Actual value = 1.64 × Experimental value

Buses : Actual value = 1.28 × Experimental value

Trucks : Actual value = 1.13 × Experimental value.

The higher fuel consumption in actual practice is due to the frequent deceleration and acceleration, stop-and-go motion, travel at speeds other than steady state, idle consumption at forced stops etc., in short the driver behaviour. However, the results demonstrate that a good deal of savings in fuel consumption can be brought about by road improvements and bettering the traffic flow conditions.

39.7. Spare Parts Consumption

The analysis of data from operators has yielded the equations given in Table 39.4.

Table 39.4. Equations for Spare Parts Consumption

S. No.	*Vehicle*	*Eqn.*	R^2
1.	Car	$\text{Log}_e\, SP = \text{Intercept} + 0.004117\, \text{Log}_e\, LK + 0.02812\, RF + 0.00005210\, RG - 0.7717\, W$ The value of intercept is in the range of – 0.1893 to + 0.8042 for the various operators.	0.530
2.	Jeeps	$\text{Log}_e\, SP = 1.069 + 0.5365\, \text{Log}_e\, LK^{**} + 0.000005271\, RG$	0.724
3.	Bus	$\text{Log}_e\, SP = 0.3096\, GA - 0.2297\, GH^{**} + 0.1287\, PA - 0.1885\, PH^{**} + 0.2781\, \text{Log}_e\, LK^{**} + 0.009078\, RF^{**} + 0.00007567\, RG^{**} + \dfrac{1.353}{W} + 0.2509\, (OHF)^{**} + 0.1104\, VCR$	0.467
4.	Truck	$\text{Log}_e\, SB = 5.656\, (PA) + 2.695\, (PH)^{*} + 10.520\, (GH)^{*} + 27.716\, (GA)^{**} + 0.01843\, LK^{**} - 1.293\, PW^{**} + 0.01235\, RF + 0.01270\, \dfrac{RG^{**}}{W}$	

[**Note** : *Denotes significant at 5 percent level.
**Denotes significant at 1 percent level]

In the above equation,

SP = Spare parts cost in paise/km (1978)

LK = Age of vehicle in thousand km

RF = Rise and fall in m/km

RG = Roughness in mm/km

W = Pavement width in m

OHF = Overhaul factor (number of overhauls during 2 years.)

GA = Zero : 1 variable, $GA = 1$ if government undertaking with average intensity of labour for maintenance, otherwise $GA = 0$

GH = Zero : 1 variable, $GH = 1$ if government undertaking with high intensity of labour for maintenance, otherwise $GH = 0$

PA = Zero : 1 variable, $PA = 1$ if private operator with average intensity of labour for maintenance otherwise $PA = 0$

PH = Zero : 1 variable, $PH = 1$ if private operator with high intensity of labour for maintenance, otherwise $PH = 0$

VCR = Volume-Capacity Ratio

The equations indicate that the spare parts consumption generally :

(*i*) increases as roughness increases.

(*ii*) increases as rise and fall or curvature increases.

(*iii*) decreases as pavement width increases.

(*iv*) increases with the age of the vehicle.

(*v*) increases with the overhauls.

(*vi*) increases as the congestion mounts up.

(*vii*) is higher for government sector than for private sector.

(*viii*) is lower where the operator has a good maintenance policy and adequate maintenance facilities.

The equations underscore the point that road improvements can bring about a good reduction in wear and tear of spare parts.

39.8. Maintenance and Repairs Labour Cost

The best equations relate the maintenance and repairs labour cost to the Spare Parts Cost, which can be obtained from known roadway and operating characteristics. The equations are given in Table 39.5.

Table 39.5. Maintenance and Repairs Labour Cost Equations

S. No.	*Category of vehicle*	*Equation*	R^2
1.	Car	$LC = 0.5498\ SP$	0.647
2.	Jeep	$LC = 0.3677\ SP$	0.774
3.	Bus	$LC = 0.4027\ SP$	0.513
4.	Truck	$LC = 0.3692\ SP$	0.837

The above equations have good R^2 values and can be reliably used. Since the labour costs are directly related to the Spare Parts costs, it is evident that just as in the case of Spare Parts costs road improvements can bring about a good saving in maintenance labour costs.

39.9. Tyre Life

The equations for tyre life are given in Table 39.6.

Table 39.6. Tyre Life Equations

S. No.	Vehicle	Eqn.	R^2
1.	Cars	$TL = 54440 - 2.418\ CV - 0.5155\ RG^* + 345.6\ W$	0.566
2.	Jeep	$TL = 45050 + 51.52\ LK - 3.953\ RG^{**}$	0.609
3.	Bus	$TL = 35030 - 980.8\ \frac{GVW}{N} - 536.8\ RF^{**} - 1.260\ RG^{**} + 2023\ W^{**} - 2963\ VCR^*$	0.509
4.	Truck	$TL = 25320\ (G) + 28757\ (P) - 2055\ \frac{GVW}{N} - 367.8\ RF^{**} - 1.016\ RG^* + 3740\ W - 6964\ VCR^{**}$	0.632

In the above equations :

TL = Tyre life of equivalent new tyre in km

CV = Curvature in deg/km

RF = Rise and fall in m/km

RG = Roughness in mm/km

W = Width of pavement in m

LK = Age of vehicle in thousand km

GVW = Gross vehicle weight in tonnes

N= Number of tyres in a vehicle

VCR = Volume/capacity ratio

G = Zero : 1 variable, $G = 1$ if government undertaking, otherwise $G = 0$

P = Zero : 1 variable, $P = 1$ if private undertaking, otherwise, $P = 0$

The equations, which have satisfactory R^2 values, indicate that tyre life generally :

(i) increases as road width increases.

(ii) decreases as roughness increases.

(iii) decreases as rise and fall increases.

(iv) decreases as the load per tyre increases.

(v) decreases as congestion builds up on the road.

(vi) is higher for private operators than for government operators.

The relationships indicate how by improving the roads economy in tyre consumption can be achieved.

39.10. Lubricants

The consumption of lubricants (engine oil, other oils and grease) is a small portion of the vehicle operating cost and is not easily amenable to accurate analysis in terms road and vehicle characteristics. All the same, some interesting results for buses and trucks presented in Tables 39.7, 39.8 and 39.9 demonstrate how the road and vehicle characteristics affect these items of cost.

Table 39.7. Equations for Consumption of Engine Oil

S.No.	*Vehicle*	*Eqn.*	R^2
1.	Bus	*EOL* = Intercept + 0.001815 *LK*** + 0.01135 *RF* – 0.02596 *W* + 0.0002155 *RF*** + 0.1634 (*OHF*)* (Intercept has values ranging from 1.21 to 6.69 for various operators)	0.483
2.	Truck	*EOL* = Intercept + 0.001479 *LK*** – 0.04183 *RF* + 0.00001849 *RG* + 0.2535 *W* 0.03765 *GVW* – 0.1986 (*OHF*) + 0.7422 *VCR* (Intercept has values ranging from 0.20 to 10.88 for various operators)	0.784
		[**Note** : *denotes significant at 5 per cent level **denotes significant at 1 per cent level] EOL = Engine Oil Consumption in litres per 1000 km	

Table 39.9. Equations for Consumption of Other Oils

S. No.	*Vehicle*	*Eqn.*	R^2
1.	Bus	*OL* = Intercept + 0.004066 *LK*** + 0.002889 *RF*** + 0.0008217 *RG*** (Value of intercept lies in the range of – 3.00 to + 5.50 for various operators)	0.400
2.	Truck	*OL* = Intercept + 0.004227 *LK** + 0.000224 *RG* – 0.2197 *W* + 0.2651 *GVW*** (Value of intercept lies in the range of – 3.37 to 23.69 for various operators)	0.761
		[**Note** : *Indicates significant at 5 per cent level **indicates significant at 1 per cent level] OL = Other O is consumption in litres per 1000 km	

Table 39.9. Equation of Consumption of Grease

S. No.	*Vehcle*	*Eqn.*	R^2
1.	Bus	*G* = Intercept + 0.001481 *LK*** + 0.03376 *RF*** – 0.3634 *W*** (Value of intercept lies in the range of 0.10 to 7.00 for various operators)	0.489
2.	Truck	*G* = Intercept + 0.001282 *LK* + 0.001928 *RF* + 0.0001030 *RG* + 0.2159 *GVW*** (Value of intercept lies in the range of – 4.024 to 23.39 for various operators)	0.824
		[**Note** : *indicates significant at 5 per cent level **indicates significant at 1 per cent level], G = Grease consumption in kg per 10,000 km	

39.11. Utilisation and Fixed Costs

As roads are improved, the vehicles can travel at increased speeds. Higher utilisation per day and per year result. Thus, the annual depreciation cost can be spread over a larger kilometerage. Similarly, the fixed costs and crew costs which are time dependent, can be spread over a larger kilometerage.

The analysis of utilisation was accomplished in two stages :

(*i*) Establishing an equation relating utilisation per day and the road characteristics.

(*ii*) Using the predicted speeds from the above equation and other additional variables, building up the utilisation equations.

The equations for vehicle speeds are given in Table 39.10 :

Table 39.10. Vehicle Speed Equations (Speed in km/per)

S.No.	*Vehicle*	*Eqn.*	R^2.
1.	Car	$V = 55.88 - 0.4969\ RF^{**} - 0.0008342\ \dfrac{RG}{W}$	0.709
2.	Jeep	$V = 59.61 - 0.2923\ RF^{*} - 0.1310\ \dfrac{RG}{W}$	0.828
3.	Bus	$V = 28.90 - 0.3102\ RF^{**} - 0.005224\ \ RF^{**} + 2.458\ W^{*}$	0.577
4.	Truck	$V = 44.78 - 0.3856\ RF^{**} - 0.00287\ \dfrac{RG}{W}$	0.482

[**Note :** * denotes significant at 5 per cent level
** denotes significant at 1 per cent level.

The various symbols have been explained earlier.

The equations prove that vehicle speeds are closely related to geometry variables, pavement width and roughness. Speeds decrease as roughness increases and increase as pavement width increases.

After obtaining the above speed equations in the first stage of regression, the second stage regression was carried out using the predicted speeds from these equations and some additional factors, in order to arrive at utilisation rates.

Table 39.11. Equations for Utilisation of Vehicles

S.No.	*Vehicle*	*Eqn.*	R^2
1.	Car	$U = 186.34\ LU + 14.94\ MU + 6.4859\ V$	0.627
2.	Bus	$U = 30.35\ MU + 206.75\ HU - 113.75\ LU + 8.011\ V$	0.558
3.	Truck	$U = 68.12\ MU + 221.35\ HU - 85.11\ LU + 5.1637\ V$	0.492

In the above equations,

U = Utilisation in km/day

V = Speed in km/hr

LK = Vehicle age in thousand km

GVW = Gross vehicle weight in tonnes

MU = Zero : 1 variable, MU = 1 if medium utilisation, *i.e.* 4 – 12 hrs/day, otherwise MU = 0

LU = Zero : 1 variable, LU = 1 if low utilisation, *i.e.* less than 4 hrs/day, otherwise LU = 0.

HU = Zero : 1 variable HU = 1 if high utilisation, *i.e.*, greater than 12 hrs/day, otherwise HU = 0.

OHF = Number of overhauls done in 2 years period.

The above equations indicate clearly how the utilisation of vehicles increases as speed increases and speeds are influenced largely by road geometry, pavement width and roughness.

Table 39.12. Percentage Utilisation during Survey Period

Sector	*Truck*	*Buses*	*Car and Jeep*
Government	52.76	84.4	69.0
Private	78.80	86.0	86.2
Overall	62.36	85.6	75.7

Thus the utilisation in km per year can be obtained from the utilisation rates km/day × 365 × percentage utilisation rates. (Table 39.12).

The utilisation in kilometres per annum provides the basis for the calculation of depreciation cost allowable per km, knowing the yearly depreciation cost. Similarly, the fixed costs per km can be calculated knowing the yearly fixed costs and the yearly utilisation in km.

Standard Tables have been presented by the author elsewhere (Ref. 19, 20, 21) for the various VOC components.

The latest updation of Road User Costs in India (Ref. 22) have yielded the following single equations for VOC for different vehicle types (Table 39.13).

Table 39.13.
Single equation for VOC for different vehicle types (Ref. 22)

Vehicle Class	Carriageway width	Equation					R^2
1. New Brand Cars (*NBC*)	(*i*) 3.0-7.0 m	$\log_e$ VOC NBC_{IT} =	1.381 (45.35)	– 0.115*W (– 24.93)	+ 0.00008300*RG (54.45)	+ 0.01302*RF (33.53)	0.91
	(*ii*) 3.0-7.0 m	$\log_e$ VOC NBC_{ET} =	0.924 (28.13)	– 0.134*W (– 26.88)	+ 0.00009256*RG (56.31)	+ 0.01411*RF (33.70)	0.92
	(*iii*) Four lane Divided	$\log_e$ VOC NBC_{IT} =	0.93 (174.9)		+ 0.00004633*RG (98.96)	+ 0.00934*RF (78.24)	0.99
	(*iv*) Four lane Divided	$\log_e$ VOC NBC_{ET} =	0.336 (52.83)		+ 0.00005624*RG (100.57)	+ 0.01023*RF (71.78)	0.99
2. Old Brand Cars (*OBC*)	(*i*) 3.0-7.0 m	$\log_e$ VOC OBC_{IT} =	1.666 (33.45)	– 0.134*W (– 17.76)	+ 0.00008789*RG (33.31)	+ 0.01145*RF (18.03)	0.80
	(*ii*) 3.0-7.0 m	$\log_e$ VOC OBC_{ET} =	1.112 (22.32)	– 0.114*W (– 19.06)	+ 0.00009774*RG (39.18)	+ 0.01251*RF (19.69)	0.83
	(*iii*) Four lane Divided	$\log_e$ VOC OBC_{IT} =	1.19 (139.2)		+ 0.00004537*RG (60.30)	+ 0.00701*RF (36.54)	0.97
	(*iv*) Four lane Divided	$\log_e$ VOC OBC_{ET} =	0.562 (72.38)		+ 0.00005521*RG (80.81)	+ 0.00820*RF (47.08)	0.98
3. Buses	(*i*) 3.0-7.0 m	$\log_e$ VOC BUS_{IT} =	2.135 (128.2)	– 0.070*W (– 27.67)	+ 0.00004553*RG (54.61)	+ .01208*RF (56.90)	0.94
	(*ii*) 3.0-7.0 m	$\log_e$ VOC BUS_{ET} =	1.705 (101.5)	– 0.073*W (28.68)	+ 0.00004718*RG (56.10)	+ 0.01225*RF (57.20)	0.94
	(*iii*) Four lane Divided	$\log_e$ VOC BUS_{IT} =	1.747 (332.9)		+ 0.00002931*RG (63.46)	+ 0.00949*RF (80.64)	0.99
	(*iv*) Four lane Divided	$\log_e$ VOC BUS_{ET} =	1.286 (243.1)		+ 0.00003128*RG (67.21)	+ 0.00970*RF (81.75)	0.99
4. Light Commercial Vehicles (*LCV*)	(*i*) 3.0-7.0 m	$\log_e$ VOC LCV_{IT} =	2.143 (342.2)	– 0.039*W (– 41.42)	+ 0.00002668*RG (85.14)	+ 0.01323*RF (165.74)	0.99
	(*ii*) 3.0-7.0 m	$\log_e$ VOC LCV_{ET} =	1.849 (318.7)	– 0.041*W (– 47.01)	+ 0.00002822*RG (97.22)	+ 0.01212*RF (163.95)	0.99

	(*iii*) Four lane Divided	$\log_e$ VOC LCV_{IT} =	1.88 (704.3)		+ 0.00001953*RG (83.11)	+ 0.01247*RF (208.24)	0.99
	(*iv*) Four lane Divided	$\log_e$ VOC LCV_{ET} =	1.548 (535.1)		+ 0.00002217*RG (87.15)	+ 0.01168*RF (179.88)	0.99
5. Two Axle Heavy Commercial vehciles (*HCV*)	(*i*) 3.0-7.0 m	$\log_e$ VOC HCV_{IT} =	2.472 (237.1)	– 0.065*W (– 41.35)	+ 0.00004121*RG (78.99)	+ 0.00992*RF (74.64)	0.97
	(*ii*) 3.0-7.0 m	$\log_e$ VOC HCV_{ET} =	2.092 (243.7)	– 0.061*W (– 46.38)	+ 0.00004047*RG (94.20)	+ 0.00938*RF (85.75)	0.98
	(*iii*) Four lane Divided	$\log_e$ VOC HCV_{IT} =	2.034 (461.2)		+ 0.00003146*RG (81.07)	+ 0.00863*RF (87.20)	0.99
	(*iv*) Four lane Divided	$\log_e$ VOC HCV_{ET} =	1.676 (385.3)		+ 0.00003227*RG (84.29)	+ 0.00871*RF (89.25)	0.99
6. Multi-Axle Commercial Vehicles (*MAV*)	(*i*) 3.0-7.0 m	$\log_e$ VOC MAV_{IT} =	2.926 (439.0)	– 0.050*W (– 49.87)	+ 0.00002969*RG (89.00)	+ 0.01443*RF (169.8)	0.99
	(*ii*) 3.0-7.0 m	$\log_e$ VOC MAV_{ET} =	2.561 (384.5)	– 0.053*W (– 52.99)	+ 0.00003039*RG (91.17)	+ 0.01337*RF (157.53)	0.99
	(*iii*) Four lane Divided	$\log_e$ VOC MAV_{IT} =	2.577 (808.9)		+ 0.00002093*RG (74.67)	+ 0.01407*RF (196.91)	0.99
	(*iv*) Four lane Divided	$\log_e$ VOC MAV_{ET} =	2.184 (885.1)		+ 0.00002224*RG (102.43)	+ 0.01289*RF (232.79)	0.99
7. Two Wheelers	(*i*) 3.0-7.0 m	$\log_e$ VOC TW_{IT} =	0.452 (8.39)	– 0.130*W (– 15.85)	+ 0.000111*RG (41.23)	+ 0.01473*RF (21.48)	0.84
	(*ii*) 3.0-7.0 m	$\log_e$ VOC TW_{ET} =	– 0.253 (– 5.19)	– 0.118*W (– 15.90)	+ 0.0001282*RG (52.54)	+ 0.01312*RF (21.11)	0.88
	(*iii*) Four lane Divided	$\log_e$ VOC TW_{IT} =	– 0.038 (– 3.97)		+ 0.00008141*RG (96.88)	+ 0.00844*RF (39.41)	0.98
	(*iv*) Four lane Divided	$\log_e$ VOC TW_{ET} =	– 0.716 (– 47.7)		+ 0.0001031*RG (78.03)	+ 0.00742*RF (22.01)	0.98

Note : The values in brackets in each equation indicate the *t*-value of the coefficients/constants.

IT = Including Tax; ET = Excluding Tax; W = Width (m); RG = Roughness (mm/km); RF = Rise and fall (m/km)

I. MULTIPLE CHOICE QUESTIONS

1. Vehicle costs include
   (*a*) direct user expenses to own and use private vehicles
   (*b*) incremental costs for mobility substitutes such as tele work.
   (*c*) Both (*a*) and (*b*)
2. Vehicle costs can be measured
   (*a*) per vehicle-mile (*b*) passenger-mile
   (*c*) vehicle-year (*d*) household-year
   (*e*) all of these.
3. Fixed costs covers
   (*a*) vehicle purchase or lease (*b*) insurance
   (*c*) registration and vehicle taxes (*d*) all of these
4. Variable costs include:
   (*a*) maintenance and repair (*b*) fuel, fuel taxes and oil
   (*c*) paid parking and tolls (*d*) all of these
5. Assuming no returns outwards or carriage inwards, the cost of goods sold will be equal to:
   (*a*) opening stock plus purchases plus closing stock
   (*b*) sales less gross profit
   (*c*) purchases plus closing stock less opening stock
   (*d*) closing stock less purchases plus opening stock

Answers

1. (*c*) **2.** (*e*) **3.** (*d*) **4.** (*d*) **5.** (*b*)

II. SOLVED QUESTIONS AND ANSWERS

Q. 1. How does vehicle operating cost vary?

Ans. Vehicle costs can vary depending on vehicle type, travel conditions, and market factors such as fuel taxes and vehicle and parking fees. Analyses vary in the scope of costs they consider. For example, when estimating automobile commuting costs, and therefore the savings from improvements to alternative modes and commute trip reduction programs, many calculators only consider fuel and parking costs. Vehicle operating costs per mile vary depending on driving patterns and conditions. Fuel consumption per vehicle-mile tends to increase at speeds over 55 mph, under 20 mph, and under stop-and-go driving conditions (Energy and Emission Reductions).

Q. 2. Where are vehicle costs are higher?

Ans. Vehicle operating costs are about 40% higher on urban arterials than highways, and costs increase proportional to travel time when congestion reduces traffic speed to 30 mph on a highway or 20 mph on an arterial. Roadway investment models provide detailed vehicle cost estimates (fuel, oil and tires, and sometimes maintenance) or various vehicle classes and road conditions.

Q. 3. What is the use of telework?

Ans. Telework often uses home office space, telecommunications equipment and services that can cost hundreds or thousands of dollars annually, but many households make such investments for personal use, and these costs can be partly offset by savings in business office space. The incremental equipment cost of telework is therefore likely to average $500-1,500 annually for a typical part-time teleworker.

Q. 4. What are the factors affecting vehicle operating costs?

Ans. The following factors affect vehicle costs (Booz Allen & Hamilton, 1999; Litman 2009; Polzin, Chu and Raman, 2008):

Vehicle type. Ownership and operating costs vary by vehicle size, class, and other characteristics. Trucks typically have much higher vehicle costs than cars.

Vehicle speed. Vehicle speed is the dominant factor affecting vehicle operating costs. Typically operating costs decrease with increasing speed to a certain point, and then begin to increase with increasing speed.

Speed changes. Changes in speed (also known as speed cycles) increase vehicle operating costs. This added cost is higher when speed cycling occurs at higher speeds.

Gradient. Grades can be either positive (uphill) or negative (downhill). Positive grades are more demanding on vehicle engines and require greater fuel consumption. This leads to an increase in operating costs. Negative grades may reduce operating costs, but may also increase wear on brakes.

Curvature. A highway curve requires a greater output of energy from a vehicle to counter the centrifugal force. This, combined with additional wear on the vehicle's tires, leads to an increase in operating costs.

Road surface. The roughness of the road surface can affect vehicle operating costs by affecting rolling resistance. Rough surfaces can reduce speed, require greater fuel consumption, increase wear on tires, and increase maintenance costs.

Q. 5. What do you mean by vehicle operating costs? Explain.

Ans. *Vehicle operating costs* refer to costs that vary with vehicle usage, including fuel, tires, maintenance, repairs, and mileage-dependent depreciation costs (Booz Allen & Hamilton, 1999). Projects that alter vehicle miles travelled, traffic speed and delay, roadway surfaces, or roadway geometry may affect travellers' vehicle operating costs, which should be considered in a benefit-cost analysis. *Vehicle ownership costs* refers to fixed costs that are not directly affected by vehicle mileage, including time-dependent depreciation, insurance and registration fees, financing, and residential parking. Projects that change per capita vehicle ownership rates, such as significant changes in the quality of alternative modes and land use accessibility, may affect vehicle ownership costs, which should be considered in benefit-cost analysis.

- Estimate changes in total vehicle miles travelled along a corridor.
- Estimate changes in vehicle travel speeds and delay due to road and traffic conditions.
- Estimate fuel consumption rates, fuel prices, and non-fuel-related operating costs.
- Calculate total changes in vehicle operating costs.
- For improvements to ride quality, such as pothole repairs and curve or grade reductions, estimate effects on vehicle wear.
- Estimate changes in per capita vehicle ownership in an area.
- Estimate average vehicle ownership costs.
- Calculate total changes in vehicle ownership costs.

III. REVIEW QUESTIONS

1. What are the components of VOC?
2. Explain the factors affecting VOC.

REFERENCES

1. Moyer R.A., Motor vehicle operating costs and related characteristics on mutreated gravel and portland cement concrete road surfaces, Proceedings of the Highway Research Board, Washington D.C. 1939.
2. Moyer R.A. and R. Winfrey, Cost of Operating rural-mail-carrier motor vehicles on pavement, gravel and earth, Iowa State College Engineering Experiment Station Bulletin No. 143, 1939.
3. Winfrey, R., Motor vehicle running costs for highway economic studies 3131 North Piedmont Street, Arlington, 1963.
4. Winfrey, R., Economic analysis for highway, International Textbook Company, Scranton, 1969.
5. Claffey, R.J., Time and fuel consumption for highway user benefit studies, Highway Research Board Bulletin, No. 276, Washington D.C. 1960.
6. Claffey, R.J., Running cost of motor vehicles as affected by road design and traffic, National Cooperative Highway Research Program Report 111, Washington D.C., 1966.
7. American Association of State Highway Officials, User benefit analysis for highway improvements, Washington D.C., 1960.
8. Dawson, R.F.F., Vehicle Operating costs in 1970, TRRL Report LR 439, Transport and Road Research Laboratory, Crowthorne, 1972.
9. Bonney R.S.P., and N.F. Stevens, Vehicle Operating Costs of trucks and Buses, Road Research Laboratory, Technical Paper No. 76, Her Majesty's Stationery Office London, 1967.
10. Adler, H.A., Economic evaluation of transport projects, Transport Investment and Economic Development, The Bookings Institution, Washington D.C., 1965.
11. De Weille, J., Qualification of road user savings, International Bank for Reconstruction and Development, World Bank Staff Occasional Paper Two, The John Hopkins Press, Baltimore, 1966.
12. Hide H., S.W. Abayanayaka, I. Sayer and R.J. Wytt, The Kenya Road Transport Cost Study: Research on vehicle operating costs, TRRL Laboratory Report 672, Transport and Road Research Laboratory Crowthorne, Berkshire, (U.K.), 1975.
13. Hodges, J.W., J. Rolt and T.E. Jones, The Kenya Road Transport Cost Study : Research on road deterioration, TRRL Laboratory Report 673, Transport and Road Research Laboratory, Crowthorne, Berkshire (U.K.), 1975.
14. Robinson, R.H. Hide, J.W. Hodges, J. Rolt and S.C. Abaynayaka, A road transport investment model for developing countries, TRRL Laboratory Report 674, Transport and Road Researh Laboratory, Crowthorne, Berkshire (U.K.), 1975.
15. Empressa Brasileira de Planejamente de Transport (GEIPOT), Research on the inter-relationship between costs of highway construction, maintenance and utilisation, Report II—Mid-term Report, Brasillia, August, 1977.
16. Kadiyali, L.R., Road User Cost Study, Objectives and methodology, Indian Highways, New Delhi, 1979.
17. Swaminathan C.G. and L.K. Kadiyali, Road User Cost Study—Vol. 44 — 1, Journal of the Indian Roads Congress, New Delhi, 1983.
18. Road User cost Study, Final Report, Central Road Research Institute, New Delhi, 1982.
19. Kadiyali, L.R., Quantifications of road user benefits and their application to Highway Economic Analysis and Highway Planning in India, Ph.D Thesis submitted to the Kakatiya University. Warangal, 1985.
20. Manual of Economic Analysis for Highway Projects, Indian Roads Congress, New Delhi, 1995.
21. Study for updating Road user Cost Data, Prepared for the Min. of Surface Transport, L.R. Kadiyali and Associates, New Delhi, 1992.
22. Reddy, T.S., et al, Updation of Road user Cost and Revised Software for Evaluation of Highway Projects, Paper No. 493, Vol. 64-2, Journal of the Indian Roads Congress, New Delhi, 2003.

40. Value of Travel Time Savings

40.1. Introduction

In the transportation sector, be it highways, railways, airways, inland waterways, ports or pipelines, considerable outlay will have to be made in India in the coming decades. The overall scarcity of resources, combined with the presence of competing demands from alternative schemes, makes it incumbent on the planner to carry out accurate economic analysis of various alternatives in order to enable the correct investment decision to be taken. In the economic analysis of transportation schemes, the emphasis in the initial stages was on the user costs, such as the vehicle operating costs. However, there are other costs borne by the transportation users which are reasonably important. One of them is the value of the travel time savings. Since travel time is quite precious, there is agreement amongst all analysts that it is highly relevant in an economic analysis of transportation schemes.

Many transportation schemes result in time saving. For example, the replacement of a road-rail level-crossing with an overbridge will result in free movement of the road traffic without getting hindered when the level-crossing gate is closed. This results in saving in time to the highway user. The construction of a flyover with grade separation in place of an existing signalised inter-section is another example where considerable reduction in travel time is possible to the highway user. When the road passing through a congested town or village reduces severely the speed of traffic and is the cause of serious accidents, a byepass is usually proposed to secure unhindered flow of through traffic. Such by-passes result in great travel time savings. Widening of narrow single-lanes roads to two lanes, improvement in geometrics, widening narrow cross drainage structures replacement of submersible causeways by high level bridges are other typical examples of schemes bringing about travel time savings.

While there can be no doubt about the importance of travel time savings, what is perhaps debatable is whether the time saved has any economic value immediately on realisation, and, if it has, how the time savings can be actually quantified into monetary terms. These are the issues on which considerable research is going on in every country.

40.2. Classes of Transport Users Enjoying Travel Time Savings

40.2.1. Travel time savings are enjoyed by different classes of transport users. They can be broadly grouped under the following heads :

1. Passengers.
2. Transport operators and owners
3. Freight consigners.

40.2.2. Passengers

The most conspicuous class of transport users who enjoy benefits due to reduction in travel time are the passengers using different transport modes such as cars, taxis, scooters, cycles, buses etc. For example, any highway improvement scheme which speeds up traffic is likely to reduce the travel time of occupants of cars, buses, taxis and scooters, cyclists and pedestrians. The modal choice of passengers is largely influenced by the monetary value they attach to travel time savings, and is governed by factors such as cost of travel by different modes, time of travel by different modes, and their income level. The monetary valuation of passengers' travel time savings should necessarily take note of these important factors.

40.2.3. Travel Time Savings Enjoyed by Transport Operators

Transport operators and owners (bus operators, truck operators, taxi operators) are direct beneficiaries of travel time savings with improved highway facilities. Bus and truck operators can run their vehicles at higher speeds and this will result in a greater utilisation of vehicles. It is generally observed that the daily utilisation of vehicles is directly related to the average speed of the vehicles in addition to other factors. A higher utilisation of vehicles will signify that fixed cost borne by the operator will be spread over a longer kilometerage. In fact, the overall cost of operation will come down when vehicles operate on improved facilities. The fixed costs also include the wages of the crew (drivers, conductors and cleaners). When transport is speeded up, the transport operator will be able to run the vehicle for a longer distance using the same crew and bearing the same expenses for the wages of the crew.

40.2.4. Time Savings Enjoyed by Freight Consigners

In the movement of freight, travel time is of considerable importance. In the case of perishable commodities, such as milk, fruits and vegetables, time is the essence of commerce. For freight consigners, speedier transportation means lesser inventory costs.

40.3. Economic Concept of Evaluation of Travel Time Savings

40.3.1. Willingness to Pay

There are two distinct concepts in the economic evaluation of travel time savings (Ref. 2). The basis of the first approach is "willingness to pay" as a value of time. The assumption is that time spent in travelling has an opportunity cost, which is a well known concept in economics. Willingness to pay for travel depends upon the individual and the situation. For example, an individual with a high income may be willing to pay for travel by air between two places whereas a person with a lower income may be content to travel the same distance by train. Thus, the air traveller attaches greater importance to savings in travel time than the rail traveller. An individual who attaches more importance to travel with an assured seat and in comfort would prefer to travel in a subscription bus (chartered bus) than in public transport bus. The particular situation facing the traveller will influence his willingness to pay to a consid-erable extent. For example, journeys made by an individual during his working hours usually encroach upon the total time in which he is productive and will thus reduce his output. On the other hand, the journeys made during leisure hours do not reduce his output but will restrict his pursuit of pleasures that are more enjoyable than travel.

40.3.2. Time as a Commodity

The second concept of value of time is that time has utility just as any other commodity and hence a cost. This concept lends itself admirably well for evaluation of time savings enjoyed by transport operators and freight consigners. The time saved in journeys can be put to fruitful use by a greater turn-round of the vehicle utilisation.

While time is undoubtedly an 'economic commodity' because people certainly place some cash value on their own time, there is no denying the fact that there is no market for time as such. An individual's demand for his own time is a drived demand in the sense that he wants time as an element of some other goods such as recreation, eating, sleeping and so on. The same applies to time which is traded in the labour market. An employer offers a wage in terms of rupees per hour, but in actual fact, he is buying the use of an employee's skill for a period of an hour rather than buying his time as such. Thus time and services are complementary economic commodities because they cannot be bought or sold separately. As such, their respective prices cannot be separately evaluated. This should in no way detract from the fact that, except for the marginal hour of the marginal worker, the wage rate paid in the market will be greater than the value which the individuals put on their own time (Ref. 3).

40.4. Certain Issues Connected with Evaluation of Time Savings Enjoyed by Passengers

40.4.1. Minor Time Savings

Perhaps the strongest criticism of the entire concept of the passenger time saving evaluation is the problem arising over minor time savings. Whereas a saving of, say, one hour, has fairly obvious benefits to the passengers since much can be accomplished in the time fruitfully, the importance of minor time savings, say, one minute or five minutes or even ten minutes, is certainly a moot point. Time saved has economic value only if it is immediately put to use on realisation. Can it be assumed that minor time savings of the order of 5-10 minutes can be immediately put to fruitful use on realisation ? Still doubtful is the concept of additivity of minor time increments. For example, can minor savings of the order of five minutes by 12 passengers be considered additive and thus, equivalent to saving of one hour by one passenger ? Though mathematically both the savings are of the same order, is the economic quantification by addition of small increments justifiable ? No simple and definite answers are available to these questions. In recent times it has become the general practice to treat time savings as additive and, therefore, accept them as significant.

40.4.2. Working Time and Leisure Time

The second issue which frequently comes up is whether to treat working time journeys and leisure time journeys in the same manner or separately. If they are to be treated separately, how is the monetary evaluation of these two distinct categories to be done ? It is commonly recognised now that the monetary value of time savings during work trip is quite different from that during leisure trip.

Work trip may occur during normal working hours or the journey may commence earlier than and may conclude later than normal working hours. In the latter case, the work trip has involved, apart from travel during working hours which is the employer's time, time spent before and after the working hours, which is the time of the individual, though ostensibly employed on work trip. Work trip also involves walking time, waiting time and what is important to transport economists, the in-vehicle time.

Research carried out abroad specifically to evaluate time has yielded consistantly higher values for time saved during walking and waiting than in-vehicle travel time. If the in-vehicle travel time is taken as 1, the walking time is generally evaluated around 2 and the waiting time around 3 (Ref. 4).

While a detailed consideration of travel time savings should necessarily make a distinction between the employer's time and other-than employer's time, one approach is to combine the two for routine analysis. This implies that the journey undertaken was performed for going to the office or business, and hence, even though the employer does not pay for the journey time, the person who undertakes the journey brackets it with his actual working time. There are others who hold that travel to or from work is a nonworking time journey. Time saving here affords more leisure to the traveller rather than more working time to the employer. If this argument is accepted, journeys made to or from work are treated as leisure journeys.

40.4.3. Time Saving in Developing Countries

Another common issue that crops up during economic analysis involving time savings is whether the concept of time saving is at all relevant in the developing countries (Ref. 5). It is often argued that there is considerable unemployment and under-employment in the developing countries and the wage levels themselves are rather low (Ref. 6). Under such circumstances, is it at all necessary to place a monetary value to travel time savings of passengers ? The answer probably is that since time savings are a distinct category of benefits they cannot be ignored totally. What can perhaps be done is to evaluate the time saving rate very carefully considering the composition of the persons travelling, including the unemployed children and housewives. Once the actual value of time is determined fairly accurately, the question of not accounting for time savings in economic analysis should not arise at all.

40.5. Methodology for Monetary Evaluation of Passengers' Travel Time

40.5.1. Possible Approaches

In literature one comes across a number of different approaches for determining the value of time of passengers. Three of the important methodologies are :

(*i*) determining the hourly income of passengers using different modes of travel and distinguishing between 'working time' and 'leisure time'.

(*ii*) An opinion survey of the passengers to elicit information as to how much extra money they are prepared to pay for reduction in travel time. This is known as the questionnaire approach, or stated preference approach.

(*iii*) determination of the time value by a study of the 'revealed preference' of passengers undertaking journeys by different modes involving different travel times and travel costs.

40.5.2. Wage Rate Approach

40.5.2.1. The most common, and at the same time elementary and simplistic, approach in the monetary evaluation of passengers' travel time is to determine the average wage rate of the traveller and treat it as the value of his time. To refine the method, distinction is made between the work time journey and the leisure time journey. Sometimes, the employer's overheads are added to the wage rate.

40.5.2.2. Since the wage rates of different categories of passengers will be varying, it is necessary to categorise the passengers into distinct groups and determine the average wage rate for each of them. The following categories seem to be distinct :

Car passengers/taxi passengers
Bus passengers
Bus drivers
Bus conductors
Truck drivers
Truck cleaner/helpers

Where cycle traffic and motorised two wheeler traffic is considerable, the average wages of these categories should also be determined. The wage rate of the pedestrian can similarly be determined if need be.

The data is generally collected by interviewing a sample passengers of the above categories and finding out their average monthly income. The sample interviewed should include non-wage earners such as the unemployed children and house-wives. Suitable account of these should be made in arriving at the average wage rate for these categories.

40.5.2.3. In applying the wage rate as a measure of value of time savings the overheads borne by the employer should be added to wages which the wage earner receives. The exact percentage of overheads that should be added to the wage rate is a matter of detailed investigation. The overheads should include the employer's contributions towards pension, insurance, and other fringe benefits. The department of Enviornment, U.K. use a figure of 18 percent as the employer's overheads (Ref. 7). Possibly, a round figure of 20 per cent may be used for Indian conditions.

40.5.2.4. The valuation of time saved for leisure purposes involves a rather empirical approach since there can be no recognised market prices for leisure. Leisure journey time consists of two distinct types : in-vehicle time when the actual journey is performed for pleasure, recreation, shopping etc. (other than for going to or coming back from work), and in-vehicle journey time when the actual journey takes place for going to or coming back from work but outside the office hours. The two can generally be clubbed together. Based on empirical studies abroad an average value of 25 percent of the hourly wage rate can be assumed for leisure time. The Department of Enviornment, U.K. assume leisure time rate is 25 percent of the wage rate (Ref. 7).

40.5.2.5. The time involved in walking and waiting is valued at a different rate than the in-vehicle time. If the in-vehicle time is taken as one, the working time is generally is taken as 2 and the waiting time is taken as 3 (Ref. 4). These values are based on European work.

40.5.3. Opinion Survey Approach

40.5.3.1. In this method, also known as the questionnaire approach, the travellers are asked to evaluate themselves the price they would pay for reduced travel of different duration. The actual rating of the travellers would depend naturally upon their wage rate. Classifying the travellers into income groups, the average value of time for each group as judged by the travellers can be worked out from the opinion elicited form a large number of travellers. The study has the obvious disadvantage that very often subjective assessment of the individuals will go into the data set and the results will be widely varying.

40.5.4. Revealed Preference Approach

40.5.4.1. Perhaps the most scientific approach in determining the value of time savings is by studying the choice people actually make when forced with a number of alternatives involving different time and cost for the journeys. The underlying principle behind the methodology is that the results will be reflecting the revealed behaviour of preference of the people and will, therefore, be very near to the reality. Choices of the following types are possible in which people trade time for money :

(*i*) Choice of mode of travel
(*ii*) Choice of route within a mode
(*iii*) Choice of destination of trip or trip frequency
(*iv*) Choice of location of home.

40.5.4.2. Choice of Mode of Travel

Several interesting studies have been carried out wherein the choice of the mode by the travellers is used as the basis for determining the time value. This is illustrated by a simple level of choice of two groups of travellers (Ref. 7, 8). Let T represent time, and C the cost of travel. The subscript A represents the preferred mode and the subscript B the rejected mode. Consider a case where the traveller had the choice of 2½ hour rail journey costing ₹ 60 or a 4 hour bus journey costing ₹ 30. Then the minimum valuation of rail traveller will be,

$$V_{R(min)} = \frac{C_A - C_B}{T_A \quad T_B} = \frac{60 - 30}{4 - 2.5} = \frac{30}{1.5} = 20 \quad \text{or} \quad ₹\ 20 \text{ per hour.}$$

For the bus traveller, the maximum valuation of time will be,

$$V_{B(max)} = \frac{C_B - C_A}{T_A \quad T_B} = \frac{60 - 30}{4 - 2.5} = \frac{30}{1.5} = 20 \quad \text{or} \quad ₹\ 20 \text{ per hour.}$$

The rail traveller's valuation must be minimum of ₹ 20, otherwise he would travel by bus and he is essentially a time saver. The bus traveller's valuation must be a maximum of ₹ 20, otherwise he would travel by rail and he is essentially by a money saver.

Using the above simple approach, Beesley (Ref. 9) sought to derive the value of time that could explain the largest number of modal choices, stratifying his sample into three different income groups. He obtained results which give the leisure time valuation to about half wage rate.

Lee and Dalvi have used a variation of the above simple concept (Ref. 8). They incorporated a national diversion charge in assessing the value of time. The diversion charge represents the change in time or price of the preferred mode that would be just enough to induce the traveller to change to the alternative mode. Using the equations given earlier and incorporating the diversion charge C_D, the travel time value for the travellers choosing the faster mode can be written as,

$$\frac{C_A + C_D - C_B}{T_B - T_A}$$

The maximum value for those preferring the slower mode can be similarly written as,

$$\frac{C_B + C_A + C_D}{T_A - T_B}$$

If suitable empirical data are available, the above expressions should make it possible to answer the following questions :

What reduction in the price of the rejected mode would induce a traveller to shift to it, what increase in journey would induce the traveller to switch away from it and what reduction in journey time would be just enough to persuade the traveller to change mode ?

Collecting data from the users of buses, trains and private cars in Manchester, Lee and Dalvi employed the values obtained by the above expressions and ran a series of regression equations to find out how various indices of quality employed as independent variables influenced the valuation of travel time. Some of their results are interesting. They found that people who preferred the train to the bus had a higher time value than those preferring the bus to the train. The latter set had a higher time value than those who travelled by bus rather than by a combination of bus and train. The in-vehicle travel time and walking and waiting times were not found to be separately important. This implies that the travellers did not attach more disutility to walking and waiting time than to the time spent travelling in vehicles. This result conflicts with the findings of other studies, where a higher value of walking and waiting time is adopted than for in-vehicle time. The average value of time savings was about 30 per cent of the gross hourly wage rate.

Since the data for the analysis carried out by Lee and Dalvi was obtained by the questionnaire approach, the analysis has its own limitations. Beesley (Ref. 9) and Quarmby (Ref. 10) use 'revealed preferences' in discriminant analysis. Quarmby developed a model incorporating seven explanatory variables to account for modal choice (Ref. 10) :

Relative overall travel time, relative excess travel time, relative cost, relative income, car demand ratio (*i.e.*, ratio of driving licences to cars in the household), the use of car for work and the ownership of cars by firms.

He found that of the various explanatory variables, travel time differences, excess travel time difference, cost difference and possi-bility of using the car for work emerged as important factors influenc-ing modal choice. He concluded that walking and waiting time could be valued between 2 and 3 times the in vehicle time and the average value of time could be put between 21 and 25 percent of the wage rate for both car and bus users.

40.5.4.3. Choice of Route

Another approach which may reflect trade-off between time and cost is where there is an ordinary road and a toll road, and latter involving some extra pavement. The toll road may facilitate faster movement and the benefit may be reduction in travel time for the user, be they passengers or vehicles. But the method can be expected to yield a reliable value of time only if the savings in time is the only or the principal benefit. But more often, it is possible that a toll road confers other benefits like reduced operating expenses, less accidents, more comfort etc. The influence of these factors will have to be eliminated if the true value of time has to be derived in this case. However, by selecting the sample appropriately, the effective price of saving time could be varied and consequently a time cost trade-off analysis could be carried out.

40.5.4.4. Choice of Destination of Trip

The assumption in this method is that trips to a given location, other things being equal, fall off with distance. Thus, the catchment area of a leisure attraction may be nationwide but the number of trips per head from a nearby location will normally exceed those from more distant ones. Trips to a given location, other things bring equal, fall off with distance. This method will allow a value of time to be obtained if the influence of generalised cost of travel on tripmaking can be successfully disaggregated into time, cost and any other relevant components. The first practical problem is to find situations where time and cost are not so highly correlated as to make their separate identification impossible. For most road trips, correlation is very high. Where leisure trips are concerned travelling may be enjoyed for its own sake and may not be traded against cost. Studies of this kind have not been very helpful for these reasons.

40.5.4.5. Housing Price Approach

Another approach to deduce value of travel time savings is to study the situation where people trade off between lower house costs or rents and lower travel costs. This is based on the premise that rents and property values in urban areas are influenced by transport or accessibility factors. Some may prefer residences in centralised area which may raise their housing costs but which may save time and transport costs. These may be savers of time and transport costs. Contrarily others may like to live in less congested surroundings and pay lower rents even at the cost of faster travel time and transport costs. This is the housing price approach. The results produced by such studies abroad are found to bear some close relationships to values derived from other approaches such as the willingness to pay approach earlier referred to.

40.6. Review of Work in India on Travel Time

40.6.1. Wage Rate Approach

There appear to have been no systematic research into the value of travel time in India till 1976. Some arbitrary figures are assumed by a number of research workers when dealing with specific economic appraisal problems. The values have been derived generally from the wage rate. The value of travel time assumed in the Indian Investigations is given below :

Table 40.1. Value of Travel Time Assumed in Economic Appraisal Investigations

Investigation	*Value assumed*	*Method of derivation*	*Reference*
Road-Rail Crossings in Delhi	Re. 0.456/hour for all categories	Monthly wage rate and 8 hour working day	Ref. 11
Road-Rail Crossings in Madras	Re. 0.5/hour for cyclists and pedestrians. Re. 0.4/hour for bus passengers.	Daily wage rate and 8 hour working day	Ref. 12
	Re. 18.75 per veh. per hour for cars. Re. 7.5 per veh. per hour for scooters. Re. 1.25 per veh. per hour for slow-traffic.	Operating cost of vehicle per hour with assumed average speed	

Level-crossing in Ludhiana	Re. 15/hour for car owners. Re. 3/hour for bus passengers. Re. 4/hour for scooterists Re. 0.75/hour for pedestrians.	Monthly income and 10 hour working day	Ref. 13

40.6.2. Revealed Preference Approach

Raghavachari and Khanna (Ref. 14) used the home interview data in Ahmedabad and developed disaggregated models incorporating mode choice and trip frequency choice. Calibrating the models, they have derived the following values of travel time in rupees per hour.

Table 40.2. Value of Travel Time in Ahmedabad City

Vehicle Owned	*Employed Value (₹/hour)*	*Self-employed Value (₹/hour)*
Car	2.71	5.90
Scooter	3.96	4.70
Bicycle	0.36	0.55
None	0.19	0.50

The values reported in the above table approximately lie between 10 per cent and 80 per cent of the average wage rate of the Indian population.

40.6.3. Results of Survey by Observing Passengers' Modal Choice and Income Under Road User Cost Study (Ref. 15)

This survey work was done on some selected routes in the country, as part of the Road User Cost Study. The data was collected according to a well structured proforma seeking information on the occupation, income, purpose of the journey (work/non-work), origin and destination of travel journey time etc.

Table 40.3 below gives the results of the analysis, including details of sample size, mean monthly income and hourly wage rate. The hourly wage rate was derived after assuming 2000 hours of working time per year. The period on which the surveys were held was 1983.

Table 40.3. Value of Time of Bus Passengers (1982)

(*Source* : Ref. 15)

S. No.	*Route*	*Sample size*	*Mean monthly income (₹)*	*Hourly wage rate (₹)*
1.	Delhi–Chandigarh (Ordinary Bus)	205 (96)	919.9	5.97
2.	Delhi–Hissar	117 (28)	717.7	4.97
3.	Delhi–Mussoorie	60 (12)	965.0	5.97
4.	Sonepat–Rohtak	48 (22)	570.6	3.42
5.	Delhi–Chandigrah	36 (21)	2142.9	12.86

Figures in brackets indicate the number of travellers on work trip.

From the results it is seen that the hourly income rate shows variations between different types of routes on which the people travel. It is seen that on the trunk routes, the average income is found to be higher whereas on the secondary or interior routes the average wage rate is lower. The Delhi–Chandigrah, Delhi–Dehradun–Mussorie and Delhi–Hissar roads are the trunk routes and the hourly wage rate of travellers on these routes is higher than on Sonepat–Rohtak routes which is a secondary and rural oriented route. It may, therefore, be concluded that the range of average hourly income of bus passengers is about ₹ 4.5–6 on trunk routes and about ₹ 3.5 on secondary routes. Adding about 20 percent to these values to account for the overheads associated with their employment a rate of ₹ 7 appears to be the average wage a bus passenger of trunk routes and ₹ 4.5 appears to be the value of bus passenger's time per hour on secondary routes. These values pertain to 1980. These values were updated in 1992 (Ref. 16). The values at 1990 prices are as under

Table 40.4. Value of Time of Passengers (1990)

S. No.	*Category*	*Value of work trip (₹/hr)*	*Value of non-work trip (₹/hr)*
1.	Bus passenger on trunk routes	27	3
2.	Bus passenger on secondary routes	10	2
3.	Car/Taxi passenger	30	4

From the data collected on these routes, it was observed that about 40 percent of the trips were work-oriented and 60 percent of the trips were non-work trips. It was also found that the 90 percent of the passenger's travel time was the employer's time for work trips.

No efforts have been made to evaluate the leisure time. For planning purposes, this value may be taken as 25 percent of the worktime. Thus, the leisure time value of bus passengers can be taken to be ₹ 1.75 per hour on trunk routes and ₹ 1.1 per hour on secondary routes.

The results of similar survey on car/taxi passengers are tabulated in Table 40.5. It will be seen that the hourly wage rate of car/taxi passengers varies from the ₹ 8 to ₹ 13.

Table 40.5. Value of Time of Car/Taxi Passengers (1982)

S. No.	*Route*	*Sample size*	*Mean monthly income (₹)*	*Hourly wage rate (₹)*
1.	Bombay–Pune (Taxi)	31 (24)	2045	12.27
2.	Dehradun–Mussoorie (Taxi)	96 (20)	1306	7.83
3.	Delhi–Karnal (Car)	17 (10)	2199	13.19

Figures in brackets indicate the number of travellers on work trip.

Adding 20 percent to the above values to account for the overheads, one can assume the hourly wage rate of car/taxi passengers to be in the range of ₹ 9–15 per hour. A round figure of ₹ 10 per hour can be assumed. The proportion or work trip to total on an overall basis for all the three routes comes to 37%. However, the ratios are different for different routes. On Bombay–Pune and Delhi–Karnal routes, where there is high business oriented traffic, work trip proportion is high. On the Dehradun–Mussoorie route, which is route of tourist importance, the proportion of leisure trip is high.

For leisure time of taxi/car passengers, value of ₹ 2.50 per hour, representing 25 percent of the value of work time, is recommended. The 1990 values are given in Table 40.4.

Based on the latest Indian studies and the observed proportion of 85% as work trips and 15% as social/leisure trips, (Ref. 17), the value of travel time of passengers at 2007 price levels has been determined (Ref. 18) and is tabulated in Table 40.6

Table 40.6. Value of Travel time of Passengers (2007)

S.No.	*Vehicle used by passenger*	*Value of time ₹/hr.*	
		Primary route	*Secondary route*
1.	Car	58.0	49.0
2.	Two wheeler	30.0	20.0
3.	(*a*) Ordinary bus	37.0	14.0
	(*b*) Deluxe bus	41.0	—

The average occupancy of vehicles may be taken as under (Ref. 18) :

Car : 4.8
Two wheeler : 1.5
Bus : 43

40.6.4. Results of Indian Study on Crew Costs

40.6.4.1. Crew Costs as Time Saving Element

One of the components of the vehicle operating costs is the cost of crew, composed of drivers, conductors and cleaners. A knowledge of crew costs is useful in two ways :

(*i*) To determine the total cost of operation of vehicles. This information is useful to the vehicle operator to fix passenger fare and freight charges.

(*ii*) To determine the value of time savings enjoyed by the vehicle operators.

The operator employs his crew for certain hours in a day and pays them wages. With highway improvement the vehicle speeds increase and hence for the same amount of wages paid to the crew, the operator can get a greater utilisation of his vehicle and a greater output from his vehicle. Thus, the crew costs per unit time represent the value of travel time saving per unit, time the operator can enjoy.

40.6.4.2. Data Collection

As part of the Road User Cost Study, data were collected from a number of vehicle operators in India on the following :

1. Number of vehicles held
2. Number of crew employed, broken up into drivers, conductors and cleaners.

3. Wages paid in one year for the different categories of crews, *viz.*, drivers, conductors and cleaners.

Table 40.7. Staff-vehicle Ratio

Crew category	*No. per bus*	*No. per truck*
Driver	2	1.5
Conductor	2	–
Cleaner	1	1

Table 40.8 summarises the wages of crew per hour, including all fringe benefits.

Table 40.8. Average Wage Rate in ₹ Per hour of Crew (1978)

Vehicle type	*Drivers (₹/hr)*	*Conductor/cleaner (₹/hr)*
Buses	4.00	3.00
Trucks	3.50	2.00

Using the crew rate from Table 40.5 and the wage rate from Table 40.7 the crew cost per day per vehicle has been constructed and presented in Table 40.9 below. The 1990 values (Ref. 16) are also indicated. Most recent values pertain to 2001 (Ref. 17).

Table 40.9. Crew Cost per Vehicle per day

Crew category	*Cost per bus per day (₹)*		*Cost per truck per day (₹)*		*Cost of crew per hour (2001) (₹)*	
	(1978)	*(1990)*	*(1978)*	*(1990)*	*Bus*	*Truck*
Driver	64	176	40	156	26	20
Conductor	48	160	–	–	21	–
Cleaner	24	40	16	40	–	10
Total	136	376	56	196		

40.6.5. Results of Indian Study on Fixed Costs of Buses and Trucks

40.6.5.1. When highway improvements are made, the vehicles can be operated at higher speed. The utilization of vehicles per day increases. As a result, the fixed charges that are incurred on the ownership of a vehicle can be stretched out over a large kilometres on improved roads than on low quality roads. A simple method of assessing the benefits enjoyed by vehicles due to time savings is to accurately find out the yearly fixed costs. The results of the research carried out on the fixed costs of vehicles as part of the Road User Cost Study in India are given here. (Table 40.9).

40.6.5.2. Constituents of the Fixed Costs

The fixed costs are generally composed of the following items :

1. Interest on capital
2. Insurance
3. Road tax
4. Other taxes
5. Registration fee
6. Garraging charges
7. Other fixed charges such as fines, tolls, octroi etc.
8. Permit charges
9. Loading and unloading charges in respect of truck only
10. Commission on booking (in respect of trucks only).

11. Overhead charges.

(*i*) Rent and water (*ii*) Clerical and managerial staff

(*iii*) Audit fees (*iv*) Advertisement

(*v*) Electricity (*vi*) Telephone and postage

(*vii*) Stationery and printing

(*viii*) Travelling expenses (*ix*) Freight and hire

(*x*) Legal fees (*xi*) Bank charges

(*xii*) Repairs and maintenance of buildings

(*xiii*) Depreciation of fixed assets

(*xiv*) Sundries

The fixed costs can be assumed to be spread over a period of 300 days in a year, which is generally the number of days operated on an average. Table 40.10 contains the summary of the fixed costs per vehicle per year and the fixed costs per day.

Table 40.10. Summary of fixed costs per vehicle (1978)

Type of Operator	*Vehicle Type*	*Cost per year per vehicle incl. taxes*	*Cost per year per vehicle excl. taxes*	*Cost per day per vehicle incl. taxes*	*Cost per day per vehicle excl. taxes*
Public Sector	Bus	78,142	47,065	260	157
Private Sector	Bus	35,098	13,545	117	45
Private Sector	Truck	68,041	59,057	248	217

40.6.6. Results of Indian Study of Value of Time Savings in Respect of Commodities Carried

As for passengers and vehicles, savings in the time have great significance in the movement of commodities. Commodities differ from one another in their characteristics. There are certain commodities like foodgrains which have vital social significance and enter into the life of everyone. Fruits, vegetables and milk are perishable to a large extent. There are items like coal and kerosene for meeting the essential fuel needs of the people and fuel and raw material needs of industries. In addition, there are various kinds of intermediate and manufactured goods. The commodities have to be moved from the production centres to points of consumption and to and from ports in the case of exports and imports. In India, road transport has a substantial share in the traffic in various commodities, particularly in the case of those commodities which have high value, high degree of perishability, and those which require door-to-door delivery. Generally, the traffic carried by road transport is found to consist of the following items :

Foodgrains, fruits and vegetables, milk, oilseeds, edible oils, mineral oils, coal, manufactured products like textiles, machinery, engineering goods, paints and varnishes, glass and glass products, pharmaceutical goods, furniture, fertilizers etc.

These are high value goods, require careful handling, safe and secure transit and are taken delivery of at the doorstep of the supplier or delivered at the doorstep of the purchaser.

To the extent the time taken for their movement can be reduced, it is possible to reduce the inventory or stock levels of these high value goods and thereby avoid locking up of capital. Road or highway improvements can bring about more certainty of supply and at less time. Theoretically it is possible to visualise the savings in the cost of holding inventories by working out the value of goods held and deriving the interest payment, thereon per hour. The hourly norm of savings can be applied to savings in transit time following highway improvements and may be reckoned as benefits.

Studies in India indicate a value of ₹ 6000 per tonne of commodity carried at 1978 prices (Ref. 15) and ₹ 20,000 per tonne at 1990 prices (Ref. 16).

The average of load of truck can be taken to be 11 tonnes. Taking the unladen weight as 5 tonnes, the average load carried comes to 6 tonnes.

With an interest rate of 15 percent per annum, and 2500 hours of operation per annum, the hourly interest charges amount to ₹ 2.16 or, rounding off to ₹ 2.00. This represents the time value of the commodities carried and can be used for rough calculations in economic analysis. At 1990 prices, it comes to ₹ 10.40.

The commonly costs at 2007 price levels based on the latest data are given in Table 40.11 (Ref. 18)

Table 40.11. Value of Commodity in transit

Type of Route	*Vehicle Type*	*Commodity holding cost ₹/day*
Primary	Light Commercial Vehicles	54.0
	Heavy Commercial Vehicles	164.0
	Multi-Axled Vehicles	308.0
Secondary	Light Commercial Vehicles	50.0
	Heavy Commercial Vehicles	137.0

I. MULTIPLE CHOICE QUESTIONS

1. Going by public transportation is worth its price compared to going by car.
   (*a*) strongly agree (*b*) agree
   (*c*) neutral (*d*) disagree
   (*e*) strongly disagree.
2. It is comfortable to go by car to work.
   (*a*) strongly agree (*b*) agree
   (*c*) neutral (*d*) disagree
   (*e*) strongly disagree.
3. It feels safe to go by car.
   (*a*) strongly agree (*b*) agree
   (*c*) neutral (*d*) disagree
   (*e*) strongly disagree.
4. It is very important that traffic speed limits are not violated.
   (*a*) strongly agree (*b*) agree
   (*c*) neutral (*d*) disagree
   (*e*) strongly disagree.
5. Measures to improve public transportation should be undertaken.
   (*a*) strongly agree (*b*) agree
   (*c*) neutral (*d*) disagree
   (*e*) strongly disagree.

Answers

1. (*b*) **2.** (*b*) **3.** (*b*) **4.** (*b*) **5.** (*a*)

II. SOLVED QUESTIONS AND ANSWERS

Q. 1. What is VTT and VTTS?

Ans. The *Value of Travel Time* (VTT) refers to the cost of time spent on transport. It includes costs to businesses of the time their employees and vehicles spend on travel, and costs to consumers of personal (unpaid) time spent on travel. The *Value of Travel Time Savings* (VTTS) refers to the benefits from reduced travel time costs.

Q. 2. Give examples of VTTS.

Ans.

- A new lane is added to a freeway, increasing traffic speeds and decreasing delays.
- A signal coordination strategy is implemented, allowing for faster travel along a particular corridor.
- An improvement to a transit system reduces waiting time or increases travel time reliability. It can also reduce travel time on a parallel highway.
- Freight and business travel time (called "paid," "on the clock," or "commercial" time) tends to have relatively high unit costs because it includes the value of driver wages and benefits, vehicles and cargo, and the need to meet delivery schedules.
- Personal travel time is usually estimated at 25% to 50% of prevailing wages, but varies by factors such as type of trip, traveller and conditions.

Q. 3. What is travel time unit cost?

Ans. Travel time unit costs tend to increase with income, and are lower for children and unemployed people (put differently, employed people are often willing to pay more for travel time savings).

- Travel time unit costs tend to increase with variability and arrival uncertainly, and are particularly high for unexpected delays during activities with strict schedules such as business travel and commuting.
- Some travel time has a low cost or positive value because people enjoy the experience, including recreational travel and errands that involve social activities.
- Under pleasant conditions walking, cycling and waiting can have low or positive value, but under unpleasant conditions (walking along a busy highway or waiting for a bus in an area that seems dirty and dangerous), costs are two or three times higher than in-vehicle time.
- Travel needs and preferences vary. For example, some people place a higher cost on time spent driving while others place a higher cost on transit travel.

III. REVIEW QUESTIONS

1. What is the concept of evaluation of travel?
2. Discuss the certain issues connected with evaluation of time savings enjoyed by passenger.
3. What is the methodology for monetary evaluation of passengers travel time.

REFERENCES

1. Draft Inception Report, Road User Cost Study in India, Central Road Research Institute, New Delhi, 1977.
2. Haney D.G., Use of two concepts of the value of time, Highway Research Road No. 12, Highway Research Board, Washington, 1963.
3. Road User Cost Manual, Special Report No. 9, Australian Road Research Board, Melbourne, 1973.
4. Harrison, A.J., The Economics of Transport Appraisal, Croom Helms Paperback.
5. Howe, J.D.G.F., Valuing time saving in developing countries, Journal of Transport Economics and Policy, Vol. X–2, 1976.
6. Winfrey R., Economic Analysis for Highways, International Text Book Company, Scranton, 1969.
7. Stubbs, P.C., W.J. Tyson and M.Q. Dalvi, Transport Economics, George Allen and Unwin, London 1980.
8. Lee N. and M.Q. Dalvi, Variations in the value of travel time, Manchester School of Economic and Social Studies, Vol. 37, 1968.
9. Beesley, M.E., The value of time spent in travelling-some new evidence, Economics, Vol. 32,1965.
10. Quarmby, D.A., Choice of travel mode for journey to work, Some findings, Journal of Transport Economics and Policy, 1967.
11. Srinivasan, N.S. and Goel, H.C., Cost Benefit analysis of road-rail grade crossing in Delhi, Journal of the Indian Roads Congress, Volume-30, 1968.
12. Muthukumaraswamy A. and Thillainayagam R., Road Rail Barrier, A traffic study of the railway lever crossings of Madras Beach-Tambaram section of Southern Railways, Journal of the Indian Roads Congress, Volume-30, New Delhi, 1969.
13. Sodhi, N.S., Level crossings in Ludhiana city—A source of traffic problem and its remedies, Indian Highways, 2(6), New Delhi, 1974.
14. Raghavachari R. and S.K. Khanna, Development of disaggregated behavioural models—A case study of an Indian city, proceedings of the Australian Road Research Board, Volume-8, Part-6, Melburne, 1976.
15. Road User Cost Study, Final Report, Vol. V, Central Road Research Institute, New Delhi, 1982.
16. Study for Updating Road User Cost Data, Prepared for Min. of Surface Transport, L.R. Kadiyali and Associates, New Delhi, 1992.
17. Reddy, T.S., et al, Updation of Road User Cost and Revised Software for Evaluation of Highway Projects, Paper No. 493, Vol. 64-2, Journal of the Indian Roads Congress, New Delhi, 2003.
18. Manual on Economic Evaluation of Highway Projects in India, Special Publication IRC : SP : 30, (Proposed Second Revision—Draft), Indian Roads Congress, New Delhi, 2008.

41. Accident Costs

41.1. Introduction

41.1.1. The steep growth of traffic on the Indian roads in recent years has caused a sharp increase in the incidence of traffic accidents. In year 2001 it has been observed that 80,000 persons were killed and 382,700 persons were injured as a result of 394,800 road accidents. (Ref. 2) Road accidents in India constitute nearly 13-14 per cent of the total deaths due to accidents of all types, and thus emerge as a major killer. (Ref. 3) An international comparison of the accident rate reveals that India has a very high fatality rate of 24 per 10,000 motor vehicles as against 3.8 in U.K., 3.2 in U.S.A. and 5.7 in Australia. (Ref. 2, 4). The situation is alarming and calls for drastic remedies from all fronts. (Ref. 5, 6)

41.1.2. Accidents cause considerable economic loss to the nation. Unfortunately, the monetary evaluation of the accident cost is rather a difficult and controversial subject. Granted so, it still becomes necessary to have an acceptable approach for the quantification of accident costs, if the enormity of the loss to the economy has to be brought home to administrators, planners, engineers and politicians. Only then can the seriousness of the calamity be grasped and measures taken up to mitigate the ghastly situation.

41.1.3. Accident costing is a complicated subject and is riddled with a number of controversies. The work done in India so far is of an extremely limited nature. The Road User Cost Study, a research project recently completed in India, provided the first opportunity for dealing with this controversial subject, through in a limited way. (Ref. 7, 8, 9, 10, 11). The results, nevertheless, should prove to be extremely valuable and handy to the transportation planner. Efforts to refine the findings should, doubtless, continue in the future.

41.2. Relevance of Accident Costing for a Developing Country

41.2.1. It has been suggested by some (Ref. 12) that for less developed countries the valuation of life is not necessary because the reduction in fatalities is likely to be of minor significance compared to the other benefits to be derived from transport investments. This argument probably is true when one deals with road investment such as new construction, realignment and resurfacing. Schemes specifically thought of for improving traffic safety will include prominantly the benefits of accidents that could be prevented. Schemes costing relatively a small amount such as simple traffic management schemes, the improvement of accident black spots, educational, enforcement and driver training programmes etc., will have to be assigned an important place in the overall traffic safety programme of developing countries. The cost-effectiveness of these schemes can be determined accurately if the accident costing can also be done correctly.

41.2.2. It has been observed that the developing countries have a very high accident rate when the number of fatalities per 10,000 vehicles is taken as the yard-stick for judgment. (Ref. 13). Unfortunately, this fact is not clearly recognised by the politicians and the administrators and there appears to be a general apathy towards the distressing situation. One of the reasons for the sorry state of affairs probably is that engineers and economists are not able to convince to politicians and the administrators in precise monetary terms the economic loss to the society caused by the accidents and how by adopting relatively simple and cheap solutions the matters can be improved to a great extent. It is for this purpose that accident costing assumes great relevance.

41.2.3. In India, for example, there has been no assured safety budget, even though considerable amount is spent on the construction and maintenance of the road. The traffic management and regulatory measures are taken up without any scientific study of the cost effectiveness. Thus, even if some budget is set apart, it cannot be assured that the most optimal solutions are always selected. Considering the acute scarcity of resources, the need for accident costing and introducing economic analysis techniques for selecting the best traffic safety measures, becomes self-evident. One of the reasons for the high rate of accidents in developing countries is lack of adequate driver training, and measures for safety education of children, pedestrians and drivers. It emphasis is given to the adoption of such simple safety measures, the total accident rate can be brought down to a significant extent at a cheap cost.

41.3. Review of Alternative Methodologies for Accident Costing

41.3.1. A number of alternative methodologies are available for accident cost. Hills and Jones-Lee have summarised the different approaches and have listed the following six (Ref. 14) :

1. The 'gross output or human capital' approach.
2. The 'net output' approach.
3. The 'life insurance' approach.
4. The 'court award' approach.
5. The 'implicit public sector valuation' approach.
6. 'Value of risk-change' approach.

41.3.2. The "gross output" approach, also known as the "human capital" approach quantifies the cost of a traffic accident as the sum of real resources cost such as vehicle damage, medical expenditure and police costs and the discounted present value of the victim's future output. Thus, the value of the prevention of an accident is defined as the avoided cost. Sometimes, a sum is added to the above cost to reflect the pain, grief and suffering of the accident victim and those who care for him or her.

41.3.3. The "net output" approach subtracts from the gross output value the amount of the victim's future consumption.

41.3.4. The "life insurance" approach treats the cost of an accident or the value of accident prevention as the sum of real resources cost and the amount for which typical individuals are willing to insure their own lives or limbs.

41.3.5. The "court award" approach considers the amounts awarded by the courts to the surviving dependents of those killed as indicative of the cost that society associates with the fatality or the value it would have placed on its prevention. To the amount of court award are added the resources cost such as vehicle damage, medical treatment and police cost, to obtain the total cost of an accident.

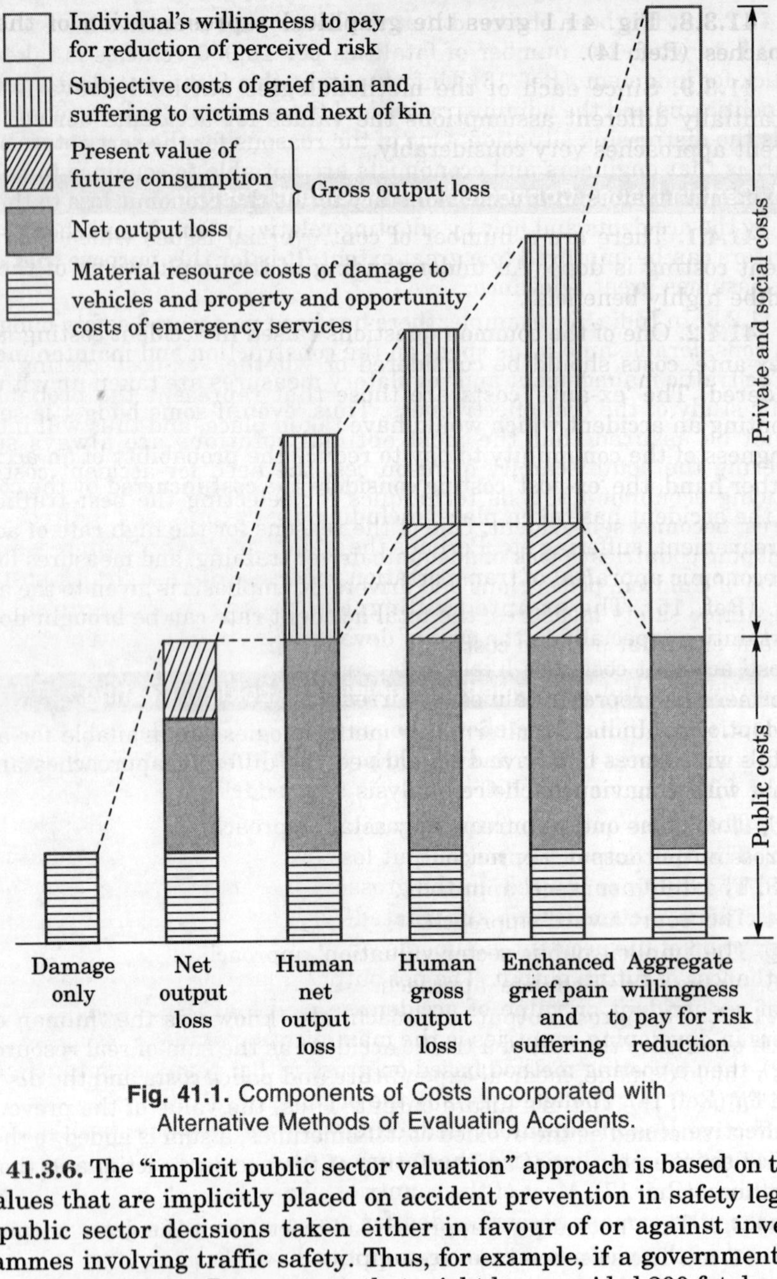

Fig. 41.1. Components of Costs Incorporated with Alternative Methods of Evaluating Accidents.

41.3.6. The "implicit public sector valuation" approach is based on the costs and values that are implicitly placed on accident prevention in safety legislation or in public sector decisions taken either in favour of or against investment programmes involving traffic safety. Thus, for example, if a government agency rejects a scheme costing Rs. one crore that might have avoided 200 fatal accidents, then the implicit public sector valuation of accident prevention must be less than ₹ 50,000 per fatality.

41.3.7. The "value of risk change" approach assumes that a typical public sector investment in traffic safety provides each indi-vidual affected with a very small reduction in the risk of involvement in a fatal accident. The value of prevention of one accident involving one fatality is defined, therefore, as an amount in aggregate that all the affected individuals in society are willing to pay for these small marginal risk reductions, both for themselves and for those whom they care.

41.3.8. Fig. 41.1 gives the graphical representation of the various approaches. (Ref. 14).

41.3.9. Since each of the methodologies outlined above is based on substantially different assumptions the values for accidents derived from the different approaches very considerably.

41.4. Certain Issues Connected with Accident Costing

41.4.1. There are a number of controversial issues which crop up when accident costing is done. An understanding of the implications of these issues would be highly beneficial.

41.4.2. One of the common questions raised in accident costing is whether the 'ex-ante' costs should be considered or whether 'ex-post' costing should be considered. The 'ex-ante' costs are those that represent the probable cost of preventing an accident which would have taken place, and thus will indicate the willingness of the community to pay to reduce the probability of an accident. On the other hand, the 'ex-post' costing considers the cost incurred by the community after the accident has taken place, including direct costs, and indirect costs such as bereavement, suffering etc. Perhaps the widely adopted procedure while dealing with economic appraisal of transportation schemes is to use the 'ex-post' accident costs. (Ref. 15). The ex-ante (willingness-to-pay) approach is riddled with uncertainties especially in the case of developing countries. In any case, since the 'ex-post' accident costs are likely to be considerably lower than 'ex-ante' values, the former costs represent almost the irreducible minimum and are recommended for adoption in India. This should not under-rate the need for research on the society's willingness to pay, and should provide a valuable area for research study dealing with behavioural choice analysis and modelling.

41.4.3. The question frequently asked in accident costing is whether to use the gross output loss or the net output loss as the basis for determining the loss caused by fatality or injured. In the gross output loss, the future consumption of the accident victim is not subtracted. In the net output loss, the future consumption of the accident victim, discounted to the present value, is subtracted from the loss of future output. The net output approach is the absolute minimum cost of an accident or value of accident prevention. If the goal of the decision makers in developing countries is the maximisation of the gross national product (GNP), then a costing method based on gross rather than net output seems to be called for (Ref. 14). Though originally the practice in U.K. (Ref. 16) was to calculate the effective loss of output by deducting the future consump-tion the current practice there is to disregard the future consumption and treat only the gross output loss. (Ref. 17). Most of the countries now adopt the gross output approach and hence the same is recommended for adoption in India too.

41.4.4. The selection of an appropriate discount rate for bringing the future loss of output to the current value is another uncertainty. Fouracre and Jacobs (Ref. 18) collected information from a number of developing countries and have found out the discount rate as follows :

South Rhodesia	6%
Ghana	7%
Thailand	12%
Great Britain (1970)	10%
Ivory Coast	10%
Turkey	5%

In view of the overall scarcity of resources in developing countries, it is recommended that an interest of 10–12% per annum be used for discounting the costs in India.

41.4.5. During the useful life spent of an individual, a normal increase in his earnings takes place. Since the increase is measured at constant prices, there is justification in assuming such an increase when calculating the future loss of output of an individual. In India the rise of per capita national income at constant prices is now around 5% and this value can be adopted.

41.4.6. Accidents involve considerable pain, grief and suffering to the victims as well as those who care for the victims. It is very difficult to quantify in precise monetary terms the additional nonresource cost reflecting such pain, grief and suffering. At the same time, it would be undesirable to neglect totally this component of the accident cost. Some notional value should be ascribed to this abstract component. The inclusion of this component is justified in a society whose goal is to maximise social welfare. Considering the seriousness of the accident problem in the country and the great human loss inflicted by it on the society, it is certainly worthwhile to include the component of pain, grief and suffering in accident costing in India. In U.K., the standard practice now is to provide for substantial increase (about 50 percent) in the accident cost to represent this component. While recognising that it is extremely difficult to quantify in monetary terms the pain, grief and suffering caused to the relatives and friends, and as such the cost values can only be judgments, it should nevertheless be accepted that ignoring the quality of life would be an extremely short-sighted and narrow outlook. Accepting the fact that the society cares for the individuals and their walfare, it is strongly recommended that in the Indian cost evaluation, a notional component be added to represent the pain, grief and suffering. The actual amount need not be as high as 50 percent used in other countries, but it could be limited to about 20 percent. It may be added here that even in the case of court awards, a small value say ₹ 5,000 to ₹ 10,000 is added to the amount of compensation to represent the notional value of pain, grief and suffering.

41.4.7. Different classes of road users have different incomes and hence the average cost per casualty of different classes of road users will vary. This is accepted as the basis for average costing in many countries. For example, the U.K. Department of Transport recommend different accident cost per casualty for pedestrians, pedal cyclists, motor cyclists, motor scooter-riders, moped riders, car and taxi occupants etc. Similarly, the differentiation of accident costing by location (rural, urban), seems desirable since the levels of income of the urban road users are different from the levels of income of the rural road users.

41.4.8. Another important issue to be sorted out in determining the loss of future output is the wage rate of the accident-victim. Two options are generally available : first, to adopt the average wage rate or earnings of the victims, and second, the national output per head of working population. The use of average wage rate or income is a simple method. However, it suffers from a disadvantage that the total output produced in a year by individuals as given by the sum of their earnings tends to be less than the total value of output of the economy because of certain unearned income elements (undistributed profits, interest, dividends etc. (Ref. 19). Reynolds (Ref. 20) employed the national income definition to measure output and determined the gross output by using the net domestic product.

41.5. Indian Studies

41.5.1. In an earlier detailed study conducted by the Central Road Research Institute, an attempt was made to determine the economic cost of road accidents in India. (Ref. 21). Data were collected on different types of accidents such as fatal, severe injury, slight injury and property damages. The data on medical expenditure, legal fees, damages to property etc. were collected. The data from the insurance companies on the road accident claims were collected and the accidents were classified according to severity, age and sex and type of vehicles involved. The details of claims made to the insurance companies and the average compensation awarded were also collected. To account for the loss of output due to death, the future income of the deceased minus the future consumption were evaluated. The future consumption was taken to be one-third of his income. The future output was calculated till the age of 55 years, which was between the average national life expectancy and the retirement age. Table 41.1 gives the average cost of various types of accidents in Delhi as calculated by this Study :

Table 41.1. Average Costs of Various Types of Accidents in Delhi

(Refer 21, 1968)

Type of Accident	*Estimated average cost in Delhi (in ₹)*
Fatal	27,805
Severe Injury	7,470
Slight Injury	870
Property damage	1,155

The above figures represent the values for the year 1968.

41.5.2. A comprehensive work on accident costing is based on data collected from some insurance companies and the Pallavan Transport Corporation (a nationalised bus undertaking) in Madras City (Ref. 22). The study considers the compensation paid by insurance companies and road transport corporations, legal fees and court expenses incurred, administrative cost of establishment, cost of medical treatment, loss of earning due to absence, loss of property damage, cost of treatment in hospitals etc. The future loss is calculated on the gross basis. The cost of various types of road accidents from the insurance companies accident sample was as follows :

Table 41.2. Cost of Various Type of Road Accidents in Madras city for 1978

(Insurance Company Accident Sample, Ref. 22)

Type of Accident	*Cost of road accidents (₹)*
Fatal	129,917
Severe Injury	35,447
Simple Injury	10,503
Property Damage	10,033

From another sample (Pallavan Transport Corporation), the study estimates the cost of various types of accidents as under :

Table 41.3. Cost of Various Types of Road Accidents in Madras city for 1978
(Pallavan Transports Corporation's Sample, Ref. 22)

Type of Accident	*Cost of road accidents* (₹)
Fatal	1,05,781
Severe Injury	67,346
Simple Injury	9,741
Property Damage	13,415

41.5.3. Road User Cost Study

41.5.3.1. The Road User Cost Study in India and its updation have investigated the cost of road accidents in India. (Ref. 25, 26). The findings are discussed below, along with the results of a recent study (Ref. 27).

41.5.3.2. Cost of fatality

Adopting an average annual income of ₹ 4,200 for all categories of total accident victims and assuming a growth of 2 percent in the income for the future years and a discount rate of 10 percent and assuming 22 years of useful services left for the average victim (54 years life expectancy minus 32 years as the average age), the present value of the loss of the future output of the fatal victim has been computed as ₹ 39,220. The notional value of pain, grief and suffering caused to the relatives and survivers of the victim has been assumed to be 20 percent of the loss of gross output. This amounts to ₹ 7,844. Adding this to the loss of future output of ₹ 39,220, the total comes to ₹ 47,064. To this the expenditure incurred on hospitalisation, medical treatment, administrative costs of the police authorities, insurance authorities, motor accident claims tribunal etc. are added.

The total cost of the fatality has been worked out as given in Table 41.4. The 1990 values (Ref. 25, 26) are also given

Table 41.4. Summary of Cost of a Fatal Accident
(Refer 25, 26)

Item	*Cost* (₹) *(1978)*	*(1990)*
1. Gross loss of future output	39,220	1,71820
2. Notional value of pain, grief and suffering	7,844	34,364
3. Hospital expenditure	140	228
4. Expenditure by courts, Tribunal and on lawyer's fee	1,600	1,692
5. Administrative expenses of Police, Insurance Companies, visits by relatives, funeral expenses etc.	1,000	2,450
Total	49,804	2,10,554

41.5.3.3. Cost of Serious Injury

The cost of a serious injury can be estimated after including the following components :

1. Expenses incurred during hospitalisation and in the post-hospitalisation medical treatment.

2. Expenses incurred in connection with the settlement of claims in court, layer's fee etc.
3. Loss of output to community due to hospitalisation of the victim and due to the any permanent disability suffered by the victim due to the accident.

The total cost of a serious injury accident has been worked out as follows :

Table 41.5. Summary of Accident Cost of Serious Injury
(Refer 25, 26)

Item	*Cost (₹)* *(1978)*	*(1990)*
1. Hospitalisation charges	7,000	10,677
2. OPD visits and hospital expenditure	300	
3. Cost of accessories like artificial limbs	500	
4. Cost of establishment of courts, tribunals and lawyers fee	1,600	1,218
5. Cost of establishment of Police, Insurance Companies and cost of visits by relatives	500	
6. Monetary value of loss of future output due to permanent disability	19,610	18,032
Total	29,510	29,927

41.5.3.4. Cost of Minor Injury

Minor injury cost include cost of medical treatment and loss of earnings. The total cost of minor injury has been worked out to be ₹ 321 in 1982 study and ₹ 1,100 in the 1992 study.

41.5.3.5. Cost of Damages of Vehicles

The cost of damages to vehicles has been summarised in Table 41.6.

Table 41.6. Summary of Cost Damages to Buses and Trucks
(Refer 25, 26)

Category of cost	*Cost in ₹ per bus* *1978*	*1990*	*Cost in ₹ per truck* *1978*	*1990*
Repair charges	4,200	9,462	4,200	14,183
Wages of crew	952	2,352	392	1,172
Fixed charges	315	4,025	1,519	2,975
Total	5,467	15,839	6,111	18,530

The cost of damages to cars has been worked out as ₹ 1,200 and the cost of damages to two-wheelers has been worked out as ₹ 300 at 1978 prices. At 1990 prices, the values are ₹ 4,700 and ₹ 1,100 respectively.

41.5.3.6. An Estimate of Total Cost of Road Accident in India

The latest data (1999) on cost of road accidents of various types is given in Table 41.7 (Ref. 27).

Table 41.7. Cost of Road Accidents in India the Year 1999

(Source : Ref. 27)

S. No.	*Accident category*	*Rate per accident (₹)*
1.	Fatal	535,489
2.	Serious injury	242,736
3.	Major injury	106,959
4.	Minor injury	18,855
5.	Damages to buses	47,100
6.	Damages to trucks	48,700
7.	Damages to cars	16,200
8.	Damages to two-wheelers	4,100
9.	Damages to three-wheelers	10,900

The total cost of accidents works out to around ₹ 7,000 crores or roughly 0.7 percent of the national income of the country (Ref. 28). In this connection, it may be mentioned that Fouracre and Jacobs found that the cost of accidents in some of the developing countries ranged from 0.3 to 1.5 percent of the gross Domestic Product as compared to 0.7 percent in the case of Great Britain (1970). (Ref. 18).

I. MULTIPLE CHOICE QUESTIONS

1. Name the traffic survey data which is plotted by means of Desire lines.
   (*a*) accident (*b*) classified volume
   (*c*) origin and destination (*d*) speed and delay. (*GATE, 2006*)

Answers

1. (*b*)

II. SOLVED QUESTIONS AND ANSWERS

Q. 1. What are accident costs?

Ans. Accident costs are one of the most high profile and highest costs of transport. This working group has examined the range of methods available for estimating accident costs and mechanisms for passing.

Q. 2. How do you estimate the marginal external costs of road and rail accidents?

Ans. The working group has focussed on estimating the marginal external cost of road and rail accidents, as requested by the high level group, with some consideration given to the costs of other transport modes. These marginal external costs are the incremental costs of an accident borne by society at large, including victims, family and friends, imposed by those who cause the accident. The relevant factors when estimating the costs are the relationship between accident risk and traffic volume, the determining the external element and the valuation of specific cost element.

Q. 3. What is marginal external costs?

Ans. External accident cost has for a long time been considered as a simple average cost, with more or less qualified assumptions on the allocation of the accident cost between external and internal components. While this lack of theoretical sound principles is serious, more serious is the fact that the underlying theory is not fully developed. Nevertheless, we do believe that enough information is available to move the transport policy a few steps further.

Marginal external costs of transport accidents are the incremental costs of an accident borne by society at large, including family and friends, and can include costs borne by the victims of an accident. The aim of an accident externality charge is to increase the efficiency of the transport system. The basic conclusion from general economic theory is emphasised; the charge should internalise the external marginal cost. In addition, internalisation based on a 'polluter pays principle' (PPP) is generally seen as a 'fair' principle. In the context of accidents the PPP may be expressed as–individual that imposes risks on others should be responsible for the extra cost they impose.

Q. 4. What is the valuation of accidents?

Ans. The valuation of accidents is a controversial matter and should probably so be. While it is evident that life is too important to value in monetary terms we know that changes in risks are a different matter. While no one would trade their life for a sum of money most people will be prepared to choose between safety equipment's with different prices and safety performance or between different ways of crossing a street compared to the saving of time. In Box 2 we briefly discuss different valuation methods. For a deeper discussion on methods we recommend the European Union COST 313 project.

Q. 5. Discuss the methods for valuation of accidents.

Ans. To value accidents we may distinguish between output-based (or human capital) definitions and social welfare (or willingness-to-pay) definitions. The output based approach that has been most widely used is the "gross output" or "human capital" definition.

The major component in these approaches is the discounted presented value of the victim's future output (income) that is foregone due to his premature death. Allowance is often made for non-market output and for various "direct" economic costs such as medical cost, cost of administration and property damage. An alternative output based concept is the "net output" definition.

This concept takes into account the fact that a person consumes a considerable amount of what he produces during his lifetime. The discounted present value of the victim's future consumption is subtracted from the "gross output". This approach provides a pure narrow economic impact for the rest of the society due to a person's premature death. The net production is approximately only 20% of the gross production for fatal accidents while the two measures are equal for light injuries. The objection to the last approach is obvious - when a person has retired the net present value will be negative. The net output approach is thus seldom used.

Q. 6. What are the components of the external marginal cost?

Ans. The external marginal cost consists of three components:

(i) System externalities: The expected accident cost to the rest of the society when the user exposes himself to risk-mainly medical and hospital costs;

(ii) Traffic volume externalities: The willingness-to-pay of the household, relatives and friends and the rest of society related to the increase or decrease in the accident risk for all other users of the same mode;

(iii) Traffic category externalities: The willingness-to-pay of the household, relatives and friends and the rest of society related to the changed accident risk in other modes of transport.

REVIEW QUESTIONS

1. Highlight the certain issues connected with accident.
2. What is the relevance of accident costing for a develoving country?

REFERENCES

1. Gopalswami, T.V., Road Accidents in India, Rates, ratios and costs RUCS Technical Paper No. 71, Fourteenth Quarterly Report, Central Road Research Institute, New Delhi, 1981.
2. Motor Vehicle Accident Statistics, Transport Research Division, Ministry of Surface Transport, Govt, of India, New Delhi. (From Internet)
3. Bureau of Police Research and Development, Accident deaths in India, (1977), Ministry of Home Affairs, Govt, of India, New Delhi, 1977.
4. World Road Statistics International Road Federation, Washington, 1987.
5. Report of the Study Group on Road Safety, Ministry of Transport, Govt. of India, New Delhi, 1972.
6. Highway Safety Workshop Lecture Notes. Indian Roads Congress, New Delhi, 1978.
7. Road User Cost Study in India-Draft Inception Report, Central Road Research Institute, New Delhi, 1977.
8. Kadiyali, L.R., Road User Cost Study in India, Objectives and Methodology Indian Highways, January 1979, Indian Roads Congress, New Delhi, 1979.
9. Swaminathan, C.G., and L.R. Kadiyali, Road User Cost Study–A major research project in transportation in India–Paper presented at the Workshop on Transportation Research, School of Planning and Architecture, New Delhi, 1982.
10. Road User Cost Study, Final Reports Vol. I to VIII, Central Road Research Institute, New Delhi, 1982.
11. Kadiyali, L.R., and A.K. Chandra, Methodology for Accident Cost Study, RUCS technical Paper No. 3, Fourth Quarterly Report, Road User Cost Study, Central Road Research Institute, New Delhi, 1979.
12. Adler, H.A., Transport investment and economic Development, (Editor: Garry Fromm) The Bookings Institution, Washington, 1965.
13. Jacobs, G.D.A., Study of road accidents in developing countries, TRRL Laboratory Report LR 732, Crowthorne, 1976.
14. Hills, P.J., and M. Jones Lee, The cost of traffic accidents and evaluation of accident prevention in developing countries PTRC Summer Annual Meeting, University of Warwick, 1981.
15. Wigan, M.R., Safety Evaluation, Some explicit and implicit considerations, Seminar on measuring social behaviour in Road Research, Australian Road Research Board, Vermont South, 1980.
16. Dawson, R.F.F., Cost of Road Accidents in Great Britain, Road Research Laboratory, Report LR 79, Crowthorne, 1967.
17. Dawson, R.F.F., Current Costs of road accidents in Great Britain, Road Research Laboratory Report LR 396, Crowthorne, 1971.
18. Fouracre, P.R. and Jacobs, G.D., Comparative accident Cost in developing countries, Transport and Road Research Laboratory Supplementary Report 206 UC, Crowthorne, 1976.

19. Goodwin, P.B., On the evaluation of human life in accident studies, Accident Analysis and prevention, Vol. 5 Pergamon Press, Oxford, 1973.

20. Reynolds, D.J., The costs of road accidents, Journal of Royal of Statistical Society, Series A, Vol. 119, Part IV, London, 1956.

21. Srinivasan N.S., Miss. D.V. Hingorani and B.M. Sharma, Economic cost of road accidents, Journal of the Indian Roads Congress, Vol. 36–2, New Delhi, 1975.

22. Natarajan, T., Cost of road accidents, M.E. Thesis, Deptt. of Traffic and Urban Engineering, College of Engineering, Guindy, Madras.

23. Chawla, K.L., Economic cost of road accidents, Journal of Transport Management, Vol. 3, No. 12, Poona, 1980.

24. Chandra, J. and T.D. Dogra, Pattern of injuries in various road users involved with different vehicles in fatal accidents, Police Research and Development, April-June, Bureau of Police Research, New Delhi, 1978.

25. Road User Cost Study in India, Final Report, Vol. V, Central Road Research Institute, New Delhi, 1982.

26. Study for updating Road User Cost Data, Prepared for Min. of Surface Transport, L.R. Kadiyali and Associates, New Delhi, 1992.

27. Evaluation of Road Accident Cost, Tata Consultancy Services, Ministry of Road Transport and Highways, New Delhi, 1999.

28. Gupta, D.P., Road Safety in India, Current Situation and Strategies Proposed, Vol. 64-3, Journal of the Indian Roads Congress, New Delhi, 2003.

42. Traffic Congestion, Traffic Restraints and Road Pricing

42.1. Congestion as a Factor in Road Traffic

42.1.1. In all situations where capacity cannot be provided for the peak demands, waiting and delay are inevitable. In road traffic especially, the peaking phenomenon is very pronounced, giving rise to congestion. Congestion is the impedance and delay imposed by one vehicle on another. For a given road, the larger the traffic volume, the greater the chances of one vehicle delaying the other and hence greater is the congestion.

42.1.2. Whenever congestion occurs the costs are affected. Delay to people, freight and vehicles is one component of the congestion cost. Another is the higher operating costs of vehicles, including fuel costs, maintenance costs, depreciation and overheads. Congestion cost is a part of the social cost, which is widely borne and shared by the society. The second of the social cost is the environmental cost, the cost resulting from the degradation of the environment.

42.1.3. Because of the seriousness of the problem of congestion an understanding of the economic effects of congestion and the methods available for its mitigation are of relevance to a traffic engineer. He should be familiar with the manner in which travel demand and cost interact with each other and the economic basis for restraining traffic which causes congestion.

42.2. Traffic Restraint

42.2.1. Since there will always be scarcity of resources, congestion is inevitable, and one has to devise ways and means of getting the maximum benefit from the existing facilities. A certain amount of restraint on traffic has to be applied before the situation is allowed to become very bad. Congestion itself acts as a restraining influence on traffic to a considerable extent ; and will induce people to make trips at off-peak periods or by alternative modes. But to allow congestion to be caused and then indirectly act as a restraining influence in preventing further congestion is not an efficient use of the resource. It fails to distinguish between the trips which are really essential to the tripmaker or the community and those that are not so essential. It just applies restraint on a first-come-first-served basis and is not a satisfactory solution. More efficient ways are, therefore, needed.

42.2.2. Methods of traffic restraint fall broadly under the following three categories (Ref. 1) :

(*i*) Physical restraint, whereby the entry of vehicles, or a particular class of vehicles is banned. Examples of such a policy are the pedestrian precincts where the vehicles are prohibited totally, reserved bus lanes where the street space is reserved exclusively for buses, prohibited entry and turns and limited access facilities. Though the method is successful on a limited scale, it is obviously unsuited for universal application. It fails to distinguish between vehicles that need to be allowed entry into an area on grounds of the value of the trip to the trip-maker or the community and those vehicles that need to be denied entry.

(*ii*) Regulatory measures, which cover a wide range of traffic regulations and parking and waiting controls. Parking restriction is known to be a powerful measure of restraint and is discussed in deal subsequently.

(*iii*) Fiscal measures, such as vehicle and fuel tax, parking charges, road pricing etc.

Buchanan (Ref. 2), in a classic study of traffic in towns, acknowledged the need for restraint of traffic and suggested three ways of directly achieving it : regulatory use by a system of permits, pricing the use of road space and parking policy. The Smeed Report (Ref. 3) has considered a number of methods for traffic restraint mainly based on the pricing system.

Some of the important methods will be discussed now.

42.2.3. Parking controls as restraint

Parking controls bring in a number of benefits : reduction in accidents, increase in road capacity, traffic restraint, preservation of the environment and so on. As regards parking traffic restraint, the relevant issue to examine is how closely do parking controls relate to road use in congested conditions. As far as regular users who make short trips for business are concerned, the number of parking occasions may well correlate highly with road use. But this may not be the case with those who park vehicles for a long duration but less frequently. Thus it would appear that encouragement of short-term parking and restraint of road use are to some extent mutually contradictory objectives. Difficulty will also be caused in respect of residents, since they may park for long periods when not contributing at all to congestion at critical times of the day. Other difficulty is that with some systems of parking control, through traffic, taxis and public transport vehicles will not be influenced. It is because of these drawbacks in the system that parking controls cannot be the final answer to traffic restraint.

42.2.4. Entry charges

A system of "entry charges" such as supplementary licensing, tolls and admission charges is one of the methods that have received consideration (Ref. 4). Supplementary licences enable a vehicle to ply in a designated area. Tolls, which are good for bridges, tunnels and highways, are probably inexpedient to deal with the urban situations. Admission charges by pre-purchased tickets are designed to give access to a relatively small are through a number of controlled entry points during the morning peak period.

The main advantage of the entry charges system is that it applies equally to parking traffic, non-parking traffic and privately parking vehicles. The disadvantages of the system are that it fails to distinguish between different amounts of road use within the licensing area, or to make any distinctions of charge on the basis of the congestion (Ref. 1). If the area is very large, the effects at the boundary will be unfair.

Singapore has successfully introduced an Area Licensing System (ALS) in 1975, according to which single occupant automobiles are required to pay supplementing licence fees to get into the city centre during restricted morning peak-periods. This was intended to make motorists change their mode and timing of trips which could then reduce the volume of traffic to the restricted city centre to a level equivalent to its off-peak period during the day.

42.3. Road Pricing

42.3.1. Introduction

Road pricing is a method of road user taxation, charging the users of congested roads according to the time spent or distance travelled on those. It is different from taxing the road users in a general way such as by annual licence charges or petrol charges, since these measures do not discriminate between those who use roads in the congested period and those who use roads in the non-congested period. The principle behind road pricing is that those who cause congestion or use the road in the congested period should be charged, thus giving the road user the choice of whether to make a journey or not.

42.3.2. Economic principles behind road pricing

An understanding of the economic rationale behind road pricing is very essential. First of all, it is necessary to be clear about the concepts of journey cost and benefit. Journey costs are made up of

(*i*) private journey costs, consisting of vehicle operating costs. Vehicle operating costs include fuel, maintenance, depreciation etc. The value of the individual's travel time is also a part of private costs.

(*ii*) congestion cost, representing the costs imposted by road users on each other, mainly in terms of extra delay :

(*iii*) environmental costs, caused by the impoverishment of the enviornment ;

(*iv*) road maintenance costs.

The benefits the road user obtains from the journey is the price he is prepared to pay in order to make the journey. As the price gradually increases, a point will be reached when the trip-maker considers it not worth performing or worth performing by other modes. This critical price would be his assessment of the benefit he derives from making the journey. At costs less than this benefit, he enjoys a net benefit equal to the difference between what he is prepared to pay and what he actually pays. This is known as the consumer surplus.

The general form of the curve relating the cost of the trip to the number of trips made is indicated in Fig. 42.1. This curve is called the demand curve.

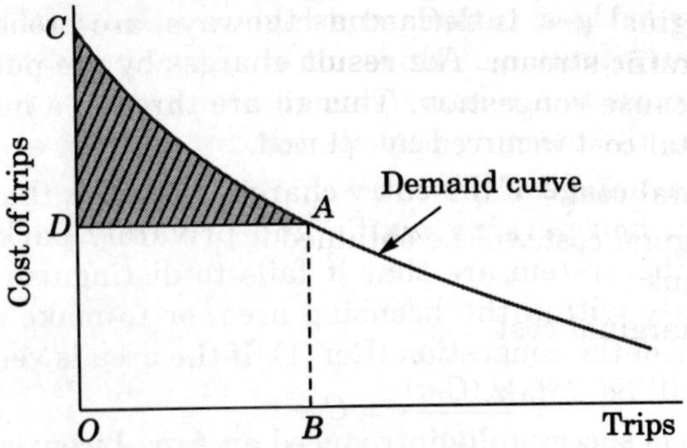

Fig. 42.1. Demand curve.

In this curve, the area *COAB* represents the absolute utility to the trip maker. The net benefit after paying for the cost of trip making is given by subtracting the rectangular area *ODAB*, and is shown by the hatched area *CDA*.

The total private costs of trip can be represented by the simple equation :

$$C = a + \frac{b}{v} \qquad ...(42.1)$$

where, C = total private costs of a trip

a = component proportional to distance

b = component proportional to speed

v = speed.

In this region where congestion is present the speed of a vehicle can be expressed as :

$$v = d - eq \qquad ...(42.2)$$

where, v = speed

d and e are constants

q = flow in veh/hour.

Substituting (42.2) in (42.1),

$$C = a + \frac{b}{d - eq} \qquad ...(42.3)$$

The private cost/flow curve represented Eq. 42.3 is given in Fig. 42.2.

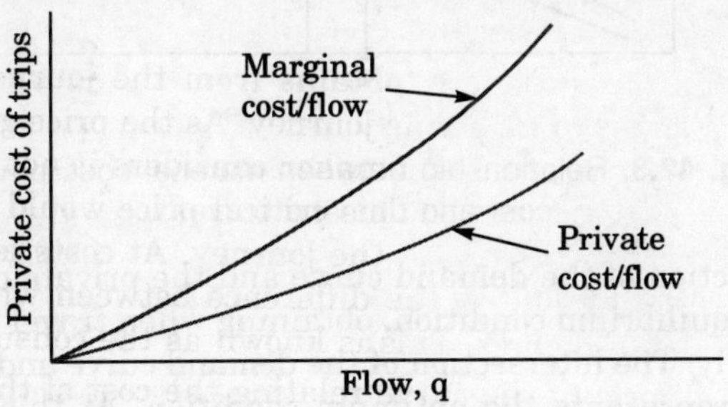

Fig. 42.2. Private cost/flow curve and cost and marginal curve.

The marginal cost is defined as the additional cost of adding one extra vehicle to the traffic stream. The result of adding one extra vehicle is to reduce the speed and cause congestion. This results in an increase in the cost of all journey. The total cost incurred by all vehicles in one hour is :

$$\text{Total cost} = C.q \qquad ...(42.4)$$

The marginal cost will be obtained by differentiating the term (42.4) with respect to q. Thus :

M, the marginal cost

$$= \frac{d\,(C.q)}{dq} = C + q.\ \frac{dC}{dq} \qquad ...(42.5)$$

From (42.2), $q = \dfrac{d - v}{e}$

Differentiating Eq. (42.1) with respect to q,

$$\frac{dC}{dq} = \frac{dC}{dv} \cdot \frac{dv}{dq} = \frac{b}{v^2}\,(-e) \qquad \text{...(42.6)}$$

From Eqns. (42.1), (42.2), (42.5) and (42.6),

$$M = a + \frac{b}{v} + \frac{(d-v)b}{v^2}$$

The above curve is also shown in Fig. 42.2. It will be seen that the marginal cost will always be greater than the private cost, the increase representing the congestion cost.

Superimposing the demand curve on the private cost/flow and marginal cost/flow curves, the position as shown in Fig. 42.3 is obtained.

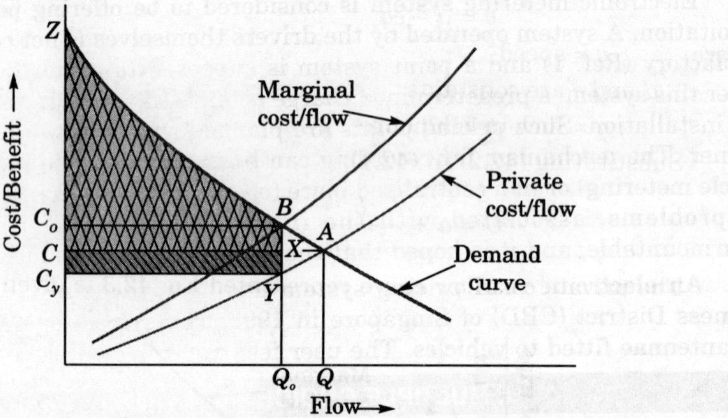

Fig. 42.3. Relationship between material cost, private cost and demand curves.

The intersection of the demand curve and the private costs curve at point A represents the equilibrium condition, obtaining when travel decisions are based on private costs only. The intersection of the demand curve and the marginal costs curve at point B represents the optimum condition. At this point the flow Q_0 corresponds to the cost C_0 which is the marginal cost as well as the value of the trip to the trip maker. The net benefits under the two position A and B are shown by the two hatched areas, ACZ and $BYCyZ$ respectively. If the conditions are shifted from the point A to B, the net benefit due to the change will be given by area $CCyYX$ minus AXB. If the area $CCyYX$ is greater than arc AXB, the net benefit will be positive. The shifting of conditions from point A to B can be brought about by imposing a road pricing charge BY. Under this scheme, private vehicles continuing to use the roads will on an average be worse off in the first place because BY will always exceed the individual increase in benefits XY.

42.3.3. Requirements of a good pricing system

The road pricing system should fulfill the following requirements (Ref. 3):

(*i*) Charges should be closely related to the amount of use made of the roads.

(*ii*) It should be possible to vary the price for different roads, at different times of the day/week/year or for different classes of vehicles.

(*iii*) Prices should be stable and ascertainable by the road users before they commence a journey.

(*iv*) Method should be simple for road users to understand and police to enforce.

(*v*) It should be accepted by the public as fair to all.

(*vi*) Payment in advance should be possible although credit facilities in certain cases be possible.

(*vii*) Equipment used should be highly reliable.

(*viii*) The system should be amenable to gradual introduction commencing with experimental phase.

(*ix*) It should be capable of being applied to the whole country.

(*x*) It should be free from fraud and evasion, both deliberate and unintentional.

Electronic metering system is considered to be offering potentialities for exploitation. A system operated by the drivers themselves is not considered to be satisfactory (Ref. 1) and a point system is supposed to avoid these drawbacks. Under this system, a predetermined charge is imposed as a vehicle crosses a road-side installation. Such pricing points are planned on the network in the desired manner. The mechanism for recording can be installed in the vehicle itself (on-vehicle metering) or in a centralised place (off-vehicle metering). Technologically, the problems, associated with the introduction of such system are not unsurmountable, and it is hoped that they will be soon introduced in future.

An electronic road pricing system has been introduced in the Central Business District (CBD) of Singapore in 1998. The system operates on devices and antennae fitted to vehicles. The user fees are collected from smart cards.

I. MULTIPLE CHOICE QUESTIONS

1. Stopping sight distance depends upon
   (*a*) total reacting time of driver (*b*) speed of vehicle
   (*c*) efficiency of brakes (*d*) all of these.
2. When speed of traffic flow becomes a zero, then traffic density?
   (*a*) attains maximum value whereas traffic volume becomes zero
   (*b*) and traffic volume both attain maximum value
   (*c*) traffic volume both becomes zero
   (*d*) becomes zero whereas traffic volume attains maximum value.

Answers

1. (*d*) 2. (*d*)

II. SOLVED QUESTIONS AND ANSWERS

Q. 1. When does traffic congestion occur?

Ans. Traffic congestion occurs wherever demand exceeds the capacity of the transportation system. This lecture gives an overview of how congestion is generated, how it can be measured or quantified; and also the various counter measures to be taken in order to counteract congestion. Adequate performance measures are needed in order to quantify congestion in a transportation system. Quality of service measures indicates the degree of traveller satisfaction with system performance and this is covered under traveller perception. Several measures have been taken in order to counteract congestion. With the evolution of society, economy and technology, the household characteristics as well as the transportation system gets affected. The change in transport system causes a change in transport behaviour and locational pattern of the system. The change in household characteristics, transport behaviour, locational pattern, and other growth effects result in the growth of traffic. But the change or improvement in road capacity is only as the result of change in the transportation system and hence finally a situation arises where the traffic demand is greater than the capacity of the roadway. This situation is called *traffic congestion*.

Q. 2. What are the effects of congestion?

Ans. Congestion has a large number of ill effects on drivers, environment, health and the economy in the following ways. Drivers who encounter unexpected traffic may be late for work and other appointments causing a loss in productivity and their valuable time. Since congestion leads to increase in travel time *i.e.*, vehicles are made to travel for more time than required which consumes large amount of fuel thereby causing fuel loss and economic loss to the drivers.

- One of the most harmful effects of traffic congestion is its impact on the environment. Despite the growing number of vehicles, cars stopped in traffic still produces a large volume of harmful carbon emissions . Increase in pollutants (because of both the additional fuel burned and more toxic gases produced while internal combustion engines are in idle or in stop and go traffic).
- Drivers who become impatient may be more likely to drive aggressively and dangerously and leads to high potential for traffic accidents.
- Negative impact on people's psychological state, which may affect productivity at work and personal relationships.
- Slow and inefficient emergency response and delivery services.
- Decrease in road surface lifetime: When a vehicle moves over the surface, the areas of contact (where the vehicles' tyres touch the road) are deflected downwards under the weight of the vehicle and as the vehicle moves forward, the deflection corrects itself to its original position.
- One beneficial effect of traffic congestion is its ability to encourage drivers to consider other transportation options like a subway, light rail or bus service. These options reduce traffic on the roads, thereby reducing congestion and environmental pollution.

Q. 3. Write a brief note on traffic congestion.

Ans. A system is said to be congested when the demand exceeds the capacity of the section. Traffic congestion can be defined in the following two ways:

- Congestion is the travel time or delay in excess of that normally incurred under light or free flow traffic condition.
- Unacceptable congestion is travel time or delay in excess of agreed norm which may vary by type of transport facility, travel mode, geographical location, and time of the day.

Q. 4. What are the types of traffic congestion?

Ans. Traffic congestion may be of two types:

- **Recurrent congestion:** Recurrent congestion generally occurs at the same place, at the same time every weekday or weekend day. This is generally the consequence of factors that act regularly or periodically on the transportation system such as daily commuting or weekend trips. Recurrent congestion is predictable and typically occurs during peak hours. It displays a large degree of randomness in terms of duration and severity.

- **Non-Recurrent congestion:** Non-recurrent congestion is the effect of unexpected, unplanned large events(road woks, accidents, special events and so on) that affect transportation system more or less randomly and as such, cannot be easily predicted.

Q. 5. What are the need and uses of congestion measurement?

Ans. Congestion has to be measured or quantified in order to suggest suitable counter measures and their evaluation. Congestion information can be used in a variety of policy, planning and operational situations. It may be used by public agencies in assessing facility or system adequacy, identifying problems, calibrating models, developing and assessing improvements, formulating programs policies and priorities. It may be used by private sector in making locational or investment decisions. It may be used by general public and media in assessing traveler's satisfaction.

Q. 6. What are the components of performance measure of a congested roadway? Explain each component.

Ans. Performance measure of a congested roadway can be done using the following four components:

- Duration,
- Extent,
- Intensity, and
- Reliability.

Duration: Duration of congestion is the amount of time the congestion affects the travel system. The *peak hour* has now extended to peak period in many corridors. Measures that can quantify congestion include:

- Amount of time during the day that the travel rate indicates congested travel on a system element or entire system.
- Amount of time during the day that traffic density measurement techniques (detectors, aerial surveillance, etc.) indicate congested travel.

Duration of congestion is the sum of length of each analysis sub period for which the demand exceeds capacity. This component measures the performance of a particular road in handling traffic efficiently i.e.,with the increase in the duration of congestion, poorer will be the performance of the transportation system. The maximum duration on any link indicates the amount of time before congestion is completely cleared from the corridor. Duration of congestion can be computed for a corridor using the following equation: For corridor analysis,

$$H = N \times T$$

Extent: Extent of congestion is described by estimating the number of people or vehicles affected by congestion and by the geographic distribution of congestion. These measures include:

- Number or percentage of trips affected by congestion.
- Number or percentage of person or vehicle meters affected by congestion.
- Percentage of the system affected by congestion.
- Performance measures of extent of congestion can be computed from sum of length of queuing on each segment. Segments in which queue overflows the capacity are also identified. This is useful for ramp metering analysis.

Intensity: Intensity of congestion marks the severity of congestion. It is used to differentiate between levels of congestion on transport system and to define total amount of congestion. It is measured in terms of:

- Delay in person hours or vehicle hours;
- Average speed of roadway, corridor, or network;
- Delay per capita or per vehicle travelling in the corridor, or per person or per vehicle affected by congestion;
- Relative delay rate (relative rate of time lost for vehicles);
- Intensity in terms of delay is given by,

$$D_{PH} = T_{PH} - T^0_{PH}$$

Reliability: Reliability is a measure of a drivers ability to accurately predict and plan for a certain travel time. The more unexpected events that occur on a roadway, the less reliable it is. Non recurrent congestion has a bigger impact on the reliability of the roadway relative to concurrent congestion. In other words, Travel time reliability is defined as the level of consistency in travel conditions over time and is measured by describing the distribution of travel times that occur over a substantial period of time. Reliability is an important component of roadway performance and perhaps more importantly, of motorists perceptions of roadway performance. The importance of measuring and managing reliability in reducing congestion is explained as follows.

- Motorists have less tolerance for unexpected delay than for expected delay
- Cost associated with unreliable travel
- Reliability is a valued service in other industries and utilities

Therefore, it is clear that reliability is the impact of non-recurrent congestion on transport system and it can be expressed as average travel rate or speed ± standard deviation or delay ± standard deviation.

Q. 7. What is marginal cost?

Ans. Marginal cost is the additional cost of adding one extra vehicle to the traffic stream. It reduces speed and causes congestion and results in increase in cost of overall journey. The total cost incurred by all vehicles in one hour (C_T) is given by:

$$C_T = cq$$

III. REVIEW QUESTIONS

1. What is meant by traffic restraint?
2. Briefly explain congestion is a factor in road traffic.

REFERENCES

1. Maycock, G., Implementation of Traffic Restraint, R.R.L. Report LR 422, Road Research Laboratory, Crowthrone, 1972.
2. Ministry of Transport, Traffic in Towns (The Buchanan and Growther Reports), H.M.S.O., London, 1963.
3. Ministry of Transport, Road Pricing : The Economic and Technical Possibilities (The Smeed Report). H.M.S.O., London, 1967.
4. Ministry of Transport, Better Use of Town Roads, H.M.S.O., London, 1967.
5. Smith, A.H., Automatic Metering System for Road Pricing, Procoeedings of the Symposium on Techniques of Improving Urban Conditions by Restraining Road Traffic, O.E.C.D., Paris, 1973.

SOME CURRENT TOPICS IN TRANSPORTATION

43. Nature of Traffic Problem in Cities

43.1. Growth of Towns

Urbanization and industrialization are two of the most important features of modern civilization. Industrialization has brought in the need to reap the benefits of economies of scale, grouping together many factories and bringing together many people, raw materials and machinery. With the present high level of technology, innovation and specialization, urbanization on a large scale is inevitable. Urbanization provides the necessary infrastructure for trade, commerce and industry. It also provides better living standards in terms of educational facilities, health care, recreation and several intellectual pursuits. Urbanization attracts the surplus labour force from the rural areas and utilises it in running the various services which are vital to the existence of towns.

In advanced countries abroad, the level of urbanization is rather high. For example, the proportion of urban population to the total in 1971 was 85.5% in Australia, 82.1% in Israel, 78.3% in U.K., 73.5% in U.S.A. and 72.2% in Japan.

India is essentially a rural oriented country. Even then, the urban population has been growing at a fast rate, as can be seen from Table 43.1.

Table 43.1

Year	*Total population (million)*	*Urban population (million)*	*Percentage of urban population*
1901	233	26	11.00
1911	246	26	10.44
1921	244	28	11.34
1931	270	33	12.18
1941	309	44	14.10
1951	350	62	17.62
1961	425	78	18.26
1971	529	107	20.22
1981	685	159	23.30
1991	846	218	25.70
2001	1020	285	27.78

India is still one of the countries having a low degree of urbanisation. The United Nations estimate show that 47 percent of the world population line in urban areas. Looking at the growth of urbanisation in other Asian countries, it is seen that countries like China and Indonesia started with low levels of urbanisation in the 1950s, but the degree of urbanisation there is now 32 and 41 percent respectively (Ref. 5). Thus, India is also poised for a big jump in the degree of urbanisation over the next few decades. Traffic engineers and transport planners have to meet this challenge.

In 2001, there were 1080 Class I cities in India with a population of one lakh and above. These included 35 cities with a population of one million and above.

With the growth of cities, pressure on land mounts up. Need for more residential buildings and office accommodation becomes acute. Office buildings tend to concentrate in the town centre, replacing existing dwelling places. High rise buildings spring up, involving great investment, to house the offices. The residences begin to be pushed outwards to the fringe of the town, increasing the separation between the places of work and residence. This gives rise to the need for transport of people in large number from the outer fringe into the inner core. The existing transport network, which was adequate for the needs of the town of the past, becomes overloaded and turns out to be incapable of handling the newly created transport demand.

The inability of the transport network to cater to the high demand often leads to relocation of activities at the edges of the town, where land is still cheap and transport capacity exists for each movement of people and goods. The town centre often starts on its road to decay, accompanied by unemployment, poverty, crime, violence and decline in educational standards. Urban sprawl, a phenomenon in American cities, stems from the dispersal of activities to the edges, and has many attendent disadvantages. Firstly, it adds to the transportation problems by increasing the distances to be travelled by people going to work, recreation, shopping or business. Secondly, unchecked urban sprawl has a great impact on primary services such as water supply, sanitary services and telecommunications, and secondary services such as restaurants, specialist shops and service industries. The advantages of compactness is lost, and the city can survive only if cheap source of oil to run the motor vehicle can be assured. The recent oil crisis has exposed the vulnerability of such development.

Whatever be the stage in the evolution of a city, unplanned growth can lead to either severe congestion in the city centre or to urban sprawl. Both of these situations bring in extreme pressure on the transport capacity.

43.2. Growth of Traffic

The private car is the most versatile transport equipment ever devised by man. It affords travel as you like facility and frees the traveller from subservience to rigid schedules of bus transport. It is speedier and more convenient. It is in fact a status symbol in the modern society. The urban higher income group are easily attracted by its virtues. In most of the developed countries, almost every family owns a car. Those who already own one may think on the second car to the family.

As to the future, it is difficult to see any new method of movement appearing in the horizon which will be a serious competitor to the motor vehicles (Ref. 1). Though the present oil crisis has given a serious jolt to the automobile industry, nothing can possibly displace the motor vehicle from the pedestal of supermacy. We will have, therefore, reckon with the inevitability of the motor vehicle on the road. The increase in the number of motor vehicles will depend upon a number of factors chief among them being the growth of family income. Other factors such as price of cars, petrol rates, insurance rates, taxation levels, measures adopted to curb the private car, demand for transport, etc. will also be significant. Considering the trend the world over, it can be predicted that the city traffic in the developing countries will grow at a very fast rate in the coming years.

43.3. Nature of the Present Difficulties in Urban Traffic Conditions

The well-known Buchanan Report (Ref. 1) deals with the nature of the present difficulties being faced by towns. Though the report is fairly old, relevance even today is unquestionable. Moreover, what developed countries experienced then is happening in the emerging economies now.

The most serious effect of increasing urbanization and accelerated growth in traffic is the severe congestion on the streets. The conditions result in concentrated traffic demand, both in time and in space. The existing street system which was evolved over the centuries was never designed to cope up with this situation. The very advantage claimed for the motor vehicle, *viz.*, its ability to provide a door-to-door service seems to be difficult to reap. It is not possible to stop the vehicle and get down at one's own will. Parking at a safe place is almost an impossibility in some areas of the city. Loading and unloading of goods by commercial vehicles is hampered by too many restrictions. Congestion results in delays and time losses. Driver stresses are caused by frustration and delays.

An inevitable result of growth of traffic has been the increase in road accidents, which take a great toll of human life every year. In many of the accidents, the pedestrians are the sufferers. Children and the aged are another group that receive severe punishment.

The deterioration of the environment due to traffic has been causing serious concern. The noise in the streets and adjoining areas has been growing up to intolerable levels. The exhaust from the vehicles pollutes the atmosphere with fumes and smell. Vibration of buildings and adjacent structures and visual intrusion are some of the other ill-effects.

43.4. Measures to Meet the Problems

The following are some of the steps that can be taken to meet the situation created by the deteriorating traffic conditions in the cities (Ref. 3, 4) :

1. Land use and city planning controls to be initiated.
2. Transportation studies to be carried out and plans for new roads and reorganisation of the existing network to be formulated.
3. Traffic restraint measures to be initiated, including :
   (*i*) Restriction on parking ;
   (*ii*) Fiscal measures such as road pricing and entry charges.
4. Promotion of public transport to be pursued.
5. Pedestrianization in congested city centres to be thought of.
6. Staggering of working hours.

The above measures are now discussed in detail.

43.5. Land Use and City Planning Controls

Traffic is a function of activities that take place in a city and is closely related to land use. The most obvious method of dealing with the deteriorating traffic situation is to impose land use and city planning controls, calculated to have an impact on traffic generation. Simple controls such as limitations on the floor area ratio of buildings and restrictions on the high rise construction can help keep a balance between land use and traffic. Functional dispersal of business, commerce, industry and commercial traffic, if enforced thoughtfully, can bring in a large amount of relief.

43.6. Transportation Studies Needed

The analysis and prediction of travel demand and formulation of plans to meet the same in the most effective manner based on transportation studies is the immediate need in most of the cities. The approach should be comprehensive and cover all aspects of urban street schemes, public transport, mass transit, traffic management, parking policies and pedestrian schemes. The methodology and techniques of these studies have been developed to a high degree and one can expect to evolve reasonably worked out answers to the problems of traffic. The plans could cover schemes for better utilization and efficient reorganization of the existing street system and transportation facilities and for new roads. New bye-passes, ring road, arterial streets and expressways can be built to give relief to the traffic congestion. Traffic management measures such as one-way streets could be implemented. Traffic control could be thought of. Measures to deal with mixed traffic, such as segregation of traffic, could be initiated.

43.7. Traffic Restraint Measures

Traffic restraint measures are becoming increasingly popular in many cities (Ref. 2). The economic theory of traffic restraint is discussed in detail elsewhere (Chapter 42). The basic idea is to impose certain measures calculated to restrain the use of roads by traffic. The main techniques that are available for this purpose are :

(*i*) Restrictions on parking ;

(*ii*) Fiscal measures such as road pricing and entry charges.

Parking facility is the basic requirement of motor vehicles. Any restriction on the place, duration and cost of parking has an immediate effect on traffic. Parking control is one of the most powerful of the direct and indirect techniques of restraining traffic from coming into the central business district. By a careful formulation of a parking policy, private car journeys to the major centres of activity in the peak period can be easily shifted to the public transport. There are many additional benefits that flow from parking controls. The safety of traffic is enhanced and better use is made of the street space for moving traffic.

Road pricing is a method whereby vehicles pay for the use of the road space and this is achieved usually by some electronic device or otherwise. It has very great potentialities if developed and implemented after sound planning.

A system of entry charges, such as licensing, tolls and admission charges, is another fiscal measure that is possible.

43.8. Promotion of Public Transport

Public transport in buses or mass transit rail facilities is one of the main planks from which the menance of traffic congestion is being fought. In view of the importance of the subject, a detailed discussion is reserved for Chapter 44.

43.9. Pedestrianisation

Complete banning of the motor vehicle from certain areas and declaring the same as pedestrian precincts has a salutory effect on the traffic situation. It increases the safety of the pedestrians. It gives them the freedom to move about and shop leisurely. It frees the area from the noise, fumes and smell. Contrary to the belief that it may affect adversely the shopping turn-over, it has led to an increase in business. By and large, the motorists, the pedestrians and the commercial interests are all agreed about the beneficial results of pedestrianisation in selected locations.

43.10. Staggering of Office Hours

Staggering of office hours is a simple and effective way to spread the duration of the peak hour and lessen the pressure on the road space. School hours can be very conveniently planned to be sufficiently away from the office hours. Even the office and industrial working hours can be staggered without detriment to the general efficiency of the activities concerned. While some modest staggering of hours is feasible and is in fact desirable, one cannot altogether avoid traffic concentrations in the peak hour, which must be accepted as a fact of life.

43.11. Promotion of Bicycle Traffic

Bicycles are an important mode of personal transport, particularly for the poor sections of the society in India. Because of their many advantages, they must be favoured and patronised. Some of the advantages are :-

(*i*) They use only human energy, and thus avoid the use of scarce oil-based energy resources.

(*ii*) They do not let out emissions

(*iii*) They are noiseless.

(*iv*) They require very small parking space.

(*v*) They are cheap to own and operate.

(*vi*) They offer door-to-door mobility.

(*vii*) They can keep the rider in good health.

(*viii*) They are safe, when they are segregated from fast traffic.

I. MULTIPLE CHOICE QUESTIONS

1. Which of the following is not one of the features of urban community
   (*a*) face to face relationships (*b*) complex life
   (*c*) materialistic (*d*) glamour in life
2. Which among the following is not a cause of growth of cities
   (*a*) trade and commerce (*b*) industrialism
   (*c*) development of transport (*d*) hygienic outlook of people
3. Which is not a feature of urban life?
   (*a*) loss of humanistic value (*b*) impersonal relationship
   (*c*) informal ties (*d*) competition
4. Conurbation means ....
   (*a*) cluster of cities and town (*b*) cities of cities
   (*c*) group of slams (*d*) none of these
5. .... refers the refurbishing or replacement of old buildings and new use of previously developed land in urban areas.
   (*a*) urban planning (*b*) urban recycling
   (*c*) urbanism (*d*) none of these
6. The movement of people from central cities to smaller communities in the surrounding area is known as ....
   (*a*) reverse migration (*b*) under urbanisation
   (*c*) suburbanisation (*d*) over-urbanisation

Answers

1. (*a*) **2.** (*d*) **3.** (*c*) **4.** (*a*) **5.** (*b*)
6. (*c*)

II. SOLVED QUESTIONS AND ANSWERS

Q. 1. In your opinion, what are the root causes of this problem and what urgent measures need to be taken to address this problem. Critically analyse.

Ans. The major reasons for the same are:

(i) Increase in number of vehicles:

(*a*) The rise of the neo-middle class, leading to increase in number of lone travellers, congesting the roads.

(*b*) Poor public transport infrastructure acting as an incentive to go for private vehicles

(ii) Poor urban transport infrastructure:

(*a*) Lack of separate ways for pedestrians, cycles, buses, etc.

(*b*) Damaged or poorly leveled or no pucca roads at all

(*c*) Frequent underground repair works (for cable wires, pipelines, etc.), damaging the road and temporary blockage of roads.

(iii) Inefficient traffic management, especially during peak hours

Q. 2. What is the concept of transit oriented development (TOD) and do you think this will address the problem of last-mile connectivity?

Ans. The concept of TOD basically means that along the (metro) corridors, since you have a means of efficient public transport, you don't need to use your private car. If you allow dense development along a metro station or a metro corridor, then the traffic is reduced, and the same space can accommodate more people. But allowing high rises along the corridors will not take care of the existing problem. If you have transit oriented development where the services, the facilities are all planned in such a way that the requirement of movement is reduced… so to that extent the use of private cars will reduce, number of trips will reduce, the requirement for a journey will reduce. So this will certainly help to the extent that the additional burden will come down, but it will not reduce the existing burden.

Q. 3. Discuss traffic composition in Indian cities.

Ans. Traffic composition in Indian cities is of a mixed nature. There is a wide variety of about a dozen types of both slow and fast-moving vehicles. Two-wheelersiand cars account for over 85% of the vehicle population in most of the metropolitan cities. They account for at least 90% of total vehicles in Ahmedabad, Bhopal, Coimbatore, Delhi, Kanpur, Lucknow, Nagpur, Vadodara, Varanasi, and Vishakhapatnam. Two-wheelers alone account for more than 80% of the total vehicles in number of metropolitan cities.

For example, during the year 2009, in Nagpur (84%), Varanasi (84%), Surat (83%), Coimbatore (83%), Madurai (82%), Bhopal (81%), Kanpur (81%), Vadodara (81%), Vishakhapatnam (81%), and Lucknow (80%), two-wheelers accounted for at least 80% of the total vehicles. Analysis of data presented in Table 3 reveals that, during the year 2009, the share of buses is negligible in most Indian cities as compared to personalized vehicles. For example, two-wheelers and cars together constitute at least 90% of the total vehicles in Ahmedabad (91%), Delhi (90%),

Lucknow (93%) and Nagpur (91%) whereas in these cities buses constitute only 1%, 0.7%, 0.3% and 0.4% respectively.

Q. 4. Why is the road safety situation so bad in so many cities of India?

Ans. The main reason for this appears to be the prevailing imbalance in modal split, inadequate transport infrastructure and its sub optimal use, and the lack of effective road-safety policies. Very few cities of India have an adequate public transport system. People rely primarily on personalized modes such as cars and two-wheelers, para-transit modes such as autorickshaws and tempos, and non-motorized transport modes such as tricycles, bicycles, and walking. In most of the cities, two-wheelers and cars account for over 90% of the motorized vehicle population whereas the share of buses is negligible in comparison to personalized vehicles. In general, the road space in cities is grossly insufficient.

Q. 5. Why do we need to encourage green modes?

Ans. Transport policy should also encourage the need for developing green modes like bicycles, cycle rickshaws, pedestrians, etc. The potential of green modes is often underestimated since they are used primarily for short distances. But, large fraction of journeys made by cars and two-wheelers are mainly for short distances say less than 6 km, a distance over which use of motor vehicle does not provide significant time advantage. Moreover, motor vehicle emissions are high for short distance travel because fuel consumption is high due to cold engine and because the catalyst is not yet working at full efficiency. Due to this reason, the use of green modes in place of motor vehicles for short distances has huge potential for pollution reduction. Green modes particularly walking and cycling have huge health benefits as well. However, to promote green modes, the safety concerns of cyclists and pedestrians have to be addressed adequately. For this purpose, there has to be a segregated right of way for bicycles and pedestrians. Apart from improving safety, this will help to improve traffic flow, increase the average speed of traffic, and reduce emissions resulting from low vehicle speed.

III. REVIEW QUESTIONS

1. What are the measures to meet the traffic problems in cities?
2. What is the promotion of public transport?

REFERENCES

1. Traffic in Towns, The Buchanan Report, Her Majesty's Stationery Office, London, 1963.
2. Techniques of improving urban conditions by restraint or road traffic, Organization for Economic Co-operation and Development, Paris, 1973.
3. National Transport Policy Committee, Planning Commission, New Delhi, 1980.
4. Dalvi M.Q. and R.C. Sharma, National Transport Policy for Urban Settlements, Seminar on Transportation Systems in Urban Settlements, Town and Country Planning Organisation, New Delhi, 1982.
5. Tenth Five Year Plan, 2002-2007, Vol. II, Planning Commissioner, Government of India, New Delhi.

44. Public Transport in Cities

44.1. Planning for Public Transport

One of the current issues pertaining to urban traffic is the role of public transport. The recent energy crisis, the steep fall in the environmental quality of the roads and the severe congestion have brought this subject to the fore. There has been a long controversy going on in almost all cities of the world on the relative role of each of the conflicting modes, *viz.,* the private car and bus. It cannot be denied that the private car is the most ideal transport vehicle devised by man. Its advantages are obvious : quicker travel, freedom from adherence to fixed timetables, possibility of carrying packages and children more comfortably and the sheer enjoyment of driving wherever and whenever one desires. At the same time, the great increase in car ownership has created problems of congestion, accidents, parking and polution. The oil crisis has made car travel extremely costly.

To improve our city streets to cater only to the growing needs of the future car travellers might not be feasible with the limited resources. For increasing the capacity of streets, whole line of buildings might need demolition. The limited space in towns might not permit the construction of expressways and superior facilities. One way of dealing with the problem is to wean the car traveller away from the car to the public transport system.

In India, the percentage of trips performed by public transport is already high as seen from Table below (Ref. 3) :

Table 44.1. Trips by Public Transport in Selected Indian Cities

City	*Average number of passengers carried by mass transport (millions)*	*Percentage of trips by public transport to total number of trips by all modes*
1. Mumbai	4.42	78
2. Kolkata	3.02	80
3. Chennai	1.28	68
4. Delhi	1.03	41

The mass transport trips are predominantly by buses, except in the case of Bombay where 53 percent of the trips are by train.

Bus transport has many advantages to offer. It can carry a lot more people for the street space it occupies. Streets can get decongested and travel can be made quicker. The dilemma of finding space for parking the cars in the urban core gets resolved. The heavy investments needed for improving the streets to cater to the insatiable demand of the private car can be postponed. Oil consumption can be reduced. Pollution can be minimised. The decay of the central business district due to congestion, absence of parking facilities and drop in business turnover can be halted. Bus transport is labour intensive and can provide jobs for the unemployed, and this is an important factor in the developing countries with high population. Buses are, in any case, the only means of travel to those who do not own a car, and to the old and the young. With this background, public transport has an important role to play.

In the more advanced countries, public transport has been losing clientale and falling into the background in recent years. The affluence of the people has caused great increase in car ownership, resulting in heavy inroads into the income of the public transport system. The baffling traffic problem has once again reactivated interest in bolstering the tottering public transport system. In a developing country like ours, the situation is different. The public transport system never lacked clientale at any time. If ever, it was a case of over-crowding and gross inadequacy in the system. The shift of car owners to the bus as a result of hike in oil prices has aggravated the crowding situation in the bus. The case for an appraisal of the policy towards public transport is stronger than ever before.

The nature of operation of bus transport in cities makes the running of the system costly. There are too many short and uneconomic routes, and it is difficult to eliminate them. Traffic congestion in the streets and frequent stopping and starting cause the fuel consumption and wear and tear of vehicles to mount up steeply. The present congested conditions in the streets bring down the speed of operation, thus increasing the time taken to travel. A higher fleet strength is needed if the speed of travel drops down. The peak-hour traffic requires the maintenance of a large fleet, a good part of which will remain idle for the rest of the day. It has, therefore, been found necessary to subsidise the public transport. This is the case in many advanced countries and the same situation obtains in India too. Subsidising public transport and fostering its growth is perhaps more desirable than facing the consequences of uncontrolled private car travel. In India, the Central Government have been providing loan assistance to the local authorities for augmenting and strengthening bus transport.

An efficient bus transport system can be built only after careful planning. The selection of bus routes should be based on detailed travel surveys. The routes should be as far as practical direct with respect to origins and destinations of passengers. Transfers are generally abhored by the passengers and should, therefore, be restricted to the minimum. The bus stops and routes should be so planned that the walking distance of the passengers from the place of residence or work to the bus stop should not be abnormally high, say 0.3–0.5 km or involving not more than 5–10 minutes walk. The frequency of service is an important consideration, and adherance to the accepted schedule is equally important. The buses themselves must be built and maintained for comfort and safety. The entry and exit doors should be carefully designed. Large windows and comfortable seats increase the pleasure of a ride. For the convenience of standing passengers adequate hand-holds must be provided. In foreign countries, where manpower is short, one-man bus operation is common. In such cases, the ticket vending should be mechanised and fare collection should be simplified. Safety of passengers is enhanced if the entry and exit doors are made to shut and open at the instance of the driver.

The streets should be planned for bus traffic. The most important requirement is the provision of recessed bus-bays to accommodate the parking bus away from the carriageway. The design aspects of bus-bays are discussed elsewhere. Bus stops should be located at convenient and safe places. Bus shelters of good design to protect the passengers from elements like sun and rain are a necessity. Exclusive bus-lanes have been tried at a number of cities to speed up bus travel. High Capacity Bus Corridors are being planned in Delhi. (Ref. 5)

Preferential treatment for buses at traffic signals by installing bus detectors has been tried in some countries with success.

Highly innovative schemes have been tried with success in some of the cities to attract passengers to public transport. These include demand-actuated bus and taxi systems (dial-a-ride service. (Ref. 2)

In congested localities, mini buses offer great advantages and need to be seriously thought of.

An example of introduction of mini buses in largely pedestrainised area is in Leeds (Ref. 1). Mini buses have also helped to serve the passengers going to or coming from Railway Stations and long distance bus terminal in Delhi.

In Delhi, the Metro rail system has succeeded in weaning the private vehicle owners to switch over to the public transport. Bangaluru city is also planning for a Metro. Kolkata already has a good Metro system built nearly 15 years ago.

44.2. Fares and Subsidies

Public transport, whether by bus or rail, in any city is a losing proposition for many reasons. The average fare level by buses is about 20 paise per km, whereas the cost of operation is generally higher. As a result, most of the public transport bus systems in the Indian cities have incurred huge losses in the past. The suburban rail transport in India is also heavily subsidised. This raises some crucial questions whether the average tax payer of the country should be made to pay for the travel of the sub-urban commuter and whether it is wise to subsidise sub-urban travel when it is clear that it will attract more people to the cities. These policy issues will have to be resolved in India in the coming future.

The fare structure within a city itself can be a powerful tool for achieving certain planning objectives. For example, a structure which favours long distance trip makers will facilitate the dispersal of residences, while at the same time increasing the load on the existing network. In general, short distance trips are made to be more costly per km than long distance trips.

I. MULTIPLE CHOICE QUESTIONS

1. All of the following are common features of cities in the developing world except:
   (*a*) few of the large cities have modern centers of commerce.
   (*b*) all have experienced massive in-migrations from rural areas.
   (*c*) most are ringed by vast squatter settlements.
   (*d*) all have populations greater than their formal functions and employment bases can support.
2. What type of transport do you use most often?
   (*a*) bus (*b*) train
   (*c*) subway (*d*) other (please specify)
3. How often do you use this mode of transport?
   (*a*) everyday (*b*) few times a week
   (*c*) once a fortnight (*d*) once a month
   (*e*) less often (*f*) other (please specify)

4. What type of ticket do you buy when using this service?

(*a*) single (*b*) return
(*c*) all day (*d*) weekly
(*e*) monthly (*f*) zone card
(*g*) other (please specify)

5. As ticketing systems are becoming more automated, do you think this is making the process easier or harder?

(*a*) easier (*b*) harder
(*c*) no change (*d*) other (please specify)

6. Urbanism is important because

(*a*) it is in cities that many modern challenges to humanity will play out.
(*b*) the pattern of urban activities affects emissions and pollution.
(*c*) the pattern of urban form affects land use and protection of ecosystems.
(*d*) all of the above.

7. Urbanism is described here as

(*a*) the space between buildings
(*b*) the connected system of public and private spaces
(*c*) big cities
(*d*) the public realm

8. A factor that is not critical to the success of urbanism is

(*a*) the level of detail at pedestrian scales
(*b*) the expressive style of buildings
(*c*) the shape and position of buildings, and how they shape public space
(*d*) the aesthetic character of buildings

9. According to the author, which of the following statements is true?

(*a*) the world is completely different today, and there are few important lessons to learn from history.
(*b*) the patterns of historical urbanism offer useful lessons today.
(*c*) we must copy the styles of historical urbanism.
(*d*) the most important thing about urban design is the excitement of a new artistic vision.

10. Garden cities were

(*a*) A human response to real problems in industrial cities.
(*b*) Based on segregation of people and functions.
(*c*) A model for modern low-density suburbs.
(*d*) All of the above.

Answers

1. (*a*)	**2.** (*a*)	**3.** (*a*)	**4.** (*a*)	**5.** (*a*)
6. (*d*)	**7.** (*b*)	**8.** (*b*)	**9.** (*b*)	**10.** (*d*)

II. SOLVED QUESTIONS AND ANSWERS

Q. 1. What are public transport modes?

Ans. Public transport modes include city buses, trolley buses, trams (or light rail) and passenger trains, rapid transit (metro/subways/undergrounds etc) and ferries. Public transport between cities is dominated by airlines, coaches, and intercity rail. High-speed rail networks are being developed in many parts of the world. Most public transport runs to a scheduled timetable with the most frequent services running to a headway (*e.g.,* "every 5 minutes" as opposed to being scheduled for any specific time of the day). Share taxis offer on-demand services in many parts of the world, and some services will wait until the vehicle is full before it starts. Paratransit is sometimes used in areas of low-demand and for people who need a door-to-door service.

Q. 2. What are the types of public transport?

Ans. Public transport consists of urban transport, regional transport and rail transport.

(i) Bus: The national government wants all buses in public transport to be clean and economical by 2020. The bus networks must also be easily accessible to people with a physical handicap.

(ii) Tram: There are trams in Amsterdam, the Hague, Rotterdam and Utrecht. They operate in these large cities because they are suitable for the transportation of large numbers of passengers. A bus is sufficient for fewer passengers.

(iii) Underground (Metro): Of the four large cities in the Netherlands, Amsterdam and Rotterdam have an underground line. The metros operate in a closed system, which does not intersect with normal road traffic. Just as with the bus and the tram, the provinces and urban regions are responsible for the safety of the underground. Moreover, the Inspectorate for Transport, Public Works and Water Management supervises the safety of the metro lines.

(iv) Regional taxi: The regional taxi is public transport on demand and provides door-to-door transport. The regional taxis also go to destinations where regular public transport does not. The price of regional taxis lies between public transport and regular taxis.

(v) Light rail: Light rail lies between the train and tram. Light rail travels over longer distances than the tram but more slowly than the train and it stops at more places. Light rail is therefore suitable for regional routes but not for long distances.

(iv) Train: The number of passengers using the train is growing. The government wishes to accelerate this growth and increase the number of trains to eventually six intercity trains and six sprinters per hour on the busiest routes. Expansion of the railway network, better travel information and Internet in the train should make the train more attractive as an alternative to the car.

Q. 3. What is car pooling?

Ans. Promoting car sharing, also known as car pooling, is when two or more people share a car and travel together. It allows people to have the convenience of the car, but at the same time helps to reduce congestion and pollution through reduction in vehicle kilometers. Car sharing may be organized through affinity groups, large employers, transit operators, neighborhood groups, or large car-sharing businesses (Singh, 2006). Car sharing provides the potential to reduce the costs of vehicle travel to the individual as well as society. In order to promote car sharing, it is important to ensure that sufficient parking places are allocated to vehicles belonging to car sharing groups at nominal or no parking fee.

III. REVIEW QUESTIONS

1. What is the planning for public transport?
2. What are the fares and subsidies?

REFERENCES

1. Baldwin, R.F., Special Contribution on Leeds Mini Bus Experiment. Proceedings of a Symposium held at TRRL, TRRL Report LR 570, Crowthorne, 1973.
2. Demand-Responsive Transportation System and Services, Special Report 154, Transportation Research Board, Washington, 1974.
3. Rao, M.S.V., Transit and Para-transit Planning-Problems and Prospects, Seminar on Transportation Systems in Urban Settlements, Town and Country Planning, New Delhi, 1982.
4. Agarwal, M.K., Urban Transportation in India, Allied Publishers, New Delhi, 1996.
5. Satindra Singh, Detailed Evaluation of Bus Rapid Transit, Delhi, M. Tech. Thesis submitted to Indian Institute of Technology, Delhi, 2009.

45. Intermediate Public Transport in Indian Cities

45.1. Introduction

Intermediate Public Transport (IPT) is a sub-set of public transport, and deals with low technology vehicles that occupy less space and provide for short distance travel. They are a part of the transport system of most of the small towns and cities in India. In fact, the three-wheeler autorickshaw, one of the IPT modes, is extremely popular even in Metropolitcan cities in India. They provide affordable travel to low-income groups. Since they are in inalienable part of towns and cities in the less developed countries, an understanding of their potentialities and problems associated with them is necessary for a Traffic Engineer.

45.2. Types of IPT Vehicles in India

A variety of IPT vehicles are found on the Indian streets. These are :

1. Motor-cycle rickshaw : These are motor cycles of Second World War vintage converted into 4-seater three-wheelers. They have disappeared from the Delhi streets. Speed is about 15 km/hr.
2. Three-wheeler auto-rickshaws : These are very popular in India and can be seen in all towns and cities. Capacity is 2-seater, besides the driver. Speed is 25 km/ hr.
3. Cycle rickshaws : These are pedal cycle vehicles with three wheels and are found in many Indian towns and cities. Seating capacity is two, besides the driver. Speed is about 10 km/hr.
4. Tonga : Horse-drawn vehicle, seen on almost any Indian town and city. Besides the driver, 2 to 4 persons can be accommodated. Speed is about 10 km/hr.
5. Six-seater auto-rickshaws : This has been recently introduced in India. There are two rows of seats with four openings. Six persons can be seated, excluding driver.
6. Trekker : A jeep-type vehicle, powered by diesel engine, having two rows in the front and two rows in the rear at right angles. Seating capacity is 9 excluding driver.
7. Matador Van : A light van, diesel operated, with a seating capacity of 8 excluding driver.
8. Matador Van : A van, diesel operated, with a seating capacity of 15, excluding driver.

45.3. Characteristics of IPT Modes

IPT modes have a useful role to play in the developing countries. Some of them are :

1. Low fuel consumption, as compared to individualised passenger car.
2. Greater manoeuvrability in congested streets.
3. Point-to-point service capacity.
4. Lower cost as compared to taxis.
5. Provision of job opportunities for the weaker sections of the society.

There are certain inherent disadvantages of the IPT modes, Some of them are not fully safe. The three-wheeler autorickshaw, for example, is unstable when turning at high speeds around sharp concerns. Some vehicles are extremely noisy and pollute the air. Research should be done to make them safe and free from pollution.

I. MULTIPLE CHOICE QUESTIONS

1. Which one of the following is an intermediate public transport?

(*a*) motor cycle rickshaw (*b*) three wheeler auto rickshaws

(*c*) cycle rickshaws (*d*) tonga

(*e*) all of the above

2. Trekker is a

(*a*) jeep type vehicle (*b*) powered by diesel engine

Answers

1. (*e*) **2.** (*b*)

II. SOLVED QUESTIONS AND ANSWERS

Q. 1. What are the characteristics of IPT modes?

Ans. IPT modes have a useful role to play in the developing countries. Some of them are:

- Low fuel consumption, as compared to individualize passenger car.
- Greater maneuverability in congested streets.
- Point-to-point service capacity.
- Lower cost as compared to taxis.
- Provision of job opportunities for the weaker sections of the society.

Q. 2. What is trekker?

Ans. Trekker is a jeep-type vehicle, powered by diesel engine, having two rows in the front and two rows in the rear at right angles. In trekker, seating capacity is 9 excluding driver.

Q. 3. Discuss the strategies for urban transport.

Ans. Strategies to meet the urban travel demand cannot succeed without the fullest co-operation of the general public. Such co-operation is best secured if the objective of any initiative is made clearly known to them so that they are able to appreciate the likely benefits of certain travel choices or sacrifices that they may have to make. Mechanisms for greater public participation in the city specific initiatives, such as the *Bhagidari scheme* in Delhi, would help bring about greater public co-operation and support. Intensive awareness campaigns that educate people on the ill effects of urban traffic congestion, especially on their health and well being could be useful. Such campaigns could also build in modules to encourage individuals, families and communities to adopt 'Green Travel Habits' that would go a long way in making travel less polluting and damaging.

Q. 4. How can you promote the use of non-motorized modes?

Ans. Yet another strategy to reduce personal motor vehicle use is to promote the use of non-motorized modes as they are 'greener' modes of travel. This can be done by investing in a segregated right of way for bicycles and pedestrians, converting crowded areas like market places into no vehicle zones, bringing about improvements in bicycle technology, providing safer parking facilities for bicycles at work places and finally promoting cycling and walking as healthy and exciting activities.

Q. 5. What is the importance of auto-rickshaw in urban transport?

Ans. The Importance of the auto-rickshaw in Urban Transport Several studies have demonstrated that IPT, and auto-rickshaws in particular, play an important role in the transport needs of Indian cities. While the former are flexible demand-based services where the passenger determines the destination, the latter are characterised by shared fixed-route services with intermediate stops for boarding and alighting. While most Indian cities have auto-rickshaws operating on the contract carriage model, Kolkata is unique in that all auto-rickshaws are required to operate on the latter model as is discussed in detail later in this report.

Q. 6. What will it take to create smart cities in India?

Ans. With the right priorities and development strategies, India's smart cities initiative has the potential to make cities more sustainable, inclusive, and safe. In June 2014, the Government of India announced its ambitious plan to build smart cities across the country. This plan will be administered by the Ministry of Urban Development, and will focus on building new smart cities and redeveloping existing urban regions with populations of over 100,000 people.

The program will include city-wide developments to improve methods for engaging citizens and addressing their needs, building capacity among city officials, and moving to e-governance through use of technology. Cities will also be asked to develop holistic city development plans with a vision for zero emissions and zero waste. They will then be provided funding to pursue strategic infrastructure projects as part of the city development plan. Cities will be asked to identify priority sectors and will receive viability gap funding from the central government. The private sector is expected to invest another INR 1000–1200 crore.

Q. 7. How is the demand for mobility services in Indian cities expected to grow? What role do taxi aggregator services like OLA play within the larger urban mobility network?

Ans. The exponential rise of OLA's service is clear evidence of the mobility service vacuum existing in India. These are exciting times and the mobility revolution is about to begin. India is on its growth trajectory, the new government and its policies are business and investment friendly, and mobility will become one of the most important factors for growth. We are not just aggregators but an online transport facilitation platform, which is a much wider term embracing all possible modes of transport that can be accommodated by our platform. We look forward to playing an important role in the growth story of India.

III. REVIEW QUESTIONS

1. What is meant by type IPT vehicle in India?
2. What are the characteristics IPT modes?

REFERENCES

1. Report of the National Transport Policy Committee, Planning Commission, Government of India, New Delhi, 1980.
2. Planning for Operation of Intermediate Public Transport, Case Study : Delhi School of Planning and Architecture, New Delhi, 1981.
3. Rao, M.S.V., Transit and Para-transit Planning, Problems and Prospects, Seminar on Transportation Systems in Urban Settlements, Towns and Country Planning Organisation, New Delhi, 1982.

46. Traffic and the Environment

46.1. Introduction

The automobile is doubtless one of the outstanding inventions of man. It has become an inalienable part of modern life and one can well imagine what life would be in the present times without this wonderful invention. Business, commerce and industry, social life of the people and their need to travel are all served by the motor vehicle. The number of motor vehicles is increasing at an alarming proportion, and there is no reason to believe that a new technology will replace the motor driven by the internal combustion engine in the near future. The oil crisis has spurred the research activity to find alternative mode of propulsion of these vehicles, but the need for the vehicles itself is taken for granted. This is the background against which one must examine the problems created by the fast growth of vehicle population. These problems have a direct impact on the environment in which we live. Unless the environment is kept healthy and safe by applying suitable corrective measures, there is a danger that the motor vehicle which is hailed as a boon to mankind might turn out to be the worst offender in defiling nature and its beauty. The traffic engineer has a great responsibility in ensuring that the road traffic does not spoil the environment in which people live and transact their day-to-day business.

46.2. Detrimental Effects of Traffic on the Environment

The following detrimental effects of traffic on our surrounding have been recognised to be of importance to merit serious study :

(*i*) Safety ; (*ii*) Noise ;

(*iii*) Air pollution ; (*iv*) Vibration ;

(*v*) Visual intrusion and degrading the aesthetics ;

(*vi*) Severance.

Safety of road users has been seriously endangered by the motor vehicle and because of its great importance this subject is dealt with separately in Chapter 17. This Chapter deals with the remaining issues.

46.3. Noise

46.3.1. Noise is unwanted sound. (Ref. 1) The increasing urbanization in cities and the growth of activities therein have contributed to the mounting volume of noise, intruding upon the quiet life and privacy of the urban dwellers. The annoyance and discomfort caused by noise can at times assume serious proportions meriting urgent attention. Noise in cities is the result of a number of activities such as road traffic, aircraft, railways and industrial and constructional works. The traffic engineer is concerned with the abatement of noise generated by road traffic, which is often the predominant source of annoyance (Ref. 3).

46.3.2. Effect of Noise

The detrimental effects of traffic noise can be considered under the following three major groups (Ref. 4) :

(*i*) Subjective effects ; (*ii*) Behavioural effects ;

(*iii*) Physiological effects.

(*i*) *Subjective effects* can be described by terms such as annoyance, disturbance, dissatisfaction, bother and noisiness. But these effects are difficult to be measured precisely, because a wide variation can exist among subjects in describing what level of sound causes them annoyance or discomfort.

(*ii*) *Behavioural effects* cover interference with sleep, speech or any general task. Disturbance in sleep caused by noise is common in high noise areas. If the noise level is high, it takes longer time to fall off to sleep and there are greater chances of awakening. Noise can cause interference in speech and enjoyment of radio and T.V. programmes. It can cause disturbance in studies and intellectual pursuits. In class-rooms considerable distraction can result in the student's mind due to continuous high level of traffic noise. The efficiency in performing general tasks involving mental activity can be impaired by noise (Ref. 4, 5).

(*iii*) *Physiological effects* are those that cause startle or fright phenomena, and can result in harmful effects on various parts of the body. At extremely high level and for long periods of exposure it may produce deafness, but this danger may not be imminent at current levels of noise in cities (Ref. 3, 5, 7).

46.3.3. Generation of Noise by Road Traffic

The generation of noise caused by road traffic can be considered under the following categories :

(*i*) Noise generated by various parts of the vehicle ;

(*ii*) Noise contributed by the interaction between the vehicle and the road surfaces ;

(*iii*) Noise dependent on the speed, flow and density of traffic.

An important source of noise generation is the vehicle itself. The elements that contribute to vehicle noise are :

- Engine (power unit), especially during acceleration ;
- Inlet ;
- Exhaust ;
- Propulsion and transmission, including gears etc. ;
- Brakes ;
- Horns ;
- Chassis ;
- Body structure ;
- Loads in the vehicle ;
- Door slamming.

Some of the facts related to noise generation by vehicles are summarised below :

(*i*) Motor cycles (Ref. 3) and scooters are generally noisier than passenger cars.

(*ii*) Because of the large horsepower of the diesel engine and the heavy loads they carry, commercial trucks are a main source of noise. (Ref. 2, 5, 6, 8).

(*iii*) As vehicles grow older and their mechanical condition deteriorates, the noise generated becomes more.

The tyre-road surface interaction is a major generator of noise. Smooth surfaces generally produce less noise than rough ones. (Ref. 5) Grooved cement concrete pavements have been particularly found to be a source of annoying noise to the neighbourhood. (Ref. 6). Wet surfaces also contribute to noise. (Ref. 5)

As the traffic volume (flow) increases, the noise level inevitably rises. Higher speeds also are known to cause higher noise level. (Ref. 6) It is believed that at lower speeds the influence of engine and transmission noise is predominant whereas at high speeds the tyre surface interaction assumes greater importance. (Ref. 8, 9) Noise levels increase during acceleration (Ref. 6).

46.3.4. Measurement of Noise Levels

The unit of measurement of noise is the decibel (dB), which is the unit of sound pressure level. In order to account for the ear's response at low and high frequencies, different weighing filters, *A*, *B* and *C* are used. *A* weighted decibel, dB (*A*), is commonly employed for measuring the relative levels of noise produced by different traffic conditions. Another useful concept is the mean sound level in dB (*A*) exceeded for a stated % of the sampling periods throughout the whole day. For instance, L_{10} and L_{50} are the mean sound levels in dB (*A*) exceeded 10 and 50% respectively, of the sampling periods throughout the whole day. The L_{10} criterion is mostly used in recent years. (Ref. 6) The instrument used for recording sound levels is the sound level meter.

People's judgment of what acceptable noise is and what is not varies. Normally, noise having a level of not more than 60 dB (*A*) will be rated by many as quiet. (Ref. 3) Noise levels of 90 dB (*A*) and above will be rated by many as extremely noisy.

In 1961-62, a survey of the noise in the London Area was carried out, and the results of the noise levels recorded at various types of streets are given below (Ref. 3), to illustrate the range of levels of noise that may be expect in urban situations.

Table 46.1. Noise Level in London Area at Locations where Traffic Noise Predominates (*Source* : Ref. 3)

Group location	*Noise level dB (A)*	
	Day	*Night*
A. Arterial roads with many heavy vehicles and buses	80–68	68–60
B. (*i*) Major roads with heavy traffic and buses (*ii*) Side roads within 15–20 yard of A or *B* group roads	75–63	61–48
C. (*i*) Main residential roads (*ii*) Side roads within 20–50 yard of heavy traffic routes (*iii*) Courtyards of blocks of flats screened from direct view of heavy traffic	70–60	54–44
D. Residential roads with local traffic only	65–57	52–44
E. (*i*) Minor roads (*ii*) Gardens of houses with traffic routes more than 100 yard distant	60–52	48–43
F. Parks, courtyards, gardens in residential areas well away from traffic route	55–50	46–41
G. Place of few local noises and only very few distant traffic noise.	50–47	43–40

46.3.5. Acceptable Levels of Noise

The standards being followed by different countries regarding the acceptable levels of noise vary. In U.S.A., the Federal Highway Administration has prescribed the following limits (Ref. 8) given in Table 46.2.

Table 46.2. Noise Limits Prescribed by the Federal Highway Administration (U.S.A.) (*Source* : Ref. 8)

	Description of area	*Exterior limits of noise dB (A)*
1.	Parks and open spaces where quietness is of primary importance.	60
2.	Residential areas, hotels, motels, schools, churches, libraries, hospitals, recreation areas, playgrounds, active sports areas and parks ; additionally, the interior design noise level in the buildings mentioned, in this category must not exceed dB(*A*) 55.	70
3.	Developed areas not included in the foregoing categories.	75

The Wilson Committee's (Ref. 2) recommendations for noise in terms of L_{10} levels are given in Table 46.3.

Table 46.3. Acceptable Noise Levels Recommended by Wilson Committee

Situation	*Noise levels dB(A)*	
	Day	*Night*
Country Areas	40	30
Suburban Areas	45	35
Busy urban Areas	50	35

These levels are for the noise inside the buildings.

From the available data it can be assumed that a level of 50 dB(*A*) is desirable, but 55 dB(*A*) is an upper limit. (Ref. 7). With these values the acceptable exterior limits of noise would be about 60—65 dB(*A*).

46.3.6. Control of Traffic Noise

46.3.6.1. Techniques available for control of traffic noise can be considered under the following headings :

(*i*) Changes in design of vehicles ;

(*ii*) Changes in tyres or road surfaces ;

(*iii*) Elimination of noisier vehicles ;

(*iv*) Modifications in traffic operation ;

(*v*) Designing streets, buildings and areas for producing less noise.

46.3.6.2. Changes in design of vehicles

Legislative measures in enforcing vehicle manufacturers to design the vehicles for a maximum noise level can prove effective if they are enforced properly. Many countries abroad have enacted such measures. As an example,

Table 46.4 gives the acceptable noise levels recommended by the European Economic Committee (Ref. 4) :

Table 46.4. Acceptable Noise Levels for Various Vehicles

(*Source* : Ref. 4)

	Class of vehicle	*Acceptable noise levels (measured at a distance of 7.5 m) dB(A)*
1.	Passenger vehicles with seating capacity for not more than 9 persons including driver.	82
2.	Passenger vehicles with seating capacity for more than 9 persons including driver and a maximum permissible weight not exceeding 3.5 tons.	84
3.	Goods vehicles with a maximum permissible weight not exceeding 3.5 tons.	84
4.	Passenger vehicles with seating capacity for more than 9 persons including the driver, and a maximum permissible weight of more than 3.5 tons.	89
5.	Goods vehicles with a maximum permissible weight of more than 3.5 tons.	89
6.	Passenger vehicles with seating capacity for more than 9 persons including driver and powered by an engine of 200 H.P. DIN or over.	91
7.	Goods vehicles powered by an engine of 200 H.P. DIN or over and having a maximum permissible weight of over 12 tons.	91

For motor cycles, the noise limits U.K. in 1970 (Ref. 4) given in Table 46.5 may regarded as indicative of trend :

Table 46.5. Noise Limits in U.K. in 1970 for Motor Cycles

	Description of motor cycle	*Noise limits (1970)*
(*a*)	Not more than 50 c.c.	77
(*b*)	More than 50 c.c. but not more than 125 c.c.	82
(*c*)	More than 125 c.c. but not more than 500 c.c.	86
(*d*)	More than 500 c.c.	86

In India, the Motor Vehicles Act (Ref. 10) makes provision for the Governments to frame rules to control the noise produced by motor cycles.

46.3.6.3. Changes in tyre/surface characteristics

Changes in tyre tread design and composition can bring about a reduction in noise level. Smoother surfaces result in less noisier roads, as already pointed out. But this requirement is against the skid resistance requirement.

46.3.6.4. Elimination of noisier vehicles

As vehicles become old, their noise level increases (Ref. 4).

Legislative measures can prevent such vehicles from being used on the road.

46.3.6.5. Modifications in traffic operations

Some of the measures that are possible under this heading are :

(*i*) Rerouting of commercial vehicles and buses away from the predominantly residential areas.

(*ii*) Providing bypasses and ring roads to siphon off the through traffic entering the towns.

(*iii*) Providing synchronized traffic signals at street crossings to reduce the stop-and-go traffic flow, and consequently eliminating acceleration noise.

(*iv*) Prohibiting blowing horns by proper signing and enforcement measures.

46.3.6.6. Designing streets, buildings and areas for producing less noise

These measures are essentially town planning measures. Narrow streets create noise conditions, a situation commonly known as the 'canyon effect'. When planning new streets, it is desirable to plan for wider streets. A related requirement is the height of buildings and environmental standards has been developed by Buchanan. (Ref. 5) With proper research and studies, this concept can be applied fruitfully to improve the noise conditions in many streets of our cities. Noise screens such as grass, trees or shrubs planted alongside the streets can serve as barriers to the propagation of sound, apart from the improvement of the general scenery they bring about.

46.4. Air Pollution

46.4.1. Pollution of the atmosphere by fumes and smell emitted by the motor vehicles makes the urban streets extremely unpleasant. With further growth of the vehicle population, the problem is bound to assume serious proportions. In special situations such as in tunnels, ventilation is a major issue.

46.4.2. Major Pollutants

The major source of the pollutant is the exhaust gas emitted by the internal combustion engine, although evaporative losses from the fuel tank and the carburretor and losses from the crank case account for some proportion of the hydrocarbons (Ref. 20). The following are the major components of the exhaust gas :

Carbon dioxide }
Water vapour }
Unburnt petrol }
Organic compounds produced from the petrol
Carbon monoxide
Oxide of nitrogen
Lead compounds
Carbon particles (smoke).

46.4.3. Effect of Pollutants

Carbon dioxide, CO_2, is the most important contributor to climate change. Under the 1997 Kyoto protocol, specific targets have been set for reducing the emissions, but the achievement of these targets involves very strict control measures, particularly in the automobile sector. Carbon monoxide, oxides of nitrogen and lead compounds are substances which are of concern from the point of view of the road user and his health. Small doses of carbon monoxide as are present in the air due to road traffic may not be a medical danger (Ref. 11), but can cause minor effects. Similarly, the concentration of nitrogen oxides due to road traffic being small, their presence does not appear to be a danger to health, but the effect of long-term exposure might cause anxiety. Lead compounds in small concentrations as are present in the city streets may not be toxic, but the increase in the concentration may ultimately affect the well-being of the residents. Smoke contains minute particles of carbon, and though by itself it is not a health hazard, it may prove dangerous in combination with some other compounds. Smog is a result of the combination of smoke and fog and can cause hazards to driving and irritation to the eyes.

46.4.4. Measures for Controlling Air Pollution

The following are some of the measures for controlling air pollution from road traffic :

(*i*) Reducing the pollutants at the source, which is the exhaust, the evaporation losses and crank case losses, by improving the vehicle design and maintenance.

(*ii*) Use of small cars instead of bigger ones.

(*iii*) Patronage of public transport system, especially tubes and railways.

(*iv*) Use of alternative fuels and methods of propulsion.

Table 46.5. Air Quality Standards in the U.S.

(*Source* : Ref. 12)

Pollutant	*Concentration limit*		
	Microgram per cubic metre	*Parts per million*	*Averaging time*
Carbon monoxide	10,000	9	8 hours[a]
	40,000	35	1 hour[a]
Photochemical oxidants	160	0.08	1 hour[a]
Hydrocarbons (methane free)	160	0.24	3 hours[a]
Nitrogen oxides	100	0.05	1 year
Sulphur oxides	80	0.03	1 year[a]
	365	0.14	24 hours[a]
	(60)	(0.02)	(1 year)
	(260)	(0.1)	(24 hours)[a]
	(1300)	(0.5)	(3 hours)[a]
Particulate matter	75	–	1 year[b]
	260	–	24 hours[a]
	(60)	–	(1 year)[b]
	(150)	–	(24 hours)[a]

Note. Secondary standards are shown in parenthesis.

a : Not to be exceeded more than once a year.

b : Geometric mean.

(*v*) Staggering work hours, to reduce peak hour traffic ;

(*vi*) institution of parking restraint to promote public transport.

(*vii*) Restraining traffic through road pricing.

(*viii*) Controlling idling engines, by insisting that engines should be stopped during traffic delays.

(*ix*) By constructing bypasses and ring roads to reduce traffic in the mid-town.

Many countries, including U.S.A. and Japan (Ref. 8) have enacted legislative measures for restricting the emission of pollutants from vehicles.

The national primary and secondary air quality standards adopted by the U.S. Environmental Protection Agency (Ref. 12) are given in Table 46.5.

In U.K. there is legislation for controlling smoke emitted by vehicles. (Ref. 11). In India, the Motor Vehicles Act (Ref. 10), makes provision enabling the State Governments to frame rules to control the emissions such as smoke, visible vapour, sparks, ashes, grit and oils. The Air (Prevention and Control of Pollution) Act of 1981 enacted by the Central Government gives necessary legal provisions in India.

46.5. Vibration

46.5.1. A vehicle moving on a road surface induces vibrations in the surroundings. These are of the following types (Ref. 13) :

(*i*) vibrations generated in the contained air ;
(*ii*) surface vibrations ;
(*iii*) underground vibrations.

On narrow streets flanked by buildings, the air contained between the buildings is vibrated when vehicles move on the streets. Such vibrations rarely cause structural damages, but may be annoying to the people. Surface vibrations are those set up on structures above the ground, whereas underground vibrations are set up in the soil mass and the foundations resting thereon.

46.5.2. Surface finish of road surfaces and its effect on generation of vibrations

Irregularities in the road surface are the major cause for vibrations. A smoothly finished surface does not generate vibrations large enough to be preceived by people or cause any damage to the structures nearby. The effects of traffic induced vibrations are cracking of plaster, rattling of doors, windows and glass panes. If the traffic is heavy and consists of a large number of commercial vehicles, the vibrations set up may induce sympathetic resonance in the buildings, causing unpleasantness to the inhabitants and even structural damage. Vibrations can be a serious problem in bridge structures giving rise to overstressing and fatigue. Vibrations can also be a problem in buildings housing scientific laboratories.

46.5.3. Levels of vibration associated with road traffic

A summary of the reactions of people and the damage to buildings at various levels of vibration normally associated with traffic is given in Table 46.6 (Ref. 14).

Table 46.6. Reaction of People and Damage to Buildings at Various Vibration Levels (*Source* : Ref. 14)

Peak Particle Velocity (in the vertical direction) mm/sec.	*Human reaction*	*Effect on buildings*
0 to 0.15	Imperceptible by people–no intrusion.	Vibrations unlikely to cause damage of any type.
0.15 to 0.3	Threshold of perception possibility of intrusion.	Vibrations unlikely to cause damage of any type.
2.0	Vibrations preceptible.	Recommended upper level the vibration to which ruins and accident monuments should be subjected.

2.5	Levels at which continuous vibrations begin to annoy people.	Virtually no risk of 'architectural' damage to normal dwelling houses.
5	Vibrations annoying to people in buildings (this agrees with the levels established for people standing on bridges and subjected to relatively short periods of vibrations)	Threshold at which there is a risk of architectural' damage to normal dwelling-house with plastered walls and ceilings. Special types of finish such as lining of walls, flexible ceiling treatment etc., would minimise 'architectural' damage.
10–15	Vibrations considered unpleasant by people subjected to continuous vibrations and unacceptable to some people walking on bridges.	Vibrations at a greater level than normally expected from traffic, but would cause "architectural" damage and possible minor structural damage.

46.5.4. Ameliorative measures

The following are some of the ameliorative measures that are possible for minimising the effect of vibrations :

(*i*) Improvement in the surface finish of the road.

(*ii*) Reducing the number of heavily loaded commercial vehicles in the badly affected zones.

(*iii*) Improvement in the suspension system of the vehicles to damp out the vibrations.

(*iv*) Provision of wider streets.

(*v*) Enforcing a certain minimum space between adjacent buildings.

46.6. Visual Intrusion and Degrading the Aesthetics

46.6.1. An imperceptible, but yet significant, effect of road traffic in the urban and rural areas has been the steady intrusion that has taken place in the visual surroundings. In the urban sphere, the motor vehicle has been competing for space for movement and it appears that buildings seem to rise from a plinth of cars (Ref. 5). The beauty of monumental and historical buildings has been mixed by the clutter of the iron and steel of these man-made machines. The tranquility and openness of parks and squares has been invaded and spoiled. Service stations, garages and petrol filling stations have sprung up along the road and added their mite to the degradation of the general scene.

46.6.2. To attract the attention of the motorist numerous signs, signals and bill-boards have sprung up all along the streets, marring the beauty of the surrounding landscape, historical landmarks and architectural masterpieces. Destruction of vegetation causing land-slides and soil erosion are some of the common bad effects (Ref. 15).

46.6.3. It is possible to guard the surroundings from the ill-effects of visual intrusion by collaborative planning amongst traffic engineers, town planners and architects (Ref. 19). The needs of traffic and the buildings they serve should be considered comprehensively and a suitable 'traffic architecture (Ref. 5) devised. The form and shape of the buildings and the requirements of the residential areas should not be dictated by traffic but should be related to it. Traffic should be subordinated to the environmental needs.

The visual ill-effects of parking can be got over by designing suitable off-street facilities for parking and enforcing regulations for on-street parking. These are dealt with in detail in Chapter 12 and 16.

It is possible to control unsightly bill-boards by suitable regulatory measures. An example of such a practice can be seen in the I.R.C. Policy on Control of Roadside Advertisements (Ref. 16).

Landscaping of rural highways can preserve and enhance the aesthetics. Proper care is needed in route location and geometric design of highways to preserve the natural beauty.

Ribbon development along highways is a result of growth of road traffic and has many ill-effects. A review of the problem and methods available for mitigation is given in the I.R.C. Special Publication No. 15 (Ref. 17). Some of the measures recommended are :

(*i*) Acquisition of adequate right of way ;

(*ii*) Control of access ;

(*iii*) Construction of bypasses around congested town ;

(*iv*) Provision of parallel service roads to meet the needs of local traffic.

(*v*) Prevention and removal of encroachments.

46.7. Severance and Land Consumption

46.7.1. Severance is the general term denoting the psychological, cultural and physical disturbance caused by a traffic facility on the neighbourhoods, the land, the society and its life styles. Highways and streets consume an enormous amount of land, and the compulsory acquisition of the some can make in roads into the homogeneous and closely knit life of a community and can cause disruption to its way of life. The amount of land devoured by streets and highways is so large that it has created problems of rehabilitation and relocation. There are other competing sectors where land would be greatly needed, for example urban slum clearance, housing and agriculture.

46.7.2. While the needs of land for streets and highways cannot be set aside and have to be reckoned with, the following guidelines can be kept it view in planning new facilities or improvement of existing ones to keep the effects of severance and land consumption as low as possible :

(*i*) Historic and cultural sites areas must be preserved.

(*ii*) Severance of school playgrounds from the school precincts should be avoided.

(*iii*) Whenever a new road is being located through a closely knit and homogenous locality the special needs of the community for the pursuit of their daily life (*i.e.*, educational medical, cultural, religious needs) should be carefully looked into and no undue dislocation caused (Ref. 15).

(*iv*) Recreational areas (parks etc.), virgin forest lands and conservation lands should not be disturbed by highway location.

(*v*) Where severance is unavoidable, suitable provision should be made for people to cross the road. Underpasses and over-bridges should be provided to mitigate the hardships.

(*vi*) The highways should be planned such that they have the least impact on the ecological system.

46.8. Evaluation Procedures

Measures to protect and improve the environmental condition cost money. At the same time, bad environmental conditions cost the society money in terms of economic loss. If scare resources have to be allotted to any scheme that scheme should be economically justifiable. Difficulties arise in assigning monetary value to benefits from schemes aimed at improving the environment. In recent times, considerable thought is being given to quantify the benefits into monetary terms, and evaluation procedures are becoming easier.

46.9. Environmental Areas

The Buchanan Report (Ref. 5) was a significant step in dissecting the problem of traffic in towns and laying open the malady as it exists today. One of the bold concepts developed in the report is of the 'environmental area'. To put it in the words of the report itself, environmental area is "an area having no extraneous traffic, and within which considerations of environment (in the specialised sense, as defined) predominate over the use of the vehicles". The areas should be safe, reasonably free from noise, and generally agreeable. Access to these areas should be planned in such a way that traffic does not degrade the environment. This implies that the environmental areas should be tied together by the interlacing network of distributory roads, into which all longer movements would be canalised without choice. A hierarchy of roads should in fact be thought of whereby the major links serving the localities should be designed for swift and efficient movement, but the minor roads planned to give direct access to the buildings. Environmental standards should be thought of for the comfort and convenience of the residents of the locality, and in order to achieve this the environmental capacity of the roads should be determined.

46.10. Situation in India

Compared to the developed countries, the traffic situation in India, is just at the bottom level.

Another happy feature in India is that urbanisation has not yet reached unmanageable proportions. The urban population in India is about 27%, whereas in other countries it is much more. The environmental pollution has already started becoming a serious problem in major cities, and road traffic is major contributor to this. The carbon monoxide level in Kolkata is reported to be 35 p.p.m. in certain streets (maximum one hour value), as against 27 p.p.m. in New York. (Ref. 18) The standard of design and maintenance of road vehicles in this country is obviously to be blamed for this sorry state of affairs. No legislative measures are in force in India to control the noise or emissions from vehicles. Visual intrusion and severance are issues which have not been given serious thought so far. Ribbon development plagues most of our highways. There is virtually no legislative measure to regulate the access to roads.

It may be mentioned that making it compulsory for buses in Delhi to use Compressed Natural Gas (CNG) instead of diesel has reduced air pollution significantly.

Under the present context, a considerable amount of work needs to be done in fostering healthy environment along our streets and highways. The I.R.C. has done some work in this regard and has brought out the following publications :

- Ribbon Development and its Prevention ;
- Control of Roadside Advertisements.

The work done as far is meagre. Further research and study needs to be channelised in the following directions :

(*i*) Determination of present noise levels in the streets of towns and cities to understand the extent of the problem.

(*ii*) Determination of the present levels of air pollution in the cities and the contribution of road traffic to this state of affairs.

(*iii*) An evaluation of the problem of vibrations of buildings, caused by road traffic.

Legislative measures need to be introduced in the following fields :

- Fixation of the noise levels of new vehicles and old vehicles.
- Fixation of the limits for emission of pollutant by vehicles.

I. MULTIPLE CHOICE QUESTIONS

1. In desire-line diagram

(*a*) width of desire-line is proportional to the number of trips in one direction

(*b*) length of the desire-line is proportional to the number of trips in both directions

(*c*) width of desire-line is proportional to the number of trips in both directions

(*d*) both length and width of desire-line are proportional to the number of tirps in both directions

2. Consider the following situations:

1. Traffic volume entering from all roads is less than 3000 vehicles per hour.
2. Pedestrian volume is high.
3. Total right turning traffic is high.
4. A road in a hilly region.

A rotary will be more suitable than control by signals, in situations listed against

(*a*) 1 and 3 (*b*) 1 and 4 (*c*) 2 and 4 (*d*) 2 and 3

3. Ratio of the width of the car parking area required at cerb for 30° parking relative to 60° parking is approximately

(*a*) 0.5 (*b*) 0.7 (*c*) 0.8 (*d*) 2.0

4. For the relationship $u = 55 - 0.44$ k, where '*u*' is the speed in kmph and '*k*' is density in vpkm, what will be the maximum flow in vph?

(*a*) 1718 (*b*) 1250 (*c*) 625 (*d*) 125

5. Consider the following statements:

Collision diagram is used to

1. study accident pattern
2. eliminate accidents
3. determine remedial measures
4. make statistical analysis of accidents

Which of these statements are correct?

(*a*) 1 and 2 (*b*) 1 and 3 (*c*) 3 and 4 (*d*) 2 and 4

6. The capacities of "One-way 1.5 m wide sidewalk (persons per hour)" and "One way Two lane urban road (PCU per hour, with no frontage access, no standing vehicles and very little cross traffic)" are respectively
   (*a*) 1200 and 2400 (*b*) 1800 and 2000
   (*c*) 1200 and 1500 (*d*) 2000 and 1200
7. The shape of the **STOP** sign according to **IRC:** 67-2001 is
   (*a*) circular (*b*) triangular (*c*) octagonal (*d*) rectangular
8. The road geometrics in India are designed for the
   (*a*) 98th highest hourly traffic volume
   (*b*) 85th highest hourly traffic volume
   (*c*) 50th highest hourly traffic volume
   (*d*) 30th highest hourly traffic volume
9. The speed and delay studies on a defined section of highway are conducted by
   (*a*) radar gun (*b*) traffic counters
   (*c*) moving car method (*d*) enoscope

Answers

1. (*c*)	2. (*a*)	3. (*c*)	4. (*a*)	5. (*b*)
6. (*a*)	7. (*c*)	8. (*d*)	9. (*c*)	

II. SOLVED QUESTIONS AND ANSWERS

Q. 1. Write a brief note on the environmental impacts and protective measures.

Ans. One particular characteristic of the environmental damage caused by road traffic is that the impact is made up of a large number of minor individual factors. Sectors of the infrastructure connected with the traffic system (refineries, fuel transportation, workshops, dumps) also make a contribution. The assumption that these cause only a negligible fraction of the environmental burden will result in road users failing to change their customary behaviour in the way that is needed.

Q. 2. What are air pollutants?

Ans. Air pollutants are not only harmful to human beings, they can also present a hazard to animals, the soil, vegetation and the climate. Remote effects (*i.e.*, remote both in time and in space) must be considered; one must also remember that combinations of air pollutants may be more harmful than the individual contaminants (synergetic effects). The principal air pollutants emitted by vehicular traffic are as follows:

- Carbon monoxide is not harmful to bloodless animals and plants, but sustained concentrations, even if small, can have an impact on both humans and other vertebrates. Compounds formed with the haemoglobin in the blood restrict the oxygen supply to the tissues, and in this way can cause disorders of the central nervous system.
- Certain hydrocarbons, resulting from incomplete fuel combustion (such as benzol), are carcinogenic among humans.
- Nitrogen oxides are irritants to human beings, and can damage the organs of respiration. In areas of high population density, nitrogen

oxides contribute to smog formation. Plants are not directly damaged by nitrogen oxides at concentrations below 200 μg/m^3 air.

- Under ultraviolet radiation, ozone is produced from nitrogen oxides and hydrocarbons. High ozone concentrations may cause smog and afflict the respiratory tract and circulation in humans. Moreover, ozone can cause serious damage to woodland and also, if present in sufficiently high concentrations, to crops such as vegetables and tobacco plants. Ethylene can have an unfavourable effect on the growth and ageing of plants. The numerous chemical interactions involved have not yet been fully explained.
- Lead and other heavy metals, which are deposited in the bones, are poisonous to humans and can cause disorders of the central nervous system. Lead aerosols are dispersed mainly over an area up to 30 to 50 meters from the roadway, while traces can be carried as far as 100 to 200 meters, depending on wind conditions. Plants absorb lead from the air by surface contact (dust deposits), or from the soil through their roots. As well as growth damage to the plant itself, follow-on effects through absorption into the food chain should also be noted (*e.g.*, milk produced from pastures near busy roads).

Q. 3. What are the measures to prevent noise pollution?

Ans. The measures to prevent noise include the following:

- regular technical inspection of vehicles,
- selection of low-noise road surfaces,
- speed restrictions in relevant areas,
- road layout and gradients optimised to help avoid unnecessary gear changes,
- reduction of through-traffic in the vicinity of town centres, residential areas, hospitals, schools and places of worship, as well as other areas sensitive to noise,
- restricted access at certain times to protected areas (e.g., heavy goods vehicles banned from inner urban areas at night),
- driver education, for lorry drivers in particular, but also for bus and coach drivers.

Q. 4. What is traffic safety? How does it play in reducing the risks of accidents?

Ans. Because of its importance, traffic safety is not usually subsumed under the heading of the "environment", being regarded as a sector in its own right. Nevertheless, traffic safety, which covers more than the prevention of accidents, is an important aspect of the quality of life, and therefore needs to be discussed here.

A vital means of reducing the risk of accidents for all groups of transport users is the imposition of vehicle speed restrictions, which differ according to locality (highways outside urban areas, main traffic routes in town and city centres, access routes etc.), and according to the state of development and condition of the road.

In addition, the accident risk can be reduced by intensive road safety training, technical inspection of the vehicles, tougher safety regulations (such as the mandatory wearing of seatbelts, driving tests administered by properly qualified persons or the banning of drink-driving). Furthermore, efforts should be made to improve the working conditions of professional drivers, particularly lorry drivers. Appropriate measures include not only requiring certain standards of comfort in the "workplace" (*e.g.* heat-shielding vehicle roofs), but also properly supervised driver working hours so as to prevent accidents due to fatigue.

Accidents involving hazardous goods transports can put not only human beings and animals at risk, but can also cause local environmental damage on a considerable scale. The same applies to the escape of hazardous substances during regular journeys. In particular, there is a risk of contamination of watercourses and groundwater.

Q. 5. Write a brief note on the environmental brief transport and traffic planning.

Ans. A transport master plan is needed, coordinated with the regional development plan, concentrating primarily on making transport more environment-friendly and less costly to the national economy as a whole.

- Decentralisation of residential, industrial and supply facilities, with the aim of reducing transport distances;
- In urban passenger transport, increasing the attractiveness of "environment-friendly combinations" (foot, bicycle, cart, bus, tram and light rail);
- In urban goods transport, deliveries by vehicles which are compatible with urban conditions (small goods vehicles with low noise and pollutant emissions), loaded to the correct capacity. In long-distance goods transport, controlled transfer to rail and ship wherever possible. This calls for effective logistical infrastructural measures to optimise the interfacing of short-distance and long-distance transport and the system interchange points between different transport media (e.g., goods traffic centres);
- For simply structured transport tasks, as in the case of raw materials projects, special transport systems such as pipelines, light railways, or cableways might be not only more environmentally acceptable but also more economical than lorries, railways or ships.

Q. 6. Discuss the assessment of environmental relevance.

Ans. Road traffic is a necessary part of the development and opening up of urban and rural areas. Increased road traffic, which is generally inevitable as infrastructure grows, has far-reaching environmental effects which can only be minimised through extensive efforts in the way of planning, administration (including informing and educating the public) and engineering.

To limit adverse environmental impact effectively, road traffic must be included within an integrated transport plan which adequately takes account of the interactions with other sectors of transport planning, urban planning, and national development planning. When making an overall evaluation of projects, it needs to be borne in mind that the environmental consequences of motorised road traffic are not all bad; in many cases it may also provide a basis for improving living conditions (*e.g.*, through improving supplies of food and medical care) or may be a necessary part of this process (construction of industrial establishments or housing). As with all planning decisions, both the positive and negative effects of road traffic projects need to be weighed up carefully on a case-to-case basis.

III. REVIEW QUESTIONS

1. What are the detrimental effects of traffic on the environment?
2. What is air polllution?
3. What is vibration?
4. Visual intrusion and degrading the aesthetics.
5. Serverance and land consumption.

REFERENCES

1. British Standards Institution, B.S. 661. Glossary of Accoustical terms, London.
2. Noise : Final Report (Wilson Committee), H.M.S.O., London, 1963.
3. London Noise Survey, H.M.S.O., London, 1968.
4. Urban Traffic Noise, Organization for Economic Co-operation and Development, Paris, 1973.
5. Traffic in towns, Buchanan Report, H.M.S.O., London, 1963.
6. Working Group on Research into Road Traffic Noise, A Review of Road Traffic Noise, RRL Report LR 357, Road Research Laboratory, Crowthorne, 1970.
7. M.E. Bult, Roads and the Environment, TRRL Report LR 441, Transport and Road Research Laboratory, Crowthorne, 1972.
8. Effects of Traffic and Roads on the Environment in Urban Areas Organization for Economic Co-operation and Development, Paris, 1973.
9. J.T. Middleton, America's Highways : where are they taking the Environment, Special Report 139, Environmental considerations in planning, design and construction, Highway Research Board, Washington, 1973.
10. The Motor Vehicles Act, 1988.
11. P.T. Sherwood and P.H. Bowers, Air Pollution from road traffic–A review of the present position, RRL Report LR 352, Road Research Laboratory, Crowthorne, 1970.
12. Highways and Air Quality, Special Report 141, Highways Research Board, Washington, 1973.
13. N. Chitharanjan and N.V. Arunachalam. The Impact of Increasing Traffic on Structures, National Seminar on Traffic Enforcement and Environment in Metropolitan Cities, Vol. II, Madras, 1972.
14. A.C. Whiffin and D.R. Leonard, A Survey of Traffic Induced Vibrations, RRL Report LR 418, Road Research Laboratory, Crowthorne, 1971.
15. A Policy on Design of Urban Highways and Artificial Streets, AASTHO, Washington, 1973.
16. A Policy of Roadside Advertisements, IRC : 46-1972,1.R.C., New Delhi, 1972.
17. Ribbon Development its Prevention, Special Publication No. 15, I.R.C., New Delhi, 1974.
18. S.C. Saxena and N.M. Surana, Transportation Modes and Environmental hazards, National Seminar on Traffic Enforcement and Environment in Metropolitan cities, Vol. II, Chennai, 1972.
19. Environmental considerations in planning and design of highways in India, Indian Roads Congress, New Delhi, 1979.
20. Rao M.N. and H.V.N. Rao, Air Pollution, Tata McGraw Hill, New Delhi, 1996.

47. Transportation and Energy

47.1. Introduction

The global energy crisis since the 1970s has posed serious problems to all sectors of the economy where energy is consumed. It has become clear that the known oil reserves have only another 25 years or so to go before they dry up. Coal, however, may last for another 200 years. The human ingenuity is being stretched to the maximum to find out new reserves of oil. Even, so, it is clear that the world appears to be doomed to live in an era of energy shortage, atleast in the near future. The impact of this will be severely felt in the transportation sector which at present is totally dependent on fossil fuel.

Transport is a vital sector of a nation's economy, and it claims a good share of the energy consumed. Globally, the nearly 40 percent of the total energy consumed was contributed by petroleum in the year 2000 (Ref. 1). The principal sources of energy in India are coal, oil, hydro-power and nuclear power. Of course, traditional sources such as animal power, animal dung, bio-gas and fire-wood occupy an important role. The country is happily endowed with a good store of coal, nearly 83,000 million tonnes, which may last a few centuries, at the current rate of consumption of 100–120 million tonnes per year. Hydro-power is also available in plenty in India and the quantity tapped so far is only 10 percent of the potential available.

In India, a large quantity of petroleum products are imported every year. In 2005-06, the import bill of crude oil and refined products was nearly ₹ 2,00,000 crores (US $ 44,000 million), (Ref. 2), representing nearly 40 percent of the country's export earnings. In that year, the total oil requirements stood at 147 mil. tonnes, the indigenous production accounting to 34 mil. tonnes, barely 23 percent of the total requirement.

Oil has come to occupy an important place as an energy source in the transportation sector because of many inherent advantages. Apart from its cheap availability so far, it is convenient to transport and handle. In the period 1991 to 2007, the consumption of petroleum products has grown at an average annual rate of 5 per cent (Ref. 2). In the road transport industry, it has so far no substitute, though efforts are continuing to run battery-powered vehicles. Technology may see a great break-through in this direction in the future, but atleast in the short term, the road transport sector has to manage with an ever-increasing shortage of oil products.

Under these grave circumstances, all possible avenues for economy in fuel consumption in the transport sector must be explored. One of the promising areas is by large scale improvements in the roads of the country. The present chapter sets forth how the various features concerning roads affect fuel consumption and how, by improving the road characteristics, sizeable reduction in fuel consumption can be easily brought about. Reference is also made about other fuel saving measures, both in the road transport and other transport sectors.

47.2. Factors Affecting Fuel Consumption of Motor Vehicles

47.2.1. Before trying to go into the details of the effect of road features on fuel consumption of vehicles, it would be appropriate to consider in total the factors affecting fuel consumption. These factors may be grouped under the following headings (Ref. 3) :

- Vehicle characteristics
- Driver characteristics
- Road characteristics
- Fuel characteristics
- Environmental conditions.

47.2.2. Vehicle Characteristics

Some of the vehicle characteristics which affect fuel consumption are briefly indicated below :

(*i*) **Age of the vehicle :** as the engine becomes old and the other moving parts get worn out, more energy is consumed in propulsion.

(*ii*) **Make :** vehicles of different makes show small variations in fuel consumption due to inherent features of engine design.

(*iii*) Horse power and engine capacity.

(*iv*) **Condition of the engine :** the condition of the engine (tuned or untuned), especially the ignition system, the spark plugs, the transmission system, the condition of the valves etc. affect fuel consumption to a large extent.

(*v*) **Tyres :** the inflation pressure of the tyres, the composition of the rubber used in the tyre, the design of the tyres itself are factors which affect fuel consumption. The tyre comes into close contact with the road surface, and the friction developed between the tyres and the road surface has to be overcome when the vehicle moves. The friction factor is governed by the tyre characteristics indicated above. It has been found by research in the Central Road Research Institute that radial tyres can bring about an economy of about 5–7% in fuel consumption of cars. (Ref. 4).

47.2.3. Driver Characteristics

The manner in which the driver drives the vehicle influences fuel consumption to a great extent. Sudden acceleration and deceleration will increase the consumption of fuel. A steady state speed, especially at optimum value, will result in minimum fuel consumption.

47.2.4. Road Conditions

The important road features which affect fuel consumptions are :

1. **Type of the surface and its roughness.** It is well known that a smooth and well maintained surface provides the least frictional resistance. Hence the fuel consumption on such surfaces will be the lowest. In India, the types of surfaces usually met with are :

(*i*) Ordinary earthen surface
(*ii*) Gravel (moorum surface)
(*iii*) Water bound macadam
(*iv*) Bituminous surface dressing
(*v*) Bituminous thin chipping carpet
(*vi*) Bituminous concrete of a superior type
(*vii*) Cement concrete.

2. **Vertical profile of the road.** When a vehicle has to negotiate an upward gradient, it has to overcome the natural forces of gravity. Hence, the steeper the upward gradient is, the greater is the fuel consumption. On the other hand, the fuel consumption when the vehicle takes a downward gradient is lower than the consumption on a level horizontal surface. This underlines the need for constructing roads with well designed vertical profile. Even though it may seem cheaper to construct roads with undulating vertical profile initially, in the long run it will be far more economical to operate the vehicles if great care is taken to select proper gradients initially.
3. **Width of the road.** The Indian roads are generally of the following widths :
   (*i*) Single lane–3.75 metres
   (*ii*) Intermediate lane–5.5 metres
   (*iii*) Two lane roads–7.0 metres.
   (*iv*) Four lane dual carriageway roads.

   The majority of the roads in India are single lane which forces the vehicles to move on the ill maintained undulating earthen shoulders whenever vehicles have to cross one another or overtake one another. The wear and tear of the vehicles and the fuel consumption mount steeply when the vehicles have to undertake such manoeuvres. This is because of the acceleration and deceleration involved and because of the rough surfaces in the shoulders. The intermediate lane roads do provide a certain amount of relief during crossing and overtaking but severely restrict the speed of the vehicles and thus cause high fuel consumption. The ideal road is a two-lane road where the vehicles can move in their respective lanes having firm and smooth surfaces. Divided carriageways are even better.
4. **Curvature of the road.** The curvature of the road indirectly affects fuel consumption because the vehicles have to slow down the speeds when entering curves and frequent acceleration and deceleration are caused. There is, however, no direct effect of curvature on the fuel consumption.
5. **Type of pavement.** It has been proved from Indian experiments that a saving of upto 14 percent in fuel is possible on concrete pavements as compared to flexible pavements. (Ref. 12)

47.2.5. Fuel Characteristics

The various properties of fuel which affect the consumption are :

1. The octane value
2. The wax contents
3. Sulphur content
4. Temperature of the fuel.

47.2.6. Environmental Factors

Some of the environmental factors which affect fuel consumption are :

(*i*) Altitude of the place : at high altitudes, there is lack of oxygen and thus the vehicles consume more fuel.

(*ii*) Ambient temperature

(*iii*) Rain or dry conditions which affect the speed of the vehicles, and thus indirectly affect the fuel consumption

(*iv*) Wind direction and velocity.

47.3. Effect of Road Condition on Fuel Consumption of Vehicles

47.3.1. An ambitious research project, entitled 'Road User Cost Study' ; has been implemented by the Central Road Research Institute (Ref. 5). This project, jointly sponsored by the World Bank and the Government of India, has cost ₹ 100 lakhs. The work on the project began in 1978 and the final results were made available in 1982. The study was updated in 1992 (Ref. 12).

47.3.2. Since fuel is one of the important components of the vehicle operating costs, this respect has given great importance in the above research project. The objective of the research project concerning fuel is to clearly understand how the fuel consumption is governed by some of the important factors (Ref. 3). Two different approaches have been followed for arriving at accurate fuel consumption experiments.

(*i*) Controlled fuel consumption experiments

(*ii*) User Cost Survey involving data collection from the actual vehicle operators.

Under the controlled experiment, vehicles belonging to the Central Road Research Institute have been run of different types of roads at different speeds and carrying different loads. The fuel consumption under steady state speeds has been noted by means of accurate fuel metres. Based on extensive data thus collected, accurate statistical relationships have been established between the fuel consumption rate and the roadway and vehicle factors which govern the same. (Ref. 6, 7, 8, 9, 12)

Under the User Cost Survey, 939 vehicles (buses, trucks and cars), belonging to different operators located in the various parts of the country have been selected and accurate data on the cost of operation of these vehicles (fuel, tyre, lubricants, spare parts, labour for repairs, depreciation etc.) have been collected. The routes on which these vehicles travelled during the data collection period, which was about 2 years, have been accurately measured. Based on this data base, accurate relationships have been established between each of the vehicle operating cost component and the factors affecting the same. (Ref. 5)

The equations developed from the controlled experiments are given in Table 47.1.

Table 47.1. Fuel Consumption Equations From Controlled Experiments (Refer 6, 7, 8, 9, 12)

S.No.	*Vehicle*	*Eqn.*
1.	Diesel Jeep	$FC = -30.83 + \frac{2258.11}{V} + 0.0242\ V^2 + 0.0012\ RG + 1.2776\ RS - 0.05647\ FL$
2.	Ambassador Car	$FC = 10.31 + \frac{1675.52}{V} + 0.0133\ V^2 + 0.0006\ RG + 1.3879\ RS - 1.0322\ FL$
3.	Premier Padmini Car	$FC = 49.84 + \frac{319.0698}{V} + 0.0035\ V^2 + 0.0019\ RG + 0.9416\ RS - 0.6769\ FL$

4.	Tata Truck	$FC = 85.07 + \dfrac{3904.64}{V} + 0.0207\,V^2 + 0.0012\,RG + 3.3281\,RF - 1.7769\,FL - 6.2396\,PW$
5.	Maruti Car	$FC = 21.85 + \dfrac{504.15}{V} + 0.004957\,V^2 + 0.000652\,RG + 1.0684\,RS - 0.3684\,FL$
6.	Light Commercial Vehicle	$FC = 63.245 + \dfrac{1615.327}{V} + 0.0245\,V^2 + 0.001525\,RG + 5.377\,RS - 3.6336\,PW$
7.	Multi-Axled Vehicle	$FC = 340.33 + \dfrac{2695.79}{V} + 0.0517\,V^2 + 0.0035\,RG + 17.75\,RS - 5.40\,FL - 51.467\,PW$

A typical diagram showing the trend in Fuel Consumption is given in Fig. 47.1. (Ref. 12)

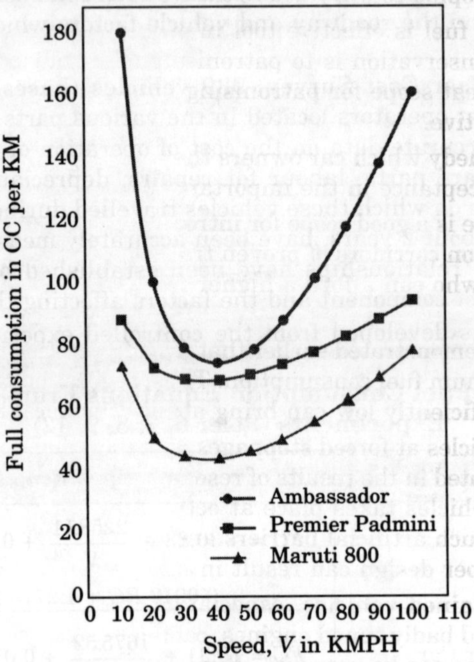

Fig. 47.1. Typical Fuel Consumption Curves.

In the above equations,

FC = Fuel consumption in c.c./per km

V = Speed in km/hr

RG = Roughness of the road, as measured by a Towed Fifth Wheel Bump Integrator in mm/km.

RS = Rise in m/km
FL = Fall in m/km
PW = Power weight ratio in kW/Tonne.

47.4. Measures for Economy of Fuel in Road Transport

Fuel consumption measures in road transport can broadly be considered under the two heads (Ref. 10) :

(*i*) Short-term measures (*ii*) Long-term measures.

Short-term measures are those intended for immediate application where results can be achieved at once. These measures include restriction on driving, rationing of fuel, increase in the fuel price, speed restrictions, car-pooling, patronage of public transport, better vehicle maintenance practice and so on.

Long-term measures are those that do not yield immediate results but involve future prospects of changing travel habits and conducting research on alternative propulsion system and fuels.

Selective restrictions on driving private cars is a widely adopted short-term measure. Banning of private car usage on Sundays and holidays has been tried successfully in countries abroad.

Increases in fuel prices has an immediate effect on private car travel, especially in a developing country where the income of car owners is not very high.

Rationing of fuel is effective tool in conserving fuel. One of the effective measures of fuel conservation is to patronise public transport. In a country like India, there is a great scope for patronising public transport by making it more effective and attractive.

A simple remedy which car owners themselves can organise is car pooling. This has gained acceptance in the important urban cities in India.

In India there is a good scope for introducing intermediate public transport systems, especially on corridors of proven traffic potential. This form of transport will cater to those who can afford a higher fare than the public transport buses and trains.

It has been demonstrated earlier that driving vehicles at the optimum speed results in the minimum fuel consumption. Thus, regulation of speeds and keeping the upper limit sufficiently low can bring about considerable fuel savings.

Idling of vehicles at forced stoppages is yet another source of fuel wastage. This has been indicated in the results of research conducted in India and presented earlier. Idling of vehicles takes place at octroi and other similar check posts in India. Removal of such artificial barriers and minimising the delay at signalised intersection by proper design can result in lesser idling by vehicles.

Badly maintained vehicles, especially the ignition system (power plug, ignition timings) and badly tuned engines, can lead to excessive fuel consumption. At worst, the increase can be upto 20%. Periodic inspection of vehicles can prevent such loss.

Changes in vehicle design can bring about considerable fuel economy. Smaller cars are the vehicles of the coming decades. Streamlined vehicle body shape can be used to minimise aerodynamic drag, which becomes critical at higher speeds and increases fuel consumption. Reduction in the weight of vehicles through the use of lighter material such as aluminium and fibre reinforced plastic is a promising possibility.

It has been established by research in USA (Ref. 13) and in India (Ref. 12) that savings in the consumption of diesel by heavy trucks of the order of 15-20 percent can be achieved by having concrete pavements in the place of flexible pavements. It has been estimated that every year a saving of nearly 0.5 mil. tonnes of diesel can be achieved in India if 14,000 km of the National Highway system is provided with a cement concrete pavement in one of the two carriageways under the four-laning programme (Ref. 14).

Alternative propulsion systems are being researched upon. It is likely that the petrol engines of the passengers cars will be converted to diesel engines in the coming decades, since the diesel engine is known to be more fuel-efficient. Efforts are continuing to make the diesel engine itself more efficient. Other alternatives such as the Wankel rotary internal combustion engine, Sterling cycle or hot air engine, Drayton cycle or gas turbine are also in the research and development stage. Electric propulsion by batteries is another alternative. Hydrogen is the fuel of the next century and continuing research is already in progress. Some of the other fuels which hold promise are liquefied petroleum gas, liquid paraffin, propane, methanol, ethanol, petrol-alcohol mixtures, bio-fuels and so on.

As a permanent long-term measure there is a need for long-term land-use transportation planning for our cities and towns. If the present system of transportation and location of offices and work centres have accentuated the transportation problem, the better remedy seems to be to relocate the residences and the work places so that transportation itself is minimized to a great extent.

47.5. Fuel Economy in Rail Transport

Rail transport is a very highly energy efficient mode. Fortunately a good system of railway communication exists in India. It is possible to improve the efficiency of this system by modernisation of track, signalling and traction equipment. The main sources of fuel for railway traction are coal, diesel oil and electricity. Since there is a large quantity of coal available in the country, and there is a large amount of untapped hydro-electric energy in the country, the most economical proposition in railway traction strategy would be to progressively go in for more electrification. It is preferable to electrify the trunk routes according to a time bound programme. A minimum target of 1,000 kilometres per annum seems to be desirable. It should be borne in mind that electricity can offer a substitute for railway traction whereas for other modes of transport and other sectors of economy such as agriculture, oil seems to be an irreplaceable source of energy at the present.

47.6. Fuel Economy in Other Modes of Transport

Coastal shipping is one of the most energy efficient modes of transport and needs to be encouraged in the country. Like any other form of water transport, the investment in line haul capacity in this mode is almost nil. Only terminal and navigational facilities are needed which can be built up at economic cost.

Measures for fuel economy in air transport consist of replacing old and obsolete air-craft by wide bodied jets. A systematic planning of the routing and fueling operations should be carried out. Since the energy spent in carrying fuel in the air-craft itself is much more than it is in transporting oil on the land, the fueling stations should be so selected that load carried by the aircraft is optimal for maximum economy.

Pipelines and inland-waterways are extremely energy efficient and should be patronised wherever there is a situation where these mode can be used.

Bullock-carts have been the traditional mode of conveyance in the country for centuries and will continue to be so even in the coming decades. It will be a wise policy to encourage this transport mode and improve its efficiency by better design of the axles, wheels and other parts. Pneumatic tyred wheels can increase the load carrying capacity of the cart considerably and at the same time prolong the life of the pavements.

47.7. The Future Outlook for Propulsion Fuel

All modes of transport currently depend upon liquid fossil fuels. At the present rate of consumption, and the future high rate of consumption in the emerging economics (China and India in particular), petroleum crude is going to last only till 2020. Natural gas is probably going to last till about the same time, or a few years further. Gas obtained from biomass, or fuel obtained from Jatropa seeds or sugarcane, is not going to change the scenario to any significant extent. Transport will have to look for alternative fuels to function in a petroleum-less world in the next 20 years or so, touted as an alternative, is costly to produce, store and handle. The only feasible alternatives are electricity from nuclear fission, nuclear fusion and solar heat. Nuclear fission cannot meet the needs for many years, as the world may run out of resources. Thus, only nuclear fusion and solar energy are the long-term viable alternatives. Electricity generated from these can charge batteries to run personalized vehicles or to run trains. More efficient batteries than the present heavy lead-acid ones will have to be found. Already alternative elements like lithium are being thought of as light-weight alternatives for the present heavy lead acid ones. Aeroplanes will have to run only on liquid fuels and will thus face great threat for their very existence when liquid petroleum reserves dry up. Perhaps some artificial liquid fuels will have to be specially manufactured to propel future aeroplanes. Or, air travel may itself be substituted by high speed (300-500 km/hr.) trains. As regards ocean-going ships, the end of diesel supply will see them to switch over to nuclear power, as some submarines already do now.

Inspite of the threat of liquid fuels getting exhausted, transport will flourish, albeit being propelled by electricity.

I. MULTIPLE CHOICE QUESTIONS

1. A drop in fuel is a substitute for

(*a*) new fuel (*b*) conventional yet fuel

(*c*) both (*a*) and (*b*) (*d*) none of these

2. Peak oil is the period when the

(*a*) minimum rate of global petroleum extraction in reached

(*b*) maximum rate of global petroleum extraction in reached

(*c*) both (*a*) and (*b*)

(*d*) none of thes

Answers

1. (*b*) **2.** (*b*)

II. SOLVED QUESTIONS AND ANSWERS

Q. 1. What are the various factors that affect fuel consumption?

Ans. The following factors which affect fuel consumption:

- Encourage/enforce less aggressive driving behavior
- Increase vehicle occupancy on substantial fraction of trips
- Reduce mileage driven per person per year
- Use biomass-based fuels to substitute for petroleum-based fuels
- More effective transportation system management
- Increase public transportation utilization
- Shift the vehicle performance/fuel economy trade-off towards lower fuel consumption
- Improve vehicle maintenance, lubricants, tire pressure, reduce parasitic loads
- Buy and use lighter weight "less big" vehicles
- Implement more efficient engine, drive train, and vehicle technologies
- Develop and implement use of hydrogen with fuel cell powered vehicles
- Use electricity with advanced batteries to reduce petroleum consumption.

Q. 2. What are the factors that affect the fuel economy of a car?

Ans. There are a lot of factors which affect the FE of a car. Some of them are in your hands, but a lot of them aren't. To top it off, doing those changes also make your car quite boring to drive.

- **Run-in of your engine:** The first 1500 kms (~1000 miles) of your car's life are the most important. Mess it up and you'll have a growling devil where you were supposed to have a placid cat. So, do that properly.
- **Carburettor tuning/FI tuning:** With a carburetted engine, it is quite easy to weak it for economy rather than power. Reduce the idling all the way down, until it is just puttering to stay alive. Then reduce the high speed setting so that it starts to strain at your country's speed limit.
- **Driving style:** Driving aggressively, engaging in traffic signal drag races, letting your teenage kids drive your car: all FE busters. Simply put: don't do it. Also, whether you spend most of your time in parking lots or on the highway also makes a difference. In general, all cars have a particular speed at which they work best. On the highway it is easier to maintain that speed. In the city, your FE figures drop by atleast 5%.

Q. 3. What are types of car?

Ans. Type of car: A small hatchback has higher FE than an SUV. So, choose a car wisely.

Remove excess weight: Extra seats, luggage and the like. If its legal in your country, remove the body work also. That'll bump up your FE at-least 20%. I'd suggest losing some weight too. If you're really desperate, that is

III. REVIEW QUESTIONS

1. What are the measures of economy of fuel in road?
2. Discuss the fuel economy in rail transportation.
3. Briefly mention the fuel economy in other mode of transport.

REFERENCES

1. Davis, S.C., Transportation Energy Data Book, Centre for Transportation Analysis, Oak Ridge National Laboratory, Oak Ridge, 2000.
2. Economic Survey, 2007-08, Government of India, New Delhi, 2008.
3. Kadiyali, L.R., Effect of Road Conditions on Fuel Economy of India Vehicles Paper presented at the Seminar or Diesel Consumption in State Transport Undertakings, Petroleum Conversation Research Association Chennai, 1982.
4. Kadiyali, L.R., et. al, A Comparative Study of Fuel Consumption with Radial *vs.* Bias Nylon Tyres, Journal of the Indian Roads Congress, Vol. 42–1, New Delhi, 1981.
5. Road User Cost Study in India, Final Report Vol. I—III, Central Road Research Institute, New Delhi, 1982.
6. Kadiyali, L.R., et. al, Fuel Consumption Experiments with Diesel Jeep, RUCS Technical Paper No. 69, Fourteenth Quarterly Report, Central Road Research Institute, New Delhi, 1981.
7. Kadiyali, L.R., et. al, Fuel Consumption Experiments with Ambassador Car and Premier Padmini car, RUCS Technical Paper No. 77, Fifteenth Quarterly Report, Central Road Research Institute, New Delhi, 1981.
8. Kadiyali, L.R. et. al, Analysis of Fuel Consumption Experiment Data on Tata Truck, RUCS Technical Paper No. 104, Eighteenth Quarterly Report, Central Road Research Institute, New Delhi, 1982.
9. Kadiyali, L.R., et al, Analysis of Fuel Consumption Data on Ashok Leyland Beaver Truck, RUCS Technical Paper No. 105, Eighteenth Quarterly Report, Central Road Research Institute, New Delhi, 1982.
10. Kadiyali, L.R., Fuel Economy in Road Transport, Paper presented at the Seminar on Transportation Systems in Urban Settlements Town and Country Planning Organisation, New Delhi, 1982.
11. Study for Updating Road User Cost Data, Ministry of Surface Transport, prepared by L.R. Kadiyali and Associates, 1992.
12. Fuel Savings on Cement Concrete Pavement, Cement Manufacturer's Association, New Delhi, 1997.
13. Zaniewsky, J.P., Effect of pavement surface type on fuel consumption, Portland Cement Association, 1989.
14. Concrete Overlays : White-topping of Roads, Cement Manufacturers' Association, New Delhi, 2005.

48. Transport Modes, Technology and Selection

48.1. Introduction

There are a number of alternative modes of transport and transport systems which can be selected to meet any given demand. Each mode and system has its own appeal and its own drawback. The transport planner must have a good idea of the characteristics of the modes and systems so that he makes a good choice. Technology of transport is undergoing a constant change, and developments are taking place very fast. Though the fuel crisis of the seventies has given a serious jolt to development of technologies based on oil as fuel, it is likely that newer fuels will be developed and newer technologies will be used in future. A glimpse into the future will be rewarding.

48.2. Transport Modes and their Characteristics

48.2.1. The important modes of transport are :

(*i*) Railways

- Surface
- Underground (Tube)
- Elevated
- Light Rail Transit (Tram)

(*ii*) Road Transport

(*iii*) Air Transport

(*iv*) Water Transport

- Coastal shipping
- International shipping
- Inland water transport

(*v*) Pipe-Lines

(*vi*) Ropeways

48.2.2. The efficiency of a transport mode can be judged in terms of the following characteristics :

1. Speed
2. Safety
3. Adequacy
4. Frequency
5. Regularity
6. Integration
7. Responsibilities
8. Comfort
9. Cheapness
10. Fuel efficiency
11. Employment generation.

Speed is the essence of transport. It reduces overall cost, reduces storage and inventroying, helps a faster turn-round with existing facilities and is of prime importance for perishable commodities.

Safety for passengers demands highest possible standards of construction, maintenance and operation. For goods, safety implies protection against theft, fire and damages.

Adequacy represents the capacity of a mode to transport people and goods.

Frequency is an important characteristic of transport. It is related to capacity, consolidation and storage. Assured frequency attracts people and goods.

Regularity is the punctuality of the service and its ability to adhere to published schedules. A regular service inspires confidence in the clientele.

Integration denotes where the mode involves transfer from one system to another or involves transaction with a single undertaking or multiple undertakings. A system with few transfers and operated under a single undertaking is the most efficiently integrated system.

Responsibility denotes the ability to meet liabilities for compensation for loss or damage. A government undertaking is generally more responsible than a private one. Usually, every system has legislation to cover regulation of transport.

Comfort is essential for passenger transport.

Cheapness is the main consideration for consumer preference. Cheapness should be calculated not only on the operating cost, but on the total resources cost, including initial facility construction, cost of maintenance of the facility and the cost of operation.

The global energy crisis has prompted transport planner to select modes and systems which are the most fuel efficient.

48.2.3. Comparative advantages and disadvantages of different modes.

In light of the criteria laid out above, the various modes can be evaluated as under:

Speed. The following table gives the speeds that are generally associated with each of the modes:

Table 48.1. Speeds of Various Transport Modes

S. No.	*Mode*	*Normal speed range in India (km/hr)*	*Remarks*
1.	Railway	50–80	(1) Speed depends on gauge and number of tracks (2) Speeds upto 250 km/hr are possible (*e.g.*, New Tokaido line in Japan)
2.	*Road transport*		
	Cars	50–80	Speed depends on pavement width and terrain. Values given are for good roads in plain terrain.
	Buses and trucks	30–60	
	Two-wheelers	25–50	
	Cycles	8–15	
	Autorickshaw	15–30	
	Bullock cart	3–5	
3.	*Air transport*		
	Boeing 747	1000	
	Boeing 707	950	
	Feeder aircraft	300–400	
4.	*Water transport*		
	Ocean liners	35–55	
	Coastal ships	15–35	
	Inland water barges	10–25	
5.	Pipe-lines	5–10	
6.	Ropeways	5–10	

It is seen that air transport is the fastest and is about 10-20 times faster than roads and railways.

Safety

Rail transport moves on single exclusive track, and as such, if adequate measures are taken to ensure safety at crossing, Railways are reasonably safe. Modern systems of electronic signal control and safety devices have further enhanced the safety of rail travel. But, if accidents take place, the death toll in the mishap is generally heavy.

Road transport is increasingly becoming unsafe with the spectacular growth of traffic on the roads. The reasons are not far to seek. The vehicles move on road space which is shared by all the vehicles and hence crossing and overtaking manoeuvres are potential hazards. There is little control on the driver requirements and vehicle maintenance, even though rules and regulations do exist to ensure safety. The enforcement often is lax.

Air transport has a good record of safety. This is because of high degree of inspection and control of materials, high standards of vehicle maintenance and medical fitness of crew. The modern instruments for landing ensure high level of safety. Air-craft, however, suffer from natural hazards such as storms, fog, lightning and bird-hits. Whenever a calamity does occur, the fatality rate is very high.

Water transport in modern times is very safe. Safety measures are highly developed and the danger from natural hazards (storms, cyclones etc.) which used to be high in olden days is very much less in the present day. The country craft used for inland waterways and coastal traffic, however, do not have safety devices and instrumental aids and are highly accident prone during storms and cyclones. Strict standards of construction and maintenance are enforced for ocean going liners and coastal ships, but the same cannot be said for country craft. Similarly, crew of ships have high standards of training, but the country craft is operated by untrained personnel.

Pipe-lines and aerial ropeways are very safe.

Adequacy

The railways can meet seasonal fluctuations in traffic, but generally capacity is wasted as a rule. Due to limitations of number of wagons and carriages that can be hauled by a train and the number of trains that can be handled on the track, the carrying capacity is restricted by gauge, number of tracks and traction device. The following rough guidelines may be noted.

Table 48.2. Capacity of Railway Track

Type of track	*Number of trains/day*
Single track	20–30
Double track, with manual block control	60–80
Double track, with centralised track control	160–200

In India, super fast express passenger trains consist of over 20 coaches and carry over 1500 passengers. Double traction engines can enhance the number of coaches that can be hauled. For urban mass transit, metros and Rapid Transit Systems can accomodate trains with a headway of 2 minutes, thus carrying upto

60,000 persons/hour/ direction. For light rail transit, capacities upto 20,000 persons/hour/ direction are possible. Exclusive busways can carry upto 40,000 persons/hour/direction. Goods trains can haul about 50-60 wagons. The pay load of wagons is in the range of 50–80 Tonnes. (The BOX car carries a load of about 55 Tonnes in India). Thus, a typical Indian goods train can haul 3000 Tonnes of goods.

The capacity of roads is governed by the pavement width and terrain. In plain terrian, the capacity of roads under Indian conditions is as below.

Table 48.3. Capacity of Indian Roads

Pavement width	*Design Capacity in PCU_S/day*
Single lane	2000
Intermediate lane	6000
Two lane	15–17,500
Four lane (divided)	35–40,000

Note : PCU signifies passenger car unit.

An average bus has a carrying capacity of 50 sitting and 10 standing passengers. A passenger car can carry 5 persons including the driver. On city streets with mixed traffic, buses can carry upto 10,000 persons/hour/direction. Beyond this range, exclusive bus lanes or LRT or Metros become necessary. An average single-unit truck, by far the most common in India, is permitted to carry a pay load of 10 Tonnes. Overloading (up to 15 Tonnes) is very common. Truck-trailer combinations can carry pay loads of 25 Tonnes easily, but have yet to become popular in India. The advantage of road transport is that it is most versatile and adaptable when considering capacity. Depending upon the load offered, any number of vehicles can be pressed into service or withdrawn.

The capacity of barges and ships are in the following range :

Table 48.4. Capacity of Barges and Ships

Small barge	:	100–400 Tonnes
Large barge	:	1000–2000 Tonnes
Coastal ships	:	3000–20000 DWT
Ocean liners		
(*a*) Crude oil	:	150,000 DWT for berths
		300,000 DWT for Single Point Moorings
(*b*) Dry bulk	:	150,000 DWT
(c) Containers	:	Feeder Vessels : 600 TEUs
		Main line Vessels : 2000-3000 TEUs

Note :

DWT : Dead Weight Tonnage

TEU : Twenty Feet Equivalent Unit

The handling capacity of ports and harbours is a major bottle neck when considering ocean going liners. The port capacity is governed by a number of factors such as (*i*) number, depth, length and layout of berths (*ii*) size of ships and the pattern of their arrival (*iii*) approach channels and locks (*iv*) port facilities (*v*) open space (*vi*) warehousing space etc. A berth occupancy rate of 67 percent is considered to be an efficient norm. The capacity of a modern berth is as follows :

Table 48.5. Handling Capacity of Port Berth

	Type of berth	*Handling capacity per year*
1.	General cargo	
	(*a*) Break bulk	60,000 T
	(*b*) Neo-bulk (steel, auto etc)	2-400,000 T
2.	Dry bulk	300,000-1,000,000 T
3.	Containers	100,000-150,000 TEUs
4.	Petroleum	1.5-6.0 million T
5.	Liquid bulk other than petroleum	80,000 T

The seating capacity of some of the modern air-craft is given below :

Table 48.6. Seating Capacity of Air-craft

	Type	*Maximum Passengers*
1.	Douglas DC–4	86
2.	Caravelle	139
3.	Boeing 707	179
4.	Boeing 737	130
5.	Boeing 747	490
6.	Airbus A-300	302

The capacity of an airport terminal depends upon a number of factors such as layout, number of docking berths, (Ref. 9) baggage handling facilities, baggage checking arrangements, security checking arrangement, number of checking-in counters, instrumentation for controlling air-craft landing and take-off, number of run-ways, area of taxiways and apron etc. The practical annual capacity of a single runway with modern facilities is in the range of 170,000 to 215,000 air-craft landings and departures per year. For detailed considerations, a standard book may be referred (Ref. 1).

The capacity of a pipe line depends upon its size and the material handled. Solids in a slurry form are increasingly being handled by pipe lines. A simple 20 dia pipe line can handle about 250 tonnes of solids per day.

Frequency and regularity

Almost all modes of modern transport offer regular, frequent and assured service. Freight transport by road in India is in private hands and is known for its better service than railways in matters of frequency and regularity.

Integration

Railways suffer from a disadvantage that a number of transfers are involved. Firstly, from the door of the consigner to the railway yard, and secondly from the destination railway yard to the door of the consignee. These transfers are generally achieved by road transport. Containerisation, which has become popular, is expected to facilitate such transfers. Railways are thus able to offer door-to-door service in India for a full container load. Piggy-back system, whereby a trailer towed by a prime-mover on road can be detached and put on a flat-bottom railway car, offers a method of overcoming the transfer door-to-door service, with no need for transfer at all. Air transport and water-transport suffer from the same drawback as the railways. In the case of sea-transport, containerisation has become very popular.

Responsibility

Railways are government owned in India and hence meet the liabilities for compensation and damage or loss. Road transport of freight, though in private hands, also offers a kind of insurance coverage. Air transport is characterised by a high degree of insurance cover.

Comfort

Passenger travel by railways, especially by sleeper berths and upper class, is generally more comfortable for long journeys than by bus. Provision of toilet facilities is absent in buses, and hence long distance by buses is tedious and inconvenient. Stoppages at convenient places for toilet facilities and refreshments help to overcome this drawback to some extent. Air travel is the most convenient form of transport.

Cheapness

Passenger travel by rail and bus in India cost about the same. For example, for long distance travel by train, the second class (sleeper) charges amount to roughly 15 paise per km (1997). By bus, the fares range from 15–20 paise per seat per km. The first class fare by train is about 65 paise per seat per km, and the air-conditioned first class costs about ₹ 1.50 per seat per km. The air-ticket cost about ₹ 3.00 per seat per km and is the costliest. As regards freight, the road transport costs about 70 paise per tonne-km. The charges levied by the Railways vary according to the type of goods. It has been recently found out (Ref. 3) that most commodity movements are cheaper by road for short hauls upto 300-350 kms by beyond this range the cost advantage lies with the railways. The high value commodities like tea, cotton textiles, cotton (raw) etc. are exception to this, where the advantage lies with road transport even for longer haulage. For perishable commodities like fruits and vegetables, road transport has comparative advantage upto 450 kms. For such commodities such as fertiliser, coal, cement, livestock, sugar and steel tubes, comparative cost advantage for roads is upto 200-300 kms.

Fuel Efficiency

Fuel has become a highly important consideration when selecting a transport mode.

Table below gives the energy efficiency of various transport modes (Ref. 3).

Table 48.7. Energy Efficiency of Various Transport Modes

(*Source* : Ref. 3)

S. No.	*Passenger*	*Energy for propulsion BTU/Pass-km*
1.	Electric traction-railway	54.6
2.	Diesel traction-railway	151.2
3.	Stream traction-railway	1445.8
4.	Diesel bus	288.7
5.	Petrol bus	526.5
6.	Average car	1909.0

S. No.	Freight	BUT/Tonne-km
1.	Electric traction-railway	84.6
2.	Diesel traction-railway	255.5
3.	Steam traction-railway	3576.9
4.	Diesel truck	1587.3
5.	Barge	328.0
6.	Pipe-line	281.7

In terms of consumption of fuel in litres, the comparison is presented in Table 48.8.

Table 48.8. Energy Intensity of Different Modes
(*Source* : Ref. 3)

	Mode Passenger	Energy requirement Litres per passenger seat km
1.	Air Bus	0.042
2.	Boeing	0.055
3.	F-27 Turbo prop	0.075
4.	HS 748 Turbo prop	0.80
5.	Car (4 seats)	0.025
6.	Bus (52 seats)	0.0048
7.	AC Railway berth	0.016
8.	Second class Railway sleeper berth	0.0031

	Freight	Litres per tonne–km
1.	Rail transport	0.0076
2.	Heavy road vehicle	0.033–0.066
3.	Light road vehicle	0.104

It is seen that the air transport is the most fuel consuming mode. In order of declining rates are road, rail inland water transport and pipe line.

Employment Generation

India is a country with a big unemployment problem. Transport is a sector which absorbs labour force in good measure. Any mode of transport that offers the best employment potential, therefore, should find favour. A recent study in India has yielded valuable results in this area.

Table 48.9. Employment Intensity of Various Transport Modes
(*Source* : Ref. 3)

S. No.	Mode	Employment potential per 100,000 rupees of investment (1977-78) (Person-year)
1.	Inland water transport operations	33.59
2.	Development of navigational channels	13.20
3.	Road construction and maintenance	27.50

4.	Road transport operation	
	Bullock cart	27.00
	Three-wheeler	17.15
	Trucks	16.95
	Buses	9.26
	Taxis	3.71
	Vehicle production	1.78
5.	Coastal shipping	
	Sailing vessels	20.00
	Other coastal shipping	2.40
6.	Railways	4.30
7.	Air transport	0.97

It is thus seen that road transport, inland water transport and coastal shipping score very high as compared to railways and air-transport in the matter of employment generation.

48.2.4. Transport Modes of India

The various transport modes discussed earlier have been developed to different levels in India.

Railways have a route length of 62,500 km and are the principal transport mode. They handle 300 billion passenger kilometres per year, which is 20 percent of the total road-rail passenger traffic. The freight traffic handled is 257 billion tonnes per annum which is 40 per cent of the total road-rail freight traffic.

Road transport has been growing very fast in the past and has been eroding into the passenger and freight traffic carried by the railways. This trend is evidenced in other countries too. The reasons are the many advantages offered by the road transport over the railways. The total road length in India is 4.0 million km, of which half is surfaced. The total length of National Highways is about 100,000 km. The number of motor vehicles in India is 26 million the annual growth rate being about 10 per cent. In addition there are nearly 15 million bullock carts in the country and a large number of cycles (about 45 millions).

India has two levels of air-lines : (*i*) International Carrier, Air India, (*ii*) Domestic Trunk Route Carrier, Indian Air Lines and some private airlines. Indian Airlines carried 8-3 million passengers in 1992. Domestic air traffic has been increasing at a rate around 10 percent per annum.

India has a coast-line of 5,660 km and hence the potentially for coastal shipping in enormous. There are 176 ports, 10 of them being major and 23 intermediate. Coastal shipping has, however, not grown as it should have been.

The total length of navigable inland waterways is 14,500 km consisting of rivers, canals, creeks, backwater and tidal inlets. Due to lack of development, only 5200 km of river water ways and 485 km of canals are suitable for mechanised crafts. There is thus a great scope of improvement in future.

There are quite a few pipelines for carrying oil and a few for carrying slurries. Aerial ropeways are few.

48.3. A Review of Transport Systems and Technology

48.3.1. The transport systems and technology have undergone great changes in the past. The invention of the wheel was the real beginning of the process. Land marks in the development are the invention of the steam-engine and invention of the internal combustion engine. These opened new forms of wheeled transport ranging from the motor car, the railway, the cycle, the motor cycle and so on. At the present, the wheeled vehicles travelling on land surface are the predominant transport system. The steam engines transformed the old ships with sails into self-propelled carriers. The air-craft also developed from the internal combustion engine.

Innovations in vehicle technology include the air-cushion vehicle which eliminates the traditional wheel. Hydrofoil boats and aircraft use airfoil to create uplift forces.

48.3.2. Developments in Personalised Vehicle Systems

The personalised motor car has grown in popularity because of great advantages it has over other forms of transport systems. But problems of fuel shortages, congestion, parking, environmental pollution etc. have hampered the further development of this system. Research is, however, continuing to develop more efficient personalised vehicle systems.

The automated highway is one such technology. It consists of a specially designed carriageway with control cables embedded in the pavement. The operations normally performed by a driver, such as speed selection and adjustment of space between successive, vehicles would be handled buy a centralised computer. It is reported that capacity 9000 vehicles/hour/lane is possible with this system.

A dual-mode system consists of a small personalised car, capable of moving on roads as well as on rail lines. When on rails they are automatically driven and controlled. The Star RR car and the Automobile are examples of this technology (Ref. 4).

Battery-powered small cars offer new power source and are likely to be popular in future.

48.3.3. Developments in Bus System

Because of many advantages, (such as lower fuel consumption, lesser congestion on the roads, lesser pollution) bus transport offers a promising future. Exclusive bus lanes have been introduced with good results in many cities. A capacity upto 40,000 seats per hour or 550 buses per lane per hour is possible on exclusive bus lanes.

Electric trolley buses (ETB) are likely to become popular in the future because of the absence of pollution and use of non-oil based fuel for propulsion.

The disadvantages of ETBs are their reduced manoeuvrability and the aesthetic intrusion caused by overhead power lines.

Chartered buses (subscription buses) are becoming increasingly popular in big cities since they cater to fixed-time commuter travel on established corridors.

Dial-a-bus is a demand responsive bus system, whereby a group of passengers can dial for a bus which operates on an exclusive bus lane.

Mini-buses can be profitably employed in congested CBDs.

48.3.4. Development in Rapid Rail Transit

Railway transit is a well-known means of rapid transit along high density corridors. They can be :

(*i*) Surface (*ii*) Underground or (*iii*) Elevated

Surface rapid rail transit serves an important role in Mumbai, Chennai and Delhi. The lines are electrified. An underground system between Dum Dum and Tollygunge has been built at Kolkata. The line is 16.4 km long and has cost about ₹ 1000 crores.

The Delhi Metro rail system, costing around ₹ 200 crores per km, is an example of how modern management practices can deliver world-class facilities to emerging economics. Bengaluru city is now planning for a 30 km long Metro corridor at a cost of ₹ 6,000 crores.

Rapid developments have taken place in the technology of rapid rail systems in the recent past. Elevated guideways with backed air-cushion vehicles hold good promise. Speeds upto 300-800 km/hr might become possible. Even in the conventional steel rail systems, technological developments are promising. The New Tokaido line of the Japanese system is an example of this, where a maximum speed of 200 km/hr is achieved. The monorail system is another variety of rapid rail transit. The ALWEG system is an example, wherein rubber tyred vehicles straddle a concrete box beam. The SAFEGE system of monorail hangs underneath the rail and has rubber tyres.

48.3.5. Light Rail Transit

Light Rail Transit (LRT) is also popularly known as the tram or the street-car. Trams used to operate on the roads of Delhi, and Mumbai where they have now been discontinued. They, however, continue to be run in Kolkata. The word light rail refers not so much to the strength of the rail, but to the lighter cars, shorter trains and methods of operation.

A resurgence in the popularity of this mode is now visible in the cities of the world. The reasons are :

(*i*) The fuel crisis has made the oil-based bus transport system costly. The tram is an electricity-based system and is thus free from the uncertainties associated with availability of oil.

(*ii*) Trams are relatively cheaper than conventional suburban rail systems.

(*iii*) Trams are free from the pollution caused by the motor vehicles.

(*iv*) Trams are suitable for pedestrian malls.

Some of the cities which have introduced the LRT are ; Boston, Buffalo, New Orleans, Cleavland, New York, Philadelphia, Pittsburg, Sacramento, San Diego, San Francisco, Montreal, Toronto Vancouver, Buenos Aires, and Oslo. A new LRT system can cost about ₹ 100 crores per Km in India. A monorail system is being built in Mumbai.

48.3.6. Magnetic Levitation (Maglev) System

Maglev is a rail system running on monorails and capable of running at as high speeds as 400 km/hr. The technology has been developed sufficiently to make the operation commercially feasible.

The technology is based on the principle of magnetic levitation. Entire trains are suspended and propelled forward by magnetic forces without touching the rail-track. A line 0.6 km long is already operational between the airport and the railway station in Birmingham (U.K.). Germany has also developed a system called Transrapid.

The advantages claimed are :

(1) Tracks require less maintenance

(2) High speeds are possible

(3) Low noise.

The cost of constructing a single track is around $ 3-4.5 million per km.

48.3.7. Para-transit

Para-transit is a new term which has come into use and is intended to cover some transport systems which are intermediate between the personal transportation system and the mass transportation system. The following systems fall under this group :

(*i*) Car-pooling (*ii*) Van-pooling

(*iii*) Dial-a-ride taxi service (*iv*) Dial-a-bus

The IPT modes have been described in detail in Chapter 45.

48.4. Need for Coordinated Development

All modes of transport have their own role to play. This should be clearly recognised in national transportation planning. No single mode should be allowed to develop at the cost of any other or to the detriment of the nation's overall economy. A coordinated approach is called for.

48.5. Multi-modal Transport Systems

Modern, day transportation covers more than one mode. For example, the goods produced in a factory are stuffed into a container and moved by road to an Internal Container Depot (ICD) or Container Freight Station (CFS). From there it is moved by rail to a port and then on it travels by ship. This system is known as multi-modal or inter-modal transport. The containers have rendered this system to flourish.

Another example of multi-modal transport are the "piggy-back" system or trailer-on-flat-car (TOFC), wherein the trailer of a truck is moved on flat rail wagons over long distances.

Konkan Railway Corporation in India is operating a "piggy-back" system, wherein trucks are moved over long distance, saving fuel, reducing congestion on the roads and relieving the trucks drivers of the drudgery of long distance driving.

The "road-trailer" is a new technology wherein the trailer of a truck can directly be moved on the rails.

The Roll-On-Roll-Off (RORO) is a system wherein the trucks are carried on ships over long distances.

48.6. An Integrated Transportation System (In Tran Sys)

An Integrated Transportation System, known by name in Tran Sys has been conceived in Colorado (Ref. 5) by Cimarron Technologies Ltd. This bold concept has the following features :

(1) All travel on the system takes place on an interconnected network of elevated guideways. Both urban and intercity transport will be covered.

(2) Maximum speed on urban links would be 100 km per hour, while on rural highways it will be 250 km per hour.

(3) The motorist drives his car to the nearest station, drives the car to a loading stall and attaches a simple electrical connection between the platform and his car. The system then checks the identity of the car to facilitate billing. The driver dials the number of the station he wants to reach. His role as a trip maker is finished and the In Tran Sys takes over and reaches his destination on the best possible route. The driver can enjoy and relax in his vehicle, and facility of viewing a TV is provided. When the destination station is reached, he takes out his car from the system's platform to the road.

(4) All types of vehicles are covered, including cars, small buses and cargo containers.

(5) The capacity is 14,400 vehicles per hour on urban tracks and 36,000 per hour on rural highways.

Though the system appears to be too futuristic, one can expect it to be operational sooner than expected, because of its safety, low cost of operation and freedom from the use of petroleum products for propulsion.

I. MULTIPLE CHOICE QUESTIONS

1. Modes of transportation are typically evaluated on the basis of all of the following criteria **EXCEPT:**
   (*a*) accessibility. (*b*) perishability.
   (*c*) frequency. (*d*) capability.
   (*e*) dependability.
2. A logistics network can best be described as:
   (*a*) the choice of rail, marine, or truck transport
   (*b*) the warehouses, production facilities, retailers, and inventory that flows between them
   (*c*) the routes taken to move goods
   (*d*) the least cost route from producer to retailer
3. Reverse logistics refers to:
   (*a*) returning trucks from deliveries
   (*b*) backward scheduling deliveries
   (*c*) movement of goods from customers back to the producer
   (*d*) transport from the producer directly to the retailer
4. FTL stands for:
   (*a*) fair transportation laws (*b*) freight tax levy
   (*c*) fixed transport lane (*d*) full truck load
5. When compared to truck transport, rail transport is:
   (*a*) less flexible but cheaper
   (*b*) less flexible and more expensive
   (*c*) more flexible but more expensive
   (*d*) more flexible and cheaper
6. The use of two or more different modes of transport in moving goods from origin to destination is known as:
   (*a*) container shipping (*b*) LTL shipping
   (*c*) double-stacking (*d*) intermodal transport
7. Consolidation warehousing refers to:
   (*a*) shipping directly from manufacturer to retailers
   (*b*) holding inventory received from suppliers in warehouses until needed
   (*c*) continuous shipment from suppliers to warehouses where goods are redirected and shipped to retailers
   (*d*) storage of full trailers in locked lots
8. Cross-docking refers to:
   (*a*) shipping directly from manufacturer to retailers

(*b*) holding inventory received from suppliers in warehouses until needed
(*c*) continuous shipment from suppliers to warehouses where goods are redirected and shipped to retailers
(*d*) storage of full trailers in locked lots

9. Direct shipment refers to:
(*a*) shipping directly from manufacturer to retailers
(*b*) holding inventory received from suppliers in warehouses until needed
(*c*) continuous shipment from suppliers to warehouses where goods are redirected and shipped to retailers
(*d*) storage of full trailers in locked lots

10. Radio frequency identification technology is expected to be used:
(*a*) to track trucks across the country
(*b*) to enhance the ability to keep accurate inventory counts in warehouses
(*c*) to eliminate the need for reverse logistics
(*d*) to lower the cost of direct shipments.

Answers

1. (*a*)	**2.** (*b*)	**3.** (*c*)	**4.** (*d*)	**5.** (*a*)
6. (*d*)	**7.** (*b*)	**8.** (*c*)	**9.** (*a*)	**10.** (*b*)

II. SOLVED QUESTIONS AND ANSWERS

Q. 1. What is the importance of choosing the correct mode of transport.

Ans. There are numerous mode of transport available at disposal of humans to meet their ever growing need for specialized and most economic mode. It becomes important to choose the most suitable mode of transport especially while working in a logistic company or transporting goods, for other purposes like travelling, which perhaps have largest share in trips, also it becomes essential to decide between various modes available.

Q. 2. Discuss the factors to consider when choosing mode of transport.

Ans. The factors to consider when choosing mode of transport include:

- **Accessibility:** This primarily deals with the ease of getting boarding a vehicle. If a mode is not easily accessible then it loses its usefulness because of its unavailability in a certain condition.
- **Cost/fare of ticket or cost of service:** Considering large number of trips taking place for different purposes and a modern capitalist economy, cost becomes one of the single most important determining factor. It can also be linked to affordability of the user. This factor is one of the most dynamic as the willingness to pay changes on the basis or urgency and need to travel/transport.
- **Fuel efficiency/carbon emission:** With the growing environmental concerns many people and companies prefer to chose a option with smaller carbon footprint. Some people prefer electric vehicle solely because of the "green" label associated with them. Fuel efficiency is closely linked with carbon emission while doing a comparative study, getting more output from same quantity of fuel is always desirable.

- **Speed:** This factor is another most important and deterministic factor depending upon the situation. Fastest mode may seem to be best option but cost associated with it makes it less lucrative, a cheap mode might result in delay and the purpose of whole trip may be defeated. In cases of emergency speed gains much more significance as compared to day to day trips.
- **Capacity/Carrying capacity:** This is another important factor especially for logistic companies, it becomes important in case of travelling in a group. While going on a trip with your complete class you cannot expect to take multiple vehicles because of safety and security concerns,a vehicle with more carrying capacity will be preferred. In case of transporting goods the concept of scale of economies is best exploited for economic profit while dealing with bulk orders.
- **Integration with other modes:** For long trips or trips which requires change of mode this factors becomes most significant for completion of the trip. In case the selected mode is not integrated with other modes than the trip will never get completed. In some cases to overcome the problem of integration a heavy cost will be incurred to complete the required trip.
- **Reliability of the vehicle/mode:** The chosen mode must be reliable. All factors will turn out to be useless if the vehicle develops some fault or is prone to develop fault. If the mode is not reliable than it will result in loss of both time and money. Reliability again becomes important on the basis of situation.
- **Comfort:** This factor is at times ignored while some makes it priority depending upon the affordability. This is essential in case of travelling and ignored in case of transporting goods. Humans undertakes trip for various purpose like work, leisure journey, picnic, shopping and thus always prefer comfort. Those who can afford to pay more are expected to get more comfort while some prioritize cost over comfort depending on their personal decision.
- **Safety:** Considering the increased cases of accidents safety becomes another important factor. Anyone will wish to ensure maximum available safety for their loved ones. In case of logistics also safety matters a lot especially while dealing with fragile material.
- **Privacy:** This factor is usually ignored and comes into mind in some cases especially when the purpose of trip is confidential like in case of business trips. All conversation taking place during the journey are required to be kept confidential. It can also be required when a person wishes to spend quality time with someone. Public transport does not provide option of privacy or if in some cases they do provide than its negligible. Private vehicles can provide high level of privacy.
- **Employment generation:** This factor is not considered for individuals and companies but is essential for government while making policies as employment generation becomes their responsibility.
- **Frequency:** A mode with higher frequency is desirable as the waiting time reduces thus saving time. Modes with good frequency helps especially in cases of delays.

III. REVIEW QUESTIONS

1. Why is there a need for coordinated development?
2. What is an integrated transport system?

REFERENCES

1. Horonjeff, R. Planning and design of airports, McGraw-Hill, New York, 1975.
2. Subramanya, K. Pipe-line transportational Status and growth, Seminar on Technological Forecasting in Transportation, Indian Institute of Technology, Kanpur, 1978.
3. Report of the National Transport Policy Committee, Planning Commission, New Delhi, 1980.
4. Ross, H.R., New transportation technology, Science and Technology Conoever-Mast Publication, New York.
5. Integrated Transport System, In Tran Sys, Cimarron Technology Ltd., From Internet.

49. Application of Information Technology in Transportation

49.1. Transport and Economic Development

Transport and economic development of a country are closely related. As the economy of a country grows, more goods are produced and people travel more. India's economy, for example, is growing at 8 percent yearly, and its road transport is growing at 10-12 percent. Air transport is growing at a rate even faster. Rail transport, water transport and pipe-line transport are also growing at a high rate. The existing stock of roads, railways, ports, airports and pipelines are proving inadequate to meet the growing demands. Capacity augmentation is proving to be very costly and time-consuming, and is soon overtaken by traffic growth. The four-laning programme under the National Highway Development Programme in India, is a example wherein no sooner the roads were four-laned, many four-lane sections now call for six-laning, or better, call for a new controlled access Expressway. This innovative approaches are needed to optimize the investments. The application of Information Technology in transport is receiving world-wide attention in this context. A new concept, known as the Intelligent Transport Systems (or ITS), is finding useful application.

49.2. What is ITS ?

Intelligent Transport Systems (ITS), also known as Transport Telemetics, are transport systems that apply modern information technologies to improve the operation of transport networks. (Ref. 1) The systems acquire vast volume of data on various aspects of transport operation (such as traffic volume, speed, headway, load carried), process them and apply the result to guide traffic, improve operations enhance safety and transport costs.

49.3. Application of ITS

ITS can cover a wide variety of application such as :

(1) Monitoring traffic flow, provide information to drivers on the congestion on the road, road closures, alternative routes, weather conditions and speeds to be observed. Advanced Traveller Information System (ATIS) gives information to highway users on traffic jams, road closures, alternative routes and weather condition.

(2) Monitoring incidents on the road, such as vehicle break-down and collisions ;

(3) Electronic collection of toll ;

(4) Intelligent Vehicle-Highway System (IVHS), in which vehicles are guided longitudinally and laterally by the use of electronic devices. The Advanced

Vehicle Control Systems (AVCS) dispense with human control of vehicles and rely on computers.

(5) Traffic control on urban streets by using information on traffic flows and adjusting the signal operations to reduce congestion and delay.

(6) Asset Maintenance Management Systems cover the data on the assets, the traffic using the asset, periodic condition survey data and use software packages to optimize maintenance interventions.

(7) Public Transport Management Systems wherein the fleet can be managed efficiently by analyzing data on vehicle location, adherence to schedules and passenger loadings. Demand-Responsive Public Transport and Taxis can also be a part of the system. The use of GIS is extensively made.

(8) Truck Transport Management Systems, wherein the data on vehicle location, breakdowns, accidents, detention etc. can be analysed and action initiated to improve the operations. The use of GIS is extensively made.

(9) Electronic Road Pricing System to decongest the city centers.

Information Technology has been successfully used for several decades now on acquiring large volume of data on travel patterns in a city, analyse the data, develop forecasting models and plan efficient transport plans.

49.4. ITS in the Various Countries

Many forms and variations of ITS are in use extensively in the developed countries. Systems like IVHS have already passed the R &D stage. Prototypes are already in operation.

Some of the systems are briefly listed below :

(1) Autoguide, introduced in U.K., which is a dynamic route guidance system, using infrared beacons (Ref. 2).

(2) Carminat (France), a vehicle-based driver information system.

(3) Advance (USA), providing dynamic route guidance, launched in Chicago.

(4) Dutch Traffic Information System (TIC).

(5) Highway 407 (Canada), which is one of the earliest all electronic toll highways.

(6) Traffic master (U.K.), giving real-time traffic information service on trunk routes in U.K.

(7) VICS (Japan), giving traffic information to motorists, using infra-red beacons, radio-wave beacons and FM radio broadcasts.

(8) SCATS (Australia), which is a traffic responsive traffic control system.

(9) Many developed countries have Electronic Toll Collection System.

China has introduced Electronic Toll Collection Systems on some of the Toll Roads.

India is modernising its highway system. Toll roads are very common. Electronic Toll Collection is being insisted upon on same of the new projects. The recently completed Mumbai-Pune Expressway has an excellent centralised traffic control system with CCTV. India can draw upon its vast reserve of technical manpower in the IT Field to find ITS applications in its expanding transport infrastructure projects. Already, on-line railway, air and long distance bus ticketing is available in some sectors. Early efforts to transplant Highway Maintenance Management Systems developed elsewhere have not proved successful. Locally developed systems which can be sustained with minimum data acquisition effort can alone succeed. The use of Area Traffic Control Systems developed elsewhere, like the SCOOT, has been tried successfully in some cities so far.

The application of ITS gives enhanced safety, higher productivity and lower user costs. It is obvious that ITS has a great potential for application, a cannot be termed anymore as too costly and using too sophisticated a technology. Many countries have successfully implemented different variations and are realising its benefits.

I. MULTIPLE CHOICE QUESTIONS

1. .... refers to deviant acts committed with the help of information technology.
   (*a*) computer crime (*b*) white collar
   (*c*) cyber crime (*d*) none of these
2. FTL stands for:
   (*a*) fair transportation laws (*b*) freight tax levy
   (*c*) fixed transport lane (*d*) full truck load
3. When compared to truck transport, rail transport is:
   (*a*) less flexible but cheaper
   (*b*) less flexible and more expensive
   (*c*) more flexible but more expensive
   (*d*) more flexible and cheaper
4. The use of two or more different modes of transport in moving goods from origin to destination is known as:
   (*a*) container shipping (*b*) LTL shipping
   (*c*) double-stacking (*d*) intermodal transport
5. Consolidation warehousing refers to:
   (*a*) shipping directly from manufacturer to retailers
   (*b*) holding inventory received from suppliers in warehouses until needed
   (*c*) continuous shipment from suppliers to warehouses where goods are redirected and shipped to retailers
   (*d*) storage of full trailers in locked lots.

Answers

1. (*a*) **2.** (*d*) **3.** (*a*) **4.** (*d*) **5.** (*b*)

II. REVIEW QUESTIONS

1. What is transport and economic development?
2. What is **ITS**?
3. Explain the application of **ITS**.

REFERENCES

1. ITS Handbook, 2000 PIARC Committee on Intelligent Transport, Artech House, Boston, London, 1999.
2. Stoneman, B., The Effect of Dynamic Route Guidance in London, TRL Report, 348, Crowthorne, 1992.

50. Public-Private Partnership in Transport Projects

50.1. Governments' Inability to Fund Infrastructure Projects

Infrastructures facilities include public utilities (power, telecommunications, piped water supply, sewerage, solid waste collection and disposal, piped gas etc.), public works (roads, railways, ports, airports, irrigation projects etc.). Infrastructure projects involve lumpy investments and long pay-back periods. Recupment of investments through a levy of user charges is often politically resisted. Hence, the government has been the main provider of infrastructure facilities in the past in all countries, till recently. With new concepts such as privatisation and deregulation gaining acceptance in the recent times, many countries are now activity pursuing the involvement of the private sector. As far as India is concerned, it has been estimated (Ref. 1) that the total investments in infrastructures would rise to 8 per cent of GDP by 2005-06, which would be ₹ 1,800 billion (US $ 50 billion). This is a huge investment, and is unlikely to be met from governments' budget or through external borrowings. The gap has to be bridged through the involvement of the private sector.

50.2. Benefits from Privatisation

There are several benefits that can be reaped from privatisation of the transport sector these are discussed below.

(1) **Additionality of resources.** It is argued that the private sector involvement will bring about additionality of resources. While this may be true if foreign direct investment (FDI) takes place, it may not be true if only domestic players are involved. Because, the domestic private entrepreneurs will raise resources from the domestic marker or borrow from banks. The government can as well do this, so, the additionality factor is not really true.

(2) **Better Management Practice.** Private sector can bring about better management practices leading to faster construction of projects.

(3) **More Innovative Design can Emerge.** The private sector can employ more innovative and cost-reducing designs than what the government had conceived.

(4) **Better Quality of Projects will Emerge.** Since the private entrepreneur has to maintain the facility for a long period, he will take care to ensure that the workmanship is of the highest quality and maintenance costs are minimal.

(5) **Loopholes in User Fee Collection will be Eliminated.** It is argued that the governments' system of user-fee collection is not always fool-proof and there are substantial leakages. The private sector is considered to be more efficient in user fee collection.

(6) **Unbundling of Infrastructure Services is Possible.** Infrastructure services comprise of number of sub-activities, and if unbundled, each activity can be separately privatised. This will reduce the burden on the private entrepreneur. For example, in the roads sector, the unbundling can be done into the following activities.

(*i*) Design of the project (*ii*) Construction
(*iii*) Maintenance

50.3. Forms of Privatisation

There are several forms of privatisation of transport infrastructure projects, as given below :

(1) BOT. Build, Operate and Transfer is one of the forms successfully adopted for National Highway Projects in India. The entrepreneur is required to build, operate and maintain the facility for a specified period, known as the Concession Period. A 20 year Concession Period is adopted for National Highway Projects. At the end of the Concession Period, the operator is expected to transfer the facility in an acceptable condition to the government. The entrepreneur is expected to recoup his investments through collection of a user fee (toll). The fee structure is so designed that a periodic increase of the fee to account for inflation is allowed.

The increase is determined by a system of price indexation, commonly the Wholesale Price Index.

Since not all projects can be viable solely on the basis of user fee collection, the government sometimes steps in to give a grant. The grant is limited to 40% of the project cost for National Highways. Competitive bids are invited fixing the Concession period and the user fee structure, and asking the bidder to quote the grant. The grant may be positive (amount given by the government to entrepreneur) or negative (amount given by the entrepreneur to the government as an up-front money to win the project).

(2) Annuity Project. The concessionaire builds, operates and transfers the project to the government after a Concession Period, using his own resources. The government makes a payment of a fixed amount at intervals (six months or a year) to cover the costs. The Concessionaire is not allowed to collect user fees, which may be collected by the government separately. In India, on National Highway works, the Annuity route is chosen when the BOT route (even with 40% grant) does not become attractive.

(3) Special Purpose Vehicles (SPV). In this route, government and the private operator and other financial entities form a separate company (SPV) which constructs and operates the facility. The user fees collected are shared by the stake-holders of the SPV in an agreed manner.

(4) Design, Build, Finance and Operate (DBFO). In the three systems described earlier, the government prepares the detailed project report (DPR) and expects the bidder to adhere to it in broad terms, though minor variations are permitted. It has been observed that the preparation of the DPR takes a minimum period of one year very often, the designs are conservative lacking in innovation and cost reducing practices. In order to cut delays and foster competitive cost-effective designs, the government can follow the DBFO route. This practice has been successfully tried in U.K., and may be tried soon on NH projects in India.

50.4. Government Incentives in India

The Government of India have recognised the following sectors as infrastructure for providing a variety of fiscal benefits :

(1) Roads, new airports, railways.

(2) Irrigation, water supply, sanitation and sewerage system. The benefits offered are :

(*i*) Five consecutive year tax holiday to be claimed within 12 years of operation. For the remaining years, 30 percent exemption is offered.

(*ii*) Concessional import duties and port charges for project related imports.

(*iii*) Exemption from Minimum Alternate Tax.

For the Roads sector, the Government of India Policy on Private Sector participation is given below :

(*a*) The policy of privatisation will be implemented by the NHAI and in exceptional cases by the State Public Works Departments.

(*b*) The Basic principles in identification of NH projects for private investments would be that the project is an approved project of the Ministry of Road Transport & Highways and it is capable of yielding adequate financial return.

(*c*) The Government will provide support by carrying out all preparatory works, such as :

- Feasibility study
- Land for right-of-way and roadside facilities
- Relocation of utility services
- Cutting of trees
- Removal of encroachments and resettlements and rehabilitation of persons affected from the projects.

Depending upon financial viability, the cost for the above activities may be recouped from the project.

(*d*) Suitable traffic support/guarantee will be provided on a case to case basis.

(*e*) Exemption will be given from import duty on identified high quality construction plant and equipment.

(*f*) Foreign Direct Investment upto 74 per cent would be permitted automatically. Beyond that, proposals would need clearance of the Foreign Investment Promotion Board.

(*g*) A five year tax holiday is available followed by a further five-year period of 30 percent exemption and the concession can be availed of in any 10 consecutive years during the first 20 years of operation. (Section 80 IA (12) (ca) of the Income Tax Act, 1961).

(*h*) Upto 40 percent may be excluded from Income Tax of the income derived by financial institutions from finance they provide for infrastructure projects.

(*i*) Exemption is available from Income Tax on the income from dividend interest on long term capital gain derived from investments in the form of shares or long term finance to any enterprise set up to develop, maintain and operate an infrastructure facility.

(*j*) Subscription to equity shares and debentures are eligible for deductions under Section 99 of the Income Tax Act.

(*k*) Real estate development can be made integral part of BOT projects to enhance their financial viability.

(*l*) NHAI can provide capital grants to the developers of the road project for project cost on a case to case basis.

(*m*) NHAI can participate upto 30 per cent of total equity of a floated SPV to develop a road project.

(*n*) The ownership of the land for the highway construction and roadside facilities would continue to vest in Government. Mortgaging and subleasing of the land for raising finances is not allowed. However, land will be given on lease to entrepreneurs.

(*o*) There will be provision of unified check barriers at the interstate borders for the use of all government authorities. Such barriers would be located outside the right-of-way with proper entry/ exit layout.

(*p*) Disputes resolution and arbitration would be under the Indian Arbitration and Conciliation Act, 1996. (It incorporates UNCITRAL provisions).

(*q*) Entrepreneurs would be protected against force majure situations including political, non political and legislative changes.

(*r*) There will be a standard concession agreement, which addresses the concerns of various stake holders in the projects.

50.5. Financial Viability of Privately Funded Projects

A Financial analysis of a typical road project is done as under (Ref. 3):

(1) A debt: equity ratio of generally 70 : 30 is considered.

(2) The Civil Construction cost to be borne by the Concession aire is found out and is spread over the estimated construction period, usually three years as under :

First year	:	30%
Second year	:	40%
Third year	:	30%

Escalation at 5% is added during the concession period.

(3) Debt financing cost at 2% is added.

(4) Interest during construction at 10% is added, considering the withdrawal schedule as given in 2 above.

(5) The interest on debt is calculated at 10% per annum.

(6) The base year traffic is found out and the future year traffic is calculated using realistic growth rates.

(7) From the projected traffic, some reduction (usually 3%) is made for non tollable traffic, such as :

(*i*) concession of local traffic

(*ii*) exempted vehicles

(8) Toll rates are considered as fixed by the National Highway Act, which are :

	₹/*Km* (*1997*)
(*a*) Car, Jeep, Van	0.40
(*b*) Light Commercial Vehicles including Mini-bus	0.70
(*c*) Truck or Bus	1.40
(*d*) Multi Axled Vehicle	2.25
(*e*) Heavy construction machinery and earthmoving equipment	3.00

The rates are escalated as per indexation of the Wholesale Price Index.

(9) Toll collection expenses, Insurance charges (at about 015% of the Project Cost), Maintenance Costs, Tax etc. are added.

(10) The profitability of a project generally judged from :

Project FIRR-Post-Tax	Minimum of 14%
Return of Equity (Equity IRR)	Minimum of 15%
Debt Service Coverage Ratio (DSCR)	Minimum 1.33
Net Present Value @ 12% discount	Positive
Pay-Back Period	Not more than 12 years for a 20 year concession Period

I. MULTIPLE CHOICE QUESTIONS

1. BOT stands for
   (*a*) building of transportation network
   (*b*) building of traffic network
   (*c*) build, operate and transfer
   (*d*) none of these
2. NHAI can participate upto of total equity
   (*a*) 30% (*b*) 20%
   (*c*) 50% (*d*) 10%
   (*e*) Operations and maintenance (*f*) Revenue

Answers

1. (*c*) 2. (*a*)

II. SOLVED QUESTIONS AND ANSWERS

Q. 1. What is public-private partnerships?

Ans. A public-private partnership (PPP) is a mutually beneficial collaboration between a public agency and a private sector entity. Through this contractual arrangement, the skills and assets of each sector are shared in delivering a service or facility for the use of the general public. In addition to the sharing of resources, each party shares in the risks and rewards potential in the delivery of the service and/or facility.

Projects with the greatest likelihood of success are those high priority projects that are clearly defined and have a demonstrated public sector commitment. Projects delivered through a PPP must allocate the risks fairly between the parties, with each sector assuming the risks that they are best able to manage.

The public agency usually assumes the project definition risk by undertaking the environmental clearance effort, assessing financial feasibility and garnering stakeholder and political commitment. The private sector can best assume the financial risk, such as project financing, construction and perhaps facility management.

Q. 2. How PPPs can work for us?

Ans. We are looking at PPPs as a means to accelerate delivery of our much-needed transportation projects. If we can use up-front private investment to finance some of the projects, we will likely see construction cost savings due to today's favorable bidding environment, efficiencies in alternative delivery and/or management strategies, congestion mitigation, earlier achievement of CO_2 emissions reductions and earlier completion of planned rail-bus-highway transportation network.

A private investor will be entitled to a reasonable return on the financing investment made in the public project. Therefore, to determine whether delivering the project through a PPP is the most financially responsible method for the public agency to pursue, we must undertake a thorough analysis and comparison of life-of-project costs to be incurred both by delivering, operating and maintaining a project as a strictly public project, and by delivering it with private financial participation.

III. REVIEW QUESTIONS

1. What are the benefits from privatisation?
2. Discuss the forms of privatisation.

REFERENCES

1. Rakesh Mohan, the India Infrastructure Report, Published by J.M. Jaine and Bros, Delhi, 1997.
2. Road Development Plan, Vision: 2021, Ministry of Road Transport & Highways, published by the Indian Road Congress, New Delhi, 2001.
3. Final Feasibility Report, Four/Six laning of Bangalore-Hoskote-Kolar-Mulbagal Section of NH4 in Karnataka, Prepared for the National Highway Authority of India by L.R. Kadiyali & Associates, New Delhi, 2006.

51. Logistics in Transport

51.1. What is Logistics ?

The concept of logistics is not new. In the military, logistics was considered as essential to win a war. It dealt with procurement, movement and supply of goods (military hardware, spare parts, food rations, POL etc.) to the armed forces to enable them to function efficiently. The concept is now gaining extreme importance in the civilian side also. To put it in plain terms, logistics is the movement of goods to obtain the best value for the money spent. Logistics is a group of activities which enables continuous movement of goods at the lowest cost to enable the product to be available to the consumer.

Activities thus include:

(1) Processing the order

(2) Management of inventory

(3) Storage and warehousing and efficient packaging

(4) Transportation at lowest cost and time

(5) Ensuring timely delivery

An excellent example of logistics is the collection, transportation, processing, packaging and delivery of milk products, the supply chain starting from a source as far as Gujarat and ending in Delhi.

Logistics cost is as high as 13 percent of the GDP in India (Based on 2001 numbers) (Ref. 1).

51.2. Multi-modal Transport as a Facilitator

For management of logistics in the modern day, multi-modal transport has become a great facilitator. Containerisation has led to an integration of various modes, particularly the road, rail and water.

Several combination of modes are possible :

Sea/Air	(Using Containers)
Road/Rail	(Example : Piggy-back System)
Road/Sea	(Example : Roll-on Roll-off)
Sea/Rail/Road	(Using Containers)

The exact combination of mode to be selected is best decided by the logistics services provider, who has the expertise in giving the most cost effective solution.

51.3. Use of Information Technology

The use of Information Technology is becoming an important part of Logistics. The exact location of a consignment can be tracked, using GPS or GIS, Radio Frequency Identification Device (RFID) or some specialized e-tracking software.

51.4. Future Prospects

It has been well summarised (Ref. 3) that the future scenario for India will be :

(1) "The potential for cost reduction and service level improvement in logistics substantial. Infrastructure development and reduction in regulation needs to take place to a large extent. The government is the main actor for this.

(2) Since complexities of coordination are large, logistics should evolve as a function in industry. The shippers are the main actors for this.

(3) Communication technology, computing technology, and problem solving technology are improving tremendously. This opportunity should be exploited in logistics management. The supply industry in logistics, as well as the shippers should act on this."

I. MULTIPLE CHOICE QUESTIONS

1. VMI stands for
   (*a*) Vendor material inventory
   (*b*) Vendor managed inventory
   (*c*) Variable material inventory
   (*d*) Valuable material inventory
2. The major decision areas in supply chain management are
   (*a*) location, production, distribution, inventory
   (*b*) planning, production, distribution, inventory
   (*c*) location, production, scheduling, inventory
   (*d*) location, production, distribution, marketing
3. Distribution requirement planning is a system for
   (*a*) Inventory management
   (*b*) Distribution planning
   (*c*) Both (*a*) and (*b*)
   (*d*) none of these
4. Reverse logistics is required because
   (*a*) Goods are defective (*b*) Goods are unsold
   (*c*) The customers simply change their minds
   (*d*) All of the above
5. 3-PL stands for
   (*a*) Three points logistics
   (*b*) Third party logistics
   (*c*) Three points location
   (*d*) None of the above
6. Logistics is the part of the supply chain involved with the forward and reverse flow of
   (*a*) goods (*b*) services
   (*c*) cash (*d*) all of the above
7. Due to small change in customer demands, inventory oscillations become progressively larger looking through the supply chain. This is known as

(*a*) Bullwhip effect (*b*) Netchain analysis
(*c*) Reverse logistics (*d*) Reverse supply chain

Answers

1. (*b*) **2.** (*a*) **3.** (*d*) **4.** (*d*) **5.** (*b*)
6. (*d*) **7.** (*a*)

II. SOLVED QUESTIONS AND ANSWERS

Q. 1. What is transportation?

Ans. Transportation is defined as the movement of people, animals and goods from one location to another. Modes of transport include air, rail, road, water, cable, pipeline and space. The field can be divided into infrastructure, vehicles and operations. Transport is important since it enables trade between people, which in turn establishes civilizations. I find it an interesting point that transportation is an enabler of civilization, but this makes sense, as it enables the ability to trade and communicate.

Q. 2. What is logistic?

Ans. According to the APICS dictionary, logistics is defined as

1. In an industrial context, the art and science of obtaining, producing, and distributing material and product in the proper place and in proper quantities.

2. In a military sense (where it has greater usage), its meaning can also include the movement of personnel.

Q. 3. What is transportation and logistics management? Do you think they are the same thing?

Ans. If you have read the above academic definitions, you will see that that after Asking "What is Transportation and Logistics Management" it seems rather easy to see the difference between the two. Transportation is the driver of logistics, but logistics is the race car driver in the seat of transportation. In fact, it's easy to see from that sentence alone, the pure difference.

Logistics requires planning, transportation is just the mode to execute the planning, when getting freight from point A to point B. Clearly, they are not the same thing, but transportation is just simply a part of logistics. When it comes to logistics, logistics executives must make further decisions beyond the mode of transportation to include:

- Packaging
- Containerization
- Documentation
- Insurance
- Storage
- Importing and exporting regulations
- Freight damage claims
- Working and collaborating with other executives within the supply chain
- Managing vendors and partners
- Responsible for mitigating risk and mitigating expenditures

III. REVIEW QUESTIONS

1. What is logistics?
2. Discuss multimodal transport as faciliatator.
3. What is the use of information technology?

REFERENCES

1. Mathur, D.S., Cost reduction in supply chain management in multi-modal transport, Proceeding of the 10th International Conference on Strategic Cost Reduction through Effective Supply Chain Management, the Chartered Supply Chain Management, the Chartered Institute of Logistics and Transport, New Delhi, 2005.
2. Blanchard S. Benjamin, Logistics Engineering and Management, Prentice-Hall Inc., Englewood Cliffs, New Jersey, 1986.
3. Raghuram, G, Logistics Management—Issues and Implications, Transportation Systems Studies, Analysis and Policy, Proceedings of the National Conference on Transportation Systems Studies, I.I.T., Delhi, 1990.

Index

M

N

O

Q

R

S